1st International Conference on Tropical Meteorology and Atmospheric Sciences (ICTMAS 2018)

IOP Conference Series: Earth and Environmental Science
Volume 303

Bandung, Indonesia
19 - 20 September 2018

Editors:

Wiwiek Setyamati
Indah Susanti
Dessy Gusnita
Eka Putri Wulandari
Iis Sofiati
Sumaryati
Tiin Sinatra
Riris Ayu Wulandari
Risyanto
Farid Lasmono
Anis Purwaningsih
Amalia Nurlatifah
Fadli Nauval
Elfira Sauvina
Fanny Aditya Putri
Sigit Kurniawan Jati Wicaksana
Christine Cecylia Munthe
Muhammad Fadhlan Putranto
Hidayatul Latifah

ISBN: 978-1-5108-9338-2
ISSN: 1755-1307

Printed from e-media with permission by:

Curran Associates, Inc.
57 Morehouse Lane
Red Hook, NY 12571

Some format issues inherent in the e-media version may also appear in this print version.

Institute of Physics
Dirac House, Temple Back
Bristol BS1 6BE UK

Phone: 44 1 17 929 7481
Fax: 44 1 17 920 0979

techtracking@iop.org

Additional copies of this publication are available from:

Curran Associates, Inc.
57 Morehouse Lane
Red Hook, NY 12571 USA
Phone: 845-758-0400
Fax: 845-758-2633
Email: curran@proceedings.com
Web: www.proceedings.com

TABLE OF CONTENTS

AIR-SEA INTERACTION

ATMOSPHERIC SCIENCES

APPLICATION OF ATMOSPHERIC SCIENCE AND TECHNOLOGY IN DISASTER, MARITIME, AND CLIMATE CHANGE

ATMOSPHERIC ENVIRONMENT AND CHEMISTRY

ATMOSPHERIC SIMULATION AND MODELLING

TROPICAL METEOROLOGY

IOP Conf. Series: Earth and Environmental Science **303** (2019) 011001 doi:10.1088/1755-1315/303/1/011001

FOREWORD

The 1st International Conference on Tropical Meteorology and Atmospheric Science (ICTMAS) 2018 has been dedicated to gather researchers, academicians, professionals and general public in sharing recent findings and knowledge on meteorology and atmospheric science of the tropics, especially over Maritime Continent to promote science and technological research on tropical meteorology and atmospheric science, and recent knowledge on the related fields. ICTMAS 2018 also aimed to strengthen international collaboration in research and education, especially on tropical meteorology and atmospheric science. It will also convey the spirit to foster a new generation of scientists in meteorology, atmospheric science, and all related discipline.

ICTMAS 2018, to which this issue of IOP Conference Series: Earth and Environmental Science (EES) is dedicated, was held on September (19-20), 2018 at the West and East Hall of Bandung Institute of Technology (ITB). The conference is co-organized by Center for Atmospheric Science and Technology (*Pusat Sains dan Teknologi Atmosfer - PSTA*) of Indonesian National Institute of Aeronautics and Space (*Lembaga Penerbangan dan Antariksa Nasional - LAPAN*) and Meteorology Department (*Program Studi Meteorologi - ME*) of Bandung Institute Technology (*Institut Teknologi Bandung - ITB*). ICTMAS 2018 is highly supported by Indonesian Ministry of Research, Technology and Higher Education through Center of Excelent (*Pusat Unggulan IPTEK - PUI*) program. The conference also supported by Meteorological, Climatological, and Geophysical Agency (*Badan Meteorologi Klimatologi dan Geofisika - BMKG*) and Agency for the Assessment and Application of Technology (*Badan Pengkajian dan Penerapan Teknologi - BPPT*). International supports to the conference are given by Japan Agency for Marine Science and Technology (JAMSTEC), Year Maritime Continent Project (YMC), Kyoto University and other institutions also contribute.

ICTMAS 2018 topics emphasize scientific research on tropical meteorology and chemistry, as well as technological advancement on atmospheric observation and modeling, and the application of those science and technology to humanity. The conference consist of 7 main topics, which are (1) Air-Sea Interaction; (2) Tropical Meteorology; (3) Atmospheric Science; (4) Environment and Atmospheric Chemistry; (5) Atmospheric Simulation and Modeling; (6) Atmospheric Observation

IOP Conf. Series: Earth and Environmental Science **303** (2019) 011001 doi:10.1088/1755-1315/303/1/011001

Technology and Instrumentation, Database, and Data Assimilation System; and (7) Application of Atmospheric Science and Technology on Disaster, Maritime, and Climate Change. The committee has received 160 research titles for both category of oral and poster presentation (76 abstracts for oral presentations and 84 abstracts for poster presentation), which come from various scientist, lecturers and students from 8 countries and 42 Institutions and University. These research papers have been reviewed by the scientific committee of ICTMAS 2018. Regarding the peer review result, 66 papers have been selected to be published in this issue, which mostly focused on meteorology and atmospheric science of the tropics, especially over Maritime Continent.

We would like to thank to all keynote speakers and invited speakers for willing to join this conference. We want to thank to the Steering Committee and Scientific Committee (Reviewer Team) which comprises several institution and country, from Indonesia, i.e, LAPAN, ITB, BMKG, BPPT, IPB, UNAND, Udayana University, BATAN, and from abroad, i.e, JAMSTEC (Japan), Kyoto University (Japan), *Universitas Kebangsaan Malaysia* (Malaysia), BoM (Australia), Chicago University (USA), UK Met Office (UK), University of California (USA), University of East Anglia (UK). We wish to thank the authors who have contributed to yield a high scientific standard to this issue. We are excited to learn your presentation in both Oral and Poster Sessions. We wish you all the very best and an enjoyable meeting. We look forward to seeing you all again in the future. We are also grateful to IOP Publishing, especially Stephanie Gill (IOP Conference Commissioning Editor), for allowing us to share some of the conference results with a large audience of colleagues through the IOP Conference Series: Earth and Environmental Science.

Dr. Trismidianto (LAPAN) and I Dewa Gede A. Junnaedhi, M.Sc (ITB)
Chairman of Organizing Committee of ICTMAS 2018

IOP Conf. Series: Earth and Environmental Science **303** (2019) 011002 doi:10.1088/1755-1315/303/1/011002

IOP Conf. Series: Earth and Environmental Science **303** (2019) 011003 doi:10.1088/1755-1315/303/1/011003

List of Committee:

Steering Committee

- Drs. Afif Budiyono, M.T. I LAPAN - Indonesia
- Dr. Widada Sulistya, DEA I BMKG - Indonesia
- Ir. Halimurrahman, M.T. I LAPAN - Indonesia
- Dr. Urip Haryoko, M.Si I BMKG - Indonesia
- Dr. Nurjanna Joko Trilaksono S.Si., M.Si. I ITB - Indonesia
- Dr. Tri Handoko Seto S.Si., M.Sc. I BPPT - Indonesia
- Prof. Mamoru Yamamoto I Kyoto Univ - Japan
- Dr. Kunio Yoneyama I JAMSTEC - Japan

Organizing Committee

Supervisor:

Dr. Lilik Slamet Suprihatin I LAPAN - Indonesia

Leader

- Dr. Trismidianto I LAPAN - Indonesia
- I Dewa Gede A. Junnaedhi, M.Si I ITB - Indonesia

Secretariat

- Nani Cholianawati, ST I LAPAN - Indonesia
- Rahmawati Rahayu, M.Si I ITB – Indonesia
- Mita Permatasari, S.Pd I ITB – Indonesia
- Fanny Aditya Putri, M.Si I LAPAN – Indonesia
- Dyah Aries Tanti, S.Si I LAPAN – Indonesia
- Fitria Novianti Sunarto I LAPAN – Indonesia
- Siti Azizah, S.Si IITB - Indonesia

Treasure

- Nur Rahmayanti, MM
- Muchtar Gunawan, SAP.

Event

- Gammamerdianti, S.Si I LAPAN - Indonesia
- Dr. Rusmawan Suwarman I ITB - Indonesia
- Haries Satyawardhana, M.Si I LAPAN - Indonesia
- Amalia Nurlatifah, S.Si., MT I LAPAN - Indonesia

- Atep Radiana, ST., MAP | LAPAN - Indonesia
- Anis Purwaningsih, S.Si | LAPAN - Indonesia
- Dita Fatria Andarini, S.Si | LAPAN - Indonesia
- Asri Indrawati, M.Si | LAPAN - Indonesia

Publication

- Edy Maryadi, MT.| LAPAN - Indonesia
- Faiz Rohman Fajary, M.Si | ITB - Indonesia
- Fandy Balbo, S.Si | ITB - Indonesia
- Farid Lasmono, S.Kom | LAPAN - Indonesia
- Listi Restu Triani, ST | LAPAN - Indonesia

Documentation

- Djoko Trianas, A.Md | LAPAN - Indonesia
- Edi Riawan, MT | ITB - Indonesia
- Shailla Rustina, S.Si., M.Stat | LAPAN - Indonesia
- Gelar Gema Gemawan, A.Md | LAPAN - Indonesia

Logistic

- Risyanto, M.Sc | LAPAN - Indonesia
- M. Ridho Syahputra, M.Si | ITB - Indonesia
- Ateng Jaelani |LAPAN – Indonesia
- Chandra Nur Ihsan | LAPAN – Indonesia
- Johan Rahmawan, SE. | LAPAN – Indonesia
- Prawira Yudha Kombara, S.Si | ITB - Indonesia
- Imam Mauli Wijaya | ITB – Indonesia

Catering

- Tirnawati, SAP | LAPAN - Indonesia
- Winasti Mahatmya. A.Md., ST| ITB - Indonesia
- Della Purnama | LAPAN - Indonesia
- Rosita | LAPAN – Indonesia

Scientific Committee (Reviewer Team):

- Dr. Tri Wahyu Hadi | ITB - Indonesia
- Prof. Dr. Ir. H. Eddy Hermawan, M.Sc | LAPAN - Indonesia
- Prof. Dr. Edvin Aldrian, B.Eng, M.Sc | BPPT - Indonesia
- Dr. Ardhasena Sopaheluwakan | BMKG - Indonesia
- Dr. Findy Renggono, B.Eng, M.Sc | BPPT - Indonesia
- Dr. Ninong Komala | LAPAN - Indonesia
- Prof. Dr. Fredolin Tangang | UKM - Malaysia
- Prof. Jun Matsumoto | TMU/JAMSTEC - Japan
- Dr. Takahiro Osawa | Japan
- Dr. Matthew C. Wheeler | BoM - Australia
- Dr. techn. Marzuki | UNAND - Indonesia
- Dr. Akhmad Faqih | IPB - Indonesia
- Dr. Rahmat Hidayat | IPB - Indonesia
- Prof. Ir. Puji Lestari Ph.D. | ITB – Indonesia
- Prof. Dr. Muhayatun Santosa | BATAN - Indonesia
- Shuichi Mori, Ph.D. | JAMSTEC - Indonesia
- Sandro Lubis, Ph.D | Chicago University - United States
- Dr. Jon Petch | UK Met Office - United Kingdom
- Dr. Piotr Flatau | Scripps Institution of Oceanography, University of California) - United States
- Prof. Adrian Matthews | University of East Anglia - United Kingdom
- Dr. Erwin Eka Syahputra Makmur, S.Si | BMKG - Indonesia
- Dr. Didi Satiadi, M.Sc | LAPAN – Indonesia
- Dr. Ibnu Fathrio | LAPAN - Indonesia
- Dr. Laras Toersilowati |LAPAN - Indonesia
- Dr. Teguh Harjana |LAPAN - Indonesia
- Dr. Asif Awaluddin | LAPAN - Indonesia
- Suaydhi, M.Sc |LAPAN - Indonesia
- Drs. Sri Kaloka P | LAPAN - Indonesia
- Drs. Mahmud | LAPAN – Indonesia
- Dr. Nurjanna Joko Trilaksono S.Si., M.Si. | ITB - Indonesia
- Dr. Trismidianto | LAPAN - Indonesia

Photograph (Documentation)

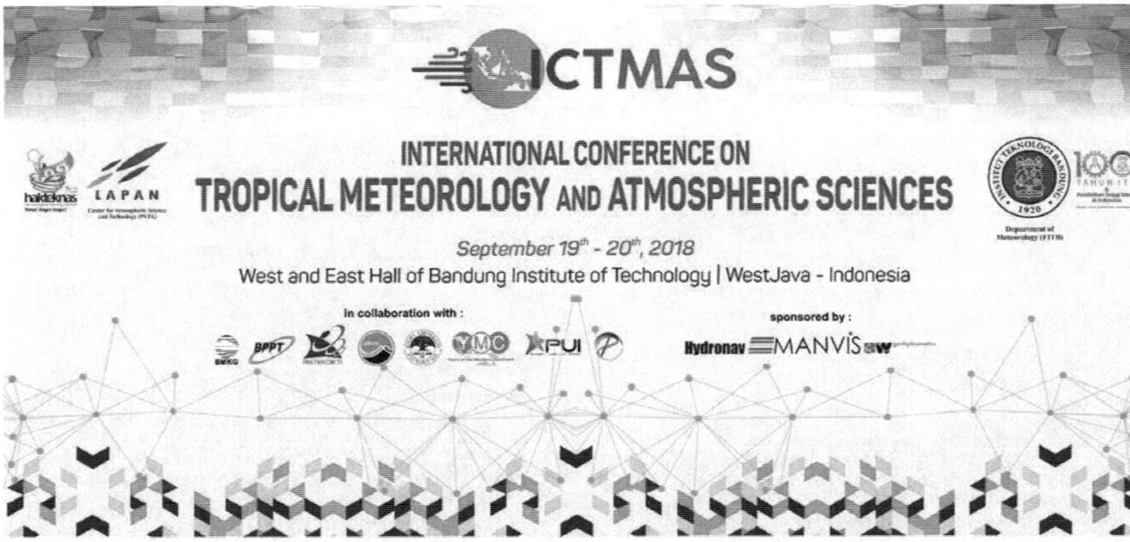

ICTMAS
INTERNATIONAL CONFERENCE ON
TROPICAL METEOROLOGY AND ATMOSPHERIC SCIENCES

ology | WestJava - Indonesia
supported by :
Hydronav MANVIS
AS

ICTMAS
INTERNATIONAL CONFERENCE ON
TROPICAL METEOROLOGY AND ATMOSPHERIC SCIENCES
ICTMAS

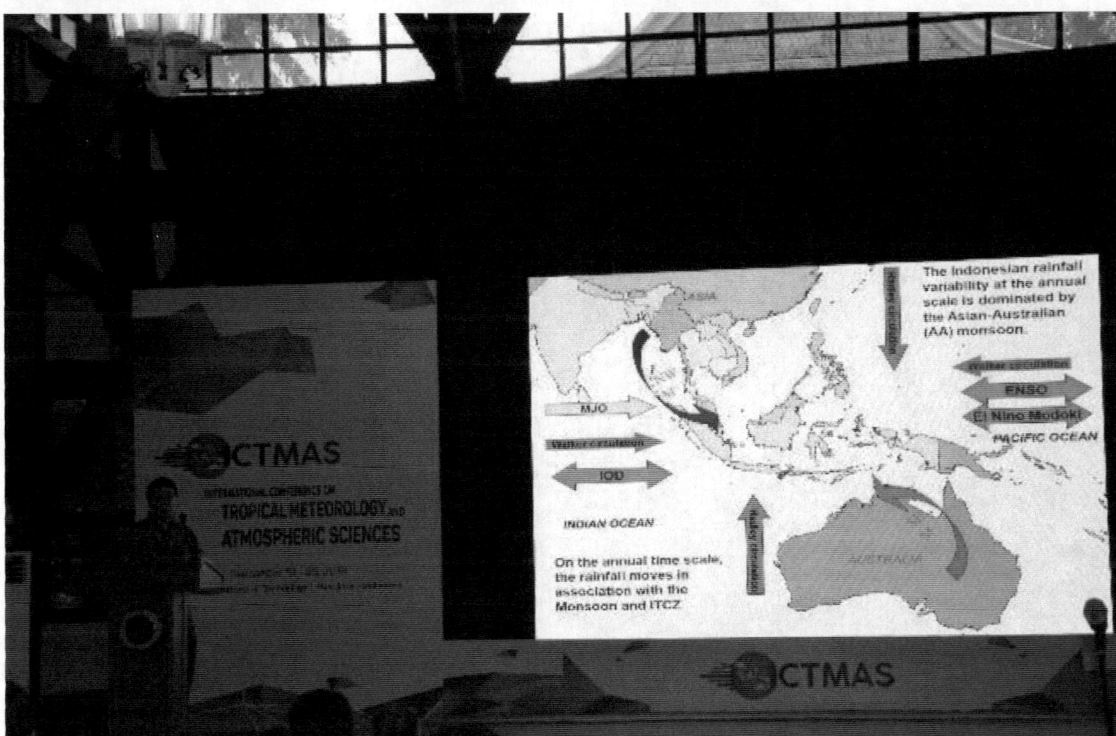
ICTMAS
INTERNATIONAL CONFERENCE ON
TROPICAL METEOROLOGY AND
ATMOSPHERIC SCIENCES
The Indonesian rainfall variability at the annual scale is dominated by the Asian-Australian (AA) monsoon.
ENSO
El Nino Modoki
PACIFIC OCEAN
MJO
IOD
INDIAN OCEAN
On the annual time scale, the rainfall moves in association with the Monsoon and ITCZ
ICTMAS

IOP Conf. Series: Earth and Environmental Science 303 (2019) 011005 doi:10.1088/1755-1315/303/1/011005

Indonesia-Australia-UK collaboration in the region of
Sumatra, Java, Lesser Sunda, and Christmas Islands
during October 2019 to February 2020 for YMC
Matthew Wheeler
CTMAS

A triggering factor of the eastward propagation of
MJO
observed in late October
during
DY2011
MOTEKI Qoosaku
CTMAS
TROPICAL METEOROLOGY AND
ATMOSPHERIC SCIENCES

IOP Conf. Series: Earth and Environmental Science 303 (2019) 011005 doi:10.1088/1755-1315/303/1/011005

IOP Conf. Series: Earth and Environmental Science **303** (2019) 011005 doi:10.1088/1755-1315/303/1/011005

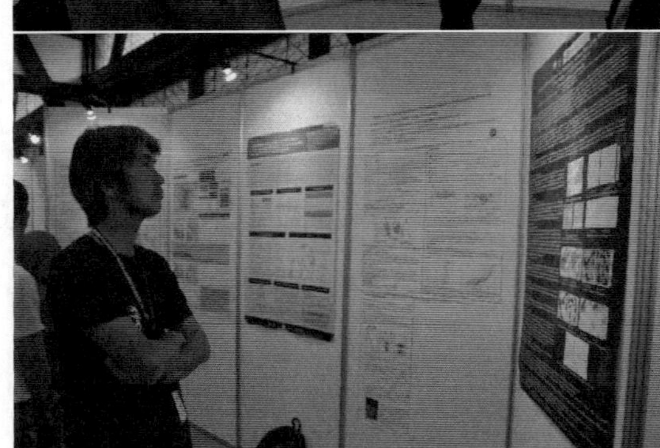

Welcome

Introduction - Background

Peer review statement

All papers published in this volume of *IOP Conference Series: Earth and Environmental Science* have been peer reviewed through processes administered by the proceedings Editors. Reviews were conducted by expert referees to the professional and scientific standards expected of a proceedings journal published by IOP Publishing.

Contrasting the evolution of moderate and extreme positive Indian Ocean Dipole events

A. Hermawanto[1,2] , Lixin Wu[1], and Mamenun[2]

[1]Physical Oceanography Laboratory/CIMST, Ocean University of China and Qingdao National Laboratory for Marine Science and Technology, Qingdao, China
[2]Center for Applied Climate, Indonesia Agency for Meteorology Climatology and Geophysics (BMKG), Jakarta, Indonesia.

andhika.hermawanto@bmkg.go.id

Abstract A recent study suggested, due to greenhouse warming, the frequency of extreme positive IOD (pIOD) events will increase in the future However, there is no information about the different character between extreme and moderate events During the period of 1979-2014, we identified 3 extreme and 4 moderate events Based on composite analysis of surface wind, sea surface temperature (SST) and temperature of the upper ocean, it is indicated there is a distinction on temporal and spatial evolution of equatorial easterly wind anomaly (EEWA) and the mean depth of 20°C isotherm (thermocline) over the eastern part of the equatorial Indian ocean The EEWA was started around June (October) and retreat in November for extreme (moderate) events The thermocline of extreme (moderate) events also showed an earlier shallowing, started in May (October) and reached the most shallow of depth around November Indicating that, during extreme events, the positive Bjerkness feedback strengthened earlier Using mixed layer heat budget analysis, it is revealed that during moderate events, the ocean dynamic process dominantly influences the cooling process of SST only in the peak phase, while during extreme events, the ocean dynamic process has dominant role in the whole process (from initial until peak phase) The early coupled process of EEWA and thermocline in shaping strong SST cooling gives time for us to monitor the developments of extreme pIOD and prepare mitigation against the impact at the peak phase However, the results presented in this study are based on limited samples from observations, thus requiring further clarification with numerical models and analysis of ensemble long integrations.

Keywords : pIOD, equatorial easterly wind anomaly, SST, thermocline.

1. Introduction

Indian Ocean Dipole is one of the dominant modes of climate over equatorial Indian ocean region, during the positive Indian Ocean Dipole (pIOD), the eastern part of the Indian ocean basin experienced negative anomalous rainfall, while the western experienced positive anomaly (Saji, et al., 1999) Indian Ocean Dipole does not affect Indian ocean basin country only but also gives serious impact on the climate over a region that is located far away from the basin (Saji & Yamagata, 2003; Hong, et al., 2008)

Figure 1 represents the distribution of anomalous rainfall during Sep-Oct-Nov (the peak phase of IOD) in the year of 2006, 1997, 1994 and 1982 We noticed that during 1994, 1997 and 2006, the area of negative rainfall anomaly over eastern equator Indian Ocean extends westward and reached on central equator Indian Ocean, while in 1982, the area of negative rainfall anomaly seems to be restricted in the eastern part of equatorial Indian ocean The area expansion of negative anomalous rainfall as one seen in 1997, is a character of rainfall during extreme pIOD (Cai et al., 2014) In a recent study by Cai et al., (2014), it was found that the number of extreme events will increase by threefold in the future as a consequence of greenhouse warming

In previous studies, the extreme pIOD events examined only as an individual case study (Webster, et al., 1999; Yu & Rienecker, 1999; Ueda, et al., 2000; Murtugudde, et al., 2000; Rao, et al., 2002; Rao, 2009; Vinayachandran, et al., 1999; Vinayachandran, et al., 2007), and some of those are the earliest studies that initiated the intensive research about IOD Since all of those studies discussed the extreme pIOD as an individual case, none of them explicitly stated the common features of extreme events Other studies which also examined the character of IOD based on many events, merely focused on the differences between positive and negative IOD (Hong, et al., 2008) or the different of pIOD with and without ENSO (Drbohlav, et al., 2007; Hong, et al., 2008; Yu, et al., 2005) or based on the term of peak time and the duration of IOD (Du, et al., 2013) or based on the strength of dipole mode index (DMI) (Saji, et al., 1999) None of them examined the different character based on the impact on rainfall during its peak phase

Annamalai et al., (2003) examined the factors that drive the initiation process of IOD, in the study, the events of moderate and extreme were grouped into one category, called strong IOD, thus the different character between moderate and extreme still have not discussed yet Cai et al., (2014) examined separately the moderate and extreme but did not examine in detail the differences in structure and time evolutions Therefore, the different character between extreme and moderate pIOD events is still an issue worth to discuss.

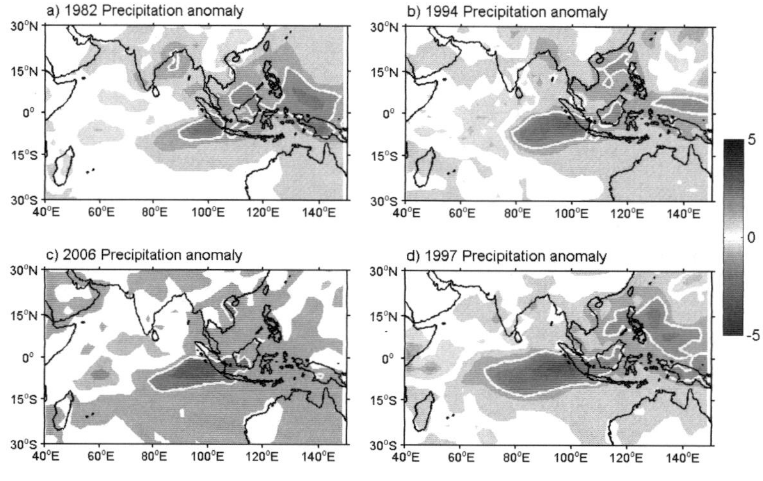

Figure 1 Precipitation anomaly (mm.day^{-1}) during peak phase of IOD (Sep-Oct-Nov) in (a) 1982, (b) 1994 (c) 2006 and (d) 1997 The year of 1982, categorized as moderate, while the years of 1994, 1997 and 2006 are categorized as extreme events Yellow enclosed contour indicates the area whose anomalous precipitation less than -1 mm.day^{-1}, to highlight the extension of significant negative rainfall anomaly during extreme events.

Since the extreme criterion in this study is based on the severity of the impact (the enhanced and vast coverage area of negative precipitation anomaly) over the eastern equatorial Indian Ocean, the influence of extreme pIOD over Indonesia as the region which is located just to the eastern equatorial Indian ocean is another point of interest In order to display the different severity of the events, using Indonesia rainfall stations observation data, the distribution of anomalous rainfall during moderate and extreme events are also demonstrated in this study

The main purpose of this study is to describe the differences in temporal evolution and structure of the two events based on the composite analysis We extend previous work by providing a comparison of all extreme and moderate events that ever happened (rather than a single case study) during 1981-2014 Therefore, from the result, we could identify the common features of the two pIOD categories Data and the composite method we used are described in section 2 In section 3, the distribution of rainfall over Indonesia is presented Evolution of the structure, analysis of heat budget and thermocline anomaly are presented in the next section Finally, the summary and discussion are presented in section 5.

2. Data and Methods

The monthly NCEP reanalysis (Kalnay et al., 1996) of wind circulation and NOAA Extended Reconstructed Sea Surface Temperature from 1948-2014 are used in this study Global Precipitation Climatology Project (GPCP) version 2.3 dataset provided by the NOAA ESRL (http://www.esrl.noaa.gov/psd) used to determine the extreme event through empirical orthogonal function (EOF) analysis In addition, we also utilized monthly rainfall data from the observed stations were provided by Indonesia Agency for Meteorology, Climatology, and Geophysics (BMKG) to show the evolution of rainfall anomaly distribution over Indonesia In order to diagnose the ocean process, we used ECMWF Ocean Reanalysis System (ORA-S3) (Balmaseda, et al., 2008) and Simple Ocean Data Assimilation (SODA 2.2.4) product of Carton, J.A and B Giese, (2008)

For anomaly calculation, the climatology is subtracted from the data series The climatology was obtained by averaging entirely available data time series, except for wind and sea surface temperature, we used the long-term monthly means of 1981-2010 dataset which available on NOAA ERSL website Nevertheless, the result does not varies much when we utilized the 1948-2013 average as the climatology.

In order to define the years of extreme events, we followed the method as used in Cai et al., (2014) by applying EOF analysis to precipitation dataset (GPCP) An extreme pIOD event is defined when the first principal component is greater than 1 s.d and the second principal component is greater than 0.5 s.d. For moderate event, we used Dipole Mode Index-DMI (Saji, et al., 1999) A moderate event was determined when the magnitude of DMI is greater than 0.75 s.d other than the extreme events From this method, we identified 1994, 1997 and 2006 as the extreme years, while 1982, 1991, 2002 and 2012 as the moderate We note that the sample size of our study is small as there are only a few extreme IOD during 1979-2013 Thus, we advise using discretion in interpreting the result from this composite analysis.

3. Distribution of anomalous rainfall over Indonesia associated with moderate and extreme events

The purpose of this section is to demonstrate the importance of studying extreme pIOD for practical implication In the previous study, As-syakur et al., (2014) examined the spatial-temporal relationship of rainfall in Indonesia during ENSO and IOD years using remote sensing dataset They found, in general (without separating the extreme and moderate) the temporal pattern relationship of rainfall with IOD and ENSO is high response during June-July-August (JJA) and September-October-November (SON) Consistent with As-syakur, A.R., et.al., (2014), as one can see in figure 2, the influence both of extreme and moderate pIOD to Indonesia rainfall during April-May is unclear (indicated with many no significant impact of rainfall) and to have significant impact started in June-July continue until August-September and October-November

During moderate years, Indonesia started to significantly experience negative anomaly of rainfall in June-July, meanwhile, for extreme events, it is started in August-September Both extreme and moderate events are terminated at the same time in November It means, time for Indonesia to experience negative rainfall anomaly is earlier and longer during moderate rather than in extreme years However, the magnitude of negative rainfall anomaly during extreme years are more severe than moderate as indicated more stations experienced negative anomalous rainfall less than -100 and -200 mm.month^{-1} (illustrated with more dots with brown and red color)

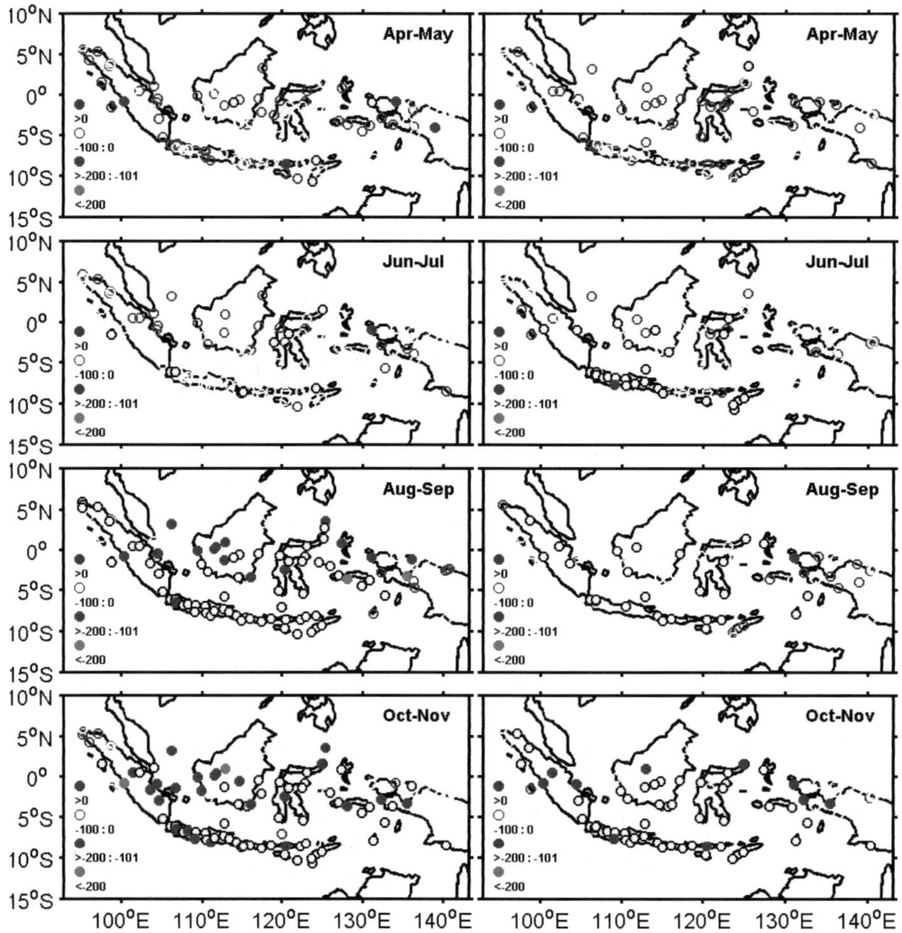

Figure 2 Composite of bimonthly rainfall anomaly during (left) extreme [1994, 1997, 2006] and (right) moderate [1982, 1991, 2002, 2012] events Filled (open) circle indicate anomalous precipitation significant (not significant) at 95% The unit is in mm.month^{-1}.

Figure 2 gives information that between moderate and extreme years, the distribution of rainfall is different Firstly, different in time to start experienced significant negative anomaly of rainfall and secondly, different in the magnitude of negative anomaly of rainfall The occurring of significantly negative rainfall anomaly in August until November gives consequences that Indonesia will experience a longer and more severe dry season than the climatology, while extreme or moderate pIOD occurred.

4. Contrasting the evolution of extreme and moderate pIOD

4.1. Wind and Sea Surface Temperature Anomaly (SSTA) pattern

In this section, we started to examine the different features of evolution between moderate and extreme pIOD derived from a composite of surface wind and sea surface temperature anomaly (figure 3 and 4) By contrasting the evolution, we identified some different features between the two events.

IOP Conf. Series: Earth and Environmental Science **303** (2019) 012001 doi:10.1088/1755-1315/303/1/012001

Strong equatorial easterlies wind anomaly (EEWA) appeared over the central Indian ocean which was started by Jun-Jul and persistently occurred until Oct-Nov during extreme events (figure 3.i) Unlike the extreme, on moderate events, the strong EEWA appeared only in Oct-Nov period (figure 3.h) This means that there was an earlier appearance of strong EEWA during extreme events In addition, the magnitude of EEWA for extreme events are stronger than moderate throughout the events (from initial to peak phase) (figure 3.a,b,c and i), while during moderate, strong EEWA only occurred during the peak phase (figure 3.d,e,f,g, and h) The coverage area of EEWA also showed something different, the EEWA extent crossing 70 °E during extreme events (figure 4.h), while during moderate, the EEWA was restricted to central of Indian ocean only (figure 4.d).

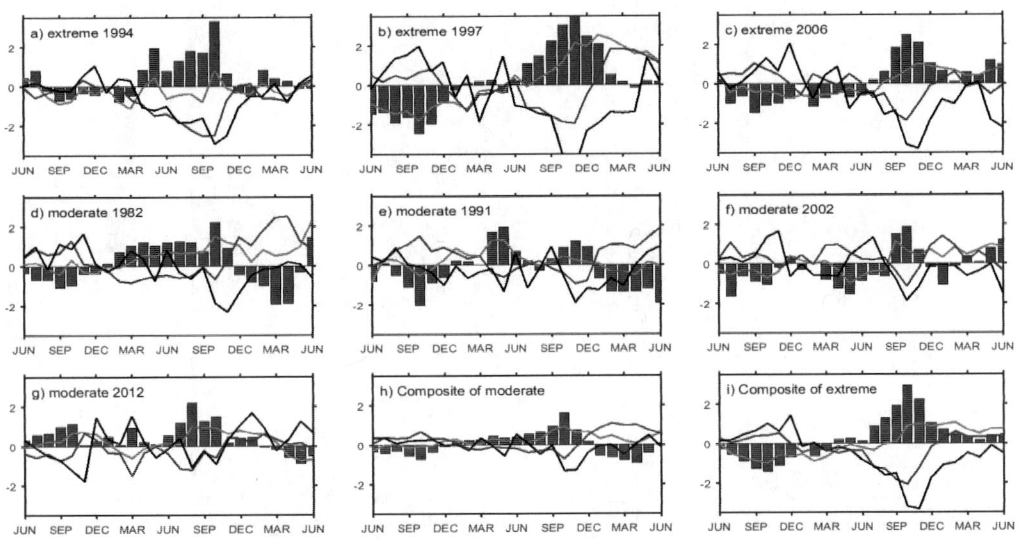

Figure 3 Time evolution of normalized DMI (index bar), a representation of zonal gradient of SST between west (IOD-W; 50°E-70°E,10°S-10°N) and east (IOD-E; 90°E-110°E,10°S-0°N) pool of Indian ocean Time evolution of normalized SSTA over IOD-W (red), over IOD-E (blue), normalized zonal wind anomaly over central Indian ocean (70°E-90°E,5°N-5°S) (black full line) (a)-(c) indicate the time evolution of extreme IOD years [1994, 1997 and 2006], (d)-(g) indicate moderate IOD years [1982, 1991, 2002, 2012], (h) the composite of all moderate events and (i) composite of extreme events.

In line with wind circulation, the spatial evolution of SSTA also showed something different During extreme years, cooling sea surface temperature over the eastern pole of Indian ocean started to active from June-July (preceded by basin-wide cooling during April-May), then continued to strengthened and extended westward along the coast of Sumatera during August-September, together with the strengthening of EEWA This cooling of SSTA reached its maximum in October-November (figure 4.h) However, during moderate events, the evolution of cooling SSTA in June-July and August-September (development phase) over eastern equatorial Indian ocean, looks there are no significant changes and then started to strengthen in October-November (peak phase)

In the other side, the basin warming over central and west of Indian ocean during moderate events looks more dominant and covered up a larger area than extreme Meanwhile during extreme events, weak warming occurred in central and west of Indian ocean It may be caused by the absence of unusual strong cold zonal advection in the initial phase (Sun, et al., 2014)

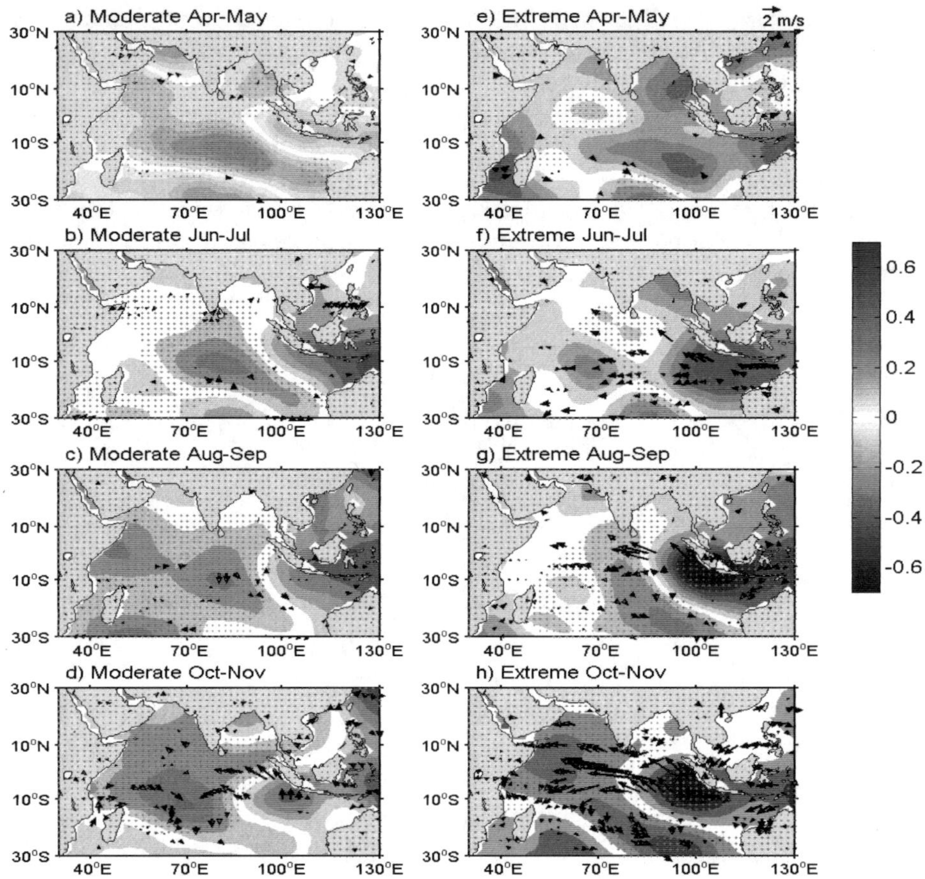

Figure 4 Composite map of bimonthly average (June-July; Augustus-September; October-November) SSTA (shading; unit: °C) and surface winds anomaly (arrow; unit: ms^{-1}) during moderate events Years of moderate pIOD are 1982, 1991, 2002, 2012 Years of extreme events are 1994, 1997 and 2006 Only the wind anomalies significant at 95% are plotted.

4.2. The ocean dynamics process of cooling SSTA over IOD-E

Since the extreme events likely more controlled by the appearance of cooling SSTA in the east pole of equatorial Indian ocean, further in this section we examined the role of the advection and surface heat fluxes for the formation of SSTA over IOD-E For this reason, a mixed layer heat budget analysis on the top 50 m depth over IOD-E was carried out The mixed layer heat balance of IOD-E can be expressed as:

$$\frac{\partial T^a}{\partial t} = -\left[\left(\frac{u^a \partial T^a}{\partial x} + \frac{\bar{u} \partial T^a}{\partial x} + \frac{u^a \partial \bar{T}}{\partial x}\right) + \left(\frac{v^a \partial T^a}{\partial y} + \frac{\bar{v} \partial T^a}{\partial y} + \frac{v^a \partial \bar{T}}{\partial y}\right) + \left(\frac{w^a \partial T^a}{\partial z} + \frac{\bar{w} \partial T^a}{\partial z} + \frac{w^a \partial \bar{T}}{\partial z}\right)\right] + \boldsymbol{Q_{SW}} + \boldsymbol{Q_{LW}} + \boldsymbol{Q_{LH}} + \boldsymbol{Q_{SH}} + \boldsymbol{R} \quad (1)$$

The variables $\boldsymbol{T}, \boldsymbol{u}, \boldsymbol{v}$ and $\boldsymbol{w}$ are potential temperature, the zonal, meridional and vertical ocean current velocities, respectively, averaged over the top 50 m of the ocean Differential operators $\boldsymbol{x}, \boldsymbol{y}, \boldsymbol{z}$ and $\boldsymbol{t}$, are along the zonal, meridional, vertical directions and time, respectively The temperature tendency ($\frac{\partial T^a}{\partial t}$) was calculated using a centered-difference approximation Superscript 'a' and overbar denote anomalous and long-term averaged quantities, respectively $\boldsymbol{Q_{SW}}, \boldsymbol{Q_{LW}}, \boldsymbol{Q_{LH}}$ and $\boldsymbol{Q_{SH}}$ represent the net downward shortwave radiation flux at the ocean surface, net upward longwave radiation flux, surface latent heat flux and sensible heat fluxes, and the sum of the fluxes will be called as heat flux ***R*** is the residual term representing either the model errors or a process other than heat fluxes and temperature advection

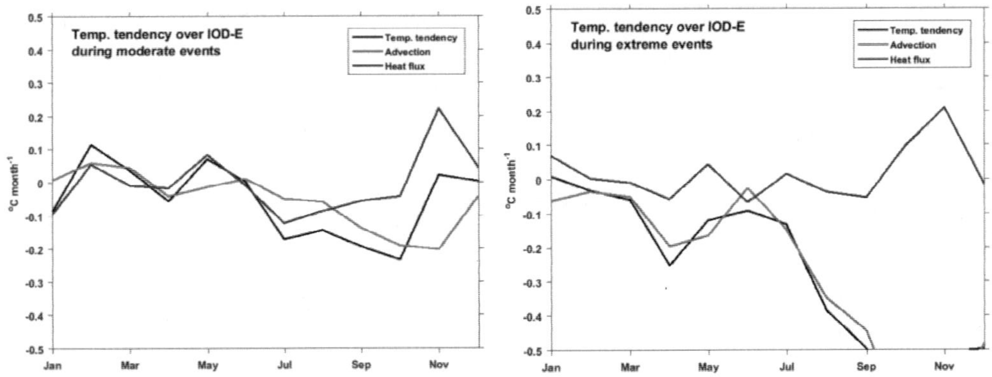

Figure 5 Evolution of mixed layer temperature tendency (black line) over east pole Indian ocean-IOD-E [90-110°E; 0-10°S] during moderate (left) and extreme (right) events Red and blue line denotes total advection (horizontal and vertical) and heat fluxes, respectively.

During moderate years, as indicated in figure 5, total advection and heat flux process, both gave significant role in the form of temperature tendency during the initial and developing phase, but during the peak phase, total advection process acted as a dominant factor Different from moderate, during extreme events, the total advection process acted as a dominant factor throughout all the phases, from initial to peak phase Furthermore, we decipher the total advection into zonal and meridional advection components By decomposing the total advection, one can find out how each of the advection components shaping the temperature tendencies

In figure 6.b, around April and May, we noticed there was an anomalous zonal advection of anomalous ocean temperature during extreme years which is stronger than moderate This can be thought that there was a strong transport of anomalous heat by the anomalous ocean current Since there were no significant wind anomalies in April and May (figure 4.e), we consider the anomalous zonal advection was driven by ocean dynamic process and was not driven by anomalous wind-driven current This matter will be discussed in the next section

Furthermore, we gave attention to peak phase in figure 6.a, one can see the magnitude of zonal advection for moderate years (illustrated in black dashed line and magenta full line), showing strong magnitude as it does in the extreme years (red, green and blue full line) But, something different was seen in meridional advection (figure 6.d), the meridional advection looks stronger than moderate, especially on the mean meridional advection of the anomalous ocean temperature component (figure 6.f)

The conditions above bring information to us that, in the development process of extreme cooling SST, which is noted as the character of extreme pIOD, it is needed not only strong zonal advection but also a combination between strong zonal and strong meridional advection If there is strong zonal (without strong meridional) advection, the pIOD only reached the moderate, but if there is strong zonal combined with strong meridional advection, the pIOD will be favor into extreme condition This situation is consistent with Weller & Cai (2014) who said that in combination with the zonal structure of temperature gradient, an anomalous meridional cross-equatorial temperature gradient in the east Indian ocean, controls the declining rainfall over the eastern equatorial Indian Ocean region

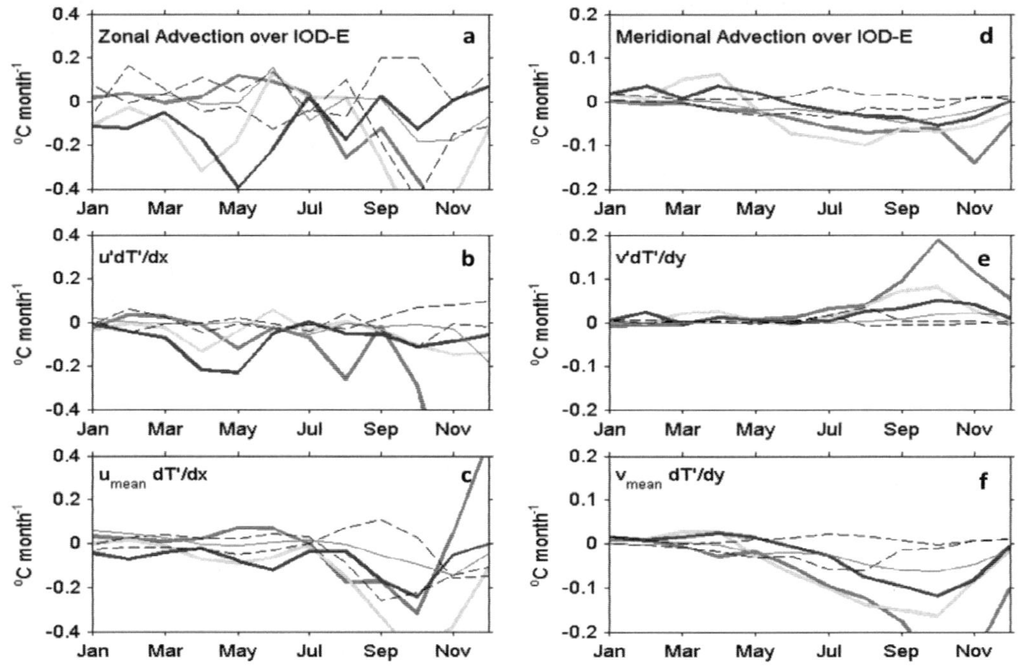

Figure 6 Evolution of zonal (left) and meridional (right) temperature advection of top 50 m depth over IOD-E (90°-110°E; 10°S-equator) during the years of pIOD events Green, red, blue and magenta denotes year of 1994, 1997, 2006 and 1982, respectively Black dashed line indicates the other pIOD moderate events Give attention to the different ordinate scale between zonal and meridional.

4.3 The thermocline (20oC isotherm depth) anomaly over IOD-E

In the previous section, we noticed that during extreme years the total advection component dominates in controlling the temperature tendency over IOD-E throughout the phases, from initial until peak Here we contrasted the condition of ocean subsurface over IOD-E during the events

Composite of 20°C isotherm depth from extreme and moderate events in figure 7 illustrates the depth of thermocline during extreme and moderate years During moderate years, the mean of thermocline depth appears to shallow (negative anomaly) in peak phase only, while during extreme years, the mean of thermocline depth started to shallow from April (initial phase) then continued and reached the maximum in November (peak phase) and finally went back to normal condition in February This situation is consistent with the evolution of wind and SSTA for extreme and moderate years (figure 4) in the previous section

If one gives attention to March, the anomalous thermocline depth during extreme and moderate events is relatively not different (in neutral position), but then in April-May, the depth of thermocline shows different progress, the thermocline going to shallow during extreme events, but it does not for moderate This situation is consistent with the appearance of basin-wide cooling of SST over east equatorial Indian ocean in April-May as indicated in figure 4.e The shallow thermocline in initial phase as indicated in figure 7, also mentioned by a previous study (Horii, et al., 2008) The study found that during May 2006, by using mooring buoy dataset, there was an appearance of negative temperature anomaly at the thermocline depth accompanied with strong westward current anomalies

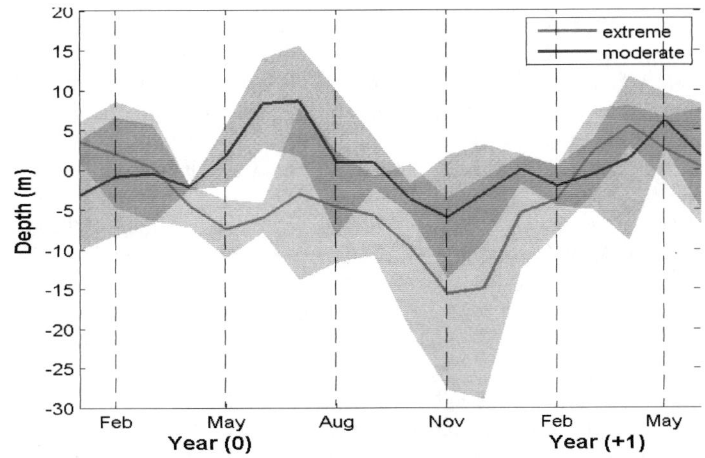

Figure 7 Mean of longitude-averaged (80°E-100°E) of 20°C isotherm depth anomaly over equator during moderate and extreme pIOD events Shading indicates the standard deviation.

5. Summary and Discussion

Motivated by Cai, et al., (2014), who stated the frequency of extreme pIOD events will increase in the future, through this study we examined and compared the spatial and temporal evolution of the structure of pIOD between extreme and moderate events The response of rainfall in Indonesia, as the region located in Indian ocean equator that experienced negative anomalous rainfall, also examined

For the anomalous rainfall over Indonesia, the influence of extreme and moderate pIOD during April-May is unclear Furthermore, during moderate pIOD, Indonesia experienced deficit rainfall earlier (June-July) than extreme events (Augustus-September) However, if we look at the severity level, rainfall during moderate events are less severe than extreme years

For the wind circulation, based on composite analysis, it indicates that there is an earlier strengthening of EEWA during the development phase of extreme pIOD This EEWA persistently occurs until peak phase, this condition is not seen during moderate events The depth of thermocline also showed earlier shallowing during the initial phase of extreme events

The early appearance of strong EEWA and shallowing of thermocline, made the Bjerkness positive feedback strengthened earlier as well, and in the end facilitated the stronger positive feedback during the next phase (peak) We expected that these are the reason why SST cooling in equatorial eastern Indian Ocean becomes stronger in extreme events than moderate.

The results presented in this study are based on limited samples from observations, due to the least number of extreme and moderate pIOD events that occurred during satellite era (1979-present), thus we realized still requiring further clarification with numerical models and analysis of ensemble long integrations.

References

[1] Annamalai H, Murtugudde R, Potemra J, Xie S P, Liu P, Wang B and Nin E 2003 Coupled dynamics over the Indian Ocean: Spring initiation of the Zonal Mode *Deep Sea Research* **II** 50 2305–2330 https://doi.org/10.1016/S0967-0645(03)00058-4

[2] As-syakur A R, Adnyana I W S, Mahendra M S, Arthana I W, Merit I N, Kasa I W and Sunarta I N 2014 Observation of spatial patterns on the rainfall response to ENSO and IOD over Indonesia using TRMM Multisatellite Precipitation Analysis (TMPA) *International Journal of Climatology* 34(**15**) 3825–3839 https://doi.org/10.1002/joc.3939

[3] Balmaseda M A, Vidard A and D Anderson 2008 The ECMWF Ocean Analysis System : ORA-S3 *Mon. Weather Rev.* 136 30188-3034

[4] Carton J A and B S Giese 2008 A Reanalysis of Ocean Climate Using Simple Ocean Data Assimilation (SODA) *Mon. Weather Rev.* 136 2999-3017

[5] Cai W, Santoso A, Wang G, Weller E, Wu L, Ashok K, Masumoto Y and Yamagata T 2014 Increased frequency of extreme Indian Ocean Dipole events due to greenhouse warming Nature **510** (7504) 254–258 https://doi.org/10.1038/nature13327

[6] Du Y, Cai W and Wu Y 2013 A new type of the indian ocean dipole since the mid-1970s Journal of Climate **26**(3) 959–972 https://doi.org/10.1175/JCLI-D-12-00047.1

[7] Hiroaki Ueda and Jun Matsumoto 2000 A Possible Triggering Process of East-West Asymmetric Anomalies over the Indian Ocean in Relation to 1997/98 El Niño *Journal of the Meteorological Society of Japan* Ser II 78(**6**) 803–818

[8] Hong C-C, Lu M-M and Kanamitsu M 2008 Temporal and spatial characteristics of positive and negative Indian Ocean dipole with and without ENSO *Journal of Geophysical Research* **113**(D8) 1–15 https://doi.org/10.1029/2007JD009151

[9] Horii T, Hase H, Ueki I and Masumoto Y 2008 Oceanic precondition and evolution of the 2006 Indian Ocean dipole *Geophysical Research Letters* **35**(3) 1–6 https://doi.org/10.1029/2007GL032464

[10] Kalnay E, Kanamitsu M, Kistler R, Collins W, Deaven D, Gandin L, Iredell M, Saha S, White G, Woollen J, Zhu Y, Chelliah M, Ebisuzaki W, Higgins W, Janowiak J, Mo K C, Ropelewski C, Wang J, Leetmaa A, Reynolds R, Jenne Roy and Joseph D 1996 The NCEP/NCAR 40-year reanalysis project *Bull Amer Meteor Soc.* **77** 437-470

[11] Lee Drbohlav H K, Gualdi S and Navarra A 2007 A diagnostic study of the Indian Ocean dipole mode in El Nino and non-El Nino years *Journal of Climate* 20(13) 2961–2977 https://doi.org/10.1175/JCLI4153.1

[12] Murtugudde R, McCreary J P and Busalacchi A J 2000 Oceanic processes associated with anomalous events in the Indian Ocean with relevance to 1997–1998 *Journal of Geophysical Research* **105**(C2) 3295 https://doi.org/10.1029/1999JC900294

[13] Rao S A, Gopalakrishna Vissa V, Shetye S R and Yamagata T 2002 Why were cool SST anomalies absent in the Bay of Bengal during the 1997 Indian Ocean Dipole Event? *Geophysical Research Letters* 29(11) 5–8 https://doi.org/10.1029/2001GL014645

[14] Rao S A, Luo J J, Behera S K and Yamagata T 2009 Generation and termination of Indian Ocean Dipole events in 2003, 2006 and 2007 *Climate Dynamics* 33(6) 751–767 https://doi.org/10.1007/s00382-008-0498-z

[15] Saji N H, Goswami B N, Vinayachandran P N and Yamagata T 1999 A dipole mode in the tropical Indian Ocean *Nature* **401**(6751) 360–363 https://doi.org/10.1038/43854

[16] Saji N H and Yamagata T 2003 Possible impacts of Indian Ocean Dipole mode events on global climate *Climate Research* **25**(2) 151–169 https://doi.org/10.3354/cr025151

[17] Vinayachandran P N, Kurian J dan Neema C P 2007 Indian Ocean response to anomalous conditions in 2006 *Geophysical Research Letters* **34**(15) 1–6 https://doi.org/10.1029/2007GL030194

[18] Vinayachandran P N, Saji N H and Yamagata T 1999 Response of the Equatorial Indian Ocean to an unusual wind event during 1994 *Geophysical Research Letters* 26(11) 1613–1616 https://doi.org/10.1029/1999GL900179

[19] Webster P J, Moore A M, Loschnigg J P and Leben R R 1999 Coupled ocean-atmosphere dynamics in the Indian Ocean during 1997-1998 *Nature* **401**(6751) 356–360

[20] Yu L and Rienecker M M 1999 Mechanisms for the Indian Ocean warming during the 1997–98 El Niño *Geophysical Research Letters* **26**(6) 735 https://doi.org/10.1029/1999GL900072

[21] Yu W, Xiang B, Liu L and Liu N 2005 Understanding the origins of interannual thermocline variations in the tropical Indian Ocean *Geophysical Research Letters* **32**(24) 1–4 https://doi.org/10.1029/2005GL024327

Thermal front condition through El Niño and Indonesian throughflow phase in southern sea of East Java and Bali on the east monsoon

A L Ahmad[1], M L Syamsuddin[1], N P Purba[2] and Sunarto[1]

[1] Departement of Marine Science, Faculty of Fisheries and Marine Science, Padjadjaran University, Jatinangor, Indonesia
[2] Marine Research Laboratory (MEAL), Padjadjaran University, Jatinangor, Indonesia

libiatyahmad@gmail.com

Abstract. The purpose of this study was to determine the effect of El Niño 2013 and 2015 in the formation of thermal fronts in the southern sea of East Java and Bali and its correlation with Indonesian Throughflow (ITF). The data that is used in this research were the SST from Aqua MODIS, Nino index of 3.4 derived from CPC NOAA, and Geostrophic current from AVISO. The used method was descriptive method with spatial and temporal analysis approach. Front can be detected in the data of SST raster image using a Cayulla Cornilon SPL 1992 algorithm with strong category of SST difference of >= 0.5 ^{O}C and with weak one of SST difference of 0.3 – 0.49 ^{O}C. The results show that thermalfront formed in the southern water of Java and Bali was the temporary front with weak and strong scale. The presence of El Niño affected the formation of thermal fronts, where it was formed during El Niño more than in normal states and El Niño affected transport ITF as well. It is because upwelling in the El Niño is stronger than at the normal condition.

Keywords : thermal front, El Niño, Indonesian Throughflow, monsoon.

1. Introduction

Southern sea of East Java and Bali is very complex because there are many different phenomena such as ITF [1,4,24], South Equatorial Current [26], Java Coastal Current [1], and Monsoon [29]. The ITF flows from the tropical Pacific Ocean to Indian Ocean via Indonesia through the Makassar Strait then out through the Lombok Strait into the Indian Ocean [4]. This water meets the water mass around java (Java Current) and results thermal fronts because of the different characteristics [3,14].

According to [9,15], *fronts* are found in the southern sea of East Java and Bali on the East Monsoon (EM). *Thermal front* is one of Oceanography phenomena that can be identified by observing the distribution patterns of SST [9]. The studies on the *front* have previously been done by several researchers such as [5,9,27], but only several studies that connected with El Niño and ITF conditions.

The changes of SST in the southern sea of East Java and Bali are influenced by the phenomenon of El Niño Southern Oscillation (ENSO) [18,20,25]. This phenomenon affects SST during El Niño or during La Niña [17]. According to [21,28] the sea temperature can be influenced by the season, the regional seawater circulation process such as warm water mass that flows from the Pacific Ocean to the Indian Ocean passing through parts of Indonesia, such as the Java Sea, and also from the natural

IOP Conf. Series: Earth and Environmental Science **303** (2019) 012002 doi:10.1088/1755-1315/303/1/012002

phenomena of ENSO. This phenomenon is a condition where there is an anomaly of SST around the central and eastern Pacific along the equator (around the coast of Peru) [23] characterized by sea surface temperature anomaly in the Niño 3.4 region (figure 1) with the changes of approximately 0.5 ° C [12].

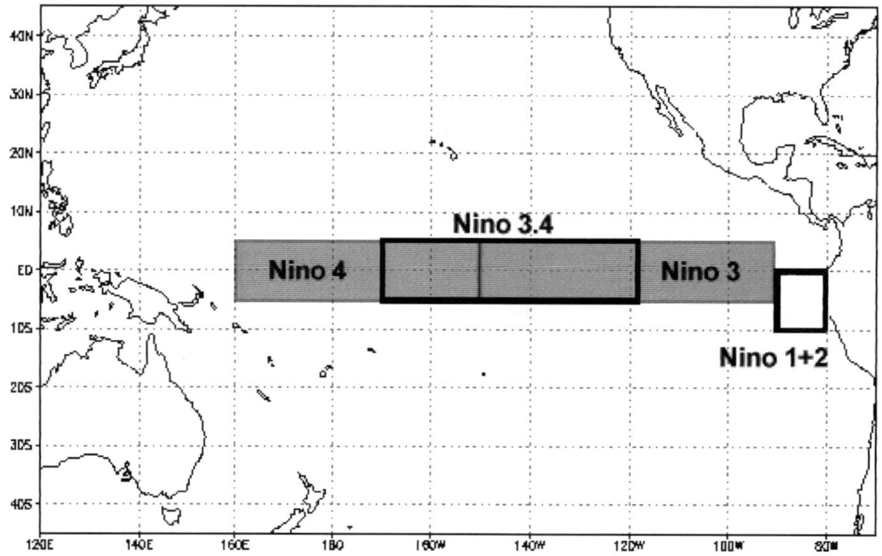

Figure 1. Niño Regions [12].

This sea surface temperature anomaly has been occurring since 1950 until now according research that conducted by [10]. It is known that the strongest ENSO period occurred in 1997 and according to the information provided by Indonesian Agency for Meteorology, Climatology and Geophysics (BMKG), the strongest ENSO also occurred in 2015. According to [13,22], it is known that ENSO had a moderate effect toward southern water of East Java and Bali during both El Niño and La Niña. These results are reinforced by research that have been done by [20] which states that El Niño causing SST of southern Java tends to be cooler. This study used the conditions in 2013 and 2015 to compare the conditions of the strongest and the weakest.

The contribution of this study was to determining the effect of El Niño in thermal fronts and ITF in the southern sea of East Java and Bali. This study focused on the effect on the number, location and the strength of *thermal front* formation in the southern sea of East Java and Bali during the period of El Niño and its correlation with ITF which occurred on the east monsoon in 2015.

2. Material and method

The region of this study was at the southern sea of East Java and Bali on the coordinate of 7°S - 16°S and 111°E - 117°E shown in figure 2 [19]. This area was an area where there was a meeting between the flow of ITF and the flow of AKS [26], there was an upwelling location [20], and the presence of boundaries with subtropical water masses [16].

The data used in this study were the data in 2013 and 2015. They were SST obtained from AquaMODIS satellites (http://oceancolor.gsfc.nasa.gov/) with a resolution of 4 km to see the distribution and the average of SST and to detect thermal front [5], Nino 3.4 index obtained from the website of CPC NOAA (http://www.cpc.ncep.noaa.gov) to see the SST anomaly [20], and the geostrophic current data obtained from AVISO website (http://motu.aviso.altimetry.fr/) to see the distribution of the current velocity [9]. The data had previously been validated by using the standard deviation.

To see the El Niño force, it can be known by observing the SST anomalies, with stages:
SST Anomalies of 0.5 - 1°C = weak,
SST Anomalies 1 - 1.5°C = moderate, and
SST Anomalies over 1.5°C = strong [2].

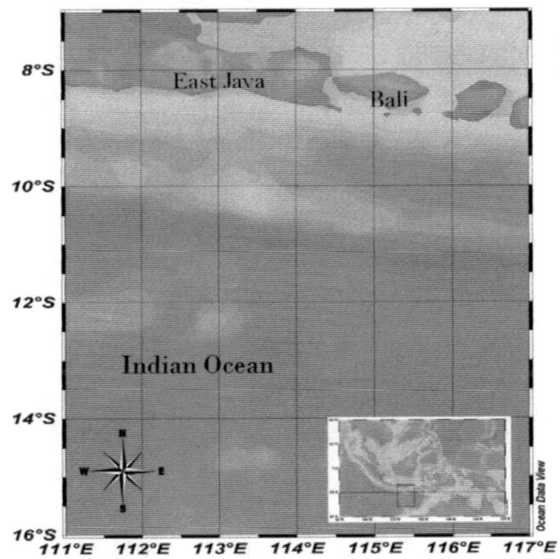

Figure 2. Research Area.

Spatial data processing to detect thermal fronts was using the Algorithms of Single Image Edge Detection (SIED) Cayulla Cornillon in 1992 and was divided into two categories [5] namely:

- Strong Front, formed due to the ST difference of > = 0.5 OC and
- Weak Front, formed due to the ST difference between 0.3 – 0.49 °C

The results of the analysis were compared to the normal state on the east monsoon in 2013.

3. Results

3.1 SST anomalies

The SST anomalies obtained from CPC NOAA were used to see the SST deviations within one month and El Niño force in that month. It can be known in table 1.

From the table above, can be seen that the El Niño force of EM in 2015 has moderate - strong force with a relatively homogeneous SST average. The SST average, when El Niño was lower than normal state, was 27.6 °C in June, 26.6 ° C in July and 26.2 ° C in August.

Table 1. The Anomaly and Average of SST (°C).

Month	Anomaly	Category	SST Average (°c)
June 2015	0.97	Moderate	27,6
July 2015	1.2	Moderate	26,6
August 2015	1.51	Strong	26,2
June 2013	-0.25	Normal	28,6
July 2013	-0.21	Normal	27,4
August 2013	-0.18	Normal	26,7

3.2 The number and size of thermal fronts

The number and the size of thermal fronts which formed in the southern sea of East Java and Bali had different numbers and sizes in each month both in normal state and in El Niño phase (table 2).

Table 2. Quantity and Large of Thermal Front.

Month and Year	Quantity of Thermal Front			
	Strong	Weak	Max (km^2)	Min (km^2)
June 2015	7	9	3640,47	49,28
July 2015	8	11	3138,99	49,28
August 2015	7	14	6738,71	49,28
June 2013	14	-	5119,95	49,28
July 2013	-	12	2629,13	49,28
August 2013	8	10	2880,59	49,28

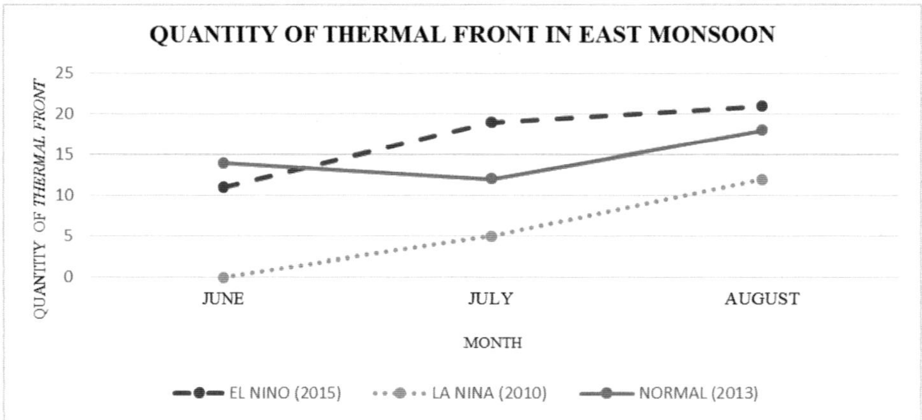

Figure 3. Quantity of Thermal Front.

It can be known in table 2 and figure 2 that thermal fronts formed in the El Niño phase have more amounts with the biggest number of 21 in August 2015 and the largest size of 6738.71 km^2 area in August 2015 compared to during normal states.

3.3 Spatial distribution of thermal fronts and currents

Thermal fronts in the southern sea of East Java and Bali were temporary fronts whose location and area changed in a certain period of time. The spatial distribution in the El Niño phase can be seen in figure 3 and in normal states it can be seen in figure 4.

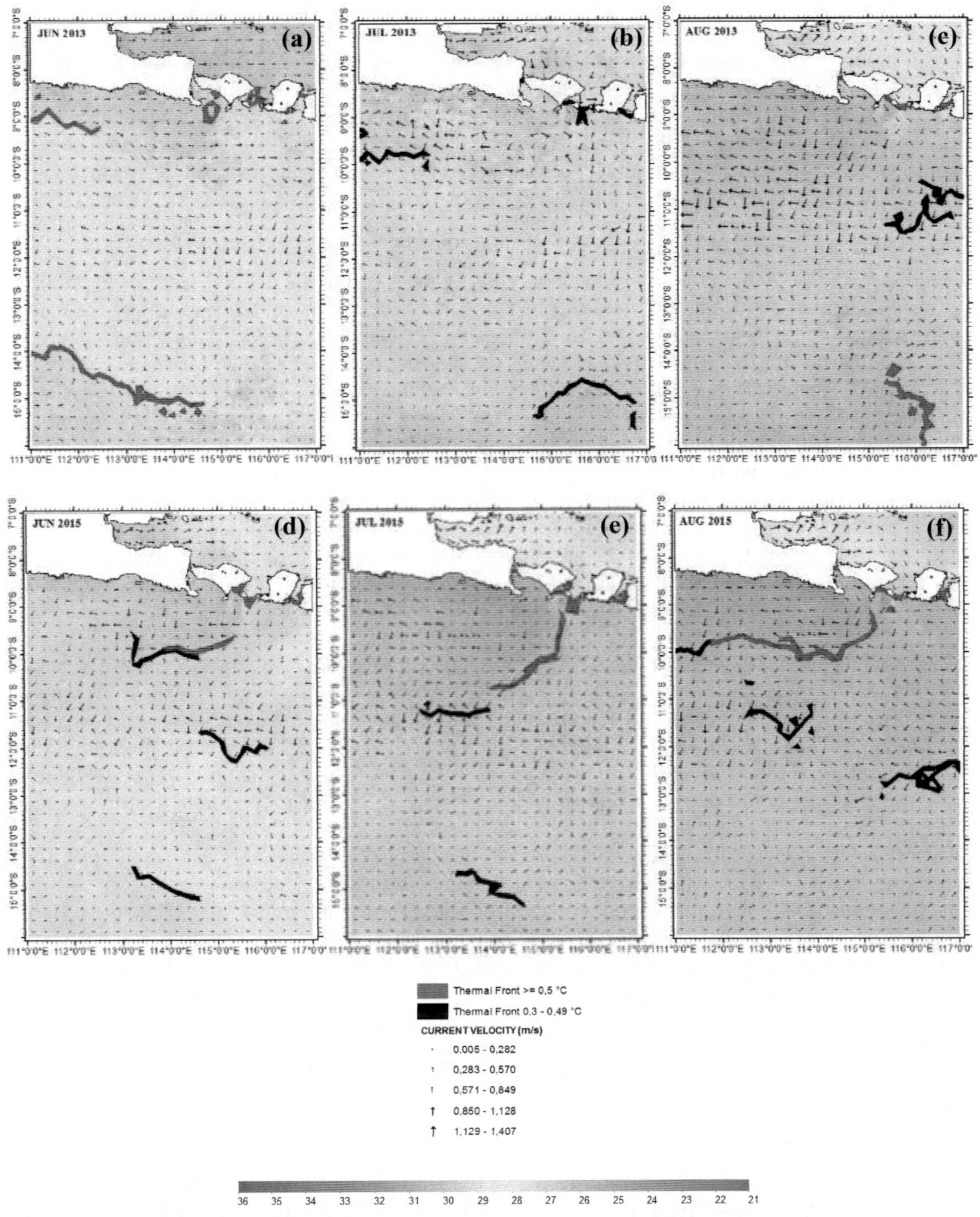

Figure 4. Spatial distribution of Current and Thermal Front (a) - (c) in normal El condition and (d) – (e) in El Niño Condition.

In six figures above, it can be known that in the area of Lombok Strait and Bali strait thermal fronts are formed due to entering flow from the northern Java and Bali. Other fronts are scattered in different areas varied over each month.

4. Discussion

The southern sea of Java and Bali in normal state in June, July, and August (Table 1) has SST average of 27.6°C and it is relatively homogeneous in each month, 28.6° C in June, 27.4° C in July, and 26.7° C in August. This corresponds to the research that has been done [8] that SST of the sea ranged between 28° C - 31° C. The homogenous SST states based on the warming rays of the sun which is at around the earth's equator [6] and because the geography of southern SST of Java and Bali was very influenced by the monsoon system so that SST is relatively higher in West Monsoon rather than EM. Besides, the SST in this area affected by some oceanographic phenomena such as ENSO and IOD (Indian Ocean Dipole) [20].

In Table 1 that the SST average of the El Niño phase had a lower value than that of the normal state. This decrease was due to the fact that El Niño was an aberration when the Indonesian SST at the time was lower than during normal states. Moreover, the decrease triggered by low SST in EM because how much the impact of El Niño depends on the conditions of SST in each region [7]. In the phase of El Niño and normal front was mostly formed in August. This caused August is the month in which the Upwelling peak occurs so that there are thermal fronts in this phase established more. These results were consistent to the statement of [20] which states that the years of 1997-2009 in August and September are Upwelling peak due to the existence of EM blowing in that month.

Many strong front forms in the El Niño period are because this phenomenon affects the oceanographic characteristics of southern Java [24]. One of them led to the high intensity of upwelling which can cause the formation of thermal fronts. According to [20], the existence of Upwelling in the southern Java affected by the positive IOD and El Niño because during this phase the mass of water at low temperatures entered more so that it strengthens the intensity of upwelling in the southern waters of East Java and Bali at the time of the SST formation had lower ones.

At the time of EM phenomenon in June, July, and August, thermal fronts formed more in the region of 8°Lat - 11 °Lat both in normal and El Niño states. In normal states, the fronts formed during the normal season are pretty much in each month. Because EM was a season that had a high intensity of ITF so that fronts in the region of influx ITF formed more and more powerful. This was consistent with the statement [9] which states that in normal condition, EM had the highest strength than any other season. According to [18] when EM current flows from east to west is in the direction of the ARLINDO transport direction from the east (Pacific) to the west (Hindia), so that it can strengthen the surface currents and the current transport flow.

During the EM with El Niño, more thermal fronts formed in the study area than when the EM in normal condition. It can be seen in Figure 3 in El Niño conditions, the number of fronts formed more than in the normal season. In normal condition, the fronts are formed around 14 to 18 while the El Niño fronts are formed around 11-21. According to [20] at the time of El Niño occurrence, there was a decrease in the volume of water moving from the Pacific Ocean to the Indian Ocean. The vacuum of the water mass will encourage the emergence of Upwelling, namely the increase in the mass of deep-sea water with the characteristics of low temperature, high salinity and rich in nutrients. According to [11] at the time of El Niño occurrence of weakened water mass transport than in normal and La Niña conditions and according to research [18] when El Niño variability of ITF transport decreased by -0.13 Sv.

5. Conclusion

Thermal fronts in the southern sea of East Java and Bali formed more during El Niño. The El Niño phenomenon affected the water mass in these waters where SST in El Niño became lower than in normal state so that it strengthens the intensity of existed upwelling having its peak in EM. Besides, the increase of Upwelling was triggered by the weakening of ITF transport causing the vacuum of water in the Indian Ocean.

Acknowledgments
The Authors would like to thanks to Padjadjaran University especially Marine Science Program, Fisheries and Marine Sciences Department, Padjadjaran University, and KOMITMEN Research Group who have supported the workmanship of this paper.

References

[1] Atmadipoera A, Molcard R, Madec G, Wijffels S, Sprintall J, Koch-Larrouy A, and Supangat A 2009 Characteristics and Variability of the Indonesian Throughflow Water at the Outflow Straits Deep Sea Research Part I: *Oceanographic Research Papers* **56** 1942-1954
[2] BMKG 2014 www.bmkg.go.id
[3] Cyr F and Larouche P 2014 Thermal fronts Atlas of Canadian Coastal Waters Atmosphere- *Ocean iFirst article* 1-25
[4] Gordon A 2005 Oceanography of the Indonesian Seas and Their Throughtflow *Oceanography* **18**(4) 14-27
[5] Hamzah R, Prayogo T, and Harsanugraha W K 2014 Identifikasi Thermal front Dari Data Satelit Terra/Aqua Modis Menggunakan Metode Single Image Edge Detection (SIED)(Studi Kasus: Perairan Utara Dan Selatan Pulau Jawa)(Thermal Front Identification from Terra/Aqua Modis Satellite Data with Single Image Edge Detection Method (Case Studies: Northern and Southern waters of Java Island) *Remote Sensing National Proceding* 2014 552-559 (In Bahasa)
[6] Harahap S A, Syamsuddin M L and Purba N P 2013 Aplikasi Penginderaan Jarak Jauh Untuk Pendugaan Hotspot Tuna Sirip Kuning Di Perairan Selatan Jawa Barat (The Aplication of Remote Sensing for Yellowfin Tuna Hotspot Prediction in the Southern Waters of Java) The Research Final Report of Grants Competition UNPAD (In Bahasa)
[7] Hasita, Firra, Zikra, Muhammad and Suntoyo 2013 Analisa Variasi Temperatur dan Salinitas Air Laut di Perairan Samudra Pasifik Akibat Pengaruh El Niño dan La Niña (Variation of Sea Water Themperature and Salinity Analysis in The Pasifik Ocean under the effect of El Nino and La Nina *Jurnal Teknik Pomits* **2** No. 2 (2013) ISSN: 2337-3539 (2301- 9271 Print) (In Bahasa)
[8] Hasyim B, Sulma S, and Hartuti M 2010 Kajian Dinamika Suhu Permukaan Laut Global Menggunakan Data Penginderaan Jauh Microwave (Study Dynamics of Global SST Using Microwave Remote Sensing Data) *Dirgantara Science and Technology Magazine* **5** No. 4 130-143 (In Bahasa)
[9] Inayah K 2015 Identifikasi Front sebagai Daerah Potensial Penangkapan Ikan Yellow Fin Tuna (Thunnus Albacares) di Perairan Selatan Jawa – Bali (Front Identification as Yellow Fin Tuna Arrest Potential in the Southern water of Java and Bali Bachelor Thesis, Marine Science, UNPAD (In Bahasa)
[10] Cane M A 2004 The evolution of El Niño, past and future Earth Planet *Sci. Lett.* 1-10
[11] Muhammad R S and Affan J M 2012 Pengaruh ENSO (El Niño and Southern Oscillation) terhadap Transpor Massa Air Laut di Selat Malaka (ENSO Influence on Marine Water Masses Transport in Malaka Strait) *Penelitian Unggulan Strategis Nasional Depok* 1**(1)**: 61-67 (In Bahasa)
[12] NCDC-NOAA 2016 Equatorial Pasific Sea Surface Temperatures Retrieved from National Oceanic and Atmopheric Administration: https://www.ncdc.noaa.gov
[13] Oktaviani N A, Jumarang M I and Ihwan A 2014 Kajian Elevasi Muka Air Laut di Perairan Indonesia pada Kondisi El Niño dan La Nina (Study of Sea Level Water the Waters of Indonesia on El Nino and La Nina Condition) *Prisma Fisika* **2** No. 1 6 - 10 (In Bahasa)
[14] Prawono and Widodo S 2005 Sistem Arus Permukaan di sekitar ARLINDO (Surface Current System around ARLINDO) In Reveals Indonesia Through Flow Jakarta: Marine and Fisheries Departement (In Bahasa)
[15] Priyanti N S 1999 Studi Daerah Penangkapan Rawai Tuna di Perairan Selatan Jawa Timur dan Bali pada Musim Timur berdasarkan Pola Distribusi Suhu Permukaan Laut Citra Satelit NOAA/AVHR dan Data Hasil Tangkapan (Study of Tuna Longline Arrest Area in the Southern Water of East Java and Bali on the East Season Based on the pattern of NOAA/ AVHR SST Satellite Data and Catch Distribution Bachelor Thesis (In Indonesian)
[16] Purba N P, Pranowo W S, Faizal I and Adiwira H 2018 Temperature – Salinity stratification in the Eastern Indian Ocean using argo float IOP Conf. Series: *Earth and Environmental Science* **162** (2018) 012010 doi :10.1088/1755-1315/162/1/012010

[17] Ruili S, Zheng L, Changlin C and Yunwei Y 2015 Interannual variability of thermal front west of Luzon Island in Boreal Winter Acta *Oceanol. Sin.* 2015 **34** No. 11 102–108
[18] Safitri M, Cahyarini S and Putri M 2012 Variasi Arus Arlindo Dan Parameter Oseanografi Di Laut Timor Sebagai Indikasi Kejadian ENSO (Indonesia Through Flow and Oseanography Parameter Variation in Timor Sea as Indicator of ENSO) *Marine Science and Technology Tropic Journal* **4** No. 2 369-377 (In Bahasa)
[19] Schlitzer R 2015 Ocean Data View http://odv.awe.de.
[20] Silalahi I A 2013 Pengaruh Indian Ocean Dipole (IOD) dan El Niño Southern Oscillation (ENSO) terhadap Variabilitas Upwelling di Perairan Selatan Jawa (The Effect of IOD and ENSO against Upwelling Variability in the Southern Java) Bachelor Thesis Marine Science UNPAD (In Bahasa)
[21] Siregar V and WAAS H J 2006 Identification of Oceanographic Parameters for Determining Pelagic Tuna Fishing Ground in the North Papua Waters Using Multi-Sensor Satellite Data *BIOTROPIA* **13** No. 1 37-48
[22] Sitompul Z and Nurjani E 2012 Pengaruh El Niño Southern Oscillation (ENSO) terhadap Curah Hujan Musiman dan Tahunan di Indonesia (ENSO Influence on Seasonal and Annual Rainfall in Indonesia) Gajah Mada University Journal (In Bahasa)
[23] Supari 2014 Sejarah Dampak El Niño di Indonesia (History Impact of El Nino in Indonesia) Retrieved from Climatology Anaysis Deputies BMKG http://www.bmkg.go.id (In Bahasa)
[24] Susanto R D, Gordon A L and Zheng Q 2001 Upwelling Along the Coasts of Java and Sumatra and its Relation to ENSO *Geophysycal Research Letters* **28** No.8 1599 - 1602
[25] Syamsuddin M L, Saitoh S I, Hirawake T, Bachri S and Harto A B 2013 Effects of El Niño–Southern Oscillation events on catches of Bigeye Tuna (Thunnus obesus) in the eastern Indian Ocean off Java *Fishery Bulletin* **111**(2) 175 - 188
[26] Utamy R M, Purba N P, Pranowo W and Suherman H 2015 The Pattern of South Equatorial Current and Primary Productivity in South Java Seas *International Conference on Environment Science and Biotechnology (ICESB)* **81** 1-5 doi: 10.7763/IPCBEE
[27] Wang Y C, Chen W Y, Chang Y and Lee M A 2013 Ichthyoplankton Community Associated with Oceanic Fronts in Early Winter on The Continental Shelf of The Southern East China Sea *Journal of Marine Science and Technology* **21** Suppl. pp. 65-76
[28] Wicaksono A, Muhsoni F F and Fahrudin A 2010 Aplikasi Data Citra Satelit NOAA-17 untuk Mengukur Variasi Suhu Permukaan Laut Jawa (NOAA-17 Satellite Data Aplication to measuring Java Sea SST Variation) *Marine Journal* **3** No. 01 70-74 (In Bahasa)
[29] Wyrkti K 1961 Physical Oceanography of Southest Asian Waters *Naga Report* Vol **02** Scripps Institute Oceanography California

Correlation and coherence analysis between sea surface temperature (SST) and surface wind in the Equatorial Western Sumatra Waters

R B Hatmaja[1], U J Wisha[2], I M Radjawane[1] and T Al Tanto[2]

[1] Earth Sciences Study Program, Faculty of Earth Sciences and Technology, Bandung Institute of Technology, Jl. Ganeca no. 10, Bandung – West Java 40135, Indonesia
[2] Research Institute of Coastal Resources and Vulnerability, Ministry of Marine Affairs and Fisheries, Jl. Raya Padang-Painan KM. 16, Bungus, Padang – West Sumatra 25245, Indonesia

Email: rahadenbagas@gmail.com

Abstract. Equatorial Western Sumatra Waters have a tremendous dynamic in ocean characteristics. It directly faces the Indian Ocean exactly located below the equator. Consequently, Equatorial Western Sumatra waters are influenced by the tropical climatic factors such as monsoons and climate variabilities, such as Indian Ocean Dipole (IOD) and Madden Julian Oscillation (MJO) that control sea surface temperature (SST) fluctuation in the Indian Ocean regions. This study aimed to review the correlation and coherence of SST distributed by surface wind in the Equatorial Western Sumatra waters. Wavelet method (cross wavelet transforms and wavelet coherence) was used to analyze the correlation and coherency between SST and surface wind. The annual variation of SST for 365 days period is the strongest event throughout the year caused by either monsoon or the changes of wind speed in the surface. Otherwise, the strongest intra-seasonal SST variation of 35 - 60 days observed from December 2012 to March 2013. The highest surface wind speed occurs in the southern and western waters. The surface wind plays a role in evoking SST distribution of 35 - 60 days period (intra-seasonal variability). Besides, the surface wind with 6 months period (semi-annual variability) influences the SST distribution which is identified only in the southern waters and the Indian Ocean regions. These conditions are possibly predicted as the influence of monsoons.

Keywords: Equatorial Western Sumatra Waters, sea surface temperature, surface wind speed, wavelet methods

1. Introduction

Western Sumatra waters have a tremendous ocean dynamic due to its geographical position located between Asia and Australia. It directly faces the Indian Ocean which exactly positioned below the equator. Consequently, water characteristics are strongly influenced by monsoons (due to pressure differences between Asia and Australia) which varies every six months [1] such as global climate variability, the Indian Ocean Dipole (IOD) [2], solar warming intensity, and equatorial current system [1]. Moreover, Western Sumatra has a unique topography consist of many islands, headlands, bays, and the convergence of two large water masses (from the Indian Ocean and the Andaman Sea).

Sea surface temperature (SST) is a significant parameter representing the water conditions which is one of the dominant parameters that can detect the climate change variability regionally globally. The significant change in SST can cause several impacts on marine ecosystems such as coral bleaching, demised biota, and many others. For example, corals will be suffering bleaching due to the increase of SST. Every 0.1°C enhancement of normal temperature results in 35 - 42% chance of coral bleaching [3,4]. That phenomenon caused by the significant change of SST reached 33 - 34°C resulted from climate anomaly occurred from April to June 2016 in western Sumatra [1].

An assessment of surface wind in the study area is significant because it has a dominant influence on the distribution of surface water masses. Therefore, it affects the distribution of the SST as well [5]. Furthermore, surface wind also plays a role in inducing upwelling near the coast and IOD cycles [2]. The combination of surface wind and SST is certainly affecting the capacity and vulnerability of coastal resources because the study area positioned below the equator, influenced by the monsoon, and IOD. Therefore, a study regarding correlation and coherence between SST distribution and the surface wind is essential. This research can be useful for local government and fisherman to determine fishing ground area based on the existing SST distribution condition, as well as climate change mitigation due to the anomaly of SST distribution. In this study, we analyzed the distribution and variability assessment of SST and surface wind speed in West Sumatra waters. This study aimed to review the correlation and coherence of SST distributed by surface wind in West Sumatra waters.

2. Data and Methods

2.1. Data

Data used in this study were gridded from reanalysis ERA-Interim daily SST and 10 m surface wind (zonal (u) and meridional (v)) for 4 years (December 2012 - November 2016) in the West Sumatra Waters (1°N – 3°S and 97°E – 101°E) which the resolution is 0.125° × 0.125°. Data were retrieved from European Center for Medium-Range Weather Forecasts (ECMWF) (http://www.ecmwf.int/en/research/climate-reanalysis/browse-reanalysis-datasets) [6].

Table 1. Observation stations coordinates

Station Point	Latitude	Longitude
Station 1 (St. 1)	0.5 °N	98.75 °E
Station 2 (St. 2)	0.5 °S	99.25 °E
Station 3 (St. 3)	1.5 °S	99.75 °E
Station 4 (St. 4)	2.5 °S	100.25 °E
Station 5 (St. 5)	0.75 °S	98.0 °E
Station 6 (St. 6)	1.75 °S	98.5 °E
Station 7 (St. 7)	2.75 °S	99.0 °E
Station 8 (St. 8)	2.5 °S	97.5 °E

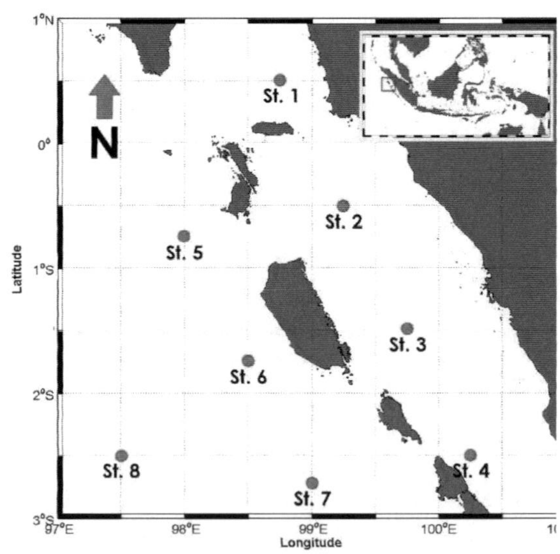

Figure 1. Research location and observation points (red pins)

IOP Conf. Series: Earth and Environmental Science **303** (2019) 012003 doi:10.1088/1755-1315/303/1/012003

Eight observation stations were selected to represent the condition of West Sumatra Waters. Station 1 - Station 4 are representing the waters between Sumatra Island and the Mentawai Islands, Station 5 - Station 7 are representing the waters near the Mentawai Islands which is directly adjacent to the Indian Ocean, and Station 8 represents the Indian Ocean. The position of these eight observation stations is shown in Table 1 and Figure 1. To obtain the wind speed value from the zonal (u) and meridional (v) components, the conversion formula was employed as follows:

$$V = \sqrt{u^2 + v^2} \quad (1)$$

Where:

V : wind speed (m/s)

u : zonal wind component (m/s)

v : meridional wind component (m/s)

2.2. Methods

To analyze the variability of SST and wind speed, spectral density is employed. The spectral density method will change the data from time series domain to frequency domain using Fourier Transform. Correlation and coherence analysis of SST distributed by surface wind was done using wavelet methods, which consist of cross-wavelet transform (XWT) and wavelet coherence (WTC) [7, 8].

2.2.1. Spectral Density. Spectral density is defined as the power of each frequency interval with amplitude as the investigated signal unit where the frequency is represented as a cyclic unit. The spectral density function $Gx(f)$ from random data $x(t)$ describes the frequency composition related to spectral density and the power of spectral density in the form of mean-square value. Mean-square value on the frequency scale off and $f + \Delta f$ may generate data filter which bypasses the frequency characteristics and calculates the mean-root during filtering processes. The mean root value will approximate the mean square value as the infinite time of observation (T), therefore the mean square value can be derived from a mathematical equation [9], as follow:

$$\psi(f, \Delta f) = \lim_{T \to x} \frac{1}{T} \int_0^T x^2(t, f, \Delta f) \mathrm{d}t \quad (2)$$

where, $x\ (t, f, \Delta f\ ) =$ The part of $x(t)$ on the frequency scale f to $f\ + \Delta f$. Spectral density function, $Gx(f)$, can be defined as the following Equation 3 below.

$$G_x(f) = 4\int_{-\infty}^{\infty} R_x(\tau) e^{-i2\pi f\tau} \mathrm{d}\tau \quad (3)$$

where, $Rx(\tau)\ =$ The self-correlation of $x(t)$ which defined as $x_T(t) = x(t), -T < t < T$

2.2.2. *Wavelet Methods.* Wavelet method is used to analyze the data in time domain and time distribution information at a certain frequency. Wavelet Transform converts data from time domain to frequency domain which the information of the phenomena at a certain frequency will be obtained. Wavelet method is a way to decompose the time series into time-frequency domain simultaneously [8, 10]. This method calculates the spectrum energy in the form of time series data. The advantages of wavelet method are detecting transient periodic fluctuations and describing complex nonlinear dynamics processes demonstrated by the interaction of disturbances spatially temporally. The wavelet spectrum describes the results of data processing from wavelet method. The y-axis shows the time of incident and the x-axis shows the incident's period or variability.

To assess the correlation between two time series data simultaneously, Cross Wavelet Transform (XWT) is used to identify relation and relative phase, as well as to analyze its covariance (Xn and Yn), as follows:

$$W^{XY} = W^{X} \cdot W^{Y*} \quad (4)$$

where:
W_{xy} : The power spectrum of XWT
W_x : The power spectrum of Xn time series data
W_y : The power spectrum of Yn time series data
(the term * shows the complex conjugate)

Wavelet coherence (WTC) is used to analyze the coherence and phase lag between two types of time series data as a function of time and frequency [8, 10]. WTC is the coherence among wavelets of those data performed to find the significant coherence with low energy level. It involves the confidence level in the calculation. It defines as follows:

$$R_n^2(s) = \frac{\left| s(s^{-1}W_n^{XY}(s)) \right|^2}{s\left(s^{-1}\left|(W_n^{X}(s)\right|^2\right) \cdot s\left(s^{-1}\left|(W_n^{Y}(s)\right|^2\right)} \quad (5)$$

Where:
$R_n^2(s)$ = the coherence between two variables
S = the spectrum values
W_n = the Wavelet Transform in n frequency.

3. Results and Discussions

3.1. Distribution and Variability of Sea Surface Temperature and Surface Wind Speed

Figure 2 shows the spatial distribution of mean SST and surface wind speed in the West Sumatra Waters for four years (from December 2012 to November 2016). The SST averages declined at Station 1 to Station 4, then terraced at Station 5, and decreased at Station 6 and 7, and terraced again at Station 8 (see Table 2), which the fluctuation is not more than 1°C. Based on its position, the SST in the south tends to be lower than in the north. It might be affected by the intense solar warming at the equator, so the value of SST is higher [11]. Otherwise, wind speed is quite random which increases during boreal summer season (June to August) and boreal autumn season (September to November). The speed ranged from 2 - 4 m/s. The average wind speed enhanced at Station 1 to Station 4, then decreased at Station 5, and eventually terraced at Station 8 (see Table 2). The wind speed closer to the equator is weak because the area near the equator is the convergence zone of trade winds well-known as *Inter-Tropical Convergence Zone (ITCZ)* [12]. Compared to the SST average in Table 2, the area with low SST tends to have strong surface wind speed due to higher air pressure in the deformation processes. Otherwise, a high temperature evokes air expansion. As a result, the wind becomes weaker inducing cloud convergence formations.

Figure 3 shows the annual variability domination of 365 days. The annual variations estimated induced by monsoon cycle. It can be affected by the changes of surface wind velocity that is stronger in the boreal summer season (June to August) and the boreal autumn season (September to November).

Besides the annual variations, there are seasonal variations of 91 to 97 days period (3 months), intra-seasonal variations over 35 to 90 days period, and semi-annual variations within 183 to 209 days period (6 months). Seasonal to semi-annual variations occurred due to the SST distribution is influenced by local phenomena such as wind, surface current, waves, *etc.* [13]. Moreover, it proves that the semi-annual variation (6 months period) is more dominant than the seasonal variations, in which it is caused by solar radiation in the equator occurred every 6 months. In general, seasonal to the semi-annual variations triggered by wind speed changing every 3-6 months (monsoon cycle).

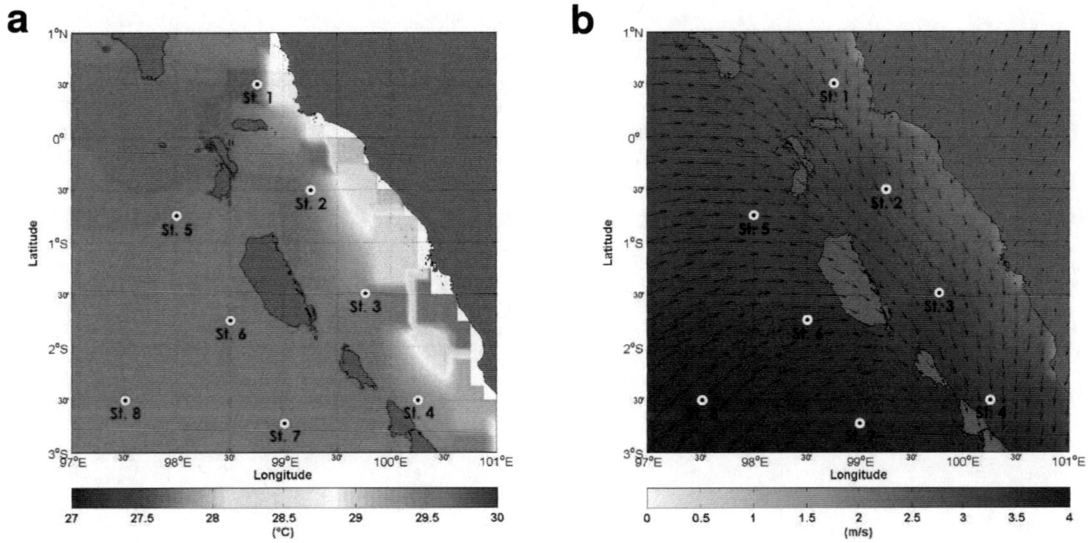

Figure 2. Mean SST (a) and surface wind speed (b) in West Sumatra Waters from December 2012 to November 2016.

Table 2. Mean SST and wind speed for four years at each station points

	St. 1	St. 2	St. 3	St. 4	St. 5	St. 6	St. 7	St. 8
Mean SST (°C)	29.74	29.42	29.34	29.33	29.60	29.54	29.49	29.52
Mean wind speed (m/s)	2.49	2.83	2.94	3.38	3.16	3.51	3.72	3.79

Figure 4 shows the spectrum density of daily variation which is very dominant, thus, the variance of wind speed in West Sumatra Waters is very high and random. Nevertheless, seasonal variations vary from 91 to 97 days, intra-seasonal variation over 35 to 90 days period, semi-annual variations within 183 to 209 days (6 months period monsoon), and annual variations of 365 days period.

The seasonal variation, which occurred for 3 months period, is caused by the change of wind speed between the boreal winter season (December to February) and the boreal spring season (March to May) and so on. The semi-annual variation shows the monsoon phenomenon between the boreal winter and boreal summer season. Those conditions caused by the fluctuations of wind speed occurred every year. In general, the dominant variation is 6-month period associated with monsoons.

3.2. Correlation and Coherency Analysis

The correlation between SST and wind speed within 32 to 60 days period (intra-seasonal variation domination) occurred in March to September 2013 which tend to be anti-phase. The annual variation correlation clearly detects for 365 days period dominating throughout 2014 until early 2015 which tend to be anti-phase as well. In addition, it correlates with semi-annual period (183 to 209 days or 6 months) occurred in October to November 2015 (anti-phase). Anti-phase term means that the wind speed leads the SST in which wind speed decreased when the SST increased. In this case, wind speed raised in the southern waters, resulting in declined SST in that region. It means that the wind distributes the warm water mass northward and westward. Otherwise, when the wind moves slowly, the warm water mass will be distributed by the other factors (currents, wave, pressure, or IOD).

IOP Conf. Series: Earth and Environmental Science **303** (2019) 012003 doi:10.1088/1755-1315/303/1/012003

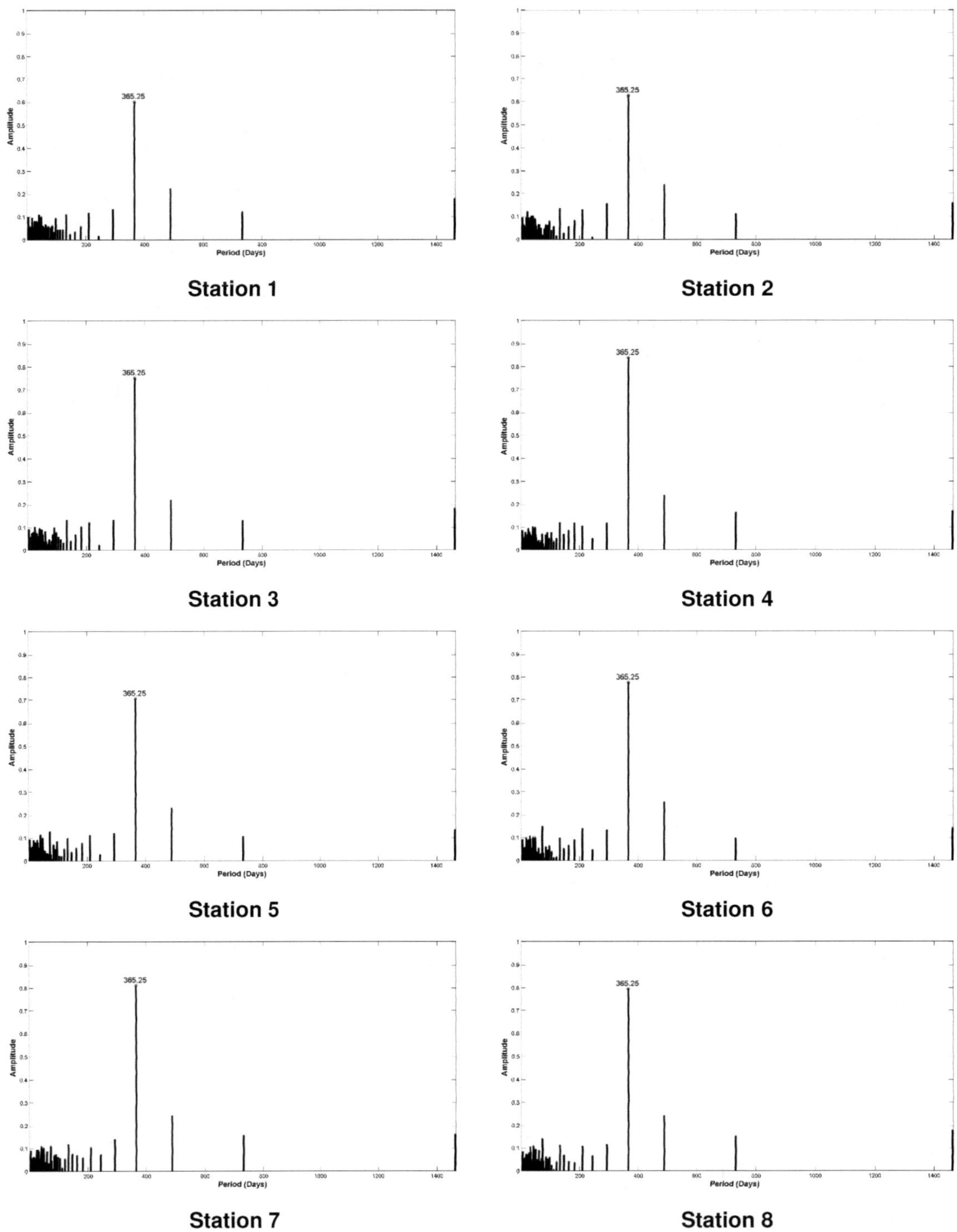

Figure 3. Spectral density of SST at each sobservation point.

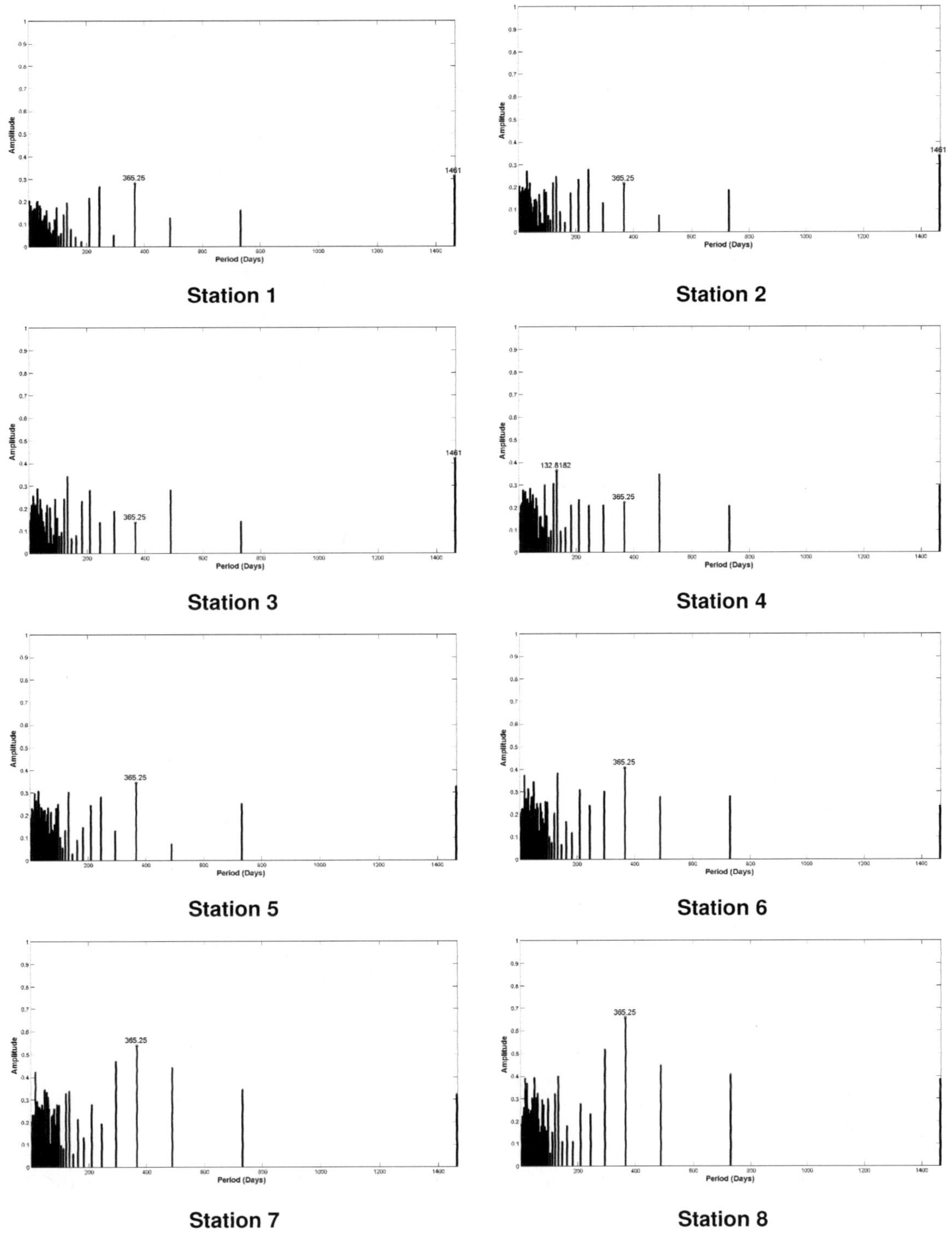

Figure 4. Spectral density of wind speed at each observation point.

IOP Conf. Series: Earth and Environmental Science **303** (2019) 012003 doi:10.1088/1755-1315/303/1/012003

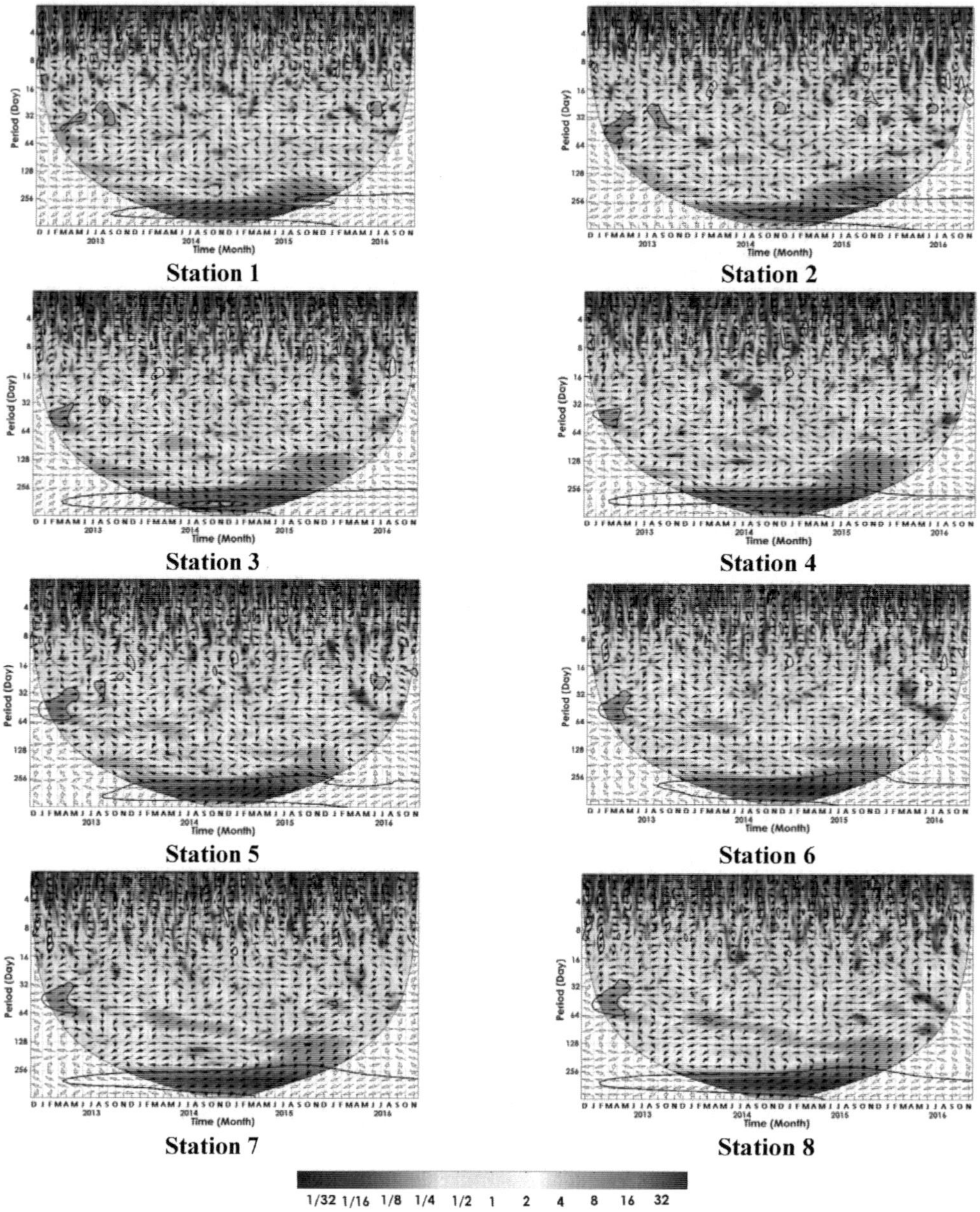

Figure 5. Cross Wavelet Transform of SST and wind speed at each observation point.

The wavelet coherence indicates the coherence between SST and wind speed with 32 to 60 days period (dominant intra-seasonal variation) occurred in March to September 2013 which tend to be anti-phase (Figure 6). Coherence in the intra-seasonal variation is increasingly strengthened at Station 4 to Station 8 in the southern waters and the area adjacent to the Indian Ocean. During March to September 2013, the coherence in the intra-seasonal period occurred from September 2015 to February 2016, only observed at Station 6 to Station 8 (stations adjacent to the Indian Ocean). A strong coherence between SST and wind speed in the southern water and the open seas (directly adjacent to the Indian Ocean) is occurred.

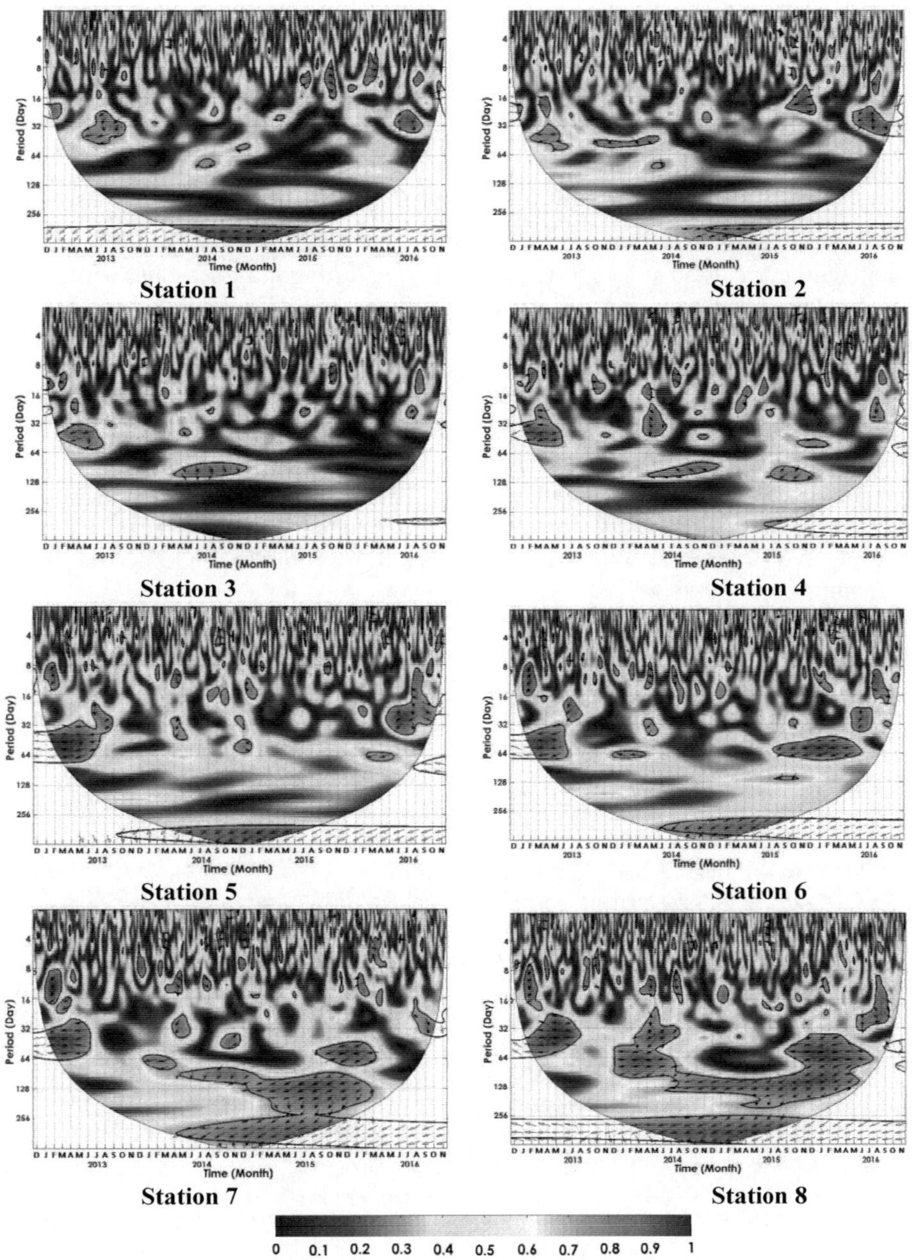

Figure 6. Wavelet coherence of SST and wind speed at each observation points.

Coherence in the annual variation is obviously detected within 365 days period that predominated in June 2014 to early 2016 tend to be anti-phase, this event only occurs at Station 5 to Station 7. Thus, there is a coherence in the semi-annual period (183 to 209 days or 6 months) from October to November 2015 that tend to be anti-phase as well observed at Station 7 and Station 8.

In general, there is a correlation and coherence between SST and wind speed, but with different both period and variation. The dominant correlation and coherence are identified in intra-seasonal variation within 35 to 90 days period. The variability of SST is caused by the changes in wind velocity and intra-seasonal variability as well [14]. Those conditions show the influence of surface wind to the SST distribution within 35 to 90 days period (intra-seasonal variation). Besides intra-seasonal variation, the

IOP Conf. Series: Earth and Environmental Science **303** (2019) 012003 doi:10.1088/1755-1315/303/1/012003

correlation and coherence during semi-annual to annual period identified are quite strong in the southern waters and open seas adjacent to the Indian Ocean indicating the limited influence of monsoon wind to the SST distribution in that region because of several other stronger influences at Station 1 to Station 3. Moreover, the location is bordered by small islands and influenced by currents, waves, tides, and upwelling.

4. Conclusion

There are correlations and coherences between SST and surface wind speed, with a different period and variation. The dominant correlation and coherence are varying during intra-seasonal with 35 to 90 days period which tend to be anti-phase. The surface wind influences SST distribution within the period approximately 35 to 90 days (intra-seasonal variation). As well as the correlation and coherence of semi-annual to the annual period that was occurred quite strong in the southern waters and the open seas adjacent to the Indian Ocean (anti-phase) indicating the limited influence of monsoon wind in that region. It's bordered by some small islands which the influences of freshwater runoff, currents, waves, tides, and upwelling take place.

This research can be useful for local government and fisherman by which it determines the fishing ground based on the existing SST condition, and climate pattern detection regarding the anomaly of SST and the surface wind. We suggest for future study to develop this information to be well implemented to the coastal communities.

Acknowledgement

Acknowledgement and gratitude are given to European Centre for Medium-Range Weather Forecasts (ECMWF) for providing the SST and wind data, and to everyone who helps, encourages, and supports us to finish this research. Cross wavelet transform and wavelet coherence software were provided by A. Grinsted (2004) [6].

References

[1] Wisha U J, T Al Tanto and Ilham I 2017 *Geomatics* **22** 21–28
[2] Saji N H, B N Goswami, P N Vinayachandran and T Yamagata 1999 *Nature* **401** 360–63
[3] Mc Williams J P, I M Côté, J A Gill, W J Sutherland and A R Watkinson 2005 *Ecology* **86** 2055–60
[4] Lesser M P 2011 *Coral bleaching: causes and mechanisms. In Coral reefs: an ecosystem in transition* (Dordrecht: Springer) p 405
[5] Stramma L P, Cornillon, Weller R A, Price J F and Briscoe M G 1986 Large diurnal sea surface temperature variability: satellite and in situ measurements *J. Physical Oceanography* **16** 827–837
[6] Dee D P *et al* 2011 *Q J R Meteorol Soc* **137** 553–597
[7] Grinsted A, J C Moore and S Jevrejeva 2004 *Nonlinear Processes in Geophysics* **11** 561–566
[8] Thomson R E and W J Emery 2014 *Data analysis methods in physical oceanography* (Oxford: Elsevier)
[9] Bendat J S and A G Piersol 2011 *Random data: analysis and measurement procedures* John Wiley & Sons Inc
[10] Bernal-Rusiel J L, Greve D N, Reuter M, Fischl B and M Sabuncu 2013 *Neuroimage* **66** 249–60
[11] Sprintall J, A L Gordon, A Koch-Larrouy, Lee T, Potemra J T, Pujiana K and Wijffels S E 2014 *Nature Geoscience* **7** 487
[12] Loo Y Y, L Billa and A Singh 2015 *Geoscience Frontiers* **6** 817–23
[13] Syamsul R, Peter D, Mulyadi W, Jurgen S, Yopie I, Taufiq I and Muhammad M 2012 *American Journal of Environmental Sciences* **8** 479–88
[14] Qu T, Du Y, Strachan J, Meyers G and Slingo J 2005 *Oceanography* **18** 50–61

The dynamics of negative Indian Ocean Dipole (nIOD) and its relation to the anomalous high rainfall in West Java Province, Indonesia

R B Hatmaja[1], A H Rusmanansari[2] and I M Radjawane[1]

[1] Earth Sciences Department, Faculty of Earth Sciences and Technology, Bandung Institute of Technology, Jl. Ganeca no. 10, Bandung – West Java 40135, Indonesia
[2] Meteorology Department, Faculty of Earth Sciences and Technology, Bandung Institute of Technology, Jl. Ganeca no. 10, Bandung – West Java 40135, Indonesia

rahadenbagas@gmail.com

Abstract. West Java Province is the most vulnerable region to hydro-meteorological disasters (i.e. flood and landslide) in Indonesia. One of the main causes of these disasters is high level of rainfall. There were many global phenomena caused an increase in rainfall level in Indonesia, one of them was the Indian Ocean Dipole (IOD), especially its negative phase. By using the Dipole Mode Index (DMI), IOD strength or intensity can be determined. Besides, the composite analysis of anomalous sea surface temperature (SST), wind, and rainfall were constructed to examine the dynamics of the negative IOD and its relation to the high rainfall level in West Java. During the negative IOD (nIOD), the significant positive (negative) anomalous SST appeared around May in eastern (western-central) of Indian Ocean. The intensity of anomalous SST in the following months reached its peak in September and accompanied by strong anomalous north-westerly wind. Furthermore, the high SST and anomalous wind increased the convection activity in the southern Java Island, and consequently, the level of rainfall in West Java increased up to 8.92 mm/day and may lead the risk of hydro-meteorological disaster.

Keywords: Indian Ocean Dipole (IOD), anomalous high rainfall, West Java

1. Introduction

Indian Ocean Dipole (IOD) is a dipole mode phenomenon with an interannual pattern which is identified by differences in sea surface temperature anomalies (SST) on eastern and western Indian Ocean, followed by wind anomalies and precipitation. There are two IOD phases, namely positive IOD and negative IOD. Positive IOD is characterized by the emergence of SST positive anomalies on the eastern side of the Indian Ocean and negative SST anomalies on the western Indian Ocean, while the opposite characteristics occur when negative IOD event [1, 2]. The IOD phenomenon has a significant correlation with global SST, rainfall, and atmospheric circulation anomalies, such as the Walker Cycle [3]. Negative IOD (nIOD) is a global phenomenon and often associated with high rainfall in Indonesia. Increased SST anomalies can increase convective cloud formation and rises the rainfall in Indonesia. In 2016, there was a nIOD event that happened from June to December with the peak in September, as well as an increase in rainfall in Indonesia [4, 5].

IOP Conf. Series: Earth and Environmental Science **303** (2019) 012004 doi:10.1088/1755-1315/303/1/012004

Along with increased rainfall in Indonesia, hydro-meteorological disasters also experienced an increase. One of the most common and most fatal hydro-meteorological disasters is landslides, especially in areas of steep hills and slopes such as in south of West Java Province [6]. In general, West Java region has a monsoonal type of rainfall because of its characteristics that are strongly influenced by the Asia-Australia monsoon system [7]. Hence, the movement of the Asia-Australia monsoon system will also be a trigger factor for high rainfall in West Java.

In interannual time scale, Indonesia's climate has been considered to be closely linked to El-Nino Southern Oscillation (ENSO) in the Pacific Ocean [7]. However, rainfall variability of West Java is significantly correlated with IOD compared to ENSO [8]. Thus, by the study of the dynamics of the IOD, especially the negative phase, and its impact on increased rainfall anomalies, it is expected to be a reference and consideration in efforts to mitigate hydro-meteorological disasters, such as landslides and floods that often occur in West Java Province, Indonesia.

2. Data and Methods

To analyze the dynamics of negative IOD (nIOD), the composite analysis will be conducted on three parameters, i.e. sea surface temperature (SST), wind, and rainfall in the area about 15°N–15°S and 40°–120°E (see Figure 1a). Data used in this research were monthly data for 30 years (1987–2017) of SST and 10 m surface wind (zonal and meridional), which are retrieved from European Centre for Medium-Range Weather Forecasts (ECMWF) (http://apps.ecmwf.int/datasets/data/interim-full-moda/levtype=sfc/) [9], and also global rainfall data, which were retrieved from Global Precipitation Climatology Project (GPCP) (https://climatedataguide.ucar.edu/climate-data/gpcp-monthly-global-precipitation-climatology-project). All those data have spatial resolution about 1° × 1° and 0.25° × 0.25° for SST, wind, and rainfall data, respectively. Dipole Mode Index (DMI) was also used to determine the IOD events years. The DMI was retrieved from National Oceanic Atmospheric Administration (NOAA) (http://stateoftheocean.osmc.noaa.gov/sur/ind/dmi.php), which is calculated using Reynolds OIv2 SST analysis [10]

Furthermore, to analyze the relationship between IOD and rainfall in West Java, the data used in this study are rainfall data of 16 rain gauges in West Java region from PUSAIR (Center for Water Resources Research & Development) (see Figure 1b) for 30 years (1987–2017), GPCP global rainfall data in the area about 5.75°S–8°S and 106.25°–108.75°E (see Figure 1b), and DMI data. Due to substantial gaps in rain gauge records, all station rainfall data will be averaged. Thus the rainfall in West Java was assumed to be homogenous. In addition, an analysis of rainfall anomalies especially in boreal autumn season was also conducted to avoid the influence of intra-seasonal phenomena such as Madden-Julian Oscillation (MJO).

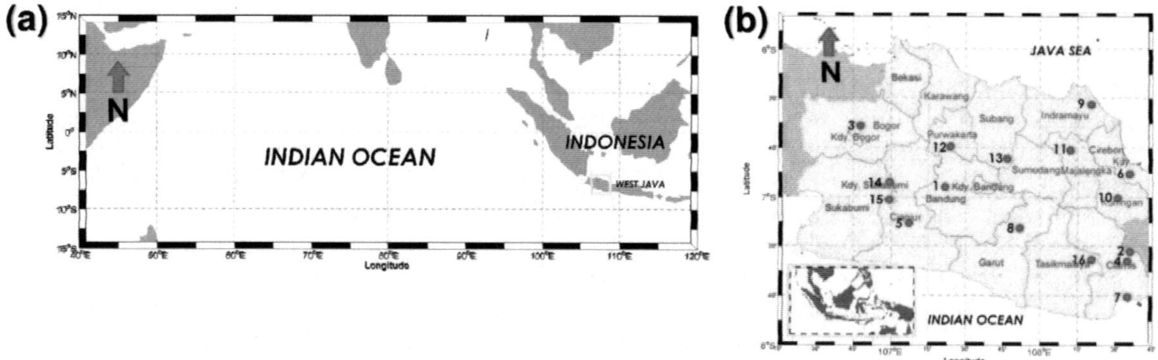

Figure 1. (a) Study area of Indian Ocean Dipole (IOD) phenomenon and (b) for anomalous rainfall analysis in West Java Province.

3. Results and Discussions

3.1. Spatial and Temporal Analysis of the Negative Indian Ocean Dipole (nIOD)

The IOD event year is defined as the year of DMI value below or above 0.5°C (one standard deviation) for two to three consecutive months during August to November [3, 11]. Based on the time series of the DMI during 1987-2016, which is presented in Figure 2, it can be seen that the years of the negative IOD (nIOD) events were in 1992, 1996, 1998, 2010 and 2016. Based on those five nIOD events, the composite analysis is conducted on three parameters, i.e. SST, wind, and rainfall, to analyse the dynamics of nIOD events in space and time.

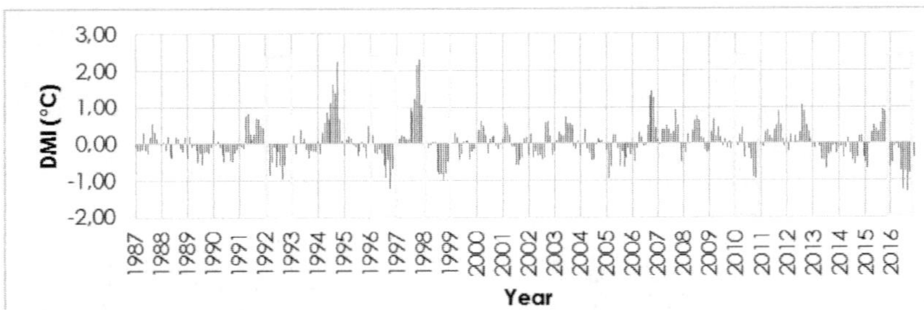

Figure 2. The time series of Dipole Mode Index (DMI) during 1987 to 2016. Negative IOD (nIOD) events are shown by yellow boxes.

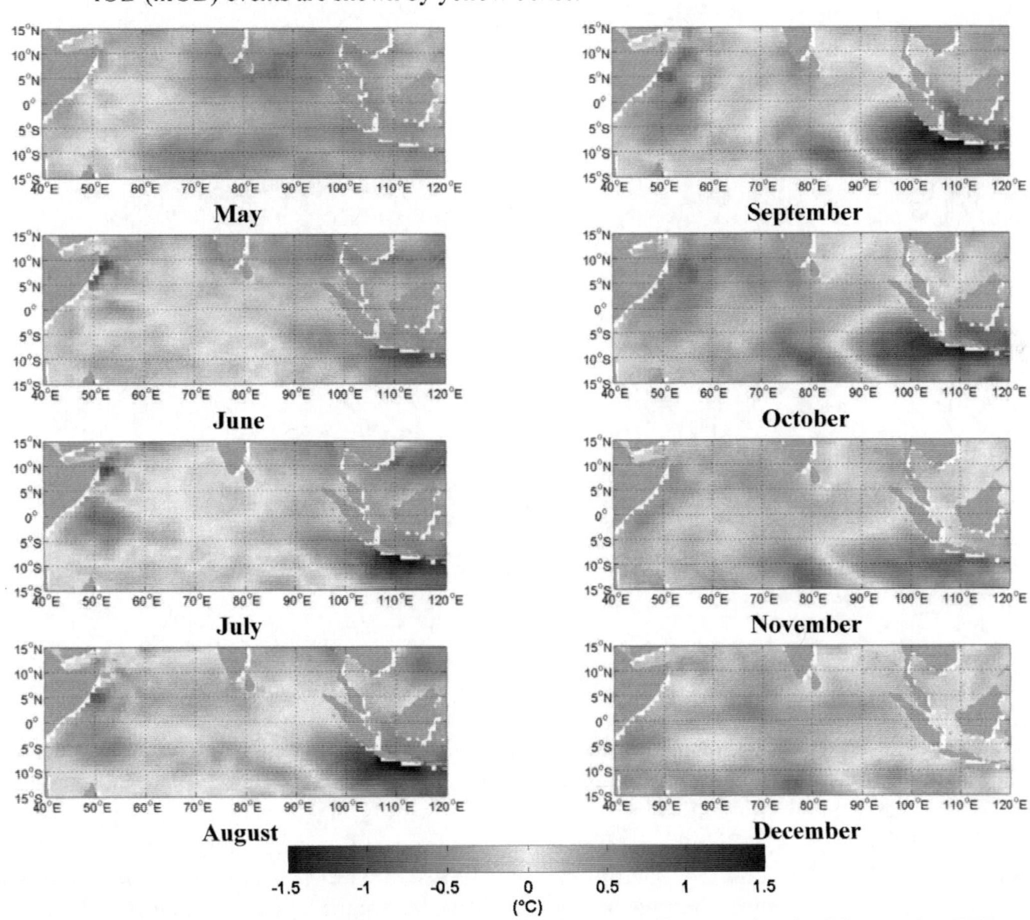

Figure 3. Composites of sea surface temperature (SST) anomalies associated with negative IOD events (based on five nIOD events) from May to December.

IOP Conf. Series: Earth and Environmental Science **303** (2019) 012004 doi:10.1088/1755-1315/303/1/012004

Figure 3 shows the dynamics of the Indian Ocean SST anomalies from May to December during nIOD event. Based on the figure, it can be seen that the dynamics of the SST anomaly begins with a positive SST anomaly on the eastern side of the Indian Ocean (0°–10°S and 90°–110°E) in May, followed by negative anomaly on the western side (10°N–10°S and 40°–60°E) about 2 months later and negative anomaly on the central side (5°–10°S and 70°–80°E) in the following month. The anomalous SST reaches the peak in September and decays in the following months [1]. During the peak of negative IOD events, the anomalous SST reached –0.88 °C on the western side and 1.63 °C on the eastern side.

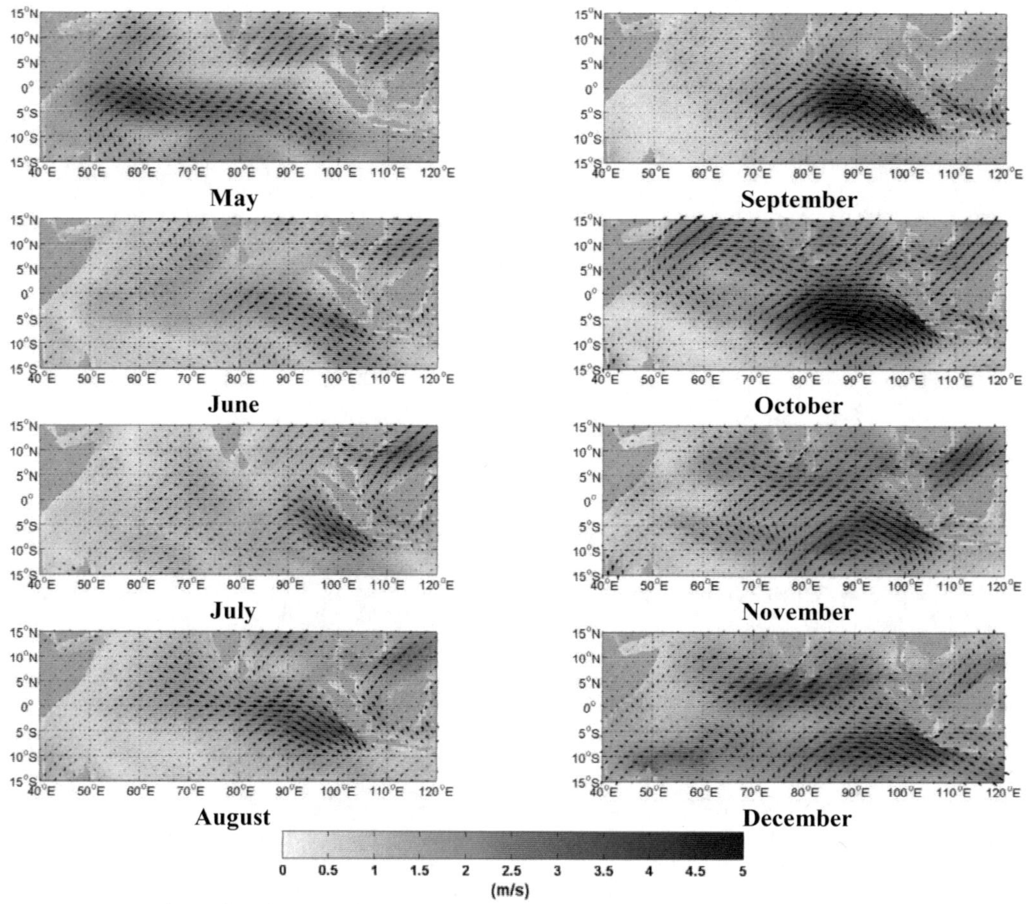

Figure 4. Composites of surface wind anomalies associated with negative IOD events (based on five nIOD events) from May to December.

Figure 4 shows the dynamics of surface wind anomalies on the Indian Ocean from May to December during nIOD event. Based on the figure, it can be seen that there is anomalous north-westerly wind, which starts to blow in June and reaches the peak in October. During the peak, the anomalous easterlies wind reaches 4.49 m/s. Spatially, the anomalous wind blows in the area about 5°N–10°S and 70°–110°E.

During nIOD event, anomalous SST and north-westerly wind are also accompanied by anomalous high rainfall in Indonesia [12]. Figure 5 shows the dynamics of rainfall anomalies from May to December during nIOD event. Based on the figure, it can be seen that there is positive rainfall anomalies in the eastern Indian Ocean (near Indonesia) and negative anomalies in the western Indian Ocean (Eastern Africa). The anomalous high rainfall tends to move eastward and become higher in the following months. Then, it reaches the peak in October up to 7.63 mm/day. It is suggested that higher

IOP Conf. Series: Earth and Environmental Science **303** (2019) 012004 doi:10.1088/1755-1315/303/1/012004

SST can trigger higher convective activities, moreover the north-westerly wind will blow the convective cloud eastward.

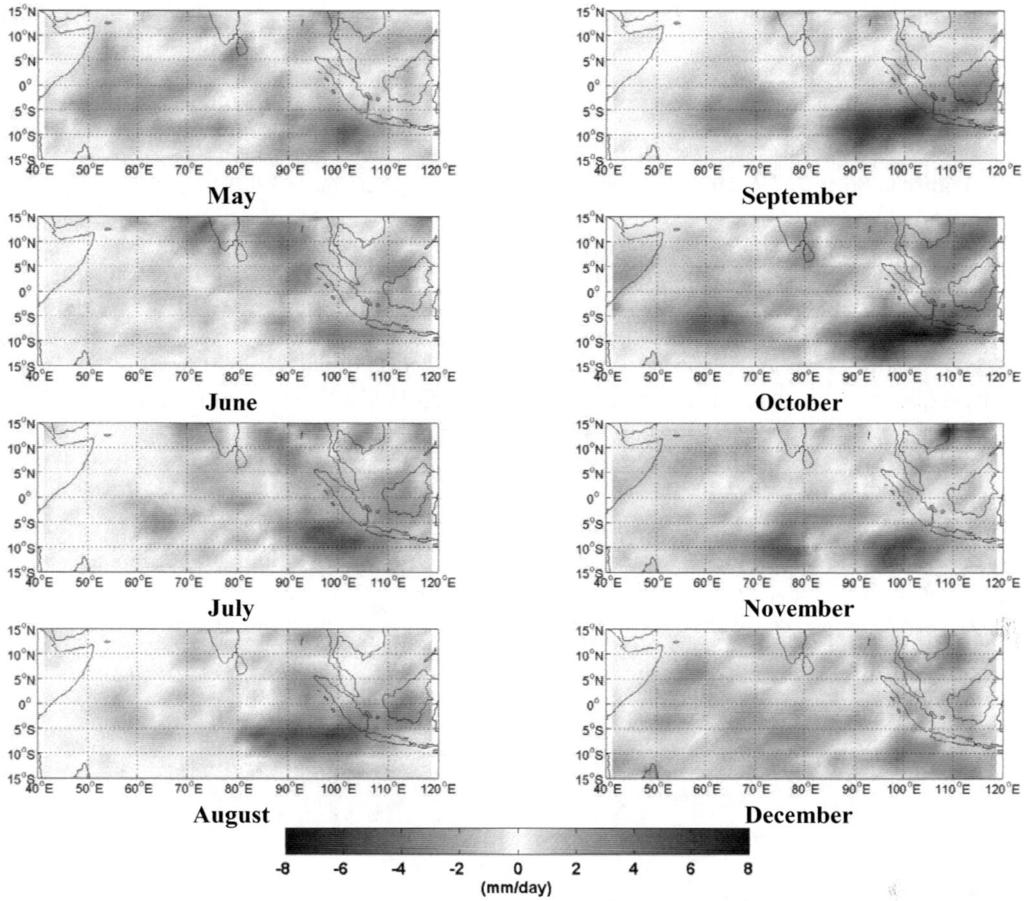

Figure 5. Composites of rainfall anomalies associated with negative IOD events (based on five nIOD events) from May to December.

3.2. The Relationship between Negative Indian Ocean Dipole (nIOD) and Anomalous High Rainfall in West Java Province

Before further analyzing the relationship between rainfall in West Java Province and the IOD phenomenon, it will be examined in advance the monthly rainfall profile or seasonal cycle of rainfall in West Java Province. Figure 6 shows the monthly rainfall profile of West Java Province. In the figure, it is seen that the highest rainfall occurred in January and February and decreased until it reached the lowest rainfall in August, then rainfall will increase until the end of the year. High rainfall in December to February due to coinciding with the west season (boreal winter) or the blowing of the northwest monsoon wind which carries moisture from the Asian Continent, while the rainfall is low in August due to coinciding with the east season (boreal summer) or the blowing of the southeast monsoon which tends to bring dry air from the Australian Continent [7, 13]. Based on this analysis, the type of rainfall in West Java Province is a monsoonal type due to its characteristics which are strongly influenced by the Asia-Australia monsoon system [7].

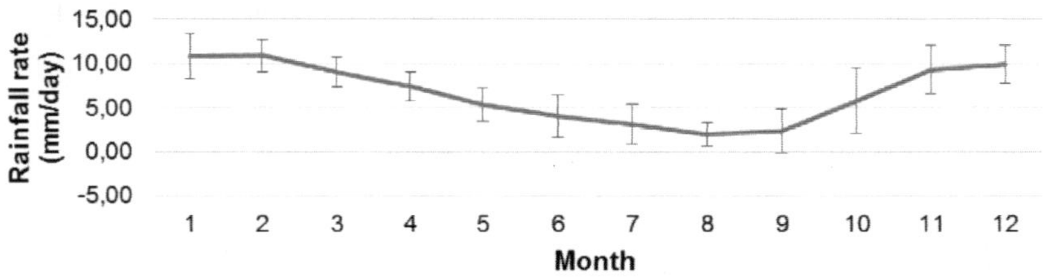

Figure 6. Profile of monthly average or seasonal cycle of rainfall in West Java Province.

If correlated linearly, the correlation coefficient between DMI and rainfall in West Java Province is very low ($r = -0.32$). This shows that the relationship between DMI and rainfall in West Java Province is not linear, so the relationship between the two is only assessed in extreme events, namely where the DMI value is below –0.5 and there is rainfall anomaly above one standard deviation (about 79 mm/month). To analyze the relationship between IOD and rainfall in West Java Province, the DMI value in October was compared to SON rainfall anomalies (September, October, and November). This is because the peak of IOD formation occurred in October. SON anomalies of seasonal rainfall are better used as a comparison than the average rainfall in October to avoid the influence of intra-seasonal phenomena, such as Madden-Julian Oscillation (MJO) [14].

Based on Figure 7 which shows the comparison between SON rainfall anomalies in West Java Province with the DMI value in October in time series, it was seen that in the years the incidence of negative IOD (DMI < –0.5) occurred an increase in extreme rainfall anomalies (above standard deviation of 79 mm/month).

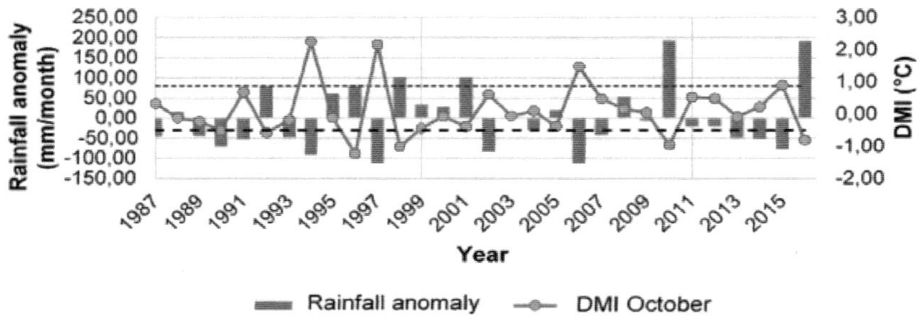

Figure 7. Comparison of SON seasonal rainfall anomalies in West Java Province with the DMI value in October from 1987 to 2016.

To further examine the relationship between DMI and rainfall in West Java Province, Figure 8 presents a scatter plot between SON average rainfall and the DMI value. In the figure, it can be seen that the lower the DMI value is below –0.5, rainfall tends to increase significantly with the coefficient of determination (R^2) of 0.57. This shows that there is a fairly strong relationship between DMI and rainfall.

Further examining the relationship between rainfall and IOD, the rainfall that occurs in a particular range of DMI (above 0.5, between 0 and 0.5, between 0 and –0.5, and below –0.5) is evenly distributed level up. Figure 9, which presents the relationship between the DMI value and the average rainfall, shows that there is a significant increase in rainfall when the DMI value is below –0.5, so it is clear that there is a significant effect of nIOD on increasing rainfall in West Java Province, which was increased averagely up to 8.92 mm/day.

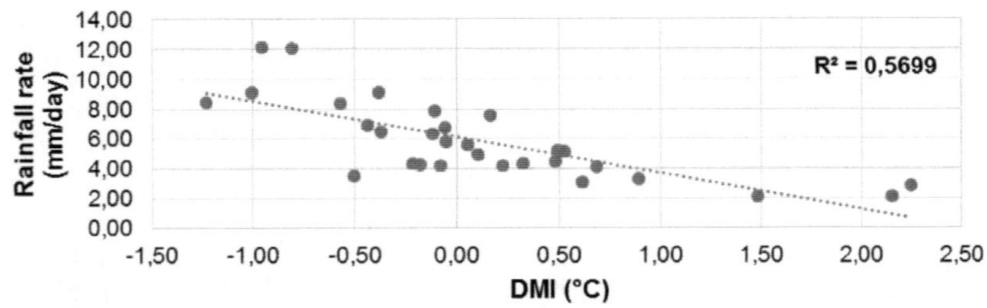

Figure 8. Scatter plot of seasonal rainfall SON (September, October, and November) in West Java Province with the DMI in October.

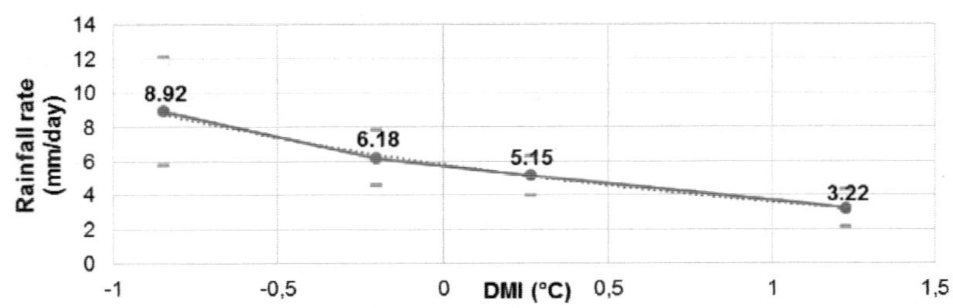

Figure 9. Relationship between DMI and average rainfall based on certain DMI ranges.

Figure 10 shows a comparison between monthly rainfall profiles with rainfall profiles when the nIOD events in 1992, 1996, 1998, 2010 and 2016, to see how the characteristics of rainfall in West Java Province when the incidence of nIODs. Based on the picture, it is seen that the rainfall in West Java Province from January to May tends not to be much different from the monthly profile (not too far above the standard deviation), but in the following months there was a significant increase in rainfall until the peak was in the month October which is the peak of the incidence of nIOD. From November to December, when the IOD has faded and disappeared, the rainfall returns to normal.

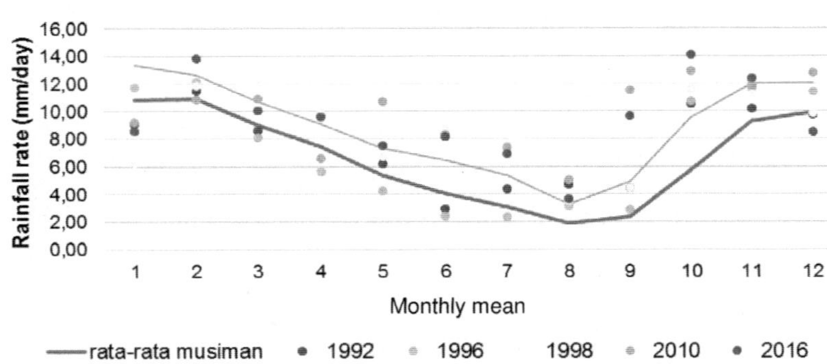

Figure 10. Comparison of monthly mean (seasonal cycle) rainfall in West Java Province with rainfall profile in West Java Province during the negative IOD events in 1992, 1996, 1998, 2010 and 2016.

4. Conclusion

Based on the composite analysis of five nIOD events, the significant positive (negative) anomalous SST appeared around May in eastern (western-central) of Indian Ocean during negative IOD event. The intensity of anomalous SST in the following months reached its peak in September and accompanied by

IOP Conf. Series: Earth and Environmental Science **303** (2019) 012004 doi:10.1088/1755-1315/303/1/012004

strong anomalous north-westerly wind. Furthermore, the high SST and anomalous wind increased the convection activity in the southern Java Island. Consequently, the average rainfall in West Java may significantly increase up to 8.92 mm/day and also lead the risk of hydro-meteorological disaster. Thus, by this study, it is suggested as reference and consideration in efforts to mitigate hydro-meteorological disasters by the government of West Java Province, particularly.

Acknowledgement
Acknowledgement and gratitude are given to European Centre for Medium-Range Weather Forecasts (ECMWF) and National Center for Atmospheric Research (NCAR) for providing the SST, wind, and rainfall data, respectively, and also to everyone who helps, encourages, and supports us to finish this research, especially Mrs. Rahmawati Rahayu from Meteorology Department of ITB.

References

[1] Saji N H, Goswami B N, Vinayachandran P N and Yamagata T 1999 *Nature* **401** 360–63
[2] Vinayacahndran P N, Francis P A and Rao S A 2009 *Current trends in science* 569–89
[3] Saji N H and Yamagata T 2003 *Climate Research* **25** 151–69
[4] Lu B, Ren H L, Scaife A A, Wu J, Dunstone N, Smith D, Wan J, Eade R, MacLachlan C and M Gordon 2018 *Climate Dynamics* **51** 89–100
[5] Lim E P and Hendon H H 2017 *Scientific reports* **7** 12619
[6] Hadmoko D S, Lavigne F, Sartohadi J, Gomez C and Daryono 2017 *Forum Geografi* **31** 1-15
[7] Aldrian E and Susanto R D 2003 *International Journal of Climatology* **23** 1435–52
[8] Nur'utami MN and Hidayat R 2016 *Procedia Environmental Sciences* **33** 196-203
[9] Dee D P *et al.* 2011 *Q J R. Meteorol Soc.* **137** 553–97
[10] Reynolds R W, Rayner N A, Smith T M, Stokes D C and Wang W 2002 *Journal of Climate* **15** 1609–25
[11] Tanizaki C, Tozuka T, Doi T and Yamagata T 2017 *Climate Dynamics* **49** 1289-304
[12] Wyrtki K 1961 *Physical Oceanography of the Southeast Asian waters* UC San Diego: Scripps Institution of Oceanography
[13] Lestari D O, Sutriyono E, Sabaruddin S and Iskandar I 2018 *Journal of Mathematical and Fundamental Sciences* **50** 257–72
[14] Black E, Slingo J and Sperber K R 2003 *Monthly Weather Review* **131** 74–94

Estimated transport of Equatorial Undercurrent observed at 90°E during the Indonesia Prima 2017 campaign

Siswanto[1], E Kusmanto[2] and Michael J McPhaden[3]

[1]Center for Marine Meteorology, The Agency for Meteorology, Climatology, and Geophysics (BMKG)
[2]Center for Oceanographic Research, Indonesian Sciences Institute (LIPI)
[3]PMEL Laboratory, National Oceanic and Atmosphere Administration (NOAA)

Abstract. Equatorial Undercurrent (EUC) plays an important role in the dynamic of the eastern Indian Ocean. EUC supplies water masses with high salinity into Indonesian waters. This article examines the EUC transport volume at 90°E across 2°S - 2°N observed on 1st – 3rd March 2017 during the Initiative on Maritime Observation and Analysis Expedition (Indonesian Prima 2017). The analysis of temperature and salinity obtained from conductivity, temperature and depth (CTD) instruments at five stations (CTD11 - CTD14) and current profiles of Shipboard Acoustic Doppler Current Profiles (SADCP) indicate the presence of high speed water column flowing the Arabian Sea High Salinity Water (ASHSW) as characterised by maximum salinity (35.15 - 35.2 PSU) in the temperature range of 18°C - 23°C. ASHSW carried by EUC from the western Indian Ocean at the upper thermocline layer had a tendency to be asymmetrically stronger to the north of the equator. The analysis shows a maximum EUC current speed of 94 cm/sec. Estimated EUC transport water masses based on the curvature area of salinity contour 35.15 and 35.2 PSU respectively result a ~3.4 Sv and a ~1.4 Sv, while at curvature area of salinity 35.00 - 35.10 PSU gave about ~8.7 Sv. The total estimated EUC mass eastward transport calculated in this study is ~13.5 Sv.

Keywords: equatorial undercurrent, Indian Ocean, SADCP.

1. Introduction

Equatorial Undercurrent (EUC) is the flow of water mass eastward that is strong in the subsurface layer in the equatorial region that occurs in the Atlantic, Indian and Pacific Oceans. There have been known two main EUC drivers namely the southeasterly trade wind which blows above the equatorial zone and the convergence of the east equatorial nonlinear currents. The convergence of east nonlinear currents toward the equator, in which caused by changes in the coriolis effects between the northern and southern hemispheres, lead to a strong subsurface current around the equator [1].

The maximum speed of EUC differs depending on the oceanic region. The maximum speed of EUC in the Pacific Ocean (often called Cromwell current) is more than 150 cm per second [2], in the Atlantic Ocean (often called Lomonosov current) with a velocity more or less similar to the Cromwell current. In the Indian Ocean, EUC is often called a strong subsurface equator jet and does not exist throughout the year. This phenomenon occurs in the winter when the northeast monsoon develops, and disappears in the summer when the southwest monsoon occurs [1], with currents between 50-100 cm/sec [6].

EUC in the Indian Ocean is already known to be a seasonal phenomenon [3]. This current has a relatively stronger magnitude in the eastern Indian Ocean [4] flows eastward and can be found in a strong thermocline layer (pycnocline) beneath the top layer (top layer mixing).

Rao et al. [5] stated the EUC presence at depths between 50 m and 150 m at the equator at a latitude range of 2°S from 61°E - 63°E during January - February 1963.

EUC is often referred to as a subsurface jet that occurs in the winter when the northeast monsoon develops, and disappears in summer during the southwest monsoon, with a current velocity of 50-100 cm/sec [6].

1.1. Observed zonal equatorial current

Observed zonal equatorial current from RAMA/ATLAS mooring buoy at 0° 90°E during 2000 - 2014 (Fig.1) shows a variation of a strong eastward current which mainly has two maximum axis speeds (>50 cm/s) namely at ~50 and ~100 m depth. The deeper presence earlier mainly during the late boreal winter, while the shallower appears later during springtime, although variation in the strength of current speed and time appearance is also observed.

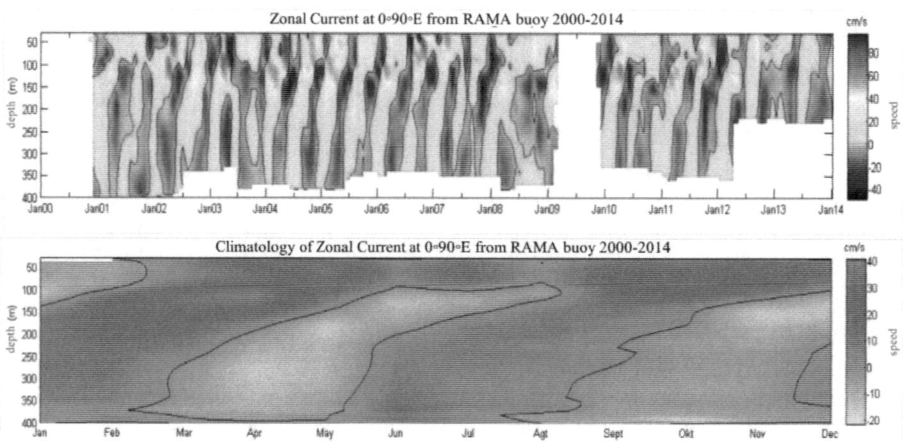

Figure 1. (upper) Observed zonal current variation from the surface to 400m depth obtained from RAMA mooring buoy at 0° - 90°E during 2000 - 2014. (bottom) The climatology means of 14 years of data. Red (blue) indicates an eastward (westward) current.

1.2. Indonesia PRIMA 2017 Field Campaign

The Indonesia PRIMA field campaign is part of a ship time BMKG - NOAA collaboration for maintaining RAMA mooring buoy (ATLAS).

The field campaign includes data survey in the marine meteorology, geophysics, oceanography, and air quality. There were 23 CTD stations and along track SADCP data extracted from the activity. This study takes the advantages of 5 CTD stations and current data represent a latitude range of 2°N - 2°S at 90°E (Fig.2).

The results of this study are expected to provide additional information about the characteristics of the EUC water mass and the volume of instantaneous water mass transport from the western Indian Ocean to the eastern Indian Ocean including Indonesian waters.

2. Objectives

This study aims to determine the characteristics of the eastern Indian Ocean water mass and the EUC current at longitude 90°E and also estimating instantaneous transport of the EUC water masses in the eastern Indian Ocean.

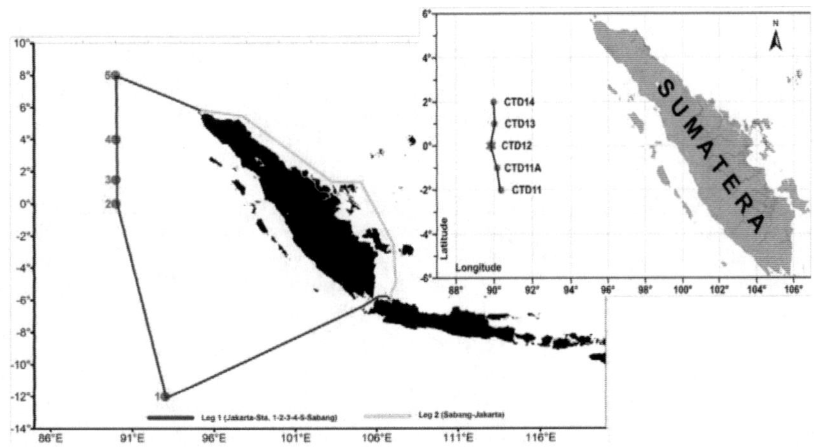

Figure 2. Full ship-track of Indonesia PRIMA 2017, 5 RAMA buoy maintenance sites including mooring buoy at 0° 90°E (red dots) and 5 CTD stations used in this study (insert). The current data was obtained from recorded along track SADCP during the field campaign.

3. Data and Methods

We use observational data from CTD11 - 14 and along track SADCP obtained from 1 - 3 March 2017 in Indian Ocean at latitude range between 2S - 2°N along longitude 90°E.
The temperature and salinity parameters are measured continuously up to 1000 m depth using SBE 911 CTD from Seabird Electronic Inc and being smoothed at 1 m resolution

SADCP 75 kHz frequency from RD Instrument measures the current parameters from a depth of 12.3 m to 500 m from the surface with a vertical bin every 5 m with 2 seconds intervals at an average 8 knots ship velocity.

The raw data is then being processed to obtain current velocity and direction, zonal current (u) and meridional current (v).

For quality control process, outlier data is eliminated by criteria of percent good below 50% and correlation coefficient below 0.8 as well as eliminating abnormal data due to a weak global positioning system signal and current velocity more than 250 cm/sec.

3.1. *Estimating the equatorial water mass transport*

Instantaneous transport estimation Q is calculated based on the mathematical sum of the gridding area of the cross section of the zonal current velocity at the equator ranging from 2°S to 2°N at 90°E longitude following the curvature area of salinity contour using the equation:

$$Q = \sum_{i=1}^{i=100} V_i . A_i \quad (1)$$

where

$$V_i = v_i \cos\theta + u_i \sin\theta \quad (2)$$

V_i is the equatorial current velocity on the ith cell whereas A_i is the cross-sectional area of the ith cell.

4. Results

The T-S diagram shows at least four different water masses (Fig.3A-B). Low-salinity water mass (33.90 - 34.80 PSU) is observed in the surface layer with sea water temperature varying between 28.0°C - 29.0°C. The second layer presents a high salinity water mass with a salinity range of 35.15 - 35.20 PSU and a temperature range of 18.0 - 23.0°C. The third layer is water mass with salinity 35.00 PSU, temperature 7.0 -11.0°C.

While the fourth deepest layer is water mass with salinity 34.70 - 34.90 PSU, the temperature range 2.0°C - 6.0°C. The mass of the water downward looks almost homogeneous.

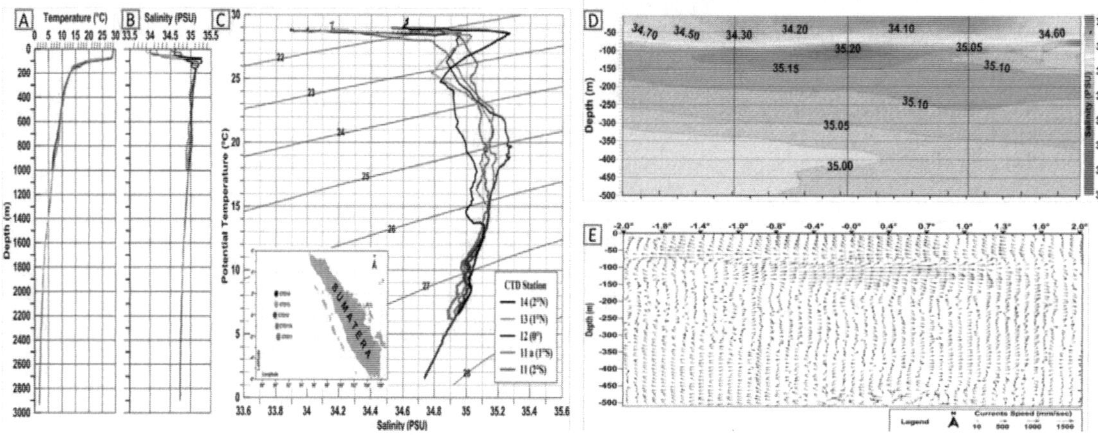

Figure 3. Vertical profile of A) temperature, B) salinity and C) T - S Diagram from CTD11--14 data. Latitudinal cross-section of (C) salinity and (D) current.

At a depth of 80m - 150m from latitude 1.2°S - 1.5°N there is a strong current with highly saline water flowing eastward with a maximum speed of 94 cm.sec^{-1} (Fig.3C-D and Fig.4). The current is characterized as an EUC current that has a stronger asymmetric tendency toward the north of the equator.

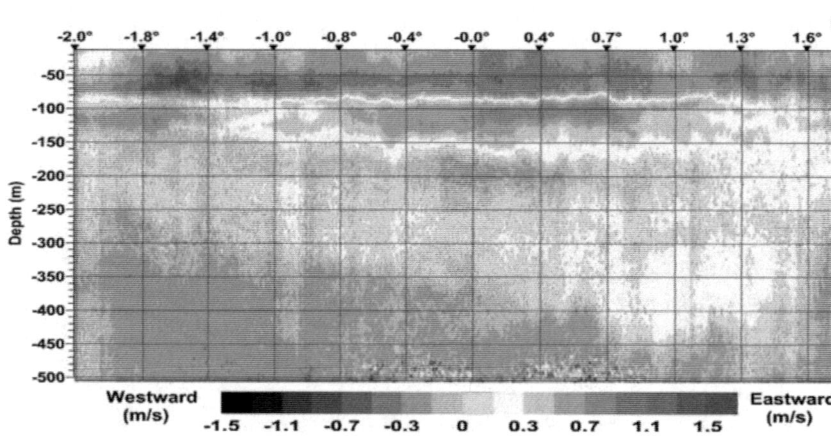

Figure 4. Meridional cross section the EUC zonal current component. Red (blue) indicates an eastward (westward) flow.

The estimated water mass transport is calculated by eastward EUC zonal current overlying the salinity contour (Fig.5A).

The salinity contour is then used to describe water mass columns that have similar characteristics, assuming the dynamics of the water mass follows the physical characteristics of seawater properties as in Fig.3B.

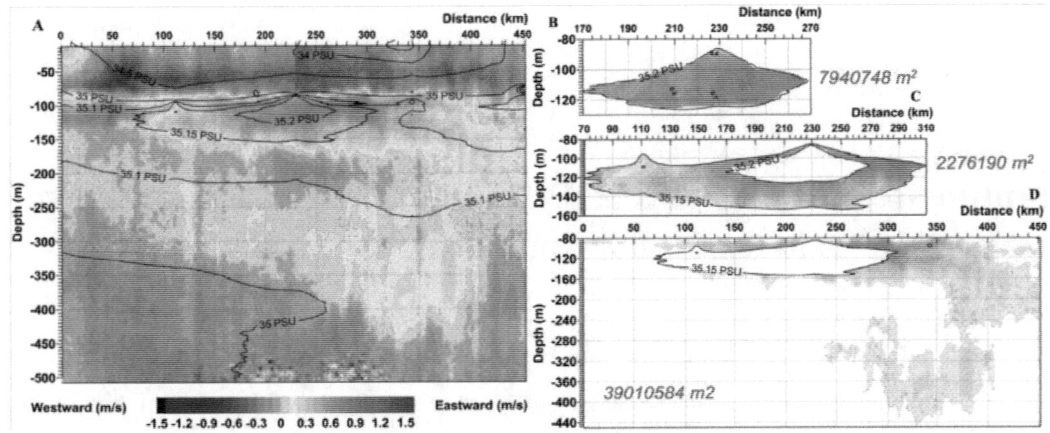

Figure 5. (A) Overlaid zonal current with salinity contour and (B) estimated transport volume based on salinity contour slices at 35.20 PSU, 35.15 PSU, and 35.00 - 35.10 PSU.

The result of estimated volumetric water mass transport based on zonal current velocity of EUC and salinity cross-sectional area defined as in Fig.5B) are 3364682 m3.s-1 ~3.4 Sv), 1377677 m3.s-1} (~1.4 Sv), and 8719226 m3.s-1 (~8.7 Sv), respectively.

Total estimated eastward transport for a high salinity range 35.0 - 35.2 PSU is ~13.5 Sv.

5. Conclusions

A strong EUC current with highly saline water flowing eastward with a maximum speed of 94 cm/sec is observed at depth of 80m - 150m from latitude 1.2S - 1.5N. This current is characterized has a stronger asymmetric tendency toward the north of the equator. The highly saline water is concluded as the Arabian Sea High Salinity Water (ASHSW). The result of estimated eastward transport based on zonal current velocity on EUC and salinity cross-sectional area of 35.0 – 35.2 PSU range is ~13.5 Sv. This result of estimated eastward transport in this study will need to be compared with other studies for forthcoming research.

Acknowledgements

We would like to thanks to BMKG, LIPI, NOAA, and RV Baruna Jaya for their available facilities to conduct this study.

References

[1] Philander S G 2008 *Encyclopedia of Global Warming and Climate Change* Sage Publications Inc 2455 Teller Road Thousand Oaks California 91320 ISBN 978-1-4129-5878-3 DOI: 10.4135/9781412963893.n238
[2] Knauss J A 1997 *Introduction to physical oceanography* Waveland Press pp. 148–151 ISBN 9781577664291
[3] Bruce J G 1973 *Equatorial undercurrent in the western Indian Ocean during the southwest monsoon* Journal of Geophysical Research volume 78 DOI: 10.1029/JC078i027p06386
[4] Knauss J A and Taft B A 1964 *Equatorial Undercurrent of the Indian Ocean* Science Jan 24 143(3604):354-6 DOI: 10.1126/science.143.3604.354
[5] Rao LVG and Jayaraman K 1968 *Vertical distribution of temperature, salinity and density in upper 500 meters of the North Equatorial Indian Ocean during the north – east monsoon period* Buletin of the National Institute of Science of India No. 38:123-148
[6] McPhaden M J 1986 *The Oceanography Report* EOS American Geophysical Union Vol 67 No. 40: 762-765

The Influence of ENSO and IOD During Mesoscale Convective Complex (MCC) to Rainfall in Indonesia

S Rustiana[1], Trismidianto[1] and H Satyawardhana[1]

[1]Center for Atmospheric Science and Technology, National Institute of Aeronautics and Space, Bandung, Indonesia

E-mail : shailla.rustiana@gmail.com

Abstract. Climate phenomena that significantly affect Indonesian rainfall to be lower (positive) and higher (negative) than normal conditions are ENSO and IOD. The other phenomenon with different time scale with ENSO and IOD is MCC that rainfall until causing storms. This study was conducted to determine the influence of ENSO and IOD during MCC to rainfall in Indonesia. The data used are monthly data of rainfall, Nino3.4-IOD index, and hourly MCC data derived from T_{BB} data of Himawari Satellite, year 2001-2015 observation. The study focused on 4 occurrence phenomena: Nino3.4 (+) IOD (+), Nino3.4 (-) IOD (-), Nino3.4 (+) IOD (-), and Nino3.4 (-) IOD (+) with the distribution of 3 longitude regions (90-105E, 106-125E, and 126-140E). The results showed that the distribution of rainfall during MCC were higher when the events of negative IOD compared to other events, especially in the western maritime of Sumatra until Kalimantan. While, in the Nusa Tenggara region there is no rainfall when positive Nino3.4 events due to positive ENSO influence from the Pacific Ocean. The MCC also give effect on interior cloud size and duration of life cycle when negative IOD events on 3 longitude regions. The effect includes the size of the interior of the cloud becomes larger and the duration of the life cycle is longer. Eccentricity interior is close to 1, means the contribution of MCC is high enough to Indonesian rainfall on negative IOD events.

Keywords : ENSO, IOD, MCC,rainfall.

1. Introduction

Indonesia's rainfall is various and dynamic mostly concentrated over the island [1]. The differences rainfall over Indonesia lies in apparent motion of the sun received in each region. Indonesia is one of the world's most convectively active regions [2]. Located between two continents (Asian and Australia) and between two oceans (Indian and the Pacific Ocean) [3] also situated at the core of the strongest monsoon region of the world, and its regional climate strongly influences both the Hadley and Walker circulations [1]. It plays an important role in the variability of the tropical and global climate system that caused climate phenomena occurred over there like El Nino/Southern Oscillation (ENSO), Indian Ocean Dipole (IOD), Madden-Julian Oscillation (MJO), Mesoscale Convective System (MCS), and Tropical Cyclone.

Rainfall in Indonesia closely related with ENSO in the Pacific Ocean. The previous study [4] said that during the early stages of 1997 El Nino, drought conditions developed across Indonesia and most of the Maritime Continent. Rainfall is normally occured over the western Pacific warm pool, whereas drier climate overall happening in the eastern and central equatorial Pacific and western Indian Ocean,

IOP Conf. Series: Earth and Environmental Science **303** (2019) 012006 doi:10.1088/1755-1315/303/1/012006

depending on the geophysical configuration of the Indo-Pacific coupled ocean-atmosphere system [5]. This region experiences higher (lower) rainfall during La Nina (El Nino) events in cased of 1997-98 [6]. The studies about the relationship between ENSO and IOD to Indonesia's rainfall have often been done. For the east Pacific El Nino type, a near-perfect linear correlation (r = 0.96) is detected between the El Niño intensity and the simultaneous IOD intensity [7]. This relatively high correlation is primarily due to the co-occurrence of the ENSO and the IOD events [8]. The occurrence of upwelling off of Java–Sumatra and the interplay of El Nino and the positive IOD generating rainfall anomalies in southwestern Western Australia [9] and certainly affected Indonesian's rainfall.

Beside ENSO and IOD, there is another phenomenon with shorter time scale (hourly) that also has an influence to rainfall distribution in Indonesia. Mesoscale Convective Complex (MCC) has an important role in the rainfall changes in the tropics, include Indonesia [10]. The contribution of MCC to total rainfall over Indonesia during fifteen years reached 20% [11]. Trismidianto [12] also reported that the largest frequency for occurrences of the MCC over Indonesia from total of 1028 MCCs during fifteen years (2001 – 2015), was found over the continental (42.32%). The previous study that conducted by Trismidiato and Satyawardhana [2], explained the influence of ENSO (El Nino and La Nina) condition reducing the number of MCC occurrences in Indonesia during June-July-August (JJA) season. Hence, the effect also reducing rainfall distribution in Indonesia. In this research, we also have seen the influence of IOD phenomenon to Indonesia's rainfall during MCC events. Because the ENSO+IOD phenomena have a strong effect to Indonesia's rainfall, so we need to determine how the ENSO+IOD phenomena influence to rainfall distribution in Indonesia during MCC occurrences.

2. Data and Methods

2.1. Data

This study utilized monthly data of Nino 3.4 from NOAA (http://www.cpc.ncep.noaa.gov/) and IOD index (DMI) data from JAMSTEC (www.jamstec.go.jp) [13] to identify ENSO and IOD phenomena during 2001 – 2015 (Figure 1). Four time occurrence of these two phenomena are chosen for comparison (Figure 2): Nino3.4(+) IOD(+), Nino3.4(-) IOD(-), Nino3.4(+) IOD(-), and Nino3.4(-) IOD(+). The selection of four occurrence phenomena based on time series data and the impact from the combination of those two phenomena (ENSO and IOD). The selected Nino 3.4 as observation data region, because based on Trenbeth's research [14], in this region ENSO events might be more precisely defined by a threshold of 0.3°C.

Analysis of spatial rainfall data provided by GSMaP data [15] from JAXA (http://sharaku.eorc.jaxa.jp/GSMaP/). The spatial distribution of rainfall from GSMaP has spatial and time resolutions 0.1°×0.1° and has time observation near real time. To identify the influence of MCC toward Indonesia's rainfall, this research utilized brightness temperature (T_{BB}) from Himawari geostationary satellites [16] that provided by the Japan Meteorological Agency (JMA) (http://weather.is.kochiu.ac.jp). In detail, the data consists of Himawari-5/GMS-5 for data from January 2001-April 2003, Pacific GOES/GOES-9 for data from May 2003-June 2005, Himawari-6/MTSAT-1R for data from July 2005 - June 2010, Himawari-7/MTSAT-2 for data from July 2010 - June 2015, and Himawari-8 for data from July 2015 - December 2015. This satellite data has spatial and time resolutions 0.05°×0.05° and one hour [2].

2.2. Methods

Identification of the four time combination of ENSO and IOD occurrences, conducted by sorting positive and negative data (Fig. 1) with a threshold of 0.3°C. Then grouped on 4 occurrences: Nino3.4(+) IOD(+), Nino3.4(-) IOD(-), Nino3.4(+) IOD(-), and Nino3.4(-) IOD(+) shown in Fig. 2. From the grouping occurrences, then we conducted the spatial analysis of rainfall distribution and becomes the reference to be associated with the MCC [2]. The MCC information used to explain the MCC characteristics over the 3 longitude regions (90-105E, 106-125E, and 126-140E) consisting of the area size interior, duration life cycle, and eccentricity interior of MCC.

3. Result and Discussion

3.1. Statistical Analysis

As mention from the previous section, to identify ENSO and IOD phenomena that influence to Indonesia's rainfall, we used Nino3.4 index and IOD index (Dipole Mode Index/DMI) data (Fig. 1). This figure explains that there are manytime when Nino3.4 index and DMI occurrence at the same time (positive on 2002, 2004, 2006, 2009, and 2015 and negative, partly in 2005, 2010, and 2013). There are also opposite conditions (Nino3.4(+) IOD(-) and Nino3.4(-) IOD(+)) at the same time, like the case on 2001, 2007, 2008, 2011, and 2013.

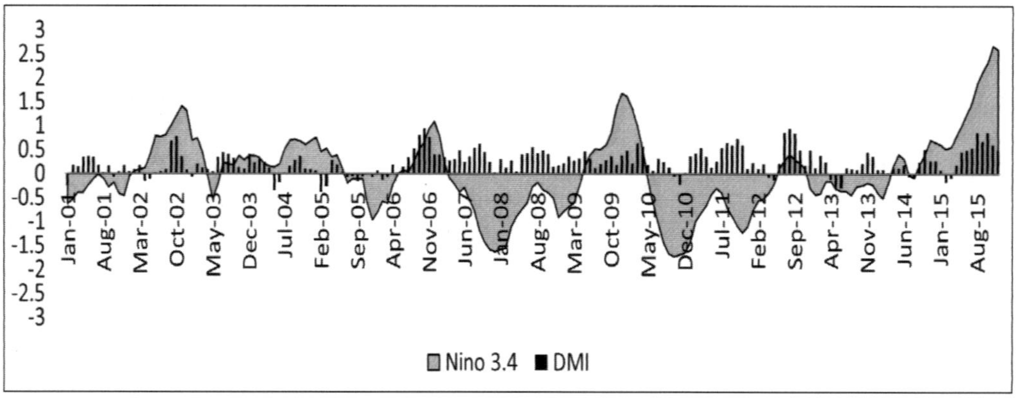

Figure 1. Time series of monthly ENSO and IOD occurrence on 2001 – 2015.

The high dataset of Nino3.4 and DMI on the boreal winter of 2014 until boreal winter of 2015 is strong El Nino phenomena that have a big impact to Indonesia's rainfall. On 2015, Indonesia has severely affected by El Nino, with marked rainfall deficits extending throughout the year. For the August-December 2015, rainfall indicates drier than average conditions over the country. On a longer perspective, Indonesia is clearly expected to experience drier than average conditions until 2016 [17].

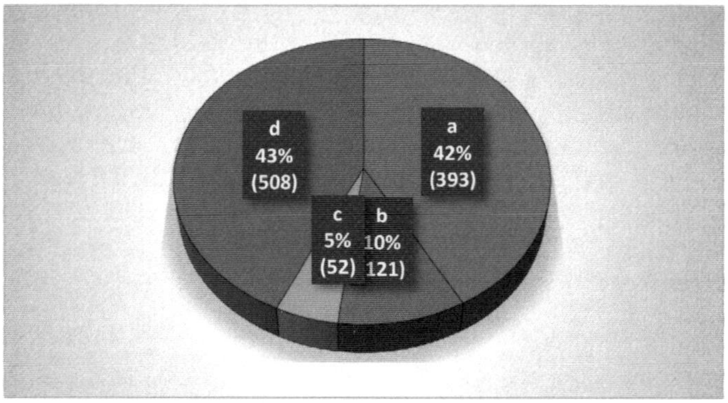

Figure 2. Percentage of ENSO and IOD combination for each time occurrence phenomena:

(a) Nino3.4(+) IOD(+),
(b) Nino3.4(-) IOD(-),
(c) Nino3.4(+) IOD(-), and
(d) Nino3.4(-) IOD(+).

The grouping 4 combinations of occurrence phenomena from the time series analysis (Figure 1) has a percentage as shown in Figure. 2. A combination of phenomena when an IOD(+) has a higher percentage than IOD (-), where a combination of Nino3.4(+) IOD(+) phenomena has 42% and a combination of Nino3.4(-) IOD(+) phenomena has 43%. Meanwhile, the combination of Nino3.4(-) IOD(-) phenomena has 5% and the combination of Nino3.4(+) IOD(-) phenomena has 10%. Hence, for

15 years of observations from the Nino3.4 and IOD index data, there are several phenomena when Nino3.4(+) and Nino3.4(-) occur with IOD(+) at the same time.

3.1.1. Analysis Area Size of MCC

To analyzed MCC occurrence more detail over Indonesia, we classify 3 region part of Indonesia from longitude large area. The western part region is 90 – 105 E, central part region is 106 – 125 E, and east part region is 126 – 140 E. For comparison, we also analyzed the total region of Indonesia (90 - 140 E) with 4 grouped combinations of occurrence phenomena.

Table 1. Interior Cloud Size Area of MCC (km^2) x10^4.

Longitude Regions	Occurrence Phenomena							
	Nino3.4(+) IOD(+)		Nino3.4(-) IOD(-)		Nino3.4(+) IOD(-)		Nino 3.4(-) IOD(+)	
	Mean	Std. Deviation	Mean	Std. Deviation	Mean	Std. Deviation	Mean	Std. Deviation
Total	15.64	11.93	14.87	11.84	15.59	13.02	14.00	9.30
90 – 105 E	19.93	15.43	19.38	15.30	19.87	17.22	17.77	11.95
106 – 125 E	12.96	7.19	10.29	3.63	13.08	9.18	12.27	7.43
126 – 140 E	12.06	7.88	13.40	9.63	11.33	4.41	11.78	5.75

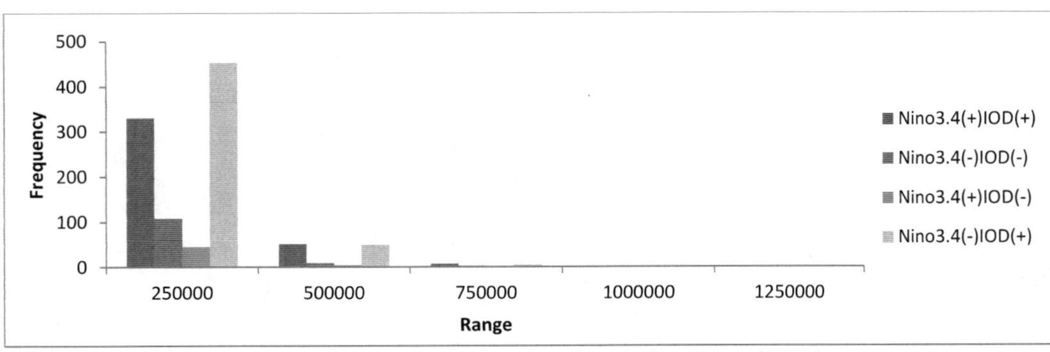

Figure 3. Histogram of Interior Cloud Size Area of MCC.

The number of MCC interior cloud size area in each group observation region and occurrence phenomena shown in Table 1. The western part region (90 -105 E) has a higher number of area sizes for each combination of occurrence phenomena than other regions. The same thing with the variation that evidenced by the higher standard deviation value than the other regions. The smallest value of standard deviation is 3.63 when Nino3.4(-) IOD(-) on 106 – 125 E region. That area makes it possible to have a lower variation of MCC because It is far from the Pacific Ocean as well as the Indian Ocean. The area size of MCC also larger when IOD(-) events both when the combination occurrence of Nino3.4(+) IOD(-) and Nino3.4(-) IOD(-) in total average around 15x10^4 km^2 compared to IOD(+) events.

Based on histogram on Figure 3, the highest MCC distribution value is on the range of 0 – 2500000 km^2 (453 frequency) in the condition of Nino3.4 (+) IOD (-).

3.1.2. Analysis Duration Life Cycle of MCC

The duration life cycle of MCC in this study define as the duration between the time of initial stage and the time of dissipation stage. The distribution frequency of the MCC duration life cycle for each combinations of occurrence phenomena shown in Table 2. The average number of duration life cycle shows around 11 hours for each region and occurrence phenomena. This number is close enough with analysis of MCC during El Nino, Neutral, and La Nina conditions over IMC that have done by a previous study [2].

Table 2. Duration Life Cycle of MCC (Hour).

Longitude Regions	Occurrence Phenomena							
	Nino3.4(+) IOD(+)		Nino3.4(-) IOD(-)		Nino3.4(+) IOD(-)		Nino 3.4(-) IOD(+)	
	Mean	Std. Deviation	Mean	Std. Deviation	Mean	Std. Deviation	Mean	Std. Deviation
Total	11.59	3.08	11.19	3.08	11.79	3.23	11.06	3.10
90 – 105 E	12.32	3.51	11.67	3.41	13.52	3.84	11.65	3.46
106 – 125 E	11.27	2.66	10.63	2.46	10.83	2.20	10.50	2.67
126 – 140 E	10.72	2.54	11.15	3.26	9.86	1.57	11.29	3.17

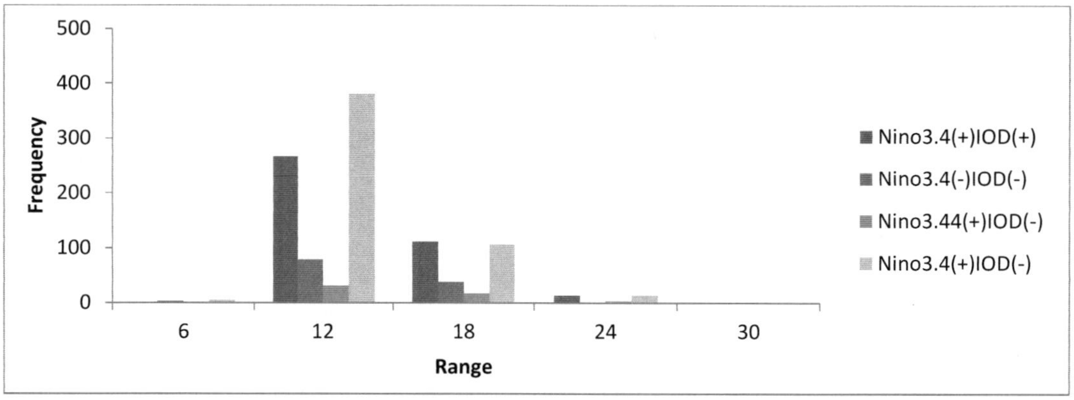

Figure 4. Histogram of Duration Life Cycle of MCC.

Same with case on Table 1 related to the interior cloud size area of MCC, the frequency of MCC duration life cycle longer in western part region (90 – 105E) than another region for each combinations occurrence phenomena, which is also proven by the magnitude of the variance of the standard deviation value. The average frequency of duration life cycle of MCC around 12 hours at this region. The combination occurrence phenomena with IOD(-) events also have longer duration life cycle than occurrence phenomena with IOD(+). Even though Lou et al. (2007) [18] said that negative IOD events do not appear to evolve into strong air-sea coupled processes in the Indian Ocean, in this study case IOD negative have strongly related with MCC occurrence. The effects of these phenomena will be explained in more detail in the next section (Figure 8). On the duration life cycle analysis, the highest MCC distribution value is on the range of 7 – 12 hour (381 frequency) shows in Figure 4. The extreme values of MCC characteristic also showed more on the Nino3.4(+) IOD(-) occurrence.

3.1.3. Analysis Eccentricity Interior of MCC

The previous study about MCS and MCC have measured the shape of the cloud-shield with calculated the eccentricity. This eccentricity is very important to distinguish the MCC with the other type of MCS. The eccentricity itself defined as the ratio of the minor axis to the major axis of the MCC best-fitting ellipse at the time of maximum extent [2]. The closer of eccentricity value to 1 means the contribution of MCC is quite high for this region and phenomenon.

Table 3. Eccentricity Interior of MCC.

Longitude Regions	Occurrence Phenomena							
	Nino3.4(+) IOD(+)		Nino3.4(-) IOD(-)		Nino3.4(+) IOD(-)		Nino 3.4(-) IOD(+)	
	Mean	Std. Deviation	Mean	Std. Deviation	Mean	Std. Deviation	Mean	Std. Deviation
Total	0.85	0.08	0.86	0.08	0.86	0.09	0.85	0.09
90 – 105 E	0.85	0.08	0.86	0.08	0.84	0.09	0.86	0.08
106 – 125 E	0.85	0.08	0.86	0.093	0.88	0.08	0.84	0.09
126 – 140 E	0.84	0.08	0.87	0.08	0.87	0.08	0.84	0.09

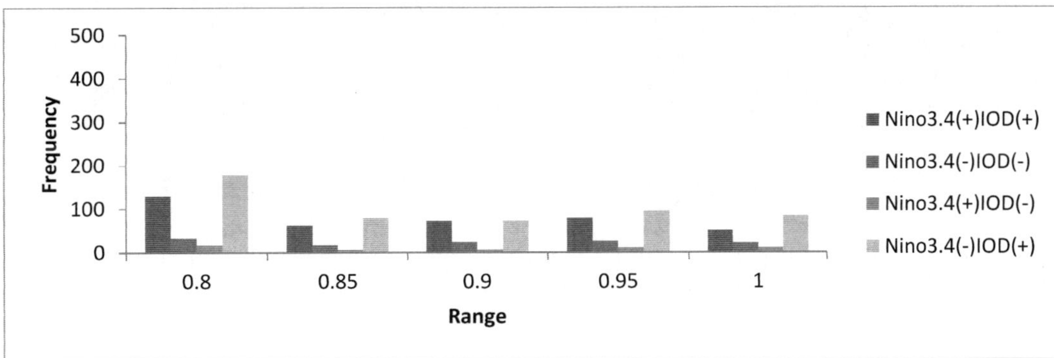

Figure 5. Histogram of Eccentricity Interior of MCC.

This study found the average eccentricity interior of MCC for all the observed regions was around 0.86, while the standard deviation value was around 0.08 (Table 3). Different from previous analysis, the analysis of this MCC eccentricity larger at central part region (106-125E). The higher value of eccentricity located around central of Kalimantan island (Figure 7), which means is located near equatorial regions. The development of the MCC near equatorial is possibly related to the ITCZ as a belt of low pressure which circles the earth generally near the equator, where the trade winds of the Northern and Southern Hemispheres come together. It is marked by convective activity which often generates strong thunderstorms over the large areas. It is most active over continental land masses by day and relatively less active over the oceans [19]. The extreme values of MCC characteristic on the eccentricity interior analysis also showed more in Nino3.4(+) IOD(-) occurrence. Based on histogram (Figure 6) analysis, the highest MCC distribution value is on the range of 0.7 – 0.8 (179) in the condition of Nino3.4(+) IOD(-).

3.2. ENSO and IOD Analysis during MCC Condition

Spatial analysis of Indonesia's rainfall for combination of ENSO+IOD occurrence phenomena (Figure 6) shows that Indonesia's rainfall is higher in Sumatra, Java, Kalimantan, and several Papua island for each occurrence phenomena. Distribution of Indonesia's rainfall when the combination of Nino3.4(-) IOD(-) occurrence phenomena is higher than other conditions. It shows with the color of rainfall distribution for all Indonesia region are green to yellow, with the average value 0.9 mm/hour near southern part of Sumatera and central part of Papua. Rainfall distribution when the combination of Nino3.4(+) IOD(-) occurrence phenomena is smaller than the other condition. The range of rainfall value is 0.1 to 0.6 mm/hour, this value smaller than combination of Nino3.4(+) IOD(+) and Nino3.4(-) IOD(+) that have average rainfall around 0.7 to 0.8 mm/hour.

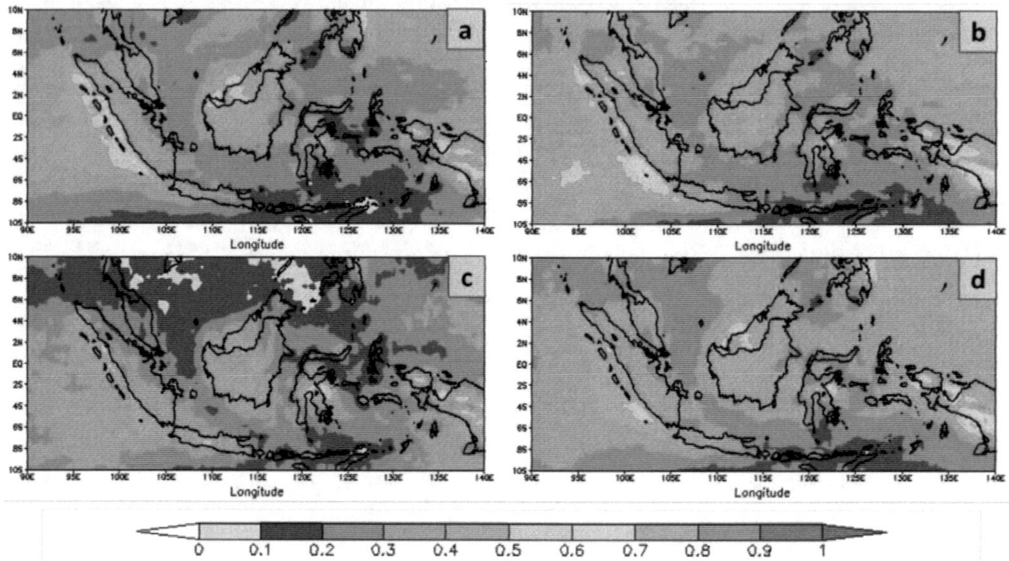

Figure 6. Rain rate distribution when ENSO and IOD event for each time occurrence phenomena : (a) Nino3.4(+) IOD(+), (b) Nino3.4(-) IOD(-), (c) Nino3.4(+) IOD(-), and (d) Nino3.4(-) IOD(+).

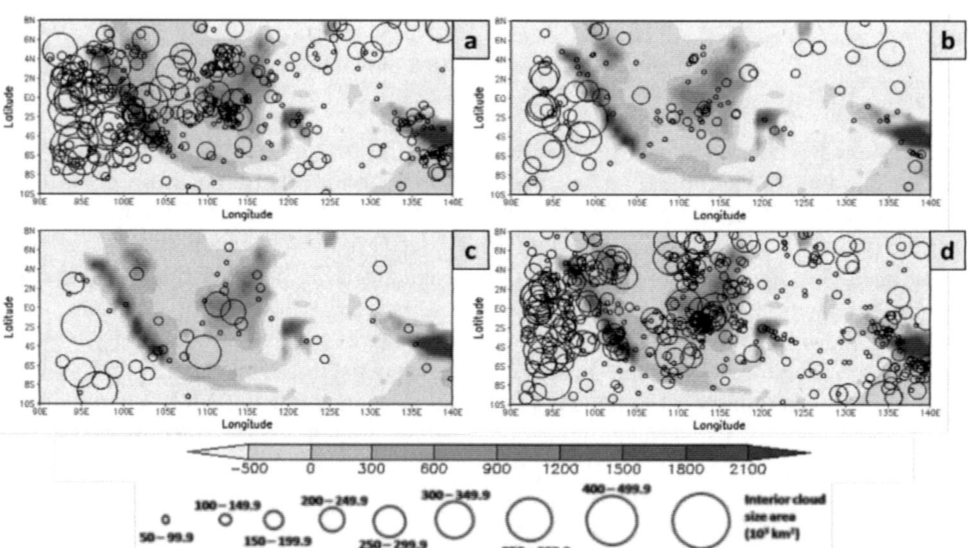

Figure 7. Frequency distribution of the MCC occurrences over Indonesia during 15 years from 2001 to 2015 when ENSO and IOD events for; (a) Nino3.4(+) IOD(+), (b) Nino3.4(-) IOD(-), (c) Nino3.4(+)IOD(-), and (d) Nino3.4(-) IOD(+). The circles shown the size and location of MCC when the mature stage at the time of the interior cloud size area of MCC reached the maximum extent (10^3 km^2). The color shaded refers to elevation which is the unit in meters (m).

Figure 7 shows the combination of ENSO and IOD phenomena influence to MCC events in Indonesia. When the combination of Nino3.4(+) IOD(+) occurrence phenomena, the MCC spread more in Sumatra and Kalimantan, while when the combination of Nino3.4(-) IOD(-), MCCs spread in the Indian Ocean and the Java Sea, and also some of MCC with small circle concentrated in Central Kalimantan. On the condition of Nino3.4(-) IOD(+), MCC spread averagely in the ocean near Sumatra and Kalimantan, also several around Papua. Then, when the Nino3.4(-) IOD(+), MCC spread almost all over Indonesian both on ocean and land.

From Figure 7, there is a condition looks like the MCC occurrences directly proportional with the percentage of combination ENSO and IOD occurrence (Figure 2), indicated by the number of cloud size interior circles. Hence, the distribution of MCC larger in the combination occurrence phenomena with IOD(+) which have percentage over 40% than IOD(-) with the percentage less or equal to 10%. Meanwhile, the impact of these combination occurrence phenomena during MCC events to Indonesia's rainfall (Figure 8) more visible than the analysis of frequency distribution of MCC (Figure 7).

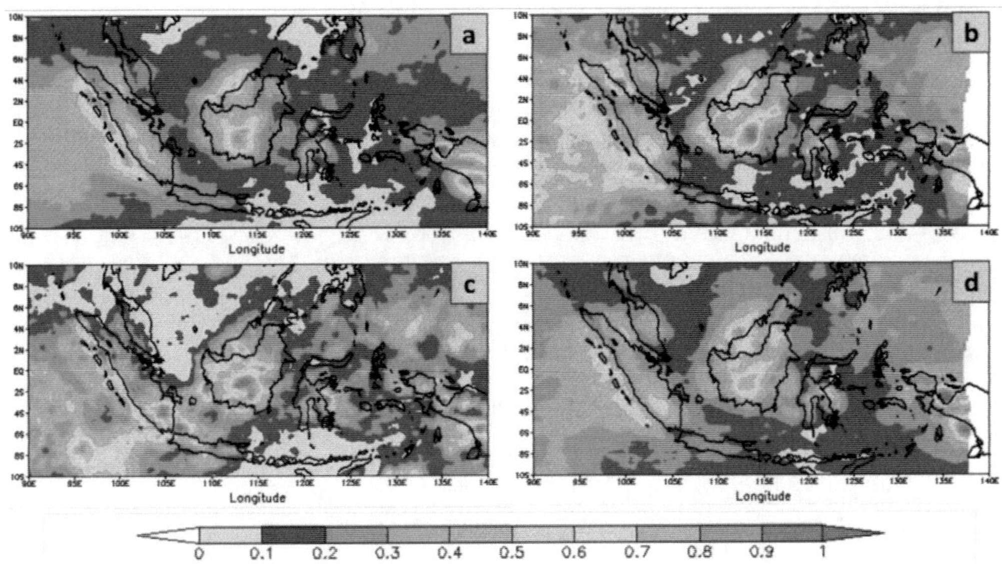

Figure 8. Rain rate distribution when ENSO IOD events during MCC for each time occurrence phenomena:(a) Nino3.4 (+) IOD (+), (b) Nino3.4 (-) IOD (-), (c) Nino3.4 (+) IOD (-), and (d) Nino3.4 (-) IOD (+).

Rainfall in Java during MCC events, not spread as much as Sumatra and Kalimantan region because the reduced of water vapor supply which become rain and producing convective clouds. As we know, the region which strongly affected by the climate phenomena in Indonesia is Java island, because it located between two continents (Asian and Australia) [1]. It appears that the largest MCCs at the time of Nino3.4(+) occurred in the southwest ocean of Sumatra (the eastern Indian Ocean), which is the most frequently formed area of large-scale MCC and the Pacific Pole (warm pole). MCC still spread both small and large in Sumatra and Kalimantan even though in the condition of Nino3.4(+), it is the same case with the previous study. The Walker's circulation shift at the condition of Nino3.4(+), possibly causing shifts and changes in the scale of MCCs further east. Hence, during the Nino3.4(-), MCCs spreads mostly in land and sea, otherwise, in Nino3.4(+), MCCs occur on land and several in the oceans [2].

4. Conclusion

Rainfall distribution during MCC event was higher when the occurrence phenomena with negative IOD (IOD(-)) compared to other phenomena, especially in the western maritime of Sumatra until Kalimantan. Meanwhile, Java, Bali, and Nusa Tenggara region there is no rainfall when positive Nino3.4 (Nino3.4(+)) occurrences, It is due to positive ENSO influence from the Pacific Ocean. The effect of MCC also seen on the larger interior cloud size area and longer duration life cycle when negative IOD events on 3 longitude regions, where the western part region (90-105E) has higher average value and variance than the other regions. The eccentricity interior of MCC is close to 1, with the average around 0.86 and standard deviation around 0.08 also when negative IOD events and the larger in the central part region (106-110E). It also supported by the results of the 3 parameter analysis that shows the highest (extreme) value occurs when negative IOD events, It means the distribution of MCC is high enough to Indonesia's rainfall when negative IOD occurrence phenomena, The next study is required to determine the monthly and seasonal analysis on the influence of the ENSO+IOD occurrence phenomena combination with the strength level from those phenomena (strong, moderate, and weak) during the MCC events.

Acknowledgments
Thank you to Japan Meteorological Agency which provides Himawari data as the most important thing to this paper. NOAA and JAMSTEC which provides Nino 3.4 data and DMI data to analyse ENSO and IOD phenomena during 15 years. JAXA which provides GSMaP rainfall data, so the Indonesia's rainfall distribution could be analysed in this study. Thank you also for all author who works on the related fields which is cited in the references, so this paper can be finished. Lastly, the authors are grateful to the National Institute of Aeronautics and Space (LAPAN), which is the institution where the author works and also as the organizer committee to this international conference (ICTMAS).

References
[1] Qian J H 2007 Why Precipitation Is Mostly Concentrated over Island in the Maritime Continent *Journal of The Atmospheric Science* Vol **65**: 1428 -1440 doi: 10.1175/2007JAS2422.1
[2] Trismidianto and Satyawardhana H 2018 Mesoscale Convective Complexes (MCCs) over the Indonesian Maritime Continent during the ENSO events *IOP Conf. Ser.: Earth Environ Sci.* **149** 012025 p. 1 – 10 doi :10.1088/1755-1315/149/1/01202
[3] Ramage C S 1968 Role of A Tropical "Maritime Continent" in the Atmospheric Circulation *Monthly Weather Review* Vol **96** No. 6: 365 – 370
[4] Hendon H H 2003 Indonesian Rainfall Variability: Impacts of ENSO and Local Air–Sea Interaction *Journal of Climate* Vol **16**: 1775 – 1790
[5] Lee H S 2015 General Rainfall Patterns in Indonesia and the Potential Impacts of Local Seas on Rainfall Intensity *Water* **7**: 1751-1769 doi:10.12063390/w7041751
[6] Gutman G, Csiszar I, and Romanov P 2000 Using NOAA/AVHRR Products to Monitor El Nino Impacts: Focus on Indonesia in 1997-98 *Bulletin of the American Meteorological Society* Vol. **81**: 1189 – 1206
[7] Zhang W, Wang Y, Jin F F, Stuecker M F and Turner A G 2015 Impact of Different El Niño Types on the El Niño/IOD Relationship *AGU Geophisical Research Letters* p. 8570 – 8576 10.1002/2015GL065703
[8] Ashok K, Guan Z, and Yamagata T 2003 A Look at the Relationship between the ENSO and the Indian Ocean Dipole *Journal of the Meteorological Society of Japan* Vol. **81** No.1, pp. 41-56
[9] Meyers G, McIntosh P, Pigot L and M Pook 2007 The Years of El Niño, La Niña, and Interactions with the Tropical Indian Ocean *Journal of Climate* Vol **70** : 2872 – 2880 doi: 10.1175/JCLI4152.1
[10] Houze R A 1977 structure and Dynamics of a Tropical Squall-Line System *Monthly Weather Review* Vol. **105** : 1540 -1567
[11] Trismidianto, Yulihastin E, Satyawardhana H, Nugroho J T and S Ishida 2017 The Contribution of the Mesoscale Convective Complexes (MCCs) to total rainfall over Indonesian Maritime Continent *IOP Conf. Ser.: Earth Environ. Sci.* **54** 012027 p. 1 – 10 doi:10.1088/1755-1315/54/1/012027
[12] Trismidianto 2018 Characteristics of the oceanic MCC, continental MCC, and coastal MCC over the Indonesian maritime continent *IOP Conf. Ser.: Earth Environ. Sci.* **149** 012024 p. 1 – 12 doi :10.1088/1755-1315/149/1/012024
[13] Cahyarini S Y and Suharsono 2015 The Influence of ENSO/IOD on SST Signal in Kendari, Southeast Sulawesi Waters: 27-year-records of Sr/Ca from Porites corals *Indonesian Journal on Geoscience* Vol. **2** No. 1: 43-51
[14] Trenberth K E 1997 The Definition of El Niño *Bulletin of the American Meteorological Society* p. 2771 – 2777
[15] Setiawati M D and Miura F 2016 Evaluation of GSMaP Daily Rainfall Satellite Data for Flood Monitoring: Case Study—Kyushu Japan *Journal of Geoscience and Environment Protection* **4** : 101-117 ISSN Online: 2327-4344

[16] Min M, Bai C, Guo J, Sun F, Liu C, Wang F, Xu H, Tang S, Li B, Di D, Dong L and Li J 2018 Estimating Summertime Precipitation from Himawari-8 and Global Forecast System Based on Machine Learning IEEE Transactions on Geoscience and Remote Sensing p. 1 – 13, DOI: 10.1109/TGRS.2018.2874950

[17] Inter-Agency Regional Analysts Network 2015 2015/2016 El Nino Event *Global Report an ACF and IRIS initiative* p. 1 – 35

[18] Luo J J, Zhang R, Bahera S K, Masumoto Y, Jin F F, Lukas R, and Yamagata T 2009 Interaction between El Niñoand Extreme Indian Ocean Dipole *J. Climate* p. 1- 46

[19] Trismidianto 2018 The Global Population of Mesoscale Convective Complexes (MCCs) over Indonesian Maritime Continent during 15 Years *IOP Conf. Ser.: Earth Environ. Sci.* **166** 012040 p. 1 – 17 doi :10.1088/1755-1315/166/1/012040

IOP Conf. Series: Earth and Environmental Science **303** (2019) 012007 doi:10.1088/1755-1315/303/1/012007

Diurnal Rainfall Propagation Relate to Cold Surge-Cold Tongue Interaction over The Northern Coast of West Java

E Yulihastin[1,3]*, T W Hadi[1], and N S Ningsih[2]

[1]Atmospheric Sciences Research Group, Faculty of Earth Sciences and Technology, Bandung Institute of Technology, Bandung 40132, Indonesia
[2]Oceanography Sciences Research Group, Faculty of Earth Sciences and Technology, Bandung Institute of Technology, Bandung 40132, Indonesia
[3]Center of Atmospheric Science and Technology, National Institute of Aeronautics and Space, Bandung 40173, Indonesia

erma.yulihastin@lapan.go.id

Abstract. Interaction of the diurnal cycle of rainfall systems between land and sea were investigated by hourly rainfall of 3B41RT TRMM dataset, daily SST dataset from OISST, 6-hourly wind and temperature data from NCEP-NCAR Reanalysis, during period of NDJFM 2000-2016. The results showed that the diurnal cycle of rainfall from sea to land and from land to sea randomly occurred on four combinations of CENS-CT, NCENS-CT, CENS-NCT, NCENS-NCT. Diurnal rainfall in northern coast of West Java for normal conditions without CENS-CT has two maximas, namely late afternoon and early morning rainfall. CENS-CT interaction changes the normal diurnal rainfall pattern into three diurnal rainfall systems: diurnal rainfall with propagate B pattern which is identified by the afternoon continous to morning rainfall and there is an offshore propagation from sea to land that occurred in the morning, diurnal rainfall with a nonpropagate pattern, namely morning rainfall which has increased intensity and duration while afternoon rainfall is minimum, diurnal rainfall with propagate A pattern which experienced rainfall from land to sea and there was a merger between the afternoon rainfall system over the land and the morning rainfall system over the sea.

Keywords : rainfall, cold surge, cold tongue, TRMM.

1. Introduction

A very strong surface wind flow across the equator is called the Cross-Equatorial Northerly Surge (CENS). Based on the research of Hattori et al. [1], the CENS shown by strong north winds (5-8 m/s) lasts from the end of December to mid-February. The CENS phenomenon which does not always occur every year in a mechanism still simplified is an extension of Cold Surge formed on the Siberian plateau [2], which is characterized by a drastic drop in temperature and pressure for several days [3]; [4]; [5]; [6]; [7]; [8]; [9]; [10]; [11]) and caused disturbances to the weather system for East Asian countries [9], such as Hong Kong and Taiwan. This Cold Surge, when passing through the South China Sea can transform into very strong north winds ([12]; [2]) and propagate south across the equator. Climatological observations of the CENS further encouraged [1] to define the location of CENS focused on strengthening winds from north or northwest from the surface of the Java Sea, Karimata Strait, and parts of the South China Sea as shown in a box that limits the region $(0-5^{o}\text{S}, 105-115^{o}\text{E})$.

IOP Conf. Series: Earth and Environmental Science **303** (2019) 012007 doi:10.1088/1755-1315/303/1/012007

On the other hand, the process that occurs at sea level in the SCS South China Sea (SCS) region is also interesting because SST variability occurs from an intraseasonal to interannual scale ([13]; [14]; [15]; [16]) and is influenced by ENSO through weakening (El Niño) and Strengthening (La Niña) of the Asian winter monsoon, respectively ([17]; [18]; [19]). In addition, during the Asian winter monsoon period, the unique features of SST have been found, ([13]; [14]; [15]), which was called by Cold Tongue (CT) ([17]; [20]; [21]; [22]) with the coldest core in the waters near Vietnam's southeast coastal [22] and show that consistently strong CT has affected the increase in monsoon activity and has an impact on the addition of evaporation. In addition, CT has also been studied to influence the distribution of diurnal rainfall between Java and the Java Sea, namely in the form of a reduction in rainfall occurring over the Java Sea and conversely an increase in rainfall over the land [21].

The CS phenomena that occur on the atmospheric surface level $(1000 - 850\ \mathrm{mb})$ and CT that occur at sea level have been studied separately ([23]; [1]; [21]). In fact, atmosphere and ocean are components that interact with each other. Therefore, it is necessary to study CT and CS by using the atmosphere and the ocean as components that interact with each other. Monitoring of the disruption of CS-CT interactions in a diurnal context in West Java during the active period of the Asian winter monsoon is interesting to study.

Previous research shows that floods that occurred in northern West Java during the DJF period in 2002, 2007([24]; [25]; [26]), 2013 [27] and 2014 were related to high rainfall which persisted from morning to afternoon. Rainfall events that cause flooding in Jakarta and the north coast of West Java can be identified as a deviation from the diurnal pattern of rainfall, which is a shift in the diurnal cycle time of rainfall from afternoon to night to morning to afternoon. So far there has been no study discussing diurnal cycle deviations in northern West Java that are thought to be related to CENS activity and reinforced by CT formation in SCS. The interaction between CS-CT is considered to influence diurnal rainfall pattern deviation in northern West Java which has not been well studied. The current research particularly interested in investigating interaction between CENS-CT and its influence to the morning rainfall peaks over the northern coast of West Java.

2. Data and Method

In this study, we identified propagation of diurnal rainfall system on the north coast of West Java by hourly rainfall data from the TRMM 3B41RT Satellite with spatial resolution of 5 km during the NDJFM period of 2000-2016. Secondly, we calculate the CENS and CT indices using meridional wind parameters and SST were carried out, respectively. The CENS index is calculated from the surface meridional wind $(925\ \mathrm{mb})$ data over the area limited by $0 - 5^{o}\mathrm{S}, 105 - 115^{o}\mathrm{E}$. Taking into account the definitions of CENS or Cold Surge which have a minimum of life cycle ([11]; [2]) and affect weather up to a duration of up to 30 days [2], the CENS index calculation using the 6-hourly wind data, the average is made into pentad or 5-daily data. By following the definition of Hattori et al. [1], the CENS index has a threshold of $5\ \mathrm{m/s}$.

Thirdly, the CT index is calculated from the SST data in an area limited by $2 - 10^{o}\mathrm{S}, 105 - 112^{o}\mathrm{E}$. Furthermore, the SST threshold for CT determination was calculated by following Koseki et al. [22] that is by calculating the average value of the SST then deducting it by doubling the data standard deviation. The result is an SST of $26.4^{o}\mathrm{C}$ as a threshold. CENS and CT incidence calculations are calculated for each month, by categorizing the events of CENS-CT, CENS without CT (CENS-NCT), CT without CENS (NCENS-CT), and normal conditions without CENS-CT (NCENS-NCT).

IOP Conf. Series: Earth and Environmental Science **303** (2019) 012007 doi:10.1088/1755-1315/303/1/012007

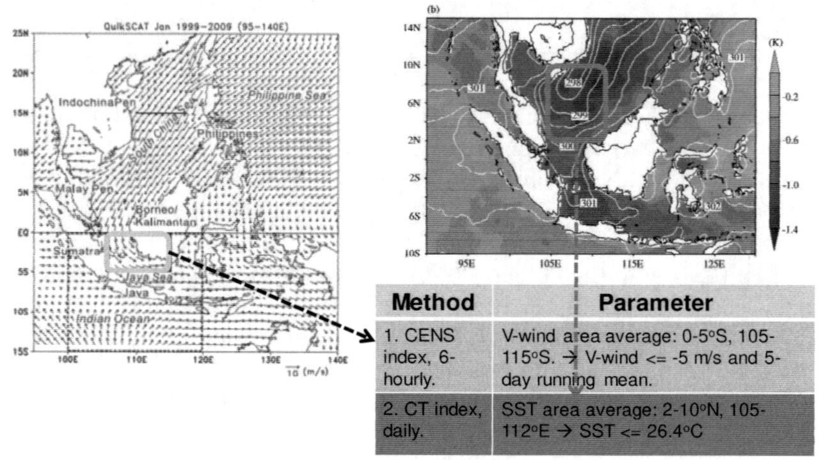

Method	Parameter
1. CENS index, 6-hourly.	V-wind area average: 0-5°S, 105-115°S. → V-wind <= -5 m/s and 5-day running mean.
2. CT index, daily.	SST area average: 2-10°N, 105-112°E → SST <= 26.4°C

Figure 1. Method to calculate CENS-CT indices with their area locations.

3. Result and Discussion

Figure 2 shows four categories of combinations between CENS-CT, CENS-NCT, NCENS-CT, NCENS-NCT during the event years in the 2000-2016 NDJFM period. Appeared during the CENS-CT events, there were at least 20 extreme conditions represented by strong northerly wind values (> 8 m/s). For the CENS category without CT (CENS-NCT), strong CENS occurs more than normal CENS. As for CT events without CENS (NCENS-CT), strong CT (< 26.4°C)occurred in almost half of all samples of NCENS-CT events.

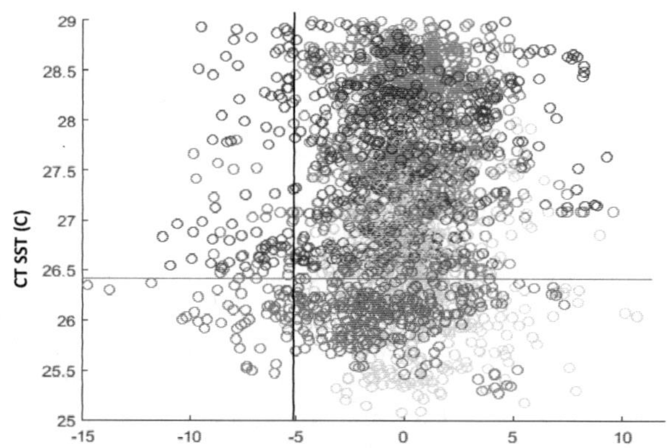

Figure 2. Scatter plot of CENS-CT (red), NCENS-CT (green), CENS-NCT (blue), NCENS-NCT (purple) during period of NDJFM 2000-2016.

Figure 3 describes the spatial formation of rainfall for four categories, namely: normal (32 months), nonpropagate (12 months), A propagate (20 months), B propagate (16 months). This shows that from 80 data, the most rainfall has a normal type with two separate rainfall systems over land and sea which are 32 months and without CENS or CT events (Figure 2-a). Meanwhile, non-propagating formations (12 months), most experienced CENS-CT (45%), both experienced CT without CENS (21.44%), CENS without CT (5%), and without experiencing both of them (15%).

For the rainfall propagation, there are two types: propagation from land to sea (A propagate), propagation from sea to land (B propagate). For A propagate, rainfall over the mainland occurs from the afternoon (1600-0100 LT) to early morning. For the B propagate, the rainfall develops over the sea from morning 0700 LT to 1300 LT. The dominant propagate type (58.3%) experienced CT, CENS-CT (25%) and without CENS or CT was 16.7%. For B propagate (16 months), most experienced CENS-CT (93.8%) and the remaining 6.2% experienced CT.

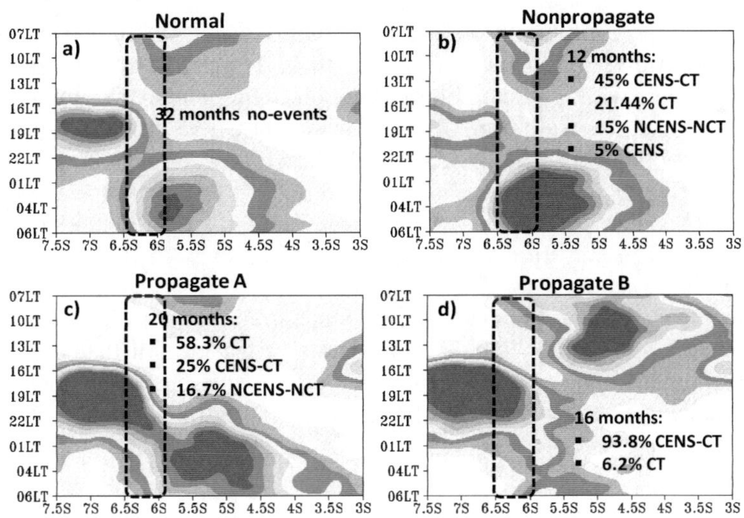

Figure 3. Spatial formation of rainfall system for (a) normal, (b) nonpropagate, (c) propagate A, (d) propogate B over the north coast of West Java.

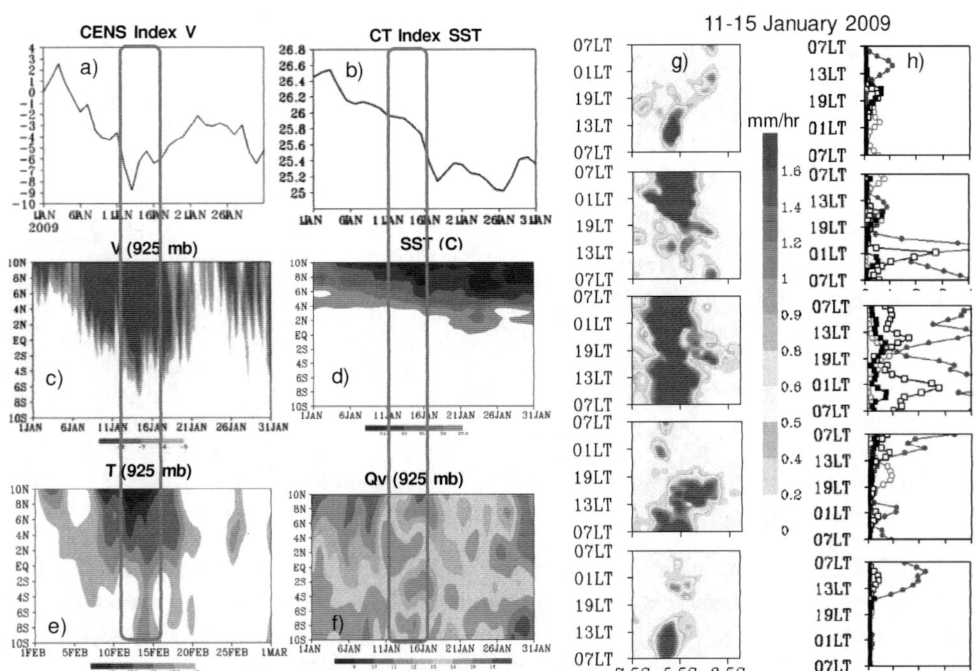

Figure 4. Rainfall evolution of the CENS-CT events during 11-15 January 2009.

Figure 4 describes the Jakarta flood events on January 11-15, 2009 which occurred simultaneously with CENS-CT. This is shown by the reinforcement (> 5 m/s) of the northerly wind through the equator (6°S) coincides with a decrease in surface temperature (Figures 3-a and 3-c). At the same time, SST has cooled ($< 26.4^{o}$C) accompanied by a front of moist and dry air masses in the north and south of the equator (2°N $- 6^{o}$S). This dry and humid air mass front has intensified the strong convection process so that heavy rainfall triggered the flood occurred during 11-15 January 2009 (Figures 3-b, d, f).

Figure 4 (g-h) is a rainfall evolution which exhibits that rainfall over the north coast of West Java occurred on 11 January from 1300 LT and persisted until 13 January 2009 (Figure 4-g). Besides that, persistent rainfall also occurs with large amplitude (blue) or high intensity (Figure 4-h).

Figure 5 explains the evolution of rainfall during CT without CENS, ie on 1-4 February 2012. It was seen on 1-4 February 2012 that SST had cooled to 2°N. In addition, there was a front between the dry and humid air masses in the region (Figure 5-f). Rainfall occur from the sea to the east on February 1 from morning to noon. On February 2, there was a spreading of rainfall from land to sea in the afternoon to night. Signals of rainfall propagation from sea to land then back to the sea are very clear in the evolution of rainfall events 1-4 February 2012.

Figure 6 describes the evolution of rain when CENS occurred without CT and coincided with flood events in Jakarta during 11-15 January 2013. It appears that the rainfall that occurred during 11-15 January 2013 did not propagate and was concentrated over the coastal area, which was preceeded by rainfall formation over the sea near the coastal. Moreover, rainfall also persists for four days in the region.

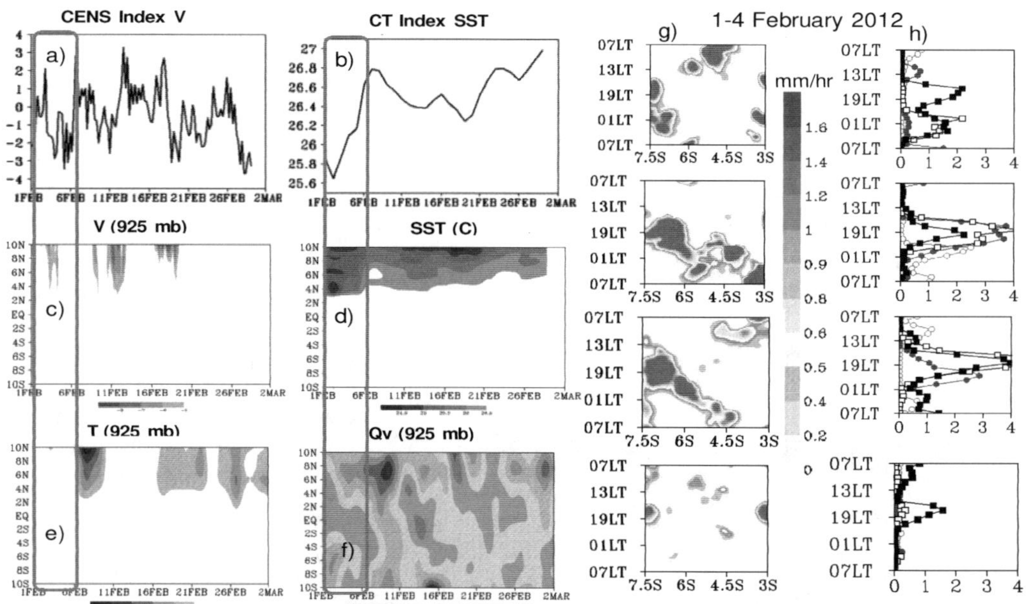

Figure 5. Same as Figure 4, but for NCENS-CT events during 1-4 Februari 2012.

Figure 7 shows the evolution of rainfall when neither CENS nor CT occurred on 1-4 January 2016. It appears that during 1-4 January, there was no strengthening of surface winds or cooling of surface temperature or sea surface temperature. In addition, the rainfall system over the land and the sea occurs separately in the afternoons (1300-1900 LT) and the morning (0100-0700 LT).

Figure 8 illustrates the synoptic conditions of the monsoon as a background wind for the four selected case studies representing: CENS-CT, NCENS-CT, CENS-NCT, NCENS-NCT. When CENS coincide with CT, it was seen that CT played a role in strengthening and increasing the intensity of the northerly wind and reducing the westerly winds over Java Island. While the CT occurs independently, the westerly wind intensify and homogeneous over Java whereas the northerly wind does not develop over the South China Sea. When CENS occurs independently, northerly develop over the South China Sea and propagate to south (Java island). However, compared to when CENS occurred with CT, the intensity of the northerly wind was weaker and the westerly wind formed strongly over Java. Under the normal conditions, without the occurrence of CENS or CT events, both the northerly and westerly wind were weakened.

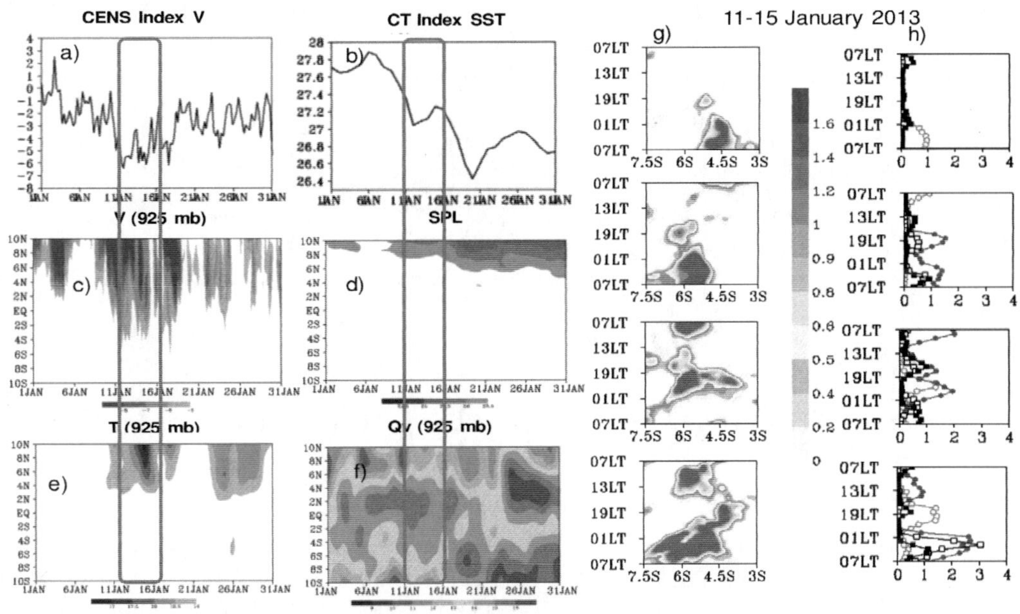

Figure 6. Same as figure 5, but for CENS-NCT events during 11-15 January 2013.

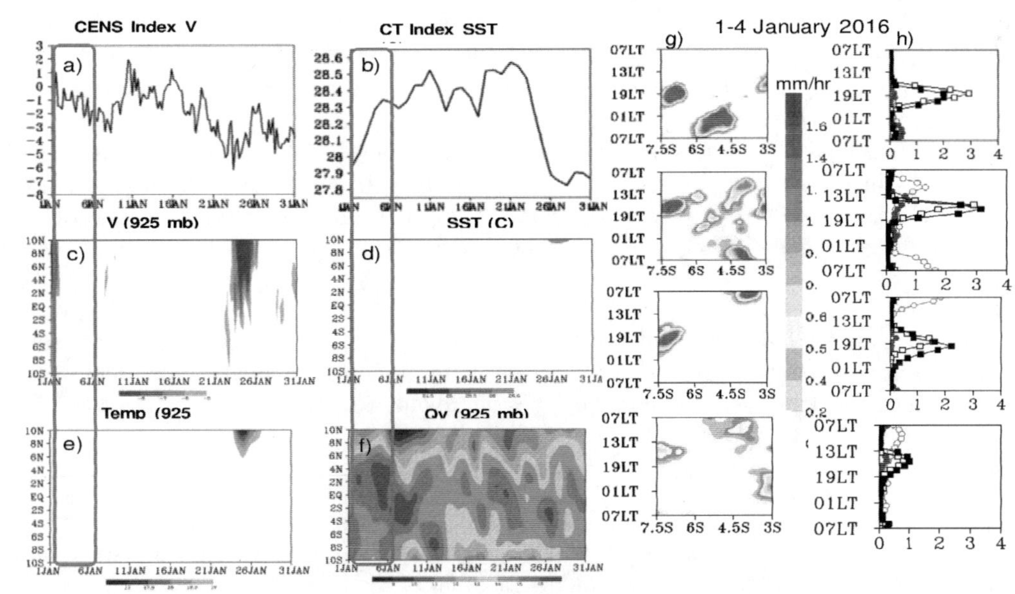

Figure 7. Same as Figure 6, but for NCENS-NCT events during 1-4 January 2016.

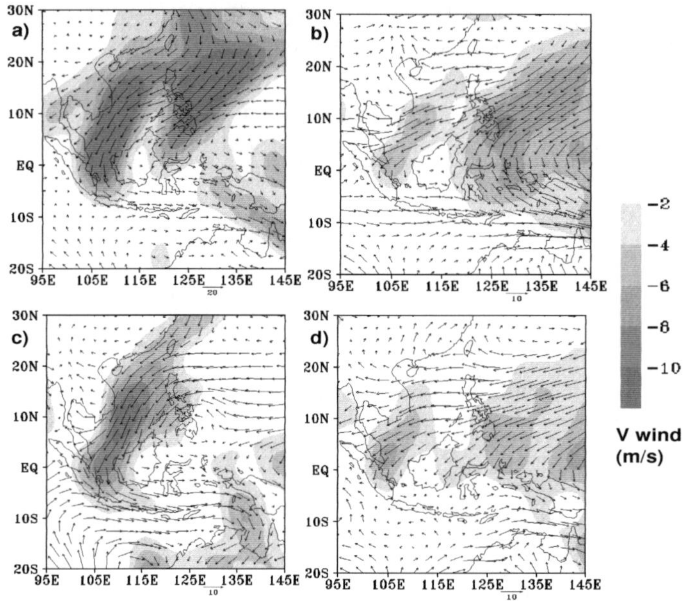

Figure 8. Spatial pattern of wind vector and meridional wind (shaded) for 4 cases: (a) CENS-CT (11-15 January 2009); (b) NCENS-CT (1-4 February 2012); (c) CENS-NCT (11-15 January 2013); (d) NCENS-NCT (1-4 January 2016).

4. Conclusion

The CT in SCS occurs almost every year during NDJFM rainy season period with duration between 1-3 months, except when there is strong El Niño occurred in 2011 and 2016. While CENS has a shorter duration, ie daily (2-30 days) and not always occur every year during the NDJFM period. CT is more influential than CENS in changing the diurnal pattern of rainfall over the northern coast of West Java. The presence of CENS is to strengthen the CT effect in extending the duration of the afternoon rainfall and increasing the intensity of the morning rainfall. Based on climatological statistics, diurnal rainfall over northern coast of West Java for normal conditions without the CENS or CT has two maxima, namely in the afternoon and in the morning. CENS-CT interaction resulted in changes in three diurnal system patterns, namely: (1) diurnal rainfall with B propagate pattern (93.8%), namely the afternoon rain experienced an extension to early morning and there was a spread of rainfall from sea to land which occurred in the morning ; (2) diurnal rainfall with a nonpropagate pattern (45%), namely morning rainfall which has increased intensity and duration while afternoon rainfall is minimum; (3) diurnal rainfall with A propagate pattern (25%) which experienced raindrops from land to sea and there was a merger between the afternoon rainfall system over the land and the morning rainfall system over the sea.

Acknowledgment

We would like to thank the Indonesia Endowment Fund for Education (LPDP) as a main sponsor of this research. This current research is part of a dissertation with the development of analytical methods in the case study section which is selected based on the four events combinations of CENS and CT.

References

[1] Hattori M, Mori S and Matsumoto J 2011 The Cross-Equatorial Northerly Surge over the Maritime Continent and Relationship to Precipitation Patterns *J. of the Met. Soc. of Japan* **89A** 27–47
[2] Compo G P, Kiladis G N and Webster P J 1999 The Horizontal and Vertical Structure of East Asian Winter Monsoon Pressure Surges *Quart. J. of The Royal Met. Soc.* **125** 29–54
[3] Lau K-M 1982 Equatorial Response to Northeasterly Cold Surges as Inferred from Satellite Cloud Imagery Month. Weat. Rev. **110** 1306-1313

[4] Chang C-P and Lau K M 1982 Short-Term Planetary-Scale Interactions over the Tropics and Midlatitudes during Northern Winter Part I: Contrasts between Active and Inactive Periods *Mont. Weath. Rev.* **111** 933-946

[5] Chang C-P, Millard J E and Chen G T J 1982 Gravitational Character of Cold Surges During Winter MONEX, *Mont. Weath. Rev.* **111** 293-307

[6] Lau K-M, Chang C-P and Chan P H 1983 Short-Term Planetary-Scale Interactions over the Tropics and Midlatitudes Part II: Winter-MONEX Period *Mont. Weath. Rev.* **111** 1372-1388

[7] Love G Cross-Equatorial Influence of Winter Hemisphere Subtropical Cold Surges *Mont. Weath. Rev.* **113** 1487-1498

[8] Yihui D 1989 Build-Up Air Mass Transformation and Propagation of Siberian High and Its Relations to Cold Surge in East Asia *Met. Atm. Phys.* **44** 281-292

[9] Chen T-C, Yen M-C, Huang W-R and Gallus Jr W A 2002 An East Asia Cold Surge: Case Study, *Mont. Weath. Rev.* **130** 2271-2290

[10] Park T-W, Ho C-H, Jeong J-H, Heo J-W and Deng Y 2015 A New Dynamical Index for Classification of over East Asia, *Clim. Dyn.*, DOI 10.1007/s00382-015-2483-7

[11] Samah A A 2016 Thermodynamic and Dynamic Structure of Atmosphere over the East Coast of Peninsular Malaysia during the Passage of A Cold Surge, *J. of Atm. and Solar-Terrestrial Phys.* **146** 58-68

[12] Lim H and Chang C-P 1981 A Theory for Midlatitude Forcing of Tropical Motions during Winter Monsoon *J. of the Atm. Sci.* **38** 2377-2379

[13] Lu S 1997 Temporal and Spatial Variabilities of the South China Sea Surface Temperature Anomaly *J. of Geoph. Res.* **102** 20937-20955

[14] Qu T, Song Y T and Yamagata T 2009 An Introduction to the South China Sea Throughflow: Its Dynamics, Variability, and Application for Climate *Dyn of Atm. and Oceans* **47** 3-14

[15] Huang E, Tian J and Steinke S 2011 Millenial-scale Dynamics of the Winter Cold Tongue in the Southern China Sea over the Past 26 ka and the East Asian Winter Monsoon *Quartern. Res.* **75** 196-204

[16] Wu R and Chen Z 2015 Intraseasonal SST Variations in the South China Sea During Boreal Winter and Impacts of the East Asian Winter Monsoon: Intraseasonal SST Variations *J. of Geoph. Res.* **120** 5863-5878

[17] Liu Q, Jiang X, Xie S-P and Liu W T 2004 A Gap in the Indo-Pacific Warm Pool over the South China Sea in Boreal Winter: Seasonal Development and Interannual Variability *J. Geoph. Res.* **109** 1-10

[18] Qu T, Du Y, Meyers G, Ishida A and Wang D 2005 Connecting the Tropical Pacific with Indian Ocean through South China Sea *Geoph. Res. Lett.* **32** 1-4

[19] Liu Q-Y, Wang D, Wang X, Shu Y, Xie Q and Chen J 2014 Thermal Variations in the South China Sea Associated with the Eastern and Central Pacific El Niño Events and Their Mechanism *J. Geoph. Res.* 10.1002/2014JC010429 8955-8972

[20] Varikoden H, Samah A A and Babu C A 2009 The Cold Tongue in South China Sea during Boreal Winter and Its Interaction with the Atmosphere *Advan. In Atm. Sci.* **27** 265-273

[21] Koseki S, Koh T-Y and Teo C-K 2012 Effects of the Cold Tongue in the South China Sea on the Monsoon, Diurnal Cycle and Rainfall in the Maritime Continent *Quart. J. of the Royal Met. Soc.* doi: 10.1002/qj.2052

[22] Thompson B, Tkalich P, Malanotte-Rizzoli P, Fricot B and Mas J 2015 Dynamical and Thermodynamical Analysis of the South China Sea Winter Cold Tongue *Clim. Dyn.* **47** 1629-1646

[23] Chang C-P, Harr P A and Chen H J 2005 Synoptic disturbance over the equatorial South China Sea and western Maritime Continent during boreal winter *Monthly Weather Review* **133** 489–503

[24] Wu P, Hara M, Fudeyasu H, Yamanaka MD and Matsumoto J 2007 The Impact of Trans-equatorial Monsoon Flow on The Formation of Repeated Torrential Rains over Java island *SOLA* **3** 93–96

[25] Trilaksono NJ, Otsuka S, Yoden S, Saito K, and Hayashi S 2011 Dependence of Model-simulated Heavy Rainfall on the Horizontal Resolution During the Jakarta Flood Event in January-February 2007 *SOLA* **7** 193–196

[26] Trilaksono NJ, Otsuka S and Yoden S 2012 A Time-lagged Ensemble Simulation on the Modulation of Precipitation over West Java in January-February 2007 *Month. Weath. Rev.* **140** 601–616

[27] Wu P, Arbain A A, Mori S, Hamada J, Hattori M, Syamsudin F and Yamanaka MD 2013 The Effects of an Active Phase of the Madden-Julian Oscillation on the Extreme Precipitation event over western Java Island in January 2013 **9** 79 – 83

Diurnal variation of stable isotopes in rainfall observed at Bengkulu for the YMC-Sumatra 2017

K Ichiyanagi[1,2], R Suwarman[3], H A Belgaman[4], M Tanoue[5], T Uesugi[1] and Warjono[6]

[1] Kumamoto University, Kurokami 2-39-1, Chuo-ku, Kumamoto, Japan
[2] Japan Agency for Marine-Earth Science and Technology (JAMSTEC), Natsushima 2-15, Yokosuka, Kanagawa, Japan
[3] Atmospheric Science Research Group, Bandung Institute of Technology, Jl. Ganesha. 10, Coblong, Kotamadya Bandung, Jawa Barat, 40132, Indonesia
[4] Agency for the Assessment and Application of Technology (BPPT), Jl. M.H Thamrin No. 8 Jakarta Pusat, DKI Jakarta 10340, Indonesia
[5] Shibaura Institute of Technology, Shibaura 3-9-14, Minato-ku, Tokyo, Japan
[6] Meteorological, Climatological, and Geophysical Agency (BMKG) Bengkulu, Jl. Raya Bandara Fatmawati Soekarno, Bengkulu, Indonesia.

kimpei@kumamoto-u.ac.jp

Abstract. Diurnal variability of stable isotopes in precipitation and its controlling factors are poorly understood, especially in the tropics. This study investigated the diurnal variability of stable isotopes in rainfall observed at Bengkulu in southern Sumatra, Indonesia. Rainfall samples were collected at 6-h intervals from 1 December 2017 to 15 January 2018 for the Year of the Maritime Continent (YMC-Sumatra 2017). The oxygen and hydrogen isotopic compositions ($\delta^{18}O$ and $\delta^{2}H$) of the rainwater varied from −15.1‰ to −1.8‰ and from −108.9‰ to −1.3‰, respectively, and the d-excess varied from 1.9‰ to 17.1‰. The observations at Bengkulu revealed that diurnal variation in rainfall amount was not significant for the entire period; however, the $\delta^{18}O$ was lighter in both the early morning and the afternoon and heavier before noon and overnight. The $\delta^{18}O$ became gradually depleted in the periods of continuous rain that occurred during 1–4 and 21–24 December. The amount of morning rain was much greater than evening rain in these periods, whereas the amount of evening rain was greater than morning rain in other periods. Based on satellite-derived CMORPH data, it is assumed that morning rain and evening rain corresponded to weak rainfall over surrounding sea region and strong rainfall over inland regions of Sumatra, respectively. The differences in moisture source and ratio of stratiform/convective rain rate between morning and evening might affect the isotopic composition of rainfall in the tropics.

Keywords : isotopes, CMORPH, stratiform, convective.

1. Introduction

Many previous studies that have discussed precipitation isotope changes in the tropics on timescales longer than a day have suggested that long-term isotopic changes might be controlled by large-scale moisture transport [1, 2, ,3]. In palaeoclimatic reconstructions, the relationships between large-scale climate variability, rainwater, and cave drip water $\delta^{18}O$ have been investigated in northern Borneo on daily–interannual timescales [4, 5]. Moreover, relationships between precipitation isotopes and precipitation amount have been determined based on daily rainfall isotopes in Papua [6]. The seasonal and intra-annual variations of rainfall isotopes have been investigated using daily rainfall isotope datasets acquired at six sites across the Indonesian Maritime Continent (IMC) [7,8]. However, the spatial and temporal resolutions of the observed rainfall isotopes were inadequate to elucidate the factors controlling the isotopic variations.

A later study investigated the high-resolution spatiotemporal characteristics of rainfall isotopes using data acquired weekly at 33 meteorological stations across the IMC [9]. The findings of these earlier studies on timescales longer than a day suggested that rainfall isotopes are affected by the Madden–Julian Oscillation, Asian and Australian monsoons, and El Niño-Southern Oscillation. However, the relationships between the the rainfall isotopes and meteorological parameters remain poorly understood, especially in the tropics where rainfall variation is predominantly diurnal. A few observational studies have conducted investigations on rainfall isotope changes on timescales shorter than a day during rain events in Sumatra [10]. Therefore, this study focused on the diurnal variation of stable isotopes in rainfall observed in the tropics and investigated the differences in the rainfall isotopes between morning and evening rain.

2. Data and Methods

Rainfall samples were collected at 6-h intervals at BMKG (Agency for Meteorology, Climatology, and Geophysics) Bengkulu meteorological station (3.87°S, 102.33°E) in southern Sumatra (Indonesia) from 1 December 2017 to 15 January 2018 during the Year of the Maritime Continent (YMC-Sumatra 2017). Overall, 63 rainwater samples were collected and the stable isotope ratios of the rainwater were measured using a cavity ring-down spectroscopy isotopic water analyser (L2120-i, Picarro Inc., Sunnyvale, CA, USA) at the Isotope Hydrology Laboratory of Kumamoto University, Japan. The stable isotopes of oxygen and hydrogen ($\delta^{18}O$ and $\delta^{2}H$) of the rainwater were expressed by convention as a deviation in parts per thousand (‰) from the Vienna Standard Mean Ocean Water (V-SMOW). The analytical errors for $\delta^{18}O$ and $\delta^{2}H$ were 0.2‰ and 0.6‰, respectively. The deuterium excess (d-excess), defined as $\delta D - 8 \times \delta^{18}O$, is a second-order isotope parameter that is sensitive to ambient conditions during the evaporation of water from the ocean surface.

Rainfall amount and air temperature data observed at 6-h intervals were acquired from the BMKG Bengkulu meteorological station. The spatial distribution of precipitation over Indonesia was produced using the Climate Prediction Center morphing method (CMORPH) rain rate (real-time version) for the YMC Domain (15°S–15°N, 80°–160°E) [11], for which the spatial and temporal resolutions were 8 km at the equator and 30 min, respectively.

3. Results and Discussions

3.1. Temporal variation

Temporal variations in $\delta^{18}O$, d-excess, and rainfall amount observed at 6-h intervals at Bengkulu throughout the entire observation period are shown in figurefigure 1. The $\delta^{18}O$ and $\delta^{2}H$ (not shown in the figure) of rainfall varied from −15.1‰ to −1.8‰ and from −108.9‰ to −1.3‰, respectively, and the d-excess varied from 1.9‰ to 17.1‰. Figure 1 shows reasonably continuous rain persisted from 1 to 4 December 2017 and from 21 to 24 December 2017. During these periods, $\delta^{18}O$ became gradually depleted from approximately −3‰ to −9‰ (hereafter, the $\delta^{18}O$ depleting periods). Sudden depletion of $\delta^{18}O$ that corresponded to periods of heavy rain occurred on 13, 14, and 27 December 2017 and on 10 January 2018. Except during the above periods, $\delta^{18}O$ showed only small variation. With reference to the temporal characteristics of $\delta^{18}O$ variation, the observation period was divided into four stages: 1–4 December (P1), 5–20 December (P2), 21–24 December (P3), and 25 December to 15 January (P4).

The local meteoric water line (LMWL) throughout the entire observation period was $\delta^{2}H = 8.03 \times \delta^{18}O + 11.6$. However, during periods P1–P4 its value was $\delta^{2}H = 8.81 \times \delta^{18}O + 15.7$, $\delta^{2}H = 7.50 \times \delta^{18}O + 8.14$, $\delta^{2}H = 8.38 \times \delta^{18}O + 16.0$, and $\delta^{2}H = 8.00 \times \delta^{18}O + 11.0$, respectively. All correlation coefficients were >0.98 with statistical significance. Both the slope and the intercept of the LMWL for periods P1 and P3 were relatively higher in comparison with periods P2 and P4. The differences in

LMWL could be attributable to processes such as evaporation in the source region, precipitation within the cloud, and evaporation of raindrops.
Generally, a "rainfall amount effect" whereby $\delta^{18}O$ values become depleted with increasing rainfall amount is found in the tropics [8]. In figure 1figure , however, the temporal characteristics of rainfall amount and $\delta^{18}O$ do not appear to match. Relationships between $\delta^{18}O$ and rainfall amount or air temperature observed at Bengkulu for the entire period are shown in figure 2figure . The correlation coefficients of these relationships were 0.02 and 0.18 for rainfall amount and air temperature, respectively; however, neither relationship was found significant.

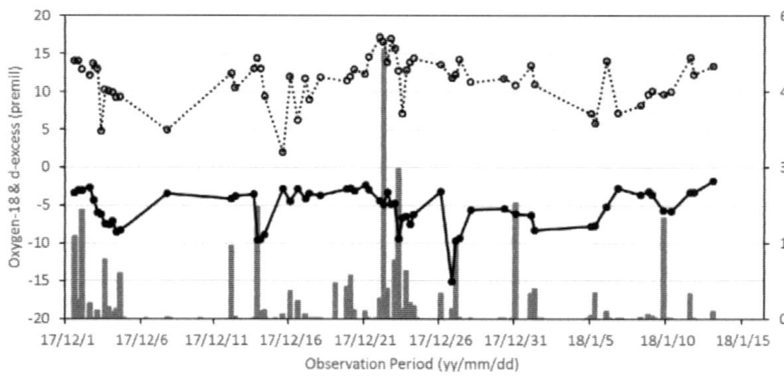

Figure 1. Temporal variations of $\delta^{18}O$, d-excess, and rainfall amount observed at 6-h intervals at Bengkulu from 1 December 2017 to 15 January 2018. Closed circles with solid line, open circles with dashed line, and bars indicate $\delta^{18}O$, d-excess, and rainfall amount, respectively.

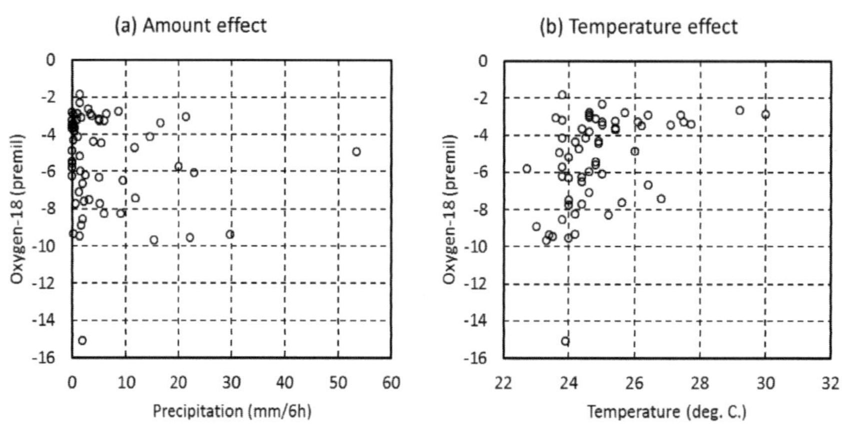

Figure 2. Scatter diagrams between $\delta^{18}O$ and (a) rainfall amount and (b) air temperature observed at Bengkulu throughout the entire observation period.

3.2 Diurnal variation
To elucidate the characteristics of the diurnal variation in rainfall isotopes, rainfall amount, air temperature, $\delta^{18}O$ weighted by the rainfall amount, and d-excess were averaged over specific 6-h intervals (i.e., 01, 07, 13, and 19 local time (LT)) for the entire period. Because the temporal characteristics of $\delta^{18}O$ were similar in the $\delta^{18}O$ depleting periods (P1 and P3) and in the other periods (P2 and P4), the meteorological and isotopic values were also averaged for these periods. Diurnal variations of rainfall amount, air temperature, $\delta^{18}O$, and d-excess are shown in figure 3figure .

Figure 3a shows that diurnal variation in rainfall amount was not evident for the entire period; however, a large difference was found between the $\delta^{18}O$ depleting periods and the other periods. The amount of morning rain (02–07 LT and 08–13 LT) was much greater than evening rain (14–19 LT and 20–01 LT) for the $\delta^{18}O$ depleting periods. Conversely, the amount of evening rain was much greater than morning rain for the other periods. Figure 3b shows clear diurnal variation in air temperature for all periods, i.e., higher during daytime (08–13 LT and 14–19 LT) and lower during night-time (20–01 LT and 02–07 LT). Remarkably, the air temperature before noon (08–13 LT) for the $\delta^{18}O$ depleting periods was much lower than that of the same time in the other periods. Heavy rainfall and low temperature before noon (08–13 LT) were assumed to prevent evaporation from the land around Bengkulu for the $\delta^{18}O$ depleting periods. From Figure 3cfigure , clear diurnal variation in $\delta^{18}O$ is evident for all periods, i.e., light values in the early morning (02–07 LT) and afternoon (14–19 LT) and heavy values before noon (08–13 LT) and overnight (20–01 LT). The $\delta^{18}O$ values before noon (08–13 LT) for both the entire period and the $\delta^{18}O$ depleting periods were found much lighter than for the other periods. Figure 3d shows that d-excess values throughout the day for the $\delta^{18}O$ depleting periods were higher than for the other periods, especially in the early morning (02–07 LT). The large difference in d-excess values indicates that the moisture source (evaporation process) and/or precipitation process within the cloud (condensation process) must have been different between these periods.

The diurnal cycle of rainfall and its regional variation over Sumatra have been investigated in earlier research [12], which revealed that convective rainfall with a broad peak between 15–20 LT predominates over the land area, whereas rainfall in the early morning, comprising almost equally of stratiform and convective types, is predominant over the surrounding sea region. To consider the diurnal cycle of rainfall and moisture source, the spatial distribution of rainfall over the IMC was investigated using satellite-derived CMORPH data.

Diurnal variations of rainfall patterns averaged over the specific 6-h intervals for the entire period are shown in figure 4. Heavy rainfall extends over the inland region of Sumatra in the afternoon (14–19 LT) and overnight (20-01 LT), while weak rainfall extends over the surrounding sea region in the early morning (02–07 LT) and before noon (08–13 LT). It is assumed that the morning rain for the $\delta^{18}O$ depleting periods and evening rain for the other periods correspond to the surrounding sea region and the inland area of Sumatra, respectively. Thus, the moisture source and the ratio of stratiform/convective rain rates between these periods might be different. These differences might affect the isotopic compositions of the rainfall.

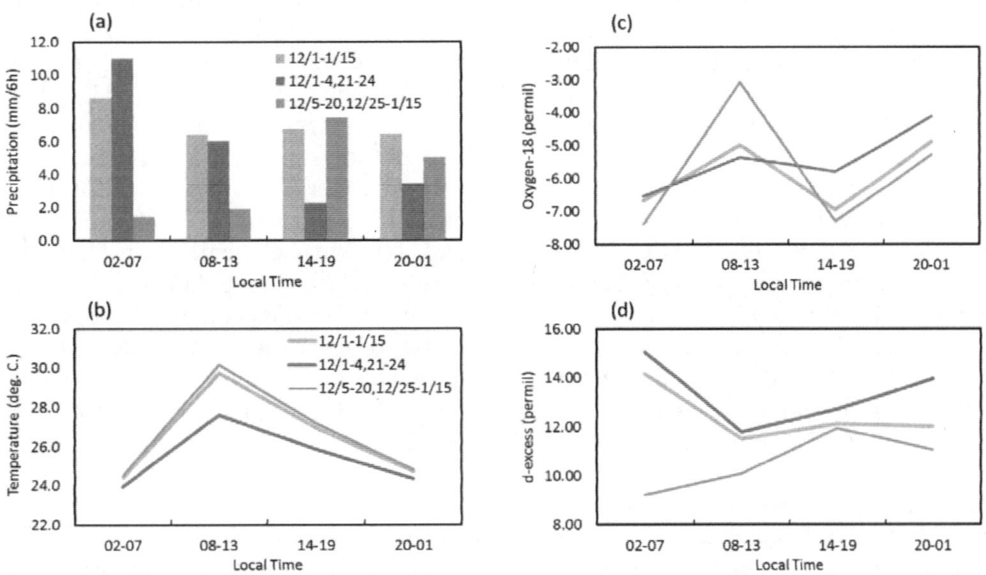

Figure 3. Diurnal variations of (a) rainfall amount, (b) temperature, (c) $\delta^{18}O$, and (d) d-excess observed at Bengkulu and averaged for each specific period. Dates 12/1-1/15 correspond to the entire period, and 12/1-4, 12/5-20, 12/21-24, and 12/25-1/15 correspond to periods P1–P4, respectively.

IOP Conf. Series: Earth and Environmental Science **303** (2019) 012008 doi:10.1088/1755-1315/303/1/012008

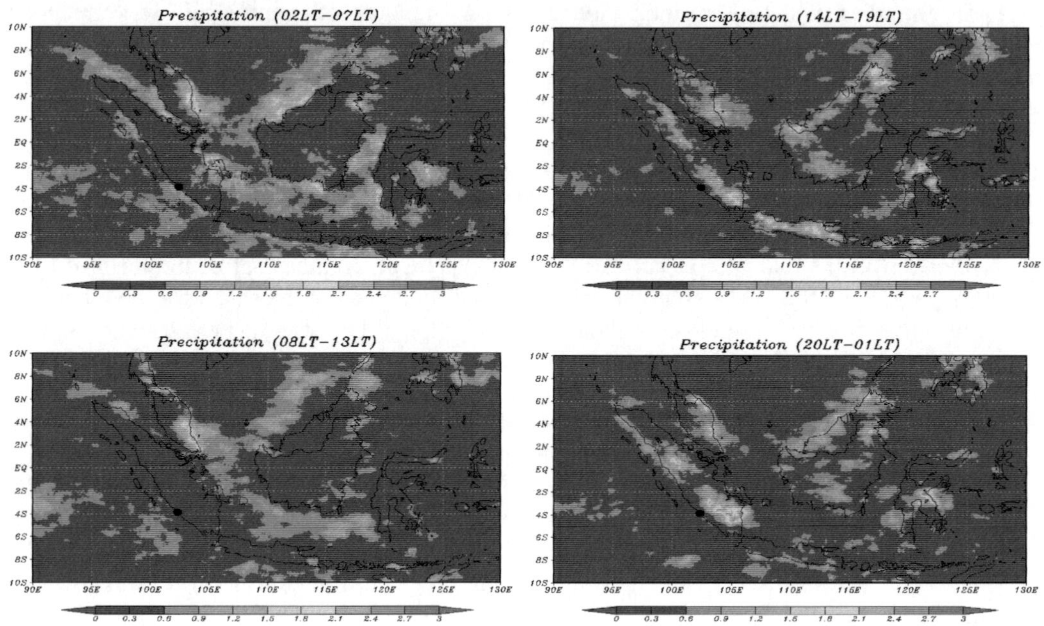

Figure 4. Diurnal variations of precipitation patterns averaged over specific 6-h intervals from 1 December 2017 to 15 January 2018. Black dot indicates the location of the Bengkulu meteorological station.

4. Concluding remarks

To elucidate the diurnal variation of stable isotopes in rainfall, rainfall samples were collected at Bengkulu from 1 December 2017 to 15 January 2018. The diurnal variation in rainfall amount was not significant for the entire period; however, the $\delta^{18}O$ was lighter in both the early morning and the afternoon and heavier before noon and overnight. Because of the temporal characteristics of $\delta^{18}O$, the meteorological and isotopic values were averaged for the $\delta^{18}O$ depleting periods (P1 and P3) and in the other periods (P2 and P4). As a result, the amount of morning rain was much greater than evening rain for the $\delta^{18}O$ depleting periods, whereas the amount of evening rain was much greater than morning rain for the other periods. The d-excess values for the $\delta^{18}O$ depleting periods were higher than for the other periods, especially in the early morning. From the average of the satellite-derived CMORPH data for the entire period, the morning rain for the $\delta^{18}O$ depleting periods and the evening rain for the other periods might have corresponded to weak rainfall over the surrounding sea region and strong rainfall over inland regions of Sumatra, respectively. The isotopic compositions of moisture and rainfall should be different depending on the moisture source and ratio of stratiform/convective rain rates. Therefore, diurnal variation of rainfall might affect the determination of the isotopic compositions of rainfall in the tropics. Further research will be needed to improve the understanding of the precipitation process using in situ observations such as rawinsondes and meteorological radar.

Acknowledgments

The authors wish to thank the staff of the BMKG (Agency for Meteorology, Climatology, and Geophysics) meteorological station at Bengkulu for assistance with precipitation sampling and for providing meteorological data. This research was supported by the Priority Organization for Innovation and Excellence, Kumamoto University, Department of Coupled Ocean-Atmosphere-Land Processes, JAMSTEC, for project research, and Japan Society for the Promotion of Science (JSPS) KAKENHI (Grant Numbers: 16H05619).

Appendix Table A1

Summary of precipitation isotope observations at Bengkulu meteorological station.

No.	Date	Time (LT)	Time (UTC)	Rain (mm/6h)	Temp. (°C)	$\delta^{18}O$ (‰)	$\delta^{2}H$ (‰)	d-excess (‰)
1	2017/12/1	1:00	18:00	-	-			
2		7:00	0:00	-	-			
3		13:00	6:00	16.4	27.7	-3.39	-13.07	14.06
4		19:00	12:00	3.7	24.6	-3.02	-10.14	14.04
5	2017/12/2	1:00	18:00	21.5	23.6	-3.05	-11.53	12.88
6		7:00	0:00	0	23.9			
7		13:00	6:00	3	29.2	-2.63	-8.95	12.11
8		19:00	12:00	0.2	24.9	-4.30	-20.67	13.70
9	2017/12/3	1:00	18:00	1.6	24.6	-5.98	-34.82	13.01
10		7:00	0:00	ttu	24.4	-6.24	-45.21	4.72
11		13:00	6:00	11.7	26.8	-7.43	-49.23	10.20
12		19:00	12:00	2.3	25.6	-7.60	-50.70	10.08
13	2017/12/4	1:00	18:00	1.3	24.6	-7.09	-46.83	9.87
14		7:00	0:00	2	23.8	-8.55	-59.22	9.19
15		13:00	6:00	9	25.2	-8.27	-56.87	9.32
16		19:00	12:00	ttu	26.2			
17	2017/12/5	1:00	18:00	0	24.4			
18		7:00	0:00	0	24.2			
19		13:00	6:00	0	30.5			
20		19:00	12:00	0	27.4			
21	2017/12/6	1:00	18:00	0	25			
22		7:00	0:00	ttu	25.6			
23		13:00	6:00	0	30.7			
24		19:00	12:00	0	28.2			
25	2017/12/7	1:00	18:00	0	26.8			
26		7:00	0:00	0	25.7			
27		13:00	6:00	0	32			
28		19:00	12:00	0.3	27.1	-3.42	-22.50	4.87
29	2017/12/8	1:00	18:00	ttu	25.3			
30		7:00	0:00	0	25			
31		13:00	6:00	0	29.7			
32		19:00	12:00	0	26.3			
33	2017/12/9	1:00	18:00	0	24.9			
34		7:00	0:00	0	24.4			
35		13:00	6:00	0	30.4			
36		19:00	12:00	0	28.4			
37	2017/12/10	1:00	18:00	ttu	26			
38		7:00	0:00	0	25.6			
39		13:00	6:00	0	31			
40		19:00	12:00	0	28.1			
41	2017/12/11	1:00	18:00	0	26.2			
42		7:00	0:00	0	25			
43		13:00	6:00	0	31.4			
44		19:00	12:00	0	28.3			
45	2017/12/12	1:00	18:00	14.4	23.8	-4.13	-20.63	12.42
46		7:00	0:00	0.4	24.6	-3.82	-20.07	10.47
47		13:00	6:00	0	29.8			
48		19:00	12:00	0	27.8			
49	2017/12/13	1:00	18:00	0	25.8			
50		7:00	0:00	0	25.1			
51		13:00	6:00	ttu	26.2	-3.49	-14.96	12.98
52		19:00	12:00	22.2	24	-9.54	-61.98	14.38
53	2017/12/14	1:00	18:00	1.4	23.5	-9.46	-62.68	12.96
54		7:00	0:00	1.7	23	-8.90	-61.80	9.37
55		13:00	6:00	0	30.8			
56		19:00	12:00	0	27.6			
57	2017/12/15	1:00	18:00	ttu	25.3			
58		7:00	0:00	0	24.7			
59		13:00	6:00	1	30	-2.87	-21.05	1.91
60		19:00	12:00	0	27.1			
61	2017/12/16	1:00	18:00	5.5	24.9	-4.44	-23.56	11.96
62		7:00	0:00	0	24.6			
63		13:00	6:00	3.5	27.4	-2.88	-16.82	6.20
64		19:00	12:00	0	26.3			
65	2017/12/17	1:00	18:00	1	24.5	-4.14	-21.37	11.72
66		7:00	0:00	ttu	25	-3.44	-18.52	8.98
67		13:00	6:00	ttu	27.6			
68		19:00	12:00	ttu	27.7			
69	2017/12/18	1:00	18:00	ttu	25.4	-3.67	-17.45	11.88
70		7:00	0:00	0	24.3			
71		13:00	6:00	0	30.6			
72		19:00	12:00	0	27.7			
73	2017/12/19	1:00	18:00	7	25			
74		7:00	0:00	0	24.2			
75		13:00	6:00	0	30.4			
76		19:00	12:00	6.4	26.4	-2.88	-11.60	11.44
77	2017/12/20	1:00	18:00	8.6	24.6	-2.76	-10.07	11.97
78		7:00	0:00	1.7	24.8	-3.09	-11.82	12.89
79		13:00	6:00	0	29.9			
80		19:00	12:00	0	28			
81	2017/12/21	1:00	18:00	1.5	25	-2.30	-6.05	12.32
82		7:00	0:00	0.2	24.6	-2.91	-8.75	14.51
83		13:00	6:00	0	30.4			
84		19:00	12:00	ttu	27.8			
85	2017/12/22	1:00	18:00	4	24.2	-4.36	-17.74	17.10
86		7:00	0:00	53.4	23.7	-4.92	-22.86	16.54
87		13:00	6:00	6	26.1	-3.26	-12.23	13.89
88		19:00	12:00	ttu	26	-4.86	-21.96	16.92
89	2017/12/23	1:00	18:00	11.6	24.3	-4.71	-22.03	15.66
90		7:00	0:00	29.7	23.4	-9.36	-62.11	12.78
91		13:00	6:00	2	26.4	-6.65	-46.15	7.05
92		19:00	12:00	9.5	24.4	-6.48	-39.02	12.84
93	2017/12/24	1:00	18:00	3.2	24	-7.49	-46.06	13.89
94		7:00	0:00	2.5	23.8	-6.19	-35.16	14.35
95		13:00	6:00	ttu	29.1			
96		19:00	12:00	0	27.4			
97	2017/12/25	1:00	18:00	0	24.6			
98		7:00	0:00	0	25			
99		13:00	6:00	0	30.4			
100		19:00	12:00	0	27.8			
101	2017/12/26	1:00	18:00	5	23.8	-3.17	-11.84	13.55
102		7:00	0:00	0	22.9			
103		13:00	6:00	0	30.4			
104		19:00	12:00	2	23.9	-15.09	-108.92	11.82
105	2017/12/27	1:00	18:00	15.2	23.3	-9.66	-65.09	12.20
106		7:00	0:00	0.3	24.2	-9.32	-60.33	14.20
107		13:00	6:00	0	29.4			
108		19:00	12:00	0	27.4			
109	2017/12/28	1:00	18:00	ttu	24.8	-5.58	-33.39	11.26
110		7:00	0:00	0	23.8			
111		13:00	6:00	0	30.7			
112		19:00	12:00	0	28.2			
113	2017/12/29	1:00	18:00	0	25.3			
114		7:00	0:00	0	24.4			
115		13:00	6:00	0	31.2			
116		19:00	12:00	0	27.5			
117	2017/12/30	1:00	18:00	ttu	25.9			
118		7:00	0:00	ttu	24.8	-5.42	-31.69	11.70
119		13:00	6:00	0	31.4			
120		19:00	12:00	ttu	27.2			
121	2017/12/31	1:00	18:00	23	25	-6.08	-37.80	10.87
122		7:00	0:00	0	24.2			
123		13:00	6:00	0	30.6			
124		19:00	12:00	0	26.6			
125	2018/1/1	1:00	18:00	5	24	-6.30	-37.02	13.41
126		7:00	0:00	6	24.2	-8.25	-55.12	10.91
127		13:00	6:00	ttu	26.8			
128		19:00	12:00	ttu	27.6			
129	2018/1/2	1:00	18:00	0	25.6			
130		7:00	0:00	0	24.7			
131		13:00	6:00	0	30.4			
132		19:00	12:00	0	28.4			
133	2018/1/3	1:00	18:00	0	25.9			
134		7:00	0:00	0	25.2			
135		13:00	6:00	0	31.2			
136		19:00	12:00	0	28.8			
137	2018/1/4	1:00	18:00	0	26.3			
138		7:00	0:00	0	25.4			
139		13:00	6:00	0	30.6			
140		19:00	12:00	ttu	28.6			
141	2018/1/5	1:00	18:00	0.7	24	-7.73	-54.85	7.00
142		7:00	0:00	5.2	24.4	-7.70	-55.90	5.71
143		13:00	6:00	0	30.4			
144		19:00	12:00	0	27.5			
145	2018/1/6	1:00	18:00	1.5	24	-5.16	-27.25	14.07
146		7:00	0:00	ttu	24.1			
147		13:00	6:00	0	29.8			
148		19:00	12:00	ttu	25.7	-2.78	-15.08	7.12
149	2018/1/7	1:00	18:00	ttu	23.8			
150		7:00	0:00	0	23.8			
151		13:00	6:00	0	32.4			
152		19:00	12:00	0	28.3			
153	2018/1/8	1:00	18:00	0	26.1			
154		7:00	0:00	0.3	25.4	-3.60	-20.68	8.13
155		13:00	6:00	0	30			
156		19:00	12:00	0.9	25.4	-3.21	-16.05	9.67
157	2018/1/9	1:00	18:00	0.5	24.4	-3.63	-18.92	10.09
158		7:00	0:00	ttu	24.5			
159		13:00	6:00	0	30.6			
160		19:00	12:00	20	23.8	-5.73	-36.23	9.61
161	2018/1/10	1:00	18:00	ttu	24			
162		7:00	0:00	ttu	22.7	-5.79	-36.28	10.00
163		13:00	6:00	0	30.8			
164		19:00	12:00	0	28.1			
165	2018/1/11	1:00	18:00	0	22.2			
166		7:00	0:00	0	23.8			
167		13:00	6:00	5	27.5	-3.27	-11.65	14.48
168		19:00	12:00	ttu	25	-3.25	-13.74	12.25
169	2018/1/12	1:00	18:00	0	23.4			
170		7:00	0:00	0	23.6			
171		13:00	6:00	0	30.7			
172		19:00	12:00	0	28			
173	2018/1/13	1:00	18:00	1.5	23.8	-1.83	-1.26	13.38
174		7:00	0:00	0	25.2			
175		13:00	6:00	0	30.6			
176		19:00	12:00	0	28.2			
177	2018/1/14	1:00	18:00	0	25.1			
178		7:00	0:00	0	25.5			
179		13:00	6:00	0	30.1			
180		19:00	12:00	0	27.5			
181	2018/1/15	1:00	18:00	0	25.1			
182		7:00	0:00	0	23.6			
183		13:00	6:00	0	32.1			
184		19:00	12:00	0	27.4			
185	2018/1/16	1:00	18:00	0	25.4			
186		7:00	0:00	0	24.8			

* Remark: ttu mean there is rain but so little that unmeasureable

References

[1] Ichiyanagi K and Yamanaka M D., 2005, *Hydrol Process* 19, 3413–3423

[2] Ichiyanagi K, Yoshimura K and Yamanaka M D., 2008, *Advances in Geosciences*, Vol. 11, 107-118. AOGS 2007 - Hydrological Science, Eds. Namsik Park et al., World Scientific Publishing Company

[3] Kurita N, Ichiyanagi K , Matsumoto J, Yamanaka M D and Ohata . T., 2009, *J. Geochem. Explor.* 102, 113–122

[4] Cobb K M, Adkins J F, Partin J W, and Clark B., 2007, *Earth Palnet Sci. Lett.* 263, 207–220

[5] Moerman J W, Cobb K M, Adkins J F, Sodemann H, Clark B, and Tuen A A., 2013, *Earth and Planetary Science Letters* E. doi:10.1016/j.epsl.2013.03.014

[6] Permana DS, Thompson LG and Setyadi G., 2016, *J. Geophys. Res. Atmos.*, 121, 2222–2245, doi:10.1002/2015JD023893

[7] Suwarman R, Ichiyanagi K,Tanoue M, Yoshimura K, Mori S, Yamanaka M D, Kurita N and Syamsudin F., 2013, *SOLA* 9, 74–78, doi:10.2151/sola.2013-017

[8] Belgaman H A, Ichiyanagi K, Tanoue M,Suwarman R, Yoshimura K, Mori S, Kurita N, Yamanaka M D and Syamsudin F., 2016, *SOLA* 12, 192–197, doi:10.2151/sola.2016-039

[9] Belgaman H A, Ichiyanagi K, Suwarman R, Tanoue M, Aldrian E, Utami A I D , and Kusumaningtyas S D A., 2017, *Hydrol. Res. Lett.*, 11(2), 92–98, doi: 10.3178/hrl.11.92

[10] Fudeyasu H, Ichiyanagi K, Yoshimura K, Mori S, Hamada - J I, Sakurai N, Yamanaka M D, Matsumoto J and Syamsudin F., 2011, *J. Meteor. Soc*. Japan. 89A, 49-59, doi:10.2151/jmsj.2011-A03

[11] Joyce R J, Janowiak J E, Arkin P A and Xie P., 2004, *J. Hydromet., 5, 487-503*

[12] Mori S, Hamada J I, Tauhid Y I, Yamanaka M D, Okamoto N, Murata F, Sakurai N and Sribimawati T., 2004, *Mon. Wea. Rev.*, 132, 2021–2039. https://doi.org/10.1175/1520-0493(2004)132<2021:DLRPMO>2.0.CO;2

The Effect of an Active Phase of the Madden-Julian Oscillation on Surface Winds over the Western Coast of Sumatra Island

P Wu[1], D Ardiansyah[2], S Mori[1] and K Yoneyama[1]

[1]*Japan Agency for Marine-Earth Science and Technology*
[2]*Indonesian Agency for Meteorology, Climatology and Geophysics*

pmwu@jamstec.go.jp

Abstract. This study examined the effect of an active phase of the Madden-Julian Oscillation (MJO) on surface winds over the western coast of Sumatra Island using surface meteorological observations, meteorological radar observations, and balloon sounding data obtained from the Years of the Maritime Continent (YMC) Sumatra 2017 field campaign. Wind gusts (sudden and large increases in surface wind speed with durations of several to tens of minutes; squall) were observed frequently at Bengkulu from late November to early December 2017 during a local active phase of the MJO. The wind gusts occurred simultaneously with heavy precipitation when the eastward propagating convection from the Indian Ocean moved over the observation site. Upper air observations showed strong northwesterly winds in the lower troposphere in the 1,000–4,000 m layer. The surface wind gusts extended from the surface up to an elevation of 560 m, with the highest wind speed occurring approximately 300 m above the ground. Small changes in surface air pressure were observed during the wind gusts. The results suggested that momentum transfer from the upper atmosphere by downdrafts was the main cause of the surface wind gusts on the western coast of Sumatra Island during an active phase of the MJO.

Keywords : MJO, surface wind, downdraft, wind gusts.

1. Introduction

Sumatra Island is located in the equatorial region, with the west of the island facing the Indian Ocean (figure 1). Convective activity over the island is generally active, with a pronounced diurnal cycle. Severe thunderstorms occur frequently over the island throughout the year. All thunderstorms have downdrafts, which cause cool air to flow out on or near the surface. A strong downdraft, which is referred to as a downburst, includes an outburst of potentially damaging winds on the ground. The sudden strong winds near the surface, that is, wind gusts, can down trees, damage buildings, and are dangerous to aviation, particularly aircraft which are on final approach or taking off.

The Madden-Julian Oscillation (MJO) [1] is the largest modulator of atmospheric intraseasonal variability in the tropics. An active phase of the MJO is characterized by an eastward progression of large areas of both enhanced and suppressed tropical rainfall, with westerly wind bursts that mainly occur between the Indian and Pacific oceans. The MJO significantly affects weather in the global tropics and subtropics. When an active phase of the MJO propagates eastwards from the Indian Ocean to the Maritime Continent, the influence of the MJO on the diurnal cycle of convection causes considerable variations in precipitation over the islands and their surroundings [2,3]. Meanwhile, convectively coupled equatorial waves (CCEWs) can significantly influence the tropical precipitation, in particular over the Indian Ocean and the Maritime continents. On average, the CCEWs can considerably affect tropical precipitation by contributing up to 16-20% of the total intraseasonal precipitation variance in the tropics [4].

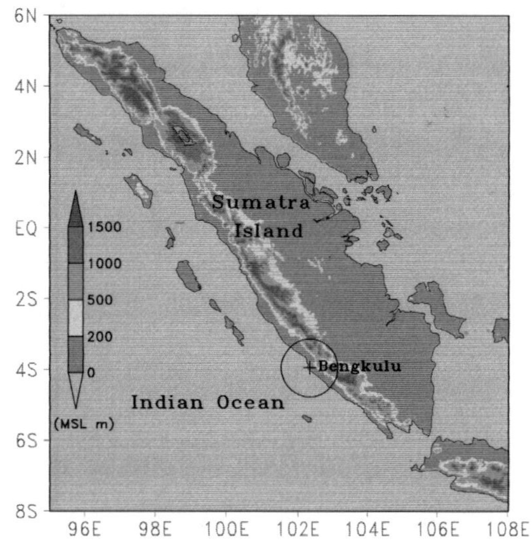

Figure 1. Topography of Sumatra Island and its surroundings, and the location of the Bengkulu Meteorological Observatory (cross). The open circles indicate the scope of the radar observations. Shading denotes terrain elevation. MSL is meters above sea level.

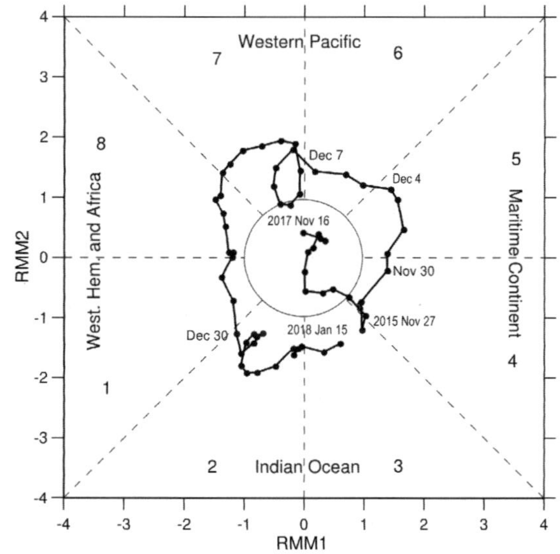

Figure 2. Phase diagrams of the Madden-Julian Oscillation (MJO) from mid-November 2017 to January 2018. The axes (RMM1 and RMM2) represent daily values of the principal components from the two leading modes. Each day within the time period is represented by a dot. The distance from the origin is proportional to MJO strength.

Wu et al. [5] investigated the effects of an active phase of the MJO on an extreme precipitation event that occurred over the western Sumatra Island in mid-December 2015 using observation data obtained from the previous Years of the Maritime Continent (YMC) field campaign. Their results showed that the leading edge of the MJO westerly wind bursts provided favorable conditions for an active phase of the MJO to work with the westward moving diurnal convection and caused torrential rain on the western coast of Sumatra Island. In this way, the MJO significantly affected the diurnal cycle and development of convection on Sumatra Island, which in turn influenced winds on or near the surface. However, because field observation data are often unavailable on the island, especially the high temporal resolution data, exactly how an active phase of the MJO influenced surface winds on the island was not well understood.

To improve our understanding of convection over the Maritime Continent and its relation to the MJO and local atmospheric circulation, we performed an intensive observation, named YMC-Sumatra 2017, on the western coast of Sumatra Island from November 2017 to January 2018 as part of the YMC field campaign. Land-based observations at Bengkulu on the western coast and ship-based observations over the sea off the western coast of the island were conducted simultaneously. A surface and upper meteorological dataset with high temporal resolution was constructed for both the land and sea from these observations. The aims of the study were therefore to examine the variations of surface winds and its relation to convection, and to clarify the effect of an active phase of the MJO on strong surface winds over the western coast of Sumatra Island using the observation data obtained from the field campaign.

2. Observations and data

Surface meteorological observations, radar observations, and balloon soundings at intervals of 3 hours were performed at Bengkulu Meteorological Observatory for 61 successive days from November 16, 2017 to January 15, 2018. The observation site was located on the coastal plain approximately 4 km from the shore (figure 1) at an elevation of 16 m above sea level. Surface meteorological variables, including atmospheric pressure, temperature, relative humidity, wind direction and speed, rainfall, and global solar radiation, were measured at 1-min intervals using an Automatic Weather Station (MAWS 201, Vaisala, Finland). Rainfall was measured by a tipping bucket rain gauge, with a sensitivity of 1 tip per 0.2 mm.

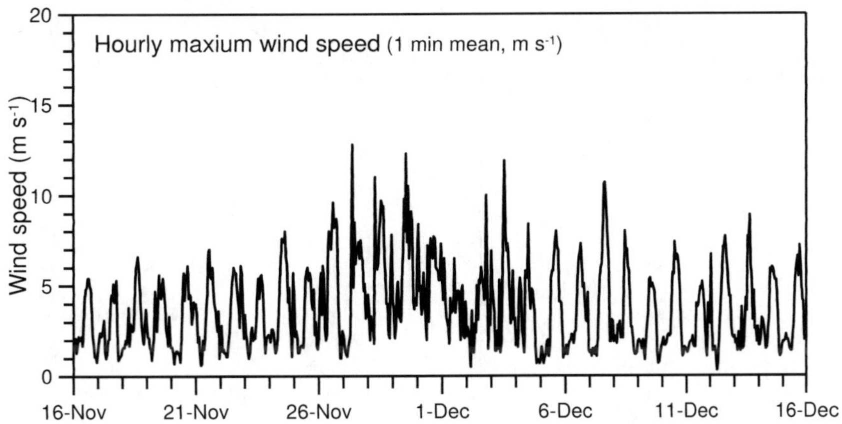

Figure 3. Time series of hourly maximum surface winds (1-minute averaged) observed at Bengkulu Meteorological Observatory on Sumatra Island from November 16 to December 15, 2017. The tick marks in the abscissa indicate 00 LT for each of the days within the time period shown.

Meteorological radar observations were conducted continuously using a C-band Doppler radar (CDR), which was operated by the Indonesian Agency for Meteorology, Climatology and Geophysics (BMKG), to obtain volume scan data at 10-min intervals with a ray and gate spacing of 1° and 250 m, respectively, at 17 elevation angles ranging from 0.5 to 40° within a 120-km range. The reflectivity data from the Constant Altitude Plan Position Indicator (CAPPI) were generated by Rainbow 5 software [6]. To assess the relative position and strength of the MJO, an all-season, real-time multivariate MJO index was used [7].

3. Occurrence of surface wind gusts on the Western Coast of Sumatra Island in an active phase of the MJO from November to December 2017

The oceanic and atmospheric features in November and December 2017 indicated mature La Niña conditions [8]. Sea surface temperatures were lower than normal across the central and eastern equatorial Pacific. Suppressed convection associated with the La Niña event persisted near the Date Line across the central and eastern Pacific, and enhanced convection over the Maritime Continent. Then, during late November and early December, enhanced convection developed over the eastern Indian Ocean and western Maritime Continent due to the development of an active phase of the MJO. The MJO was constructively interfering with the La Niña base state, resulting in enhanced convection over the eastern Indian Ocean and the Maritime Continent. Subsequently, enhanced convection shifted eastward across the Maritime Continent and the western-central Pacific in December.

The MJO index [7] showed a rapid increase in the MJO amplitude from November 25–27, 2017 (figure 2). On November 27, the index had an amplitude greater than 1.0, with a phase of 4 (the enhanced convective phase was centered on the western Maritime Continent). Although the position of the MJO showed little eastward propagation from November 27–29, in the subsequent five days from November 30 to December 4, an eastward propagation of the MJO over the Maritime Continent indicated a steady eastward movement of the robust MJO signal. Thereafter, the MJO index continued with amplitudes greater than 1.0 in December and propagated smoothly across the western Pacific and the Western Hemisphere.

IOP Conf. Series: Earth and Environmental Science **303** (2019) 012009 doi:10.1088/1755-1315/303/1/012009

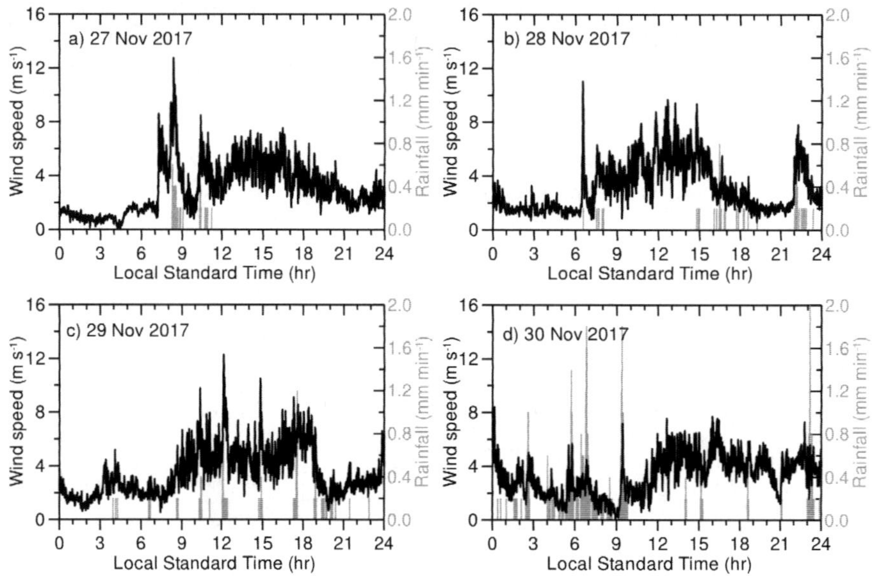

Figure 4. Time series of wind speed and precipitation observed at Bengkulu Meteorological Observatory on Sumatra Island on November a) 27, b) 28, c) 29 and d) 30, 2017 at 1-min intervals.

The time series of hourly maximum surface winds (1-min averaged) at Bengkulu on the western coast of Sumatra Island during the period from November 16 to December 15, 2017 is shown in figure 3. The surface winds were weak prior to the local active phase of the MJO until November 25. Conversely, during the local active phase of the MJO from November 27 to December 4, 2017, surface winds with maximum wind speed greater than 10 ms^{-1} were observed frequently. The surface winds showed a clear diurnal variation in wind speed throughout the period. In general, the maximum surface wind occurred in the afternoon and the minimum wind in the early morning. The strong daytime wind was largely due to the greater temperature differences between the sea and land surfaces during the day compared with at night [9]. The sea breeze on the western coast of the island in the afternoon was much stronger than the land breeze in the early morning [10]. Furthermore, the daytime enhancement of the wind speed was also partly attributable to the diurnal variation of the vertical momentum transport by turbulent fluxes, as it was much easier for fast-moving air above to mix down to the surface during the daytime [11].

4. Occurrence of precipitation and its effect on surface wind gusts

The 1-min mean wind speed and precipitation for the 4 consecutive days from November 27–30, 2017 when the MJO enhanced convective phase was centered near the western Maritime Continent are shown in figure 4. The variations of the 1-min mean winds showed that wind speed consistently displayed short time fluctuations. It was more windy during the daytime than at night. On each of the days, sudden strong surface winds, or wind gusts, occurred several times. The duration of the surface wind gusts had short time periods of 8 to 30 minutes. The World Meteorological Organization (WMO) defined the term "squall" as a sudden wind speed increase of 8 ms^{-1} or more to a speed of greater than 11 ms^{-1}, lasting for 1 min or longer in duration. Squalls occurred on November 27, 28, and 29, as the wind speed increased suddenly by more than 8 ms^{-1}, and wind speed of greater than 11 ms^{-1} were observed on these days. The surface wind speed increased simultaneously with the start of rainfall, and the highest wind speed of the wind gusts occurred during heavy precipitation at the observation site.

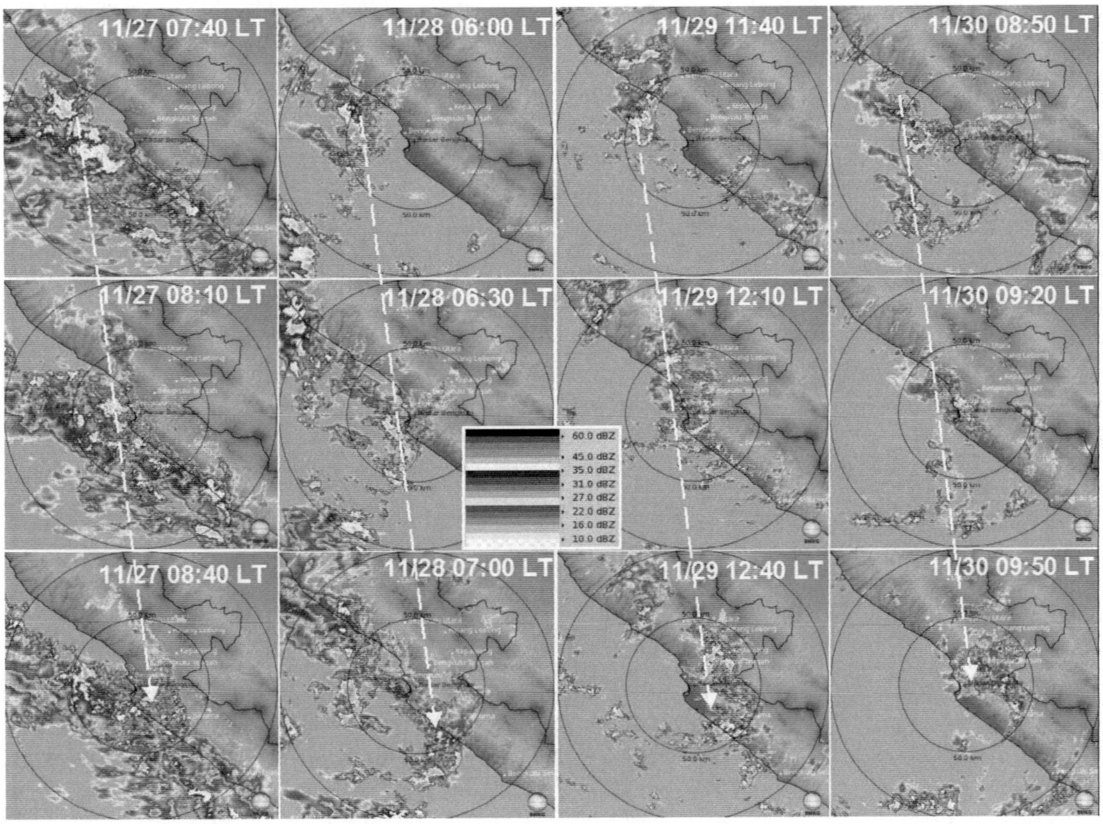

Figure 5. Reflectivity data (in dBZ) from the Constant Altitude Plan Position Indicator (CAPPI) at 2.0 km altitude obtained from the Indonesian Agency for Meteorology, Climatology and Geophysics (BMKG) Bengkulu C-band Doppler radar (CDR) on November 27, 28, 29 and 30, 2017, with the end times (local time) of the volume-scan in the panels. The CAPPI is a radar display that gives a horizontal cross-section of data at a constant altitude.

The reflectivity data from the Constant Altitude Plan Position Indicator (CAPPI) at 2.0 km altitude, obtained from the Bengkulu CDR radar for the times of the strongest wind gusts on each of the four consecutive days from November 27–30, 2017, are shown in figure 5. The middle panels show CAPPI at the times when the surface winds began to increase during the wind gusts on each day. The upper panels and lower panels show CAPPI at 30 min before and after the surface winds began to increase, respectively. Thirty min before the strong surface winds were observed at the observation site, large areas of convection appeared over the eastern Indian Ocean about 30 km to the west of the radar observation site, moving southeastward toward Sumatra Island. Then, a high reflectivity area passed over the observation site during the times when the strong surface winds were observed. The convection became weaker as it moved over the island compared to when it was over the ocean.

The eastward propagation of an active phase of the MJO often causes strong northwesterly winds in the lower troposphere over Sumatra Island [5]. Upper-air sounding data have shown that strong upper westerly to northwesterly winds 20-25 ms^{-1} persisted over the island during the active phase of the MJO from November 27–30, 2017 (shown in Section 5). From radar observations, the migration direction of the precipitation system was almost the same as the northwesterly winds in the lower troposphere. The propagation speed of the precipitation system was estimated as approximately 20 m s^{-1}. It has been reported that the propagation of mesoscale convective complexes (MCSs) is mainly determined by the mean flow of the cloud-layer (850-300 hPa) and the development of new convective cells [12]. The above results suggested that the movement of the mesoscale convective systems from the Indian Ocean was steered towards the west coast of Sumatra Island by the upper northwesterly winds in the cases being studied, which were caused by an active phase of the MJO. Surface wind gusts were then observed when the eastward propagating convection moved over the observation site.

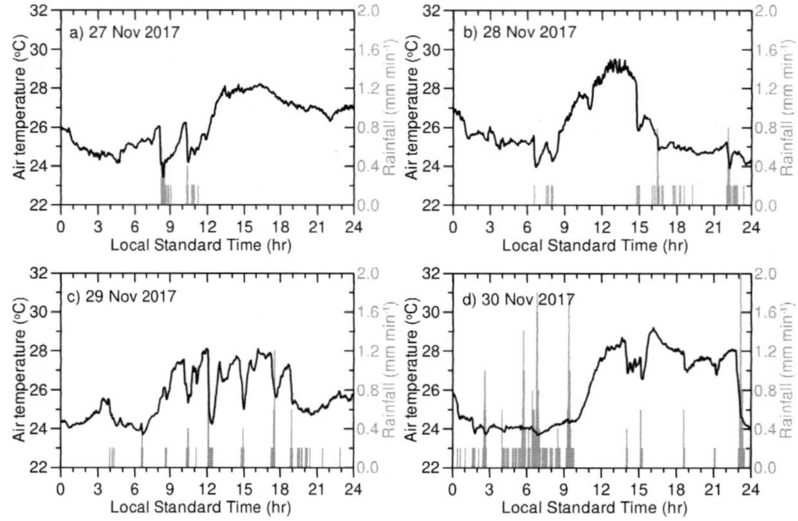

Figure 6. Time series of the surface air temperature and precipitation observed at Bengkulu Meteorological Observatory on Sumatra Island on November a) 27, b) 28, c) 29 and d) 30, 2017 at 1-min intervals.

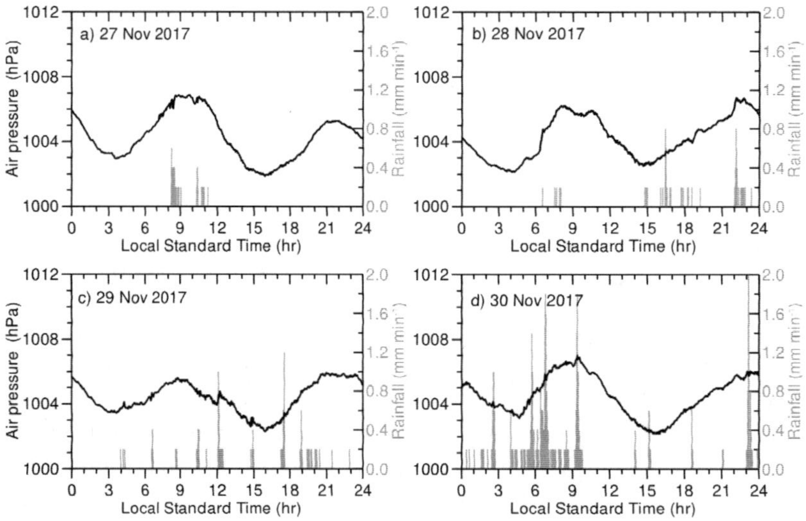

Figure 7. Time series of surface air pressure and precipitation observed at Bengkulu Meteorological Observatory on Sumatra Island on November a) 27, b) 28, c) 29 and d) 30, 2017 at 1-min intervals.

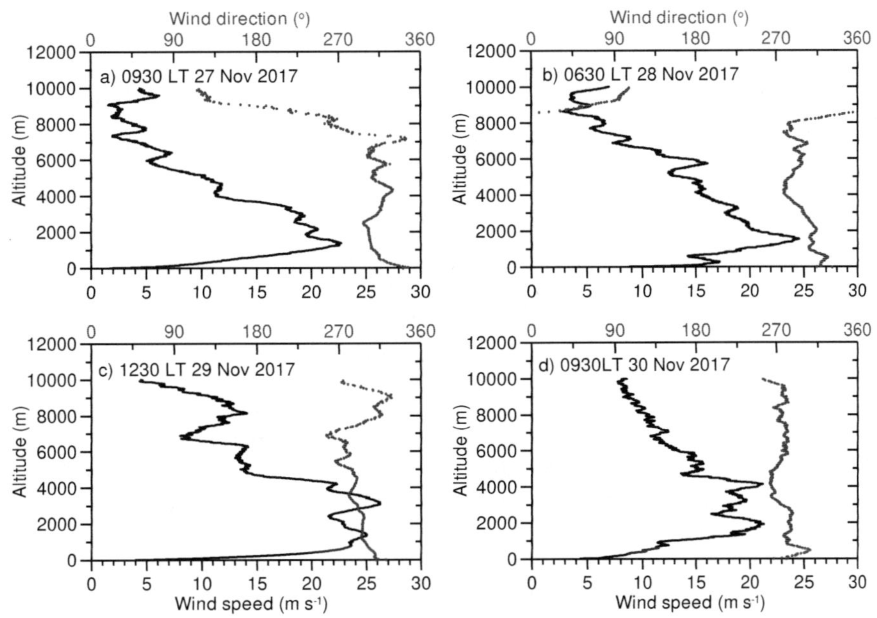

Figure 8. Profile of winds obtained from upper-air soundings at Bengkulu on November a) 27, b) 28, c) 29 and d) 30, 2017, with the balloon-release times (local time) in the panels. Wind speed is labeled along the bottom of the abscissa, while wind direction in degrees is labeled along the upper part of the abscissa.

5. Surface temperature and pressure variations, and vertical distribution of the wind of a wind gust

Continuous surface meteorological observations and balloon soundings at intervals of 3 h were performed at Bengkulu Meteorological Observatory during the YMC-Sumatra 2017 field campaign. As the wind gusts were caused by the downdraft of convection, other meteorological elements on the surface were also affected by the downdraft. In this section, we examined the variations in surface temperature and pressure, and the vertical distribution of the wind gusts.

The time series of surface air temperatures (1-min intervals) and precipitation for the four days from November 27–30, 2017 are shown in figure 6. The surface temperatures show large diurnal variations on these days, with a maximum of approximately 28°C in the afternoon and a minimum of approximately 24°C in the early morning. The diurnal cycle in surface air temperature on the island was caused by daytime solar radiation, even during a local active phase of the MJO [9]. In figure 6, abrupt decreases in surface air temperature occurred frequently, which were all accompanied by precipitation. For the four cases of the strongest surface wind gusts on each day, decreases in surface temperature varied from 0.2 to 3.9°C. Surface air temperature in the early morning on November 28 and 30 was relatively low close to 25°C, and decreases in air temperature of less than 1.0°C were observed during the wind gusts. The reason for the small changes in air temperature was probably the lower surface air temperature. Decreases in surface air temperature by downdraft on the western coast of Sumatra Island only reached a minimum of approximately 24°C [9] as the warm ground heated up the air near the surface. Furthermore, because of the lower air temperature, a layer of cold air near the surface before precipitation most likely acted as a "cushion" and weakened the downdraft near the surface [13], resulting in smaller decreases in the surface air temperature.

The time series of surface air pressure (1-min intervals) and precipitation for the four days from November 27–30, 2017 are shown in figure 7. The variations in surface air pressure were characterized by dual maxima in the morning and evening, with minima after midnight and in the afternoon. The diurnal and semidiurnal variations of pressure were the surface representation of a diurnal tide of the entire atmosphere due to the warming of the upper atmosphere by the Sun [14]. Other than the large diurnal variations from the atmospheric tide, the sudden abrupt increases in pressure coincided with precipitation at the observation site. However, only small changes in surface air pressure of approximately 0.5-0.7 hPa occurred in the four cases of the strongest surface wind gusts on each day from November 27–30, 2017. In comparison to the cases shown above in figure 4, 6 and 7, which occurred during strong upper northwesterly winds in an active phase of the MJO, similar gusts of wind accompanied by a large decrease in surface temperatures occurred on November 22 and 24, 2017 (not shown). At this time widespread convection passed over the observation site in an inactive phase of the MJO with weak winds at the lower troposphere, whereas the wind speeds of the wind gusts were much weaker than those in the active phase of the MJO (figure 3). The above results suggested that the surface wind gusts that were observed on the western coast of Sumatra Island in an active phase of the MJO were mainly caused by momentum transfer from the upper atmosphere to the surface by downdrafts, rather than by a pressure gradient that resulted from the temperature contrast between the cold pool and air surroundings.

Figure 8 shows the wind profiles of the strongest surface wind gusts for each of the four days from November 27–30, 2017, which were obtained at the nearest balloon-release times (within 90 min) to the occurrence of the surface wind gusts. The wind speeds increased rapidly from the surface to approximately 1,000 m elevation, with maximum wind speeds (>20 ms^{-1}) occurring in the 1,000-4,000 m layer. The northwesterly wind directions remained steady from near the surface up to approximately 8,000 m altitude in the lower troposphere. Above this elevation, the wind direction changed to an easterly, consistent with the passage of an active phase of the MJO. Surface winds measured on November 27, 28 and 30 (figure 8a, c, d) were weak because the balloon soundings were not carried out during the occurrence of the surface wind gusts.

On November 28, radiosonde began to ascent at 0630 LT when a surface wind gust occurred at the observation site. One min later, the largest surface wind speed of 11.0 ms^{-1} was measured at 0631 LT at the surface. The wind speed profile (figure 8b) shows that the gust winds extended from near the surface up to an elevation of 560 m, with the largest wind speed (> 17.0 ms^{-1}) of the gust occurring in the 230–303 m layer.

As shown previously in section 4, on the morning of November 28, the precipitation systems propagated southeastward from the Indian Ocean to Sumatra Island. The surface wind gusts occurred when the precipitation systems passed over the observation site. In general, the highest winds in a surface wind gust in a downburst occur closer to the splashdown point near the surface when strong downdraft reaches the ground. In the present study, the highest winds of the winds gust on November 28 were not observed on the surface, but approximately 300 m from the ground. This is most likely due to the strong northwesterly winds in the lower troposphere associated with an active phase of the MJO and the fast propagation of the precipitation systems. Momentum transfer from the upper atmosphere to the surface by downdrafts caused the wind gusts near the surface. However, the surface exerts a frictional drag on the air flowing just above it, forcing the winds near the surface to slow down. As a result, the highest winds occurred close to 300 m above the ground, rather than near the surface. To reveal the detailed structure and formation mechanism of the surface wind gusts, further studies based on an analysis of the C-band Doppler radar (CDR) data and other observation data obtained from the field campaign are needed.

As previously mentioned, the oceanic and atmospheric features in November and December 2017 indicated mature La Niña conditions. Meanwhile, the Indian Ocean Dipole (IOD) remained near neutral during November to December 2017. Sea surface temperatures were close to normal across the tropical Indian Ocean; seas to the west of Sumatra Island were warm, as usual. Westerly winds prevailed in the lower troposphere. The active phase of the MJO was constructively interfering with the La Niña base state, resulting in enhanced convection over the Maritime Continent region. The effect of MJO on the development of convection and precipitation over Sumatra is expected to be highly associated with the nature of MJO itself in modulating total precipitable water (TPW/column water vapor) in the tropics, in particular during MJO phases 3 and 4, and seasonal variations of TPW modulated by the MJO are maximized in the tropics during boreal winter [15]. Therefore, it may be that the ENSO, IOD and monsoon large-scale atmospheric circulations during November to December 2017 provided favorable conditions for the development of heavy rainfall and strong surface wind gusts on the western coast of Sumatra Island.

6. Summary

We successfully performed an intensive observation, named YMC-Sumatra 2017, on the western coast of Sumatra Island from November 2017 to January 2018 as part of the YMC field campaign. During the field campaign, MJO activity took place from late November to December 2017. This study examined the variations in the surface winds and the effect of an active phase of the MJO on the occurrence of surface wind gusts over the western coast of Sumatra Island using the observation data obtained from the field campaign.

Wind gusts, or sudden and large increases in the surface wind speed with durations of several to tens of min (squall), were observed frequently at Bengkulu from late November to early December 2017 in a local active phase of the MJO. The occurrences of surface wind gusts were accompanied by heavy precipitation, when an eastward propagating convection from the Indian Ocean associated with an active phase of the MJO moved over the observation site. Upper-air observations showed strong northwesterly winds (>20 m s^{-1}) in the lower troposphere in the 1,000-4,000 m layer. The gust winds extended from near the surface up to an elevation of 560 m, with the highest wind speed occurring approximately 300 m above the ground. Decreases in surface temperature varied from 0.2 to 3.9°C and small changes in air pressure were observed during the wind gusts. The results suggested that momentum transfer from the upper atmosphere by downdrafts was the main cause of the surface wind gusts on the western coast of Sumatra Island in an active phase of the MJO.

Acknowledgments

Many thanks to Mr. Warjono, the head of Meteorological Station of Fatmawati Soekarno Airport in Bengkulu, Indonesia, for his strong support in the observations of this study. The authors are grateful to the reviewers for their constructive comments and suggestions. We are indebted to the Ministry of Research, Technology and Higher Education (RISTEKDIKTI), Agency for the Assessment and Application of Technology (BPPT), and Japanese Embassy for their highly understanding and support to conduct the observations.

References

[1] Madden R A and Julian P R., 1994, *Mon. Wea. Rev.* **122,** 814-837
[2] Peatman C Matthews A and Stevens D., 2014, *Q. J. R. Meteorol. Soc.* **140,** 814-825
[3] Birch C Webster S Peatman C Parker D Matthews A Li Y and Hassim M., 2016, *J Climate* **29,** 2471-2492
[4] Lubis S W and Jacobi C., 2015, *Int. J. Climatol.* **35,** 1465-1483. doi:10.1002/joc.4069
[5] Wu P Ardiansyah D Yokoi S Mori S Syamsudin F and Yoneyama K., 2017, *SOLA,* **13,** 36-40
[6] Selex, 2010: *Rainbow 5. Products and Algorithms, Release 5.31.0* Selex SI GmbH Neuss pp 442
[7] Wheeler M and Hendon H., 2004, *Mon. Wea. Rev.* **132** 1917-1932 Deser C and Wallace J., 1990 *J Climate,* **3,** 1254-1281
[8] Deser C and Wallace J., 1990, *J Climate,* **3,** 1254-1281
[9] Wu P Mori S and Syamsudin F., 2018, *Progress in Earth and Planetary Science* **(2018)5:4** https://doi.org/10.1186/s40645-017-0160-7 Accessed 10 Jan 2018
[10] Wu P Hara M Hamada J Yamanaka M D and Kimura F., 2009, *J. Appl. Meteor. Climatol.* **48,** 1345-1361
[11] Fujibe F., 1985, *J. Meteor. Soc. Japan,* **63,** 52-59
[12] Corfidi S Merritt J and Fritsch J., 1996, *Wea. Forecasting,* **11,** 41-46
[13] The University of Illinois at Urbana-Champaign *Microbursts* http://ww2010.atmos.uiuc.edu/ Accessed 4 Aug, 2008
[14] Haurwitz B., 1956, *New York University Press* **2(5)** 1-36
[15] Fathurochman I Lubis S W and Setiawan S., 2017, *IOP Conf. Ser.: Earth Environ. Sci.* **54** 012034. doi:10.1088/1755-1315/54/1/012034

IOP Conf. Series: Earth and Environmental Science **303** (2019) 012010 doi:10.1088/1755-1315/303/1/012010

Characteristic of anabatic wind in Bandung basin observed by AWS

P Y Kombara[1], I D G A Junnaedhi[1,2], and E Riawan[1,2]

[1]Weather and Climate Prediction Laboratory, ITB, Indonesia
[2]Research Group of Atmospheric Science, ITB, Indonesia

pyudha@meteo.itb.ac.id

Abstract. A field campaign has been done for the wet season in 2015 to identify the pattern of local circulation such as anabatic wind at Bandung basin. In this field campaign, the Automatic Weather Stations (AWS) were installed at seven points from the north until the south region of Bandung basin. The result of this research defines the pattern of anabatic wind in Bandung basin into three categories: the pattern on weak convection, strong convection, and strong convection with rain event. On weak convection condition, after 12.00 LT, the anabatic wind tended to move to the north region of Bandung basin. The wind still moved to the north and south region, on strong convection and strong convection with rain event condition after 12.00 LT. However, after 15.00 LT the wind on strong convection with rain event changed to become northward. On strong convection and strong convection with rain event, the pattern of anabatic wind was influenced by the convective activity in the north and south region of Bandung basin. Whereas on weak convection condition, the pattern of anabatic wind was not influenced by the convective activity.

Keywords : AWS, convection, anabatic wind,

1. Introduction

Bandung is one of the areas in West Java, Indonesia which has a complex topographic form. Since the region is surrounded by mountains, it makes Bandung area resembled the shape of a basin, so-called the Bandung basin [1]. The land area that has basin shape will have a local circulation in the form of anabatic and katabatic wind [2]. The existence of local circulation in the Bandung basin has been investigated in the previous studies. Based on the previous research, Bandung basin has a local circulation in the form of anabatic wind that has a semidiurnal pattern [3,4,6]. The prior researches were conducted during the dry season [3] and wet season [4] in 2012 used the SODAR instrument. While in the subsequent research, surface observations by using three AWS placed in the northern part of the Bandung basin has succeeded identify anabatic wind patterns [6]. This research proved the existence of the anabatic wind circulation which was stated in the previous research [3,4].

The previous research still has some weaknesses, which the observation tools used and the number of observation points were still limited so that the existing wind patterns could not be properly described. Therefore, to complement the previous research, the other similar research needs to be carried out on the same basis with the additional observation points so that better observations and wind patterns in the Bandung Basin can be described properly.

2. Data and Method

The method of this study is the surface observation by conducting a campaign using AWS (Automatic Weather Station) which was installed at seven points in the Bandung basin during the wet season to define local wind patterns. This campaign was conducted from mid-December 2014 until the end of January 2015. However, the data used for processing was only six days due to incomplete data on the other days. The dates of the selected days are January 5th,6th,9th,10th,11th, and 12th, 2015. Observation of local circulation such as anabatic wind circulation requires several weather parameters such as temperature, pressure, humidity, the direction and speed of wind that must be observed simultaneously and follow the instruction from WMO [8]. In addition, observations were made in several places to get more comprehensive weather data. The wind which was defined in this study is the local wind that cuts the valley, so observations were made by referring to the ideal sketch of Stull [5] in the form of a basin as shown in figure 1.

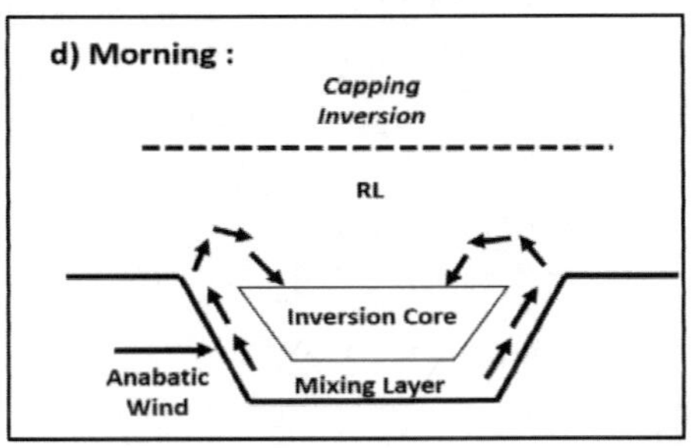

Figure 1. Ideal sketch of Stull [5] for local circulation in basin-shaped areas (reproduces from Stull [5])

Cross section and location of the spread point of the observation are shown in figure 2. Observations of this study used three AWS of Davis Vantage Pro, two AWS of Davis Vantage Vue and two MAWS (Meteorology-ITB Automatic Weather Station).

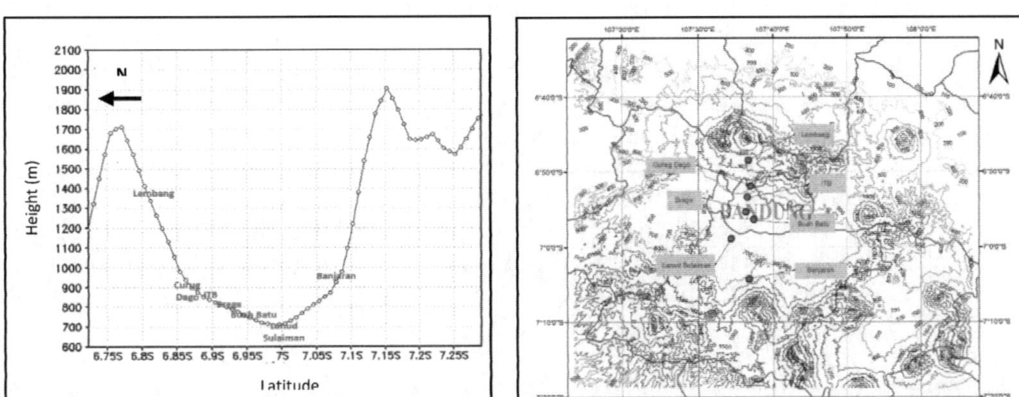

Figure 2. Cross section of observation point (left) from north to south and point distribution (right)

The next step is grouping the days to determine the convective conditions occurred during observation by using satellite imagery of MTSAT IR1 data. The convective condition should be determined because during observation the convective conditions were different so that the data must be grouped. Day grouping is done by processing the temperature of black body (Tbb) data on MTSAT IR1 satellite images with the following calculations:

IOP Conf. Series: Earth and Environmental Science **303** (2019) 012010 doi:10.1088/1755-1315/303/1/012010

$$CI = 253 - TbbIR1 \quad (Tbb\ IR1 < 253) \tag{1}$$
$$CI = 0 \quad (Tbb\ IR1 > 253) \tag{2}$$

The threshold value of Tbb-IR1 of 253 K [2] is used in this research. After being processed with the above calculation, the results were overlaid with rainfall data to determine which days belong to the strong convective day category accompanied by rain (figure 3). Therefore, the convective conditions can be divided into three categories: weak convective days, strong convective days, and strong convective days accompanied by rain events.

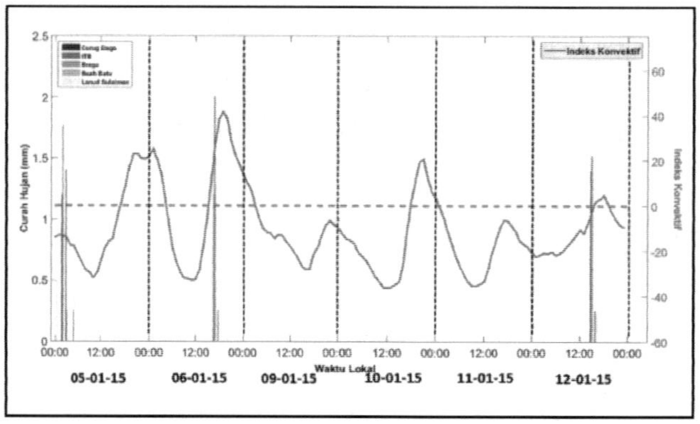

Figure 3. Convective index plot of MTSAT IR1 data overlaid with rainfall to determine convective conditions

The results of determining convective conditions are as follows:
Weak Convection (WC): 9th, 11th January 2015
Strong Convection (SC): 5th, 10th January 2015
Strong Convection with rain event (SCR): 6th, 12th January 2015

After being grouped into three convective conditions, the data went into the processing stage. In the first stage, temperature, air pressure and wind were converted to become virtual potential temperature (θv), Mean Sea Level Pressure (MSLP) and meridional winds (V Wind). In the second stage, the data of virtual potential temperature (θv), Mean Sea Level Pressure (MSLP), and meridional winds (V) were interpolated using linear interpolation method. The interpolation process aims to see diurnal or semi diurnal patterns as seen as an hovmoller image. After being interpolated, the next step was compiling the data and calculating the anomaly. The example of results is shown in figure 4. The existing results were then analyzed spatially and temporally.

3. Result

3.1 Weak Convection

At 00.00 to 07.00 LT in Lembang, Lanud Sulaiman, and Banjaran there was a **θv** anomaly whose value was negative, while in Curug Dago, ITB, Braga, and Buah Batu, the values were positive. This condition made the air temperature in Lembang and Banjaran was cooler and the air density was large so that the air parcels became heavier. This condition caused the air parcel tended to descend down the slope to the valley. In the valley, there was also a low-pressure center that attracted airflow from Lembang and Banjaran. When looking at the V-wind anomaly (figure 5c) the condition was consistent, the wind from Lembang moved southward while from Banjaran moved northward. In the range between 00.00 - 07.00 LT was called as the katabatic wind regime.

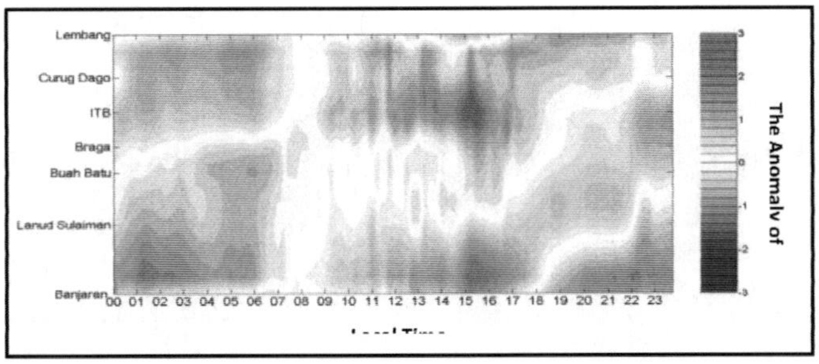

Figure 4. An example of linear interpolation results for the θv parameter

The warming from the sun began to occur at 07.00 LT and it warmed the air around the valley wall. The condition of θv anomalies in Lembang, Sulaiman Lanud, and Banjaran area showed positive values so that the air density was smaller. Whereas, in Curug Dago, ITB, Braga, and part of Buah Batu area, the anomalies were negative. But for the case of anabatic wind, the difference in air density was not enough to move the air, so the MSLP anomaly condition was also reviewed (figure 5b). There was a low-pressure center in the area of Lembang, Lanud Sulaiman, and Banjaran and on the contrary, the high pressure center existed in parts of Curug Dago, ITB, Braga, and Buah Batu. Since the wind will move from high to lower pressure [9], there should be the wind that moves towards Lembang and Banjaran.

When reviewing the anomaly of V-wind, there was a wind that moved to Banjaran at 07.00-12.00 LT and to Lembang at 09.00-17.00 LT so that the wind conditions were in accordance with the theory and in this span of time it is called as an anabatic wind regime. Based on the theory [5], for anabatic winds there should be wind moving from the center of the valley to the two valley walls in the afternoon at maximum insolation. But for the weak convection conditions, this did not happen. The existence of the wind that moved towards the two valley walls only occurred until 12.00 LT. After that, the wind tended to move to one of the valley walls that was only to the north or Lembang. This was presumably because there was an influence from the position of the sun during the wet season which was in the south of equator.

At 6:00 to 18:00 LT, radiative cooling of the surface began to occur because of the sunset. Radiative cooling in the mountain cooled the surrounding air, causing air to move down the slope towards the valley [5]. Radiative cooling caused air θv to decrease so that the atmospheric conditions tended to stable again. Therefore, in this span of time, the circulation turned into a katabatic wind regime again.

3.2 Strong Convection

In the SC condition, katabatic wind regime occurred at 00.00 - 07.00 LT, as in the WC condition. In the morning at 07.00 LT warming from the sun began to occur and warmed the air around the valley wall first. At 10.00 LT, the θv anomalies in the Banjaran region and part of Lanud Sulaiman showed positive values so that the air density became lighter which caused the air parcels tended to move upwards. However, Curug Dago, ITB, and Braga areas still showed negative θv anomaly values. The difference in the anomalous value of θv between the valley floor and the valley wall generated the pressure difference (figure 6b). When referring to the theory [5], during the day there is an anabatic wind moving from the bottom of the valley to the two valley walls. In the SC condition, the anomaly pattern of meridional wind was in accordance with the theory [5]. At 10.00 - 16.00 LT there were positive and negative of V-wind anomalies (figure 6c) at the same time, which showed that there was a wind moving to north and south.

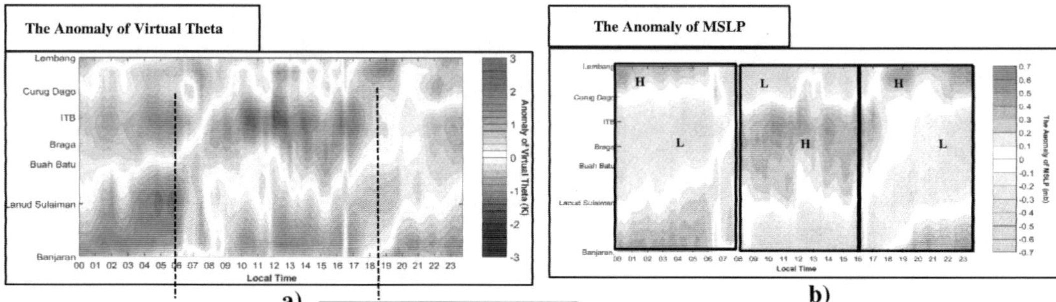

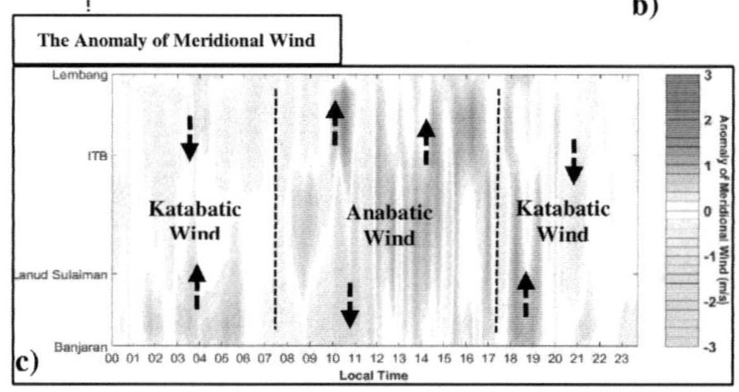

Figure 5. The anomaly of Virtual Theta (a), The anomaly of MSLP (b), and The anomaly of meridional wind (c) in WC Condition

This suitability was because in the Bandung basin during the wet season, there was a strong influence of convective activity on both valley walls. However, the anabatic wind regime that moved to the south occurred more strongly than the anabatic wind that moved to the north. This was due to the influence of stronger convective activity in the south. In the north, there was also the influence of convective activity, but the effect was not as strong as in the south. At 5:00 LT heating from the sun weakened and cooling of the surface of the valley walls began to occur. As in WC conditions, at night the katabatic wind regime returned. The negative value of θv anomaly in both valley walls (Lembang and Banjaran) caused the air density to become heavier so that the air parcel moved down the slope towards the valley floor.

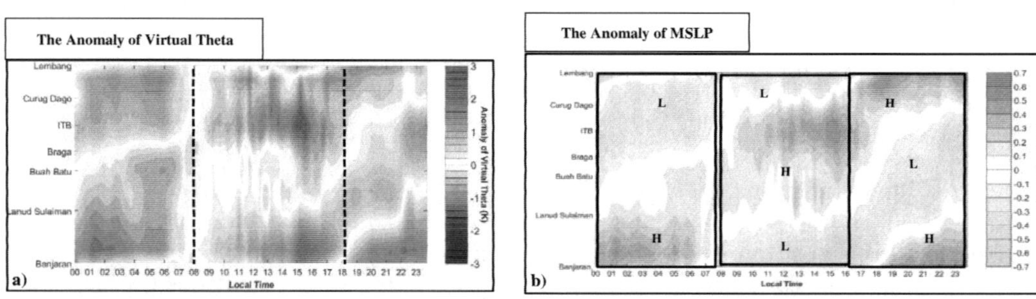

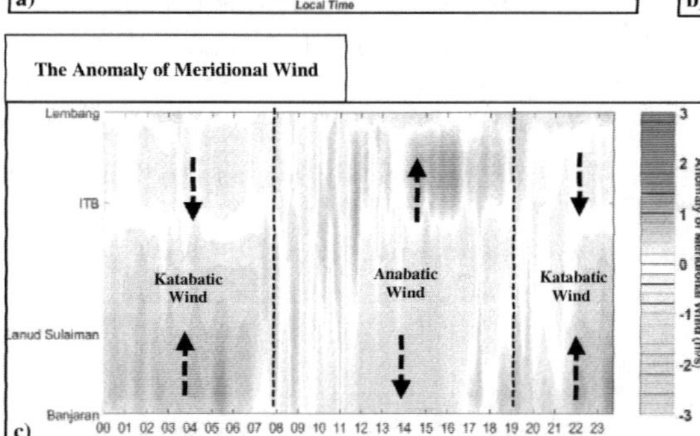

Figure 6. The anomaly of Virtual Theta (a), The anomaly of MSLP (b), and The anomaly of meridional wind (c) in SC Condition.

3.3 Strong Convection with Rain Event

In the conditions of SCR at the time of 00.00 - 07.00 LT there was a katabatic wind regime. In the morning, there was heating from sunlight that warmed the air near the valley wall. As in the condition of SC, there was also an anabatic wind regime in the SCR condition which was in accordance with the theory [5]. The positive θv anomaly condition in Lembang and Banjaran areas made the air parcel lighter so that the region became a low-pressure center (figure 7b). Then, there was an airflow that moved from the valley to the north and south of the basin (figure 7c) as in the condition of SC. However, in the SC condition, after 3:00 LT the anabatic wind that moved into Banjaran area suddenly weakened and disappeared. The wind direction was turned towards Lembang. This tendency was similar to the WC condition which occurred after 12.00 LT. The change in wind direction was due to a positive θv anomaly (figure 7a) which was strong in Lembang region. The positive θv anomaly caused very low-pressure anomalies (figure 6b) which attracted airflow to Lembang region.

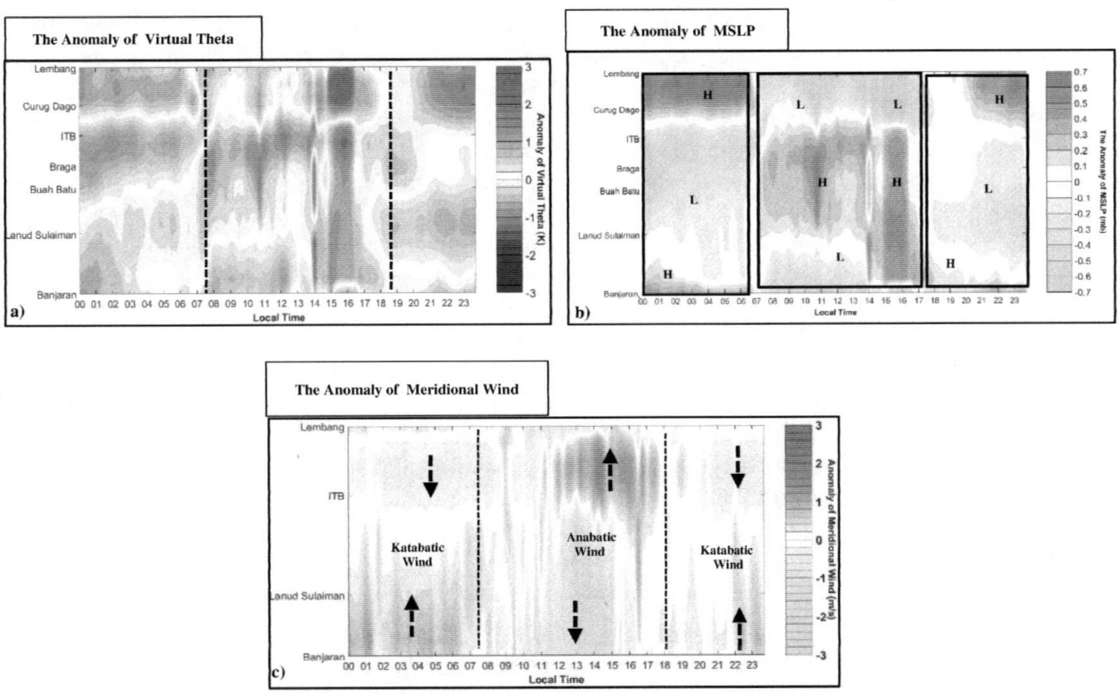

Figure 7. The anomaly of Virtual Theta (a), The anomaly of MSLP (b), and The anomaly of meridional wind (c) in SCR Condition.

In addition, at that time in the Bandung basin there were rain events (figure 8a and figure 8b). When the rain stopped, the convective activity has weakened. However, this weakening only occurred in Banjaran region so that anabatic winds that moved into this region has also weakened. The weakening of convective activity in Banjaran caused a change in the pattern of anabatic wind movement in the afternoon of the SCR condition. At night (18.00-00.00 LT) when there was no more solar warming, radiative cooling occurs. Similar to the WC and SC conditions, the katabatic wind regime occurred in the SCR conditions at night. The weight of the air density caused the air mass flew down the slope to the valley which was called katabatic wind (figure 7c).

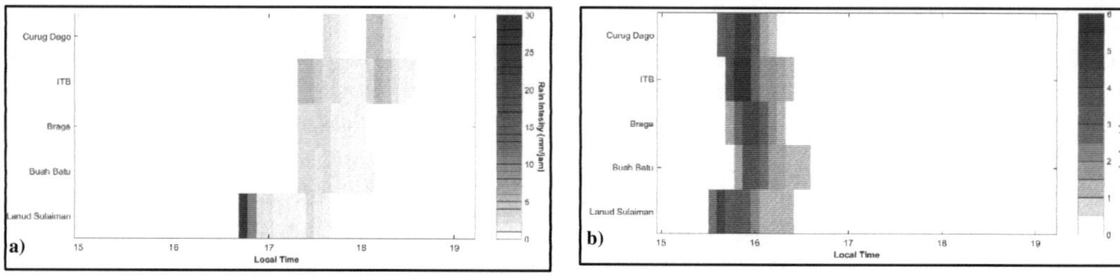

Figure 8. Rain rate on 6 January 2015 (a) and 12 January 2015 (b) in SCR condition.

3.4 The Pattern of Local Wind in Bandung Basin on Several Convective Conditions

Cross-valley winds during the day are stronger than at night [7]. Figure 9 shows the scheme of the local wind pattern in each condition and its development with respect to time. The wind speed is shown in the figure as an arrow symbol with different colors. Cross-valley winds in the afternoon showed an anabatic wind. Cross-valley winds during the day or anabatic wind was stronger only in a few conditions. These conditions were WC at 10-12 LT, SC at 13-18 LT, and SCR at 10-15 LT. Furthermore, when comparing the anabatic wind speed between the southern and northern parts, the southern part had a higher speed except for WC conditions at 13.00-18.00 LT and SCR conditions at 16.00-18.00 LT. Stronger anabatic wind speeds in the southern part showed a stronger influence of convective activity in the south when the condition of SC and SCR.

Generally, in the wet season, local wind patterns in the Bandung basin can be divided into three categories, those are Weak Convection (WC), Strong Convection (SC), and Strong Convection with Rain event (SCR) as shown in figure 9. At the WC conditions, the local wind in the Bandung basin begins with katabatic wind at 00.00 - 06.00 LT which blows from Banjaran and Lembang to the center of Bandung Basin or the valley floor. Then in the morning until the evening, an anabatic wind blows coming from the middle of the basin to Lembang and Banjaran. However, these conditions only last until 12.00 LT. Then the anabatic wind tends to blow into Lembang area and there is a weakening of the anabatic wind that blows towards Banjaran. In the evening, the katabatic wind returns from Banjaran and Lembang to the center of the basin.

At the SC condition, at night there is a katabatic wind regime from Banjaran and Lembang to the center of the basin as in WC condition. Furthermore, during the day, it turns into an anabatic wind regime from the center of the basin which blows to Lembang and Banjaran. But unlike the WC condition, the condition of the SC anabatic wind that blows to Lembang and Banjaran lasts until late afternoon. Then in the evening, it changes back to the katabatic wind regime.

As in the WC and SC conditions, the wind pattern in SCR condition begins with a katabatic wind regime and changes in the afternoon to an anabatic wind. Anabatic wind pattern of the SCR condition is similar to the SC condition, except in the afternoon. The anabatic wind blows from the center of the basin to Lembang and Banjaran areas until 15:00 LT. Thereafter, there is an anabatic wind which blows into Banjaran. At that time, anabatic wind tended to move to Lembang. Then in the evening, it changed again to become a katabatic wind regime.

4. Conclusion

Based on the analysis of AWS observation data on the seven stations in the wet season, the following conclusions are obtained:

1) Anabatic winds of the Bandung basin during the wet season have three patterns in three conditions (WC, SC, and SCR). At the WC conditions, anabatic wind after 12.00 LT tends to move towards Lembang after previously moving towards Lembang and Banjaran. In the conditions of SC and SCR after 12.00 LT, it still moves towards Lembang and Banjaran, but for SCR at 15.00 LT, its direction changes to Lembang.

2) In WC conditions, the convective activity does not affect anabatic wind patterns. On the contrary, for the conditions of SC and SCR convective activity influences anabatic wind patterns in the Bandung Basin.

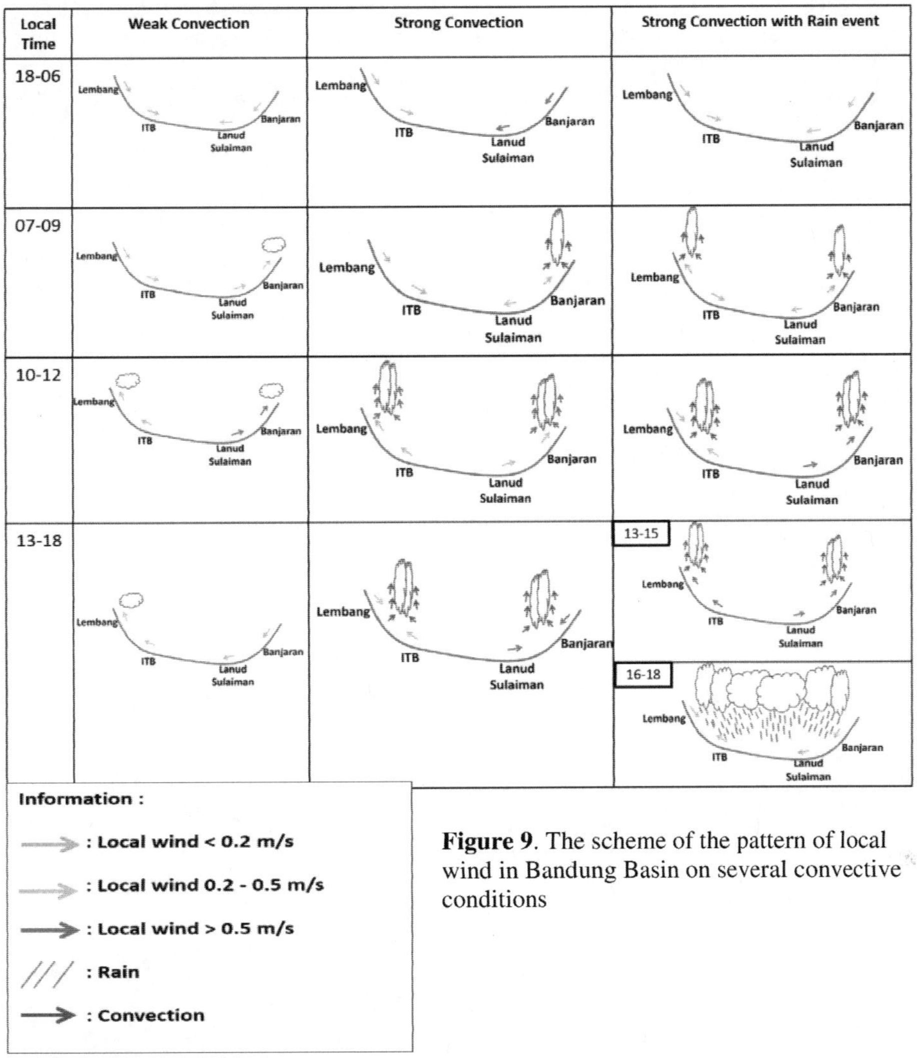

Figure 9. The scheme of the pattern of local wind in Bandung Basin on several convective conditions

Acknowledgment

The authors gratefully acknowledge the contribution to WCPL-ITB and crew for valuable assistance while this observation was done.

References

[1] Abidin, Hasanuddin Z, Andreas, Heri, Gamal M and Darmawan, D., 2006, Land Subsidence Characteristics of Bandung Basin (Indonesia) between 2000 and 2005 as Estimated from GPS Surveys. (*Munich: Landslide Control and Monitoring Survey*)

[2] Yang L, Wang Z, Chu Y, Zhao H and Tang M., 2014, *Adv. Atmos. Sci.* **31**: 1386. https://doi.org/10.1007/s00376-014-3165-9

[3] Megatroika A., 2013, The Analysis of Local Wind at Bandung Basin Based on Data of SODAR Observation and The Result of WRF Model Simulation (*Bandung: Department of Meteorology, ITB*)

IOP Conf. Series: Earth and Environmental Science **303** (2019) 012010 doi:10.1088/1755-1315/303/1/012010

[4] Ramadhan M., 2014, The Analysis of Local Wind of Bandung Basin in Wet Season Based on Data of Sodar Observation, AWS, and MTSAT. (*Bandung: Department of Meteorology, ITB*)
[5] Stull R B., 1988, An Introduction to Boundary Layer Meteorology (*Boston: Kluwer Academic Publisher*)
[6] Sugiarto and Aldi Nursepta Nirwana, 2014, The Analysis of Local Wind at North Mountains of Bandung Basin Based on The Data of AWS Observation (*Bandung: Department of Meteorology, ITB*)
[7] Whiteman C D., 2000, Mountain Meteorology Fundamental and Application (*Oxford: Oxford University Press*)
[8] WMO, 2008, Guide to Meteorological Instruments and Methods of Observation (*Geneva: World Meteorological Organization*)

A study of gravity wave activities based on intensive radiosonde observations at Bengkulu during YMC-Sumatra 2017

T Kinoshita[1], R Shirooka[1], J Suzuki[1], S Ogino[1], S Iwasaki[2], K Yoneyama[1], U Haryoko[3], D Ardiansyah[3] and D Alyudin[3]

[1] Japan Agency for Marine-Earth Science and Technology, Japan
[2] National Defence Academy, Japan
[3] Indonesian Agency for Meteorology, Climatology and Geophysics, Indonesia

Email: t-kinoshita@jamstec.go.jp

Abstract. Gravity waves are generated from local regions such as convection, topography, jet stream and fronts. It is important to understand their activities because the quasi-biennial oscillation and general circulation in middle atmosphere are affected by their activities. However, the scale of gravity waves is too broad to understand their characters and activities in detail. The purpose of this study is to clarify three-dimensional dynamical circulation driven by gravity waves around the equatorial region. The Campaign of Years of Maritime Continent-Sumatra (YMC-Sumatra) has been operated since 2017 to expedite the progress of improving understanding and prediction of local multi-scale variability of the maritime continent weather-climate system and its global impact. We use intensified radiosonde observation data at Bengkulu during December 2017 and examine the relation between gravity wave activities and background state. The results showed that the heat flux of the gravity waves is proportional to the vertical shear of the background zonal wind. From the relation between gravity wave energy and Richardson number, it is suggested that the gravity waves having large amplitude modify the background state around 15 km.

Keywords : gravity wave, radiosonde, YMC, maritime continent.

1. Introduction

A quasi-periodic oscillation of the equatorial zonal wind between easterlies and westerlies with a mean period of about 28 months, which is widely known as the quasi-biennial oscillation (QBO), exists in the equatorial middle atmosphere. There is a general circulation, which transports the tracer gas from equator to pole regions, in the middle atmosphere. It is well known that atmospheric gravity waves play an important role in the QBO and the general circulation. Sato and Dunkerton [1] showed that disturbances having periods shorter than 3 days have sufficient amplitudes to drive the QBO by estimating the absolute value of the vertical momentum flux from the radiosonde observation data in Singapore. Kawatani et al. [2] examined a contribution from plane equatorial waves, extratropical Rossby waves, and internal gravity waves to the driving of the QBO using the gravity-wave resolving general circulation model (GCM) data. It was found that the eastward-propagating equatorial waves including Kelvin waves contribute up to ~25%−50% to the driving of the QBO under the eastward wind shear of the QBO, and the inertia-gravity waves contribute to the rest. On the other hand, the westward-propagating equatorial waves including Rossby-gravity waves contribute to ~10% to the driving of the QBO under the westward wind shear of the QBO.

Extratropical Rossby waves propagating into the equatorial region contribute to ~10%–25%, whereas inertia-gravity waves with zonal wavelengths ≤ 1000km are the primary contributors to the westward shear phase of the simulated QBO. Thus, it is found that the QBO is driven by not only Equatorial Kelvin waves, mixed Rossby-Gravity waves but also inertia-gravity waves. Note that the role of convectively coupled equatorial waves in modulating the QBO is still unclear [3,4]. On the stratospheric general circulation, which is called "Brewer-Dobson circulation" [5,6], Okamoto et al. [7] revealed that the Brewer-Dobson circulation associated with gravity waves including both orographic and non-orographic components is dominant in the summer hemisphere upwelling part (about 20◦S) of the winter hemisphere cell, although the overall structure of the residual mean flow is primary determined by the Rossby waves, especially planetary waves. On the origin of gravity waves, since they are generated from local regions such as convection, topography, jet stream and fronts, it is difficult to understand all gravity wave activities affecting the QBO and the general circulation. Moreover, the scale of gravity waves is too broad to understand their characters in detail.

The maritime continent in the tropical area is one of the regions having the largest rainfall and one of the major equatorial atmospheric convection center. Since strong convections generate various scales of gravity waves, high frequency observing of atmosphere and ocean around maritime continent is important to understand the relation between gravity wave activities and background state including QBO and Brewer-Dobson circulation. To expedite the progress of improving understanding and prediction of local multi-scale variability of the maritime continent weather-climate system and its global impact, a two-year field campaign project named "the Campaign of Years of Maritime Continent-Sumatra (YMC-Sumatra)" has been operated since 2017. We conducted enhanced radiosonde soundings with every 3 hours at Bengkulu from November 2017 to January 2018.

Based on the above considerations, the purpose of this study is to examine the gravity wave activities during December 2017 by using intensified radiosonde observation data. Section 2 introduces the observation data. Observed results are shown in section 3. The summary and discussion are given in section 4.

2. Observation data and analysis

The intensified radiosonde (Meisei ims100) observations of horizontal wind velocities and temperature were performed at Bengkulu (3.86S, 102.34E) during December 2017 at a time interval of 3 h. Although the original data sampling time intervals are 2 s, corresponding to about 10 m vertical intervals, to make further analysis easier, we used data with a vertical interval of 50 m made by linear interpolation.

3. Observed results

The obtained components are plotted in figure 1 as time-height cross sections of observed zonal wind, meridional wind, and vertical derivative of temperature during December 2017. At first glance of the zonal wind, the eastward flow was observed in the lower troposphere and larger eastward flow from 1 to 5 in December 2017 may be associated with MJO. In the upper troposphere, the zonal wind was almost westward. The eastward flew from 17 to 22 km and westward flew at altitudes higher than 22 km were QBO. Next, although the meridional wind has a complicated structure in the troposphere, there are many wave-like structures whose phases propagate almost downward in the stratosphere. We think these structures are meridional component of gravity waves. The vertical derivative of temperature is easy to diagnose the tropopause (the negative value expresses troposphere and the positive value expresses stratosphere) and the tropopause height is about 17 km. However, from 9 to 30 in December 2017, the positive value can be seen around 15km. This temperature inversion layer corresponds to the eastward wind.

Figure 2 shows the time-height cross sections of disturbances of zonal wind, meridional wind, and temperature during December 2017. The disturbance is defined as the deviation from the monthly mean. The wave-like structures can be seen in the zonal wind, meridional wind and temperature disturbances in the stratosphere (at altitudes higher than 20 km). From the temperature inversion layer to the tropopause (15-19km), the zonal wind and temperature disturbances are in-phase. Since this correspondence is not seen in meridional wind disturbance, this structure is thought to be due to Kelvin waves and will be examined near future.

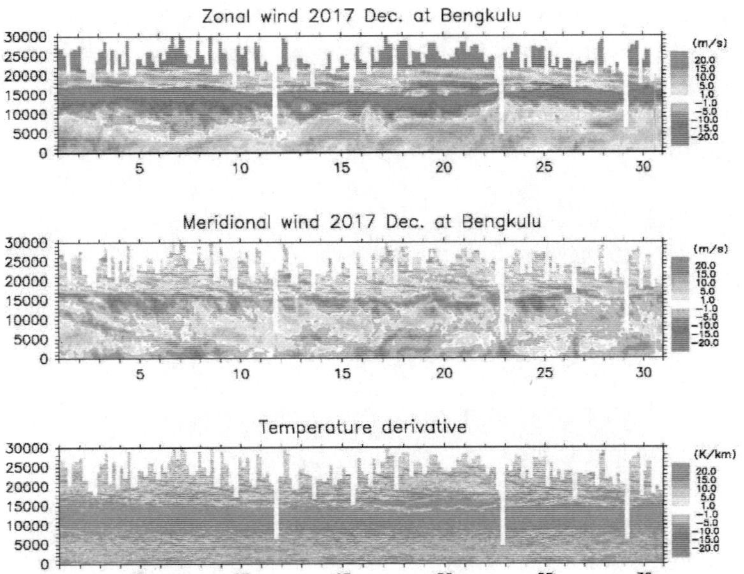

Figure 1. Time-height cross section of zonal wind, meridional wind and vertical derivative of temperature at Bengkulu during December 2017.

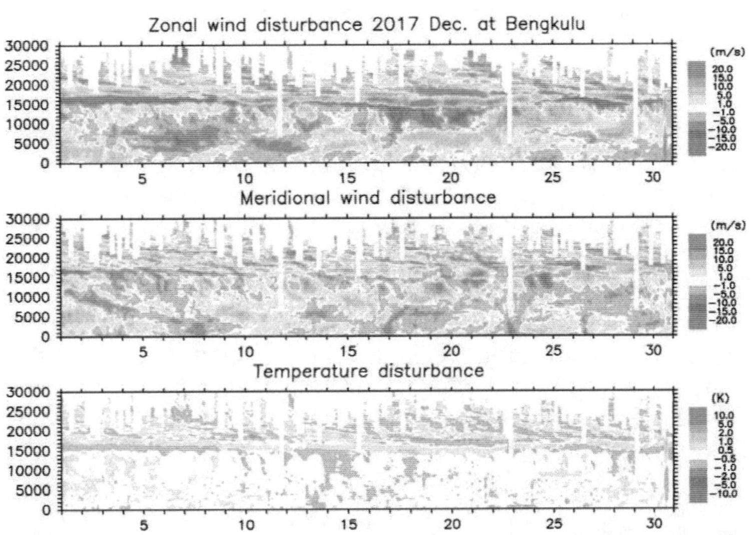

Figure 2. Time-height cross section of zonal wind, meridional wind and temperature disturbances during December 2017.

Next, the period-height cross section of zonal and meridional wind disturbances and temperature disturbances are shown in figure 3. The zonal wind disturbances having wave period range of about 5 days from lower to middle troposphere and about 10 days in the upper troposphere are dominant. At altitude higher than 10km, the disturbances having wave period longer than 10 days cannot be seen in the meridional component. Since the equatorial Kelvin wave does not have meridional component, these disturbances having wave period longer than 10 days include Kelvin waves. The meridional wind disturbances having wave period from 1 to 5 days are dominant. Since the inertial period is about 7.4 days at Bengkulu, we think these meridional wind disturbances include inertia gravity waves. The temperature disturbances having wave period of 1 day are dominant in the troposphere and stratospheric temperature disturbances have broad wave periods. To focus on gravity wave activities, we use a band-pass filter with periods from 1 to 10 days for zonal wind, meridional wind and temperature disturbances. Note that the band-pass filter with vertical wavelength is not used in this analysis to investigate gravity wave activity around the strong vertical shear of background state zonal wind although we calculate the vertical wavelengths of gravity waves by using Space-Time Spectral analysis (not shown). The background state is defined as the component having a period longer than 10 days.

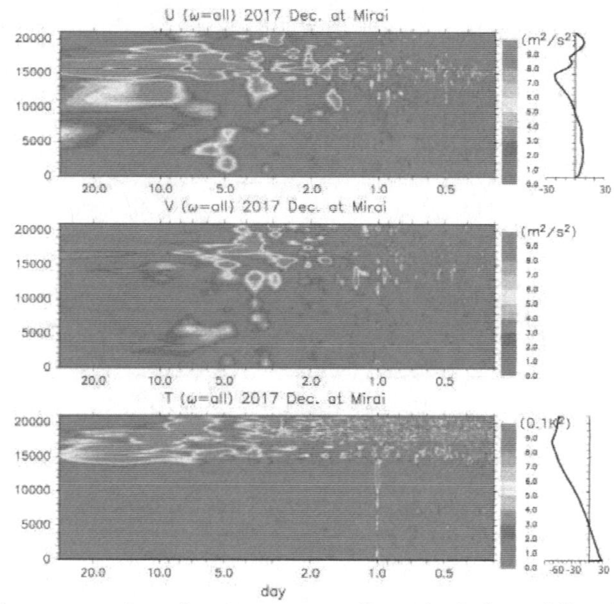

Figure 3. Period-height cross section of power spectra of zonal wind, meridional wind and temperature disturbances. The line plots are monthly mean zonal wind (top) and temperature (bottom).

Figure 4 shows the time-height cross section of heat flux of disturbances having wave period range of 1-10 days ($\overline{u'T'}$: u' and T' are zonal and temperature disturbances and the bar expresses the background state) and the vertical derivative of zonal wind having period longer than 10 days. The vertical shear of background state of zonal wind is positive/negative at altitude lower/higher than 20 km which is the centre of QBO westerly. It is found that the relation between gravity wave heat flux and background state of zonal wind vertical derivative is in-phase. Dunkerton [8] and Sato and Dunkerton [1] showed that the covariance of temperature and zonal wind components of slowly varying two-dimensional gravity waves satisfies the following relation:

$$\overline{u'T'} = \frac{N\bar{T}}{2gk|\hat{c}|}|\overline{u'w'}|\bar{u}_z, \tag{1}$$

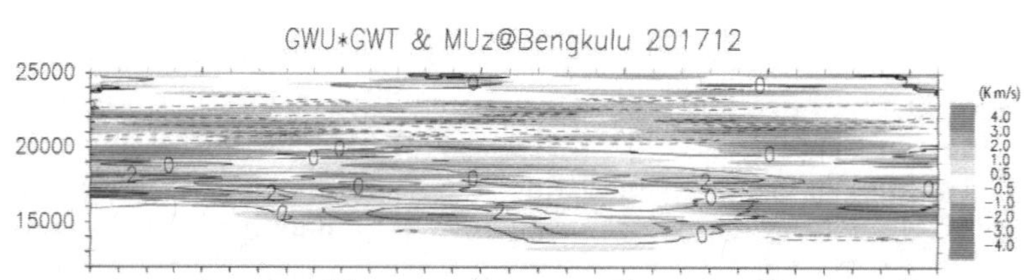

Figure 4. Time-height cross section of heat flux of disturbances having wave period range of 1-10 days (colors) and background state of zonal wind vertical derivatives (contours).

where N is the Brunt-Väisälä frequency, g is the acceleration of gravity, k is the horizontal wavenumber which is taken to be positive without loss of generality, $\bar{T}$ is the background state of temperature, w is vertical wind velocity and $\hat{c}$ is the intrinsic phase velocity. Thus, it is suggested that the disturbances having wave period range of 1-10 days have the character of two-dimensional gravity waves ($\overline{u'T'}$ is proportional to $\bar{u}_z$). Next, the disturbance energy ($\frac{1}{2}\left(\overline{u'^2} + \overline{v'^2} + \frac{g\overline{\theta'^2}}{\bar{\theta}\,\bar{\theta}_z}\right)$, where θ is the potential temperature) and background zonal wind are shown in Figure 5. It is found that the disturbance energy is large around 15 km where the large vertical shear of zonal wind exists. It is also found that the large disturbance energy region (figure 5) corresponds to the temperature inversion layer (figure 1). To examine the background situation where the large disturbance energy exists, the Richardson number ($\mathrm{Ri} \equiv \left(\frac{g}{\bar{\theta}}\frac{\partial \bar{\theta}}{\partial z}\right)\left(\frac{\partial \bar{u}}{\partial z}\right)^{-2}$) is plotted in figure 6.

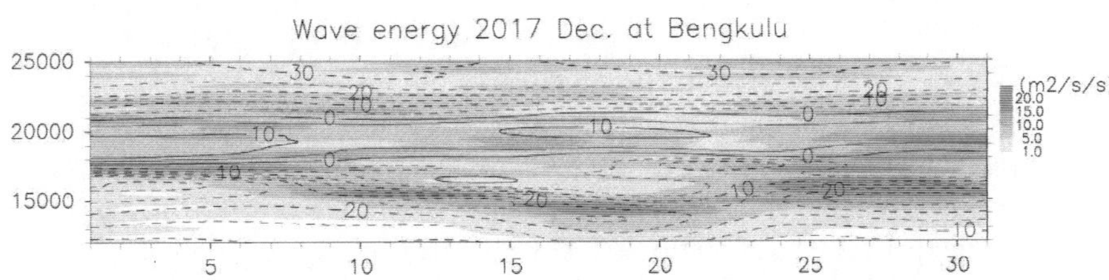

Figure 5. Time-height cross section of the disturbance energy (colours) and background state of zonal wind (contours).

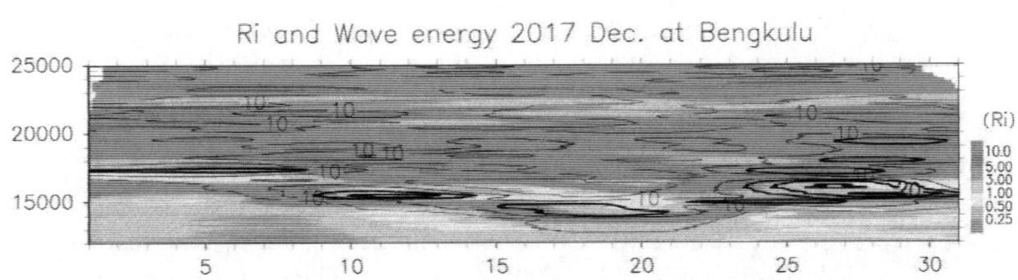

Figure 6. Time-height cross section of Richardson number (colours) and the disturbance energy (contours). The bold contours are larger than 15 m^2s^{-2}.

Note that Ri < 1 is required for instability if the relative vorticity of the mean flow vanishes [9]. When we focus on the large disturbance energy seen around 15 km from 10 to 15 and from 25 to 30 in December 2017, it is found that the region of Ri < 1 has occurred after the disturbance energy is enhanced. Thus, it is suggested that the wave having large amplitude modifies the background state around 15 km. On the other hand, the regions of small Richardson number below 15 km can be seen when the large disturbance energy exists around 15 km. From a comparison between 10-15 days and 25-30 days, the larger disturbance energy corresponds to the smaller Richardson number. These results suggest that the disturbances having wave period range of 1-10 days (gravity waves) are generated by the instability seen below 15 km, but the further analysis is needed to understand what makes the instability region.

4. The summary and discussions

In this study, the gravity wave activities from upper troposphere to lower stratosphere during December 2017 were investigated by using the intensified radiosonde observation data at Bengkulu. From the frequency analysis, the disturbances having wave period range of 1-10 days was defined as gravity waves. As the result of examining the relation between the gravity waves and the background state, it is found that the heat flux of the gravity waves is proportional to the vertical shear of the background zonal wind. The relation between gravity wave energy and Richardson number suggest that the gravity waves having large amplitude modify the background state around 15 km. It is also suggested that the gravity waves around 15 km are related to the region of small Richardson number below 15 km. On the generation mechanism of gravity waves, orographic forcing such as mountains [10], convective activities [11,12], instability associated with Potential vorticity gradients [13,14] and spontaneous generation due to jet and fronts [15,16] can be thought. The region of small Richardson number below 15km observed in this study corresponds to the temperature inversion layer, which is thought to be due to Kelvin waves. Thus, although the observed gravity wave energy enhancements are thought to be due to Kelvin waves and modified convective activities associated with Kelvin waves, the further discussion is needed to understand the generation mechanism of gravity waves.

Finally, this study showed the initial results of examining relation between gravity wave activities and background state. Near future, we will focus on the temperature inversion layer and examine the generation mechanism of gravity waves observed in this campaign. Moreover, we will use observation data of other stations to examine the observed gravity wave character (e.g., wavelength and intrinsic frequency).

Acknowledgments
The authors deeply appreciate strong supports in the observation of this study by Mr. Warjono, the head of Meteorological Station of Famawati Soekarno Airport in Bengkulu and the local staffs. We thank the anonymous reviewer for providing constructive comments. Also, we are indebted to the Ministry of Research, technology and Higher Education (RISTEKDIKTI), Agency for the Assessment and Application of Technology (BPPT), and Japanese Embassy for their highly understanding and support to conduct the observations. The GFD-DENNOU library was used for drawing figures. This study was supported by a Grant-in-Aid for Young Scientists No. 18K13616 of the Japan Society for the Promotion of Science (JSPS).

References

[1] Sato K and Dunkerton T J., 1997, *J. Geophys. Res.*, **102**, 26 247–26 261
[2] Kawatani Y, Sato K, Dunkerton T J, Watanabe S, Miyahara S and Takahashi M., 2010, *J. Atmos. Sci.*, **67**, 963–980, doi:10.1175/2009JAS3222.1.
[3] Yang G -Y, Hoskins B J and Slingo J M., 2011, *J. Atmos. Sci.*, **68**, 839–862, doi:10.1175/2010JAS3514.1
[4] Fathullah N Z, Lubis S W and Setiawan S., 2017, *IOP Conf. Ser.: Earth Environ. Sci.*, **54**, doi:10.1088/1755-1315/54/1/012032
[5] Brewer A W., 1949, *Quart. J. Roy. Meteor. Soc.*, **75**, 351–363
[6] Dobson G M B., 1956, *Proc. Roy. Soc. London.*, **236A**, 187–193
[7] Okamoto K, Sato K and Akiyoshi H., 2011, *J. Geophys. Res.*, **116**, D10 117, doi:10.1029/2010JD014953
[8] Dunkerton T J., 1995, *J. Meteorol. Soc. Jpn.*, **73**, 747-755
[9] Holton J R., 2004, *Elsevier Academic Press*, 529 pp
[10] Eliassen A and Palm E., 1961, *Geofys. Publ.*, **22(3)**, 1–23
[11] Tsuchiya C, Sato K, Alexander M J and Hoffmann L 2016 *J. Geophys. Res. Atmos.*, **121**, 7641–7651, doi:10.1002/2015JD024463
[12] Alexander M J, Ortland D A, Grimsdell A W and Kim J -E 2017 *J. Atmos. Sci.*, **74**, 2701os. Sci.D024463s.ergy in stationary mounta44.1
[13] Lubis S W, Omrani N, Matthes K and Wahl S 2016 *J. Atmos. Sci.*, **73**, 2509–2528, https://doi.org/10.1175/JAS-D-15-0189.1
[14] Yasui R, Sato K and Miyoshi Y 2018 *J. Atmos. Sci.*, **75**, 3635–3651, https://doi.org/10.1175/JAS-D-17-0337.1
[15] Plougonven R and Snyder C 2007 *J. Atmos. Sci.*, **64**, 2502–2520, doi:10.1175/JAS3953.1
[16] Kalashnik M V and Chkhetiani O 2017 *J. Atmos. Sci.*, **74**, 293–308, https://doi.org/10.1175/JAS-D-16-0134.1

The evaluation of drought indices: Standard Precipitation Index, Standard Precipitation Evapotranspiration Index, and Palmer Drought Severity Index in Cilacap-Central Java

F Balbo[1], R A Wulandari[2], M R R Nugraha[3], A Dwiandani[3], M R Syahputra[1] and R Suwarman[1]

[1] Atmospheric Science Research Group, Bandung Institute of Technology, Indonesia
[2] Department of Magister of Environment Engineering, Bandung Institute of Technology, Indonesia
[3] Department of Meteorology, Bandung Institute of Technology, Indonesia

fandy.balbo@meteo.itb.ac.id

Abstract. The Cilacap Regency-Central Java, is the largest agricultural area with a high level of drought vulnerability. This study aims to evaluate three indices in quantifying a drought condition in Cilacap Regency and its application in forecasting. The indices that were used in this study are Standard Precipitation Index (SPI), Standard Precipitation Evapotranspiration Index (SPEI), and Palmer Drought Severity Index (PDSI). The comparison between SPI and SPEI shows that there is no significant difference in terms of determining the drought severity. SPEI should be used when there is a temperature difference more than 2°C in 30 years. PDSI may provide more frequency of drought event and a good result in indicating the effects of drought on agricultural productivity. We used Climate Forecast System Version 2 (CFSv2) to predict drought severity by SPI. The result of model prediction shows that there is no significant improvement in accuracy before and after statistical bias correction. The prediction can be done on three months (lead3) before initial planting.

Keywords : SPEI, PDSI, CFSv2, drought.

1. Introduction

Cilacap Regency, located in Central Java, has the highest level of drought hazard [1], due to its high population. The region is also being one of the main rice resources for Indonesian food resilience [2], where its agriculture area covers 30% of the total area. Consequently, the drought hazard may affect the availability of rice production in complying the food needs in Cilacap.

There are various methods that are often used for the drought hazard monitoring, such as Standard Precipitation Index (SPI) [3,4], Standard Precipitation Evapotranspiration Index (SPEI) [5], and the Palmer Drought Severity Index (PDSI) [6,7,8], which can be estimated by long monthly precipitation data. SPI has several advantages. Firstly, SPI is a simple method since it only calculates based on precipitation data. Secondly, the calculation of SPI can be done on the different timescales, as for example one month (SPI-1), 3 months (SPI-3), 6 months (SPI-6), and 12 months (SPI-12) [3]. In 2010, however, there was a study that introduced a new drought index, namely the Standard Precipitation Evapotranspiration Index (SPEI) [5], computing both precipitation and evapotranspiration data. Moreover, there is Palmer Drought Severity Index (PDSI) method [6,7,8] that uses precipitation, temperature, and available water capacity (AWC) as input data, thus the parameters of water balance can be calculated, such as runoff, recharge, evapotranspiration, and coefficient of soil moisture loss. Therefore, the aim of this study is to evaluate SPI, SPEI, and PDSI as drought indices in Cilacap, henceforth to be used in the drought prediction that uses rainfall prediction data from Climate Forecasting System Version 2 (CFS v2).

2. Methods

Data used in this study are precipitation, temperature, land cover, soil, and rice productivity. Monthly precipitation and temperature data were provided by Indonesian Meteorological, Climatological, and Geophysical Agency (BMKG) from 1982-2012 (31 years). Land cover and soil data were provided by Plantation and Forestry Service of Cilacap Regency, while rice productivity data was provided by Center of Statistic Agency (BPS Cilacap) from 1997-2012 as annual data.

SPI only needs precipitation data and its calculation refers to [3,4]. The calculation uses gamma distribution to attain the cumulative probability of precipitation as in equation (1) and equation (2).

$$G(x) = \int_0^x g(x)dx = \frac{1}{\hat{\beta}^{\hat{\alpha}}\Gamma(\hat{\alpha})}\int_0^x x^{\hat{\alpha}-1}e^{-x/\beta}dx \tag{1}$$

$$with\ \hat{\alpha} = \frac{1}{4A}\left(1+\sqrt{1+\frac{4A}{3}}\right),\ \hat{\beta} = \frac{\bar{x}}{\hat{\alpha}}\ , and\ A = \ln(\bar{x}) - \frac{\sum \ln(x)}{n} \tag{2}$$

The function above is only able to be used for $x > 0$. Thus when $x = 0$ (or no rainfall event), then the cumulative probability will be approached by the comparison between the number of no rainfall events with the number of total data (q).

$$H(x) = q + (1-q)(G(x)) \tag{3}$$

$H(x)$ is the cumulative probability with the calculation of zero rainfall event. The value of $H(x)$ is then being transformed to the standard normal random variable value with zero average values and variance equals to one. Z is the value of SPI that is determined by the value of cumulative probability with the approximation of [9] as in equation (4) and equation (5) [4].

Z for $0 < H(x) \leq 0.5$ is

$$Z = SPI = -\left(t - \frac{c_0 + c_1 t + c_2 t^2}{1 + d_1 t + d_2 t^2 + d_3 t^3}\right) \quad where\ t = \sqrt{ln\left(\frac{1}{(H(x))^2}\right)} \tag{4}$$

For $0.5 < H(x) \leq 1.0$ is

$$Z = SPI = \left(t - \frac{c_0 + c_1 t + c_2 t^2}{1 + d_1 t + d_2 t^2 + d_3 t^3}\right) \quad where\ t = \sqrt{ln\left(\frac{1}{(1-H(x))^2}\right)} \tag{5}$$

With the constant values of $c_0 = 2.515517,\ c_1 = 0.0802853,\ c_2 = 0.010328,\ d_1 = 1.432788,\ d_2 = 0.189269,$ and $d_3 = 0.001308$ [4].

The SPEI and SPI calculations are not far different, where the SPEI calculation refers to [5] and both indices characteristics are the same. The basic difference lies in the data and distribution used. The distribution of the three-parameter log logistic is used to standardize the D value to calculate the SPEI as in equation (6). D is the value of precipitation minus by evapotranspiration (P-E). Evapotranspiration is estimated by Thornwaite [5], that has proven the log logistic distribution is good to use and more coherent, especially at very low D values. In addition, no value is found below the origin parameter of this distribution.

$$F(x) = \left[1 + \left(\frac{\alpha}{x-\gamma}\right)^{\beta}\right]^{-1} \tag{6}$$

Where $\alpha, \beta,$ and γ are scale, chape, and origin parameters. Then the $F(x)$ value will be transformed to the standard normal random variable value with zero average values and variance equals to one. Z is the value of SPI that is determined by the value of cumulative probability with the approximation of [9] as in equation (7) and (8) [5].

Z for $0 < P \leq 0.5$ is

$$Z = SPEI = \left(W - \frac{c_0 + c_1 W + c_2 W^2}{1 + d_1 W + d_2 W^2 + d_3 W^3}\right) \quad where\ W = \sqrt{-2\ln(P)} \tag{7}$$

For $0.5 < P \leq 1.0$ is

$$Z = SPEI = -\left(W - \frac{c_0 + c_1 W + c_2 W^2}{1 + d_1 W + d_2 W^2 + d_3 W^3}\right) \quad where\ W = \sqrt{-2ln(1-P)} \tag{8}$$

with the constant values of $c_0 = 2.515517,\ c_1 = 0.0802853,\ c_2 = 0.010328,\ d_1 = 1.432788,\ d_2 = 0.189269,$ and $d_3 = 0.001308$, and $P = 1 - F(x)$ [5].

The PDSI calculation refers to [7,8,10]. The Palmer method is started by the calculation of water balance, using precipitation and temperature data. Calculation of soil moisture capacity is attained by dividing the land into two layers and assuming 25 mm of water can be stored in the soil surface layer (Ls). The bottom layer (Lu) has water capacity that depends on the soil characteristics. Potential evapotranspiration (PE) is generally calculated using the Thornthwaite method [5]. Evapotranspiration occurs when PE (Potential Evapotranspiration) > P (Precipitation), where P is monthly precipitation.

As part of the water balance calculation, the Palmer method calculates evapotranspiration (ET), recharge (R), runoff (RO), loss (L), potential recharge (PR), potential loss (PL), and potential runoff (PRO). Here Palmer assumes that runoff will occur if and only if the moisture content in the two soil layers has reached field capacity. From the results of the PE, PR, PL and PRO calculations, the value will be determined using equation (9) from the four climate constants, namely the evapotranspiration coefficient (α), the charging coefficient (β), runoff coefficient (γ), and water loss coefficient (δ) as in equation (9).

$$\alpha_j = \frac{\overline{ET_j}}{\overline{PE_j}}, \quad \beta_j = \frac{\bar{R}_j}{\overline{PR_j}}, \gamma_j = \frac{\overline{RO_j}}{\overline{PRO_j}}, and\ \delta_j = \frac{\bar{L}_j}{\overline{PL_j}}, where\ j = 1, \ldots, 12 \tag{9}$$

Overbar shows the fact that the coefficients are calculated using the monthly average value (j). These coefficients are used to calculate the differences in d for other months between actual precipitation and "CAFEC" (Climatically Appropriate For Existing Conditions) as in equation (10).

$$d = P - \hat{p} = P - \alpha_j PE + \beta_j PR + \gamma_j PRO - \delta_j PL \tag{10}$$

Definition of $\hat{p}$ is analogous to a simple water balance. Moisture anomaly index (Z) is defined as in equation (11)

$$Z = K_j d \tag{11}$$

with *Kj* is a weighted factor that defined as in equation (12)

$$K_j = \frac{17.67\widehat{K}_j}{\sum_{i=1}^{12} \bar{D}_i \times \bar{K}_i}, where\ j = 1, \ldots, 12 \tag{12}$$

with $\overline{Dj}$ is the average of the absolute value d for the month j. The purpose of the weighted factor is to manage the decrease of normal precipitation, therefore it can be compared to different areas for different months. Moisture anomaly index (Z) shows the relative decrease of weather, in month and at a specific location, from the average humidity conditions in the related month. Palmer has evaluated Z at two study areas in Iowa and Western Kansas. Then the final equation to calculate the level of drought as in equation (13).

$$X(i) = 0.897X(i-1) + \frac{Z(i)}{3} \tag{13}$$

with X(i) is the value of PDSI for month i.

However, the values of those indices are quantified in different categories as shown in table 1, thus the categories equalization is needed. The equalized values will be explained in the next section. Rice productivity data was detrended in order to eliminate the data trend and then to find out its relationship with SPI and PDSI.

We also evaluated CFSv2 data performance in predicting the drought with several lead times, by correcting it using statistical downscaling and bias statistical correction toward BMKG station data [11].

Table 1. SPI and PDSI Categories (unequalized) [3] [4].

SPI	Category	PDSI	Category
≥ 2,00	*Extremely wet*	≥ 4.00	*Extremely wet*
1,50 to 1,99	*Very wet*	3.00 to 3.99	*Very wet*
1,00 to 1,49	*Moderately wet*	2.00 to 2.99	*Moderately wet*
-0,99 to 0,99	*Near normal*	1.00 to 1.99	*Slightly wet*
-1,00 to -1,49	*Moderately dry*	0.50 to 0.99	*Incipient wet spell*
-1,50 to -1,99	*Severely dry*	0.49 to (-0.49)	*Near normal*
≤ -2,00	*Extremely dry*	-0.50 to (-0.99)	*Incipient drought*
		-1.00 to (-1.99)	*Mild drought*
		-2.00 to (-2.99)	*Moderate drought*
		-3.00 to (-3.99)	*Severe drought*
		≤ -4.00	*Extreme drought*

3. Results and Discussion

3.1 *SPI and SPEI*

Figure 2 denotes three months of the SPI, SPEI, and PDSI from 1982 to 2012. There are no significant differences between SPI and SPEI in quantifying a drought. Both of them show the drought in the same episodes, namely 1982, 1983, 1995, 1997, 2002, 2007, 2008, and 2012. The relationship between SPI and SPEI is expressed in linear equations (14), which shows that both indices have the same classes in quantifying a drought, thus the index categories equalization is not required.

$$SPEI = 0.9426\ SPI - 0.0001 \tag{14}$$

However, the two indices differences may be seen in the level of severity. The severity of SPI is worse than SPEI. When SPI reaches the extreme level (index ≤ -2), SPEI only reaches the moderate level. The variance of the P-E value is lower than the rainfall values, thus the results of the SPI and SPEI methods have different variances. In this case, it turns out that SPI has a larger data range than SPEI (see in methods), therefore SPI is able to define drought up to extreme levels. Another case that we can see is in 2008, where SPI and SPEI described the duration of drought for 6 months length. The index given by SPI has more extreme values for 4 months, while SPEI only able to provide a drought index in the moderate and severe categories.

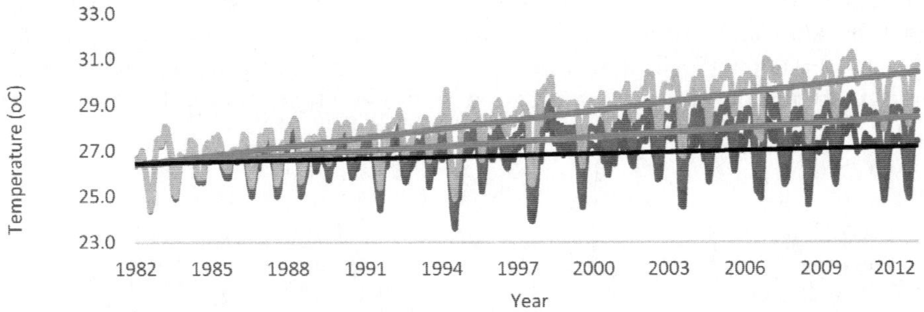

Figure 1. Synthetic data with the temperature increase of 2°C and 4°C, and its real data in 30 years.

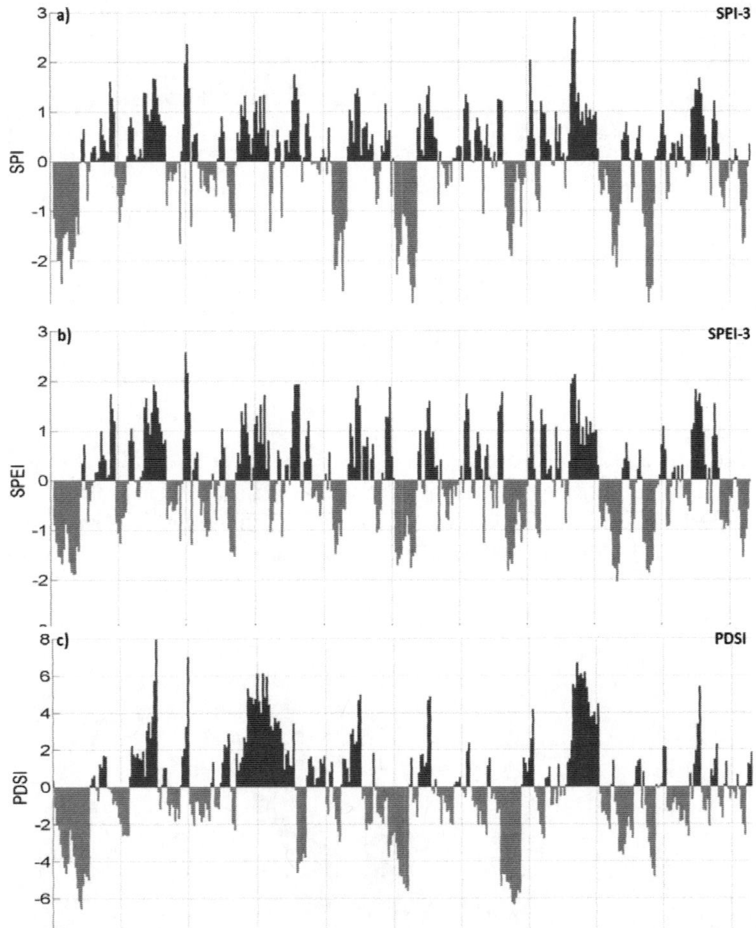

Figure 2. Time series of drought indices, a) Standard Precipitation Index (SPI), b) Standard Precipitation Evapotranspiration Index (SPEI), and c) Palmer Drought Severity Index (PDSI). SPI-3 and SPEI-3 imply the SPI and SPEI that calculate in three months time scale.

It can be concluded that SPEI is able to describe the drought severity in terms of intensity and duration when there is an increase in temperature. Meanwhile, SPI is not able to compute the effects of temperature increase to the drought severity. This certainly makes the use of the SPI method less optimal in drought analysis in the area affected by increasing temperature. However, the temperature in Cilacap Regency has increased about 0.7°C in 30 years, thus there is no significant difference between SPI and SPEI in quantifying a drought. In addition, equation 1 shows that the relationship between SPI and SPEI is SPEI ≈ SPI.

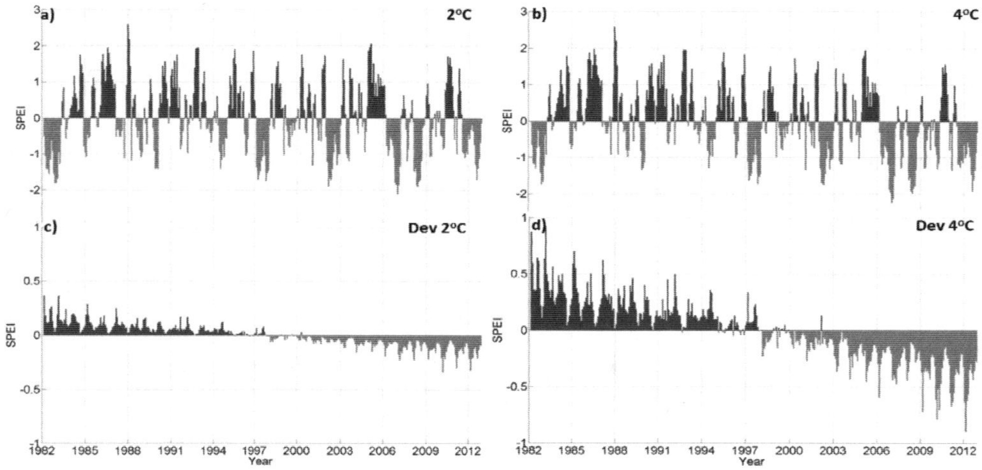

Figure 3. SPEI and the deviation of SPEI, **a)** SPEI with the temperature increase scenario of 2°C, **b)** SPEI with the temperature increase scenario of 4°C, **c)** The deviation of the SPEI 2°C scenario, and **d)** The deviation of the SPEI 4°C scenario towards the original SPEI.

3.2 *SPI and PDSI*

It can be shown that the drought categories of PDSI are different with SPI and SPEI (figure 2). Consequently, the drought categories equalization between SPI and PDSI is required. Figure 4 shows a relationship between PDSI and SPI that can be expressed in equation (15). The equation can be simplified by PDSI ≈ 1.4 SPI, thus the SPI and PDSI categories can be equal as shown in table 2.

$$PDSI = 1.395SPI - 0.1309 \tag{15}$$

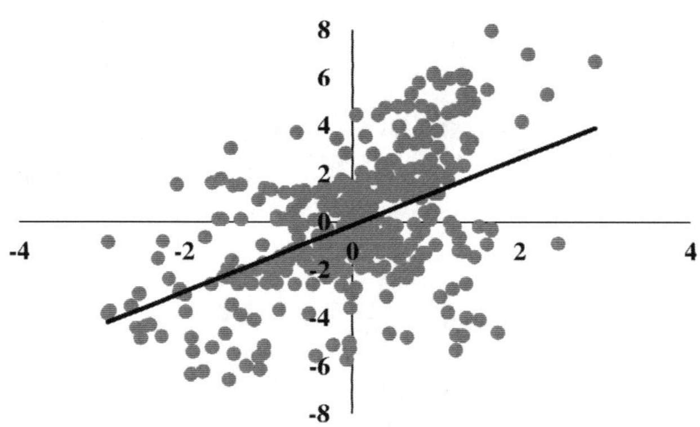

Figure 4. The relationship between Palmer Drought Severity Index (PDSI) and SPI. *y*-axis is PDSI and *x*-axis is SPI.

Table 2. SPI and PDSI categories (equalized)

Category	SPI	PDSI
Extremely wet	≥ 2,00	≥ 2,80
Very wet	1,50 to 1,99	2,10 to 2,79
Moderately wet	1,00 to 1,49	1,40 to 2,09
Near normal	-0,99 to 0,99	-1,39 to 1,39
Moderately dry	-1,00 to -1,49	-1,40 to -2,09
Severely dry	-1,50 to -1,99	-2,10 to -2.79
Extremely dry	≤ -2,00	≤ -2,80

One of the differences between SPI and PDSI is the frequency of drought event (table 3). In consequence, it can indicate that PDSI is able to quantify more in the drought events frequency rather than SPI. In figure 5, there is low rice productivity when the number of drought events is high. SPI shows the decline in rice productivity rate when there are values of SPI in 1997, 2001, 2002, 2003, 2006, 2008, and 2012. Where the declines that are shown by PDSI are in 1997, 1999, 2000. 2002, 2003, 2006, 2008, 2011, and 2012. In conclusion, both SPI and PDSI are good to illustrate the relationship between rice productivity and drought event. However, in particular case, namely 2007 and 2010, PDSI shows drought events when the rice productivities are high. These are due to the temporal resolution of rice productivity data which is not good enough (it is yearly data) thus there is a need for better data resolution to understand the drought in further studies. However, it still can be concluded that both SPI and PDSI can be useful as drought indices in the agricultural sector.

Table 3. The frequencies of drought events in *extreme, severe,* and *moderate* categories for SPI and PDSI.

Category	SPI	PDSI
Extreme Dry	16	49
Severe Dry	15	22
Moderate Dry	28	38
Sum	**59**	**109**

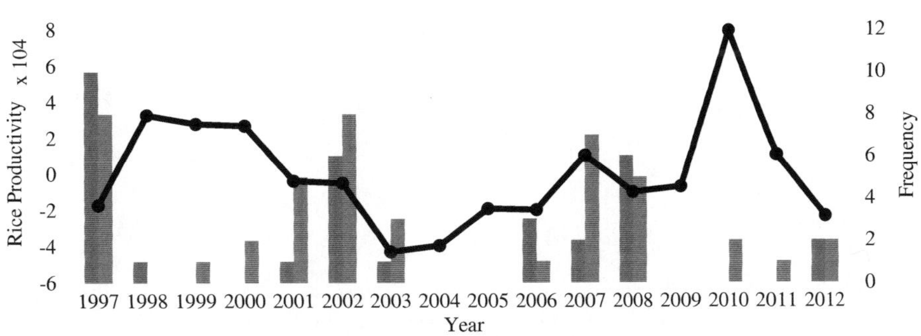

Figure 5. The frequencies of extreme, severe, and moderate drought occurrences for rice productivity. Blue bars and red bars show SPI and PDSI, respectively. Black line shows detrended rice productivity data.

3.3. Experiment using CFSv2 data to predict the drought with SPI

We chose SPI to predict the drought rather than two other indices, to be used in the experiment with CFSv2 data in this section, due to the simple calculation process (it requires only precipitation data).

Precipitation in SPI uses CFSv2 output as prediction data. Figure 6 shows CFSv2 data before and after being corrected in our experiment. CFSv2 data underestimates (red lines) when it is compared to the observation data (blue lines), thus statistical downscaling and bias correction are applied to CFSv2 output for correction (green lines). Predictions for the next 1 month, 2 months, and 3 months are expressed by lead 1, lead 2, and lead 3, respectively.

SPI-3 was calculated using corrected CFSv2 data (CFSv2-C) and raw CFSv2 (CFSv2-Raw) for each lead, and SPI-3 which was calculated with CFSv2-C and CFSv2-Raw, later called as SPI3-C and SPI3-Raw, respectively. Figure 7 shows SPI3-C and SPI3-Raw boxplot diagrams in 2001-2010. A value below zero is a negative SPI value. There are missing values in SPI3-C and SPI3-Raw, namely in 2004 (for lead 4 and lead 2) and 2008 (for leads 2 and 1). SPI3-C and SPI3-Raw do not have any significant differences. Those are because the SPI calculation process uses normal distribution and depends on historical rainfall data. Thus, CFSv2 data can be used to calculate SPI without data correction process, where in this case is with the statistical bias correction.

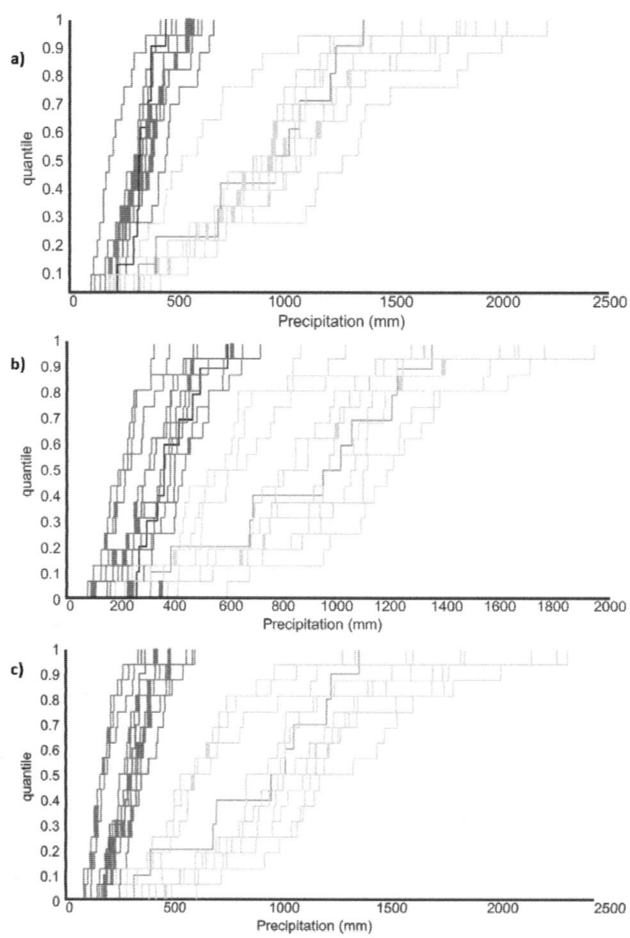

Figure 6. CDF (Cumulative Distribution Function) with bias correction testing period plots, (a) lead3; (b) lead2 and (c) lead1. Red lines and green lines show CFSv2 data before and after correction, respectively. Blue lines show observation data.

IOP Conf. Series: Earth and Environmental Science **303** (2019) 012012 doi:10.1088/1755-1315/303/1/012012

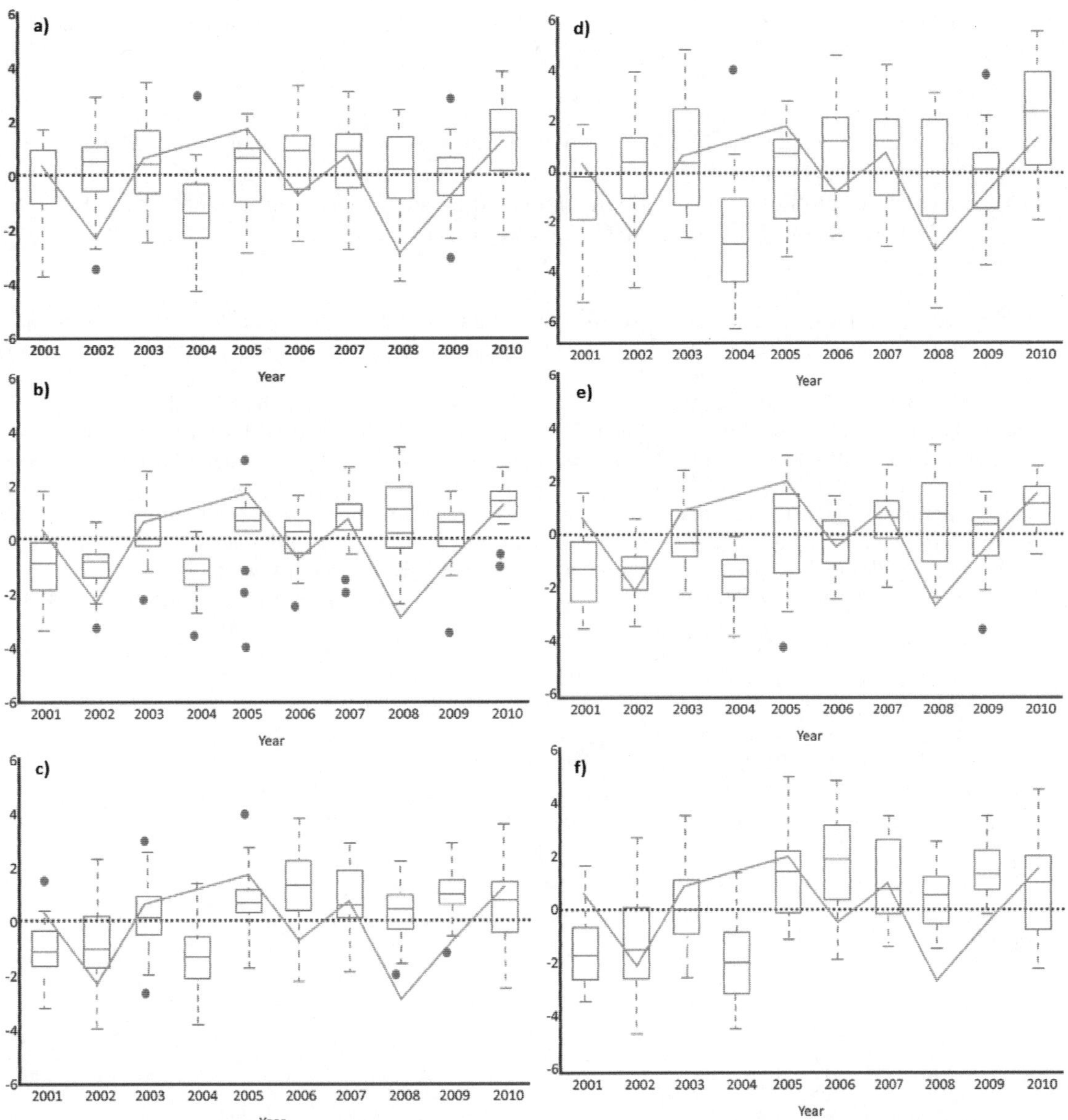

Figure 7. SPI boxplots from CFSv2 data in May, a) SPI-3 Correction lead3, b) SPI-3 Correction lead2, c) SPI-3 Correction lead1, d) SPI-3 Raw lead3, e) SPI-3 Raw lead2, f) SPI-3 Raw lead1.

4. Conclusion

We have evaluated three indices, namely SPI, SPEI, and PDSI in quantifying a drought. Those three indices are able to identify the drought in Cilacap-Central Java. There are no significant different characteristics of SPI and SPEI in quantifying a drought in the present time. The SPEI will significantly different with SPI if there is an extreme change in climatological temperature (e.g. an increase more than 2°C/30 years). SPI and PDSI are both able to identify the drought in the agricultural sector. However, PDSI overestimates when quantifying rice productivity.

We chose SPI to be used for drought prediction because of the ease in data processing. The CFSv2 rainfall model is capable enough to forecast the drought by SPI, in lead3, lead2, and lead1. There are no significant differences between corrected CFSv2 data and raw CFSv2 data. Thus, the precipitation of CFSv2 does not need to be corrected to calculate SPI.

Acknowledgment
This research is partly funded by Research, Community Service and Innovation Program (P3MI LPPM) Bandung Institute of Technology. The authors would like to thanks to Indonesian Meteorological Climatological and Geophysical Agency (BMKG), Plantation and Forestry Service of Cilacap Regency, and Center of Statistic Agency of Cilacap (BPS-Cilacap) for providing data.

References

[1] BNPB., 2013, Indeks Risiko Bencana Indonesia (*Bogor: Direktorat Pengurangan Risiko Bencana*) p 292
[2] BPS,, 2013, Cilacap in figures *2013* ed IPDS Section (Cilacap: *Badan Pusat Statistik Kabupaten Cilacap*) p 119
[3] Hayes M J., Wilhite D A and Vanyarkho O V., 1999, *Bull. Amer. Meteor. Soc.* **80** 429–438
[4] Cacciamani C., Morgillo A., Marchesi S and Pavan V., 2007, Monitoring And Forecasting Drought on A Regional Scale: Emilia-Romagna Region Methods And Tools For Drought Analysis and Management *(Water Science and Technology Library vol 62)* ed V P Singh (Dordrecht: Springer) chapter 2 pp 29-48
[5] Vicente-Serrano S M, Begueria S and Lopez-Moreno J I., 2010, *Journal of Climate* **23** 1696-1718
[6] Vicente-Serrano S M, Begueria S, Lorenzo-Lacruz J, Camarero J J, Lopez-Moreno J I, Azorin-Molina C, Revuelto J, Moran-Tejeda E and Sanchez-Lorenzo A., 2012, *Earth Interactions* **16** 1-22
[7] Palmer W C., 1965, Meteorological Drought (*Washington D.C : U.S. Department of Commerce*) vol 45 p 6-49
[8] Jannah N., 2015, Drought Analyze in Sub-Basin Slahung, Ponorogo, Using Palmer Drought Severity Index Thesis (*Malang: Brawijaya University*) p 4-9
[9] Abramowitz M and Stegun I A., 1965, Handbook of Mathematical Function,with Formula, Graphs, and Mathematical Tables, *Dover Publication* pp 1046
[10] Wells N and Goddard S., 2004, *J. of Climate,***17,** 2335-2351
[11] Inomata H, Takeuchi K and Fukami K., 2011, *J. Japan Society of Civil Enginers* **67** 247-252

IOP Conf. Series: Earth and Environmental Science **303** (2019) 012013 doi:10.1088/1755-1315/303/1/012013

The characteristics of ozone concentration over the maritime equatorial stratospheric region derived from the quasi-biennial oscillation

P Wulandari[1], E Hermawan[2] and H Halide[1]

[1]Departement of Physics, Geophysics Study Program, Hasanuddin University, Jl. Perintis Kemerdekaan Km. 10, Makassar 90245, Indonesia
[2]Center for Atmospheric Science and Technology of National Institute of Aeronautics and Space Republic of Indonesia, Jl. Dr. Djundjunan No. 133, Bandung 40173, Indonesia

E-mail: wulandarip14h@student.unhas.ac.id

Abstract. The characteristics of ozone concentration [O_3] over the Maritime Equatorial Stratospheric region until now are not yet fully understood well due to the limitation of data observations. However, there are many techniques to derive ozone concentration data; from satellite, in-situ observation, modeling, etc. We have investigated, then found a good agreement between the characteristics of ozone concentration and Quasi-Biennial Oscillation (QBO). This study mainly concerned with the analysis of the stratospheric zonal wind from Singapore's QBO monthly data at 100-10 hPa layers. Total ozone, ozone mass mixing ratio, and stratospheric temperature were considered in the analysis. Predominant peak oscillation in each layer was investigated by using the spectral technique. The result shows that a strong signal of zonal wind oscillation pattern was evident at 30 hPa layer. As the altitude decreases, zonal wind oscillation decays. Box-Jenkins final model contained 28 months of a seasonal component $(5,0,4)(0,1,0)^{28}$ and a percentage error of 46.6%, whereas multiple linear regression obtained a quite good correlation score, which was about 0.55.

Keywords : ozone, stratospheric temperature, zonal wind oscillation, Box-Jenkins.

1. Introduction

The study of the equatorial stratospheric region which came to be known as quasi-biennial oscillation or QBO firstly discovered by Ebdon [1] and Reed et al. [2]. This phenomenon is characterized by easterly and westerly winds, which originates from above 30 km and propagates downward through the stratosphere at the speed of 1 km per month. The alternating wind regimes repeat at varying intervals of 22 to 34 months, with an average period of ~28 months [3]. The maximum amplitude of QBO mostly associated with the westerly wind rather than the easterly, and its symmetry toward the equator follows the Gaussian distribution. Even though this phenomenon occurs in the tropics area, but it also influences the stratospheric flow from pole to pole. Therefore, QBO particularly impacts the atmospheric dynamics and changes the circulation of chemical constituents such as O_2, H_2O, and CH_4. Furthermore, it also affects the distribution and transport of trace gases, which may be a factor that causes stratospheric ozone depletion [3].

As proposed by Holton and Lindzen [4], the mechanism of QBO is the result of interaction between the mean flow and the equatorial wave equation mode. They are the first that modeled the QBO based on the planetary equatorial wave propagation, known as Kelvin waves and Rossby-gravity waves. Hence, the QBO only occurs in the equatorial region that caused by these waves, and formed in the region of about 12^0N - 12^0S [5]. This area provides considerable energy from the release of latent heat in large-scale convective clouds to form planetary scale-waves and push them down to the lower stratosphere.

The QBO in the total ozone (TOZ) amount firstly reported by Funk and Garnham [6]. A large number of studies describe possible connections between this phenomenon and the interannual variation of ozone [7,8]. These studies mainly reported that the QBO influences TOZ amount at about 1–2%, with a high correlation ($r > 0.7$) based on the equatorial zonal wind, especially at 30 and 50 hPa.

Additionally, there is evidence suggesting that the variation of TOZ at ±10° latitude is nearly in phase with the zonal winds at 30 hPa. Interannual variability of equatorial TOZ is dominated by the QBO that appears after removing the annual cycle of original data [3]. Moreover, the clear signals in temperature and QBO are revealed in the tropics. The equatorial temperature anomalies associated with the QBO in the lower stratosphere are in the order of ±4K, which maximize at near 30–50 hPa. Holton and Tan [5] have proven that QBO will satisfy thermal wind equations if the vertical shifts of zonal wind (wind shear) associated with the temperature in the stratosphere. Based on this result, we then investigate how the zonal wind QBO and stratospheric temperature anomalies are related to the TOZ in the equatorial region. Meanwhile, we thought that the ozone concentration which is TOZ, could be derived from predicted wind QBO over the maritime equator region.

2. Data and Method

2.1 Data

To study the ozone variations, we used the monthly total ozone columns (TOZ) dataset of a MERRA-2 satellite from 1988 to 2018, obtained from https://disc.gsfc.nasa.gov/datasets/. While the zonal wind data of this study were taken from Singapore's QBO dataset that available since January 1979 until present. This dataset is observed by regular radiosonde measurements of daily vertical wind profiles at the levels of 100 to 10 hPa. Moreover, stratospheric temperature data from NOAA NCEP-NCAR CDAS-1 reanalysis were also used. This dataset was accessed through the IRI operational website, online data repository and analysis tools. Both zonal wind and stratospheric temperature data were on a month-by-month basis. As a review, the ozone mass mixing ratio data were included in the analysis.

2.2 Method

In this present study, we applied spectral analysis techniques, hereafter referred to power spectral density and wavelet. Time series data was decomposed into time-frequency space, then variability dominant modes and variations modes in time would be determined as well [10]. Power spectral density function shows the strength of each variation (energy) as the function of frequency, mostly uses to maintain the magnitude of frequency variations. This method is useful to determine a level in which stratospheric zonal wind has similar characteristics with ozone. As similar to power spectral density, wavelet analysis is known as an effective tool for analyzing local variations of power within a time series by decomposing it into time-frequency space [10]. This method has been used in the previous study e.g. the tropical convection [11], and the El Niño–Southern Oscillation (ENSO) [12,13].

The ozone model was developed to predict the future monthly of TOZ by using zonal wind QBO and temperature data, especially at 30 hPa. As a univariate method, ARIMA (Autoregressive Integrated Moving Average) model was used to derive ozone concentration data from the QBO. This method divided the dataset into a training set for model learning and a testing set for validating an appropriate ARIMA model. For the detail of this method, please refer to Box and Jenkins [14]. In this case, the total ozone column is a single layer property, where an extracted data typically represents the equatorial region in 0° latitude and 100^oE longitude. TOZ anomaly was obtained by subtracting monthly dataset with the average of TOZ from January 1988 to January 2018. Finally, the selected ARIMA model was then compared to the multiple regression model that formed from zonal wind QBO and temperature data.

3. Results and Discussion

Figure 1 shows the PSD of monthly zonal wind data for period of January 1988 – January 2018 which starts from 100 to 10 hPa's layers. From this figure, it is clearly shown that pre-dominant peak oscillation is about 28 month that related to the Quasi-Biennial Oscillation (QBO) phenomenon.

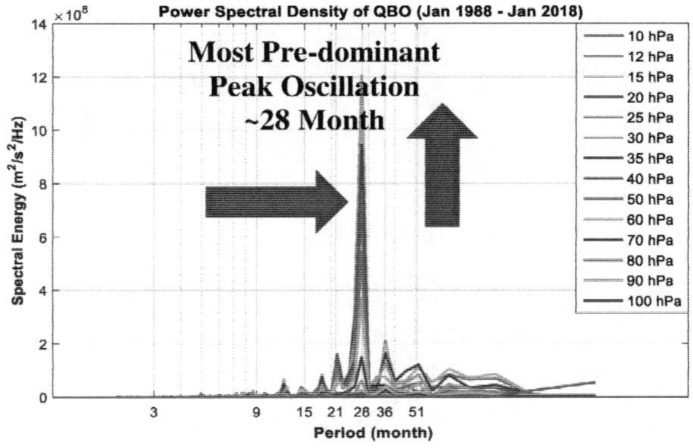

Figure 1. The Power Spectral Density of QBO for period of January 1988 to January 2018.

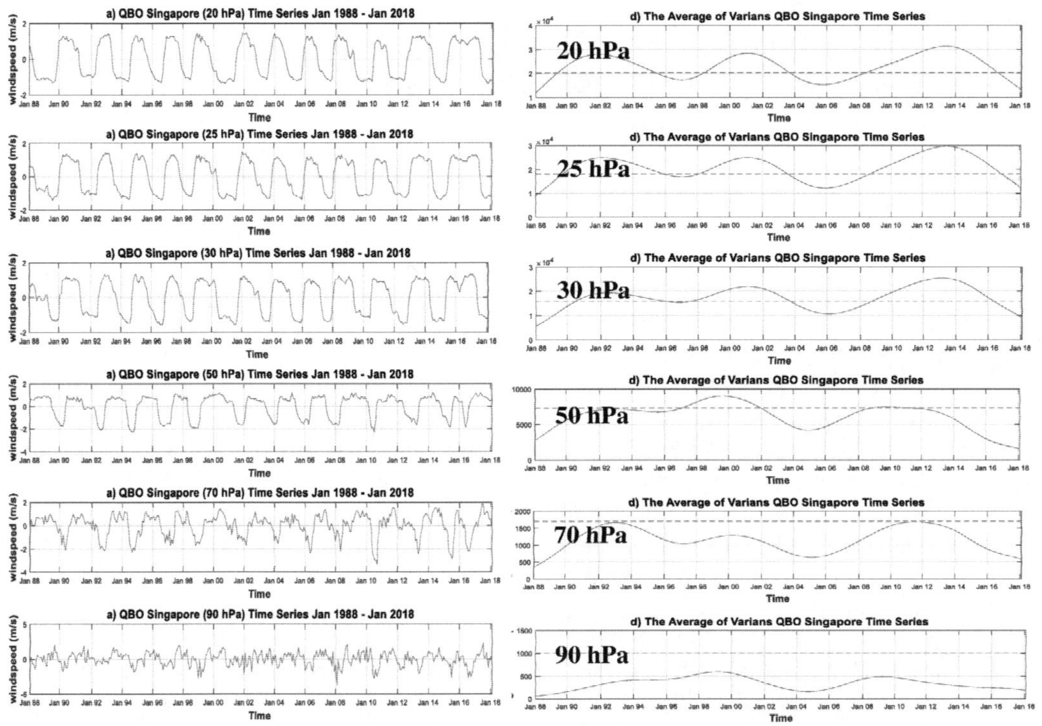

Figure 2. Time series of zonal wind (left-hand panel) and the average of varians (right-hand panel) at 20, 25, 30, 50, 70, and 90 hPa layers for period of January 1988 to January 2018.

From figure 2, in the left panel, we can see a similar oscillation pattern at the upper layer of the stratosphere (20 – 30 hPa), whereas inconsistency appears at 50 hPa until the bottom layer. The same result is also displayed by the average of variance for each layer as shown in the right panel. A periodic easterly and westerly phase of wind is clearly shown in the layer of 20-50 hPa. Along with a decrease in the altitude, zonal wind oscillation decays by the stratospheric disturbance. It can be seen from the random oscillation of QBO signal that turns up in the bottom layer.

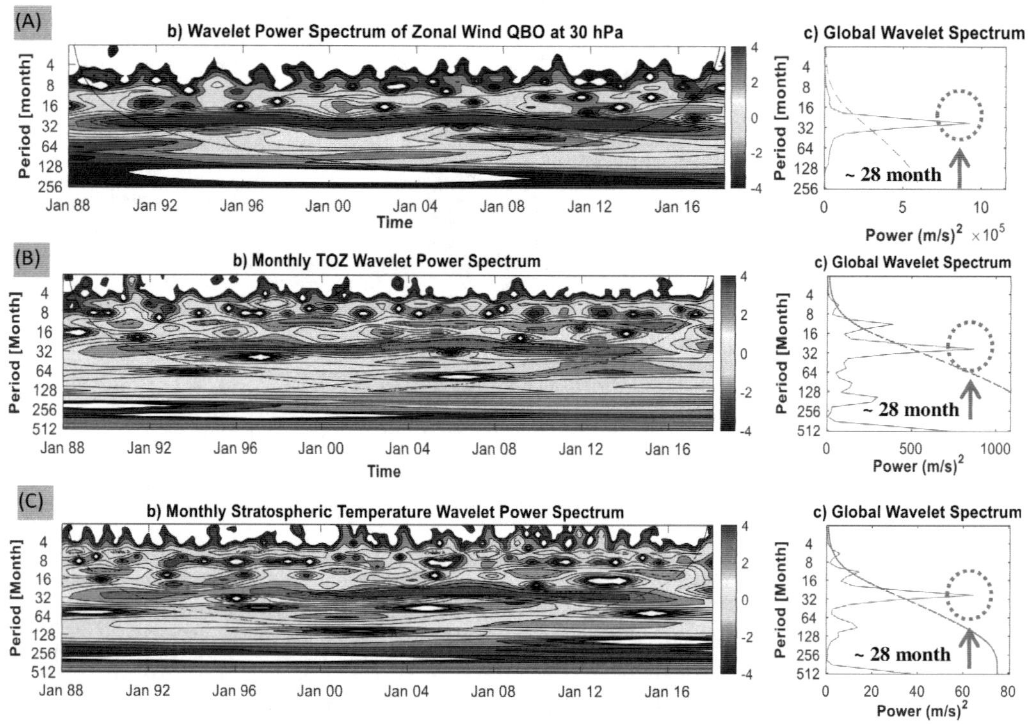

Figure 3. Wavelet analysis of (A) zonal wind QBO at 30 hPa (B) total ozone column and (C) stratospheric temperature at 30 hPa for period of January 1988 to January 2018.

Since our analysis only focussed on 30 hPa layer, we applied the wavelet technique to investigate when the strong signal has occurred. Then, we obtained the pre-dominant peak oscillation of 30 hPa was still 28 month. The same method was also used for monthly TOZ and stratospheric temperature. From the global wavelet spectrum, we saw that pre-dominant oscillation of ozone was around 28 months, with two other oscillation in 6 and 12 months. It means that the variation of ozone concentration and temperature at the stratospheric layer were affected by the wind QBO, Annual Oscillation (AO) and Semi-Annual Oscillation (SAO) as shown in figure 3.

Then, we concentrate to analyze the comparison between TOZ against QBO and temperature anomalies over the maritime equatorial stratospheric region, started from January 2001 to December 2017 as shown in figure 4. From here, we can see a good pattern of both parameters. Ozone concentration in the stratosphere is a combination of the transport mechanism and photochemical process at the layer of ~30 – 20 hPa [15]. Figure 4(a) shows that there is a difference of ozone concentration when the westerly and easterly phase of QBO occurred. Generally, it increases in the westerly phase and conversely, decreases in the easterly phase. Thereby, the oscillation of zonal wind at 30 hPa layer influenced the ozone fluctuation. The transport mechanism in the stratosphere was not only associated with the wind regimes, but also the temperature. A negative anomaly of temperature described the low

IOP Conf. Series: Earth and Environmental Science **303** (2019) 012013 doi:10.1088/1755-1315/303/1/012013

of temperature at the easterly phase of QBO and related to the descending rate of ozone which resulted in ozone depletion as shown in figure 4(b). In other words, a negative anomaly of TOZ was associated with negative anomaly of temperature, hence the ozone absorbed solar and infrared radiation. The stratospheric layer with slight ozone concentration will cause a limited amount of absorbed radiation, so the temperature is low.

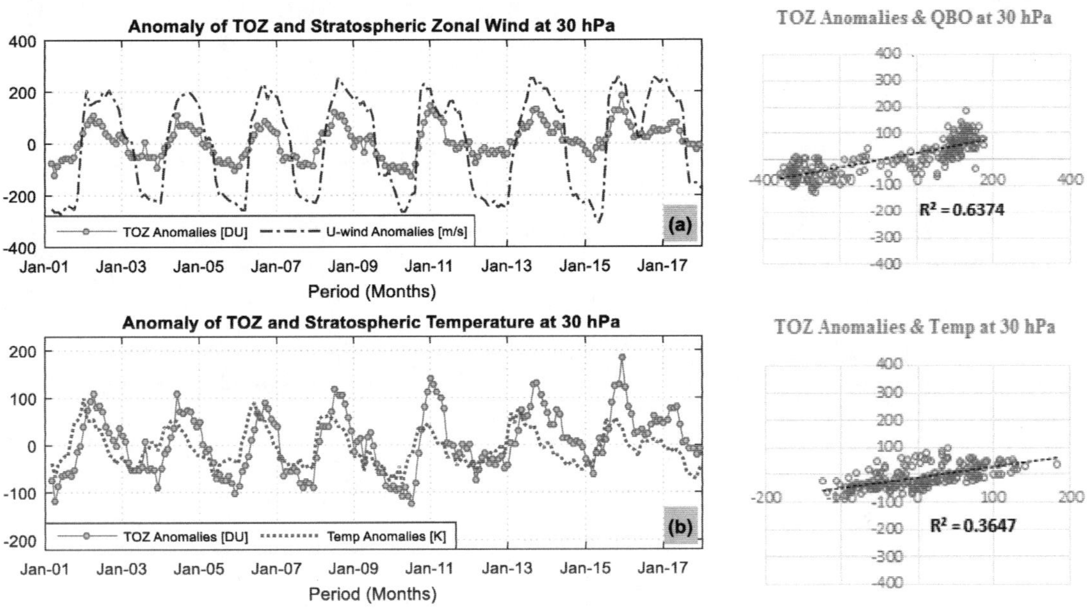

Figure 4. Time series comparison of TOZ anomalies against (a) stratospheric zonal wind and (b) temperature at 30 hPa during the 2001-2017 period (left panel) and scatter plot for (a) and (b) (right panel).

Please note, the similar result is also obtained by Hermawan [10] that compared QBO and ozone anomalies from AIRS satellite over Kototabang, West Sumatera. Hereafter, we made the scatter plot between TOZ anomalies with QBO and stratospheric temperature at 30 hPa layer as shown in the right panel of figure 4. We found a good agreement based on the coefficient determination, which is 0.64 to QBO and 0.36 to temperature. It means that 64% of TOZ anomaly can be explained by the QBO phenomenon.

Finally, to derive ozone concentration from Quasi-Biennial Oscillation data, we used Box-Jenkins or so-called ARIMA method. The first step of the method is the identification of stationarity of the data by time series plot, autocorrelation functions (ACF) and partial autocorrelation functions (PACF) of the whole dataset. Then, the next step is the determination the order of AR and MA processes from the PACF and ACF patterns which useful to consider the initial model. Based on the least square criteria, the model parameters were calculated. A diagnostic test consisted of the significance parameter and residual white noise test was applied to the initial model (right panel of figure 5). If the model passed the test, it will be used to forecast ozone concentration for the next period. As in the spectral analysis above, the QBO contains seasonal variation which oscillates around 28 months. Therefore, Box-Jenkins final model was SARIMA $(5,0,4)(0,1,0)^{28}$ with model equation, $Y_t = 0.3Y_{t-1} + 1.11Y_{t-2} + 0.2Y_{t-3} - 0.87Y_{t-4} + 0.16Y_{t-5} + e_t - 1.09e_{t-1} + 0.36e_{t-2} + 1.29e_{t-3} + 0.44e_{t-4} - e_{t-28} + 1.09e_{t-29} - 0.36e_{t-30} - 1.29e_{t-31} - 0.44e_{t-32}$ and percentage error of 46.6%

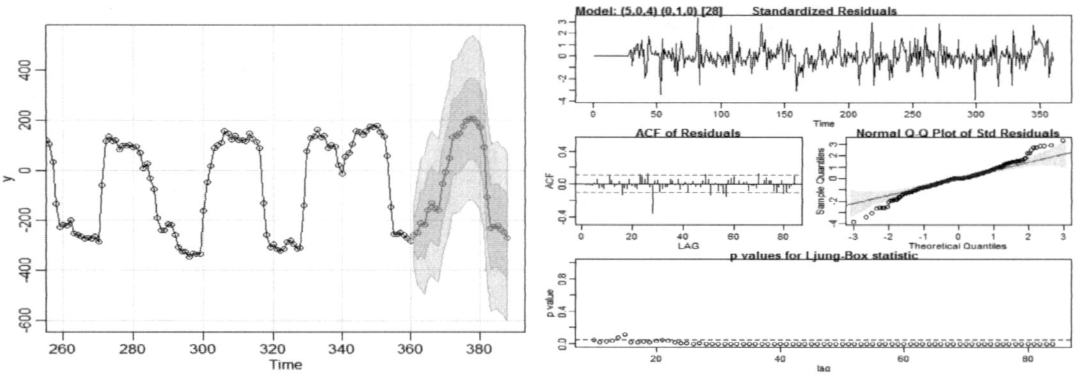

Figure 5. The forecast model of QBO at 30 hPa layer (left panel) and the result of diagnostic testing (right panel) from January 1988 to December 2017.

We have divided the QBO into 2 dataset, which were a training and a testing set. For model validation, comparison between the data observation and model has been done. The dataset of testing started from April 2008 to December 2017, were validated for each 28 months as shown in figure 6 below.

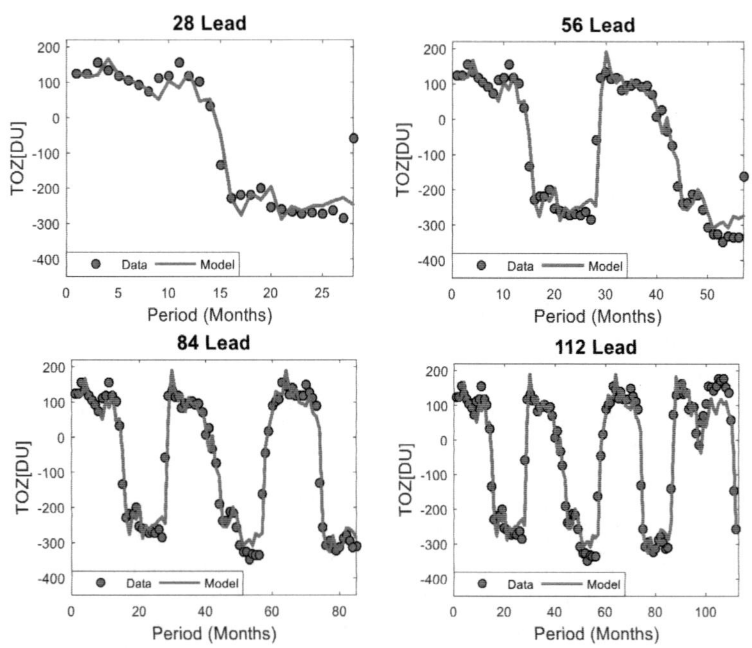

Figure 6. The model validation of SARIMA $(5,0,4)(0,1,0)^{28}$

If we include the temperature at 30 hPa in the analysis, then developed a multiple linear regression model (figure 7) to derive ozone concentration data, we obtained the correlation score was 0.55. Meanwhile, the model with 2 predictors is still good enough to explain the ozone concentration.

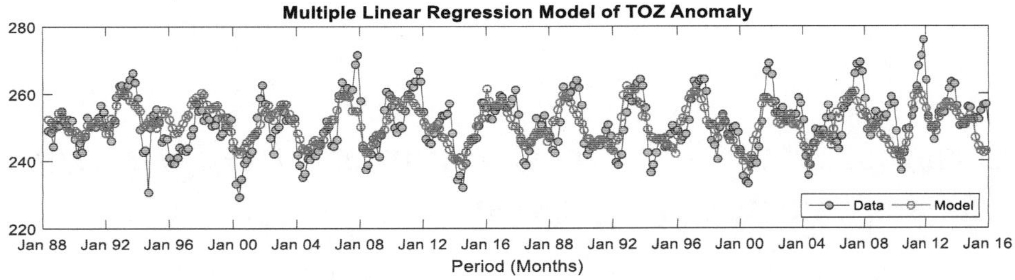

Figure 7. Comparison of TOZ with multiple linear regression model

4. Conclusion

The behavior of ozone concentration at the equatorial region is important to be investigated due to the ozone depletion. Since equatorial TOZ is well correlated with the zonal wind and stratospheric temperature, in this present study we have investigated the QBO and the temperature to derive ozone concentration data. The pre-dominant peak oscillation of QBO from Power Spectral Density (PSD) and wavelet is around 28 months, as the same as temperature oscillation. The result shows that zonal wind oscillation pattern was evident by a strong signal at 30 hPa layer. Zonal wind oscillation decays by the decreasing of the altitude. As the results of the TOZ comparison to zonal wind and temperature anomaly for a period of January 2001 to December 2017, the coefficient determinations (R^2) were obtained about 0.64 and 0.36, respectively. Both parameters look similar to each other. For this reason, we developed a model prediction of ozone concentration using Box-Jenkins and multiple linear regression (MLR). Box-Jenkins final model contains seasonal component of 28 month $(5,0,4)(0,1,0)^{28}$ with the model equation is $Y_t = 0.3Y_{t-1} + 1.11Y_{t-2} + 0.2Y_{t-3} - 0.87Y_{t-4} + 0.16Y_{t-5} + e_t - 1.09e_{t-1} + 0.36e_{t-2} + 1.29e_{t-3} + 0.44e_{t-4} - e_{t-28} + 1.09e_{t-29} - 0.36e_{t-30} - 1.29e_{t-31} - 0.44e_{t-32}$ and percentage error of 46.6%, whereas MLR obtained a fairly good correlation score, that was about 0.55.

Acknowledgments

We wish to thank the NASA GSFC, NOAA NCEP-NCAR, and Institut für Meteorologie Freie Universität Berlin, who have provide the TOZ of MERRA-2 Satellite, temperature, and Singapore's QBO data available to access online. We also express our gratitude to Hasanuddin University for supporting this study.

References

[1] Ebdon R A., 1960, *Q. J. R Meteorol. Soc.* **86.** pp. 540-542
[2] Reed R G, Campbell W J, Rasmussen L A, Rogers D G., 1961, *J. Geophys. Res.* **66** pp.813-818
[3] Baldwin M P, Gray L J, Dunkerton T J, Hamilton K, Haynes P H, Randel W J, Holton J R, Alexander M J, Hirota I, Horinouchi T, Jones D B A, Kinnersley J S, Marquardt C, Sato K, Takahashi M., 2001, 1972 *American Meteorological Society* **29** 1076-1080
[4] Holton J R, Tan H C., 1980, *Journal of Atmospheric Science* **37** 2200-2208
[5] Funk J P, Garnham G L., 1962 *Tellus* **14** 378–382
[6] Camp C D, Roulston M S, Yung Y L., 2003, *J. Geophys. Res.* 108 (D**20**) 4643
[7] Kinnersley J S, Tung K K., 1998, *J. Geophys. Res.* 94 **11**,559–11,571
[8] Eddy Hermawan 2018, *IOP Conf. Ser.: Earth Environ. Sci.* **149** 012064
[9] Torrence C, Compo G P., 1998 *Bull. Am. Meteorol. Soc* **79** 61-78
[10] Gu D, Philander S G H., 1995, *J. Climate* **8** 864–876
[11] Wang B, Wang Y., 1996, *J. Climate* **9** 1586–1598
[12] Gamage N, Blumen W., 1993, *Mon. Wea. Rev.* **121** 2867–2878
[13] Box G E P, Jenkins G M., 1976, *Time series analysis: forecasting and control* San Fransisco: holden-day
[14] Brasseur G P, Orlando J J, Tyndall G S., 1999, *Atmospheric chemistry and global change* New York, NY, USA: Oxford University Press

Comparison of Heat Flux Observation at Three Different Locations around Bandung, West Java

I D G A Junnaedhi[1], K Muchtar[1], S H S Herho [1], P Y Kombara[1] and F R Fajary[2]

[1]Atmospheric Science Research Group, Faculty of Earth Sciences and Technology, Bandung Institute of Technology
[2]National Disaster Management Authority

dewa108@meteo.itb.ac.id

Abstract. We observed heat flux profiles at three different locations around Bandung, West Java. Heat flux values were calculated using the covariance method with the vertical flow and potential air temperature fluctuation parameters. Observations have shown diurnal patterns in the three observation locations. Increasing heat flux during morning time flux on all three location shows similar pattern, but decreasing heat flux on the afternoon varies between locations. Urban residential area of Dago shows the most distinct pattern, with a rapid decrease in the afternoon and also after midnight. We also compared observed heat flux data with estimate values using the Bulk Parameterization method. We found that the estimated heat flux was not able to produce values in accordance with the observation data. Wind shear factor and inadequate implementation of the Bulk Parameterization method were thought to causing poor estimation of surface heat flux in Bandung. Improvements in this method, especially on the bulk transfer coefficient and surface skin temperature, are necessary to get better estimation result.

Keywords : heat fluxm temperature, bulk transfer, surface skin temperature.

1. Introduction

Atmospheric boundary layer is part of the atmospheric layer that receives direct influence from activities on the earth's surface [1]. Changes in physical parameters of the earth's surface also change the transfer of surface heat and moisture, which in turn affect the structure of the atmospheric boundary layer. An important parameter that most influences changes in the structure of the atmospheric boundary layer are the phenomenon of air mass exchange, which is represented in the energy equilibrium equation, and turbulence parameters [1]. Therefore, the phenomenon of energy transfer represented through surface flux parameters is one of the important things in the study of atmospheric boundary layers.

Surface observation data from observation stations, such as wind; air temperature; etc., used as initial input in parameterization for flux calculations using various numerical weather prediction (NWP) models. There are various types of parameterization of heat fluxes used by NWP, one of which is Bulk Parameterization. This parameterization method has been widely used in various types of NWP and is usually combined with other parameters, such as the Penman-Monteith approach and the multi-source approach [2].

Because of the lack of availability of data on heat flux observations used as the main reference, research on the characteristic of heat flux in tropical regions are still very limited. Therefore, it is necessary to study heat flux in the tropics and their response to differences in surface characteristics in each observation area. This study aims to analyze heat fluxes at three different points in Bandung, West

Java. This research was conducted on June 16, 2016, until July 21, 2016. In addition, we also estimate heat fluxes using Bulk Parameterization and compared it with the observed data.

2. Data and methods

Data used in this study are weather observation data from three different locations in Bandung. The first point located in the Dago area, with coordinates 6 ° 52 '17.3 "S, 107 ° 37' 8.9" E and an altitude of 837 m.a.s.l. (figure 1). This location considered as a representation of the urban residential area of Bandung with dense settlements and little vegetation (figure 2a). The second location is at Sulaiman Air Force Base (Lanud Sulaiman or Lanud) with coordinates 6° 58' 53.1 "S, 107° 34' 25.49" E and an altitude of 688 m.a.s.l (figure 1). The second location was chosen because of flat and wide surface characteristics, with a homogeneous surface cover in the form of grass (figure 2b). The third location is in Manoko plantation with coordinates 6° 48' 25.02 "S, 107° 36' 46.62" E and height of 1312 m.a.s.l. (figure 1). This location was chosen because of the hill characteristics that represent the morphology of the North Bandung region (figure 2c).

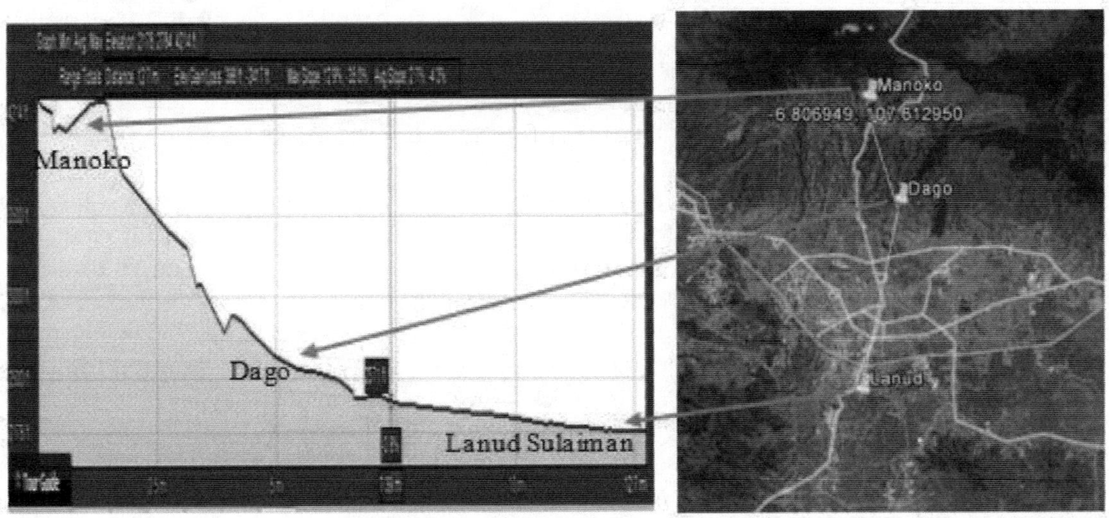

Figure 1. Altitude (left) and map (right) of observation locations, namely Dago, Lanud Sulaiman and Manoko

(a) (b) (c)

Figure 2. The appearance of the observation locations in a) Dago, b) Lanud Sulaiman and c) Manoko

There are three types of instruments used in the observation, namely Young Model 81000 Ultrasonic Anemometer, Davis Vantage Vue Automatic Weather Station (AWS), and Davis Vantage Pro Automatic Weather Station (AWS). Data obtained from the ultrasonic anemometer measurements are vertical wind velocity data (w) and virtual air temperature (T_v). To remove the effect of different pressure altitude, T_v should be converted into a virtual potential air temperature (θ_v) [1]. This could be done using Poisson's equation (1), by changing T with T_v, thus resulting in θ_v instead of θ. P_0 is the reference pressure, which usually is set to 1000 hPa, and P is station pressure which we take from the AWS measurement. Apart from P, we also obtain other parameters from AWS measurement, including air temperature (T), relative humidity (Rh), and horizontal wind velocity (u, v). In order to compare result from ultrasonic anemometer and AWS equally, we also convert the temperature measurement of AWS into virtual potential air temperature (θ_v). The conversion was done by calculating potential temperature (θ) using equation (1), and then convert the result into virtual potential air temperature (θ_v) using equation (2) [1]. r on equation (2) is the unsaturated mixing ratio which could be calculated from Rh, T, and P using formulas that could be found in [3], in the Annex 4 part.

$$\theta = T\left(\frac{P_0}{P}\right)^{0.286} \tag{1}$$

$$\theta_v = \theta(1 + 0.61r) \tag{2}$$

For each site, the AWSs were installed on two different height. The Vantage Pro AWS was installed at two meters above the ground and the Vantage Vue AWS was placed at 10 meters above the ground. This configuration is needed because the Bulk Parameterization estimation method requires air temperature data at two different heights. Installation configuration of these instruments in this study can be seen in figure 3.

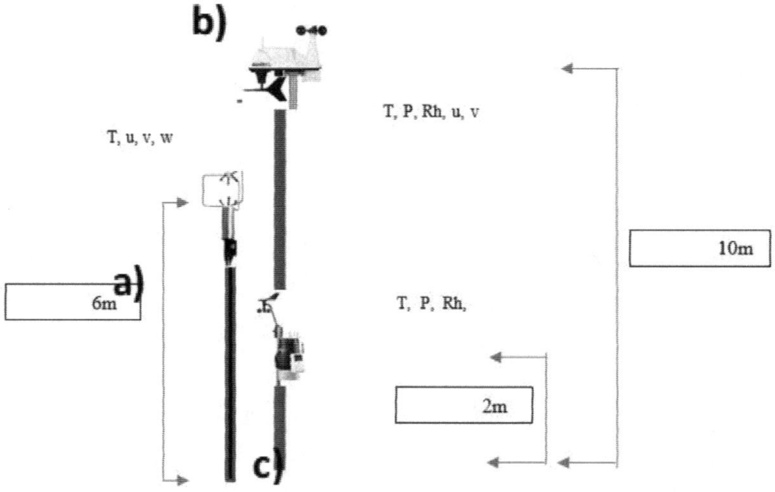

Figure 3. Instruments configuration used in observation activities. a) RM Young 81000 Ultrasonic Anemometer. b) Davis Vantage Vue AWS. c) Davis Vantage Pro AWS.

The first step to calculate sensible heat fluxes are to obtain vertical flow fluctuation (w') and air temperature fluctuation (T') [1]. However, to remove the influence of altitude on air temperature, the air temperature parameter (T) is replaced by potential air temperature (θ), so the fluctuation parameter used is the potential air temperature fluctuation (θ'). The values of w' and θ' can be calculated using equations (3) and (4). The overbar on these equation denotes the time average over a certain time window. In this study, we choose 30 minutes time window for averaging, just like [4], to account for most of turbulent eddies spectrum [1].

$$w' = w - \overline{w} \tag{3}$$

$$\theta' = \theta - \bar{\theta} \tag{4}$$

After obtaining the fluctuation value of the two parameters, the heat flux can be calculated using a covariance method, as shown in equation (5), where C_p is the air heat capacity at constant pressure and ρ is the dry air density [1].

$$Qh = C_p\, \rho\, \overline{w'\theta'} \tag{5}$$

However, in this study, the parameters of heat flux used are kinematic sensible heat flux, where the heat capacity and the air density are removed from the calculation [1]. Since in this experiment we do not have any fast response temperature sensor (e.g. thermocouple), we use the air temperature recorded by the ultrasonic anemometer instead. As previously explained, the ultrasonic temperature measurement is actually a virtual temperature, or more precisely is an acoustic virtual air temperature, which highly affected by moisture in the air. Although slightly different, but these acoustic virtual temperature could be considered as virtual temperature as the ultrasonic anemometer (i.e. RM Young 81000) has already corrected for crosswind velocity contamination [5]. Thus equation (5) was modified into equation (6), where θ is replaced by θ_v , and the heat capacity and air density was removed. Conversion from virtual temperature into virtual potential temperature could be done using equation (1) as previously explained.

$$Qh_k v = \overline{w'\theta_v{}'} \tag{6}$$

The usage of virtual temperature (or ultrasonic temperature) to measure heat flux may yield an over or underestimate results, compared with using actual or absolute temperature. As pointed out by [6], a correction is necessary to get a good heat flux measurement from ultrasonic anemometer. This will require a fast response hygrometer or an H_2O gas analyzer to estimate the Bowen Ratio; or a net radiation observation to completing the energy balance equation [6]. However, research by [4] found that for horizontal wind speed that was less than 8 m/s, heat flux measurement using ultrasonic temperature have a good agreement with measurement using thermocouple, with variation less than 30 W/m-2 or less than 0.02 $K.m.s^{-1}$ in kinematic form. Our observation show that the horizontal wind speed on all sites were less than 8 m/s, thus we consider that the heat flux measurement was quite accurate to represent the heat flux exchange over the study area.

$$(\overline{w'\theta'})_{est} = -C_H\, \overline{U}(z)[\bar{\theta}(z) - \theta_0] \tag{7}$$

$$\left(\overline{w'\theta_v'}\right)_{est} = -C_H\, \overline{U}(z)\left[\overline{\theta_v}(z) - \theta_{v0}\right] \tag{8}$$

To estimate surface heat flux using Bulk Parameterization method, we used formulation defined in equation (7), adapted from [1] and [2]. All data to calculate these estimates comes from AWS, both on 2 m (Vantage Pro) and 10 m (Vantage Vue). Since we used virtual potential temperature for measurement of heat flux, the equation (7) was modified into equation (8). $\overline{\theta_v}(z)$ is the virtual potential temperature on 10 m height, while is the virtual potential temperature on 2 m height. $\overline{U}(z)$ is the average horizontal velocity at an altitude of 10 meters from the ground. We use 30 minutes averaging for all averaged variables to match the heat flux measurement from ultrasonic anemometer. CH on the equation is called bulk transfer coefficient or sometimes called Stanton number. The value of C_H used in this study is 1.5×10^{-2} based on trial and error to get a comparable value of heat flux estimation.

During observation, several rain event took place on the observation site. Since rain could induce large scale error on ultrasonic anemometer measurement, we discard all data during the rain event. Thus, we focus this study on clear non-rainy days.

3. Results and discussion

3.1. Diurnal pattern analysis

We process and display wind data from observations of an ultrasonic anemometer, Vantage Pro AWS, and Vantage Vue AWS in the form of wind rose graphs for general wind conditions. The wind rose graphs produced from the data of the three instruments did not show significant differences, so we will only display the wind data obtained through ultrasonic anemometer measurements (figure 4). The wind rose graph plot shows that in the Dago observation area, the wind blows from the west, east and south directions (figure 4a). This shows that there is no dominant wind in the Dago observation area. The maximum wind speed in Dago ranges from 2 - 2.5 m/s. Meanwhile, the dominant wind that flows over Lanud Sulaiman came from the northeast with a tendency towards east with a maximum speed of 7 m/s (figure 4b). Whereas over Manoko the dominant wind blew from the southeast with a maximum value range between 3 - 3.5 m/s (figure 4c).

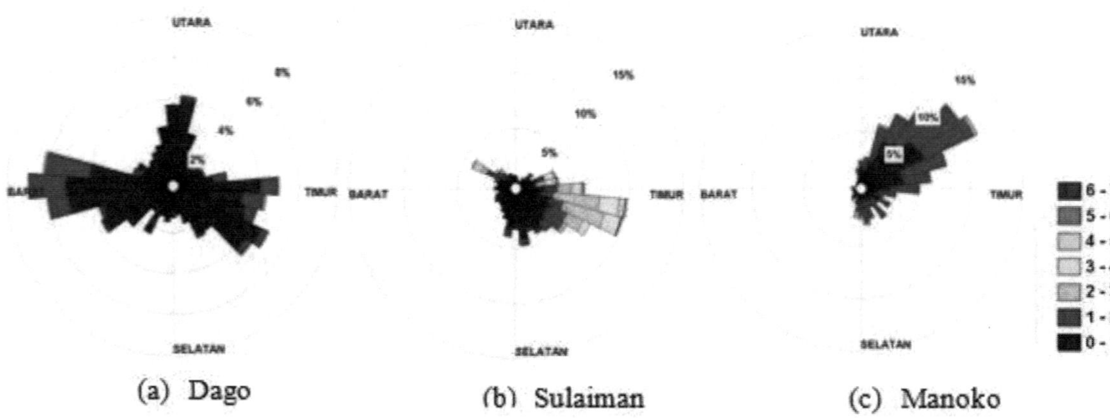

Figure 4. Wind rose graphs in (a) Dago, (b) Lanud Sulaiman, and (c) Manoko which were obtained from observations of ultrasonic anemometers.

Wind over Dago site was highly affected by numerous building surrounding the observation site. Those buildings causing distortion to the wind field and decreasing the wind speed. In Manoko, wind flows more freely than in Dago, so there is no significant change in direction. However, the condition of the observation area in Manoko, which is a plantation with many vegetation covers and hilly contour characteristics, also influences the wind speed in the observation area. Sulaiman observation area is a very wide open space. The flowing wind does not get many obstacles, so the wind speed in this region was greater than the other two observation locations.

Figure 5 shows the values of u', v', w', and θ' in Lanud Sulaiman. We chose to display the observations in this observation area because the characteristics of the location were in accordance with the criteria for the area of observation of heat flux, namely the location of the observation with an open state and without many obstacles, and homogeneous surface conditions. From these data, it can be seen that the fluctuations of the four variables above are influenced by diurnal patterns. During the daytime, the fluctuations of the four parameters are very high (shown in a red circle), while at night the fluctuations are relatively lower than during the daytime (indicated by a black circle).

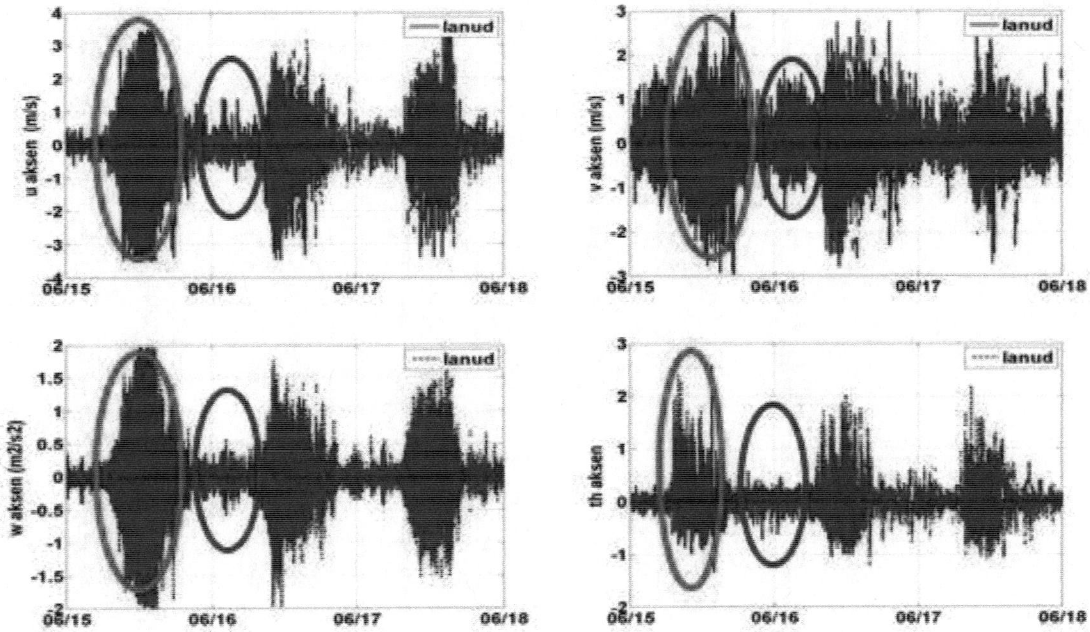

Figure 5. $\boldsymbol{u'}$, $\boldsymbol{v'}$, $\boldsymbol{w'}$, and $\boldsymbol{\theta'}$ data which were observed by the ultrasonic anemometer on Lanud Sulaiman. The red circle shows fluctuations during the day, while the dark circles show fluctuations at night.

In figure 6 we show the histogram of frequency of occurrence of w', θ', and $w'\theta'$ in all three observation locations. The frequency occurrence of w' in all sites tend to be balance between positive and negative. The same thing also happen for the frequency occurrence of θ'. However, the frequency occurrence of $w'\theta'$ Dago, Lanud Sulaiman, and Manoko tends to be more positive. This indicated a more active occurrence of heat flux transfer from ground to the atmosphere above it [1]. It was an expected result as positive heat flux was highly supported by sun radiation during the day, while negative heat flux was just depended on surface temperature, which may not have big discrepancies with the air above it.

All days composite value of kinematic sensible heat flux over three locations in Bandung was presented in figure 7. Comparing this results with [7], we found that our observation was comparable with their results. The kinematic sensible heat flux observation in Bandung ranges from -0.025 to 0.15 $K.m.s^{-1}$, while the range of kinematic sensible heat flux from [7] varies between -0.02 to 0.2 $K.m.s^{-1}$. The diurnal pattern shown in figure 7 also shows similarities with the diurnal pattern observed in [7]. So it can be concluded that the measurement of the heat flux that we are conducting in Bandung was in accordance with the theory that we referred.

Figure 7 also shows a clear distinction in the diurnal pattern of heat flux on the three locations. Manoko exhibits more rapid decrease of heat flux on the afternoon, while on Dago and Lanud Sulaiman the decreasing rate were much slower. In the morning, all sites show an increase of heat flux in a comparable rate among sites. In the night, all sites also exhibit negative values which comparable one to another, although Dago site shows slightly lower value. This means that during nighttime, in Dago, the surface took a little bit more heat from the atmosphere, compared to the other two places. Nevertheless, all site shown to have similar diurnal pattern of heat flux.

IOP Conf. Series: Earth and Environmental Science **303** (2019) 012014 doi:10.1088/1755-1315/303/1/012014

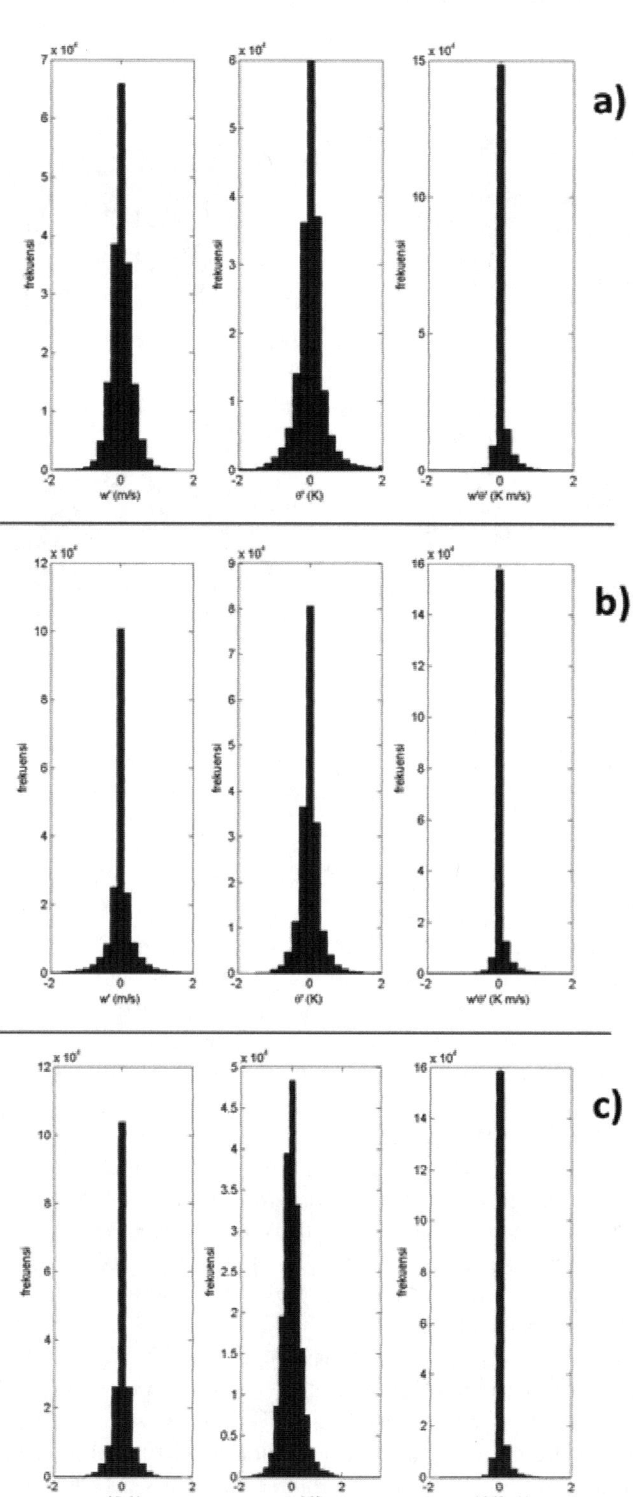

Figure 6. Histograms of $\boldsymbol{w'}$, $\boldsymbol{\theta'}$, and $\boldsymbol{w'\theta'}$ for a) Dago, b) Lanud Sulaiman and c)Manoko.

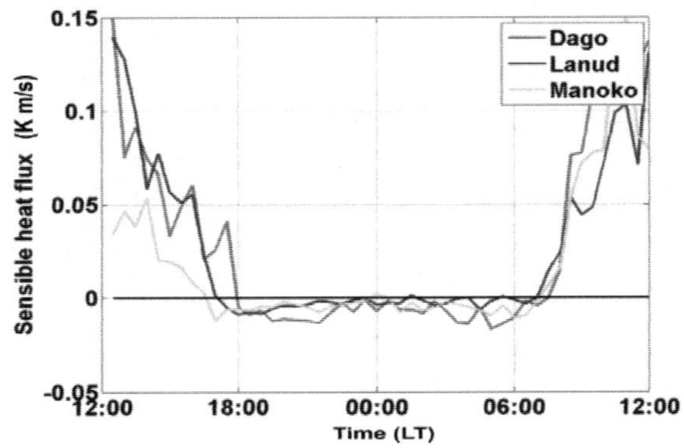

Figure 7. All observation day composite of kinematic sensible heat flux observed over three locations in Bandung.

3.2. Comparison of the heat flux values in the three observation locations.

Figure 8 shows the results of the measurement of heat flux on a sunny day in the three observation locations. It appears that the heat flux in Dago (which is marked by a red line) has the fastest increase in value compared to the other two observation locations. Heat fluxes at Manoko (green line) and Lanud Sulaiman (blue line) reach their peak faster than Dago. The peak already reached around 10:30 AM in Manoko and around 11 AM in Lanud Sulaiman, but in Dago the peak was reached after 12 AM. The heat flux in Manoko and Sulaiman experienced a relatively slow decline after reaching its peak and experienced an increase again after 3 PM. By 5 PM, the value of the heat flux in Manoko and Sulaiman changed to negative. While in Dago, the pattern was completely different. After it reaches the peak, the heat flux was decreasing rapidly and reach negative value just after 3 PM.

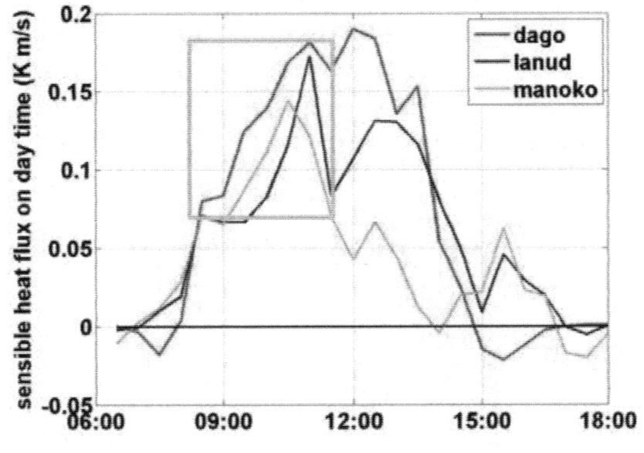

Figure 8. Heat fluxes observed during sunny days. The yellow box shows a significant increase in the heat flux values towards midday.

Based on observational data on the two observation locations, it can be concluded that heat flux during the clear night is more inclined to have a negative value (figure 9). The heat flux in Sulaiman has a lower value compared to Dago at 6 PM until 9 PM. Around 10 PM, there was a fluctuation in the value of heat flux in both observation locations. After midnight. the heat flux in Dago experienced a very significant decline, this condition continued until it reached a positive value at 6 AM. The lowest value of heat flux during the clear night in Dago is -0.023 $K.m.s^{-1}$, while the lowest value in Sulaiman is -0.0117 $K.m.s^{-1}$.

IOP Conf. Series: Earth and Environmental Science **303** (2019) 012014 doi:10.1088/1755-1315/303/1/012014

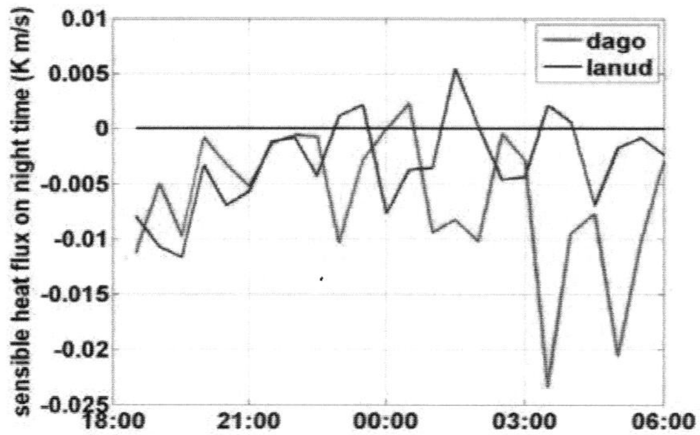

Figure 9. Heat flux values during the clear night over Dago and Lanud Sulaiman. Data over Manoko was not used due to long rainfall over the site during night time.

3.3. Comparison of heat flux observation and estimation results

Observed heat flux values are shown by the red line and the estimation results are shown by the green line in figure 10. It can be seen in figure 10a, that the estimation results in Dago are in line with the observations, but tend to take place in different time frames. At 9 AM to 1 PM the estimated heat flux results tend to stagnate at 0.1 $K.m.s^{-1}$. Observation results showed that the heat flux in Dago reached its peak around 12 PM., while the peak of estimation occurred around 2 PM. These show the time lag between the results of the observation and the estimation results until the heat flux reaches its peak. After reaching the peak point, both heat flux values decrease rapidly.

Comparison of the two heat flux values in Lanud Sulaiman (figure 10b) shows a different pattern when compared to the pattern in Dago. The heat flux from the estimation results is very slow to increase its value when compared to the observational flux. The estimated heat flux is still negative until 2 PM., then it experiences a very rapid increase until it reaches its peak, and it returns down very quickly. Comparison of heat flux estimation and observation results in Manoko (figure 10c) produced a better pattern compared to the pattern in Lanud Sulaiman. The heat flux estimation results in Manoko show a pattern that is in line with the observations. The estimated heat flux tends to fluctuate around the value of 0.5 $K.m.s^{-1}$ until 1:30 PM, while the observations show a continuous increase until it reaches its peak at 11 AM. The estimated heat flux reaches its peak at 3 PM.

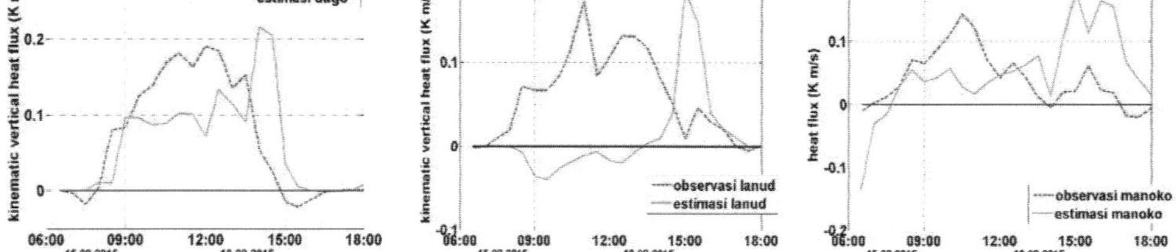

(a) Dago (b) Sulaiman (c) Manoko

Figure 10. Comparison of the daytime observed (red line) and estimated (green line) heat flux values in (a) Dago, (b) Lanud Sulaiman, and (c) Manoko.

Compatibility of the comparison results in three observation locations is thought to be influenced by the heat flux formation mechanism. The observation location that has a buoyancy formed mechanism as in Dago generally has a greater alignment between the estimation results and the results of its observations. Wind shear factors also affect the accuracy of the heat flux estimation. Wind shear is the wind velocity difference at two and 10 meters above the ground at each observation location. It can be seen in figure 11, that the wind shear in Sulaiman is very high with a maximum value at noon reaching 2.46 $m.s^{-1}$, while the wind shear value in Dago is much smaller. So it can be concluded that wind shear is inversely proportional to the estimation results accuracy.

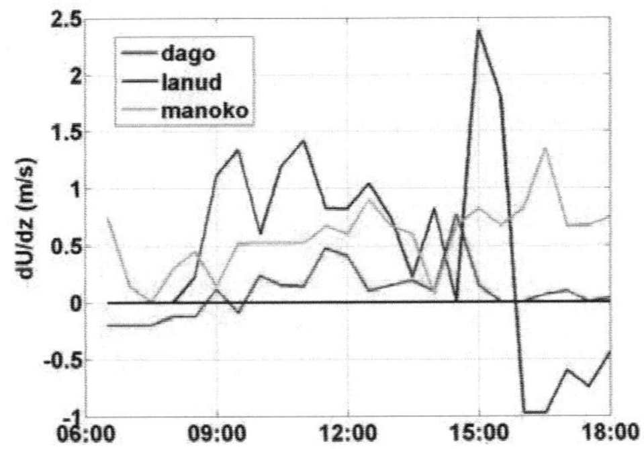

Figure 11. Wind shear charts in three observation locations.

Figure 12 shows the heat flux plots in the three observation locations at night. Observed heat flux values are shown by the red line and the estimation results are shown by the green line. There are differences in the pattern of heat flux results from estimations and observations in the Dago region. The estimated heat flux tends to have a value of 0 $K.m.s^{-1}$, only at 02:30 and 05.30 there is a slight change in values to -0.01 and 0.012 $K.m.s^{-1}$, throughout the observation period. This is different from the results of observations that show negative values throughout the observation period. At Sulaiman, the heat flux from the estimation is positive from 6 to 9 PM., while the observation data shows the opposite. After 9 PM, the estimated heat flux shows a value of 0 $K.m.s^{-1}$, but the observation data shows a negative value. The estimation results of heat flux in Manoko showed a significant difference compared to the observation data. This large difference is seen in terms of its magnitude, the value of the observed heat flux ranges in the range of 0 to -0.005 $K.m.s^{-1}$, while the estimated value can even reach -0.19 $K.m.s^{-1}$. The positive heat flux at the beginning of the night in Dago and Sulaiman occurs because of the effect of the time lag from heat flux during the day. As mentioned earlier, the ability of estimation results to approach the observation pattern is influenced by the well-mixed mechanism. At night, there is no buoyant production in the process of heat flux formation, so the comparison between the observation results and the estimation results is much different. It seems clear that there is an inability of the estimation results to approach observation results, especially during night time. The absolute correlation between estimated value and the observed value was less than 0.3, indicating a very weak relation between them.

Based on our results, the Bulk Parameterization method seem unable to produce good estimate of sensible heat flux over Bandung, either during day time nor night time. However, further investigation has found out that our implementation of the method seems inadequate. According to [2], the bulk transfer coefficient (C_H) need to be carefully calculated using roughness lengths for momentum and heat. The surface skin temperature (θ_0) also need to be measured or estimated correctly and could not simply be estimated using air temperature [1]. By implementing these improvements, further research could be conducted to give a better estimation on the surface sensible heat flux over Bandung.

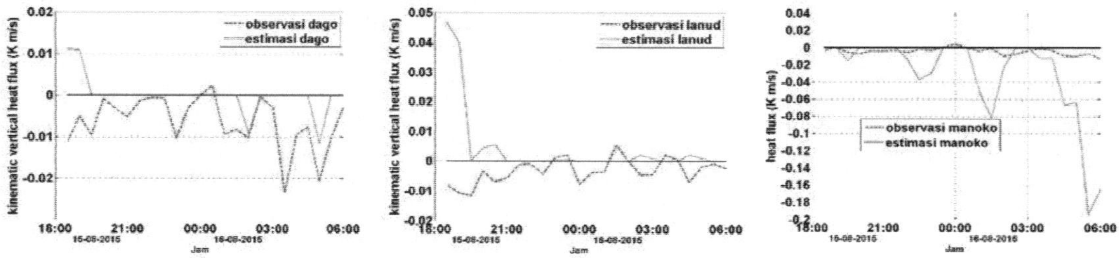

(a) Dago (b) Lanud (c) Manoko

Figure 12. Comparison of the night time observed (red line) and estimated (green line) heat flux values in (a) Dago, (b) Sulaiman, and (c) Manoko.

4. Conclusion

Surface sensible heat flux observation over three locations around Bandung has shown a diurnal pattern that match the pattern from other research. In the morning, the heat flux over all location were increasing in relatively same rate. However, the decreasing rate of heat flux during afternoon shows varying results among three locations, with urban residential area (Dago) show most rapid decrease. In the night time, all location shows mostly negative value of heat flux. Dago area also shows more rapid decrease in flux value, especially after midnight.

Estimation of heat flux using Bulk Parameterization method over three sites on Bandung shows a poor results. The estimated values were barely matched the observed heat flux. Wind shear over the sites was found to have a strong influence on the estimation results, but the major error may be caused by simplification of the bulk transfer coefficient and the surface skin temperature. A more careful and detailed research must be carried out to gain a better estimation of surface sensible heat flux over the study area.

Acknowledgement

This research was supported by Program Penelitian, Pengabdian kepada Masyarakat dan Inovasi ITB (P3MI-ITB) for year 2017.

References

[1] Stull R B 1988 *An Introduction to Boundary Layer Meteorology* (Dordrecht ; Boston: Kluwer Academic Publishers)
[2] Mahrt L and Vickers D 2004 Bulk formulation of the surface heat flux *Boundary-Layer Meteorology* **110** 357–79
[3] World Meteorological Organization (WMO) 2014 *Guide to Meteorological Instruments and Methods of Observation* (WMO)
[4] Burns S P, Horst T W, Jacobsen L, Blanken P D and Monson R K 2012 Using sonic anemometer temperature to measure sensible heat flux in strong winds *Atmospheric Measurement Techniques* **5** 2095–111
[5] Kaimal J C and Gaynor J E 1991 Another look at sonic thermometry *Boundary-Layer Meteorol* **56** 401–10
[6] Schotanus P, Nieuwstadt F T M and De Bruin H A R 1983 Temperature measurement with a sonic anemometer and its application to heat and moisture fluxes *Boundary-Layer Meteorol* **26** 81–93
[7] André J C, De Moor G, Lacarrère P and du Vachat R 1978 Modeling the 24-hour evolution of the mean and turbulent structures of the planetary boundary layer *J. Atmos. Sci.* **35** 1861–83

IOP Conf. Series: Earth and Environmental Science **303** (2019) 012015 doi:10.1088/1755-1315/303/1/012015

Investigation of the regeneration of precipitating convective cloud in basin topography area

R Fitriani[1,2], N J Trilaksono[2,3] and T W Hadi[2,3]

[1]Master Program in Earth Science, Faculty of Earth Sciences and Technology, Institut Teknologi Bandung, ID
[2]Weather and Climate Prediction Laboratory, Faculty of Earth Sciences and Technology, Institut Teknologi Bandung, ID
[3]Atmospheric Science Research Group, Faculty of Earth Sciences and Technology, Institut Teknologi Bandung, ID

rachmy.f@gmail.com

Abstract. The occurrence of precipitation in basin topography area at Bandung can be generated from repeatedly formed convective clouds in one diurnal cycle, which also is detected as multiple rain peaks during a day. In satellite observations, the formation of a new convective cell after dissipation stage of the predecessor cell around or in the same location indicates a regeneration process. The daily convective activities were analyzed during February–May 2017 to understand the regeneration of precipitating convective clouds in Bandung area. The rainfall events were obtained from cumulative distribution function (CDF) at the highest percentages of daily precipitation. This study proposed regeneration index (RegI) as initial identification for convective cloud regeneration from daily rainfall data with multiple peaks. The cases that obtained through the RegI were then verified by satellite data. Both cases at 27 February and 1 May 2017 show that the regeneration of convective cloud was detected in the study area. The precipitating convective clouds initiate in the north Bandung while regenerate and propagate southward. These cases suggest that the highest percentages of daily rainfall in basin topography area can be generated by multiple convective clouds.

Keywords : precipitation, cumulative distribution function, convective clouds, RegI.

1. Introduction

Indonesia Maritime Continent (IMC) is known to have a strong convective activity, which involved in the formation of precipitating convective cloud [1]. The convective activity in Bandung region has a single diurnal peak related to valley wind circulation and produces local scale convection [2]. The heavy rainfall events in Bandung at daytime may cause flash flood event with 150 minutes duration as stated by Agency of Meteorology, Climatology and Geophysics (BMKG) on local news Tempo.co (available online at: http://nasional.tempo.co/read/814715/bmkg-ingatkan-banjir-bandung-bisa-terulang). These events also yield material losses and affect human activities in the study area [3,4]. During diurnal convective cycle, convective clouds regeneration was spotted frequently, which is also indicated on rainfall parameter in other cases at the same region by diurnal observation and numerical simulation [2].

Previous studies detected the formation of regenerative convective clouds that lead to heavy rain in IMC region, associated with external forcing such as monsoon intrusion [5] or intraseasonal oscillation [6]. At extratropical region, regeneration process in convective cloud formation occurs when cold pool

IOP Conf. Series: Earth and Environmental Science **303** (2019) 012015 doi:10.1088/1755-1315/303/1/012015

– wind shear interacts with relatively strong intrusion [7]. Meanwhile, in Bandung region the rainfall area that affected by convective cloud has local coverage [3,4,8]. The formation of convective cloud in basin area of Bandung is suspected to occur repeatedly during diurnal active phase due to wind valley circulation [2]. Therefore, this study investigates the evolution of repeated convective cloud formations during diurnal cycle which is used to understand regeneration process and characteristics of precipitating cloud formation events in Bandung region. The regeneration of precipitating convective cloud is analyzed from rainfall data and Himawari 8/9 satellite images during February to May 2017.

2. Data and Method

The precipitating convective cloud life cycle in Bandung region was observed by surface and satellite imagery. Convective cloud activity was monitored by surface observation stations via rainfall parameter. Observation of rainfall data during February-May 2017 was obtained from Weather and Climatology Prediction Laboratory ITB (WCPL ITB) and *Balai Besar Wilayah Sungai Citarum* (BBWS Citarum). The rainfall data in this study were spread across Bandung region as in figure 1 with coordinates of each station listed in table 1. The coordinates of Meteorological station ITB was measured by WCPL ITB and other stations were measured by BBWS Citarum.

The identification of convective cloud regeneration by rainfall was detected from multiple peaks during a day. Prior to the convective cloud identification, the cumulative distribution function (CDF) of daily rainfall in each station was calculated to obtain threshold values at highest 20% (P80), 15% (P85), 10% (P90) and 5% (P95). The rainfall threshold for each cumulative probability was served as initial identification. Rainfall threshold value for all stations was calculated through three methods: mean, median, and single station. Single station method (Meteorological station ITB) was used to determine whether the rainfall data from one station could represent the Bandung region. The cases were identified when daily rainfall in all stations surpasses the threshold value for each method. Daily time series of rainfall from all stations with 5 minutes time step were used to identify the rain peak. This peak was identified when the distance between each peak was above 30 minutes [9] and its value was greater than 0.5 mm from rainfall modes during observation.

The convective cloud regeneration was represented by regeneration index (RegI). This index is proposed to quantify regeneration identification in the study area. The RegI was obtained from array number of observation stations with the multiple rain peaks divided by an array number of observation stations with the single rain peak. This calculation is shown in equation (1).

$$RegI = \frac{[Station_{peaks\geq 2,c=1} \quad Station_{peaks\geq 2,c=2} \quad \cdots \quad Station_{peaks\geq 2,c=n}]}{[Station_{peak=1,c=1} \quad Station_{peak=1,c=2} \quad \cdots \quad Station_{peak=1,c=n}]} \tag{1}$$

Station $_{peaks\geq 2}$ is total stations that detect multiple rain peaks for a day, Station $_{peak=1}$ is total stations that detect single rain peak for a day, and n is total cases that surpass rainfall threshold during observation period. The RegI has a maximum limit (Station – 1) and minimum limit $(\text{Station} - 1)^{-1}$, with Station is total observation stations in study. This index is used to locate local-scale rain events when similar peak value is detected at 30-70% observation stations or equal to RegI scale at 0.4-2.5 for this study. Calculation of RegI is an approach to find rainfall characteristics based on its coverage and the number of rainfall event detected in one day. However, an analysis of convective cloud regeneration that relies only on rainfall data may have weakness hence verification using satellite data is needed.

The Himawari 8/9 satellite image was used to observe convective cloud growth at the time of the event through black body temperature of IR channel at 10.4 μm ($T_{10.4}$) and IR 12.4 μm ($T_{12.4}$). Convective cloud activity can be observed from index of black body temperature (I_{TBB}) in Kelvin [10] [11]. I_{TBB} is obtained from:

$$I_{TBB} = 255 - T_{10.4} \qquad T_{10.4} < 255 \tag{2}$$

$$I_{TBB} = 0 \qquad T_{10.4} \geq 255 \tag{3}$$

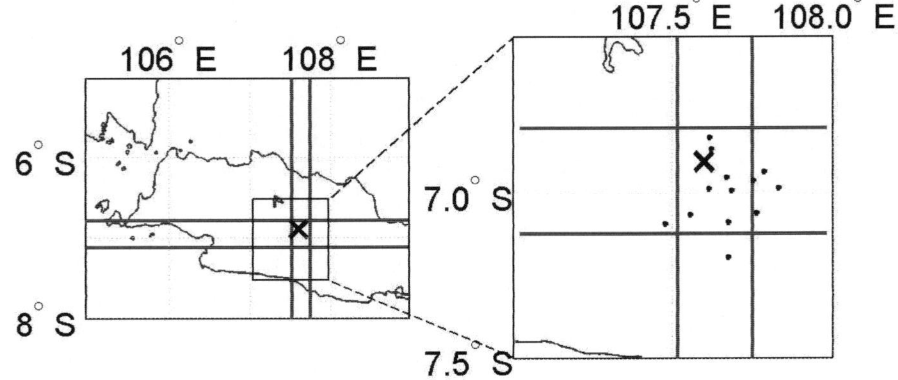

Figure 1. Bandung region as the domain of the study. Black cross depicts the location of meteorological station ITB, while black dot signs are other stations located in Bandung region. The area between vertical (horizontal) lines on map is time-latitude (longitude) cross section.

Table 1. Surface observation stations coordinates in Bandung region.

Location	Coordinates	
	Latitude (°S)	Longitude (°E)
Ancolmekar	7.0853	107.6772
Cicalengka	6.9813	107.8300
Cidurian	6.9490	107.6716
Cisondari-Pasirjambu	7.0903	107.4817
DagoPakar	6.9848	107.6164
Dayeuhkolot	7.0632	107.5559
Hantap	6.9298	107.7871
Jatiroke-Cikuda	7.1919	107.6769
Kertasari	6.8266	107.6177
Meteo-Lembang	6.8618	107.6245
PasehCipaku	7.0566	107.7638
Rancaekek	6.9586	107.7539
Sapan	6.9899	107.6873
Meteorologi ITB	6.9000	107.6000

Identification of convective clouds on IMC uses blackbody temperature threshold at 250 K [10], but in addition, this threshold shows weak correlation with rainfall intensity [12]. This study used threshold value as in equation (2) and (3) to obtain more convective cloud coverage.

The convective cloud identification is performed using an algorithm [13] that relies on data from these channels. Convective cloud detection is obtained from subtraction of IR 10.4 μm and IR 12.4 μm ($T_{10.4\text{-}12.4}$). In this study, identification of convective cloud was obtained by trimming convective cloud detection with I_{TBB}'s outer lines. Both I_{TBB} and convective cloud identification was used to analyze convective cloud activity. The analysis of the convective cloud activity of each case was observed through time-longitude variation with latitude averaging at 6.8°S - 7.12°S and time-latitude variation with longitude averaging at 107.52°E - 107.75°E as shown in figure 1.

The convective cloud evolution in each case was reanalyzed to understand precipitating convective cloud regeneration patterns. The I_{TBB}'s time-latitude and longitude variation plot was used to observe convective activity. Convective cloud identification from $T_{10.4-12.4}$ was calculated to find convective core. The core was identified when the result of convective cloud value was less than 2. Rainfall temporal distribution at meteorological station ITB was used along with time-latitude and longitude variation of convective cloud core to ascertain whether the convective cloud core actually produces rain. The time-latitude and longitude variation plot of satellite data was used to observe convective cloud. Based on these data, the convective cloud regeneration was identified and analyzed thoroughly.

3. Results

3.1. Identification of precipitating convective cloud by rainfall data

Daily rainfall data from all observation stations in Bandung region during February-May 2017 were analyzed using cumulative distribution function (CDF) as shown in figure 2. High intensity precipitation is produced by convective cloud [14] therefore the values above threshold represent the highest percentage of daily rainfall data. The rainfall threshold was obtained from CDF of all stations using mean, median and single station (ITB) methods. This threshold value can be seen in table 2.

The case is confirmed when rainfall data exceeds the threshold from the cumulative probabilities at P80, P85, P90, and P95. The threshold for each method has similar value at highest 20% of daily rainfall (P80), whereas the values start to differ when cumulative probability increases up to 5% highest rainfall event (P85-P95). The daily rainfall at highest 5% (P95) has threshold value 33.5-35.7 mm day^{-1}. The cases that meet the criteria were calculated using equation (1) to derive regeneration index (RegI) as shown in table 2. The RegI values in table 2 are between the range of 0.4-2.5 which show that rainfall with high intensity in Bandung region during this period have local coverage with single or multiple peaks.

The regeneration indices from P80 to P90 have varying values. The RegI determined by the mean and median methods give a range values of 1-2.5 that dominated by local rainfall with multiple peaks. Different results are shown by RegI values for single stations method, which is dominated by local rainfall with single peak (0.4-1). However, P95 values for all methods show that rainfall detected at highest 5% of intensity have local character with multiple peaks. The identification of cases conducted with this method gives the same result, which is 27 February 2017. All cases that meet criteria for P95 are shown in figure 3 as composite of hourly rainfall distribution at each station. This composite plot shows the rainfall distribution of each station, variation and diurnal pattern. Some stations on the figure also detect multiple rain peaks for a day.

Heavy rainfall event at 27 February 2017 has the highest rainfall intensity (70 mm day^{-1}) at ITB meteorological station. This event was also detected by most ground stations with intensity 5-56 mm day^{-1}. Figure 4 shows a convective activity for a day. The I_{TBB} values higher than 50 that predominate in figure 4a illustrate a strong convective activity during the day. The convective cloud identification was obtained from $T_{10.4-12.4}$. This result is shown in shading contour in figure 4b and recognized as convective cloud when the calculated value is near to zero.

Table 2. Daily rainfall threshold value and regeneration index (RegI) in Bandung region.

Probability	Rainfall threshold (mm day^{-1})			Regeneration index (RegI)		
	Mean	Median	Single station (ITB)	Mean	Median	Single station (ITB)
P_{80}	10,38	10,80	10,60	1,02	1,23	0,84
P_{85}	16,04	16,75	15,50	1,39	1,61	0,77
P_{90}	22,97	21,30	21,60	1,64	1,51	0,71
P_{95}	34,50	33,50	35,70	2,00	1,03	1,12

IOP Conf. Series: Earth and Environmental Science **303** (2019) 012015 doi:10.1088/1755-1315/303/1/012015

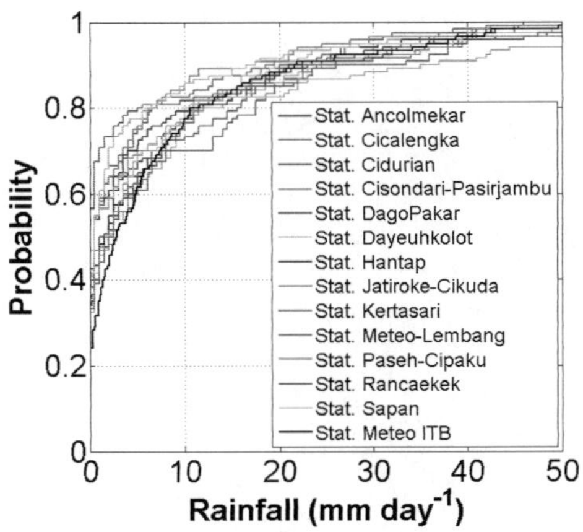

Figure 2. Cumulative distribution function of daily rainfall during period of February-May 2017 at observation stations in Bandung region.

In figure 4 there are two formations of convective clouds detected in northern Bandung after 12.00 local time (LT). Convective cloud formed at 6.8°S was regenerating up to three times while propagating to southern part of Bandung, meanwhile, other convective cloud formed at 6.5°S grew in situ then extended southward. Figure 5 shows a horizontal plot of the convective cloud core in the study area. The convective cloud observed in figure 5a is the first cloud formed on this case; this cloud has reached mature stage at 13.00 LT with a radius of ~10 km. This convective cloud then weakened an hour later as shown in figure 5b. The new convective cloud formed with larger radius (>20km) at southeastern part of predecessor cell was also detected in figure 5b. This cloud formation was also detected in figure 4b as the first regeneration of convective cloud in figure 5a. Meanwhile, the convective cloud at northern part in figure 5b was observed as another initial convective cloud formation at 6.5°S in figure 4b. The rainfall in this case at 15.00 LT was generated from the first regeneration of convective cloud above Bandung region (6.9°S - 7°S) that weakened when intersects with other convective clouds that aroused in the northern part. These cloud formations were also detected in figure 5b. The formation of regenerative convective clouds on the same area was also indicated in [8] where local rainfalls occur in a consecutive manner while propagating southward. This case shows that the weakening of convective cloud activity during regeneration process is related to precipitation that detected on the stations. During active period, convective cells are repeatedly formed until the convective activity decays at night.

3.2. Analysis of regenerative precipitating convective cloud patterns on diurnal scale

The precipitating convective cloud activities in these case study show patterns associated with the initial convection of cloud formation, duration of activities, cloud propagation, and lag time of new convective cell formation. The convective clouds here either have regenerative formation or long duration with strengthening and weakening activities. Another example of cloud regeneration case is shown in figure 5, where the convective cloud activity exhibits multiple peaks in consecutive period. The regeneration pattern of precipitating convective clouds analysis used the cases from the highest 10% of daily rainfall (P90) which yielded 15 events. Analysis of satellite observations combined with surface stations shows that most cases observed also have regenerating convective cloud formation.

One of the cases of regeneration pattern with P90 occurred on 1 May 2017. In this event, ITB meteorological station has detected rainfall with the intensity of 21.9 mm day^{-1} and other four stations detected 6-24.5 mm day^{-1}. The rainfall on 1 May 2017 was detected by fewer observation stations as opposed to the previous case. However, the convective cloud regeneration pattern is also presented here as shown in figure 5.

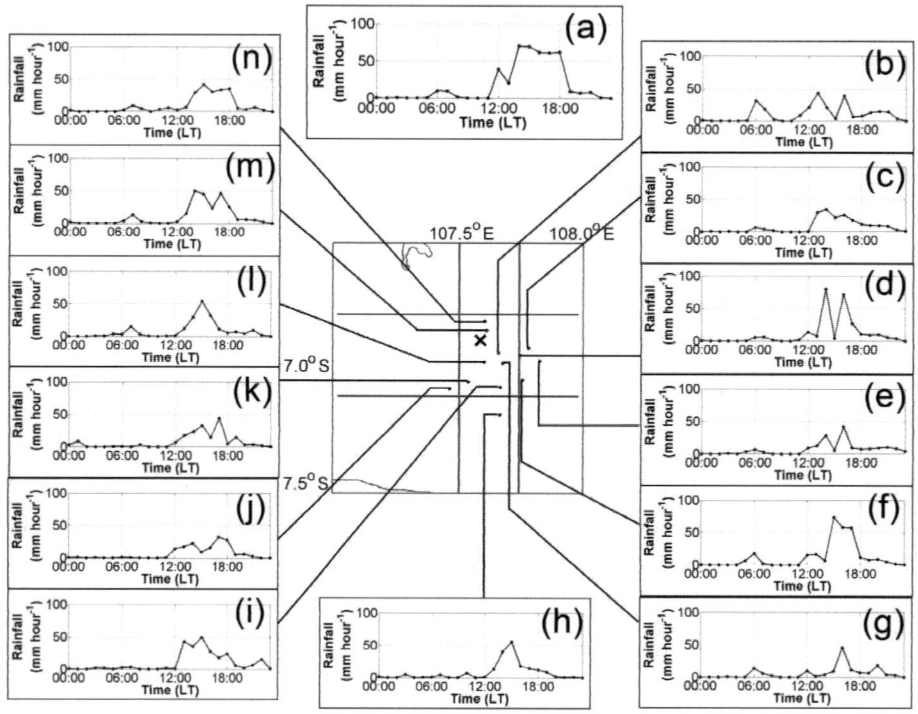

Figure 3. Time series of diurnal rainfall composite for P95 during February-May 2017 at (a) meteorological station ITB, (b) Cidurian, (c) Jatiroke-Cikuda, (d) Rancaekek, (e) Cicalengka, (f) Paseh-Cipaku, (g) Sapan, (h) Kertasari, (i) Ancolmekar, (j) Cisondari-Pasirjambu, (k) Hantap, (l) Dayeuhkolot, (m) Dagopakar and (n) Meteo-Lembang.

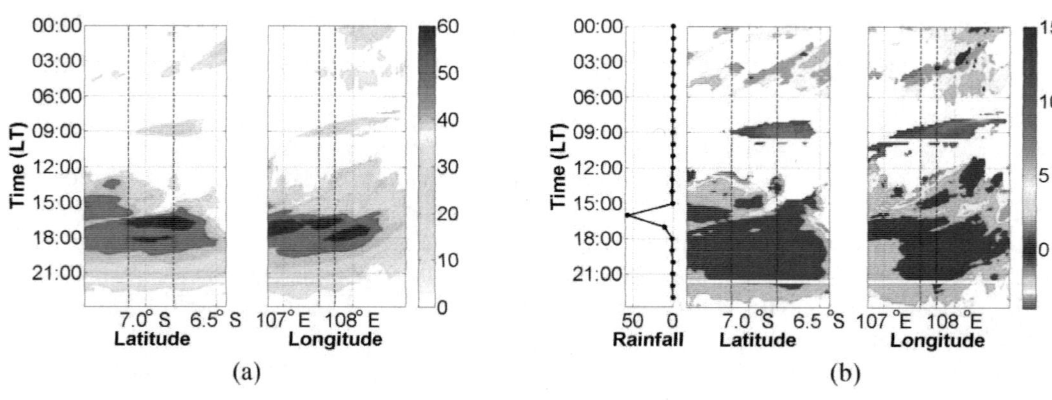

Figure 4. (a) (left) Time-latitude cross sections and (right) time-longitude cross sections of I_{TBB} on 27 February 2017. (b) (left) Time series of precipitation at meteorological station ITB. (middle) Time-latitude cross sections and (right) time-longitude cross sections of convective cloud identification on 27 February 2017. The study area is shown in the region between thick dashed lines in both (a) and (b).

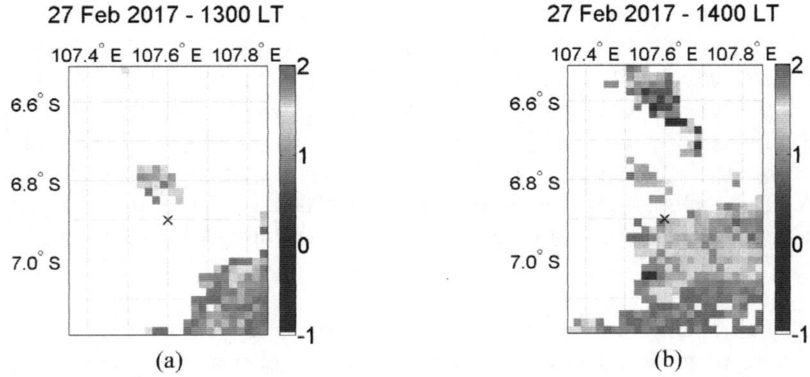

Figure 5. Horizontal distribution of convective cloud core on 27 February 2017, (a) 13.00 LT and (b) 14.00 LT. The cross sign is location of meteorological station ITB.

The convective activity shown in figure 5a exhibits an increase after 13.00 LT then weakened at night. The peak of convective activity around the observed area was detected three times as indicated by increasing I_{TBB} value in figure 5a, which lasted for 1-2 hours and then weakened. The convective cloud cores activity is shown in figure 5b. The contour has higher possibility identified as convective core when the values were close to zero. This figure also shows that the convective cloud cores were mostly detected in northern part of Bandung.

Convective activity started at 13.00 LT in northern and southern part of Bandung as shown in figure 5. The cloud evolution in this case was more prominent in northern part. The first cloud formed in the northern Bandung has been developed in situ for an hour. The precipitation of convective cloud was generated when the first convective cloud activity weakening in the study area, similar to the previous case on 27 February 2017. The precipitating convective cloud was identified by comparing the rainfall time series and the convective core activity in figure 6b. The rapid formation of new convective cloud in figure 5b was detected from the southern part of the preceding cloud and propagated southward with the convective core in the north. The new convective cloud formation on 6.6°S was also detected at 15.00 LT, approximately one hour after the first regeneration was detected. This second regeneration of convective cloud propagated southward then weakened at 17.00 LT. On the same location, a new convective cloud was generated and developed in situ then reached dissipation stage at night.

Convective clouds based on satellite imagery in figure 5 have radius 10-20 km with 1-3 hours lifespan. Similar characteristics of precipitating convective cloud in Bandung region were also shown by radar observations [3,4]. The time duration of new convective cloud needed to form relative to its predecessors varies between 30-120 minutes, but the new convective cloud formed in the south of predecessors have a shorter time lag than the new convective cloud that formed in northern part.

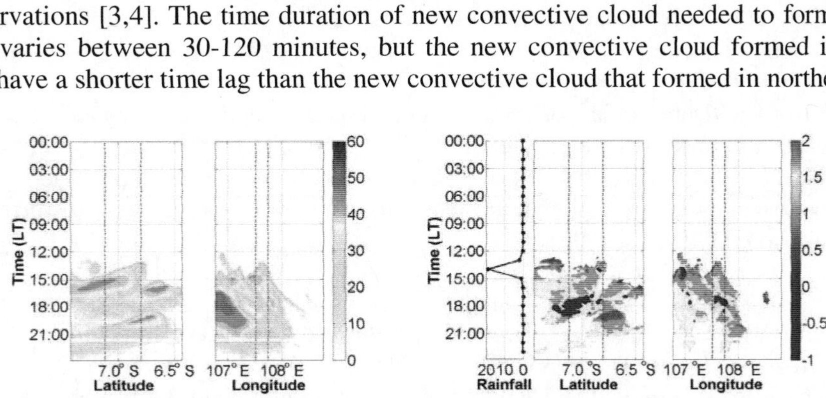

Figure 6. As in figure 4, but for 1 May 2017. (b) (middle) Time-latitude cross sections and (right) time-longitude cross sections of convective cloud core identification on 1 May 2017.

The convective cells that recently formed tend to be at southern part of predecessor cells with a linear regeneration pattern as shown in figure 6. This formation of linear shape at single regeneration of convective cloud can be explained by convective cell life-cycle [15]. The cold air mass from the northern basin meets the wind from southern Bandung producing forced updraft that strengthens positive buoyant force and forms convective cloud cell. Both cold air mass and valley wind are important factors in convective formation in Bandung region based on previous study [2]. Whereas the formation of subsequent convective cells is the result of circulation decay at the front of convective cloud predecessors that raises mixed layer with high potential temperature to the top layer [2].

3.3. Discussion

The formation of precipitating convective cloud in this study mostly started from northern part of Bandung, which corresponds to other studies that showed the location of initiation cloud formation at 6.7°S - 6.9°S. [3,4]. This result is in contrast to the previous studies that concluded a precipitating convective initiation in Bandung began at southern part and propagated northward [2]. The convective clouds that undergo regeneration can propagate southward, or develop in situ then extend southward, although it is possible that the regeneration of convective cloud formed in northern part. This difference of convective initiation may occur since convective cloud activity in this study is restricted to events with heavy rainfall, therefore there is a possibility that the convective activity generated outside criteria yields different result.

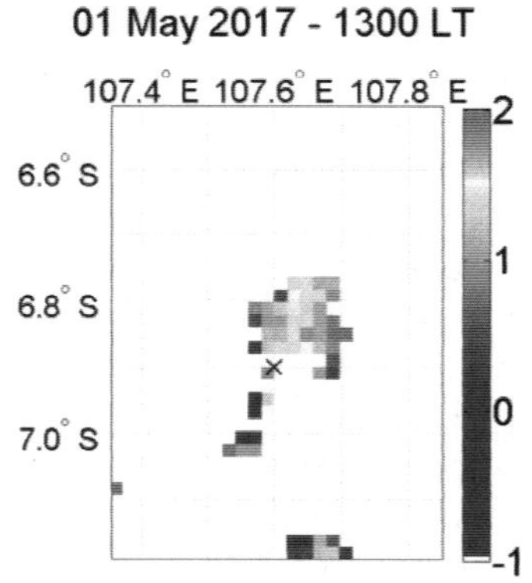

Figure 7. As in figure 5, but for 1 May 2017, 13.00 LT.

The precipitating convective clouds in the study area are not always formed by regeneration, it can also generate by decay-enhance convective activity or stratiform cloud. However, the event that affected by regenerating convective cloud has dominated most events. The rainfall events observed in this study mostly resulted from a decrease in convective cloud activity in diurnal active phase. This result corresponds to previous study [10] that also exhibits the frequency of precipitating cloud in February to March is dominated by deep convective cloud. This cloud type also reaches peak activity faster than other types. Some cases of precipitation that are affected by other clouds such as stratiform type are also detected despite its scarcity.

4. Conclusion

Daily rainfall with high intensity is dominated by multiple rain peaks indicating the presence of precipitating convective clouds. The regeneration of precipitating convective cloud was identified by RegI which proposed in this study. The RegI used high temporal resolution of rainfall data as a proxy to diurnal convective activity. The results of RegI during observation period denote that most of high intensity rainfall has a local coverage with single or multiple peaks. The cases obtained from the RegI were verified with satellite imagery, two sample cases for P95 at 27 February 2017 and P90 at 1 May

2017 were presented on this study. Both cases show the repeated convective cloud formation in the study area, therefore RegI can be used for the initial analysis to identify regeneration of precipitating convective cloud. By using satellite observation, the case on 27 February 2017 was shown to have stronger diurnal convective activity than the other cases. The weakening of convective activity in study area was sometimes accompanied by precipitation. Meanwhile, the daily rainfall observed had a higher maximum value and detected by more stations than the daily rainfall on 1 May 2017.

Both cases show the initial location of convective cloud formation in north Bandung (6.5°S - 6.8°S). The precipitating convective cloud in the study area has a maximum radius of 10-20 km with duration of 1-2 hours. These convective clouds also tend to propagate southward. Convective cloud regeneration is detected in both cases with time lag of new convective cloud formation to its predecessor varies up to 120 minutes. The duration of time lag is also related to location of new convective clouds relative to its predecessor, where new convective cloud in the southern part is formed faster than in the northern part. Heavy rainfall events in Bandung region are dominated by convective clouds that form repeatedly; hence daily rainfall observed is an accumulation of precipitating convective cells.

This study gives different conclusion compared to the previous study that denotes the convective initiation started from southern part of Bandung and propagate northward. This difference may occur since the case observed in this study only restricted to daily rainfall with the highest percentages. This study shows that the convective activity in Bandung region can produce regenerating convective cloud during diurnal active phase. Further analysis for this regeneration pattern is necessary to understand its contributing factors.

Acknowledgments

Authors thank Balai Besar Wilayah Sungai Citarum (BBWS Citarum) for providing rainfall data. We are greatly appreciated Japan Meteorological Agency (JMA) for providing Himawari 8/9 satellite data. Himawari 8/9 gridded data are distributed by Center for Environmental Remote Sensing (CEReS), Chiba University, Japan, available online at http://www.cr.chiba-u.jp/databases/GEO/H8_9/FD/. This research was supported by Institut Teknologi Bandung through Research, Community Service, and Innovation Program (P3MI) ITB 2018.

References

[1] Ramage C., 1968, *Mon. Wea. Rev.* **96** 365-70
[2] Oigawa M, Matsuda T, Tsuda T and Noersomadi, 2017, *J. Met. Soc. Japan* **95** 261-81
[3] Nugroho G A, Sinatra T, Trismidianto and I Fathrio, 2018, *IOP Conf. Ser.: Earth Environ. Sci.* **149** 012058
[4] Sinatra T and Nugroho G A., 2018, *IOP Conf. Ser.: Earth Environ. Sci.* **166** 012039
[5] Wu P, Hara M, Fudeyasu H, Yamanaka M D, Matsumoto J, Syamsudin F, Sulistyowati R and Djajadiharja Y S., 2007, *SOLA,* **3** 93-6
[6] Shibagaki Y, Shimomai T, Kozu T, Mori S, Fujiyoshi Y, Hashiguchi H, Yamamoto M K, Fukao S and Yamanaka M D., 2006, *Mon. Wea. Rev.* **134** 1682-96
[7] Tuttle J D and Carbone R E., 2004, *Mon. Wea. Rev.* **132** 192-207
[8] Dewi E S and Trilaksono N J., 2018, *AIP Conf. Proc.***1987** 020050
[9] May P T and Ballinger A., 2007, *Mon. Wea. Rev.* **135** 82-92
[10] Renggono F, Hashiguchi H, Fukao S, Yamanaka M D, Ogino S -Y, Okamoto N, Murata F, Sitorus B P, Kudsy M, Kartasasmita M and Ibrahim G., 2001, *Ann. Geoph.* **19** 889-97
[11] Nakazawa T., 1988, *J. Met. Soc. Japan* **66** 821-39
[12] Yuter S E and Houze R A., 1998, *Quart. J. Roy. Meteor. Soc.* **124** 53-99
[13] Yasuhito S, Hiroshi S, Takahiro I and Akira S., 2017, *Meteorological Satellite Center Technical Note,* no 62, pp 19-37.
[14] Dai A 2001 *J. Climate* **14** 1112-28
[15] Fovell R G and Tan P H 1998 *Mon. Wea. Rev.* **126** 551-77

Preliminary result of characteristic of convective cloud development observed by x-band meteorological radar in bandung basin

R A Syaraswati[1,3], N J Trilaksono[2,3] and G A Nugroho[4]

[1]Undergraduate Program in Meteorology, Faculty of Earth Sciences and Technology, Institut Teknologi Bandung, ID
[2]Atmospheric Science Research Group, Faculty of Earth Sciences and Technology, Institut Teknologi Bandung, ID
[3]Weather and Climate Prediction Laboratory, Faculty of Earth Sciences and Technology, Institut Teknologi Bandung, ID
[4]Center for Atmospheric Science and Technology, Indonesian National Institute of Aeronautics and Space, ID

rima.syaraswati@gmail.com

Abstract. The initial, mature, and dissipation stage changes in a convective cloud development have a short period of time so that the development stage analysis is needed by using data that has a high temporal resolution. The data that used are product CAPPI-V and CAPPI-Z from X-Band meteorological radar and also BBWS Citarum rainfall data from 13 stations that scattered in the area of Bandung basin. The selection of rainfall days is done to take the areas that have a strong convective system, and the result is that Sapan area has higher rainfall than 12 other stations on April 27th, 2017. The precipitation is caused by convective clouds. The initial stage of convective clouds begins with the existence of convergence phenomena in the afternoon and at 10 minutes before the rainfall event. It is also when there are maximum rainfall events on April 27th, 2017 which can indicate the formation of convective clouds. After the formation, the development stage occurs when there is an increase in reflectivity values on the radar and then convective cloud size starts to grow. When convective cloud size is dissipating, there is an identification of a decrease in reflectivity of the previous value in the evening. There is a new convective initiation next to the cloud that is in the dissipation stage and then joins to develops in a short time for about 30 minutes.

Keywords : convective, cloud, rainfall, reflectivity.

1. Introduction

Convective activity can generate the convective clouds that have the potential to cause rainfall event with clear diurnal intensity [1]. Mountainous area has complex topography that can influence diurnal intensity of convective activity in them that caused by valley winds [2]. Fovell and Tan [3] divide the stage of convective cells based on the presence of convective activity: cell formation, cell strengthening, and dissipation. Leary and Houze [4] also divided the convective cloud development stage which consists of the formation stage, strengthening convection, mature stage, and dissipation. Then Anggoro and Pramujo [5] statistically quantified the stage of convective cloud development in Bogor, Indonesia and found that convective clouds had a longer time to develop at the mature stage of 80 minutes. Nishiwaki et al. [6] analyze convective cloud structures that develop in mountainous areas in the North Kanto, Japan using X-Band meteorological radar. Jackson et al. [7] also apply the same radar instrument to see the evolution of convective clouds in southwestern England. Identification using radar X-band is an option because X-Band radar measurements provide more detailed information on the structure of precipitation in it [8].

One of the triggers of a strong convective system producing convective clouds is convergence. This convergence was analyzed statistically by the model simulation by [9] because the link between convergence and convection was not good enough in global model, with the result that it will produce a lot of bias. On the other hand, convective clouds are small-scale phenomena then it is difficult to be simulated or even predicted the development of the convective cloud model [10]. Therefore, the main purpose of this study is to determine the characteristic of convective cloud development using X-band radar observation is needed to provide the detailed information in the observed area so that it can develop the reliability of numerical weather prediction. The preliminary result of characteristic of convective cloud development stage is from the initial, mature, and dissipating in the Bandung basin in April 2017. The data used and the methodology followed for the radar data product processing are explained in the following section.

2. Data and Methods

The stages of the development of convective clouds studied in this research are in the Bandung basin area in April 2017 with coordinates 7.93°S-6.13°S and 106.86°E-108.66°E as in figure 1. However, before analyzing the stages, the selection of rainfall days will be conducted in this research. The type of rainfall that chosen is convective rainfall type, because the localized convergence occurs and resulting the large vertical velocities in convective rainfall type [11]. From the 13 rainfall stations that owned by BBWS Citarum that scattered in Bandung basin (figure 1), rainfall data will be taken which shows that rain occurred in all stations in the range of April 2017. Then the data will be percentage and we will take the dates that have the highest percentage of rainfall event in all stations. These rainfall data have 10 minutes temporal resolution. Data can be downloaded via http://bbwscitarum.com. After obtaining this date, it is necessary to validate whether this rainfall event is due to convective cloud.

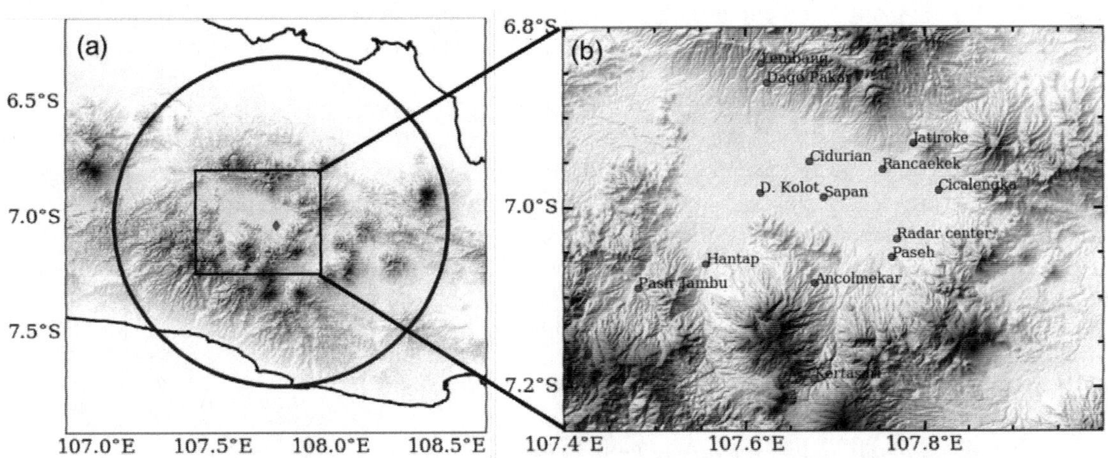

Figure 1. Radar coverage area (a) and distribution of rainfall stations in Bandung basin (b).

The next step is to identify the stage of convective cloud development using radar, after the convective clouds, which produce rainfall are identified. Identification of convergence and convective clouds will be examined using X-Band radar products in the form of Constant Altitude Plan Position

Indicator (CAPPI). CAPPI is a horizontal cut of a PPI volume scan at multiple elevation angles. The number of angles depends on the range and height of the CAPPI that want to produce [12]. In this study, 20 elevations from the range 0°-45° in radar PPI scanning will be used. The CAPPI radar is from the observational campaign that held by National Institute of Aeronautics and Space, Indonesia in April 2017. The center point of the X-Band radar is in Majalaya, with coordinate 7.035°S and 107.769°E. Then to identify the convergence, we use CAPPI-V that can show the radial velocity value on the radar. This CAPPI-V product has an image format with the extension. V, therefore the color definition of each pixel in this product will be done. The CAPPI-V pixel color definition according to [13] is:

$$\text{Velocity (ms}^{-1}\text{)= -14,346+ } \left(\frac{14{,}346\text{-(-14,346)}}{255}\right) \text{pixel value} \quad (1)$$

The threshold of radial velocity is +14,346 ms^{-1} and -14,346 ms^{-1}. If there is a pixel value of CAPPI-V that exceeds that threshold, then that value will not be defined. If the radial velocity value has an opposite value in which a positive value means away from the radar and a negative value means approaching the radar, then in an area where there is an opposite value converge, it indicates the occurrence of convergence. Convergence occurs when two opposing radial velocity values meet each other. CAPPI-V height for convergence identification is 2 km and 3 km, because at height of 1 km, the mountains will block the signal that captured by the X-Band meteorological radar. This obstacle is called ground clutter. The next step is identifying the convective cloud using reflectivity in CAPPI-Z which can show the reflectivity of the radar. CAPPI-Z pixel color definition is:

$$\text{Reflectivity (dBZ)= -32+ } \left(\frac{95{,}5\text{-(-32)}}{255}\right) \text{pixel value} \quad (2)$$

Clutter at height of 1 km can be removed in CAPPI Z (reflectivity) by using the Gabella clutter filter [14] module in the [15], so analysis at height of 1 km can be solved for radar reflectivity parameters without being blocked by clutter that identified on the radar. The removal of ground clutter at the height of 1 km has a threshold of $tr_1 = 9$, $np = 10$, and $tr_2 = 2.9$. Nishiwaki et al. [6] define the convective echo as a radar that has a reflectivity of more than 8 dBZ with an area of about 3.75 km. If there is a convective echo with a reflectivity of more than 17 dBZ, then the reflectivity is defined as a convective cell. This convective cell on the mature stage that have 37 dBZ reflectivity can be detected by their movements and can result a rainfall to form [16]. Convective initiation is defined by echo that have a reflectivity of 35 dBZ by [17].

Convective cloud development and dissipation stage will be monitored by defining the grid area based on that maximum reflectivity. After the grid area has been determined, the next step is averaging the area of the grid so that it produces one value at one height and one time. Then to average the next height at the same time, the grid area at that level refers to the initial height. There will be 11 values of reflectivity in one grid area. When time increases and reflectivity moves, it is necessary to redefine the area by looking back at the maximum reflectivity value and determining the grid area again. The results of the levelling reflectivity coverage area will be obtained in the form of time series analysis, namely time height. This information can be used to analyse when the cloud develops and dissipates at that time.

3. Result and discussion

3.1. Rainfall days selection based on stations data

The selection of rainfall days during April was taken from 12.00 - 18.00 LT, because the convective cloud was formed at 12.00 and weakened after 18.00 local time. It turns out that there are rainfall events every day in the study area in April 2017. Therefore, a filtering is needed using a percentage analysis of rainfall events to find out on what date the rainfall occurred with a large area. This percentage analysis will take the maximum value, with the result that the rain producing convective clouds in the entire station coverage.

From the bar diagram in figure 2 shows that 13 BBWS Citarum rainfall stations that scattered in the Bandung basin recorded that on April 22 and 27 there was rainfall event. From these two dates, on April 27th at the Sapan station recorded high rainfall values than the 12 other stations (table 1). This difference can be associated as a powerful meteorological phenomenon in the Sapan area, for example convergence, or an intense updraft that forms convective clouds in it. Rainfall on April 27th tends to predominantly move from South to North when viewed from the start time of the station that recording the rainfall events.

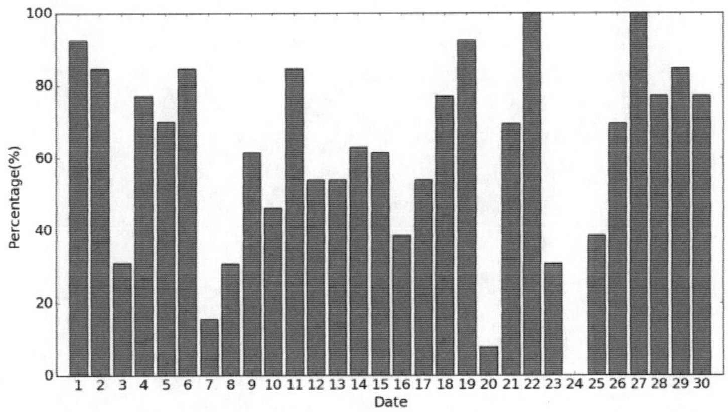

Figure 2. Bar plot of the percentage of rainfall events in April 2017 in 13 stations in Bandung basin

Table 1. Rainfall data (mm) from 13 stations that scattered in Bandung basin on 27th April 2017

Station	Rainfall value (mm)
Ancolmekar	22.5
Cicalengka	6
Cidurian	32
Pasir Jambu	9.5
Dago Pakar	2.5
Dayeuh Kolot	1
Hantap	3.5
Jatiroke	8
Kertasari	37
Lembang	1
Paseh	6
Rancaekek	22.5
Sapan	58

3.2. Analysis of convergence phenomena

The first convergence on April 27th exists at 12.30 LT (figure 3). Rainfall event in the Sapan area occurred at 13.30 LT. Before rainfall occurs (13.20 LT), there is a convergence around Sapan when viewed from the radial speed on the radar both at 2 km and 3 km (figure 4a and 4d). But when there is maximum rainfall (13.40 LT), there is a convergence that decreases speed detected on the radar.

Different things occur at the height of 3 km, where at this height convergence occurs only during the rain event. The radial velocity value is not greater than 2 km height when it rains.

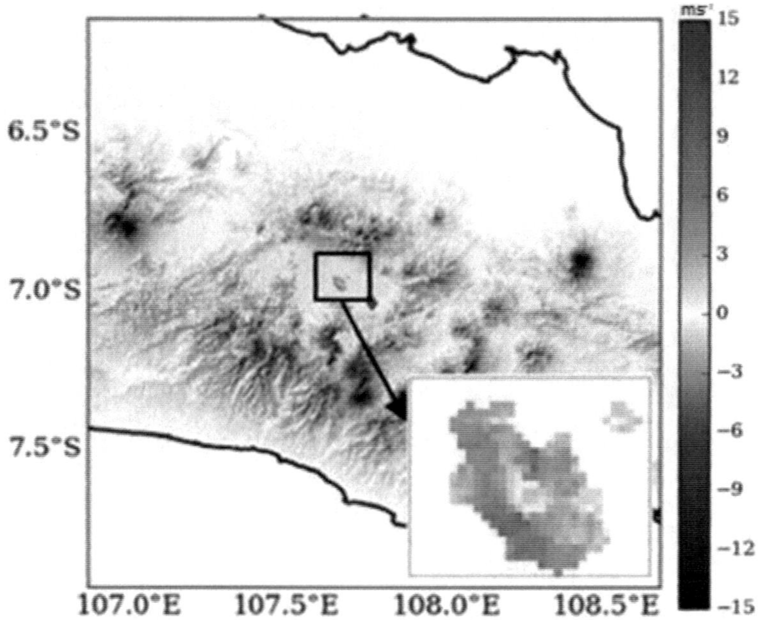

Figure 3. Spatial plot radial velocity of CAPPI-V at height 2 km at 12.30 LT

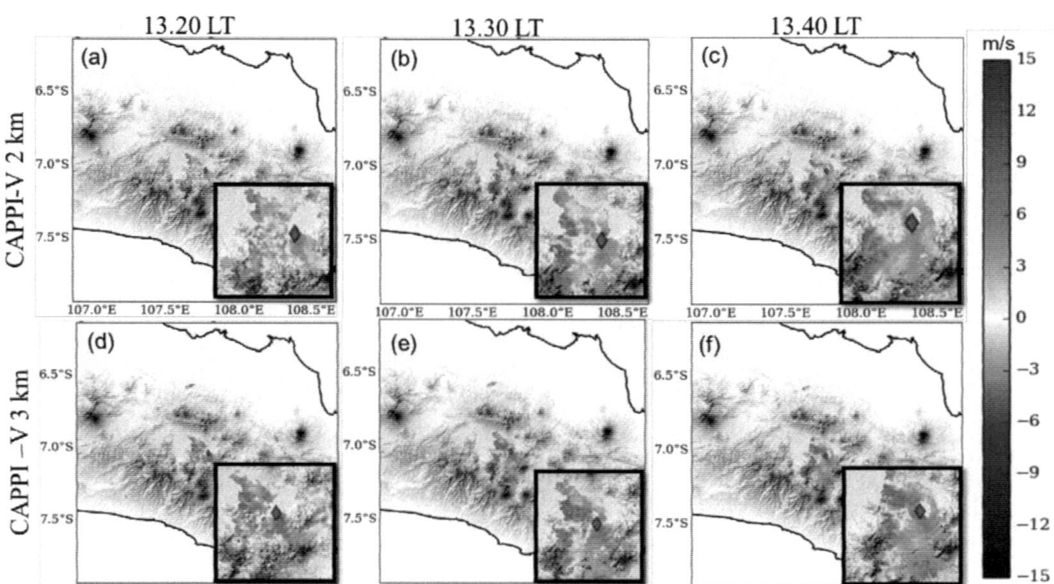

Figure 4. Spatial plot radial velocity of CAPPI-V at height 2 km (a,b,c) and 3 km (d,e,f)

3.3. Development and dissipation stage of convective clouds analysis

By monitoring the reflectivity movements, the analysis can be carried out using a time height plot as in figure 5 to see the development and dissipation process of the convective cloud. The time height is done by averaging the grids around the maximum reflectivity and follows the movement of the cloud, therefore it will produce the mean reflectivity of cloud. When it is related to the rainfall value recorded at the Sapan station, the developing convective cloud does not cause rainfall. This means that radar does not directly measure rain droplets, because the captured reflectivity is not necessarily immediately identified by the occurrence of rainfall. The increase in rainfall value at 13.30 LT was followed by a decrease at the mean value of radar reflectivity at the height of 5 km, then continued to decline until it reached a value of 16 dBZ in the event of maximum rainfall in the first peak (13.40 LT). However, there was a condition where at 13.50 LT, there was a significant decrease in rainfall for about 8 mm from the previous rainfall peak, then the rainfall increased again until it reached the second peak (14.10 LT) and decreased until 14.40 LT.

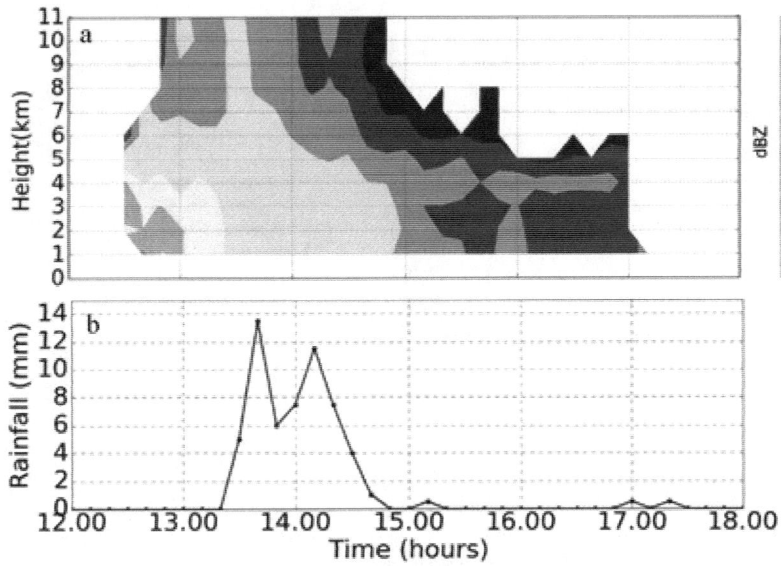

Figure 5. Mean reflectivity time height (a) and time series (b) plot of CAPPI-Z at 27th April at Sapan station

There is no difference at the mean reflectivity at 13.40 - 13.50 LT at the height of 1 - 7 km, although the process of decreasing reflectivity still occurs at 8 km height. But there was an increase in reflectivity at 14.00 LT at the height of 1-3 km, even though at that time the detected rainfall had increased. When rainfall reaches the second peak at 14.10 LT, the reflectivity decreases, and the height of the cloud begins to decrease. From the description above, the increase in rainfall value after experiencing a previous decrease is also supported by the identification of decreased reflectivity.

When analyzed with vertical cross section of reflectivity, the convection is initiated on the plain at 12.30 LT. At this time, there is a strong reflectivity of up to 40 dBZ at 3 km height and there are also other convective cells that have a value more than 20 dBZ. This reflectivity increased at 12.40 LT. The reflectivity, which originally had a value of 40 dBZ is strengthened to have a value of 48 dBZ. Strong reflectivity is detected in two places, at 1 and 3 km height. Increasing of the reflectivity value and identification of this wider area can be associated as convective cloud growth. This is consistent with the statement that if the higher reflectivity value identified on the meteorological radar, there is an indication of convective clouds which will potentially produce rainfall [18].

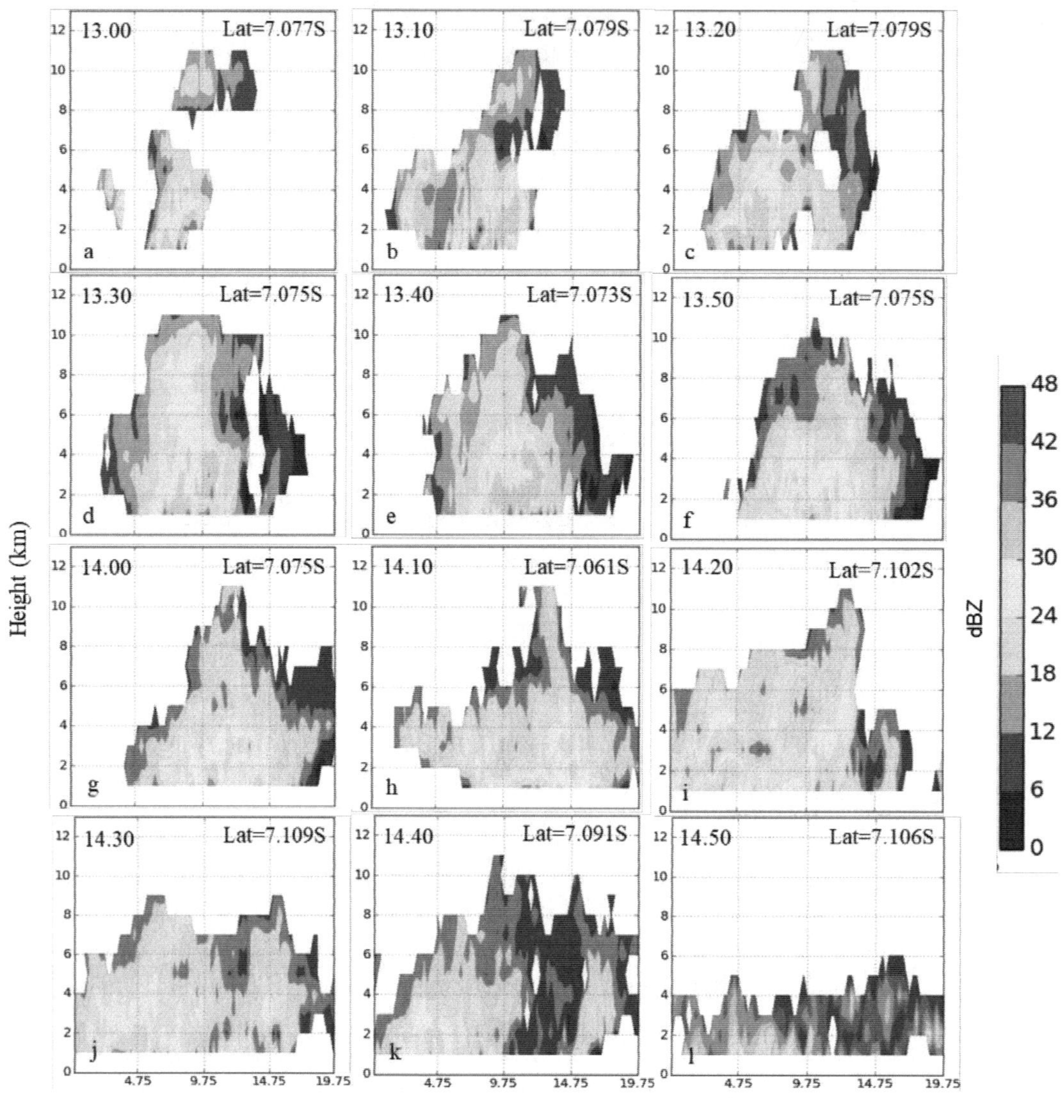

Figure 6. Vertical cross section plot of CAPPI-Z radar reflectivity. The position longitude that used from this v-cut is 107.64E – 107.81E

The convective cloud then continues to grow to a height of 11 km and has a large area, followed by a reflectivity value of about 30 dBZ, which increases to a height of 10 km (figure 6c). But at 13.30 LT, there was a significant decrease in reflectivity at low altitude (2-4 km), although a reflectivity of 27 dBZ was still detected up to a height of 10 km. This decrease in reflectivity is supported by the occurrence of rainfall event at that time in the Sapan area. This shows that the convective clouds identified are included in the vertical cloud type and cause rain (small cumulonimbus cloud). High reflectivity values are found at an altitude of 1 km, which indicates that the reflectivity is moving down. Then when maximum rain occurs at the first peak (figure 6e), the decrease in reflectivity is quite clearly identified, it can be seen that the reflectivity that previously had a value of 27 dBZ at 10 km height is dropped significantly to a height of 5 km. There is still a high reflectivity value of 40 dBZ at an altitude of 1 km, which indicates that rain is still occurring in this area and clouds enter the dissipation stage.

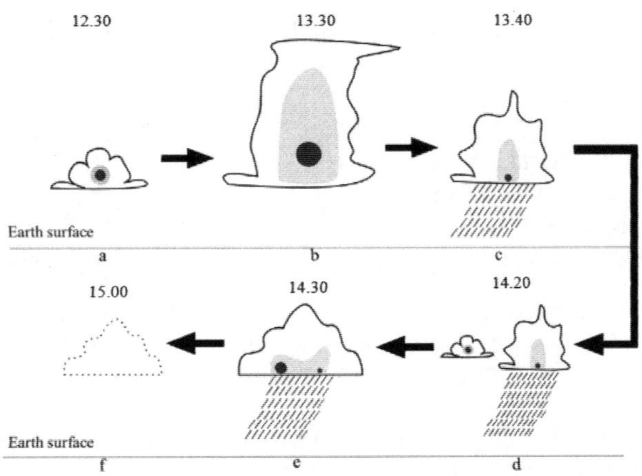

Figure 7. Schematic illustration of characteristic of convective cloud development in Bandung basin (LT). The gray color is the higher reflectivity and black point is the strongest reflectivity.

When the rain event reaches the second peak, a new reflectivity is identified to the west of the convective cloud that is being dissipated. It is indicated that there is convective cloud regeneration at this time. This phenomenon is likely the same that explained by [19] about the regeneration of diurnal cycle of convective cloud formation. Here, the cloud begins to develop followed by an increase in reflectivity but within a period of 30 minutes, shorter than the previous convective cloud. This makes convective clouds have a fairly short development time because rainfall event is still recorded when the second convective cloud grows, therefore the clouds enter the dissipation stage more quickly. The stage of dissipation in clouds causes a significant decrease in reflectivity until the convective echo is not identified is occurred at 14.50 LT.

The previous vertical analysis can be simplified in schematic illustration as shown in figure 7. In the initial stage, there is a high reflectivity value which is included as convective cells (figure 7a), then the convective cell strengthens, so that the reflectivity increases so that the size of the cloud enlarges to mature stage (figure 7b). When the rain occurs, a decrease in reflectivity occurs which indicates that the cloud has been dissipated (figure 7c). However, when it entered the stage of dissipation a new convective cloud initiation is placed next to the cloud which is in the dissipation stage and then joins and develops in a short time (figure 7e). The dissipation stage ends when no reflectivity is identified in the area, which indicates that the cloud is no longer exist.

4. Conclusion

The characteristic of cloud convective development like initial, mature, and dissipation stage in the Bandung basin can be identified using X-Band meteorological radar. It has an initial stage which can identify the existence of convergence phenomena at 12.20 LT and at 10 minutes before the rainfall event (13.20 LT) also when there are maximum rainfall events (13.40 LT and 14.10 LT) on April 27th, 2017 which can indicate the formation of convective clouds. After the formation, the development stage occurs when there is an increase in reflectivity values on the radar and then convective cloud size starts to grow at 12.30-13.30 LT, and when there is a rainfall event, the convective cloud size is dissipating, and there is an identification of a decrease in reflectivity of the previous value at 14.10-17.30 LT, therefore it is entered the dissipation stage. Convective cloud initiation occurs next to the cloud which is in the dissipation stage and then joins and develops in a short time.

Acknowledgement

Rainfall data in this study was obtained from BBWS Citarum and X-band meteorological radar data obtained from an observation campaign conducted by the Center of Atmospheric Science and Technology of the National Institute of Aeronautics and Space in 2017. This research was supported by

Institut Teknologi Bandung through Research, Community Service, and Innovation Program (P3MI) ITB 2018.

References

[1] Sato T and Kimura F *2005* Diurnal cycle of convective instability around the central mountains in Japan during the warm season *J. Atmos. Sci.* **62** 1626–36

[2] Oigawa M, Matsuda M, T Tsuda and Noersomadi 2016 Coordinated observation and numerical study on a diurnal cycle of tropical convection over a complex topography in West Java, Indonesia *J. Meteor. Soc. Japan* **95** 261–281

[3] Fovell R G and Tan P-H 1998 The temporal behavior of numerically simulated multicell-type storms part II: The convective cell life cycle and cell regeneration *Mon. Wea. Rev.* **126** 551–577

[4] Leary C A and Houze Jr R A 1979 The structure and evolution of convection in a tropical cloud cluster *J. Atmos. Sci.* **36** 437–457

[5] Anggoro M D and Pramujo B 2018 Study of convective cloud lifetime and movement using radar image and ECMWF *Int. J. Adv. Sci. Eng. Inf. Techno* **8** 988–997

[6] Nishiwaki N, Misumi R, Shimizu S, Maesaka T, Iwanami K, Sakurai N, Maki M, Suzuki S, Kato A and Yamaji A 2013 Behavior and structure of convective clouds developing around a mountainous area observed by stereo photogrammetry and ka-band and x-band radars: case study of Northern Kanto, Japan *J. Meteor. Soc. Japan* ***91*** 609–626

[7] Jackson R, French J R, Leon D C, Plumer D M, Trapp S L, Blyth A M and Korolev A 2018 Observations of the microphysical evolution of convective clouds in southwest United Kingdom Atmos. Chem. Phys. Discuss. *(Preprints)*

[8] Lengfeld K, Clemes M, Münster H and Ament F 2014 Performance of high-resolution x-band weather radar networks – the PATTERN example *Atmos. Meas. Tech.***7** 4151–4166

[9] Birch C E, Marsham J H, Parker D J and Taylor C M 2014 The scale dependence and structure of convergence fields preceding the initiation of deep convection *Geophys. Res. Lett.* **41** 4769–4776

[10] Kanada S, Nakano M, Hayashi S, Kato T and Nakamura M 2011 Reproducibility of maximum daily precipitation amount over japan by a high-resolution non-hydrostatic model SOLA **4** 105–108

[11] Tjasyono and Bayong H K 2007 Mikrofisika Awan dan Hujan Indonesia National Meteorological, *Climatological and Geophysical Agency* p 199-200

[12] Geçer, C 2005 Training Course on Weather Radar Systems Module D: Radar Products and Operational Applications WMO and CIMO p 24

[13] Sinatra T and Noersomadi 2015 Pemanfaatan Transportable Radar Cuaca Doppler X-Band Untuk Pengamatan Awan *Berita Dirgantara Vol.* **16** No. 2 Desember 2015p 91-97

[14] Gabella M and Notaprieto R 2002 Ground clutter characterization and elimination in mountainous terrain *Proceeding of ERAD* 305–311

[15] Heistermann M, Jacobi S and Pfaff T 2013 Technical note: an open source library for processing weather radar data (wradlib) *Hyydrol. Earth Syst. Sci* **17** 863–871

[16] Kober K and Tafferner A 2009 Tracking and nowcasting of convective cells using remote sensing data from radar and satellite *Meteorologische Zeitschrift* **1** 75–84

[17] Mecikalski J R and Bedka K M 2006 Forecasting convective initiation by monitoring the evolution of moving cumulus in daytime goes imagery *Mon. Wea. Rev.* **134** 49–78

[18] Steiner M, Houze Jr R A and Yuter S E 1995 Climatological characterization of thre e dimensional storm structure from operational radar and rain gauge data *J. Appl. Meteor* **34** 1978–2007

[19] Fitriani R 2018 Regenerasi awan konvektif penghasil hujan di wilayah Bandung *Thesis of Master Graduate of Earth Science Bandung Institute of Technology* p 40

Identification of Mesoscale Convective Complex (MCC) Over Bangka Belitung Area

R W Saragih, A H Hestiwi, F N Indah Sari and PA Winarso

State College of Meteorology Climatology and Geophysics Jakarta, Indonesia

rinosaragih14@gmail.com

Abstract. Mesoscale Convective Complex (MCC) is a mesoscale cloud convective system characterized by large, long-lived and quasi-circular pattern. The existence of MCC may trigger extreme weather events such as heavy rains which cause floods and landslides around the center of interest area. This paper aims to identify the presence of a storm cloud system which is observed as the MCC. For that purpose, this study used Himawari-8 satellite infrared channel data. Case of the study is the rain event in Bangka Belitung on 8 February 2016 which recorded 183.9 mm rainfall a day. Satellite data was processed using image processing software by applying MCC cloud system identification algorithm based on its physical characteristics, i.e. peak cloud temperature, coverage area, central point of cloud, cloud eccentricity and ellipse fitting, and lifetime of the cloud system. The final result of this study is able to yield the coordinates of latitude and longitude of MCC location, which are MCC cloud systems appeared around Bangka Belitung. This result confirmed that the occurrence of heavy rain on February 8th, 2016 in Bangka Belitung was caused by the presence of MCC around the interest area.

Keywords : MCC, quasi-circular, floods and landslides.

1. Introduction

Mesoscale Convective System (MCS) is a complex convection system that grows up into a larger scale of convection system and normally lasts for several hours or more [1]. Mesoscale Convective Complex (MCC) is one of the MCS characterized by large, long-lived and quasi-circular pattern. Generally, MCC is nocturnal and tends to cause severe weather such as heavy rain, strong wind, and thunderstorm in a large coverage which may cause floods even landslide around a point of interest. This group of convective system is triggered by low level jet causing larger cumulonimbus coverage (hundred kilometers in approximation) and has more than three hours lifetime cycle [2].

Researchers have shown that the MCSs produce huge precipitation at the Midwest and the Great Plains when growing season [3]. Previous research on South Afrika [4] and on South America [5] mentioned that the impact of MCC are quite dangerous, which include heavy rain, strong winds, hail, tornadoes, and may result in flooding if it occurs for a long enough duration. The rainfall effects of MCC can be found not only on MCC area but also on its surrounding area [6]. On February. 8th and 9th, 2016 in Bangka Belitung region, there were heavy rains with an intensity of 183.9 mm a day. Rain on February 8th 2016 occurred with a long duration and had caused flooding in the region, therefore it can be expected that on the day of the heavy rain there was an MCC phenomenon. This paper aims to identify and confirm the existence of a storm cloud system that is monitored as an MCC cloud system using Himawari-8 satellite infrared channel data. Identification of the existence of MCC is done by processing the satellite data using image processing software and algorithms that are built based on the characteristics of MCC referring to the results of research conducted [7].

2. Data and Methods

The area that became the focus of the research is Bangka Belitung and its surroundings with geographical location at 104^0 BT and -1^0 LS to 106^0BT and -3^0LS. This area covers the entire Province of Bangka Belitung.

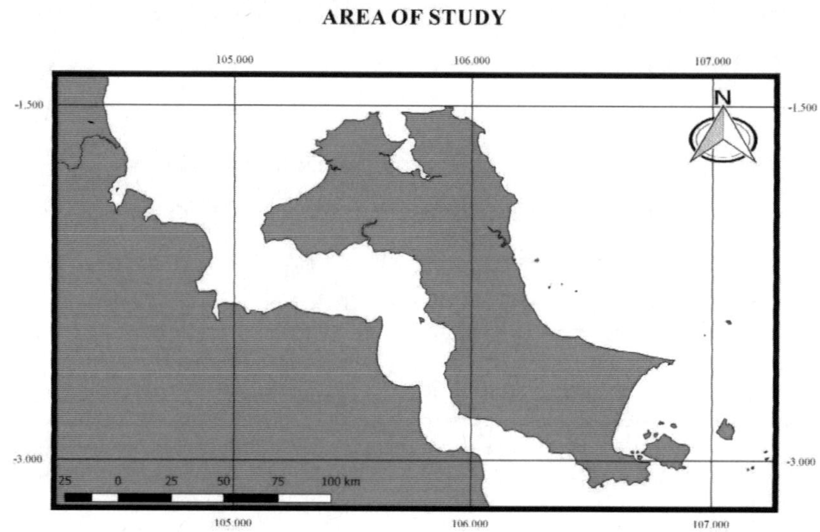

Figure 1. Map of Research Area

The data used in this study were Himawari-8 infrared channel (IR1) satellite imagery on 8 February 2016 from Kochi University website (hourly data) and for verification, rainfall data were taken from Pangkal Pinang Meteorological Station observation on 8 February 2016 and 9 February 2016. The method used in the data processing process to identify the existence of MCC is by processing satellite data using image processing software and algorithms that are built based on the characteristics of the MCC cloud system [7,8,9]. The methods are :

1. First, satellite data in portable gray map format (PGM) with a scale of 0-255 were converted to cloud top temperature data in Kelvin.
2. Determine the blanket clouds with the characteristics of the MCCs [7]. The value of the cloud peak temperature <241 K signed as a coverage cloud (SA) [7,8] and 217 K as a cloud core (IA) [7,8]. Temperature which every grid met the requirement will be regarded as 1 or 0 (changing data into binary data form)
3. Find the area that meets the temperature requirements (procedure 2, the pixel that has a binary value "1") by counting the number of pixels that are connected to 4 surrounding grid. Select area requirements by selecting SA area that has an area of $\geq$ 100,000 km^2 [7,8] or if it is converted into pixels, there are 3305 pixels, because Himawari-8 satellite data has a resolution of 5 km, and IA area that has $\geq$ 50,000 km^2 [7,8] or around 1652 pixels.
4. Find the center point of the selected area from the procedure 1-3 where the center point is the center of mass of the interconnected area, calculated by this formula [10].

$$X_0 = \frac{\sum_1^N Xi}{N} \; and \; Yo = \frac{\sum_1^N Yi}{N} \quad (1)$$

Where:

Xi	: pixels to-i at X-axis
Yi	: pixels to-i at Y-axis
X_0 and Yo	: *center*
N	: total of pixels

5. Determine the MCC based on the eccentricity value using the Machado threshold method with a value of Mad is 0.7 [7], and added with the Ellipse Fitting method with a limit of ≤ 0.8 to capture cloud clusters that have relatively sloping shapes [8].
6. Determine and identify MCC after the conditions in the procedure 1-5 above are fulfilled based on the active life of the cloud system for at least 6 hours.

3. Results

On February 8th, 2016, Pangkal Pinang Meteorological Station reported that moderate to heavy rainfall had occurred in Bangka Belitung area from February 8th, 2016 at 9:00 UTC until February 9th, 2016 at 07.00 UTC. Accumulated 24-hours rainfall on February 8th, 2016 was 183.9 mm. To find out the cloud system that was formed on that day, verification of the causes of rain was done using Himawari-8 satellite image data. On the same day, the area of the cloud that spread above Bangka Belitung region was observed and almost formed a circle pattern that lasted for more than six hours starting at 21:00 UTC around Bangka Belitung area.

Table 1. The output of satellite data processing by Image Processing Software and MCC characteristic algorithm

Month	Day	Time	Core				Coverage			
			Eccentric -ity	Pixel	Lon	Lat	Pixel	Lon	Lat	Eccen tricity
2	8	21	0,865	5478	105,5	-1,04	8708	105,4	-0,70	0,666
2	8	21	0,878	2213	108,9	-5,24	9745	109,3	-5,36	0,876
2	8	22	0,979	8426	92,50	-2,63	27226	93,36	-2,95	0,795
2	8	22	0,948	2905	92,73	-12,39	7529	93,52	-12,70	0,789
2	8	22	0,912	5723	105,5	-1,00	17469	106,8	-2,73	0,72
2	8	23	0,980	8367	93,95	-5,10	28342	93,33	-2,84	0,882
2	8	23	0,985	3225	92,81	-12,58	7941	93,67	-12,72	0,733
2	8	23	0,994	5737	105,7	-1,08	16359	106,9	-2,38	0,636
2	9	0	0,949	7432	93,84	-5,29	28347	93,14	-2,96	0,856
2	9	0	0,910	3137	92,90	-12,72	8014	93,62	-12,75	0,61
2	9	0	0,965	6164	105,7	-1,02	10003	105,5	-0,65	0,638
2	9	1	0,926	2362	91,93	-3,58	28304	93,01	-2,70	0,863
2	9	1	0,993	2962	93,04	-12,85	6073	92,81	-13,02	0,471
2	9	1	0,882	6635	105,8	-1,01	11102	105,6	-0,70	0,649
2	9	2	0,920	2156	93,03	-13,18	5792	92,72	-13,14	0,534
2	9	2	0,927	6507	105,9	-1,06	17953	107,3	-2,29	0,727
2	9	3	0,954	2056	92,90	-13,41	5308	92,74	-13,19	0,68
2	9	3	0,919	6162	106,4	-1,23	19388	107,2	-2,09	0,817
2	9	4	0,971	6030	93,63	-6,65	15573	93,55	-6,03	0,95
2	9	4	0,946	2069	92,89	-13,52	4791	92,86	-13,37	0,684
2	9	4	0,934	6014	106,7	-1,42	20411	107,4	-2,25	0,816
2	9	5	0,988	6003	93,54	-6,63	15142	93,56	-6,171	0,958
2	9	5	0,968	1969	92,94	-13,64	4734	92,96	-13,236	0,714
2	9	5	0,993	6072	106,9	-1,58	26946	107,8	-1,036	0,766
2	9	6	0,998	6059	93,10	-6,71	15765	93,52	-6,352	0,956
2	9	6	0,985	1871	92,99	-13,73	4738	93,18	-13,20	0,645
2	9	6	0,805	6086	106,9	-1,41	28088	107,7	-1,05	0,803
2	9	7	0,883	6042	106,9	-1,22	22459	106,7	0,48	0,818

The characteristics of cloud life cycle and cloud cluster shape patterns observed were identified as the characteristics of MCC cloud clusters. The identification of MCC cloud events around Bangka Belitung region as a cause of rain was done using Himawari-8 satellite image data, which were processed by using image processing software with an algorithm adapted to MCC cloud characteristics. This process produced several identifications of cloud clusters with blanket and cloud core area in accordance with MCC characteristics. These results can be seen in table 1 below.

Based on table 1, MCC has been detected since February 8th, 2016 at 21.00 UTC until February 9, 2016 at 07.00 UTC. The largest coverage of cloud cluster occurred at 06.00 UTC which covered 28088 pixels or approximately equal to 842.640 km^2 and the widest core cloud occurred at 01.00 UTC, covered 6635 pixels or 199.050 km^2. However, not all cloud cluster that fulfilled the characteristics of MCC's cloud coverage and core cloud could be categorized as MCC, because some of the observed cloud clusters did not grow and last more than six hours.

One of the cloud clusters that has active life cloud system more than six hours is displayed as yellow boxes in the table, which then identified as the MCC because its clusters fulfilled all the characteristics of MCC. This cloud cluster lasts from 21.00 UTC until 07.00 UTC, which mean the cloud cluster that caused heavy rainfall in Bangka Belitung start to grow at 21.00 UTC. The result of the identification can be seen in figure 2.

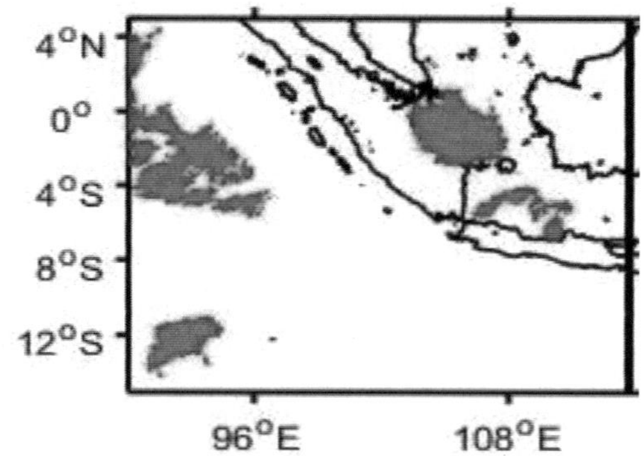

Figure 2. Satellite image of cloud cluster at 21.00 UTC

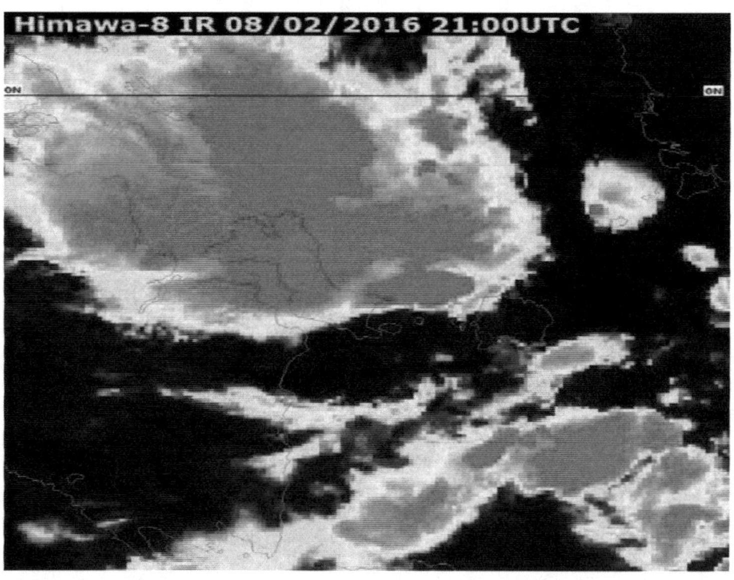

Figure 3. Himawari-8 satellite image on February 8th, 2016 at 21.00 UTC which is the beginning of MCC over Bangka Belitung which caused heavy rainfall on February 8-9, 20

From figure 2 and figure 3, it can be seen that not all cloud clusters that have the coverage more than 100.000 km^2 will also have the core more than 50.000 km^2. There are only a little number of cloud clusters that fulfilled the MCC's characteristics. Figure 2 is a Himawari-8 satellite image at 21.00 UTC which shows a cloud cluster with cloud coverage and core cloud that fulfilled the characteristics of MCC, where the pattern almost forms a circle, around Bangka Belitung.

Furthermore, to identify the cloud system that already fulfilled the MCC's characteristic, which located in 105.615 E, 0.703 S and 105.884 E, 1.011 S, its life cycle should be analyzed. The life cycle of the MCC can be seen in figure 4.

a

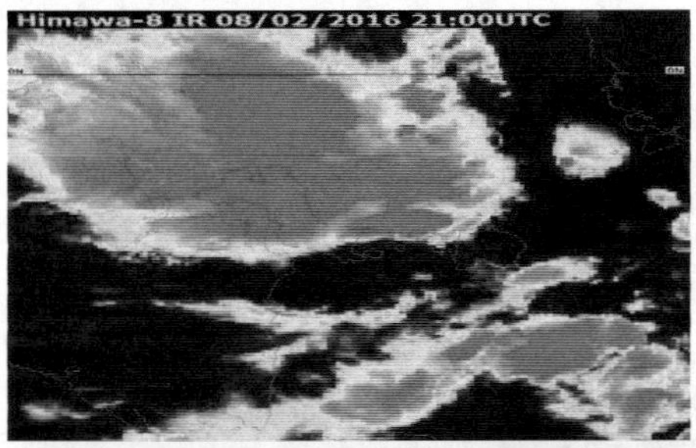

b

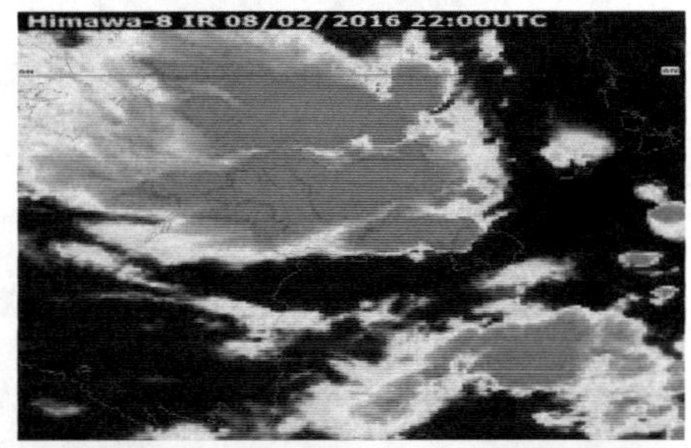

c

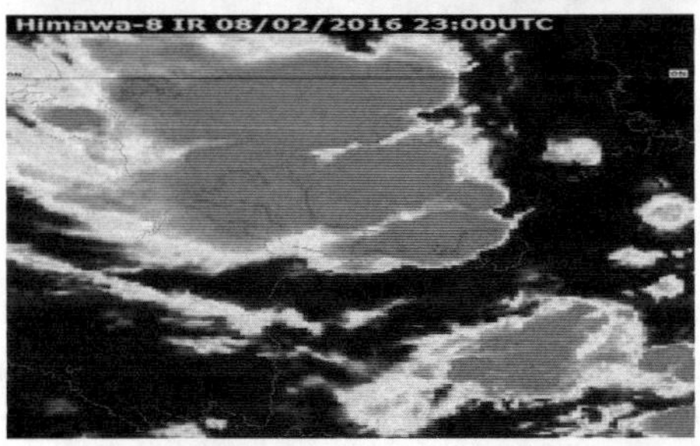

d

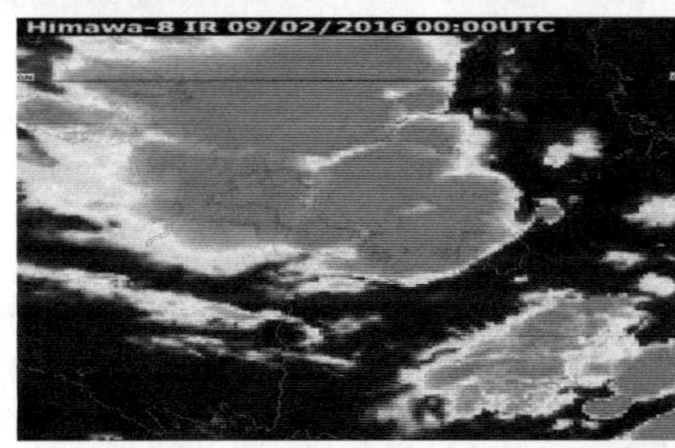

e

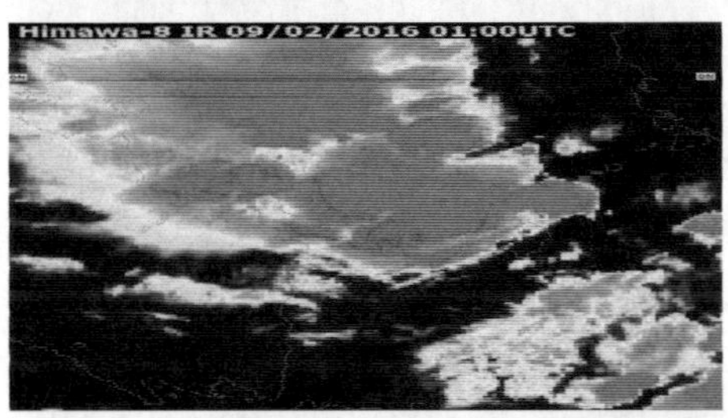

f

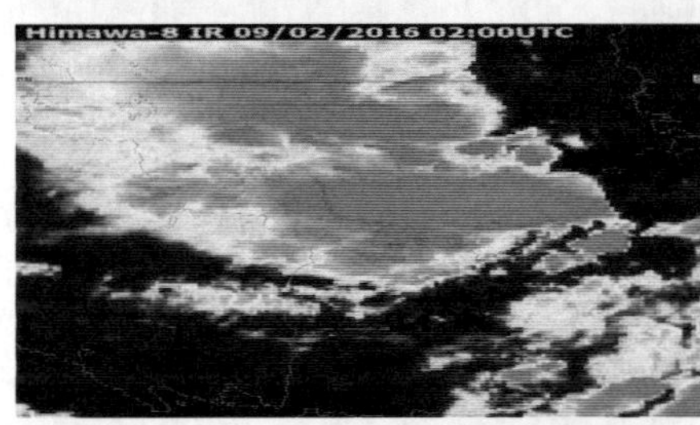

g

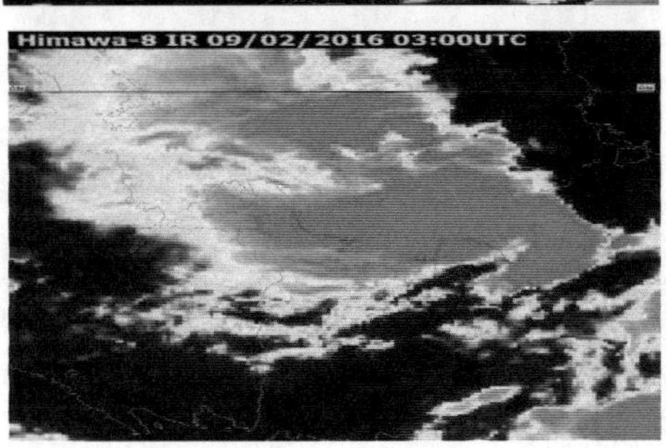

h

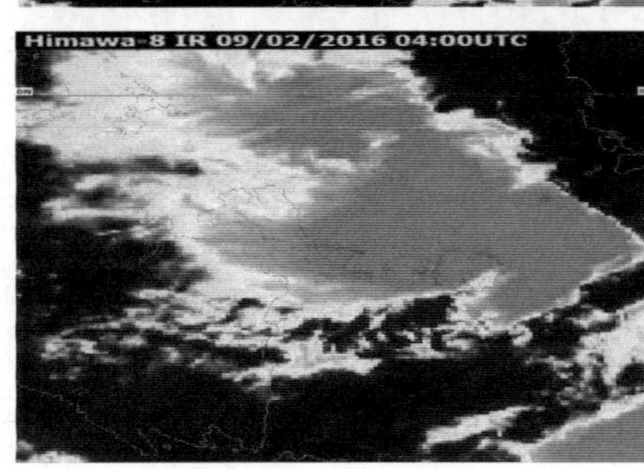

i

j

k

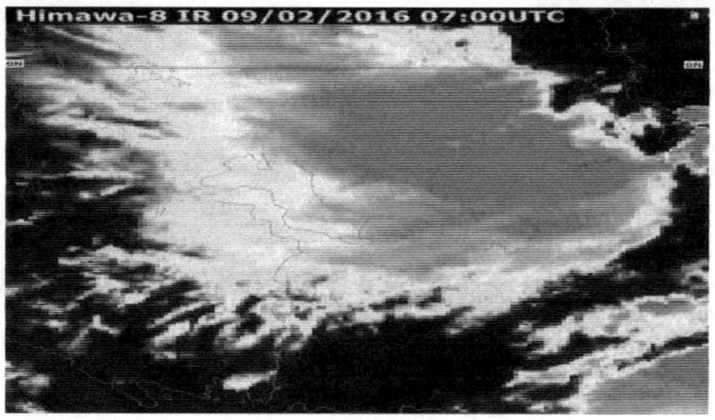

Figure 4. Output of Image Processing Software and satellite image of Himawari-8 at (a) 21.00 UTC (b) 22.00 UTC (c) 23.00 UTC (d) 00.00 UTC (e) 01.00 UTC (f) 02.00 (g) 03.00 (h) 04.00 (i) 05.00 (j) 06.00 (k) 07.00

Figure 4 shows that cloud cluster of MCC continued to develop from 21.00 UTC on February 8th, 2016 until its deployment on February 9th, 2016 at 07.00 UTC. The circle pattern was formed on February 8th, 2016 at 21.00 UTC (figure 3a), the development and the growth phase occurred since 22.00 UTC until 07.00 UTC on February 9th, 2016.

The MCC's location which being monitored from satellite image is then verified using synoptic data for every three hours using rainfall data from Pangkal Pinang Meteorological Station. The rainfall data can be seen in table 2 as below.

Table 2. 3-Hourly rainfall data from Pangkal Pinang meteorological station [11]

Date/Hour	03.00	06.00	09.00	12.00	15.00	18.00	21.00	00.00	Total Rainfall
8	4,5	53,8	56,5	47,3	1,3	1,1	3,4	16,0	**183,9**
9	22,8	0,8	TTU	0,0	0,0	0,0	0,0	0,0	**23,6**

Table 2 shows that rain had occurred even before 21.00 UTC on February 8th, 2016, but the cloud cluster that showed the formation of MCC has not been seen on the satellite processed data. Rain still continues until 07.00 UTC on February 9th, 2016 but the cloud cluster that forms a circle pattern has begun to split based on Himawari-8 satellite data which was the MCC's dissipation period.

4. Conclusion

Based on the processing of infrared channel of Himawari-8 satellite data using image processing software and algorithm that has been adapted to the characteristics of MCC, it can be shown that MCC cloud clusters have been formed with a life cycle more than six hours over Bangka Belitung area. This has verified that moderate to heavy rainfall with thunderstorms occurred on February 8th, 2016 from

21.00 UTC until February 9th, 2016 at 07.00 UTC which caused a total 24-hours rainfall of 183,9 mm over Bangka Belitung area. This condition is consistent with the previous research [4,5,8] which stating that MCC contributes significantly to rainfall almost throughout the MCC area. It is also proved that the presence of MCC as a cloud system that has a life cycle more than six hours with the possibility of integrating with other cloud system is related to the occurrence of heavy rainfall in a long duration of time around its system as happened in Bangka Belitung area.

Acknowledgements
We gratefully thank Dr. Paulus Agus Winarso as a mentor for this research.

References
[1] Orlanski I., 1975, *Bul. Amer. Meteorology Society*, **56,** 527-530
[2] Ismanto H., 2013, *Megasains*, **4,** No 2, 74-81.
[3] Ashley W S, Mote T L, Dixon P G, Trotter, S L, Powell, E J, Durkee, J D and Grundstein A J., 2003, Distribution of Mesoscale Convective Complex Rainfall in the United States (*Semanticsscholar*)
[4] Blame R C, dan Reason C J C., 2011, *American Meteorology Society,* **25,** 753-766
[5] Durke J D, Mote T L, and Shepherd J M., 2009, *American Meteorology Society,* **22,** 4590-4605
[6] Trismidianto, 2012, Study of Mesoscale Convective Complexes in the Indian Ocean to Weather Pattern above Sumatra (*Magister Thesis, FITB ITB*)
[7] Maddox R A., 1980, *Bull. Amer. Meteor. Soc.,* **61** 1374-1387.
[8] Ismanto H., 2011, Characteristics of Mesoscale Convective Complexes over Maritime Continent (*Magister Thesis, FITB ITB*)
[9] Miller Dand Fritsch J M., 1991, *American Meteorology Society*, **119,** 2978-2992
[10] Carvalho L M V and Jones C., 2001, *J. of Applied Meteorology American Meteorology Society* **40**, 1683-1701
[11] Trismidianto, Yulihastin E, Satyawardana H, Nugroho J T and Ishida S., 2017, *IOP Conference Series Earth and Enviromental Series*, **54,** 012027

Analysis of ultraviolet index, ultraviolet B insolation, and sunshine duration at Bandung in year 2017

S. Hamdi

Center for Atmospheric Science and Technology, National Institute of Aeronautics and Space, Bandung, Indonesia

saipulh@yahoo.com

Abstract. Ultraviolet index, ultraviolet B insolation, and sunshine duration in 2017 have been analyzed at Bandung, Indonesia (106.6 W and 6.9 S). The daily maximum ultraviolet B insolations have been grouped in ten-daily sequence following BMKG's method in determining of dry/rainy season. Ultraviolet index classified following small modification WHO's classification. Sunshine durations are clustered in 3 levels of exposure. It is found the average of sunshine duration in range 4.1-8 hours/day has 50% occurrence in 2017, meanwhile ultraviolet index more than 10 have been recorded in 46% occurrence in 2017, even of average of maximum ultraviolet B insolation is about 1.22 W/m^2, and maximum intensity on 2017 is 2.17 W/m^2. Both of ultraviolet index and maximum ultraviolet B insolation have a similar pattern on 2017 and the minimum value is on aphelion.

Keywords : ultraviolet, sunshine duration, aphelion.

1. Introduction

As a main source of energy in the environment, solar emits energy in the radiative electromagnetic wave and known as solar radiation. Solar electromagnetic wave is a polychromatic wave which its content is very wide spectrum of wavelength i.e. ultraviolet spectrum. The wavelength of ultraviolet spectrum is 100-400 nm. This spectrum is divided to be three sub_spectrums: ultraviolet A (315-400 nm), ultraviolet B (280-315 nm), and ultraviolet C (290-400 nm). The ultraviolet B radiation is used in this paper, even of only small percentage from wide spectrum but has significant effects to environment ecosystem, human health, and some processes in atmosphere [1]. Some medical effects to human are increasing of risk of skin cancer [2,3], cataracts [4,5] and human immune system [6,7].

In the other word, global radiation is also defined as incoming solar radiation (insolation). The sum of time for which the direct solar radiation exceed 120 W/m^2 during a given period is defined as sunshine duration [8]. Sunshine duration is measured in one cycle of the length of day (hours/day) or time duration of solar at horizon. By using sunshine duration data at LPD Sumedang on period 1999-2013, it is found the distribution of sunshine duration at Sumedang is depend on the season [9], it is mostly 0-1 hour/day on DJF, 6-7 hours/day on MAM, 9 hours/day on JJA, and distributed smoothly on SON. At cloudy condition, humid, and rainy all day sunshine duration mostly very short, as a characteristic of tropical rain region.

The International Commission on Non-Ionizing Radiation Protection (ICNIRP) in collaborative work with WHO and UNEP has recommended using of Solar Global Ultraviolet Index in order to estimate and conveying information to open public simply about ultraviolet irradiance level which can reach earth's surface [10]. Ultraviolet index is calculated quantitatively by multiplying UVER (W/m^2) with factor 40. The calculated value is presented in integer that is very close to. Rahayu and Komala [11] reported ultraviolet index of 2008-2010 reach maximum value as 13.4 by project research at Bandung using AURA OMI satellite data, especially in September 2009. Ultraviolet index as value as 13.4 is categorized in extreme and very high risk for human life corresponded to cataract and skin cancer symptoms. Medical effects of skin burned and erythema has been reported by using ultraviolet erythemal radiation (UVER), and defined as amount of solar irradiance at surface by standard response curve adopted by Commission Internationale d l'Eclairage (CIE) in 1987 [12]. Study about medical effects of ultraviolet irradiance generally correlated to Minimum Erythemal Dose (MED) and has close relationship with type of human skin.

Incoming solar radiation is influenced widely by some environmental factors, for example stratospheric ozone and atmospheric aerosol. Change of these factors will be the main factor for variation of global ultraviolet radiation and direct-diffuse component in clear sky condition [13]. Except by water vapor, ultraviolet B spectrum also absorbed strongly by atmospheric ozone layer, with the result the persistence of ozone is very important in extinction of incoming solar ultraviolet radiation. By research of ultraviolet index at Medan by using satellite AURA data year 2005-2012, it has been reported ultraviolet index varies between 8 to 13 and has a good correlation with stratospheric ozone layer thickness [14]. Seasonal correlation of ultraviolet index shows a negative correlation with coefficient correlationbetween 0.66 to 0.92. This means increasing of stratospheric ozone concentration will decrease the ultraviolet index, and vice versa.._It is found good correlations among UVER irradiance, total ozone column, water vapor, aerosol, and solar zenith angle [15]. The aim of this paper is to analyze the ultraviolet index classification and ultraviolet insolation, and sunshine duration classification at Bandung-Indonesia in 2017.

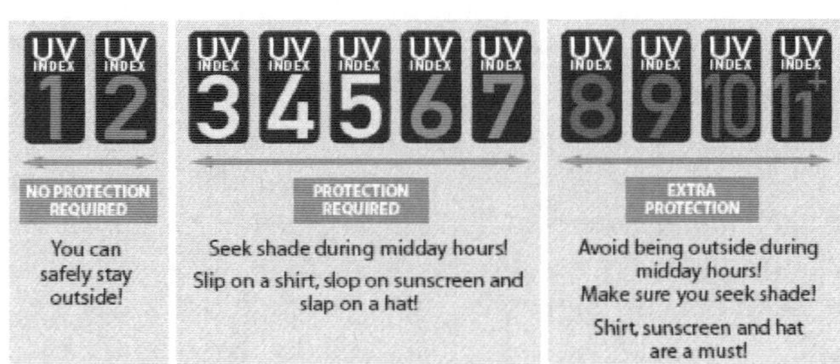

Figure 1. Ultraviolet index by WHO

2. Method

In this paper, three kinds of data have been used i.e. solar ultraviolet B insolation, ultraviolet index, and sunshine duration data. Solar ultraviolet B insolation is measured by using pyranometer MS-212W which passing solar wavelength 285-315 nm. Insolation of ultraviolet B data is recorded every 10 minutes using Data Logger Hioki LR8402-20. Sunshine duration is measured by using MS-093 sensor, recorded simultaneously by the same data logger. Both of sunshine duration sensor and pyranometer are produced by Eko Instrument Co, Japan, and installed on rooftop 4th floor of PSTA LAPAN, about 20 meters from land surface, and altitude of Bandung is 720 meters mean sea level. Measuring of ultraviolet index also using ultraviolet sensor integrated into automatic weather station (AWS), produced by Davis. The type of Davis's AWS is Vantage Pro2 Plus and records the data every 15 minutes automatically. AWS is placed at same location too.

The daily maximum ultraviolet index has been calculated for every day-data and found 365 ultraviolet indexes for Bandung in 2017. Ultraviolet index of 2017 then classified following WHO's classification modified. After re-classified, there are only 4 classes, because of ultraviolet index 3-5 and

6-7 are grouped to be one class. By reason of measured ultraviolet index has one decimal then range of classification has been revised. This revision can be seen in table 1.

Tabel 1. Ultraviolet index classification.

UV Index (1)	Categories (2)	Classifications (3)	Descriptions (4)
0-2	Low	≤ 2.5	Low risk for public.
3-5	Moderate	2.6 to 7.9	Potentially risk if not protected
6-7	High		Need protection to avoid sun burned
8-10	Very high	8 to 9.9	Avoid direct exposure, use glass and shirt
11-15	Extreme	> 10	Protect your body

The daily maximum ultraviolet B insolation has been grouped in ten daily sequence following BMKG's method in determining of dry/rainy season [16]. By using this method the daily maximum insolation grouped for every 10 days and average in 10 days have been calculated. There are 37 values of an average of 10-days daily maximum ultraviolet B insolation in year 2017. Meanwhile, the different technique has been used for calculating sunshine duration data. Sunshine durations are clustered in three levels of exposure, i.e. 0-4 hours, 4.1 – 8 hours, and 8.1 - 12 hours.

3. Result and Discussion

Data processing for ultraviolet index, ultraviolet B insolation, and sunshine duration are shown in figure 2, 3, and 4. Figure 2a shows the trend of daily maximum of ultraviolet index of Bandung in 2017 and figure 2b shows monthly ultraviolet index classification in 2017 after clustered into four classes. The high ultraviolet index on wet season DJF (December January February) correspond to the nearest distance between solar and earth, causes higher solar intensity on top of atmosphere relatively. Besides of changing of distance between solar and earth, the variability of ultraviolet index depends on atmospheric constituent also, i.e.: cloud and water vapor content, aerosol, and stratospheric ozone. Changing of cloud cover is locally component, but stratospheric ozone is generally. Ozone attenuates ultraviolet irradiance strongly in the region 280-315 nm (ultraviolet A band) and affects weakly on region 315-400 nm (ultraviolet A band). Moreover, by referring to figure 2b, it is shown that the extreme value of ultraviolet index occurred every day on January and decreased gradually in the next two months. High amount of water vapor content on DJF cannot absorb solar radiation effectively and causes ultraviolet index in extreme class persistently.

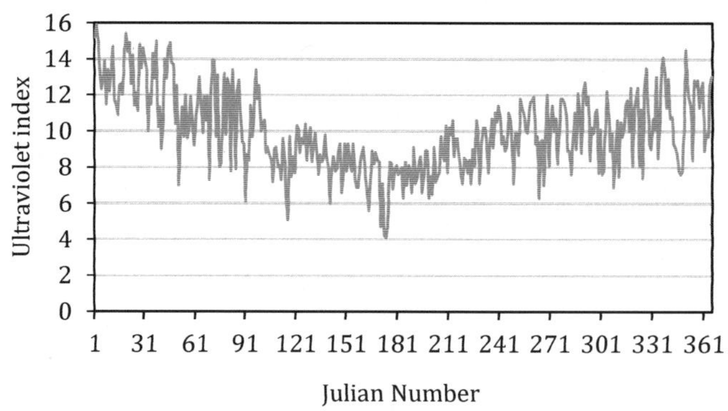

Figure 2a. Daily maximum of ultraviolet index at Bandung in 2017 has very high value especially when distance between earth and sun become short relatively (perihelion) on December, and low ultraviolet index value occurred when longest distance (August).

Figure 2b shows distribution of ultraviolet index at Bandung. In January 2017, all of the days have maximum ultraviolet index more than 10, and amount of days decreased gradually to several days in July, then increased again until December. By comparing to figure 2a, number of days with ultraviolet index less than 2.0 is not found in 2017, while in the middle of year 2017 distribution of days in both second class (3 – 7) and third class (8 – 10) are balanced. The percentage of days-event in 2017 with index ≥ 10 is 46%.

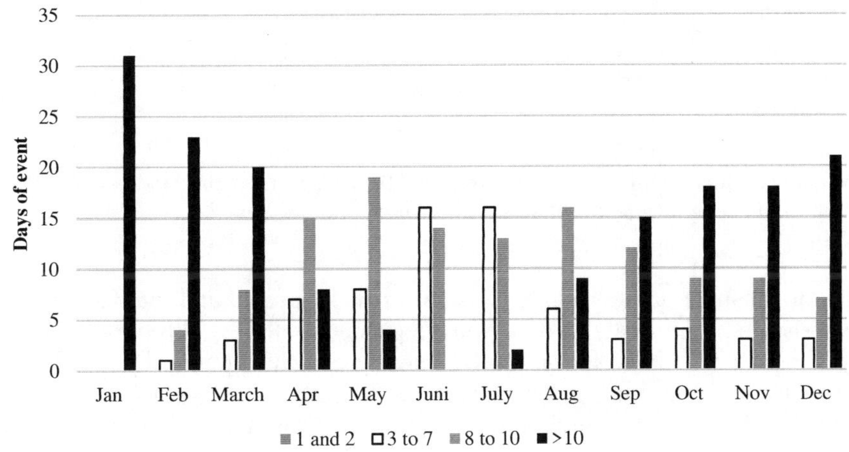

Figure 2b. Classification of monthly ultraviolet index in 2017 after grouped into four classes in Tabel 1.

10-days daily maximum ultraviolet B insolation in 2017 is shown in figure 3. By data analysis technique explained in advance the 10-days ultraviolet B maximum intensity follows trend of yearly maximum-minimum intensity, and have a higher intensity at the beginning of year correspond to perihelion, whereas has lower intensity at the mid of year. In general, the average of 10-days ultraviolet B insolation in the range 0.8-1.8 W/m^2 rated lower and can be concerned to ultraviolet B absorption by atmospheric constituents especially stratospheric ozone layer. Average of ultraviolet B insolation is 1.22 W/m^2, and clearness index is 0.3-0.7 [17], both of these have mutual relationship. The lower 10-days ultraviolet B insolation correspond to turbid atmosphere event most of the strongest absorber of solar ultraviolet B radiation is stratospheric ozone layer. Besides of ozone layer, another atmospheric constituents contribute to attenuation of incoming solar B radiation are aerosol and water vapor. The attenuation is caused by both absorption and scattering mechanisms. In addition, increasing of carbondioxide and water vapor content in stratosphere will cause infrared radiative cooling effects. However, absorption of ultraviolet B radiation by ozone layer will counteract the effects [18,19]. In correlation with absorption of aerosol, some research have found negative correlation between ultraviolet flux to aerosol optical depth (AOD) [20,21,22]. Moreover, reducing of incoming both solar UVA and UVB radiation in a big city can be correlated to loading of heavy pollutant to atmosphere [23,24].

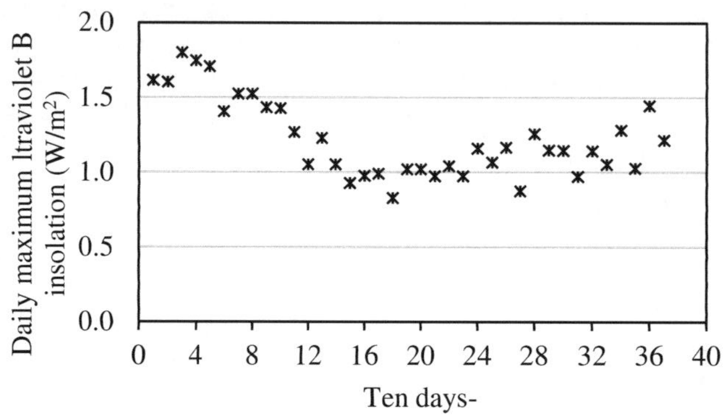

Figure 3. Average of 10-days daily maximum ultraviolet B insolation (Watt/m^2). Days in 2017 are grouped by every 10 days and average value every 10 days are shown as one mark.

The statistics of sunshine duration in 2017 is shown in figure 4a and the distribution of sunshine duration event is shown in figure 4b. In figure 4a, sunshine duration is clustered to be 3 classes: short (0-4 hours), medium (4.1-8 hours), and long (8.1-12 hours). Sunshine duration in January, February and November are mostly short duration less than 4 hours/day. These correspond to weather condition which mostly rainy and cloudy. Sunshine duration in other months (March to October) mostly dominantly by medium duration (4.1-8 hours/day). Because of pseudo orbit of solar, on September 23rd and March 21st the solar will be on equator exactly, this also causes duration or length of day time and night time become similar. Location of Bandung over southern hemisphere causes longer duration of day time around beginning of October and beginning of March. In fact, sunshine duration in March and October are not dominantly by longer duration but have similarity for short and medium duration in March and medium duration in October. This can be presumed with high atmospheric water vapor content. Water vapor is a major atmospheric constituent in absorb of incoming solar radiation.

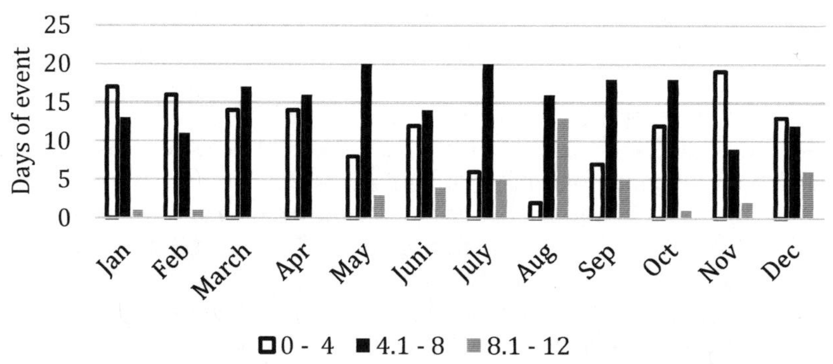

Figure 4a. Statistics of monthly sunshine duration in 2017 at Bandung after clustered to be 3 classes.

In general, sunshine duration in 2017 is mostly in medium duration class of 4.1-8 hours/day as amount 184 days-event or 50% of data, and secondly in short duration class as amount 140 days-event or 38%, and remaining in long-duration class as amount 41 days-event or 12%. Mostly duration of sunshine is 4-5 hours/day and occurred as 54 days (14.8 %) in 2017. By comparing with results found in LPD Sumedang then sunshine duration in DJF are mostly 1 hours/day, 7 hours/day in MAM, 9 hours/day in JJA, and distributed evenly in SON [9]. In that 15 years observation (1998-2013) the trend of increasing of sunshine duration from 7 hours/day to be 10 hours/day have been found. Sunshine duration in Bandung (2017) is shorter than sunshine duration in Sumedang. Sunshine duration only explains about time accumulation of incoming solar radiation in a day, but not accumulation of energy nor intensity. Sunshine duration depends on minimum intensity of incoming solar radiation 120 W/m^2. In small solar zenith angle, sunshine duration is not affected widely by changes of atmospheric constituents, but increasing of atmospheric turbidity in large solar zenith angle contribute to decreasing of sunshine duration. Work in South Asia by using Net Downward Shortwave Radiation (NDSWR) satellite data showed the decreasing of sky clearness average -0.05 $W.m^{-2}y^{-2}$ has been occurred in period 1979-2004 [25], and can be corresponded to solar dimming phenomena with average decrease as amount -0.54 $Wm^{-2}y^{-2}$. The decreasing can be correlated to increasing of optical depth caused by increasing of atmospheric aerosol content. Off course, the decreasing of sky clearness is not directly correlate to shorter sunshine duration. Only in very serious decreasing of sky clearness will affect directly to sunshine duration.

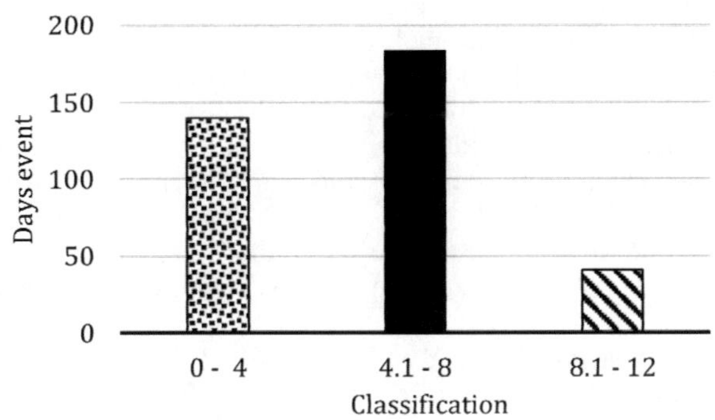

Figure 4b. Distribution of sunshine duration events in 2017 after clustered to be 3 classes.

4. Conclusion

After taking analysis of sunshine duration data, it has been found that Bandung was exposure by solar radiation as long as 4.1-8 hours /day average almost in part of 2017 and less than 10% was exposured up to 12 hours/day. Mostly duration of sunshine is 4-5 hours/day and occurred as 54 days (14.8 %) in 2017. Although daily maximum ultraviolet B insolation lower relatively (average 1.22 W/m^2) but not similar to ultraviolet index case. The ultraviolet index in 2017 at Bandung reached extreme value ($\geq$10) as amount 46%, and it was not found daily maximum ultraviolet index less than 2.

Acknowledgments

Many thanks are delivered to Center of Science and Technology of Atmosphere for permission to using the data for this work, and especially be delivered to Bpk. Drs. Sri Kaloka for long discussions and suggestions.

References

[1] Thomas P, Swaminathan A and Lucas R M 2012 Climate change and health with an emphasis on interactions with ultraviolet radiation: a review *Glob. Change Biol.* **18** 2392-2405
[2] Diffey B L 1998 Ultraviolet radiation and human health *Clin. Dermatol.* **16** 83-89
[3] Gallaghar R P and Lee T 2006 Adverse effects of ultraviolet radiation: a brief review, Prog. Biophys *Mol. Biol.* **92** 119-31
[4] Roberts J 2001 Ocular phototoxicity *J. Photochem. Photobiol.* **B**(64) 136-43
[5] Lonsberry B B, Wyles E, Goodwin D, Caesar L and Lingel N, ed Bartlett J and Jaanus S 2008 Diseases of cornea *clinical Ocular Pharmacology* Butterworth-Heineman Pub.
[6] Norval M 2001 Effects of solar radiation on human immune system *J. Photochem. Photobiol.* **B**(63) 28-40

IOP Conf. Series: Earth and Environmental Science **303** (2019) 012018 doi:10.1088/1755-1315/303/1/012018

[7] Norval M 2006 The mechanisms and consequences of ultraviolet-induced immunosuppression *Porg. Biophys. Mol. Biol.* **92** 108-18
[8] WMO 2008 Guide to meteorological instruments and methods of observation 8th ed.
[9] Sumaryati and Hamdi S 2014 Distribusi lama penyinaran matahari di LPD Sumedang (6.91LS dan 107.84 BT) LAPAN *Proc. Pertemuan Ilmiah XXVIII Himpunan Fisika Indonesia Jateng & DIY*
[10] WMO, UNEP, ICNIRP and WHO 2002 Global solar UV index: A practical guide *Watch* **95** WMO/TD no. 625 Geneva Switzerland
[11] Komala N and Rahayu S A 2011 Karakteristik indeks ultraviolet (UV) dan ozon di Bandung menggunakan data Ozone Monitoring Instrument (OMI) *Proc. Seminar Nasional Sains Atmosfer dan Antariksa* Bandung
[12] McKinley A F and Diffey B L 1987 A referrence spectrum for ultraviolet induced erythem in human skin *CIE Journal* **6** 17-22
[13] Alexander K 2008 Aerosol optics light absorption and scattering by particles in the atmosphere (Berlin : Springer)
[14] Komala N 2013 Analisis variasi temporal ozon dan indeks ultraviolet Kota Medan *Proc. Seminar Penelitian Masalah Lingkungan di Indonesia* (Buku) Bandung
[15] Bilbao J, Roman R, Yousif C, Mateos D and Miguel AD 2014 Total ozone column, water vapour and aerosol effects on erythemal and global solar irradiance in marsaxlokk, Malta *Atmospheric Environment* **99** 508-18
[16] Dayantolis W, Supeni A and Ripaldi A 2016 Penentuan normal musim hujan di Indonesia berdasarkan frekuensi curah hujan dasarian *Megasains* **7** 25-32
[17] Hamdi S 2015 Bilangan kebeningan atmosfer dan aplikasinya dalam ilmu lingkungan atmosfer *Berita Dirgantara* **1** 17-22
[18] Canada J, Pedros G, Lopez A and Bosca J 2000 Influences of the clearness index for the whole spectrum and of the relative optical air mass on UV solar irradiance for two locations in the Mediterranian Area, Valencia and Cordoba *J. Of Geophys. Res.* ***25*** 885-96
[19] Liu H, Hu B, Zhang L, Zhao X, Shang K, Wang Y and Wang J 2017 Ultraviolet radiation over China : spatial distribution and trends *Renewable and Sustainable Energy Reviews* **76** 1371-83
[20] Singh S, Lodhi N K, Mishra A K, Jose S, Kumar S N and Kotnala R K 2018 Assessment of satellite-received surface UVA and UVB radiation by comparison with ground-measurements and trends over Mega-City Delhi *Atmospheric Environment* **188** 60-70
[21] Fountoulakis I, Bais A F, Fragkos K, Meleti C, Tourpali K and Zempila M M 2016 Short- and long-term variability of spectral solar UV irradiance at Thessaloniki, Greece: effects of change in aerosols, total ozone and clouds *Atmos. Chem. Phys.* **16**(4) 2493-2505
[22] Anton M, Gill J E, Fernandez-Galvez J, Lyamani H, Valenzuela A, Foyo-Moreno I, Olmo FJ and Alados-Arboledas L 2011 Evaluation of the aerosol forcing efficiency in the UV erythemal range at Granada, Spain *J. Geophys. Res.* **116** D20214
[23] Wai K M, Peter K N, Lam KS 2015 Reduction of solar U V radiation due to urban high-rise buildings - a coupled modelling study *PloS One* **10**(8)
[24] Czerwinska A E, Kszyscin J W, Jarowlawski J and Posyniak M 2016 Effects of urban agglomeration on surface-UV doses: a comparison of Brewer measurements in Warsaw and Belsk, Poland, for period 2013-2015 *Atmos. Chem. Phys.* **16**(2) 13641-51
[25] Kambezidis H D, Kaskaoutis D G, Kharol S K, Moorthyd K K, Satheesh S K, Kalapureddy M C R, Badarinath K V S, Shama A R and Wild M 2012 Multi-decadal variation of the net downward shortwave radiation over south Asia *Atmospheric Environment* **50** 362-72

Trend of rainfall over Indonesian major lakes from tropical rainfall measuring mission data

Hidayat
Research Center for Limnology, Indonesian Institute of Sciences, Cibinong Science Center - Botanical Garden, Cibinong 16911, Indonesia

hidayat@limnologi.lipi.go.id

Abstract. Establishing patterns of natural variability of water resources in lakes is important for water management. However, the availability of reliable data often becomes obstacles, especially in ungauged regions. Rainfall products from Tropical Rainfall Measuring Mission (TRMM) potentially fill this gap. The aim of this study is to assess the trend of rainfall over Indonesian major lakes using TRMM rainfall estimates to investigate the natural variability of rainfall over the lakes. TRMM 3B43 monthly data over the period of 1998–2017 obtained from Giovanni data portal maintained by NASA were used to assess the trend of rainfall over Indonesian major lakes. Rainfall correlation with El Niño-Southern Oscillation (ENSO) variability was also investigated using Southern Oscillation Index (SOI) and Multivariate ENSO Index (MEI). Trend analysis was carried out using Mann-Kendall test. Rainfall depths over Lake Laut Tawar in Sumatra tend to gradually decline, which is confirmed by the Mann-Kendall test. Result of the test of rainfall over all other investigated lakes on Kalimantan, Sulawesi, Papua, and Java indicate no trends except for that of Lake Paniai on Papua, which indicates an increasing trend.

Keywords : ENSO, TRMM, rainfall

1. Introduction

Establishing patterns of natural variability of water resources in lakes is important for water management in the present and future time of increasing water demand and climate variability. However, the availability of reliable data including those on rainfall often becomes obstacles, especially in ungauged remote regions. The availability of satellite-based meteorological data including that of rainfall products from Tropical Rainfall Measuring Mission (TRMM) potentially fill this gap. TRMM precipitation estimates have been available since the satellite was launched in 1997. TRMM combines observations from multiple satellites, as well as gauge analyses where available with a resolution of 0.25° x 0.25° in a latitude band covering 50° N to 50° S [1]. TRMM rainfall estimates have been reported to be valuable as an alternative source of rainfall information in hydrological studies [2,3].

Indonesia has at least 840 natural lakes, 735 ponds and 162 reservoirs [3] spreading over Indonesian Archipelago, most of them are fed by rainfall as source of water. The aim of this study is to assess the trend of rainfall over Indonesian major lakes using the TRMM rainfall estimates to investigate the natural variability of rainfall over this important fresh water resources. The representative lakes on each major islands of Indonesia and their brief descriptions are given below.

1. Sumatra:
   a) Lake Laut Tawar (Aceh). Lake Laut Tawar is located in the Gayo Highland near Takengon of Central Aceh Regency. This lake has important roles for local Gayo community as source of freshwater for domestic, agriculture, industry, and fishery uses [5].
   b) Lake Toba (North Sumatra). Lake Toba is the largest lake in Indonesia that has large potentials of water resources [6]. It has been used for electricity generation through the Sigura-gura hydro-power plant with a capacity of 286 MW, for tourism, also for fishery production and source of drinking water for domestic use.
   c) Lake Maninjau (West Sumatra). Lake Maninjau, is socially important to local community for domestic uses and recreational function, economically as source of water for irrigation, fishery (both open waters and aquaculture fishery activities), tourism, and hydro-power generation with a capacity of 68 MW. Ecologically, the lake plays its role in controlling groundwater balance, as habitat for aquatic organism, as well as controlling micro-climate [7].
2. Kalimantan
   a) Sentarum Lakes complex (West Kalimantan). The Sentarum Lakes complex is an assemblage of floodplain lakes in the upper Kapuas River system of Kapuas Hulu Regency. The lakes area is declared as a Ramsar site due to its unique tropical wetland ecosystems rich in biodiversity of typical aquatic as well as terrestrial flora and fauna [2]. The Kapuas lakes are important for local communities, not only as a source of water for domestic purposes, but also to sustain the livelihoods of people, especially in the open water fishery subsector.
   b) Mahakam cascade lakes: Semayang, Melintang, Jempang Lakes (East Kalimantan). The Mahakam cascade lakes are part of the middle Mahakam River floodplain system that include some 30 shallow lakes located in the Kutai Kartanegara and Kutai Barat Regencies. The lakes are the core of inland fisheries in East Kalimantan and is considered one of the most productive freshwater fisheries in south-eastern Asia [8].
3. Sulawesi
   a) Malili lakes complex: Matano, Towuti, Mahalona Lakes (South Sulawesi). The Malili Lakes are the only hydrologically connected ancient lake system in the world located in the central eastern part of Sulawesi that include three major lakes, Matano, Mahalona, and Towuti, and two smaller lakes, Wawantoa and Masapi [9]. These tectonic lakes are biodiversity hotspots.
   b) Lake Tempe (South Sulawesi). This floodplain lake experiences hydrological variation during dry and wet seasons. Its area is highly influenced by fluctuating water level, ranging from 280–430 km^2 with a maximum depth of 9 m during extreme high water level to only about 10 km^2 remain inundated during low water period with a maximum depth of 1.5 m [10].
   c) Lake Poso (Central Sulawesi). Lake Poso is a large (323 km^2; 450 m deep)[9] tectonic lake located at about 502 m asl. The lake drains northward via the Poso River into the Gulf of Tomini.
4. Papua
   a) Lake Sentani (Papua). Lake Sentani is located on the South of the City of Sentani at about 70–90 m asl in Jayapura Regency. The lake has an area of 93.6 km^2 that is part of the Cyclops Nature Reserve [11]. Its source of water is from direct precipitation and and from at least 14 mountain rivers debouching into the lake.
   b) Lake Paniai (Papua). It is an ancient tectonic lake (160 km^2 area; 54 m deep) located near the backbone of Central Papua Island mountain range at 1.742 m asl [12]. On the south of this lake, lies a smaller sister lake, Lake Tage (24 km^2 area; 52 m deep).
5. Java
   a) Citarum cascade dams: Saguling, Cirata, and Jatiluhur Dams (West Java). The three dams are interconnected reservoirs of the Citarum river located in West Java. Outflow from Saguling Dam goes to Cirata Dam, outflow from Cirata Dam goes to Jatiluhur Dam and

outflow from Jatiluhur is used for many purposes, including electricity generation, domestic consumption, irrigation water for agricultural uses and flushing Jakarta canals [13].

b) Gajah Mungkur Dam (Central Java). Gajah Mungkur or Wonogiri Dam is located in Wonogiri Regency, which was built with the main purpose for flood control that often occurs due to the overflowing of Bengawan Solo river in the rainy season [14]. It is also used as source of freshwater for drinking water, irrigation, and hydropower.

2. Methodology

In this contribution, TRMM 3B43 version-7 monthly data product over the period of 1998–2017 obtained from Giovanni data portal (https://giovanni.gsfc.nasa.gov/giovanni/) maintained by National Aeronautics and Space Administration were used to assess the trend of rainfall over Indonesian major lakes. Thirteen major lakes on five major Indonesian islands were selected. Table 1 shows the bounding box of the lake catchments used for the area-averaged TRMM data download.

Rainfall correlation with El Niño-Southern Oscillation (ENSO) variability was investigated using Southern Oscillation Index (SOI; http://www.cpc.ncep.noaa.gov/data/indices/soi) and Multivariate ENSO Index (MEI; http://www.esrl.noaa.gov/psd/enso/mei/mei.html). Normalized monthly precipitation (P_N) was taken as $P_N = (P - \overline{P}) / \sigma_P$, where $\overline{P}$ is mean and σ_P is standard deviation, which were plotted along with SOI time-series. Trend analysis was carried out using Mann-Kendall test, a non-parametric monotonic trend detection [15] that test of the null hypothesis (H_0) of trend absence in the time series, against the alternative of trend (H_1). The result of the test is returned in H=0 indicates a failure to reject the null hypothesis at the α significance level. H=1 indicates a rejection of the null hypothesis at the α significance level.

Table 1. The bounding box (decimal degree) of the representative Indonesian lakes (covering also the catchment of the lakes) used for the area averaged TRMM data download

Lake Name	West bound	South bound	East bound	North bound
Lake Laut Tawar	96.82	4.50	97.03	4.66
Lake Toba	98.50	2.25	99.15	3.00
Lake Maninjau	99.90	-0.42	100.27	-0.20
Sentarum Lakes Complex	111.96	0.56	112.30	1.10
Mahakam Cascade Lakes	115.50	-0.50	116.70	0.50
Malili Lakes Complex	121.10	-3.01	121.78	-2.92
Lake Tempe	119.74	-4.25	120.25	-3.90
Lake Poso	120.39	-2.18	120.75	-1.72
Lake Sentani	140.24	-2.76	140.76	-2.49
Lake Paniai	136.15	-4.15	136.51	-3.74
Citarum Cascade Dams	107.14	-7.18	107.70	-6.38
Gajah Mungkur Dam	110.73	-8.06	111.05	-7.72

3. Results and discussion

Rainfall patterns in Indonesia are generally influenced by ENSO and the Indo-Pacific air-sea interaction. Twenty years of rainfall data length used in this study is actually less than the recommended data length of 30 years. However, the current analysis can be regarded as a modest proxy to rainfall characteristics over Indonesian major lakes. The analysis of rainfall trends over Indonesian major lakes are presented herein for the five Indonesian major islands.

3.1. Sumatra

Average annual rainfalls over Lake Laut Tawar, Lake Toba, and Lake Maninjau are 2600, 2745, and 2670 mm, respectively. Figure 1 a, b, c show the average TRMM 3B43 monthly rainfall estimates over Sumatran lakes plotted along with those of El-Niño (2015) and La-Niña (2000) years. The three lakes were affected by ENSO although with different intensity. For example, during the 2015 El-Niño year, the three lake basins generally experienced below average rainfall depth except for January, June, and August for Laut Tawar and March, April, June and November for Maninjau with above average rainfall. Another example with the opposite condition was that during the La-Niña year 2000, Lake Laut Tawar experienced a highly above average rainfall for nearly the whole year followed by Lake Toba with an extremely high rainfall depth in September and Lake Maninjau with a high rainfall depth in November but with also below average rainfall depth during the first half of the year. ENSO variability affects rainfall pattern in Sumatran lake basins with varied intensity among places and times, for example, using only data in August Rainfall–SOI correlation coefficients were respectively 0.25, 0.46, and 0.29 for Lake Laut Tawar, Lake Toba, and Lake Maninjau (Fig. 1 d). MEI–rainfall depth correlation coefficients for August were respectively 0.00, -0.32, and -0.09 for Lake Laut Tawar, Lake Toba, and Lake Maninjau. Figure 2 a shows the variation of normalized monthly rainfall over Lake Toba plotted along with SOI time series from January 1998 through December 2017. During certain periods, rainfall rates over this lake perfectly follow the pattern of SOI such as those in early 1998, late 2000, and late 2016 indicating a very strong correlation, while during other periods they show no clear relation such as those in late 2005, early 2007, and early 2014.

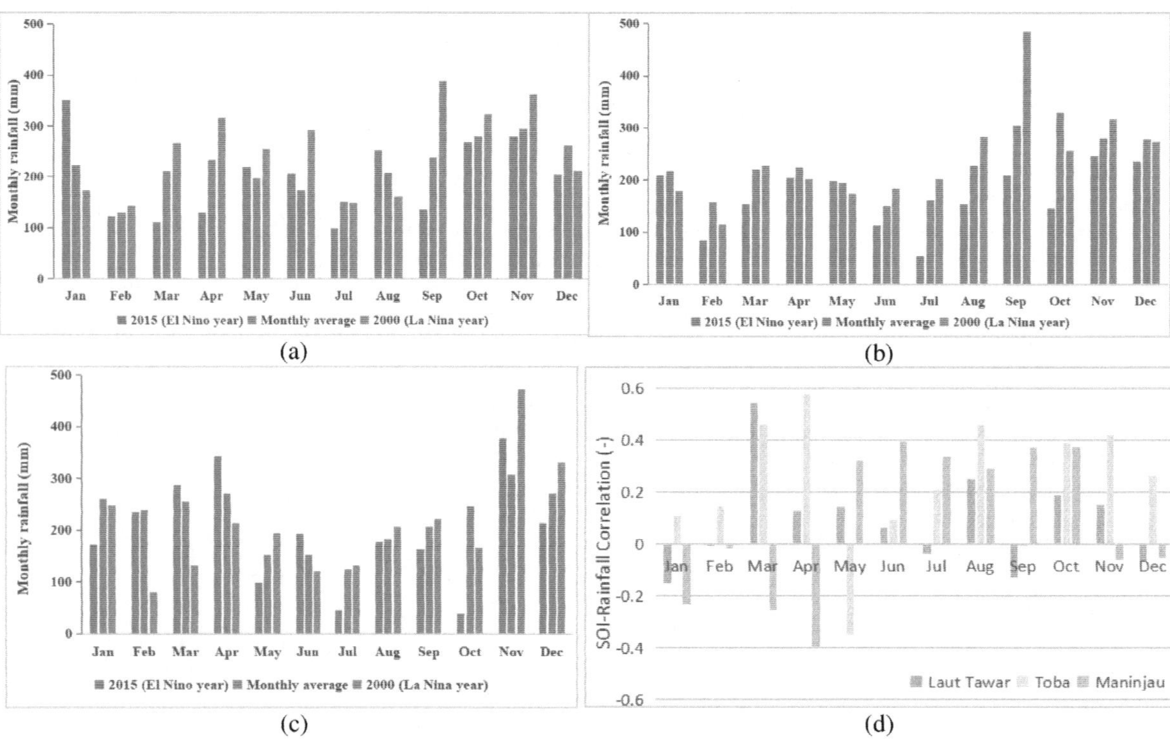

Figure 1. Average TRMM monthly rainfall estimates along with those during an El Niño (2015)/La Niña (2000) years over Lake Laut Tawar (a) Lake Toba (b), Lake Maninjau (c), and variation of SOI–rainfall correlation values throughout the year

Rainfall depths over the Sumatran Lake basins show a gradual decline from year to year. Figure 2 b shows the accumulated annual rainfall over the lake basins with their corresponding linear trend lines. This trend is inline with previous research [7,16]. The decline of rainfall depth over Toba Lake basin has caused the decrease of water levels in the lake [16]. Similar trend was found by other team [7] who analysed rainfall data from rain gauges in the catchment area of Lake Maninjau for the period 1994–2000. They found that rainfall depth over Maninjau Lake basin shows a decreasing trend that needs to be anticipated in terms of the sustainability of the lake's function for water supply, hydro-power generation, tourism, as well as its ecological function. The declining trend is also in agreement with previous research [5], who found that water levels of Lake Laut Tawar has decreased about 1 to 2 m. They reckon that this issue was due to global warming effect and deforestation in the catchment area of the lake, which in turn has caused the decline in water resources around the lake and the decrease of fishery production of about 80% during the period of 1989 through 2008. The result of the Mann-Kendall test only confirms the trend of rainfall over Lake Laut Tawar.

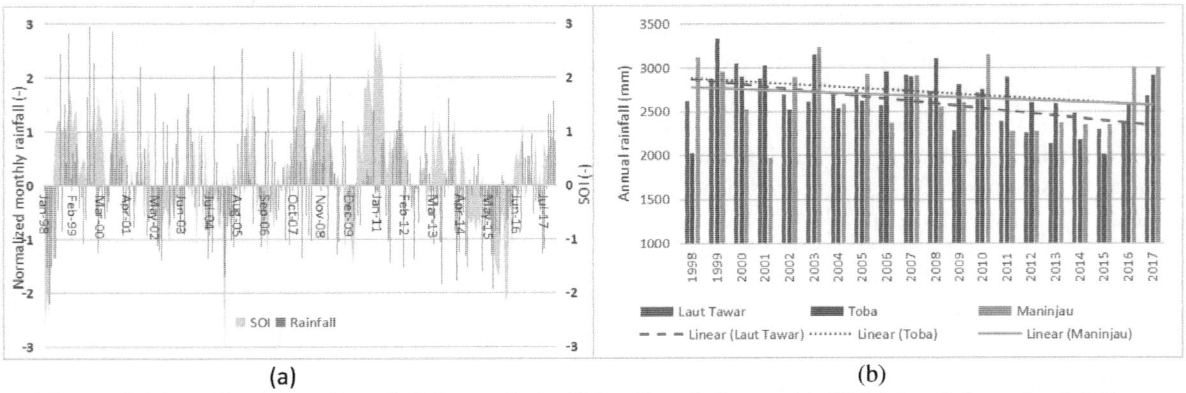

Figure 2. Normalized monthly rainfall over Lake Toba plotted along with SOI (a) and Annual rainfall over Lakes Laut Tawar, Toba, and Maninjau (b)

3.2. Kalimantan

Both the Sentarum Lakes complex and Mahakam Cascade Lakes are flood plain lakes. Hydrological regime is the key factor determining the ecological function and biodiversity in flood plain lake. Average annual rainfalls over Sentarum Lakes complex and Mahakam Cascade Lakes are 3680 and 2770 mm, respectively. Figure 3 a and b show the average TRMM 3B43 monthly rainfall estimates over those lakes.

Using the entire 20 years data series, SOI-rainfall depth correlation coefficients were 0.14 for Sentarum and 0.25 for Mahakam Cascade Lakes, while those using only August were 0.68 for Sentarum and 0.70 for Mahakam Lakes. Figure 3 c shows SOI–rainfall correlation throughout the year. MEI–rainfall depth correlation coefficients were -0.19 for Sentarum and -0.27 for Mahakam Cascade Lakes, while those using only August were -0,54 for Sentarum and -0.65 for the Mahakam Lakes. These numbers indicate that rainfall over the Mahakam Lakes are more strongly affected by ENSO. Figure 4 a shows the time series of normalized monthly rainfall over Mahakam Lakes plotted along with SOI. Figure 4 b shows the annual rainfall over the Sentarum and Mahakam lake basins. Mann-Kendall test results indicate no trend of rainfall over the selected lake basins of Kalimantan.

3.3. Sulawesi

Average annual rainfalls over Malili lakes complex, Lake Tempe, and Lake Poso are 3010 and 2235, and 3075 mm, respectively. Figure 5 a, b, c show the average TRMM 3B43 monthly rainfall estimates over those lakes. Using the entire 20 years, SOI–rainfall depth correlation coefficients were 0.16 for Malili, 0.21 for Tempe, and 0.20 for Poso. These correlation values are varied throughout the year as shown in Figure 5 d, which also indicates that rainfall over the three Lakes are equally affected by ENSO. The highest correlation of SOI and rainfall over the catchments of lakes in Sulawesi of about 0.7 were found during the period of July through November. Using the entire data series, MEI–rainfall depth correlation coefficients were -0.08 for Malili, -0.17 for Tempe, and -0.13 for Poso. Similar pattern (opposite signs) with those of SOI (Figure 5 d) was found when MEI-rainfall analysis in monthly basis was carried out. Figure 6 a shows the variation of normalized monthly rainfall over Lake Tempe plotted along with SOI time series. Figure 6 b shows the annual rainfall over the Malili,

Tempe, and Poso lake basins. Mann-Kendall test results indicate no trend of rainfall over those lakes of Sulawesi.

3.4. Papua

Average annual rainfalls over Lake Sentani and Lake Paniai basins are 2010 and 3870 mm, respectively. Figure 7 a and b show the average TRMM 3B43 monthly rainfall estimates over those lakes. SOI–rainfall depth correlation coefficients were 0.05 and 0.11 for Lake Sentani and Lake Paniai respectively. Throughout the year, correlation between SOI and rainfall over Paniai Lake basin were generally higher than those over Lake Sentani basin with the highest value of about 0.6 for July (Figure 7 c). Using the entire dataset, MEI–rainfall depth correlation coefficients were -0.15 and -0.14 for Lake Sentani and Lake Paniai respectively, while analysis on monthly basis results in similar patterns as those of Figure 7 c except for the opposite signs. Figure 8 a shows the variation of normalized monthly rainfall over Lake Paniai plotted along with SOI time series from January 1998 through December 2017. Figure 8 b shows the annual rainfall over the Sentani and Paniai lake basins. Result of the Mann-Kendall test indicates there is a trend of rainfall over Lake Paniai basin (increasing).

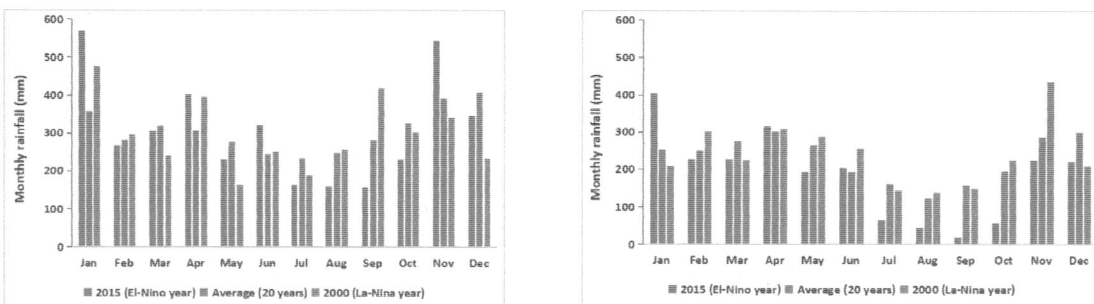

(a) (b)

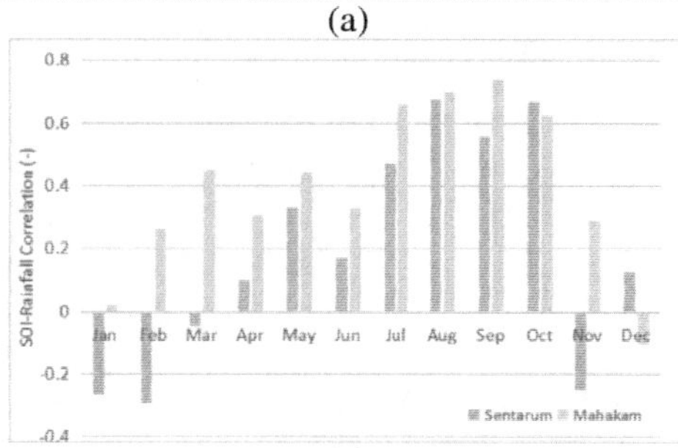

(c)

Figure 3. Average TRMM monthly rainfall estimates along with those during El Niño (2015) and La Niña (2000) years over Sentarum Lakes Complex (a), Mahakam Cascade Lakes (b), and variation of SOI–rainfall correlation values throughout the year (c)

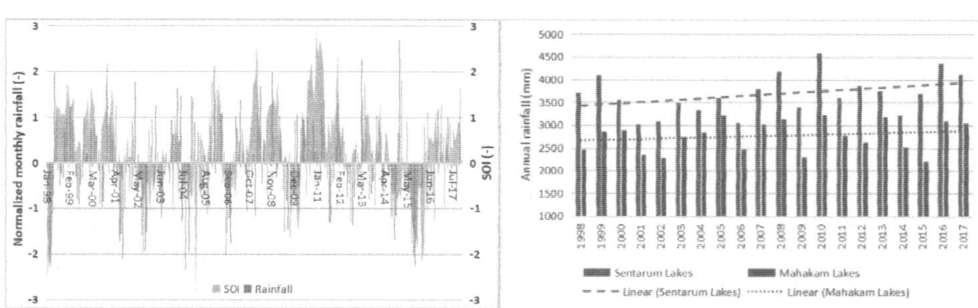

(a) (b)

Figure 4. Normalized monthly rainfall over Mahakam Lakes plotted along with SOI (a) and annual rainfall over the Sentarum and Mahakam Lakes (b)

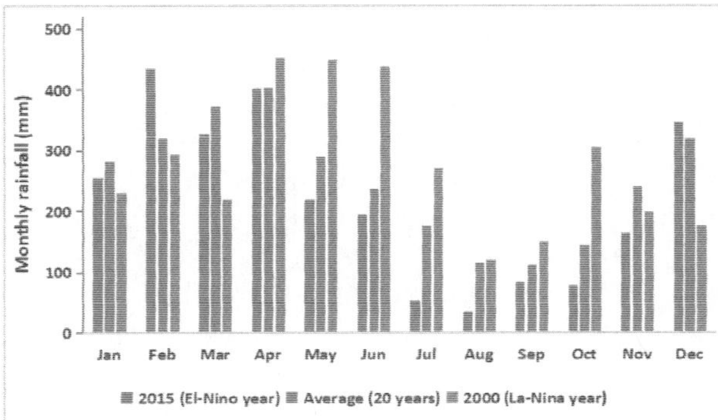

(a)

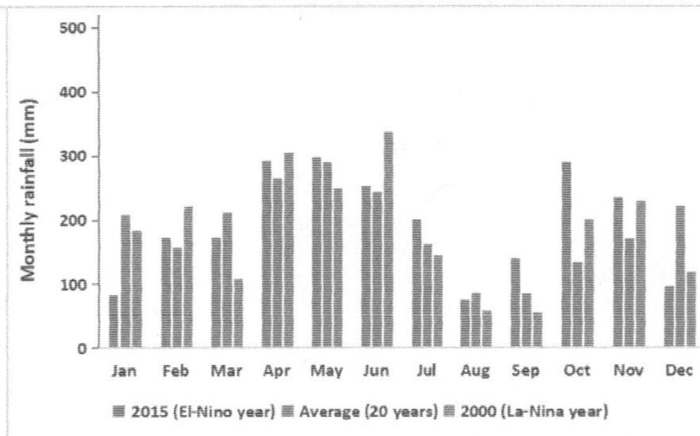

(b)

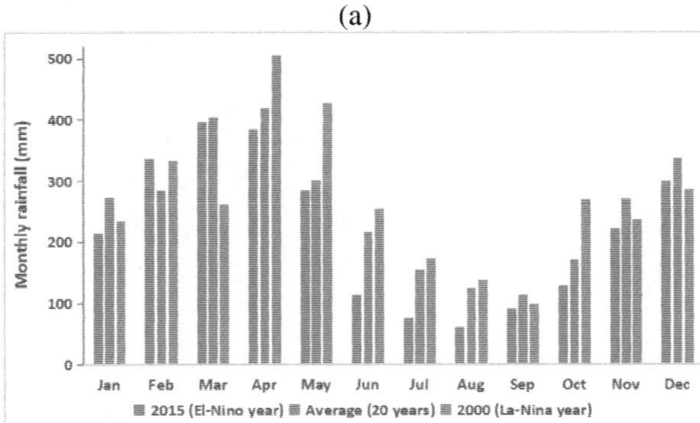

(c)

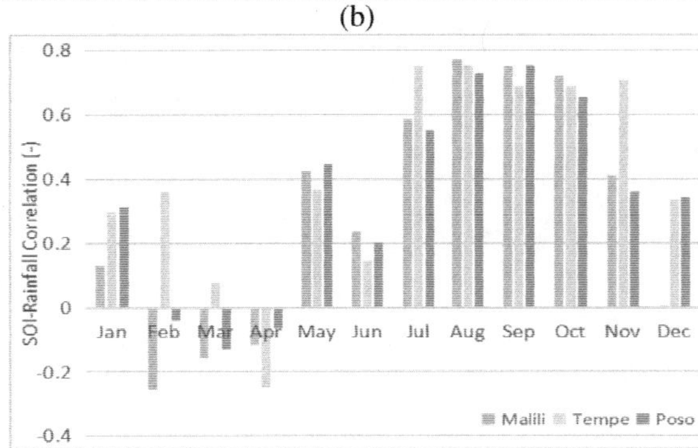

(d)

Figure 5. Average TRMM monthly rainfall estimates along with those during an El Niño/La Niña years over Malili lakes complex (a) Lake Tempe (b), Lake Poso (c), and variation of SOI–rainfall correlation values throughout the year (d)

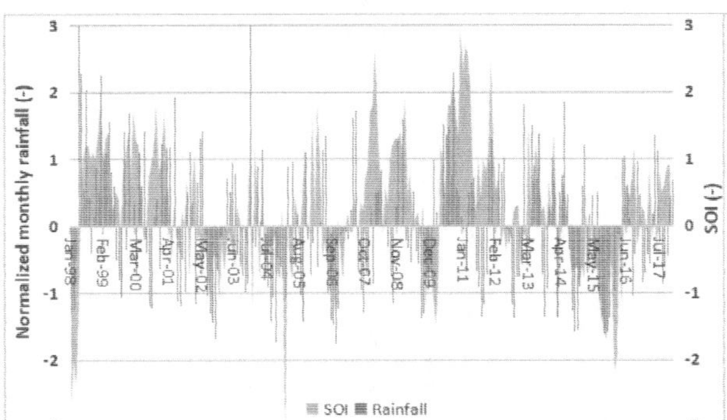

(a)

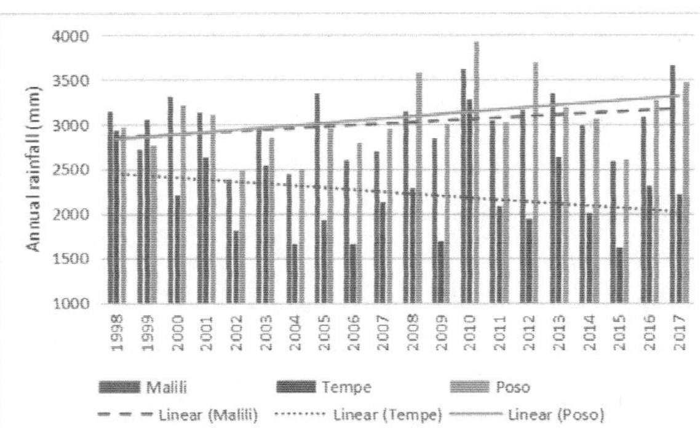

(b)

Figure 6. Normalized monthly rainfall over Lake Tempe plotted along with SOI (a) and annual rainfall over the Malili, Tempe, and Poso lake basins with their corresponding trend lines (b)

3.5. *Java*

Average annual rainfalls over Citarum Cascade Dams and Gajah Mungkur Dam basins are 2795, and 2300 mm, respectively. Figure 9 a, and b show the average TRMM 3B43 monthly rainfall estimates over the basins of those man-made lakes. Using the entire 20 years, SOI–rainfall depth correlation coefficients were 0.08 for Citarum and 0.16 for Gajah Mungkur. Throughout the year, the highest correlation of about 0.7 was found in June (Fig. 9 c). Using the entire dataset, MEI–rainfall depth correlation coefficients were -0.11 for Citarum and 0.20 for Gajah Mungkur, while results of analysis on monthly basis are shown in Figure 9 d. Figure 10 a shows the variation of normalized monthly rainfall over Gajah Mungkur basin plotted along with SOI time series. Figure 10 b shows the annual rainfall over the basins of Citarum Cascade Dams and Gajah Mungkur Dam. Mann-Kendall test results indicate no trend of rainfall over those dams of Java.

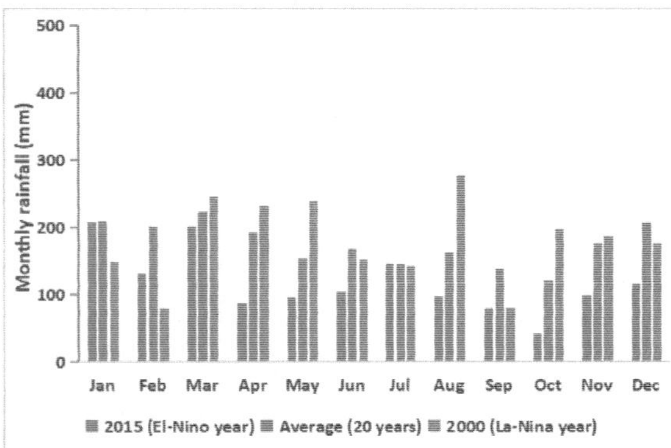

(a)

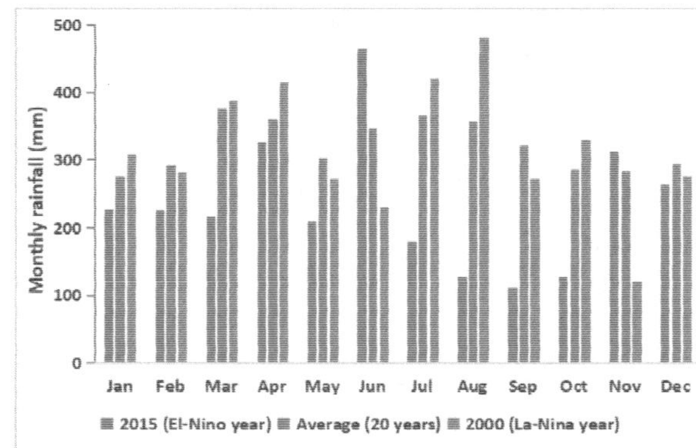

(b)

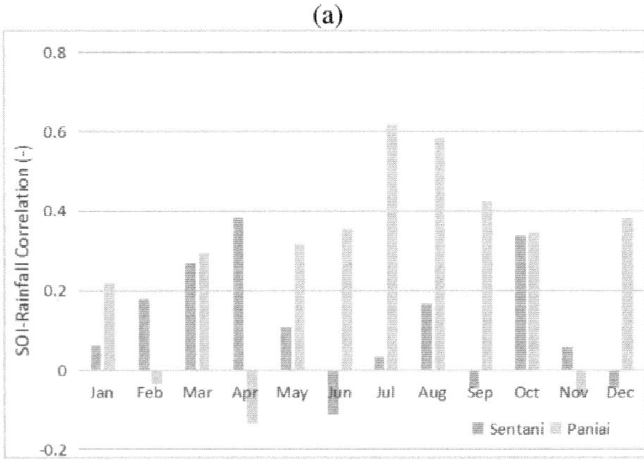

(c)

Figure 7. Average TRMM monthly rainfall estimates along with those during an El Niño (2015) and La Niña (2000) years over Lake Sentani (a), Lake Paniai (b), and variation of SOI–rainfall correlation values throughout the year (c)

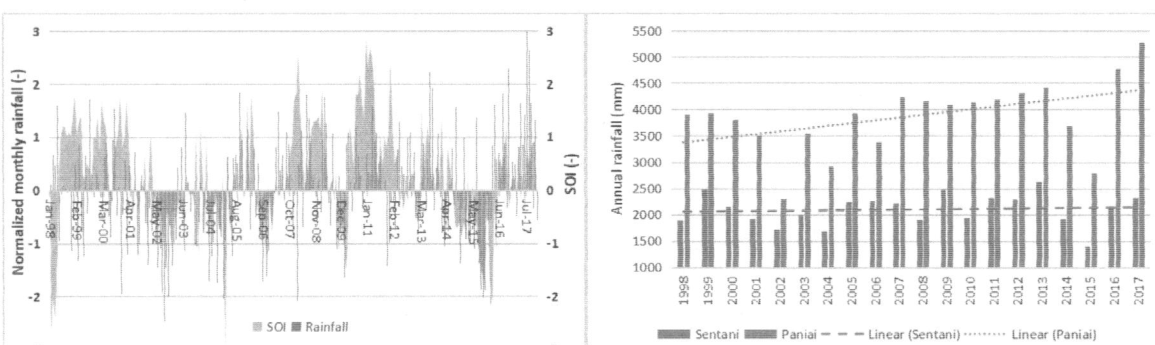

(a) (b)

Figure 8. Normalized monthly rainfall over Lake Paniai plotted along with SOI (a) and annual rainfall over the Sentani and Paniai lake basins with their corresponding linear trend lines (b)

4. Conclusion

Rainfalls over Indonesian major lakes are generally affected by ENSO variability with varied correlation values throughout the year. This study that based on 20 years of available TRMM data agrees the declining trend of annual rainfall over Sumatran lakes basins (Laut Tawar, Toba, and Maninjau) found by previous site-specific respective researches. However, the Mann-Kendall test only confirms that of Laut Tawar. Result of Mann-Kendall test of rainfall over all other investigated lakes in Java, Kalimantan, Sulawesi and Papua indicate no trends except for that of Lake Paniai (Papua), which indicates an increasing trend.

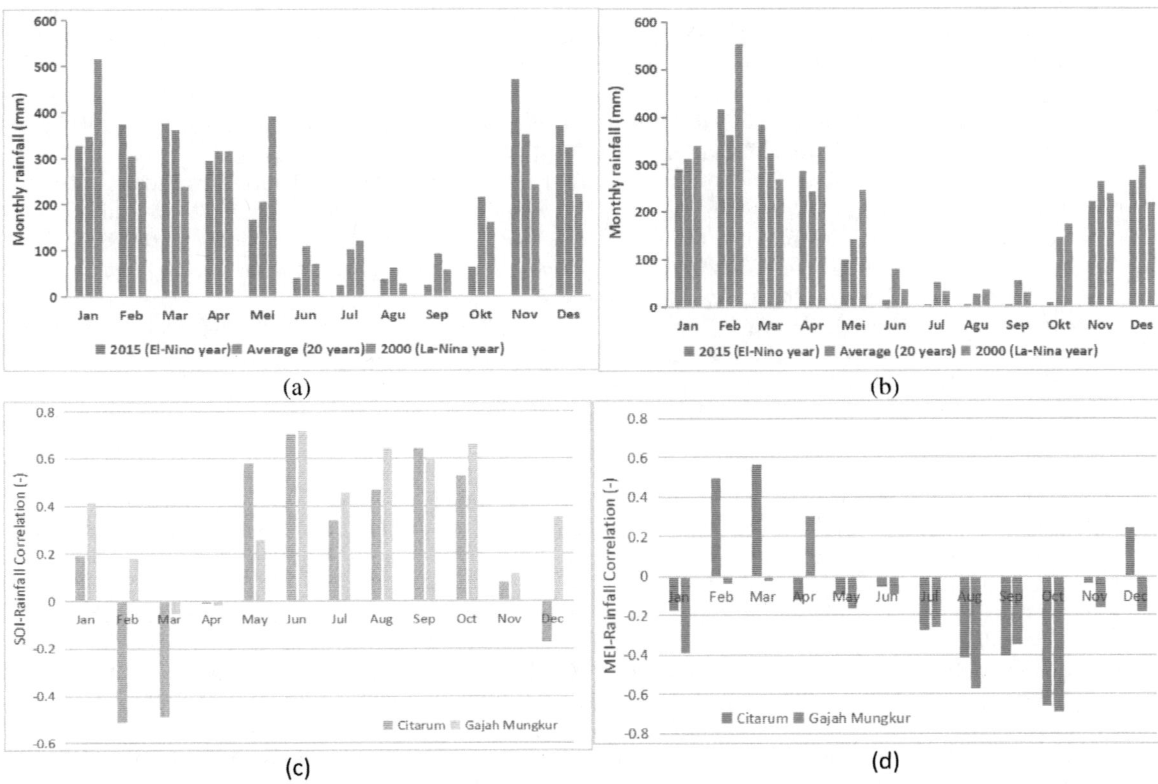

Figure 9. Average TRMM monthly rainfall estimates along with those during an El Niño (2015)/La Niña (2000) years over Citarum Cascade Dams (a), Gajah Mungkur Dam (b), SOI–rainfall correlation throughout the year (c), and variation of MEI-rainfall correlation values throughout the year (d)

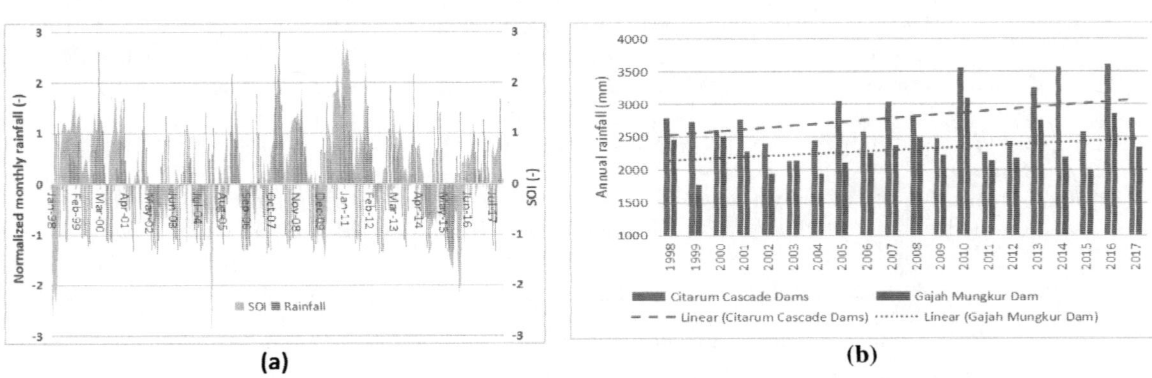

Figure 10. Time series of normalized monthly rainfall over Gajah Mungkur Dam plotted along with SOI (a) and annual rainfall over the Citarum Cascade Dams and Gajah Mungkur Dam basins (b)

References

[1] Huffman G J, Adler R F, Bolvin D T, Gu, G, Nelkin, E J, Bowman, K P, Yong Y, Stocker E F and Wolff D B 2007 The TRMM Multi-satellite Precipitation Analysis (TMPA): Quasi-global, multi-year, combined-sensor precipitation at fine scales *J. Hydrometeor.* **8** 38-55

[2] Hidayat H, Teuling A J, Vermeulen B, Taufik M, Kastner K, Geertsema T J, Bol D C C, Hoekman D H, Haryani G S, Van Lanen H A J, Delinom R M, Dijksma R, Anshari G Z, Ningsih N S, Uijlenhoet R, and Hoitink A J F 2017 Hydrology of inland tropical lowlands: the Kapuas and Mahakam Wetlands *Hydrol. Earth Syst. Sci.* **21** 2579–94

[3] Su F, Hong Y and Lettenmaier, D P 2008 Evaluation of TRMM multiscale precipitation analysis (TMPA) and its utility in hydrologic prediction in the La Plata basin *J. Hydrometeor.* **9** 622- 40

[4] Komite Nasional Pengelolaan Ekosistem Lahan Basah 2004. Strategi Nasional dan Rencana Aksi Pengelolaan Lahan Basah di Indonesia. Kementerian Lingkungan Hidup p 153 Available at http://wetlands.or.id/PDF/buku/Buku%20NSAP%202004.pdf

[5] Muchlisin Z A, Siti Azizah M N, Rudi E and Fadli N 2009 Danau Laut Tawar dan Beberapa Permasalahannya. Proc. *Workshop Selamatkan Danau Laut Tawar*, Takengon 21-22 November 2009. Accesed on 29 Aug 2018 from internet site at: https://www.researchgate.net/publication/268309345

[6] Lukman 2013 *Danau Toba: Karakteristik Limnologis dan Mitigasi Ancaman Lingkungan dari Pengembangan Karamba Jaring Apung* Jakarta: LIPI Press

[7] Fakhrudin M, Wibowo H, Subehi L, and Ridwansyah I 2002 Karakterisasi Hidrologi Danau Maninjau Sumbar. Proc. *Seminar Nasional Limnologi: Menuju Kesinambungan Pemanfaatan Sumberdaya Perairan*, Puslit Limnologi LIPI, Bogor, 22 April 2002 65 – 75

[8] MacKinnon K, Hatta G, Halim H, and Mangalik, A 1996 *The ecology of Indonesia series – The ecology of Kalimantan*, Oxford University Press

[9] Vaillant J J, Haffner D G, and Cristescu M E 2011 The Ancient Lakes of Indonesia: Towards Integrated Research on Speciation *Integrative and Comparative Biology* **51** (4) 634–43 doi:10.1093/icb/icr101

[10] Toruan R L and Setiawan F 2017 Hydrological regimes and zooplankton ecology at Tempe Floodplains, Indonesia: preliminary study before the operation of the downstream barrage In book: *Trends in Asian Water Environmental Science and Technology* doi: 10.1007/978-3-319-39259-2_p 10

[11] Fauzi M, Rispiningtati and Hendrawan A P 2014 Kajian Kemampuan Maksimum Danau Sentani dalam Mereduksi Banjir di DAS Sentani J. Teknik Pengairan, **5** (1) 42–53

[12] Hehanussa P E , Haryani G S and Tjiptasmara 2005 Two ancient lakes in Paniai and Tage, the central mountains range, Papua, Indonesia *Proc. 11th World Lake Conference, Kenya* vol 2 253-58

[13] Boer R, Dasanto B D, Perdinan and Marthinus D 2012 Hydrologic Balance of Citarum Watershed under Current and Future Climate in W. Leal Filho (ed.) Climate Change and the Sustainable Use of Water Resources, Climate Change Management, DOI: 10.1007/978-3-642-22266-5_3, Springer-Verlag Berlin Heidelberg

[14] Hidayani F, Sardjono Y, Fandeli C, Rukmini A R 2017 Creative Environmental Energy Technology assesssment Hydroelectric Power Plant (Case Study of Wonogiri Reservoir) *Indonesian Journal of Physics and Nuclear Applications* **2** (3) 101-10

[15] Burkey and Jeff 2006 A non-parametric monotonic trend test computing Mann-Kendall Tau, Tau-b, and Sens Slope written in Mathworks-MATLAB implemented using matrix rotations. King County, Department of Natural Resources and Parks, Science and Technical Services section. Seattle, Washington. USA. Matlab code available at https://www.mathworks.com/matlabcentral/fileexchange/11190-mann-kendall-tau-b-with-sen-s-method-enhanced

[16] Tanakamaru H, Kato T and Takara K 2004 Water balance analysis and water level simulation of Lake Toba, Indonesia. *Proc. 2nd Asia pacific Association of Hydrology and Water Resources Conf.* **2** 107-15

Disaster mapping as decision support system to decrease abrasion impact due to climate change in Bantul Coastal Area

M Naufal[1] M Nandini,[1]Y A Rahmanul D W A Najib,[1] P B Kusumaningrum,[1] M I Ahyar[1] and D R Hizbaron[2]

[1] Bachelor Student of Faculty of Geography, Universitas Gadjah Mada, Yogyakarta, 55281, Indonesia
[2] Lecturer of Faculty of Geography, Universitas Gadjah Mada, Yogyakarta, 55281, Indonesia

naufalmuhammad@mail.ugm.ac.id

Abstract. Global climate change affects the physical condition of coastal areas, one of which is the rise in wave height. The increase in wave height allows higher volume of carried substrate on the swash and backwash motions. These events have an impact on increasing the abrasion intensity, specifically on the coastal areas which consist of sand material (beaches). Coastal area of Bantul Regency is one of the areas exposed to abrasion because it is directly facing the Indian Ocean. Application of atmospheric science and technology is needed to reduce the impact of losses caused by abrasion. Disaster mapping is an effort to reach that expectation. This research aimed (1) to identify abrasion susceptibility, and (2) to analyse social vulnerability due to abrasion in Bantul coastal area. The assessment was done by GIS-based spatial-multi criteria analysis to process physical and socio-cultural aspects in regards of abrasion, which the analytical unit was the coastal area between Opak and Progo River estuaries. This research resulted in susceptibility map and social vulnerability map as references for coastal mitigation directives supporting the resilience of region and community towards climate change.

Keywords : climate change, abrasion, susceptibility, vulnerability.

1. Introduction

The coastal area is defined as the area between highest tide shoreline and the boundary of landform genesis change. Coastal dynamics can be caused by natural and artificial factors. One of the natural causes of coastal dynamics is the abrasion phenomenon. Abrasion is a destructive process that is affected by sea waves and currents.

Global climate change is hypothesized as a factor in the emergence of various coastal disasters [1], one of which causes a rise in wave height [2] and further causes intensive coastal abrasion. Hydrodynamic factors of coastal genesis in the form of sea waves, tidal waves, and sea currents would practically be intensified due to global warming, thus affect the swash and backwash cycles on the shore. It consequently leads to the more eroded substrate and following shoreline changes.

The coastal area of Bantul Regency is covered by sand substrate. It geographically faces Indian Ocean, which allows continuous accumulation of sandy materials on the beach. As seen from the coastal typology, the coastal area with such conditions could be classified as marine depositional coast [3]. As an exposed area, Bantul coastal is basically susceptible to coastal hazards, and thus the regions' elements would be vulnerable. Reference [4] has studied Depok Coastal as a sample of Bantul coastal area and stated that the coastal is relatively susceptible to disasters in the form of tsunami, abrasion, landslide, and mass movement.

Beside from the shore hydrodynamic factors, abrasion in coastal area of Bantul is also influenced by the reduction of sand deposit transported by Opak River and Progo River. Reference [5] stated that the sand mining activities along those rivers cause the deposition process on the beach part of Bantul coastal. Therefore, this research is focussing on the coastal area bounded by Opak and Progo River estuaries, which are Samas, Pandansari, Goa Cemara, Kwaru, Baru, and Pandansimo. The coastal area is located in the southern part of the districts of Sanden and Srandakan. These regions have the most severe abrasion impacts, which can be seen from its building damages, specifically food stalls and settlements [6]. The abrasion susceptibility is caused by the geophysical aspect of the coastal area, which consists of morphodynamic and hydrodynamic sub-aspects. The morphodynamic factors causing to abrasion are elevation, the area distance to shoreline, morphology, and slope; while the hydrodynamic factors are wave height, tidal wave height, and sea current intensity.

Abrasion poses a serious threat to the area including the people who live there, as the coastal area had become a space to settle and do economic activities to its various resources [7]. The response of social environment in experiencing hazards or disasters could be defined as social vulnerability. Social vulnerability is conditions that are determined by social factors or processes that can increase the vulnerability of a community on the impacts of harm [8]. The Asian Disaster Preparedness Center (ADPC) stated that the social vulnerability assessment is based on its factors including perceptions of the risks and views of the cultural aspects in community, religion, ethnicity, social interaction, age, gender, and poverty [9]. Factors causing to social vulnerability of abrasion are population density, sex ratio, poverty ratio, disability ratio, and dependency ratio [10].

An effort of coastal mitigation is needed to support appropriate policy-making of conserving coastal environment including geophysical and socio-economic aspects. Along with that urgency, this research was conducted to identify the coastal area susceptibility and social vulnerability to abrasion, as the phenomenon has massively damaged the coastal ecosystem in Bantul Regency. Construction of susceptibility and social vulnerability maps based on multi-criteria analysis is the exact attempt to understand the levels of possibility to survive from coastal hazards and disasters. The maps would consequently allow deciding actions to mitigate based on the regional status in terms of abrasion or other coastal hazards. The assessment then followed up by alternatives of attempts to rehabilitate as well as to conserve coastal environment towards the irresistible abrasion impacts, by the applicable efforts to improve the coastal area both geo-physically and socially.

2. Data and Method

The susceptibility and social vulnerability mapping of this research were conducted based on quantitative design. The studied area is the coastal area of Bantul Regency, Daerah Istimewa Yogyakarta Province, Indonesia. The specific area was then selected based on the worst existing condition caused by abrasion, so to construct the analysis unit, i.e. (1) the beach in the whole coastal area between Opak River Estuary and Progo River Estuary for the susceptibility assessment, and (2) each coastal area of Samas, Pandansari, Goa Cemara, Kwaru, Baru, and Pandansimofor the social vulnerability assessment. Figure 1 shows the study area.

2.1. Data Collection

The data consist of main spatial data attributes and also supporting data of coastal characteristics as evidence. Primary data were collected from (1) geomorphological survey, (2) spatial data interpretation, and (3) structured interview to people who live in research area; while secondary data were collected from institutional sources, such as Meteorological, Climatological, and Geophysical Agency and Alaska Satellite Facility.

The main data used to construct susceptibility map were adapted from the Underlying Physical Susceptibility Model (UPSM), which had been applied in to the coastal area of Scotland [11]. The parameters were then sorted to be applied in this research, considering the local coastal characteristics in Bantul Regency. The susceptibility assessment was mainly based on elevation and proximity to open coast (area distance from shoreline) parameters, and supported by coastal morphodynamic (morphology and slope) and hydrodynamic (sea wave height and current intensity) aspects.

The social vulnerability is based on parameters of population density, sex ratio, poverty ratio, disability ratio, and dependency ratio [10], thus the main data were total population, area, total men and women, impoverished people, disabled people, children under-15-years-old and elderly people above-65 years old; and were supported by people's perception of abrasion phenomenon. Those were collected by structured interview census to 35 heads of household as respondents. All the collected data for this research are shown in table 1.

Table 1. Data Source, Place, and Year

Data	Source	Place	Year
Area Susceptibility			
Elevation[a]	DEM processing	Study area	2010
Distance to shoreline[a]	DEM processing	Study area	2010
Morphology[b]	DEM processing and field survey	Study area	2018
Slope[b]	DEM processing and field survey	Study area	2018
Wave height[b]	Meteorological, Climatological, and Geophysical Agency	Bantul Shore	2016
Sea current intensity[b]	Meteorological, Climatological, and Geophysical Agency	Bantul Shore	2016
Social Vulnerability			
Total population[a]	Census	Study area	2018
Men and women[a]	Census	Study area	2018
Impoverished people[a]	Census	Study area	2018
Disabled people[a]	Census	Study area	2018
Elderly people (>65 y.o.)[a]	Census	Study area	2018
Children (0-14 y.o.)[a]	Census	Study area	2018
Perception to abrasion[b]	Census	Study area	2018

[a]Main data.
[b]Additional-supporting data.

2.2. Data Processing

The collected data were put into geodatabase. Both susceptibility and social vulnerability maps were constructed by GIS-based spatial multi-criteria analysis as provided by ArcMap 10.3software. The multi-criteria analysis is basically done by weighting the main spatial data attributes according to their influences on coastal abrasion. The parameters' influence was determined based on literature study.

2.3. Data Analysis

The data were analyzed by descriptive quantitative method. As for the susceptibility map, both elevation and distance parameters were considered to have an equal influence to abrasion; while as for the social vulnerability map, the different weights were applied in parameters of social vulnerability which was based on the agreed national reference of Indonesian National Agency for Disaster Management. The research procedure is showed by a workflow diagram in figure 3.

IOP Conf. Series: Earth and Environmental Science **303** (2019) 012020 doi:10.1088/1755-1315/303/1/012020

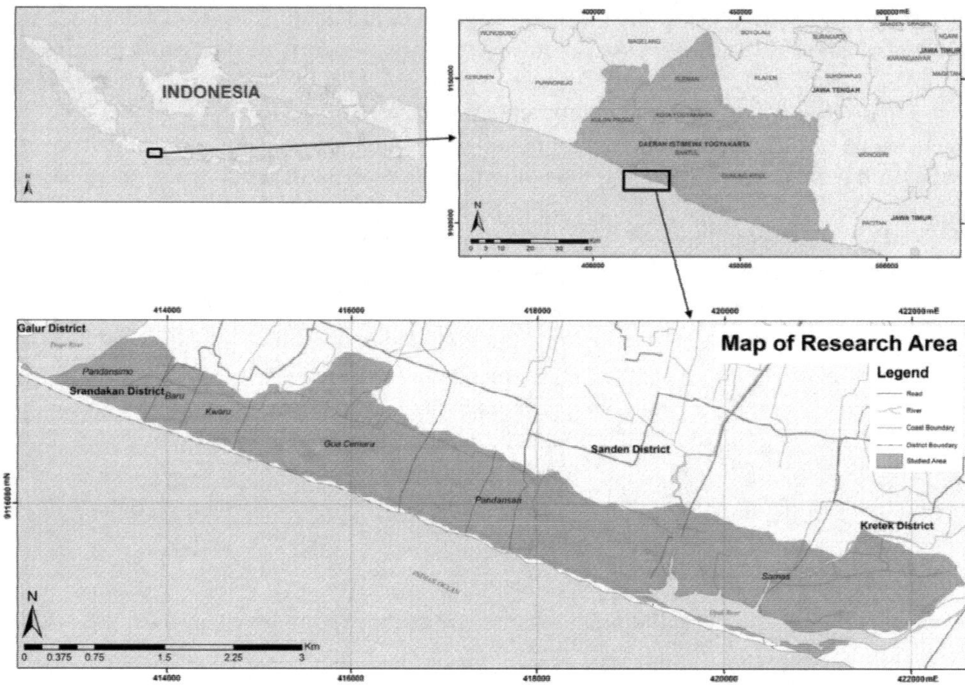

Figure 1. Map of research area

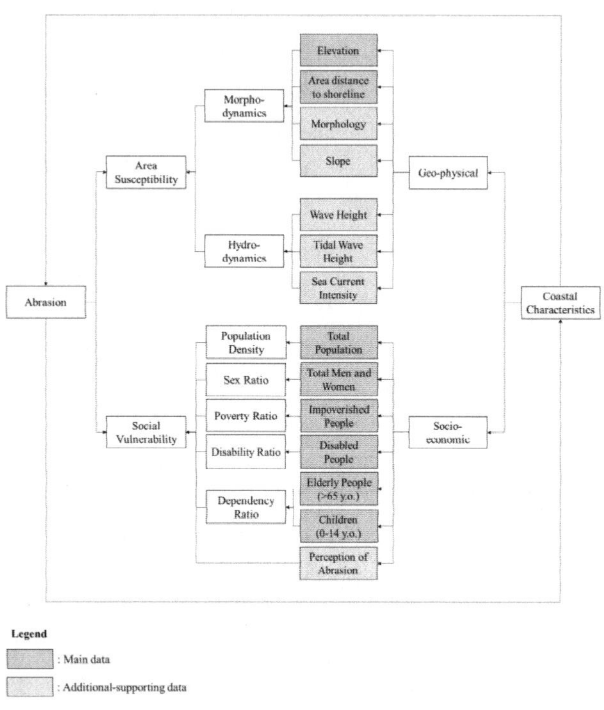

Figure 2. Collected data of coastal characteristics

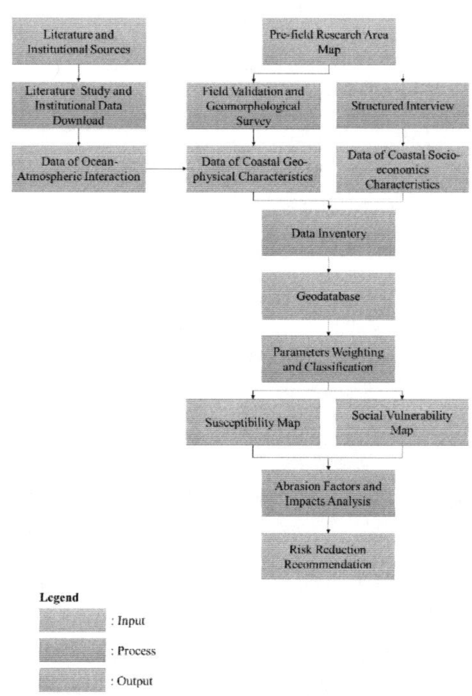

Figure 3. Research procedure

3. Result and Discussion

3.1. Abrasion as an Impact of Coastal Dynamics

Reference [12] stated that coastal area development could be influenced by complex geomorphic processes. Bantul coastal area is classified as marine depositional coastal typology which is characterized by flat to undulating reliefs, sandy material, and marine processes (near the sea) and aeolian processes (near land) [13]. The coastal typology thus causing coastal erosion and sedimentation as dominant geomorphic processes that works in Bantul coastal area. The coastal area of Bantul Regency, specifically the area that is bordered by Opak River and Progo River estuaries has the main substrate in the form of sand materials (sandy shore), so that it has a large beach.

Coastal area is associated with morphodynamic, hydrodynamic, and aeolian processes [14]. Furthermore, the broader factors that affect coastal dynamics has been described, including astrodynamic, aerodynamic, hydrodynamic, morphodynamic, geodynamic, ecodynamic, and anthropodynamic [15]. There are interactions between those processes that affect coastal conditions [5].

Sandy substrate could easily be carried away through geomorphic processes by water energy (forming beach) or wind energy (forming sand dunes). Hydrodynamic factors on the shore in the form of extreme sea waves and currents at certain times could affect the coastal ecosystem. The sea wave dynamics in Bantul in 2016 is shown in figure 4, with the highest wave reaching 5 meters in June while the lowest waves in between 1 and 1.5 meters from December to February. Wave height fluctuation is influenced by wave generator factors such as the velocity, duration, and direction of wind; and fetch length [16]. The extreme weather condition would increase the region's threat to abrasion damages, along with the morphodynamic factors.

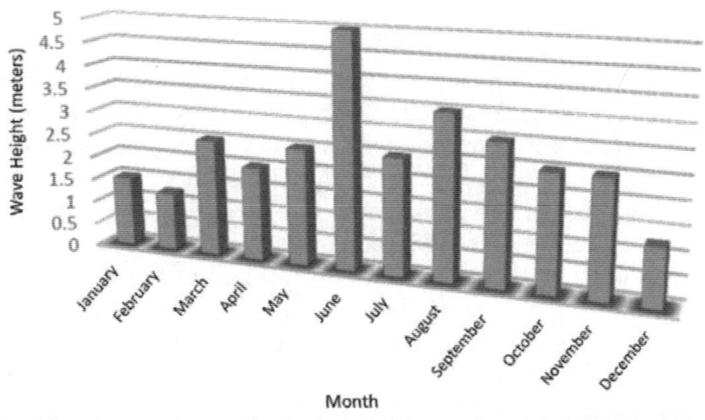

Figure 4. Wave height dynamics in Bantul shore, 2016 (Meteorological, Climatological, and Geophysical Agency, 2016 as quoted in [18])

Global climate change that occurs has increased hydrodynamic intensity in coastal areas. Air temperature on land and sea that has increased due to global warming affects to rising wind velocity and practically followed by wave height [5]. The situation causes intensive abrasion in the study area.

In terms of geomorphology, sand deposits in the beach area of Bantul come from transported volcanic eruption materials from Mount Merapi by Opak River and from Mount Sumbing by Progo River. However, sand mining activities that continue until 2017 in the middle part of Opak River and Progo River cause the reduction in sediment supply of beach. Thus, increasing abrasion intensity in the coastal area of Bantul is an accumulation of suppressing factors in the upstream, middle, and downstream parts of Opak and Progo River.

Abrasion as a result of the coastal dynamics causes damages. Physically, the damages could be seen from worst level in coastal areas of Kwaru, Pandansimo, Samas, and Goa Cemara respectively, including disturbed building structure and land cover pattern. Those coastal areas experienced shoreline retreat due to intensive abrasion until 2013. Table 2 describes the risk elements that were damaged by abrasion.

Table 2. Physical Damages Caused by Abrasion[18]

Coastal Area	Year	Damages
Kwaru	2009	Infrastructure facilities, SAR viewing post, shrimp ponds
	2010	Tourism infrastructure
	2011	Public infrastructure and tourism facilities
	2012	*Casuarina equisetifolia* vegetation
	2013	51 buildings were lost, 25 houses were exposed, *Casuarina equisetifolia* vegetation
Pandansimo	2006	6 houses, 2 stalls, 1 market stall
	2010	Infrastructure facilities owned by fishermen
Samas	2013	20 damaged/missing buildings including houses and public infrastructure
Goa Cemara	2013	*Casuarina equisetifolia* vegetation

3.2 Area Susceptibility to Abrasion

The beach in coastal area of Bantul Regency basically has a high level of susceptibility to shoreline changes, because it contains deposits of unravelled/unconsolidated sandy material and is directly facing high energy of sea wave and current from Indian Ocean [19]. The abrasion susceptibility assessment shows that the lower elevation and distance values would affect the higher level of susceptibility to abrasion, or it could be stated that abrasion susceptibility is inversely proportional to the parameters of elevation and distance from shoreline. The condition causes the graded level of susceptibility as follows: (1) shore area (at elevation of 0-25 meters above sea level (m.a.s.l.) and distance of 0-100 meters from shoreline) is highly susceptible to abrasion; (2) beach area at elevation of 26-35 m.a.s.l.and distance of >100-400 meters from shoreline is moderately susceptible to abrasion; and (3) beach area at elevation of >35 m.a.s.l.and distance of >400-500 meters from shoreline is lowly susceptible. The spatial pattern of susceptibility level to abrasion on shore to beach area in Bantul coastal is shown in figure 5.

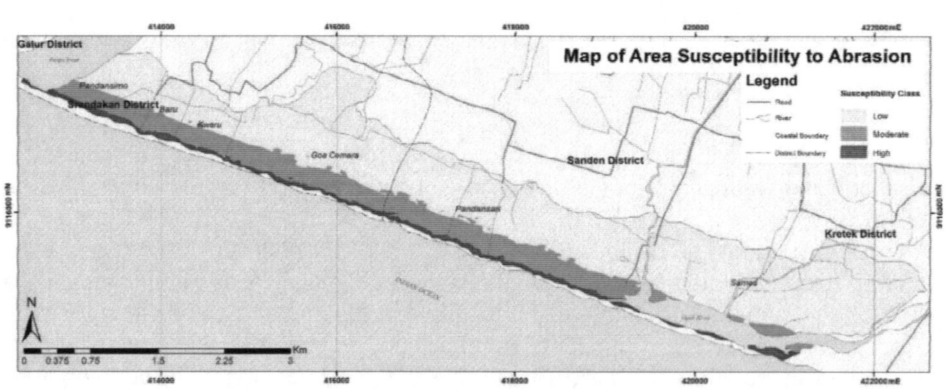

Figure 5. Map of area susceptibility to abrasion in part of coastal area of Bantul Regency

Abrasion is a geomorphic process in the form of erosion in coastal areas that could cause shoreline changes. Changes to the coastline have greater susceptibility if the slope condition is steeper [20]. Relief condition in each coastal area indicates the susceptibility to abrasion horizontally among all the coastal areas, as it contains its certain material and geomorphic process. As seen from beach relief, the coastal areas of Samas and Kwaru are considered to have higher abrasion susceptibility than other coastal areas. It is based on their shore areas that are relatively steeper and wider than other coastal areas.

3.3. Social Vulnerability to Abrasion

Social vulnerability is the level of social inability to the face any forms of hazard [17], which indicates the condition of population that potentially experiencing disadvantages due to hazards that may occur [20]. The level of social vulnerability is determined based on the parameters of population density, sex ratio, age ratio, disability ratio, and poverty ratio [10]. The analysis is focused on the coastal areas of Samas and Kwaru, as only the two coastal areas that have domiciled residents. Thus, the coastal areas of Pandansari, Goa Cemara, Baru, and Pandansimo are practically have zero-value of social vulnerability or classified as not socially vulnerable to abrasion.

Kwaru Coastal Area is categorized in high level of social vulnerability while Samas Coastal Area is categorized in moderate level (figure 6). The assessment result (table 3) shows that the parameters of population density and poverty ratio are the ones that highly influencing social vulnerability differences in the coastal areas of Samas and Kwaru. Samas Coastal Area has a population density of 13 people per square kilometre, while the Kwaru Coastal Area has 86 people per square kilometre. Samas Coastal Area has larger population but larger area, and thus the lower population density than Kwaru Coastal Area. On the contrary, Kwaru Coastal Area certainly the densest coastal area, and thus have the highest level of social vulnerability because the population density parameter is considered to be the most influencing factor. Kwaru Coastal Area experienced narrowing 150 meters of its area in 2013, which the continuation may increase the social vulnerability index for various coastal hazards and disasters.

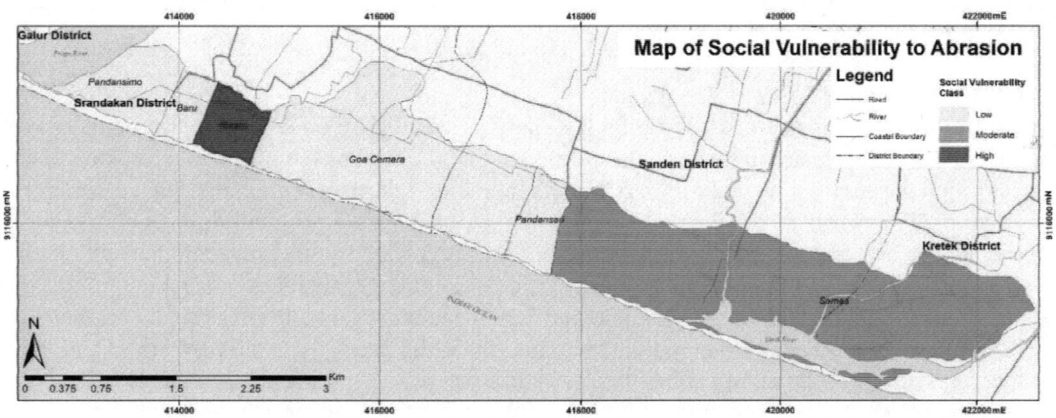

Figure 6. Map of social vulnerability to abrasion in part of coastal area of Bantul Regency

The poverty condition of the population is assessed based on variable of individual income per month. Between the two compared coastal areas, Samas has the worst condition, as the poverty ratio reaches 195.24. It means that there will be 195 poor people if there were 100 non-poor people, thus there might be 95 people who become dependency expenses to the coastal area. The main livelihoods in Samas Coastal Area are fishermen, merchants, and dry land farmers; those are the activities which commodities are seasonally fluctuate. Meanwhile, the residents in Kwaru Coastal Area are in middle social class, which main economic activities are more diversified from traditional farming (including fishing), such as processed seafood seller, shrimp pond farmers, and civil servants.

Table 3. Social Vulnerability Levels of Abrasion Phenomenon.

Coastal Area	Parameter					Social Vulnerability Index	Social Vulnerability Class
	Population Density	Sex Ratio	Poverty Ratio	Disability Ratio	Dependency Ratio		
Samas	0.6	0.3	0.3	0	0.3	1.5	Moderate
Pandansari	0.6	0.1	0.1	0	0.1	0.9	Low
Goa Cemara	0.6	0.1	0.1	0	0.1	0.9	Low
Kwaru	1.8	0.3	0.1	0	0.3	2.5	High
Baru	0.6	0.1	0.1	0	0.1	0.9	Low
Pandansimo	0.6	0.1	0.1	0	0.1	0.9	Low

The gap of tourism development among coastal areas is the main issue causing social inequalities. Kwaru Beach is considered to have greater appeal in terms of tourism than Samas Beach, thus affecting the local people's preference to build up a small business (usually in the form of food stalls) or become tour guides. Those kinds of business have become the main sources of income for most people in the area, as well as the additional sources of income for the existing fishermen and shrimp farmer communities.

3.4. Coastal Defence Recommendation as Disaster Risk Reduction

The analysis of susceptibility and social vulnerability to abrasion in part of Bantul coastal area is the basis for mitigation efforts in the area due to abrasion or any coastal hazards that triggered by extreme hydrodynamic. Thus, it is important to plan the applicable mitigation as an attempt to reduce the potential risks. Based on the results of susceptibility and social vulnerability analysis, the recommended mitigation is consequently need to be area-specific, as each coastal area has different urgency in terms of abrasion phenomenon.

3.4.1. Highly Susceptible Coast to Abrasion.

The susceptibility to abrasion in part Bantul coastal is assessed based on coastal topographic condition. The highly susceptible area requires structural defence efforts, as it is the most effective one. The area mitigation requires retaining wave buildings to decrease the higher wave impacts. This effort certainly requires further studies related to the engineering aspects of wave-retaining buildings. Otherwise, beach reclamation could be done to recover the lost deposit that is eroded by sea wave and current.

The non-structural effort that might be done as coastal abrasion mitigation is renewing the coastal conservation zones periodically, based on the repetition of sea wave and current cycle that affects coastal area of Bantul; so as to evaluate the ecological function of the area. As the follow up to the zonation renewal, reforestation of *Casuarina equisetifolia* vegetation as natural breakwater should also be done, specifically at the nearest exposed area to sea wave and current (shore and front beach).

3.4.2. High-Socially Vulnerable Coast to Abrasion.

The main social problems faced by coastal areas of Samas and Kwaru are respectively poverty and population density. The effort to handle social vulnerability in both coastal areas is focused on the two strategic issues. The empowerment of coastal communities can be aimed in prior to people who live in Samas Coastal Area which are relatively underdeveloped as seen from a whole development perspective. The program that can be implemented is training for workforce and opening local small business that funded by government as an opportunity to promote local coastal commodities. The program is expected to improve the perception of community towards tourism conduciveness in Samas Coastal Area. Coastal commodities such as shrimp ponds, salt ponds, and seafood stalls could be great sources of income, as it is proven in Kwaru Coastal Area. Furthermore, the development of tourism sector towards ecotourism is also recommended to develop an environment-responsive activity as well as to improve regional economy.

Meanwhile, Kwaru Coastal Area has its potential human resource, with higher level in life quality so that the region develops more rapidly. The progress of tourism sector still needs to be balanced with community awareness of abrasion threat, especially with the condition of Kwaru Coastal Area which has historically been the worst affected. Tourists in Kwaru Coastal Area could also be given the responsibility to monitor the coastal ecosystems in line with governmental acts. Thus, the preservation of coastal function could be carried out in a participatory manner by utilizing the human resource potential in Kwaru Coastal Area.

One of the main issues related to intensive abrasion of the beach area is the massive sand mining along the rivers. Thus, the area rehabilitation in each coastal should include programs of sand mining restriction along the strategic line of beach deposit transportation. The strict enforcement, restriction, and periodical monitoring of sand mining activities along the river to the coastal are urgent to be implemented, as the sand mining is consistently done with less consideration of the environment threshold. The proposal could be applied initially along Opak River and Progo River lines, as the routes of beach sediment of Samas, Pandansari, Goa Cemara, Kwaru, Baru, and Pandansimo Coastal Areas. Thus, the proper management involving multi-institutions and multi-administrative regional governments needs to be optimized.

4. Conclusion

The findings in this research could be concluded as follows. Abrasion susceptibility is inversely proportional to the parameters of elevation and distance from shoreline, as in the studied area it resulted in three sub-areas with graded susceptibility levels: (1) shore area at elevation of 0-25 m.a.s.l. and distance of 0-100 meters from shoreline is highly susceptible to abrasion; (2) beach area at elevation of 26-35 m.a.s.l. and distance of >100-400 meters from shoreline is moderately susceptible to abrasion; and (3) beach area at elevation of >35 m.a.s.l. and distance of >400-500 meters from shoreline is lowly susceptible. The social vulnerability to abrasion is primarily influenced by population density and poverty ratio. The coastal area of Kwaru and Samas are respectively have high and moderate level of social vulnerability, as Kwaru Coastal Area is the densest area among the studied area while Samas Coastal Area has dependency expenses; while coastal areas of Pandansari, Goa Cemara, Baru, and Pandansimo are not socially vulnerable to abrasion as there are no domiciled people.

The mitigation to coastal abrasion could be done by structural and non-structural efforts. Structurally, the highly susceptible area requires retaining wave buildings to decrease the higher wave impacts, and also reclamation to recover the beach part as to rehabilitate its function in facing the front shore hydrodynamic. The non-structural effort could be in the form of conservation zones renewal of coastal area periodically, following up the local sea waves and currents cycle; and completed by reforestation of *Casuarina equisetifolia* to optimize its function as a natural breakwater.

Attempting to improve the coastal community capacity to abrasion, it requires the empowerment of people in underdeveloped coastal areas, with some alternatives programs related to workforce training and developing small business which highlights the local coastal commodities, so that they would be stable under the threats of abrasion damages. The coastal areas which have population density problem could use its potential human and economic resources to balance the development of tourism with community awareness of abrasion threat, in actions that involve community, government, and tourists to monitor the coastal ecosystem as well as the river ecosystem which carries out deposit supply for it.

Acknowledgement

The authors would like to thank to all respondents in the cosastal area of Samas, Pandansari, Goa Cemara, Kwaru, Baru, and Pandansimo for the necessary information that had been given as the input of this research; The Meteorological, Climatological, and Geophysical Agency of Indonesia for providing data as the input of this research; and also Mrs. Dyah Rahmawati Hizbaron for the supervision during this research.

References

[1] Hoque M A, Phinn S, Roelfsema C and Childs I 2016 Modelling tropical cyclone hazards under climate change scenario using geospatial techniques IOP *Conf. Ser. Earth Environ. Sci.* **47**, pp. 1–7
[2] Masselink G and Russell P 2013 Impacts of climate change on coastal erosion *Mar. Clim. Chang. Impacts Partnersh. Sci. Rev.* November pp. 71–86
[3] Shepard F P 1972 Submarine Canyons *Earth-Science Rev.* **8** pp 1–12
[4] Subarjo P and Ario R 2015 Uji kerawanan terhadap tsunami dengan sistem informasi geografis (SIG) di pesisir Kecamatan Kretek, Kabupaten Bantul, Yogyakarta *J. Kelaut. Trop.* **18** no. 2 pp 82–97
[5] Sunarto M A, Marfai and Setiawan M A 2014 Geomorfologi dan Dinamika Pesisir Jepara (Yogyakarta: UGM Press)
[6] Hakim B A, Suharyanto and Hidajat K 2012 Efektifitas Penanggulangan Abrasi menggunakan Bangunan Pantai di Pesisir Kota Semarang *Prosiding Seminar Nasional Pengelolaan Sumberdaya Alam dan Lingkungan* pp 122–128
[7] Fuad M A Z and Faiz M DA 2017 Automatic detection of decadal shoreline change on northern coastal of Gresik , East Java-Indonesia IOP *Conf. Ser. Earth Environ. Sci.* **98** pp 1–10
[8] UNISDR 2005 Hyogo Framework for Action 2005-2015: Building the Resilience of Nations and Communities to Disasters Hyogo
[9] Jaswadi, Rijanta R and Hadi P 2012 Tingkat kerentanan dan kapasitas masyarakat dalam menghadapi risiko banjir di Kecamatan Pasar Kliwon Surakarta *Maj. Geogr. Indones.* **26** no. 1 pp 119–148
[10] Indonesian National Agency for Disaster Management 2012 General Guidelines for Assessment of Disaster Risks **2**
[11] Fitton J M, Hansom J D and Rennie A F 2016 A national coastal erosion susceptibility model for Scotland Ocean *Coast. Manag* **132** pp 80–89
[12] Nuraini F 2016 Geomorfometri Lubang Deflasi pada Lorong Angin Alami Gumuk Pasir Aeolian di Pesisir Parangkusumo Kabupaten Bantul Daerah Istimewa Yogyakarta (Yogyakarta: Universitas Gadjah Mada)
[13] Khakhim N, Soedharma D, Mardiastuti A, Siregar V P and Boer M 2008 Analisis preferensi visual lanskap pesisir Daerah Istimewa Yogyakarta untuk pengembangan pariwisata pesisir menuju pada pengelolaan wilayah pesisir berkelanjutan Forum Geogr **22** no. 1 pp 44–59
[14] Pintado E G and Jackson D 2017 Geomorphometry in coastal morphodynamics *Geophys. Res. Abstr.* **19**
[15] Sutari C A T, Siswanti E and Jiwaningrat Y 2015 Antropodinamik Tambak Udang Sebagian Pesisir Kabupaten Bantul *Bul. Geomaritime* **1**
[16] Coastal Hydraulic Laboratory 2000 Coastal Engineering Manual Part I-IV. Washington DC: Department of the Army, US Army Corp of Engineers
[17] Widodo M M and Hizbaron D R 2017 Kerentanan sosial ekonomi di wilayah kepesisiran terhadap dampak gelombang tinggi di Kecamatan Srandakan, Kabupaten Bantul *J. Bumi Indones.* **6** no. 1
[18] Marine Affairs and Fisheries Agency of Bantul Regency 2013 Data Kelautan, Pesisir dan Pulau-Pulau Kecil Bantul
[19] Saputro G B, Marchiavelli M I C, Ibrahim F and Maulana E 2017 Identification of typology related to the coastal line changes in Bantul IOP *Conf. Ser. Earth Environ. Sci.* **54** pp 1–8
[20] Hizbaron D R, Baiquni M, Sartohadi J and Rijanta R 2012 Urban vulnerability in Bantul District, Indonesia—Towards Safer and Sustainable Development Sustainability **4** pp 2022–2037

Rice plantation projection based on scenario representative concentration pathways (RCP) 4.5 in West Java Province

N M Hidayat, L Agustina, D R Wibawanty and A Safril
School of Meteorology Climatology dan Geophysiscs, South Tangerang, Banten, Indonesia

nmanarul@gmail.com

Abstract. Climate change is characterized by the changing of weather and climate elements in the earth, such as temperature and precipitation over a long period. The changing of these climate elements will affect the water balance then affect the agricultural management system, especially in timing and cropping patterns. This study aims to explain the projection of climate change result in the future and its relation with the changing of planting time and crop patterns of rice in West Java Province based on Representative Concentration Pathways (RCP) 4.5 scenario from 2006 until 2040. The data used in this study were observation data of temperature and rainfall in period of 2006-2015 from 39 rainfall posts with the height data of each rainfall post. The analysis conducted in this study was based on Thorntwaite and Matter water balance models by using observation data (2006-2015) as the baseline and future periods (2021-2030 and 2031-2040). Air temperature and rainfall data in period of 2006-2015 were used to correct the model data. Calculation of groundwater availability was calculated for each representative of Oldeman climate type. Furthermore, the rice planting schedule can be known when the condition of the level of groundwater availability reaches field capacity. The results of the analysis show changes in the Oldeman climate type and the level of groundwater availability due to changes in air temperature and rainfall parameters resulting from projections on current conditions. This caused a change in the planting schedule. The planting schedule will start early when the climate type turns wet, so the condition of the level of groundwater availability increases. The planting schedule in September was seen more broadly in the East and West parts of West Java in the period of 2021-2030 compared to the period of 2006-2015 and 2031-2040.

Keywords : plantation, climate change, temperature, precipitation

1. Introduction

The evaluation results by the Ministry of Agriculture in 2015 showed that the fundamental problems predicted to still be faced by the food crop sub-sector in the period of 2015-2019, covering aspects such as environmental damage, climate change, infrastructure, facilities, land, and water. Plants will produce high production if given enough water at the right time [1].

West Java Province is geographically located in 5°-50'-7°50' south latitude and 104°-48'-108°48' east longitude. This province is a center of rice production that has the highest rice production rates compared to other provinces [2]. Based on data from Jawa Barat Pertanian (2011), several regions in West Java experienced a shift in rice planting periods such as in Indramayu. Information on planting calendars in this region is divided into three growing seasons, namely the first planting period (November), the second planting period (April), and the third planting period (August).

Suriadi [3] said that the rainfall pattern of West Java during the last two decades (1991-2007) had significant changes. Therefore, adaptation and mitigation efforts are needed. One effort is to establish information about the groundwater availability for plants and provide an overview of planting schedules that are suitable for future climate conditions.

The view on future climate change can be obtained with Representative Concentration Pathways (RCP). RCP has four scenarios, namely RCP 2.5, RCP 4.5, RCP 6.0 and RCP 8.5. The use of scenarios is not aimed at predicting the future, but to get a better understanding of the uncertainties and alternatives of the future [4]. The results of climate projections provide information about the projectional changes in groundwater availability, in order to determine the planting schedule especially for rice in West Java.

This study aimed to determine the pattern of rice planting schedules by comparing the current period (2006-2015) with future periods (2021-2030 and 2031-2040) using projection data. The parameters used were temperature and rainfall. Both were used for the analysis of future groundwater availability so that the projected planting schedule can be determined. The results of the climate projection provided an overview of climate conditions until 2040. This can be one step to deal with climate change, by establishing planting patterns and planting calendars based on climatic conditions, especially for rice in West Java Province so as to minimize crop failure due to climate change and to obtain maximum production.

2. Data and Method

The research data was based on the monthly rainfall observation data from 2006 to 2015 in 39 observation points consisted of coperating rainfall posts and BMKG stations in West Java Province. The air temperature was obtained based on the height of the place using the Braak formula [5]:

$$T_h = T_{h_0} - \left(\frac{0.6}{100} h\right) \tag{1}$$

with:
T_h = temperature at the rainfall post (°C)
T_{h_0} = temperature at the reference station (°C)
h = height difference of rainfall posts asl (meter)

Numerous GCMs simulation runs have been implemented by various countries as part of the WCRP's Fifth Coupled Models Inter-comparison Project (CMIP5) in which the simulation outputs can be downloaded from its web portal. For SEACLID/CORDEX SEA, member countries within the Southeast Asia region use GCM model simulation outputs archived at the Abdus Salam International Centre for Theoretical Physics (ICTP), Trieste, Italy. Model data for RCP4.5 scenario projection get from the center of climate change in BMKG. The data projection RCP4.5 needs correction to be equalized with observational data. The correction was calculated based on the difference between observation data and model data. Model and observation data must have the same period. Corrected rainfall can be calculated using the following equation [6]:

$$CH_{model\_kor} = CH_{mod} \; x \; \frac{\overline{CH}_{obs}}{\overline{CH}_{mod}} \tag{2}$$

where,
CH_{model_kor} = model monthly rainfall after correction
CH_{mod} = model rainfall before correction
$\overline{CH}_{obs}$ = observed average rainfall in baseline period
$\overline{CH}_{mod}$ = model average rainfall in baseline period

Whereas for corrected air temperature can be calculated by the following formula (Weiland et al., 2010):

$$T_{mod\_kor} = T_{mod} + (\bar{T}_{obs} - \bar{T}_{mod}) \quad (3)$$

T_{mod_kor} = monthly air temperature after correction
T_{mod} = monthly air temperature before correction
$\bar{T}_{obs}$ = observed average of monthly air temperature in baseline period
$\bar{T}_{mod}$ = model average of monthly air temperature in baseline period

Determination of Oldeman climate types for each period was done using corrected model data [7]. This was based on the number of wet months and dry months respectively with the following criterias:

Table 1. Classification critera for Oldeman climate types

Main types	Respective wet month	Sub Division	Respective dry month
A	>9	1	<2
B	7-9	2	2-3
C	5-6	3	4-6
D	3-4	4	>6
E	<3		

Reference: (Oldeman, 1982)

The next step was to determine the planting schedule based on the level of groundwater availability. This was to find out the favorable planting period. To find out the water availability for plants, it was necessary to know evapotranspiration values calculated by the method of Thronthwaite and Matter [8] as follows:

a. Calculating monthly heat index (i):

$$i = \left(\frac{t}{5}\right)^{1.514} \quad (4)$$

t = monthly average temperature

b. Calculating total monthly heat index (I):

$$I = \sum_{Jan}^{Des} i \quad (5)$$

c. Calculating standard ETP :

$$ETP = 16\left(\frac{10t}{I}\right)^{a} \quad (6)$$

ETP = monthly average standard ETP (mm)

A $= 675 \times 10^{-9} I^3 - 771 \times 10^{-7} I^2 + 1792 \times 10^{-5} I + 0.49239$

d. Correction of standard ETP uses day length (for latitude 0, 1 day = 12.1 hours) and the number of days per month = 30 days, thus:

$$ETP = \left(\frac{X}{30}\right)\left(\frac{Y}{12}\right) ETP_{standard} \quad (7)$$

IOP Conf. Series: Earth and Environmental Science **303** (2019) 012021 doi:10.1088/1755-1315/303/1/012021

X = number of days in a month
Y = day length in hours
The day length table was obtained from secondary data in "Crop Evaporation"[9].

e. The column for air temperature was filled according to the results of average temperature estimation in each rainfall station using the Braak formula that has been done before.

Furthermore, the evapotranspiration value was used to determine the value of groundwater availability for plants. Groundwater availability was calculated using the method of Thorntwaite and Matter land water balance on a monthly basis. Data used were rainfall, water content at field capacity level (KL) and water content at permanent wilting point (TLP). Steps in water balance calculation were as follows:

1. The column for rainfall (CH) was filled with CH data on a monthly average or CH with certain probability.
2. The column for potential evapotranspiration (ETP) was filled with the results of the ETP estimation using the equation by Thorntwaite & Matter (1957)
3. The column for CH-ETP was filled with the difference between rainfall and plant
4. evapotranspiration.
5. The column for accumulation potential of water loss (APWL) was filled with negative results in step 3, accumulated month by month as Accumulation Potential of Water Loss (APWL) value and then filled in the corresponding column.
6. The column for groundwater content (KAT). First, field capacity (KL) was determined. The filling of the KAT column began at the first month of APWL based on the formula calculation as follows:

$$KAT = KL \text{ x } k^{|APWL|} \tag{8}$$

where:

k $= p_0 + p_1/KL$
p_0 = 1.000412351
p_1 = -1.073807306
KL = field capacity
|APWL| = absolute value of APWL

The column for KAT at the first month, where CH-ETP is positive, was filled with (KAT = last KAT + CH-ETP), so on until the KAT = KL value is reached. Since that month, during the excessive raining, the KAT value was constant, equal with KL. From the results of KAT, the index value or water requirement criteria for plants can be found as follows:

$$ATS = \frac{KAT-TLP}{KL-TLP} \, x \; 100\% \tag{9}$$

Where :

ATS : percentage of available groundwater
TLP : permanent wilting point
KAT : groundwater content
KL : field capacity

Table 2 describes the percentage of ATS by the level of groundwater availability. ATS results are divided into 3 classes, as follows:

Table 2. Percent of water available (% ATS)

Percentage of ATS	Level of groundwater availability
<40	Low
40-60	Medium
>60	Sufficient

3. Results and Discussion

3.1 Correction results of RCP 4.5 scenario model data

RCP 4.5 scenario data consisted of air temperature and rainfall data which still have significant errors. To minimize this error, the RMSE statistical test was performed. The smaller the RMSE value, the better the performance and accuracy of this model data, thus made it proper for research use. The correction results showed that the model data has lower RMSE value compared to the model data before correction. This can be seen in table 3 and table 4.

Table 3. RMSE value of rainfall model data

No	Rain Post	RMSE Before Correction	RMSE After Correction
1	Darmaga	206.74	159.32
2	Depok	199.55	180.57
3	Gunung mas	262.83	190.31
4	Montaya	144.44	138.1
5	Cibeet	189.5	149.15
6	Cibereum	123.6	121.66
7	Empagra	161.73	147.5
8	Jatiseeng	185.05	106.27
9	Batujaya	266.89	151.57
10	Krangkeng	202.24	76.63

Table 4. RMSE value of air temperature model data

No	Rain Post	RMSE Before Correction	RMSE After Correction
1	Darmaga	1.15	0.25
2	Depok	0.53	0.38
3	Gunung mas	0.48	0.35
4	Montaya	3.05	0.4
5	Cibeet	1.81	0.39
6	Cibereum	0.59	0.4
7	Empagra	1.37	0.38
8	Jatiseeng	0.65	0.25
9	Batujaya	0.65	0.36
10	Krangkeng	0.63	0.32

3.2 Changes in Oldeman climate types

Spatially, the regions with B2 and B3 climate types extended in the central part during the period of 2021-2030 compared to 2006-2015 and 2031-2040. B2 and B3 climate types have sufficient amount of rainfall for the needs of rice plants up to two planting times and one planting times of secondary crops. This was because monthly average rainfall in the central part of West Java Province was higher in that period. Meanwhile, during 2031-2040 the regions with D3 and E climate types in the northern coast were more extended than in the period of 2006-2015. This condition was expected because in that period, monthly average rainfall in the rainfall posts located in the northern coast had fewer wet months and more dry months. The distribution of Oldeman climate types in West Java Province in each period can be seen in figure 1, figure 2, and figure 3.

3.3 Groundwater availability

In each Oldeman climate type in West Java Province, one representative rainfall post was chosen and its groundwater availability was calculated. Air temperature and rainfall data from direct observation and model data projection were used as input in the calculation of groundwater availability, thus the condition of groundwater availability in normal period (2006-2015) and projection period (2021-2030 and 2031-2040) were obtained.

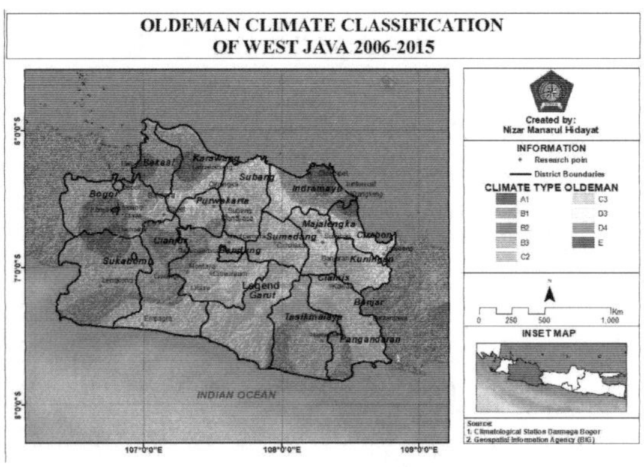

Figure 1. Oldeman climate types in West Java Province in the period of 2006-2015.

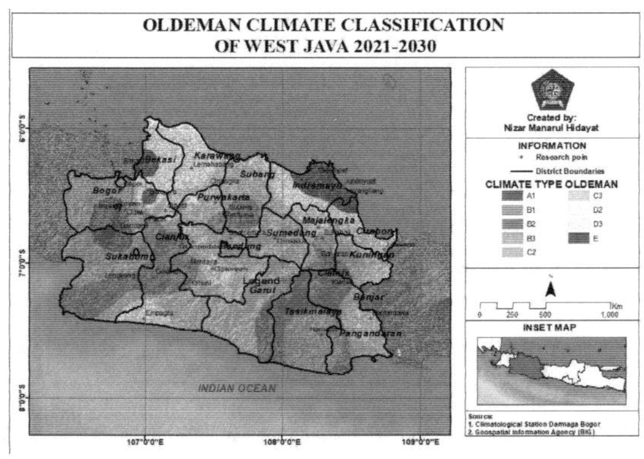

Figure 2. Oldeman climate types in West Java Province in the period of 2021-2030.

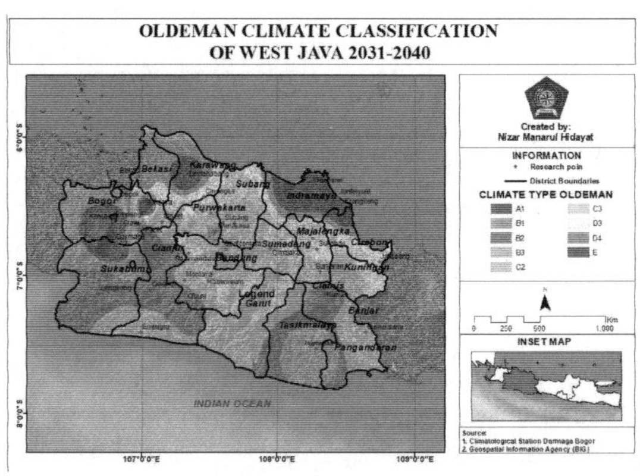

Figure 3. Oldeman climate types in West Java Province in the period of 2031-2040.

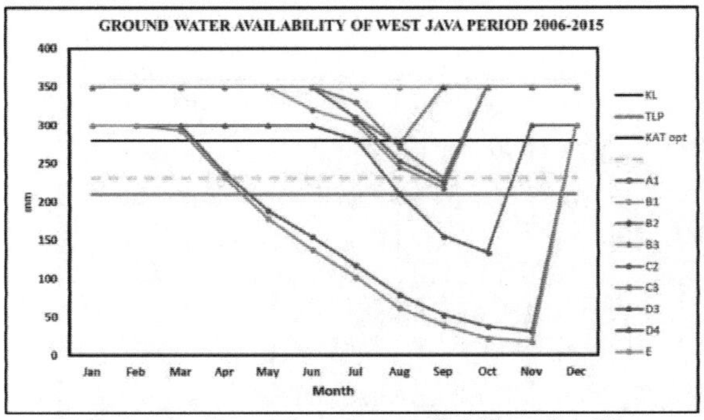

Figure 4. Groundwater availability in West Java Province in the period of 2006-2015.

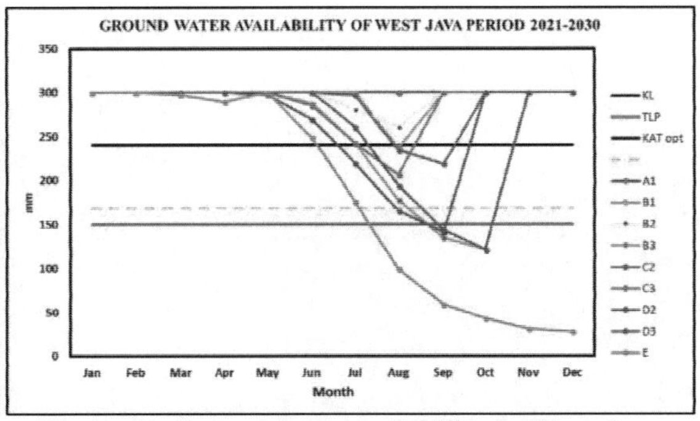

Figure 5. Groundwater availability in West Java Province in the period of 2021-2030.

In the period of 2006-2015 (figure 4), the A1 and B1 climate types almost reached field capacity in all months. However, there was a decline in B2 climate type in August, although insignificant, because in the next month it was above the permanent wilting point and rose progressively to be able to fulfill the need for rice plantings. As for D4 and E climate types, the groundwater availability levels decreased from May to November. For this reason, the regions with D4 and E climate types could not be planted with rice.

The value of groundwater availability during 2021-2030 in A1, B1, and B3 climate types (figure 5) reached field capacity in most months, but in B1 climate type began to decline in July and August. The decline was still above the optimum KAT; thus, it was still sufficient for rice planting throughout the year. As for E climate type, the value of groundwater availability was very low so it was not sufficient for rice planting.

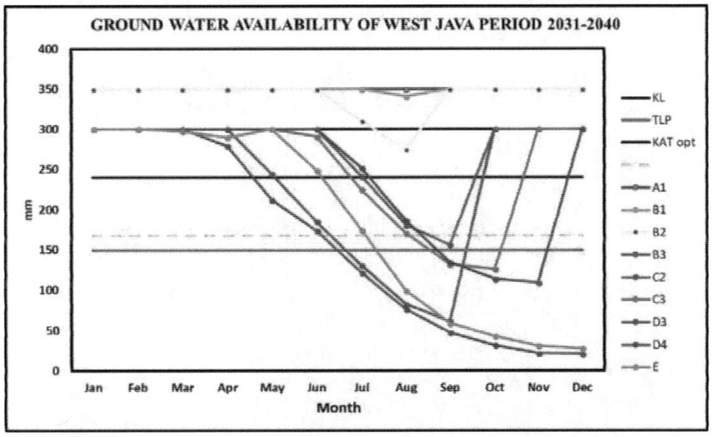

Figure 6. Groundwater availability in West Java Province in the period of 2031-2040.

The value of groundwater availability in the period of 2031-2040 (figure 6) in A1, B1, and B2 climate types reached field capacity in most months. When compared with the previous period, the period of 2031-2040 showed a significant decline in June to September. Whereas in D4 and E climate types, the value of the groundwater availability were very low, thus insufficient for rice planting.

3.4 Changes in planting schedules
Based on the calculation of groundwater availability in the baseline period, the first, second and third rice planting schedule for each period were obtained.

3.4.1 Changes in the first rice planting schedule

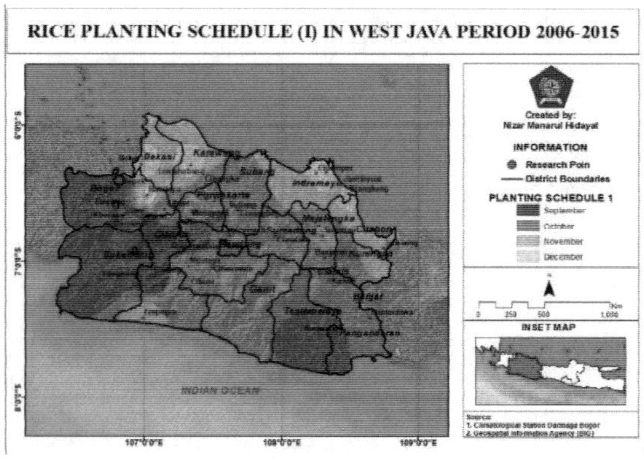

Figure 7. The first rice planting schedule in the period of 2006-2015.

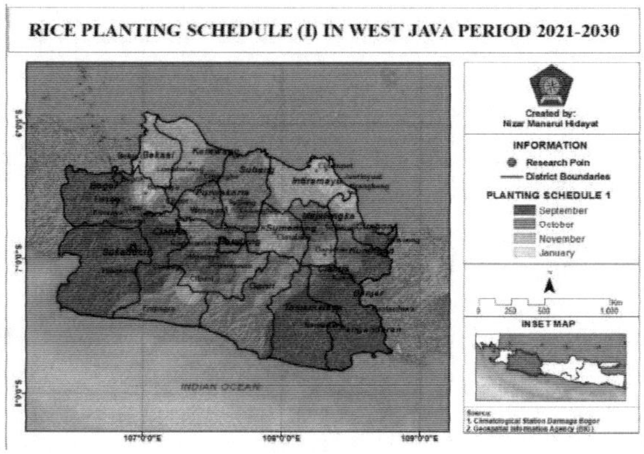

Figure 8. The first rice planting schedule in the period of 2021-2030.

Based on figure 7, it can be seen that during the period of 2006-2015, in the central part of West Java, most began to plant rice in October, while small portions in the Bekasi, Indramayu, and Karanganyar planted rice in November. Whereas, the northern coastal region had a planting schedule in December.

Figure 8 showed a change in the first planting schedule in the central part of West Java. In the period of 2021-2030, the first rice planting schedule in central West Java tended to begin in September, earlier than in the baseline period (2006-2015). The western part of West Java generally had a planting schedule in September that extended to Bandung Regency. In the northern coastal region of West Java, the planting schedule was later, which was in January.

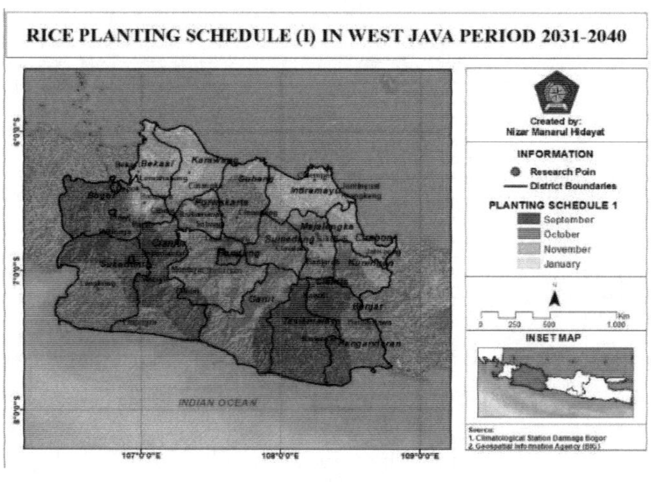

Figure 9. The first rice planting schedule in the period of 2031-2040.

The period of 2031-2040 (figure 9) had a similar pattern to the baseline period (2006-2015). It is shown in the spatial illustration where in most parts of the central West Java, the first rice planting occurred in October. However, the western part of West Java experienced a narrowing areas.

3.4.2 Changes in the second rice planting schedule

Changes in the second planting schedule were seen in the central part of West Java. In the period of 2006-2015, the central part of West Java began the second rice planting in February and a small part began in January. During the baseline period (2006-2015) and near period (2021-2030) it was divided into 4 areas, namely areas with planting schedules in January, February, March, and areas with no planting schedule. The difference was seen in the middle period (2031-2040) that only has three areas, namely the areas with planting schedules in January, February, and areas with no planting schedule. Spatially, the second rice planting schedule can be seen in figures 10, 11 and 12.

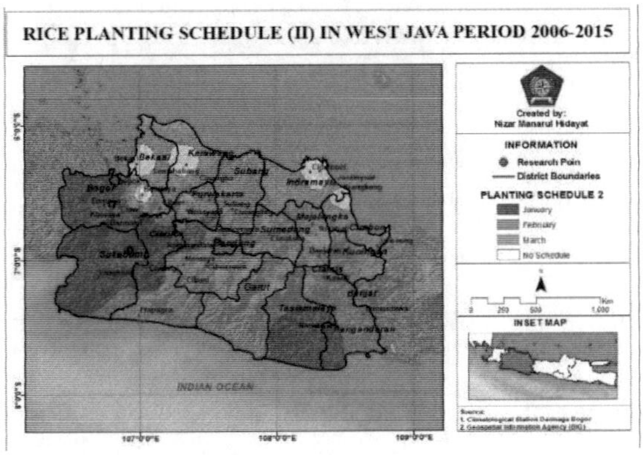

Figure 10. The second rice planting of schedule in the period of 2006-2015.

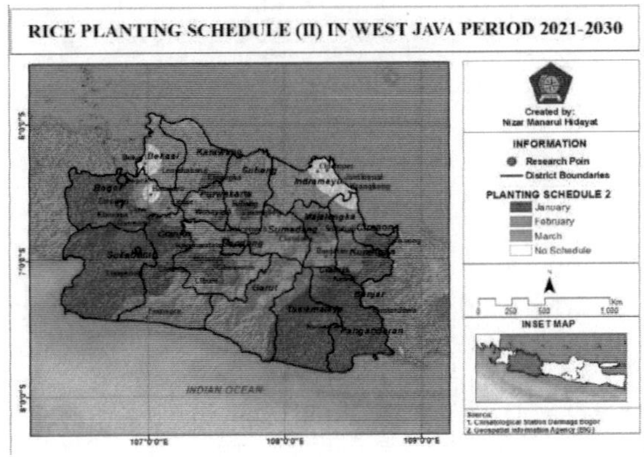

Figure 11. The second rice planting of schedule in the period of 2021-2030.

3.4.3 Changes in the third rice planting schedule

In general, during the third planting season of the baseline period (figure 1) the entire region of West Java Province had two planting areas, namely the area with planting schedules in May and June. This was due to the deficient value of groundwater availability, so that it could not reach three planting seasons. In the baseline period (2006-2015) the regions with planting schedule in May were almost the same in each period, covering a small part of Bogor, Sukabumi, and a small part of Tasikmalaya Regency.

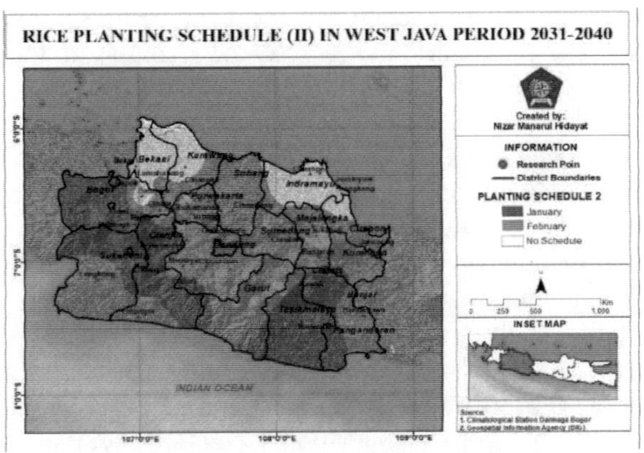

Figure 12. The second rice planting schedule in the period of 2031-2040.

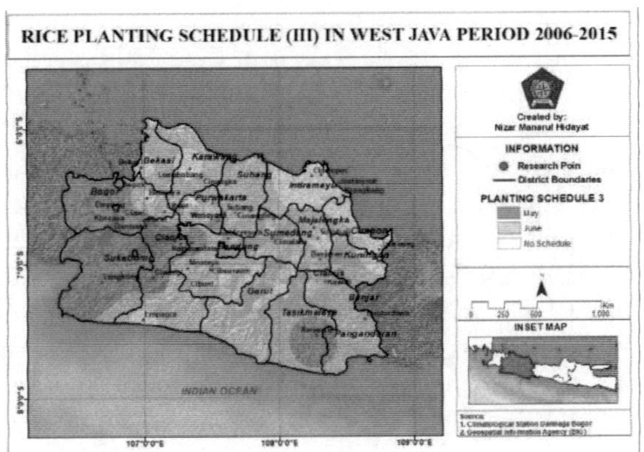

Figure 13. The third rice planting schedule in the period of 2006-2015.

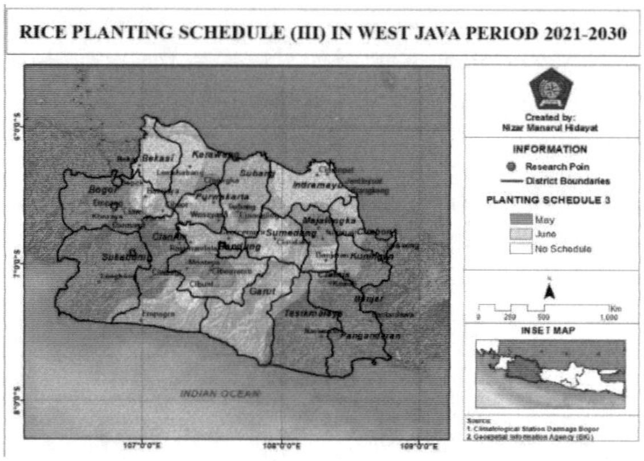

Figure 14. The third rice planting schedule in the period of 2021-2030.

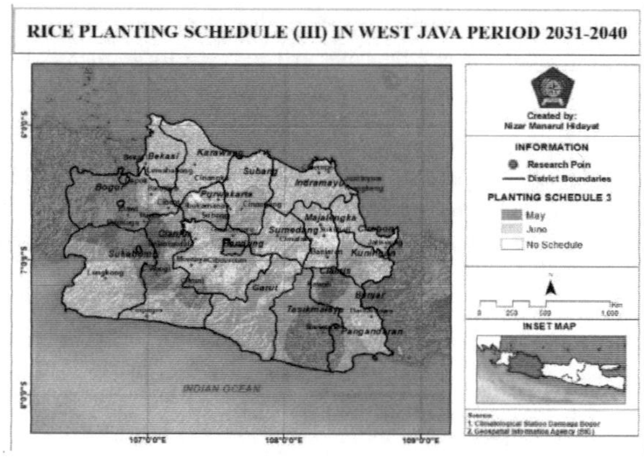

Figure 15. The third rice planting schedule in the period of 2031-2040.

In the period of 2021-2030 (figure 14) the region with the start of planting in May extended to the central and eastern parts, which was different from the previous period. There also several areas that did not have planting schedules, namely in parts of Indramayu, Bekasi, Sumedang, and the eastern part of Bandung. Figure 15 showed the third planting schedule in the middle period. It can be seen clearly that the northern coastal region did not have planting schedule. Whereas the western part of West Java such as Bogor, Cianjur, and parts of Sukabumi Regency still had planting schedules which fell in May. Spatially, the area that did not have planting schedule extended to the middle part.

4. Conclusion

Based on the results of this research, it can be concluded that the projection of Oldeman climate types had insignificant changes compared to the baseline period (2006-2015). In the period of 2006-2015 to 2031-2040, the Oldeman climate types consisted of 9 types. Regions with B2 and B3 climate types were more extended in the central part during the period of 2021-2030 than in the period of 2006-2015 and 2031-2040. Meanwhile, in the period of 2031-2040, there was an extension of regions with D3 and E climate types in the north coast.

The groundwater availability in the central part of West Java Province was generally categorized as "sufficient" in all periods. Meanwhile, regions with "low" to "medium" level of groundwater availability were more spread in the northern coast. Of the three periods that have been analysed, the largest region with "low" groundwater availability occurred in September during 2006-2015, since September is the peak of dry season.

The change in air temperature parameters and the projected rainfall to the baseline period (2006-2015) caused changes in the rice planting schedule. Changes in the first planting schedule were evident in the central part and the north coastal region of West Java. During the planting season of the 2006-2015 period, generally the rice planting began in October in the western part and November in the north coast. In the period of 2021-2030, the start of planting was the same as in the period of 2006-2015, which was in October. Meanwhile, in some parts of the northern coastal region, the start of planting in November changed to January. In the period of 2031-2040, the first planting in the central part of West Java began in October while in the north coastal region began in January. The second planting in the central part of West Java during the period of 2006-2015 generally began in February, during 2021-2030 generally began in January and during 2031-2040 began in the same month as in 2006-2015 (February). The start of the third planting schedule only existed in some parts of Bogor, Sukabumi, Cianjur, Garut, Bandung, Tasikmalaya, Ciamis, some parts of Subang, Banjar, and Pangandaran, which was in May.

5. Acknowledgments
We gratefully acknowledge the Dramaga Bogor Climatology Station for providing and making available their observation data, and special thanks to three anonymous reviewers for their most insightful and constructive comments.

6. References
[1] Tjasyono B H K 2004 *Klimatologi* (Bandung: Institut Teknologi Bandung)
[2] BPS Data Produksi Padi Provinsi Jawa Barat 2015 jabar.BPS.go.id (Accessed: 30 November 2017)
[3] Suriadi A B 2010 Perubahan Iklim dan Ketahanan Pangan di Jawa Barat *Majalah Ilmiah Globe* **12** 48-56
[4] IPCC 2013 *Climate Change 2013: The Physical Science Basis* Cambridge: Cambridge University Press
[5] Rusdiana O and Amalia R F 2012 Kesesuaian Lahan Pintus Merkusii Jungh et de Vriese Bekas Tegakan Tectona grandiss Linn *Jurnal Silvikultur Tropika* **3** 174-81
[6] Weiland F C, Van Beek L P H, Kwadijk J C J and Bierkens M F P 2010 The Ability of A GCM-Forced Hydrological Model to Reproduce Global Discharge Variability *Hydrology and Earth System Science Journal* **14** 1595-621
[7] Oldeman L R and Frere M 1982 *A Study of the Agrometeorology of the Humid Tropics of South East Asia* (Geneva: World Meteorological Organization)
[8] Thornthwaite C W and Mather J R 1957 *Instruction and Tables for Computing Potential Evapotranspiration and the Water Balance* (New Jersey: Drexel Institute of Technology)
[9] Allen R G, Pereira L S, Raes D and Smith M 1998 Crop Evapotransporation – Guidelines for Computing Crop Water Requirements *FAO Irrigation an Drainage Paper 56.*

Local Climate Zone classification for climate-based urban planning using Landsat 8 Imagery (A case study in Yogyakarta Urban Area)

Y F Pradhesta[1], E Nurjani[2] and B I Arijuddin[1]

[1]Geography Faculty, Gadjah Mada University, Yogyakarta
[2]Department of Environmental Geography, Gadjah Mada University, Yogyakarta

yayang.fajar.p@mail.ugm.ac.id

Abstract. Urbanization of the last decade contributed to the urban morphological changes in Yogyakarta Urban Area (YUA). These conditions have an impact on local climate zone variations. There is an urgent need for enhanced climate-based urban planning. This study aims to: 1) Classify the Local Climate Zone (LCZ) in YUA using the World Urban Database Access and Portal Tool (WUDAPT) method, and 2) develop climate-based urban planning recommendations of the classification results. WUDAPT method used in this research based on the concept of LCZ framework by utilizing Landsat 8 imagery. LCZ classification of the whole area is determined using random forest classifier algorithm by SAGA GIS. The results of this study indicate that there are 12 LCZs in YUA, consisting of 6 types of LCZ based on building type and 6 LCZs based on land cover type. The highest increase of LST are in LCZ 3 and LCZ 5. Consequently, there's a change in diurnal temperature from low to medium. Therefore, the thermal comfort aspect define by four components of LCZ is important in YUA urban planning strategy.

1. Introduction

Over the past three decades, Indonesia has experienced a substantial urbanization. The number of Indonesian people living in urban areas has increased from 42% in 2000 to 49.8% in 2010 [1]. Based on National Development Planning Agency (Bappeda), 66.6% of Indonesian population will live in urban areas by 2035. The level of urbanization for some provinces, especially in Java and Bali are already higher than in all of Indonesia. The highest urbanization rate with percentage above 80% is in four provinces in Java; DKI Jakarta (100%), West Java (89.3%), Banten (84.9%), and Yogyakarta Special Region (84.1%) [2].

In the Special Region of Yogyakarta, a high level of urbanization goes according to its function as a student and tourist city. Therefore, population growth and urban activities are also increasingly high, demand greater land requirements. Land use change in the term of urban physical development become undeniable. Yogyakarta City experienced massive acceleration of physical development in urban areas due to the urbanization process. From 2006 to 2009, the changes of open land into settlements and built land reached 442.61 ha [3].

The characteristics of urban development during the urbanization process are identical with extensive paving, intensive physical development both vertically and horizontally, accompanied by emissions from various pollutants that have a major influence on changes in the natural conditions of urban areas. As a result, several environmental problems such as Urban Heat Island (UHI), pollution and air quality degradation will directly affect the lives of urban residents. The vulnerability to disasters will be higher along with the high concentration of population, infrastructure and socio-economic activities, especially the vulnerability to climate disasters. If mitigation is not immediately carried out, this condition can lead to the emergence of various climate disasters [4].

1.1. The need of applying the LCZ classification system in Yogyakarta Urban Area

In mitigating climate disasters and undesirable changes during the urbanization process, every urban area in the world has a different capacity and response. This is due to the differences in the background of political culture, economic base, and social conditions of the population. Likewise in climate-based urban planning. Developed countries and developing countries have significant differences in spatial planning in urban areas. In supporting the planning process, developing countries generally have limited availability of urban data, while in developed countries urban data is easily accessible by open source. Therefore, standardization of the information is needed regarding urban data that can describe urban landscape, globally implemented in all urban areas both in developed or developing countries. This standardization is then used as a guide in the study of urban climate and planning throughout the world.

Although there are plentiful of previous studies have revealed the impact of urban physical development on the local climate, the findings of the study have limited influence on the urban planning process [5][6]. This is because there are obstacles between researchers and planners, specifically the differences in the availability of information at different scales [7]. In a climate perspective, the information needed to determine differences in the spatial characteristics of the region's climate in urban areas are the form of the city and the function of the city. The form of the city is described through land surface cover, building construction material, and urban morphology. The function of the city is described based on its function, for example as a residential, commercial building, or industrial area. Ideally, both information are spatially-temporally collected and analysed. In the final stage, the right recommendations were found to overcome changes due to the urbanization process.

Yogyakarta is one of the city in a tropical country with a high level of urbanization, the role of LCZ in determining LST variations in YUA is interesting. The higher the level of urbanization, the higher the growth of urban areas. This will also increase LST. The threat of stress, thermal discomfort, increased energy demand because the use of air conditioning has a direct impact on the lives of residents of the YUA region. The threat is not only for now, but also in the future.

1.2. LCZ Mapping Method

The World Urban Database and Portal Tool (WUDAPT) is a community-based project collecting data in urban areas around the world using the Local Climate Zone (LCZ) scheme. LCZ is a region with similar land surface cover, building construction materials, structures and human activities with a range of 100 meters to several kilometres on a horizontal scale [8]. Tools and materials used in the LCZ classification can be obtained free of charge (open source), namely Landsat, Google Earth Pro, and SAGA GIS. LCZ has been widely applied, developed and modified in urban areas in various countries, both developed and developing countries.

Developed countries usually apply LCZ level 1 (detail) and level 2 (very detailed), while in developing countries it is better to apply LCZ level 0 because of the limited data availability and government policies [9]. There are 17 types of LCZ, including ten built types (LCZ 1-10), and seven land cover types (LCZ A-G) [10]. The selected training areas used by SAGA GIS [11] to automatically classify the pixel using random forest classifier algorithm [12].

Considering that Yogyakarta is one of the cities in a tropical country with a high level of urbanization, the role of LCZ in determining the local climate of the urban area of Yogyakarta and its surroundings is quite interesting. The final results of this study are to classify the LCZ in Yogyakarta Urban Area and its surrounding using the World Urban Database Access and Portal Tool (WUDAPT) method and develop climate-based urban planning recommendations of the classification results.

2. Data and Methodology

2.1. Study Area

This research was conducted in urban areas of Yogyakarta City and the expansion of the surrounding area (7°32'24.30"S - 8°2'50.02"S and 110°6'31.63"E-110°39'1.77"E) which was influenced by urbanization, precisely inside the ringroad area. This boundary is used to determine Region Of Interest (ROI). Based on the WUDAPT protocol, the ROI should contain the entire urbanized area with a buffer of about 20 km around this area. The ROI should not be smaller than 50 km in each direction. The study area in the research that has been described above is called Yogyakarta Urban Area (YUA) and its surroundings.

2.2. Data

Landsat 8 Level 1 image with 30 m spatial resolution downloaded from the United States Geological Survey (USGS) page. Since Landsat 8 was launched in 2013, not all the images at each recording time are good and right to use. Those which have less cloud coverage are better. The presence of the clouds and their shadows can cause misclassification and false detection of land cover [13].

Table 1. Description of Landsat images that were used in this study

Image Acquisition Date	Sensor	Bands for LCZ Classification	Thermal Bands
18 May 2017	Landsat 8 OLI/ TIRS	1-7	10
5 May 2018	Landsat 8 OLI/ TIRS	1-7	10

To determine changes in LCZ and LST spatial distribution for one year, the images used are Landsat 8 on the recording as in the table at path 120 and row 65. This satellite images were preferred due to the less cloud coverage leading to more intisified results of this LCZ and LST mapping [14].

2.3. LCZ mapping and validation

2.3.1. LCZ Mapping method

Basically, the LCZ classification requires detailed urban morphological data. However, these data are often difficult to obtain due to the limitations of technology and government policies, especially in developing countries such as Indonesia. This study follows the World Urban Database and Access Portal Tools (WUDAPT) methods proposed by Bechtel et al. [15] using SAGA GIS to do LCZ classification. The LCZ classification mapping workflow starts with determining the Region of Interest (ROI) in Google Earth Pro. After ROI was determined, the next step is to digitize the training area. In the process of digitizing the training area, aspects of the shape and function of the city must be considered, then categorize into one of the 17 categories of LCZ according to the type of building and land cover [16]. Users only need to digitize several training areas in each category, because the overall area is classified by the random forest classifier algorithm based on the specified training area. The more training areas, the more accurate the classification results.

2.3.2. LCZ Classification

In the process of digitizing the training area, aspects of the form and function of the city must be considered. Below (table 2) are 17 LCZ categories according to building type and land cover type.

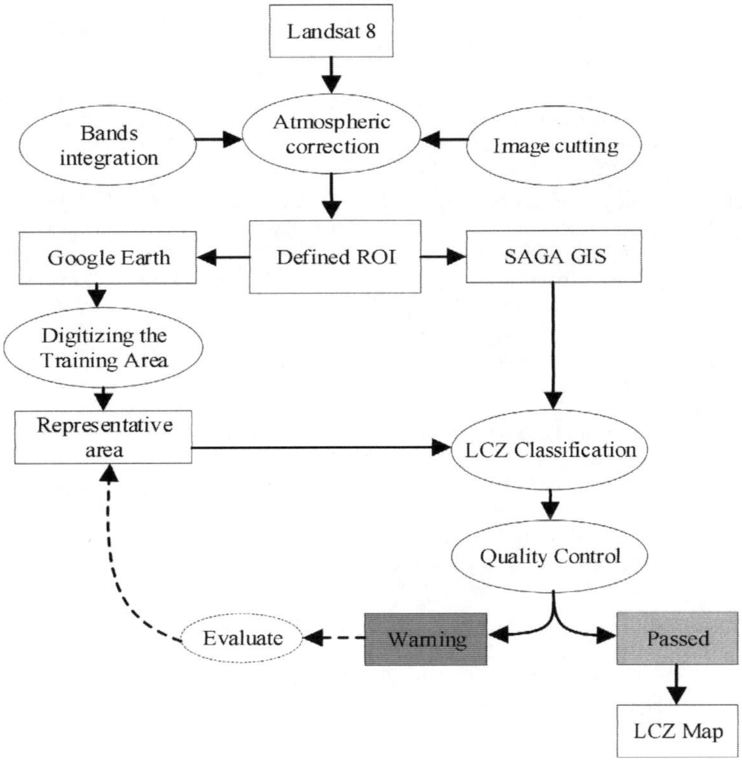

Figure 1. Research flowchart

Table 2. LCZ Classification
(Source:Steward & Oke (2012)

Building Type	Definition	Land Cover Type	Definition
LCZ 1: Compact high-rise 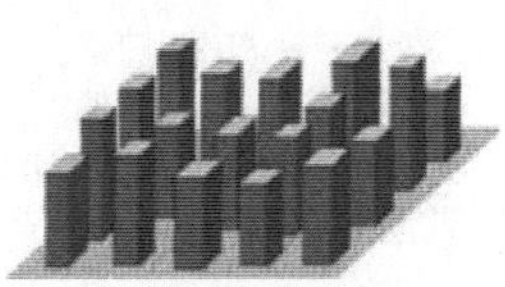	- tightly packed buildings to tens of stories tall - few or no trees - little or no green space - concrete, steel, stone and glass construction materials - small diurnal temperature range	**LCZ A: Dense trees** 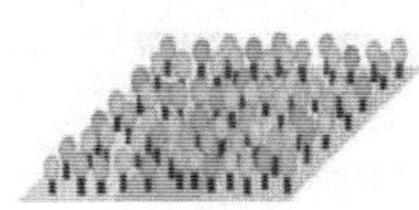	Heavily wooded landscape of deciduous and/or evergreen trees. Land cover mostly pervious (low plants). Zone function is natural forest, tree cultivation, or urban park.
LCZ 2: Compact mid-rise 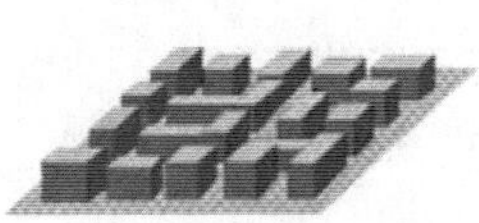	- tightly packed buildings of 3 to 9 stories tall - few if any trees - little or no green space - stone, brick, tile and concrete construction materials - small diurnal temperature range	**LCZ B: Scattered trees** 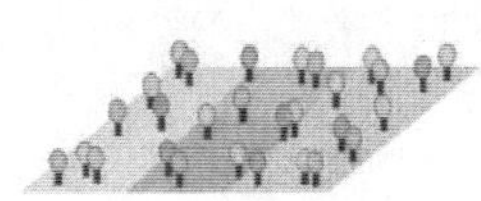	Lightly wooded landscape of deciduous and/or evergreen trees. Land cover mostly pervious (low plants). Zone function is natural forest, tree cultivation, or urban park.
LCZ 3: Compact low-rise 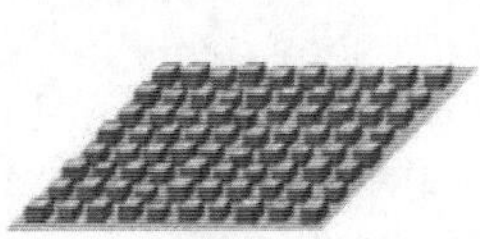	- tightly packed buildings of 1 to 3 stories tall - few or no trees - little or no green space - concrete, steel, stone and glass construction materials - diurnal temperature range is medium	**LCZ C: Bush, scrub** 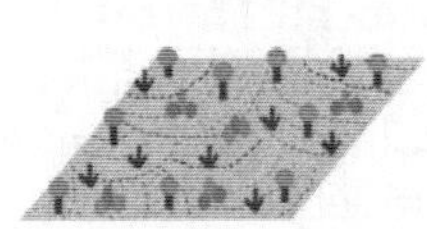	Open arrangement of bushes, shrubs and short, woody trees. Land cover mostly pervious (bare soil or sand). Zone function is natural scrubland or agriculture

LCZ 4: Open high-rise 	- openly arranged buildings tens of stories tall - abundance of trees and pervious cover (low plants) - concrete, steel, stone and glass construction materials - small diurnal temperature range	**LCZ D: Low plants** 	Featureless landscape of grass or herbaceous plants/crops. Few or no trees. Zone function is natural grassland, agriculture, or urban park.
LCZ 5: Open mid-rise 	- openly arranged buildings of 3 to 9 stories tall - abundance of trees and pervious cover (low plants) - concrete, steel, stone and glass construction materials. - medium diurnal temperature range	**LCZ E: Bare rock or paved** 	Featureless landscape of rock or paved cover. Few or no trees. Zone function is natural desert (rock) or urban transportation.
LCZ 6: Open low-rise 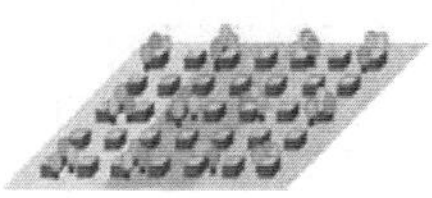	- openly arranged buildings of 1 to 3 stories tall - abundance of trees and pervious cover (low plants) - wood, brick, stone, tile and concrete construction materials	**LCZ F: Bare soil or sand** 	Featureless landscape of soil or sand cover. Few or no trees or plants. Zone function is natural desert or agriculture.
LCZ 7: Lightweight low-rise 	- lightweight building materials of 1 to 2 stories tall - few or no trees - land cover hard-packed - large diurnal temperature range	**LCZ G: Water** 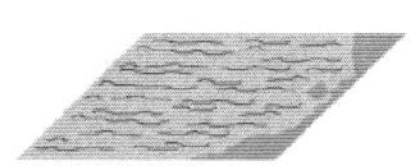	Large, open water bodies such as seas and lakes, or small bodies such as rivers, reservoirs and lagoons.
LCZ 8: Large low-rise 	- large, openly arranged buildings of 1 to 3 stories tall - few if any trees - land cover mostly paved - medium diurnal temperature range		
LCZ 9: Sparsely built 	- sparse arrangement of small or mid-sized buildings - natural setting - abundance of pervious cover - large diurnal temperature range		
LCZ 10: Heavy industry 	- Low-rise and mid-rise industrial structures (towers, tanks, stacks) - few or no trees - land cover mostly paved or hard-packed - metal, steel and concrete construction materials		

Variable Land Cover Properties

Variable or ephemeral land cover properties that change significantly with synoptic weather patterns, agricultural practices, and/or seasonal cycles.

b. bare trees
Leafless deciduous trees (e.g. winter). Increased sky view factor. Reduced Albedo.

s. snow cover
Snow cover > 10 cm in depth. Low admittance.

d. dry ground
Parched soil. Low admittance. Large Bowen ratio. Increased albedo

w. wet ground
Waterlogged soil. High admittance. Small Bowen ratio. Reduced Albedo.

2.3.3. Validation of LCZ

Choosing the right method and representative training area is very important for LCZ classification and subsequent accuracy assessment. The selection of training area polygons is very dependent on the knowledge of local experts. Everyone has various cognitive interpretations of the urban landscape, this can significantly affect the results of classification and accuracy. Bechtel et al. [17] shows this in the LCZ HUMINEX (Human Influence Experiment). Validation in this study was carried out by reviewing the results of the LCZ generating from SAGA GIS by random forest classifier algorithm. Random forest classifier algorithm chosen due to its high accuracy and computational performance

[18]. Validation is done to get a good distribution, therefore the role of local expertise here is highly important. Checking the LCZ results is done by validating the results of the LCZ mapping with the existing conditions in Google Earth Pro to obtain a good LCZ distribution.

2.3.4. LST retrieval

Remote sensing is a method commonly used for analysis and visualization of LST in a relatively wide area of observation due to its ability to provide tight spatial data with good accuracy and wide area coverage [19]. Previous research uses a single channel algorithm to retrieve LST from satellite image data because it is relatively simple and has high accuracy [20]. In this study, LST acquisition was obtained using SAGA GIS. Here are the equations in LST calculations.

$$LST = TB_{10} + C1(TB_{10} - TB_{11}) + C2(TB_{10} - TB_{11})^2 + C_0 + (C_3 + C_4W)(1-X_{\varepsilon}) + (C_5 + C_6W)\,\Delta_{\varepsilon} \quad (1)$$

Where the $C_0 - C_6$ are *Split Window Coefficients*, *TB* is the number of the thermal band (K), $X\varepsilon$ is the number of LSE mean of thermal band, *W* is the atmospheric water vapour content = 0,013 and $\Delta\ \varepsilon$ is the difference between LSE.

3. Results and Discussion

3.1. Result

Based on the classification results, Yogyakarta and its surroundings consist of 13 LCZs. The description is at the point below.

3.1.1. LCZ Maps

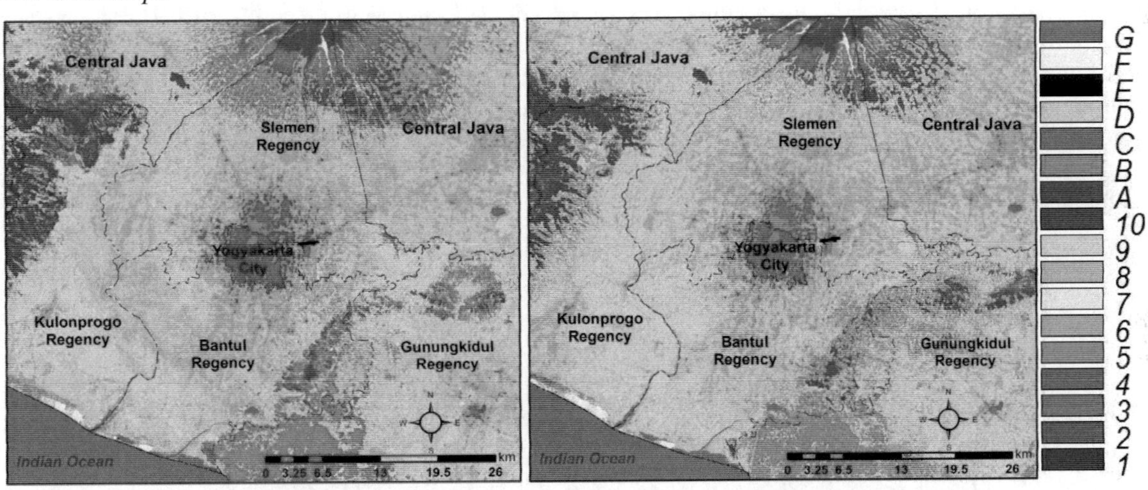

(a) 2017 (b) 2018

Figure 2. LCZ Maps Yogyakarta and its surrounding area

On the map, it may not seem obvious that there is a difference if seen at a glance. But more thoroughly, there are clear differences. The following (table 3 and table 4) are the description.

Table 3. LCZ percentage in Yogyakarta and its surrounding area 2017

LCZ 2017	Area (%)		
	Yogyakarta City (Administrative)	Inside the ring road	Outside the ring road
LCZ 2	4.66%	0.5%	0.1%
LCZ 3	76.71%	35.3%	7.0%
LCZ 5	8.58%	11.9%	2.3%
LCZ 6	4.95%	27.3%	34.9%
LCZ 8	1.69%	7.2%	3.1%
LCZ 9	0.12%	2.6%	17.2%
Σ Building Type	**96.72%**	**84.82%**	**64.5%**
LCZ A			0.3%
LCZ B	0.11%	1.4%	2.3%
LCZ C	1.43%	2.5%	1.9%
LCZ D	0.31%	6.5%	27.4%
LCZ E	0.59%	0.2%	0.6%
LCZ F	0.81%	4.3%	3.0%
LCZ G	0.03%	0.3%	0.3%
Σ Landcover Type	**3.28%**	**15.18%**	**35.5%**

Table 4. LCZ percentage in Yogyakarta and its surrounding area 2018

LCZ 2018	Area (%)		
	Yogyakarta City (Administrative)	Inside the ring road	Outside the ring road
LCZ 2	4.76%	0.49%	0.06%
LCZ 3	70.23%	29.17%	5.94%
LCZ 5	8.35%	10.14%	1.56%
LCZ 6	11.18%	38.09%	39.84%
LCZ 8	2.25%	5.48%	2.36%
LCZ 9	0.20%	2.36%	12.59%
Σ Building Type	**96.97%**	**85.72%**	**62.36%**
LCZ A			0.05%
LCZ B	0.09%	0.88%	1.67%
LCZ C	1.48%	1.50%	1.11%
LCZ D	0.38%	9.85%	32.71%
LCZ E	0.37%	0.17%	0.62%
LCZ F	0.61%	1.71%	1.30%
LCZ G	0.10%	0.16%	0.18%
Σ Landcover Type	**3.03%**	**14.28%**	**37.59%**

Generally, Yogyakarta and surrounding areas have 13 LCZs classes consisting of 6 LCZs based on building type, and 7 LCZs based on land cover. Urban areas (administrative and inside the ring road) do not have LCZ A in the form of dense tress. Dense trees are only in the outside of ring road area. Based on the table above, the three areas are dominated by LCZ based on the building type. In 2017, the developed area of Yogyakarta City increased by 0.25% from 96.7% in 2017 to 96.97% in 2018. Conversely, the type of natural land cover decreased by 0.25% from 3.28% in 2017 it became 3.03% a year after that.

3.1.2. LST Maps

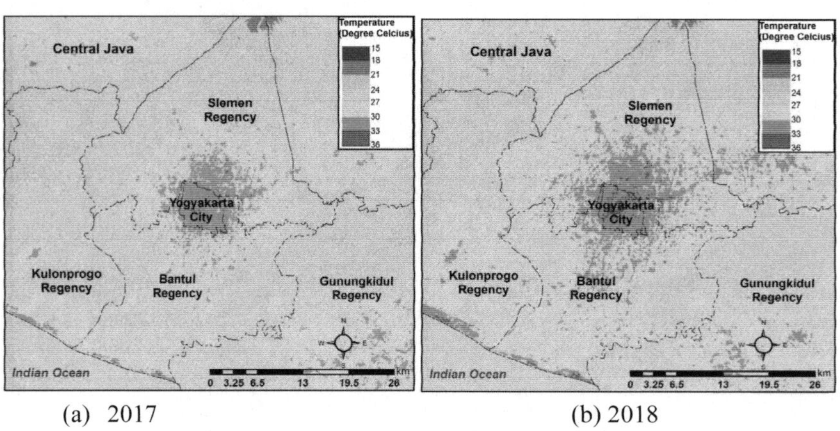

(a) 2017 (b) 2018

Figure 3. LST Maps of Yogyakarta Urban Area and its surrounding

In just one year, Yogyakarta and surrounding areas experienced a temperature increase. From figure 3, the Urban Heat Island (UHI) phenomenon is also clearly visible. By using LCZ and LST schemes, this can be further investigated in the future studies.

Table 5. LST in each area

Area	LST 2017	LST 2018	Temperature Increase (°C)
Yogyakarta City (Administrative)	32.1	32.7	0.6
Inside the ring road	30.6	31.3	0.7
Outside the ring road	28.8	29.7	0.9

3.1.3. LCZ and LST Map in Yogyakarta City Administrative and Inside the Ring road

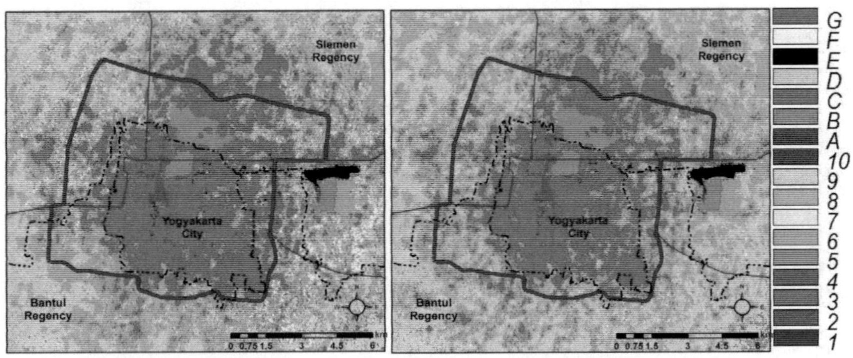

(a) 2017 (b) 2018

Figure 4. LCZ maps in Yogyakarta City Administrative and Inside the Ring road

This part is the focus of analysis in urban areas of Yogyakarta (administrative and in the ring road) without the area outside the ring road. If analyzed in depth only in urban areas, namely the administrative area of Yogyakarta and inside the ring road of Yogyakarta, LCZ percentage can be observed in the following table.

Table 6. LCZ percentage in Yogyakarta City Administrative and Inside the Ring road

	LCZ Classes	Area (%) Yogyakarta City (Administrative) 2017	2018	Inside the Ring road 2017	2018
Building Type	Compact midrise (LCZ 2)	4.66%	4.76%	0.49%	0.53%
	Compact low-rise (LCZ 3)	76.71%	70.23%	29.17%	35.26%
	Open midrise (LCZ 5)	8.58%	8.35%	10.14%	11.91%
	Open low-rise (LCZ 6)	4.95%	11.18%	38.09%	27.29%
	Large low-rise (LCZ 8)	1.69%	2.25%	5.48%	7.24%
	Sparsely built (LCZ 9)	0.12%	0.20%	2.36%	2.58%
	Total of Building Type	**96.72%**	**96.97%**	**85.72%**	**84.82%**
Land Cover Type	Dense trees (LCZ A)	0.00%	0.00%	0.00%	0.00%
	Scattered trees (LCZ B)	0.11%	0.09%	0.88%	1.45%
	Bush, scrub (LCZ C)	1.43%	1.48%	1.50%	2.47%
	Low plants (LCZ D)	0.31%	0.38%	9.85%	6.50%
	Bare rock or paved (LCZ E)	0.59%	0.37%	0.17%	0.20%
	Bare soil or sand (LCZ F)	0.81%	0.61%	1.71%	4.29%
	Water (LCZ G)	0.03%	0.10%	0.16%	0.26%
	Total of Land Cover Type	**3.28%**	**3.03%**	**14.28%**	**15.18%**

In the administrative area of the city of Yogyakarta, it can be seen that >96% is in the form of buildings, and the rest is natural land cover. While inside the ring road, >84% are also in the form of buildings and the rest is natural land cover. There were no dense trees in the two regions. If we look at the LCZ distribution pattern with the LST distribution map in Yogyakarta and its surroundings, LCZ has a relationship with LST. The more intense the LCZ by type of building is, the higher the LST will be. For more details, let's look at the following table.

Table 7. LST in YUA (Administrative and Inside the Ring road)

LCZ	LST of Yogyakarta Administrative Area		Temperature Increase	LST Inside the Ring road		Temperature Increase
	2017	2018		2017	2018	
Compact midrise	32.4	33.1	**0.6**	32.3	32.9	**0.7**
Compact low-rise	32.3	33.0	**0.7**	31.8	32.6	**0.9**
Open midrise	31.6	32.4	**0.8**	30.9	31.7	**0.8**
Open low-rise	30.6	31.7	**1.1**	29.9	31.0	**1.0**
Large low-rise	30.9	31.5	**0.6**	30.3	31.0	**0.7**
Sparsely built	29.2	29.6	**0.4**	28.3	29.2	**0.9**
Scattered trees	29.2	29.2	**0.0**	28.6	29.3	**0.7**
Bush, scrub	31.9	32.5	**0.6**	30.0	31.1	**1.0**
Low plants	29.9	29.7	**-0.3**	28.5	29.1	**0.6**
Bare rock or paved	32.7	33.2	**0.5**	31.9	32.4	**0.5**
Bare soil or sand	31.1	31.6	**0.5**	29.6	30.6	**1.1**
Water	29.7	30.7	**1.0**	28.8	30.4	**1.6**
Average	32.1	32.7	**0.6**	30.6	31.3	**0.7**

Based on the table, the highest temperature increase in the administrative area of Yogyakarta City is LCZ 5 or Open Midrise (0.8°C) and LCZ 3 or Compact Low-rise (0.7°C). While in the ring road range, the highest increase is at LCZ 3 (0.9°C) and LCZ 5 (0.8°C). The distribution of the two LCZs can be seen in figure 4.

3.2. Discussion

The graph below (figure 5) shows how LST is in each LCZ, in urban areas of Yogyakarta. Above has also been explained that the most significant increase in LST is LCZ 3 and LCZ 5. LCZ is based on four main components of urban landscape, namely the height of roughness feature, packing of roughness, surface cover around roughness feature, and thermal admittance of materials. Therefore, climate-based urban planning can be planned based on LST and LCZ. Through this concept, climate-based urban planning can be carried out in the LCZ region which has high temperatures by observing and mitigating through the components of the LCZ compiler.

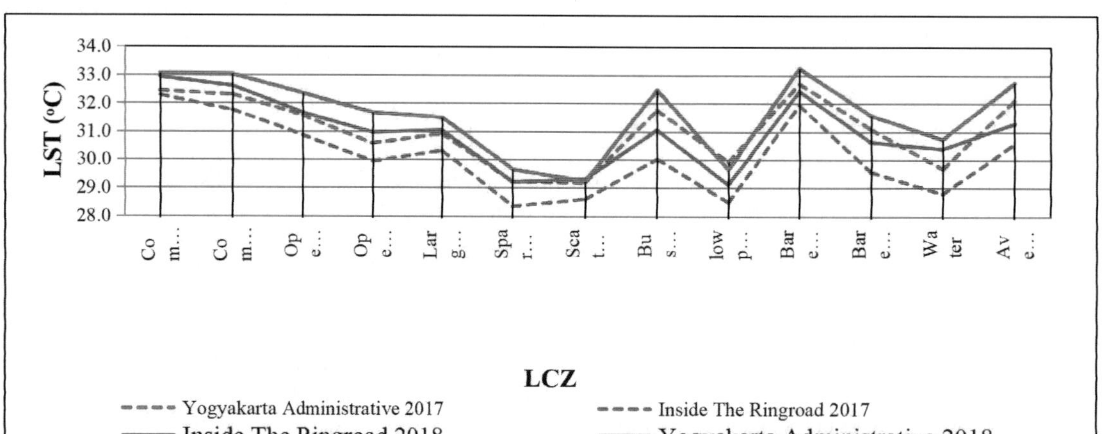

Figure 5. Graphic of each LST in LCZ in Yogyakarta Urban Area 2017-2018

4. Conclusion

Yogyakarta Urban Area have 12 LCZs consist of 6 LCZs based on building type and 6 LCZs based on landcover type. The highest increase of LST occurs in LCZ 3 (Compact low-rise) and LCZ 5 (Open midrise). LCZ 3 has characteristics; tightly packed buildings of 1 to 3 stories tall, few or no trees, little or no green space, concrete, steel, stone and glass construction materials, diurnal temperature range is medium. While LCZ 5 has characteristics openly arranged buildings of 3 to 9 stories tall, abundance of trees and pervious cover (low plants), concrete, steel, stone and glass construction materials, medium diurnal temperature range. Climate-based urban planning can be done by adding greenspace, replacing material that absorbs heat more throughout the area, spread out LCZ 3 and LCZ 5 according to those in the map. This will be quite effective because any area that needs the climate-based urban planning has been displayed on the map.

Acknowledgments

The authors appreciate the support of the Department of Environmental Geography, Gadjah Mada University. This research is part of the final project to get a master's degree in geography post-graduate program, so we would like to express our gratitude to our teachers at the Faculty of Geography for giving stimulation to this research idea and useful suggestions on the initial draft of this paper; M. Pramono Hadi and Retnadi Heru Jatmiko. We thank the anonymous reviewers and editors for contributing their time and attention to helping us build our research to be more accurate, clear and relevant.

References

[1] UNFPA 2013 Urbanization, demographics, and adaptation to clmate change in Semarang, Indonesia *Urbanization and Emerging Population Issues Working Paper 11* (London: UNFPA-Human Settlements Group IIED)
[2] Bappenas 2013 *Indonesian Population Projection 2010-2035* (Jakarta: Publications with Bappenas, BPS and UNPF)
[3] Wibowo S T and Suharyadi R 2012 Application of object-based image analysis (OBIA) for detection of landuse change using alos AVNIR-2 Image *Jurnal Bumi Indonesia* vol **1** (Yogyakarta: UGM University Press) p 130-138
[4] UN Habitat 2012 Cities and climate change initiative: tool series *Developing Local Climate Change Plans: A Guide for Cities in Developing Countries* (Nairobi: UNON Publishing Services Section)
[5] Ng E 2012 Towards Planning and practical understanding of the need of meteorological and climatic information in the design of high-density cities: a case-based study of Hong Kong *Int. J. Climatol* vol **32** (New York: Wiley-RMetS) p 582-598
[6] Ren C, Ng E and Katzschner L 2011 Urban climatic map studies: a review *Int. Journal Climatol* vol **31** (New York: Wiley-RMetS) p 2213-2233
[7] Zheng Y, Ren C, Xu Y, Wang R, Ho J, Lau K and Ng E 2018 GIS-based mapping of Local Climate Zone in the high-density city of Hong Kong *Urban Climate* vol **24** (Amsterdam: Elsevier) p 419-448
[8] Stewart I D and Oke T R 2012 Local climate zones for urban temperature studies *Bulletin of the American Meteorological Society* vol **93** p 1879-1900.
[9] Bechtel B, Foley M, Mills G, Ching J, See L, Alexander P J, O'Connor M, Albuquerque T T A, Andrade M F, Brovelli M A, Das D, Fonte C C, Petit G, Hanif U, Jimenez J, Lackner S, Liu W, Perera N, Rosni N A, Theeuwes N and Gál T 2015 *CENSUS of cities: LCZ classification of cities (Level 0): Workflow and Initial Results from Various Cities* Int. Conf. on Urban Climate 12th (Toulouse: Meteo France)
[10] Conrad O, Bechtel B, Bock M, Dietrich H, Fischer E, Gerlitz L, Wehberg J, Wichmann V and Böhner J 2015 System for Automated Geoscientific Analyses (SAGA) v**2.1.4.** *Geoscientific Model Development*
[11] Bechtel B and Daneke C 2012 Classification of local climate zones based on multiple earth observation data *IEEE J. Sel. Top. Appl. Earth Obs. Remote Sens.* vol **5** p 1191–1202.
[12] Stewart I D and Oke TR 2012 Local climate zones for urban temperature studies *Bulletin of the American Meteorological Society* vol. **93** p 1879–1990
[13] Zhu Z and Woodcock C E 2012 Object-based cloud and cloud shadow detection in Landsat imagery *Remote Sens Environment* vol **118** p 83-94
[14] Dong L P, Shen S H and Jiang ZH 2014 Urban heat island change and its relationship with urbanization of urban aglomeration in Yangtze River Delta in past decade *Trans. Atmos. Sci.* vol **37** (Amsterdam: Elsevier) p 146-154

[15] Bechtel B, Alexander P J, Böhner J, Ching J, Conrad O, Feddema J, Gerald M, See L and Stewart I D 2015 Mapping local climate zones for a worldwide database of the form and function of cities *ISPRS International Journal of Geo-Information* vol **4** (1) p 199-219

[16] Stewart I and Oke T R 2009 A new classification system for urban climate sites *Bulletin of the American Meteorological Society* vol **90** (7) p 922-923

[17] Bechtel B, Demuzere M, Sismanidis P, Fenner D, Brousse O, Beck C, Van Coillie F, Conrad O, Keramitsoglou I, Middel A, Mills G, Niyogi D, Otto M, See L and Verdonck M-L 2017 Quality of crowdsourced data on urban morphology : the Human Influence Experiment (HUMINEX) *Urban Sci.* 2017 vol **1** p 15

[18] Liaw A and Wiener M 2002 Classification and regression by random forest *R News* vol **2** (3) p 18-22

[19] Lo C P, Quattrochi D A and Luvall J C 1997 Application of high-resolution thermal infrared remote sensing and GIS to assess the UHI effect *International Journal of Remote Sensing* vol **18** (London: Taylor & Francis Group) p 287-304

[20] Jiménez-Muñoz J C and Sobrino J A 2003 A generalized single-channel method for retrieving land surface temperature from remote sensing data *Journal of Geophysical Research: Atmosphere* vol 108 (New York: Wiley Online Library)

Determining best East Java monthly rainfall projection using spatial-based validation

R R Mahmudiah[1,2], A Kurniawan[1,2] and E E S Makmur[3]

[1]Malang Climatological Station, Malang, 65152, Indonesia
[2]Indonesian State College of Meteorology Climatology and Geophysics, Perhubungan Street 5, Pondok Aren, South Tangerang, 15221, Indonesia
[3]Research and Development Center, Meteorology Climatology and Geophysics Agency, Angkasa I Street 2, Kemayoran, Jakarta, 10720, Indonesia

rikharizki.mahmudiah@gmail.com

Abstract. In production of climate change information, determining the best projection is a very important step. Missed projection can lead to loss of public trust in climate change. In some previous studies, most analyses were conducted on point approach due to the lack of observational data covering large spatial areas. In this study, rainfall data from 197 observation points are used to get correction coefficients by comparing it to monthly rainfall historical model from 1972-2005. Area of study is 110.89°E–116.27° E and 8.78°S–5.04°S. Two RCP scenarios (4.5 and 8.5) under CSIROMK3.6 RegCM4 BMKG SEA-Cordex are compared to get its ensemble weighting factor. Validation parameters used are mean absolute error and quartiles of error. Shifting correction is introduced in this study with some evidence based on visual analysis and high correction coefficients from the unshifted one. It is also shown that bilinear resampling gives better result than nearest neighbour. Ensemble weighting factors are then set 0.4 for RCP 8.5 and 0.6 for RCP 4.5. In next 10 years (2017-2026), dry condition average rainfall condition is expected to happen between May and October.

Keywords: Projection, rainfall, correction coefficient, bilinear, shifting correction

1. Introduction

Projection is information about how far climatic parameter will change. It is calculated using greenhouse gases (GHG) scenarios, global climate models (GCMs), and regional climate models (RCMs). The variation of its possibilities is so high due to uncertainty carried by its modelling process. Determining or choosing the best one is a very important step. Climate change itself refers to statistically significant variation in either the mean state of the climate or in its variability, including for periods extended (typically decades or longer). Climate change may be due to persistent anthropogenic changes in the atmosphere of land-use. The change is clear in global level but the complexity is increasing drastically for more local area [1].

In Indonesia, rainfall is more important than temperature. Indonesian season also determined not by temperature but rainfall. Seasonal prediction is then focused only on rainy-dry season and its properties such as its start, length, and rainfall amount compared to normal condition [2].Rainfall is also the primary driver of streamflow in tropical regions [3]. Because of it, this climate change study is also will be focused on rainfall.

BMKG (Indonesian meteorology climatology and geophysics agency) also contributes to the South-East Asia Coordinated Regional Climate Downscaling Experiment (SEA-Cordex). In this framework, there are two Representative Concentration Pathways (RCP) used. RCP 8.5 is assumption that condition similar toA2R SRES is happen or business as usual is carried out [4] while RCP 4.5 is assumption that realistic adaptive and mitigative policies are agreed so a constant population after 2050 [5]. GCM used is CSIROMK3.6 [6] and RCM used is RegCM4 [7]. There is possibility to identify the best projection due to that it is been at least 10 years after the last historical year (2005) has passed.

East Java Province is the provincial administration with the biggest in number of local government under it. 38 local governments are under its administration. For making comprehensive climate information for those stakeholders, Malang Climatological Station collects and maintains at least 197 rainfall data each month. This number is also the highest among other province. It is interesting to investigate climate change in term of monthly rainfall projection in this area (spatial-based). So, the aim of this study is determining best East Java monthly rainfall projection using spatial-based validation.

2. Methodology

2.1. Area of Study and Data

This study is focused in East Java Province which is astronomically bordered by 110.89°E–116.27° E and 8.78°S–5.04°S.197 points used in this study are the same points used for operational monthly climate monitoring. In figure 1, the dots from right image are the 197 observation points and the black filled area shows the area position inside maritime continent.

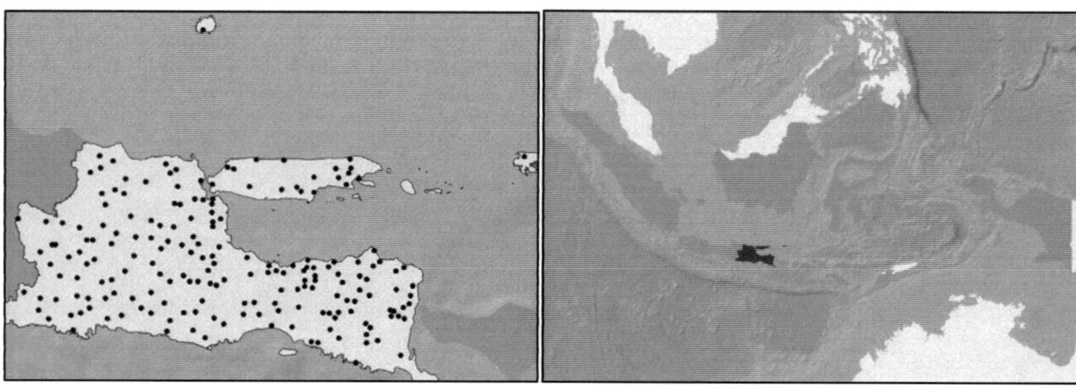

Figure 1. Area of the study.

The rainfall data are collected by Malang Climatological Station. The data is not available at all time. Each observation point began to operate at different times. Data that is not available or suspicious is removed and denoted as NA values. The figure 2 shows the data availability. A rising pattern happened near 1981, 1991, and 2001 due to regular update of climate data normal. Availability reaches above 80% after 1990.

Projection data used in this study are monthly rainfall projections output of CSIROMK3.6 RegCM4 models ran by BMKG with RCP 4.5 and 8.5 in the period for 2006-2099 (94 years). The model has historical data from 1970-2005. This data is so popular among students of Indonesian State College of Meteorology Climatology and Geophysics. It is being used as under graduate research material and also be one of data used in climate change analysis course.

In addition, polygon shapes file from Malang Climatological Station also used in shifting detection. The shapes file is under standard of National Geospatial Agency (BIG). Shuttle Radar Topography Mission (SRTM) data [8] is also used to create elevation contour.

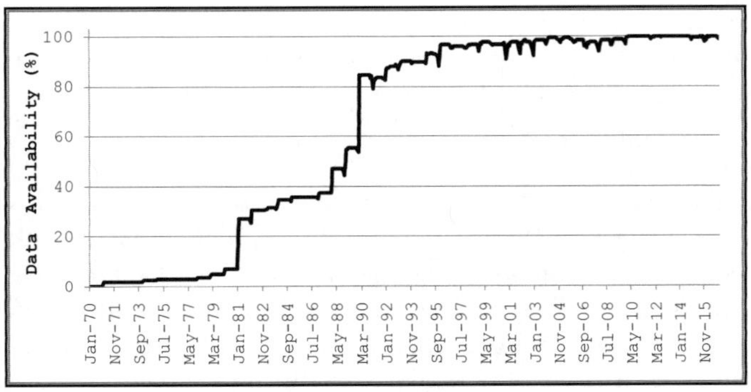

Figure 2. Data availability

2.2. Method

Because this study orientation is spatial through some area not points, the first thing to do is making sure that the model data is spatially correct. The historical data is then mapped using Grid Analysis and Display System (GrADS) with shaded output. Default colours level and Indonesian National Standard no. 8196:2015 aboutcolours level of monthly rainfall mapping issued by Badan Standardisasi Nasional (BSN) are used. Contour of elevation using 1000 m as interval is also used in the mapping process.

In investigating spatial correctness of the model, correction coefficient (CC) is used. CC is similar to multiplicative factor used in climate change bias correction study [9]. CC is calculated for every available time if there is observation data at the time. The historical model data is extracted (P_{mod}) then used as divisor to observation data (P_{obs}) at same point and month (p.m.). Model output having zero value can make some error in CC calculation. 0.01 is then used as substitution for zero value due to possibility of error from division by zero. The result will be 197 x 12 = 2364 CC for every extraction method.

$$CC_{(p,m)} = \frac{\sum_{y=1970}^{y=2005} P_{obs(p,m,y)}}{\sum_{y=1970}^{y=2005} P_{mod(p,m,y)}} \quad (1)$$

Extraction method used in this study is nearest neighbor (NN) and bilinear interpolation (BL). NN will take one nearest model point from observation coordinate so the possibility of same value between close points is relatively higher. On the other hand, BL uses four nearest points to calculate some value in observation coordinate [10].

The CC is then multiplied with 2006-2016 average monthly rainfall from projection model for both RCP 4.5 and RCP 8.5. To determine the better one, mean absolute error (MAE) is used due to its uncomplicated interpretation [11]. MAE is calculated from 2364 error values. Quartile1 (Q1), median (Med), and quartile 3 (Q3) from those error values are also calculated.

$$MAE = \frac{1}{197} \sum_{n=1}^{n=197} |P_{obs(p,m)} - P_{mod(p,m)} * \mathrm{CC}_{(p,m)}| \quad (2)$$

The plan was actually only choosing between NN and BL so that the one with lower error will be used as extraction method. Due to its spatial orientation, this study interpolates the best CC with inverse distance weighting (IDW) [12] with parameter power =1.5, point used in interpolation = 9, and resolution = 0.01^0. Resampling the average projection model data from period 2017-2026 with the best extraction method to resolution 0.01 is also done so the resolution is same with CC raster. Ensemble projection using RCP 4.5 and RCP 8.5 is done by weighting. The weighting factor is got from percentage ratio of when and where RCP 4.5 is better than RCP 8.5.

IOP Conf. Series: Earth and Environmental Science **303** (2019) 012023 doi:10.1088/1755-1315/303/1/012023

3. Result and Discussion

This first step of the method is factually unconsidered before, but plotting experience near Madura islands made it is clear that there is something wrong about the historical data. The first case that is found is Jan 1972 historical rainfall as shown in figure 3. For making sure that it is not the only case, the first step of method is done.

The result of mapping shows that using GrADS default colours, at least there is 32 cases that Madura islands rainfall is look-like being shifted to North-East. Using SNI colours template, 33 cases is found. Justification that the model is wrong cannot be done in this study but the fact shows that there is enough evidence that there is shifting in coordinate of model. So, for this study, using some basic intuitive visual analysis, it is decided that there is possibility to correct it by shifting the historical data model to South-West one resolution each. The example of change due to shifting correction (SC, this term will also will be used as shifting corrected) can be seen in Figure 1.

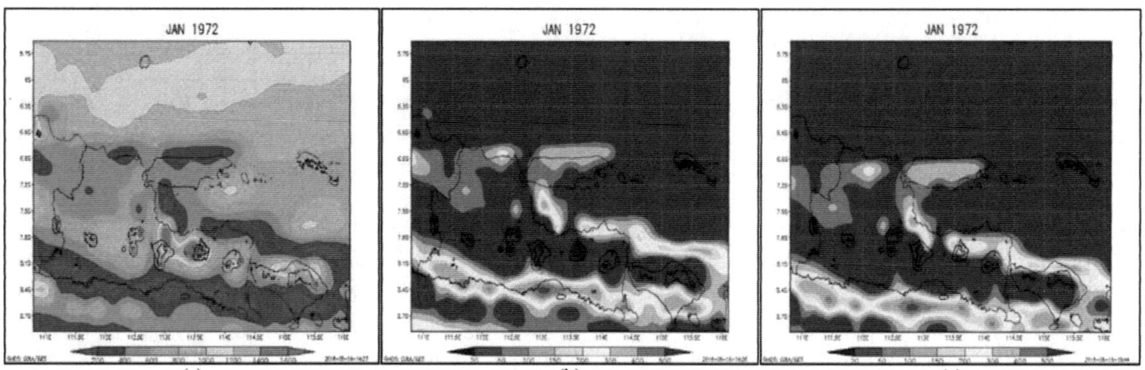

(a) (b) (c)

Figure 3. Mapping of rainfall historical data: (a) GrADS default colours, (b) SNI colours, and (c) shifting-corrected data.

The impact of SC can also be analysed using CCs that are produced. Lower CC is produced when SC is done. Figure 4 shows an example of how the high CC values near south coast of East Java is decreased when SC is applied. It encourage some principle that differently from continental area [13] or global scale [9], maritime continent climate change study needs more attention to its spatial data precision .This study can also give more detail of some bias that may exist from previous study with larger scale [14]. CC in figure 4 is calculated using NN extraction method. It can be stated that maybe there is some other mechanism that makes it better in corrected but at least by this one jump to South-West, the CC is more logically accepted.

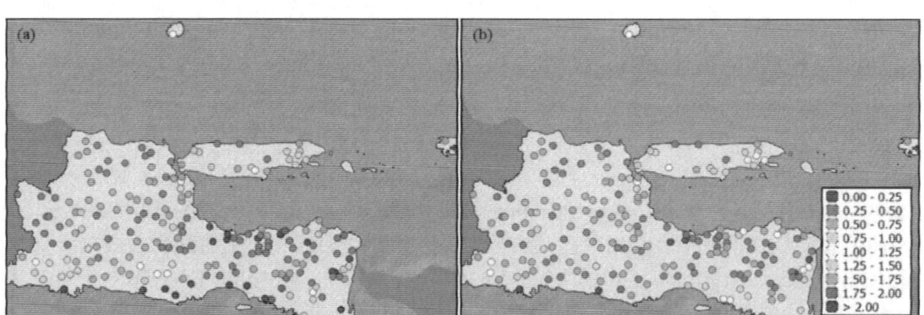

Figure 4. CC for March of (a) before SC and (b) afterSC.

For further study and practical use in course of climate change analysis, the extraction script for R Statistics user that can be used is as follows. The raster package used in the script is written by Hijmans et al [15].

```
>library(raster)
>SC=raster("...\csiromk3.6-hist-pr.nc")
>res(SC)
>cor_lon=-1*res(SC)[1]
>cor_lat=-1*res(SC)[2]
>SCm=extent(SC)
>SCm[1]= SCm [1]+cor_lon
>SCm[2]= SCm [2]+cor_lon
>SCm[3]= SCm [3]+cor_lat
>SCm[4]= SCm [4]+cor_lat
```

Then apply "SCm" as the new extent before extracting historical model.

Other impact of using SC to CC can also be shown in figure 5. In this figure, CC is distributed each month. CC of NN without SC shows values as high as 100. It means that in some case, when rainfall model value is 1 mm, the observation is above 100 mm. Very high CC is not acceptable because if CC itself shows the ratio between model and reality captured in observation. Furthermore, using BL method as extraction method can also decrease high CC values. It can be decided that to determine best projection using no SC will give bad result.

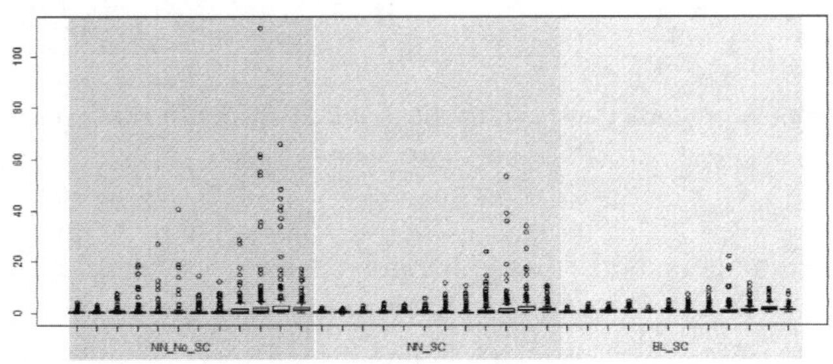

Figure 5. Correction coefficients boxplot for rainfall data.

Although CC from BL method does having lower CC, it does not mean the best projection will be come from BL. Using MAE, the error from model projection in 2006-2016 is calculated. From table 1, BL method's superiority can be shown. Its MAE is lower than NN method. Errors' Q1, Med, and Q2 of BL are also lower than NN method. It can be confirmed that BL is better than NN for extracting the model data.

Table 1. Errors from shifting corrected NN and BL methods.

	Nearest Neighbour (NN)		Bilinear (BL)	
	RCP 4.5	RCP 8.5	RCP 4.5	RCP 8.5
MAE	195.07	217.66	73.52	70.77
Errors Q_1	18.36	24.62	15.51	17.62
Errors Med	55.05	72.18	39.61	45.79
Errors Q_3	138.06	179.01	90.13	98.04

IOP Conf. Series: Earth and Environmental Science **303** (2019) 012023 doi:10.1088/1755-1315/303/1/012023

It is simple that BL is better than NN but choosing the best between RCP 4.5 and RCP 8.5 cannot be differentiated clearly. MAE of RCP 8.5 is lower than RCP 4.5 but 3 quartile parameters show the opposite. Because of its inconsistency, the ensemble method is then used. For making better ensemble, weighting is needed but the weighting factor is calculated need some justification.

In this study, how better RCP 4.5 than RCP 8.5 is then examined using both spatial and temporal approach. Spatially, conditions that RCP 4.5 better than RCP 8.5 cannot be stated clearly. 58.01% from 197 points states that RCP 4.5 is better. It means that average percentage of condition from 12 months at the point is 58.01%. For example, at a specific point, if condition RCP 4.5 is better than RCP 8.5 is 7 month so the percentage at the point will be 7 divided by 12 or around 58%. Percentages from all 197 points are then averaged. From temporal perspective, it is also unclear which RCP is better. RCP 4.5 is better in all month except August, October, November, and December. Using same averaging approach with the spatial separation, 57.92% conditions show RCP 4.5 is better. Figure 6 shows both temporal and spatial (RCP 4.5 oriented).

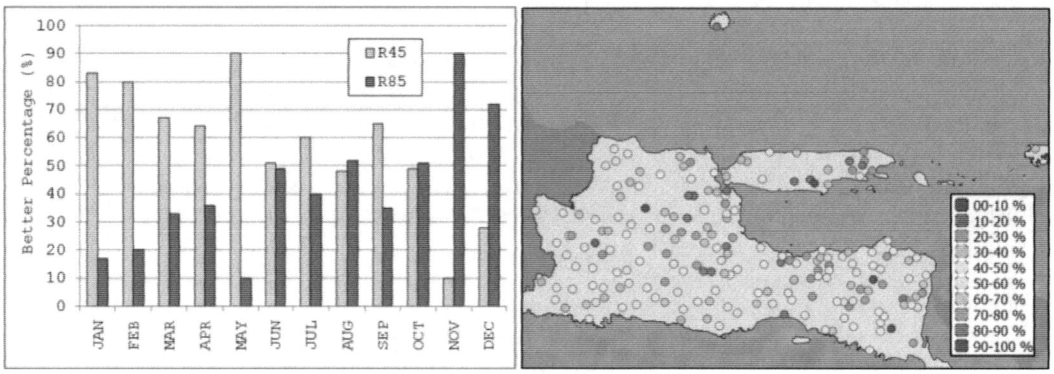

Figure 6. Temporal (left) and spatial (right) distribution of condition that RCP 4.5 better than RCP 8.5.

In projection, separated model based on month can be used. For example, RCP 4.5 will be used for eight months and the rest will use RCP 8.5. Climate change is a global system so that the impact to regional or more local one can be separated like that. From the ratio of spatial and temporal that RCP 4.5 better than RCP 8.5 is around 60%, the ensemble weighting in this study is decided to be 0.6 for RCP 4.5 and 0.4 for RCP 8.5. Other weighting can be taken but these possibilities will not be done in this study.

The CC from BL with SC is then interpolated using System for Automated Geoscientific Analyses (SAGA) [16]. The interpolation results then multiplied as raster with resampled RCP 4.5 projection. Same procedure is carried out for RCP 8.5. The RCP 4.5 raster is then multiplied with 0.6 and the RCP 8.5 raster with 0.4. These two rasters then combined by raster calculator for every month. So, the result can be seen in figure 7.

Period of next 10 years is used in this so that if there is some urgency, stakeholder can use the information. Longer period will only decrease practical benefit of projection. It is projected that in next 10 years, average condition from November till April in mountainous area will be wet. Topography still plays a big role in determining rainfall distribution. Based on the result, peak dry condition will happen between July and September. Relatively wetter condition will back in November. May till October is expected as dry period for almost East Java. Further analyses can be done by making raster from this study as input.

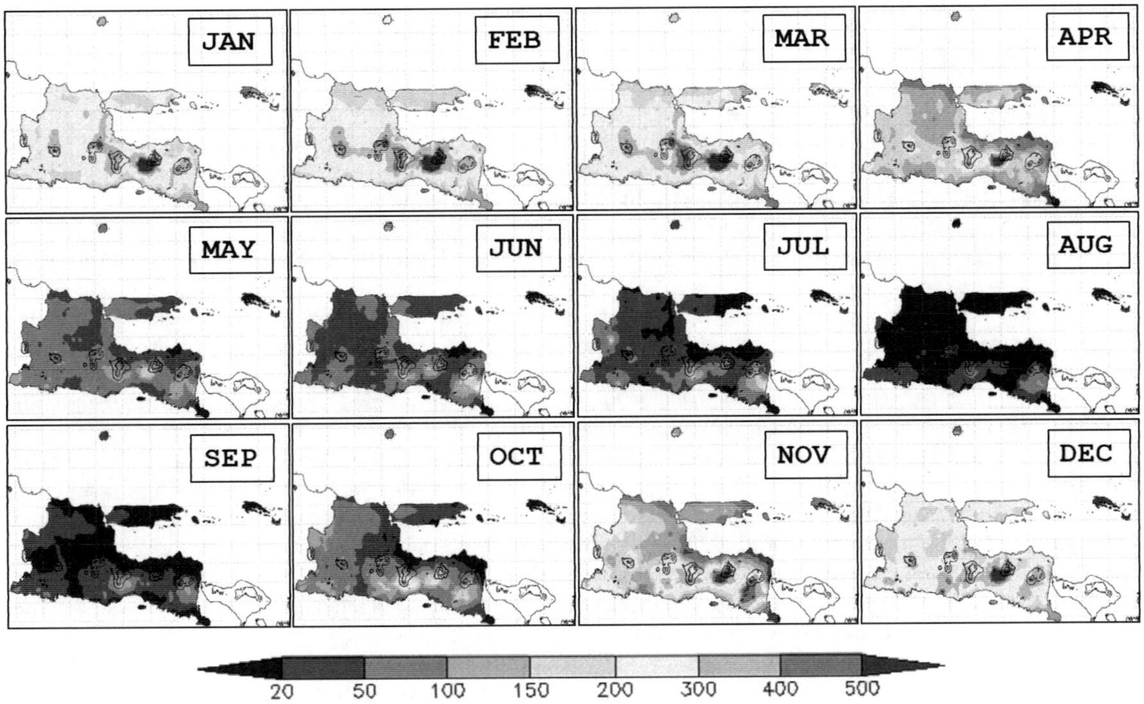

Figure 7. Average monthly rainfall projection 2017-2026.

4. Conclusion

The first aim of this study is factually determining best projection, but the side result give wider perspective. This study brings new standard of pre-testing in spatial correctness of model before using it. Correction coefficient (CC) is proposed not only as corrector but also parameter to determine correctness of model's coordinate. Although bring no good answer, this study at least shows that extraction using bilinear is better and demonstrate the possibility of making ensemble weighting factor justification from spatio-temporal error analysis. In the nutshell, dry condition for next 10 years (2017-2026) is expected to happen between May and October.

Acknowledgments

We would like to thank Malang Climatological Station and Indonesian State College of Meteorology Climatology and Geophysics (STMKG) for supporting this study by providing the data and opportunities.

References

[1] IPCC 2014 *Climate Change 2014: Synthesis Report In Contribution of Working Groups I, II and III to the Fifth Assessment Report of the Intergovernmental Panel on Climate Change* ed Pachauri RK and Meyer LA (Geneva: IPCC) p 151
[2] BMKG 2012 *Verifikasi Prakiraan Iklim Indonesia* (Jakarta: BMKG)
[3] Yanto, Livneh B, Rajagopalan B and Kasprzyk 2017 Hydrological model application under data scarcity for multiple watersheds, Java island, Indonesia *J. of Hydrology: Regional Studies* **9** 127-139
[4] Riahi K, Rao S, Krey V, Cho C, Chirkov V, Fischer G, Kindermann G, Nakicenovic N and Rafaj P 2011 RCP 8.5—a scenario of comparatively high greenhouse gas emissions *Climatic Change* **109** 33–57

[5] Thomson A M, Calvin K V, Smith S J, Kyle G P, Volke A, Patel P, Delgado-Arias S, Bond-Lamberty B, Wise M A, Clarke L E and Edmonds J A 2011 RCP4.5: A pathway for stabilization of radiative forcing by 2100 *Climatic Change* **109** 77–94

[6] Jeffrey S, Rotstayn L, Collier M, Dravitzki S, Hamalainen C, Moeseneder C, Wong K and Syktus J 2013 Australia's CMIP5 submission using the CSIRO Mk3.6 model *Aust. Meteor. Oceanogr. J.* **63** 1-13

[7] Giorgi F, Coppola E, Solmon F, Mariotti L, Sylla M B, Bi X, Elguindi N, Diro G T, Nair V, Giuliani G, Turuncoglu U U, Cozzini S, Güttler I, O'Brien T A, Tawfik A B, Shalaby A, Zakey A S, Steiner A L, Storda F, Sloan L C and Brankovic C 2012 RegCM4: model description and preliminary tests over multiple CORDEX domains *Clim. Res.* **52** 7-29

[8] Yang L, Meng X and Zhang X 2011 SRTM DEM and its application advances *Int. J. Remote Sensing* **32** 3875-3896

[9] Ngai S T, Tangang F and Juneng L 2016 Bias correction of global and regional simulated daily precipitation and surface mean temperature over Southeast Asia using quantile mapping method *Global and Planetary Change* **149** 79-90

[10] Teegavarapu R S V, Meskele T and Pathak C S 2012 Geo-spatial Grid Based Transformations of Precipitation Estimates Using Spatial Interpolation Methods *Computers & Geosciences* **40** 28–39

[11] Willmot C and Matsuura 2005 Advantages of the Mean Absolute Error (MAE) over the Root Mean Square Error (RMSE) in assessing average model performance *Clim. Res.* **30** 79-82

[12] Yanto, Livneh B and Rajagopalan B 2017 Development of gridded meteorological dataset over Java island, Indonesia 1985-2014 *Sci. Data* **4** 170072

[13] Ahmed K F, Wang G, Silander J, Wilson A M, Allen J M, Horton R and Anyah R 2019 Statistical downscaling and bias correction of climate model outputs for climate change impact assessment in the U.S. northeast *Global and Planetary Change* **100** 320-332

[14] Siew J H, Tangang F T and Juneng L 2014 Evaluation of CMIP5 coupled atmosphere–ocean general circulation models and projection of the Southeast Asian winter monsoon in the 21st century *Int. J. Climatol.* **34** 2872-2884

[15] Hijmans R J, Etten J V, Sumner M, Cheng J, Bevan A, Bivand R, Busetto L, Canty M, Forrest D, Ghosh A, Golicher D, Gray J, Greenberg J A, Hiemstra P, Karney C, Mattiuzzi M, Mosher S, Nowosad J, Pebesma E, Lamigueiro O P, Racine E B, Rowlingson B, Shortridge A, Venables B and Wueest R 2017 *Geographic Data Analysis and Modeling* https://cran.r-project.org/web/packages/raster/raster.pdf accesed September 2018

[16] Conrad O, Bechtel B, Bock M, Dietrich H, Fischer E, Gerlitz L, Wehberg J, Wichmann V and Böhner J 2015 System for Automated Geoscientific Analyses (SAGA) v. 2.1.4 *Geosci. Model Dev.* **8** 1991-2007

IOP Conf. Series: Earth and Environmental Science **303** (2019) 012024 doi:10.1088/1755-1315/303/1/012024

Exposure of dryland Farming to the ENSO Phenomenon in Kebumen District

T I Utami, T Handayani, Kuswantoro

[1]Department of Geography, Faculty Mathematics and Natural Science, University of Indonesia

trindahutami@gmail.com

Abstract. The impact of the ENSO phenomenon causes a shift in the pattern of rainy and dry season. This will affect the agricultural activities Kebumen regency, crop failure and decreased productivity are few of the results of such impacts. This study aims to find the exposure pattern of dryland agriculture in Kebumen regency towards ENSO phenomenon and its association with maize productivity. The Data used is the daily rainfall period of 1986-2016 from 32 observer rain stations. There are five parameters to identify exposures are early deviation and duration of a dry season, rainfall of the maize, the maize harvest, the number of rainy days in maize harvest and the method used is scoring and overlay. The pattern of high exposure to dryland agriculture tends to be in mountainous areas. The El Nino Period of 2015 does not result in dryland agriculture Significantly being exposed due to low exposure, while the La Nina period results in moderate to high exposure cleaning Because land does not require too much water. In the El Nino period of 2015, maize productivity tends to rise 43% from normal conditions. In the La Nina period in 2010, the productivity of maize dropped by 63% Compared to normal conditions.

Keywords: El Nino, ENSO phenomenon, Exposure, La Nina, Dryland.

1. Introduction

The geographical position of Indonesia is located in the tropics, between Asia and Australia, between the Pacific Ocean and the Indian Ocean, as well as passed the equator, consists of the island and archipelago stretching from west to east, there are many straits and bays, causing parts of Indonesia prone to the phenomenon of changes in weather /climate. Indonesia climatic conditions affected the ENSO phenomenon that comes from the eastern region of Indonesia (Central Pacific Equator / Nino 3.4).

El Nino is a global phenomenon of ocean-atmosphere interaction system that is characterized by the warming of sea surface temperatures in the equatorial Eastern Pacific (Nino 3.4) or sea surface temperature anomalies in the area of positive (hotter than the average). La Nina is the opposite of El Nino events ie reduction in sea surface temperature in the equatorial region of the Pacific Ocean from its normal temperature [1], the El Nino and La Nina are both associated with the Southern Oscillation, therefore this phenomenon is known as ENSO. El Nino's impact in Indonesia depends on the condition of Indonesian territorial waters. El Nino will reduce the amount of rainfall in Indonesia when the seawater temperature is relatively cold. However, if the temperature of seawater in Indonesia is relatively warm, El Nino does not affect the rainfall in Indonesia. Meanwhile, La Nina will tend to cause heavier rainfall than the normal condition in Indonesia.

Climate variability associated with El Nino events and La Nina has a significant impact on agricultural land. The natural phenomenon El Nino causes droughts and drought in agriculture, while

IOP Conf. Series: Earth and Environmental Science **303** (2019) 012024 doi:10.1088/1755-1315/303/1/012024

La Nina could increase rainfall and extend the length of harvest season and can lead to crop failure. ENSO events are often accompanied by a prolonged drought due to the decrease in the amount of rainfall, including Java. The agricultural sector is very vulnerable to climatic aberrations due to indicate a shift in the rainy season or drought and changes in rainfall patterns that affects the shifting of planting and harvesting of food commodities [2]. The agricultural sector in Kebumen regency is one of the main driving sectors of the regional economy because it is a staple food buffer in Central Java. Kebumen district has 65 632 ha of dry land agriculture area (49.4%), which means having a sizeable area [3]. Kebumen is also a producer of various such as maize. The average productivity of maize in Kebumen 2015 72.33 Ku / ha increased significantly when compared to 2014 which reached 68.90 Ku / ha. However, the productivity of maize in Kebumen can bedecreased due to a shift in the pattern of the rainy season and dry season (4%), which means having a sizeable area [3].

ENSO phenomenon provides a significant impact in Kebumen, Central Java, mainly in a shift of the pattern between the rainy season and dry season. It affects the agricultural activities in the district of Kebumen, including crop failure, reduced productivity and changing planting time. Exposure to dry land farming will occur along with the ENSO phenomenon in Kebumen. Therefore the research needs to be done as to the benefits of this research can anticipate the impact of the ENSO phenomenon will result in the loss of agricultural land Kebumen. This study has two research questions are:

1. What is the pattern of dryland farming exposure to the ENSO phenomenon in Kebumen?
2. What is the impact on the productivity of maize towards the ENSO phenomenon in Kebumen?

2. Methods

2.1. Study Area

Kebumen Regency is one of 35 districts/ municipalities in the province of Central Java. Administratively Kebumen is divided into 26 districts, 449 villages and 11 urban villages with conditions some areas are coastal and hilly areas, while most of the lowland [3]. The total area of Kebumen amounted to 133.749 hectares, or by 1,337, 49 km^2. Kebumen Regency of Wonosobo and Banjarnegara regency in the north, Purworejo in the east, Banyumas and Cilacap districts in the west and is adjacent to the Indian Ocean in the south (figure 1).

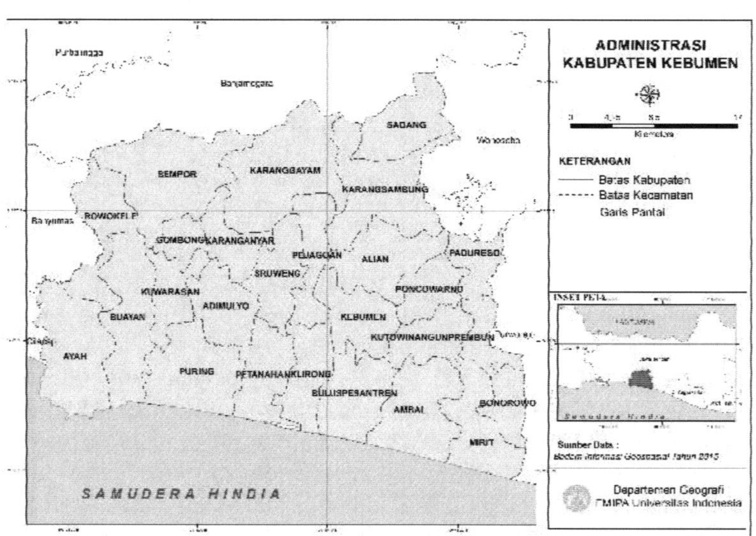

Figure 1. Study Area in Kebumen District

2.2. Materials

The secondary data required in this study include daily rainfall data for 30 years in the period 1986 to 2016 from the Department of Natural Resources and Energy and Mineral Kebumen. DEM data from the USGS used to create a map of altitude and slope, land use data and administration of BIG in 2015, ONI index of the web http://origin.cpc.ncep.noaa.gov/products/analysis_monitoring/ensostuff/ONI_v5.php maize production data from the CPP and the Department of Agriculture Kebumen.

2.3. Data Processing

The method used in this research is to process tabular data and statistics data, as well as the processing of spatial data. Tabular data processing is to determine the deviation at the beginning of the dry season, the duration of the dry season, the average intensity of precipitation planting, the harvest rainfall and number of rainy days in the period as El Nino deviate 2015 and La Nina in 2010 using standard deviation formula in each 32 rainfall stations.

$$S = \sqrt{\frac{\sum_{i=0}^{n}(x_i - \bar{x})^2}{n-1}}$$

S = Standard Deviaton
xi = sample data
x = average
n = amount of data

Determining the class value of the deviation of the five parameters, by calculating the number of classes (k) determined by k = 1 + (3,3) log n, where n is the number of rainfall stations. Determine the fifth class interval parameter. To get the class interval (I) on the values of the parameters of exposure, using this formula I = R / k. The value of R is the difference between the largest data value and the smallest data value, while k is the number of classes that have been predetermined. The five parameters are the deviation of the beginning of the dry season, the duration of the dry season, the rainfall of the planting season, the rainfall during the harvest period, and the number of days of rain during the harvest, then each overlayed to become a map of the level of exposure to the ENSO phenomenon, which is stated in the following formula (Level of Exposure = P1+P2+P3+P4+P5). Scoring results are divided into three classes, namely low, moderate, and high exposure (table 1).

Table 1. Exposure Scoring Levels to the El Nino and La Nina Period

Interval La Nina	Level of Exposure	Score	Interval El Nino
< 7	Low	1	<2
8 s.d 15	Moderate	2	3 s.d 5
> 16	High	3	>5

Source : Data Processing, 2018

Spatial data processing are mapping the area of irregularities with Thiessen polygon method using software Arc.GIS 10.1, and mapmaking exposure area and intersect with overlay techniques using software Arc.GIS 10.1.

2.4. Analyze Data

In a study on the patterns of exposure to dry agricultural land to the ENSO phenomenon which aims to answer the problem of how exposure pattern affects the dry agricultural land to the ENSO phenomenon in the district of Kebumen using spatial analysis to see the formation of the exposure pattern within the research area resulting from the parameters of climate change based on the level exposure regions.

To answer the second question, how the ENSO phenomenon to agricultural activities by using comparative analysis between exposure maps of agricultural land to the ENSO phenomenon with land productivity values described by chart or table to determine the relationship. ENSO phenomenon considered as a factor, while the result of the productivity of
maize is considered as a result of factors.

3. Results and Discussion

3.1. Deviations Rainfall and Drought

Deviations seasons and rainfall is a parameter that can be measured to see climate change as El Nino and La Nina. ENSO phenomenon (El Nino Southern Oscillation) can affect the variability of annual rainfall. Total rainfall may be higher when the La Nina, and can be lower when the El Nino. Climate change caused by the ENSO phenomenon will lead to changes in annual climate patterns such as the late start of the rainy season and dry season. The years of the occurrence of El Nino, La Nina, and abnormal conditions can be seen in figure 2 below.

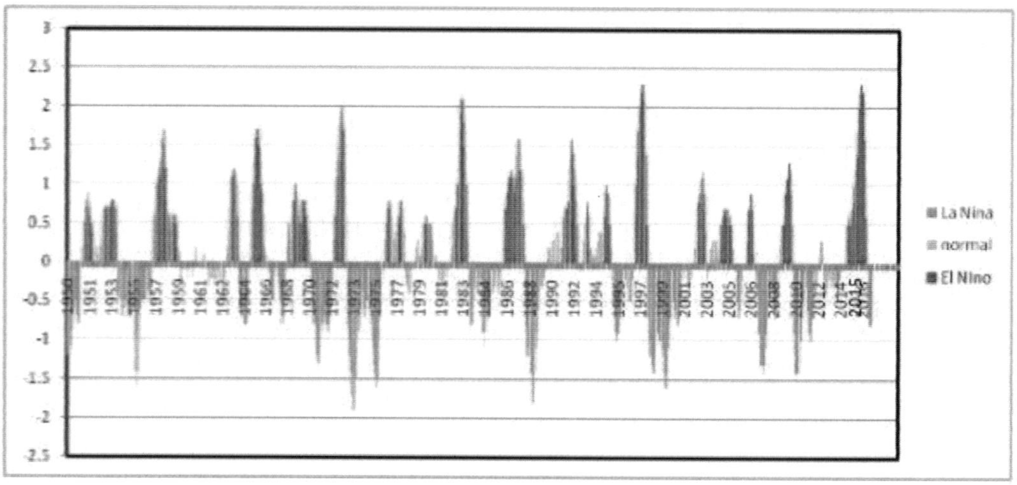

Figure 2. Occurrence Graphic of El Nino year, La Nina and Normal Condition.
Source : Data Processing, 2018

By looking at the chart above, a strong El Nino happened in 1997 with the ONI index of more than +2, El Nino moderate and weak in the years 2003, 2005, 2009, and in 2015 El Nino tends to be moderate to strong. A strong La Nina years, namely in 1973, 1988 and 1999. La Nina happened in 2007 and 2010. In the occurrence of La Nina which is in 2010 resulted in higher average annual precipitation in Kebumen. In the occurrence of El Nino, namely 1997, 2002 and 2015, the average annual rainfall tends to be low. This proves that the impact of the ENSO phenomenon on rainfall in Kebumen. The intensity of the rain tends to rise or lower than the average normal conditions could result in an impact on several agricultural crops.

Deviations of the season happened in 2010 that in La Nina phenomenon happened and in 2015 when El Nino in comparison with the normal conditions in the years 1986-2016. This is evidenced by the existence of irregularities early dry season when the La Nina is back 3 dasarian of normal conditions. Deviations duration of the dry season happens more quickly with reduced duration of as much as 13 dasarian the period of La Nina and going on longer with increasing 1 dasarian on the El Nino period of normal conditions (table 2).

Table 2. Deviations of Dry Season In the period of El Nino, La Nina and Normal Condition

Information	Year			Irregularities in El Nino	Irregularities in La Nina
	On average (1986 - 2016)	2015 (El Nino)	2010 (La Nina)		
Early Drought	Dasarian May 3rd	Dasarian May 3rd	Dasarian June 2nd	No irregularities	Backward 3 dasarian
Duration Drought	16 dasarian	17 dasarian	3 dasarian	I grew dasarian	Reduced 13 dasarian

3.2 Cultivation of Maize in Kebumen

Maize farmers in Kebumen usually use dry agricultural land or paddy fields and carry out a rotation of seasonal crops planted with rice. Types of maize that grown in Kebumen majority is hybrids maize, due to a high price for corn farmers. Maize plant growth is strongly influenced by one of the climatic factors, that is rainfall. Increase or decrease in production of corn is determined also by the intensity of the rainfall. The corn crop is cultivated in the dry season or when the climate is relatively dry because these plants tend not to require a lot of water. Based on interviews with farmers corn, it can be concluded that the time is used for planting to harvest (table 3).

Table 3. The Pattern of Corn Planting in Kebumen

Corn Cultivation time	Month									
	Jan	Feb	March	April	May	June	July	Agt	Oct	Nov	Dec
Planting season											
The harvest											

3.3 Exposure Levels Against Regional ENSO phenomenon

Climatic deviation caused by the ENSO phenomenon causes aberrations and irregularities season rainfall intensity. Changes in precipitation can be used as an indicator to determine the value and grade five parameters. To obtain the final value of exposure to the ENSO phenomenon, by using the scoring method for the fifth parameter, namely the deviation of early dry season, the duration of the dry season, rainfall during the planting season, rainfall during harvest and rainy days of harvest. Deviations early dry season, duration of the dry season and rainy days of the harvest maize do not deviate significantly when the El Nino period. El Nino 2015 and La Nina 2010 phenomenon do not have too much significant effect on the third parameters.

Deviation of rainfall in the growing season of maize during the 2015 El Nino does not deviate significantly if compared to the period of La Nina in 2010. This is because maize is a plant that did not require water at planting time, if the excess water in the growing season it can cause some damage, including bulih growth on maize stalks that will affect the quality of maize. Deviation of rainfall during the growing period of La Nina 2010 was dominated by a class 2 aberrant deviation 69-137 mm dominate in the region ramps up rather steep. Grade 6 aberrant deviation >343 mm, which is potentially damaging maize crops in the area are steep in the period of La Nina in 2010. Distribution of classes of rainfall deviations during El Nino and La Nina maize planting periods can be seen in figure 3 and 4.

Deviations of rainfall in the harvest of maize in the period of El Nino is not the irregularities. This is because at the time of the maize harvest does not need water, dry conditions are needed in maize. Dry conditions needed to dry the maize kernels and maize stalks are then able to generate income for maize farmers. Deviations average rainfall the harvest of maize in the period of La Nina in 2010 had a very significant impact in Kebumen. This is proven by the deviation of rainfall in the highest class is class 6 deviated >130mm dominate (figure 4). In the event of a La Nina 2010, maize crops failed and provide impacts to the farmers because the produced cannot be sold to the collectors.

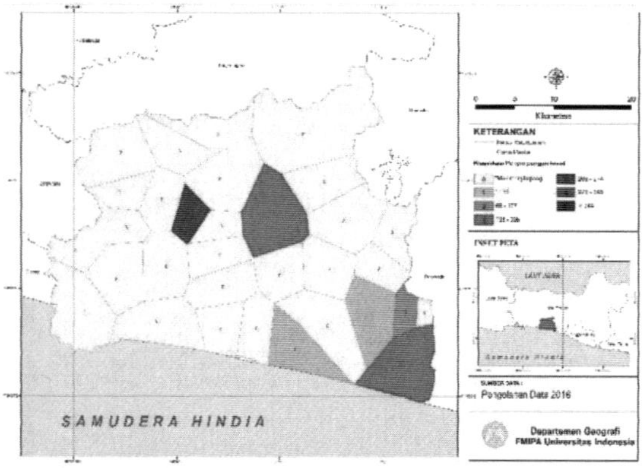

Figure 3. Deviation of Rainfall Maize Cropping Period El Nino 2015 in Kebumen

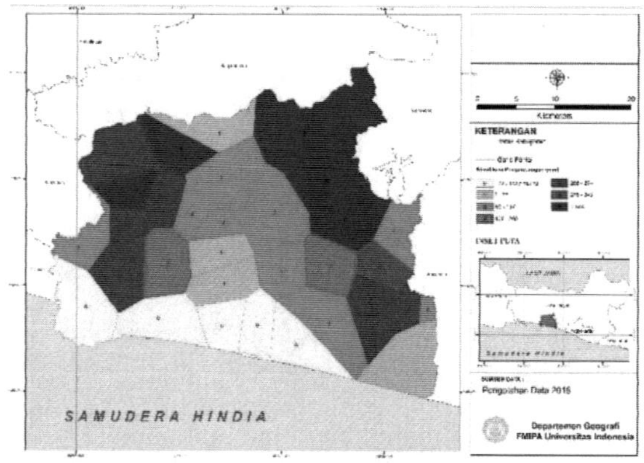

Figure 4. Deviation of Rainfall Maize Cropping Period La Nina 2010 in Kebumen

3.4 Exposure Pattern of Dryland Agriculture to the ENSO phenomenon

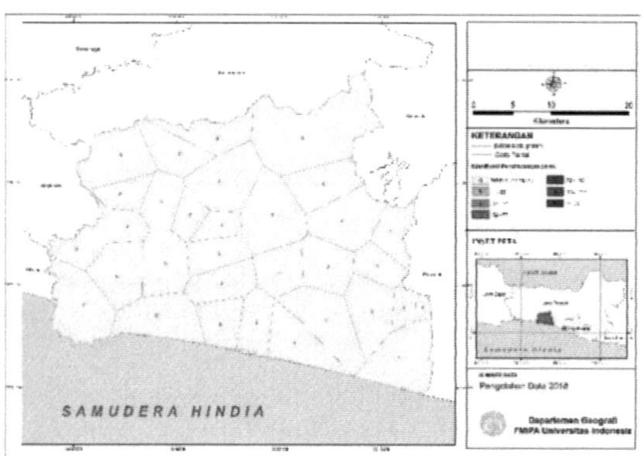

Figure 5. Rainfall Irregularities Corn at Harvest Time Period El Nino 2015 in Kebumen

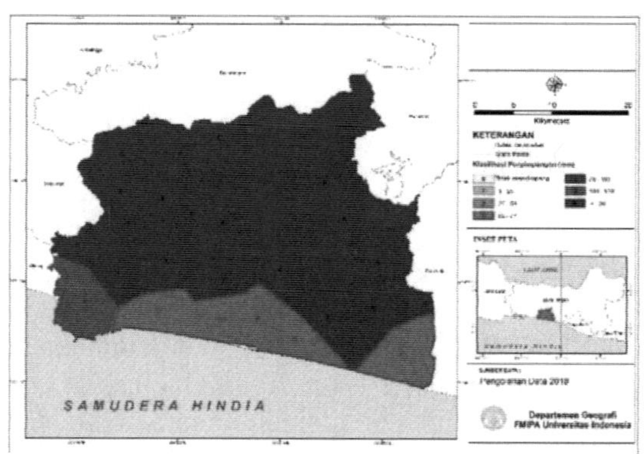

Figure 6. Rainfall Irregularities Corn at Harvest Time Period La Nina 2010 in Kebumen

Pattern exposure of dryland farming on the ENSO phenomenon in Kebumen obtained from combining variable deviation early dry season, the duration of the dry season, precipitation growing season, rainfall harvest, and rainy day harvest has classified the value of exposure to long fetches using the scoring method to map farmland , Mapmaking exposure processed by methods intersect with ArcGIS 10.1 software. Agricultural land belonging to the dry land agriculture in Kebumen include rainfed areas, moor, dry forest, mixed farms, and fields / huma. Dryland farming are mostly located in the northern district and a small portion is located in the coastal Kebumen. Dryland agriculture, in general, produce commodity crops, fruits, vegetables, and forest products. The spatial pattern of dryland farming exposure to the weather phenomenon El Nino and La Nina can be seen in figure 7 and 8.

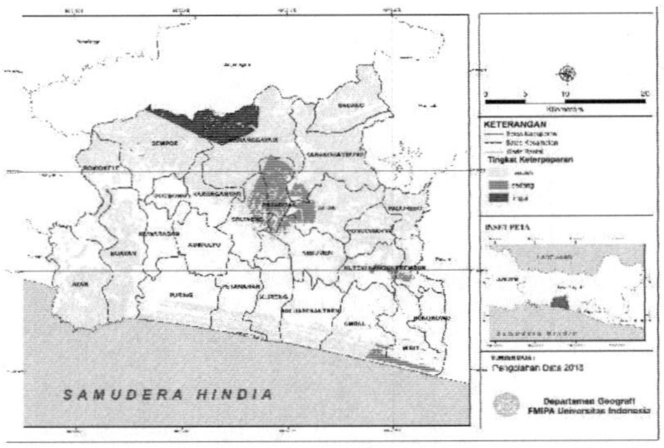

Figure 7. Exposure Pattern of Dryland Agriculture in the Period of El Nino 2015 in Kebumen

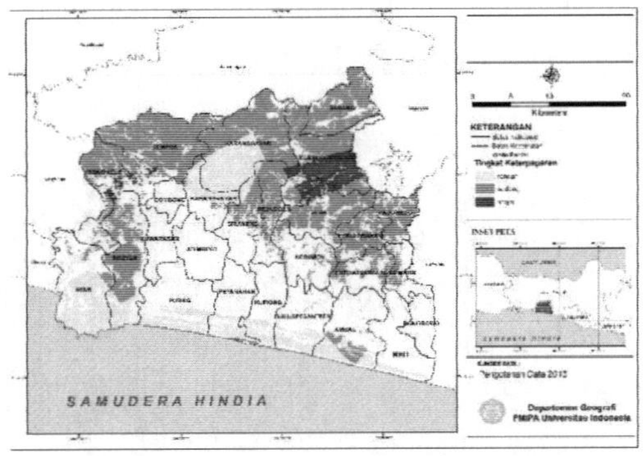

Figure 8. Exposure Pattern of Dryland Agriculture in the Period of La Nina 2010 in Kebumen

Farming dry land exposed by El Nino in 2015 was dominated by exposure to low levels. This is evidenced by the low exposed area has the greatest general 85% of the total area of agricultural land is dry and dominated by a mix of garden land. Dryland agriculture in the period of La Nina in 2010 was dominated by a moderate level of exposure. 67% of dryland farming area has a moderate exposure that dominates in the garden mix.

3.5 Exposure Dryland Agriculture Linkages Against ENSO phenomenon With Maize Productivity

According to interviews with farmers indicated that the presence of the La Nina phenomenon in maize production can be decreased drastically due to high rainfall in the growing season led to crop failure. However, at the time of occurrence of El Nino in 2015, maize production does not decline, but even tends to rise due to the decreasing of rainfall in the growing season. This can be evidenced by the average productivity of maize in figure 9.

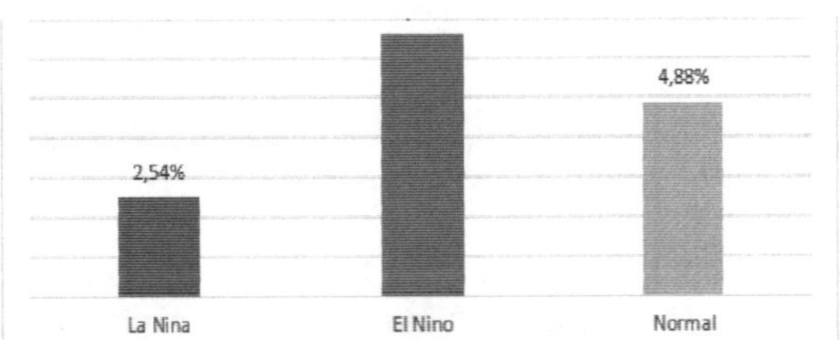

Figure 9. Percentage of Average Productivity Maize Period El Nino, La Nina and Normal in Kebumen

On dryland farming exposure exposed to low, productivity gains in the period 2015 El Nino tends to be higher than normal and La Nina. The productivity of maize in coastal areas tend to have high productivity, as in the district Ambal, Mirit, Puring, Petanahan, Klirong. This is because the rainfall received is less than the maize plants that grow in mountainous areas so that productivity in coastal regions tend to be higher. The highest productivity in Sub Petanahan with El Nino conditions are low at 8.10 tonnes / ha and the lowest is exposed District of La Nina Buluspesantren low at 2.11 tonnes / ha (figure 10).

Exposures exposed dry land farming El Nino was only one sub-districts and 13 sub Pejagoan La Nina being exposed. In this moderate exposure, the productivity of maize decreased compared with the productivity of maize exposed La Nina is low. The productivity of the largest and smallest are in Sub Pejagoan, with the largest exposure was that El Nino 6.67 tons/ha while the smallest with La Nina was namely exposure of 1.22 tonnes/ha (figure 11).

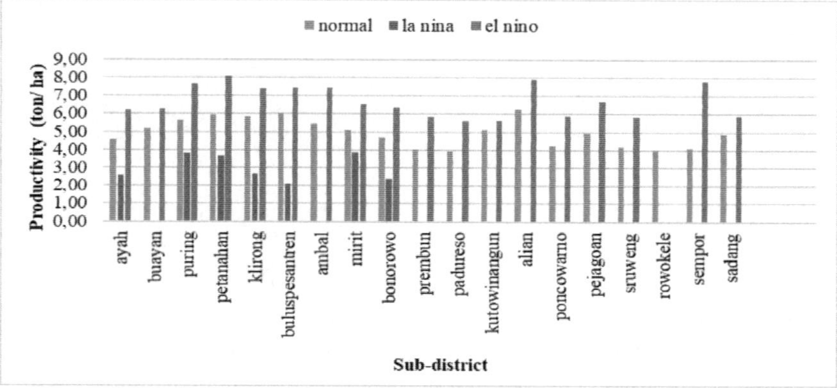

Figure 10. Productivity of Maize with Low Exposed of ENSO Phenomenon

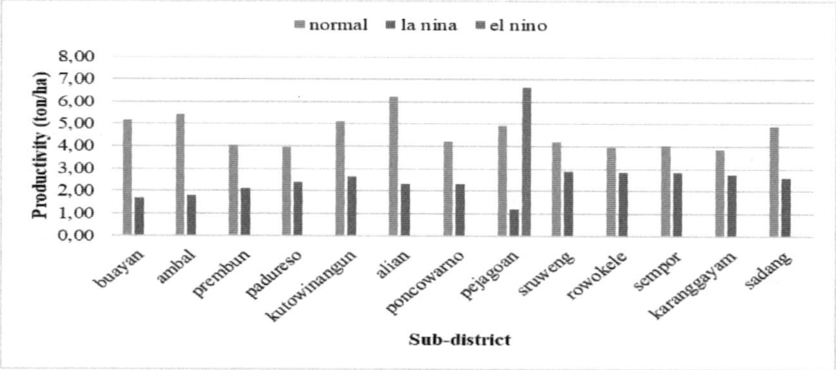

Figure 11. Productivity of Maize with Moderate Exposed of ENSO Phenomenon

Exposures dryland farming Heavily exposed only 2 sub-districts and sub-district Karangsambung Karanggayam. High exposure in the period of La Nina causes corn productivity is only 1.80 tonnes/ha compared to the normal time is 4.63 ton/ha and in the period of El Nino is 6,13 t/ha (figure 12). The corn crop is exposed to high located in the district and sub-district Karangsambung Karanggayam which is a mountainous area and has a harvested area is not too broad because corn can not grow well than in coastal areas.

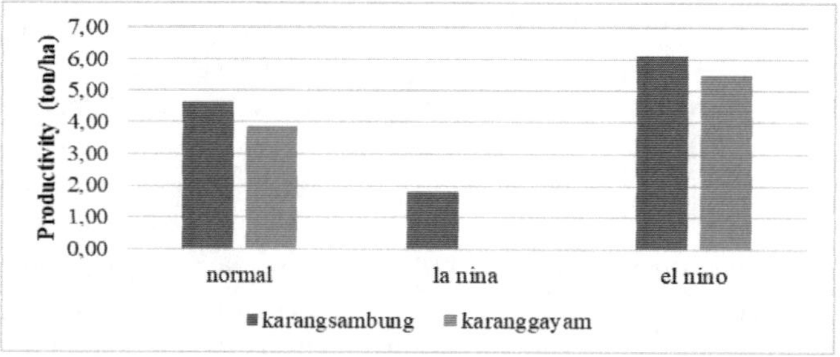

Figure 12. Productivity of Maize with High Exposed of ENSO Phenomenon

4. Conclusion

Exposure patterns exposed dry land farming tends to be high in the mountains. El Nino period in 2015 did not result in dryland farming significantly exposed as predominantly low exposure, while La Nina period resulted in moderate to high exposure because of dryland farming does not need much water. Exposure of dryland agriculture in Kebumen is dominated by low exposure to the El Nino period that dominates in the garden mix, higher exposure in the period of El Nino is in the northern part of the district which is dominated by dry land agriculture fields and forest types. In the period of La Nina dominated by the exposure being distributed in the central part to the northern districts with altitude region 50-500 meters above sea level and dominates in dryland farming types mixed farms. Low exposure in the period of La Nina distributed in parts of the coastline is dominated by rainfed and high-exposure is in the northeastern part of the district that dominates in the garden mix.

El Nino and La Nina have an impact on the productivity of the maize crop in Kebumen. In the period 2015 El Nino, maize productivity tends to rise 43% from normal condition. In the period of La Nina in 2010 significantly lower maize productivity is decreased 63% from normal condition.Maize productivity declined in La Nina conditions in 2010 caused by presence shift in the early dry season, reduced the duration of the dry season, the average rainfall planting and harvests maize deviate above the normal range, and the number of rainy days the harvest of maize distorted.

Acknowledgement

The research work reported in this paper was funded by PITTA Grant No. 2326/UN2.R3.1/HKP.05.00/2018 Universitas Indonesia (UI) and Thank you to directorate Research and Community Service Universitas Indonesia.

References

[1] Prabowo M and Nicholls 2002 *The Impact of Southern Oscillation and El Nino in Indonesia* Department of Primary Industries Queensland, Australia

[2] Irawan B 2006 El nino and la nina climate anomaly phenomena: long-term trends and their effects on food production *Forum of Agro Economic Research* **24** (1) 28-45

[3] Central Agency of Statistics, Kebumen Regency 2015 www.kebumenkab.bps.go.id accessed on October 10, 2017

[4] IPCC 2001 Climate Change 2001 : Impacts, Adaption, and Vulnerability In Contribution of Working Group II to the Third Assessment Report of the Intergovernmental Panel on Climate Change, Australia

[5] ____ 2007 Climate change 2007: Impacts, adaptation and vulnerability. In: Contribution of Working Group II to the fourth assessment report of the intergovernmental panel on climate change. Cambridge: Cambridge University Press

[6] ____ 2014 Fifth Assesment Report A Special Report of Working Groups I and II of the Intergovernmental Panel on Climate Change (IPCC): Cambridge University Press, Cambridge, UK and New York, NY, USA, 555-64

IOP Conf. Series: Earth and Environmental Science **303** (2019) 012025 doi:10.1088/1755-1315/303/1/012025

Prediction of Seawater Flooding Hazard on Settlement Areas in Padang City as a Climate Change Impact using GIS and Remote Sensing Technology

AY Nofrizal, H Rahman and M Hanif

Geography Science, Padang State University, Padang 25132, Indonesia

adenannofrizal23@gmail.com; henzrahman12@gmail.com

Abstract. Recently, climate change is becoming the most popular issue in the 20th century. One impact of climate changes is seawater flooding. It occurs in a lowland of coastal area. One of coastal cities that will have impact of sea level rise is Padang City. This is because part of Padang's morphology is a lowland and constrained by the sea. This research aims to analyse and to predict hazard potential of seawater flooding, especially on settlement areas. Statistic model used was GIS and Remote Sensing. DEM classification, representing altitude of Padang City, was used to predict sea level rise. Land change modeler is used to predict settlement areas development, and to calculate impact of flooding on settlement areas. The result showed that there is a sea level rise of 2.8 m, causing seawater flooding in 2500. It also predicts that settlement will cover area of 22,522 ha in 2500. The settlement area impacted by seawater flooding was 1,322 ha. A subdistrict had the highest risk of being drown was Koto Tangah with affected area of 1,026 ha, while the lowest one was Lubuk Begalung with affected area of 6 ha.

Keywords : climate change, flooding, settlement area, GIS and Remote Sensing.

1. Introduction

Climate change is becoming the most popular issue of the 20th century and various conferences in the world are held to discuss climate change, ranging from cause, prevent and impact from climate changes. Increase in fossil fuel burning and land use change have emitted, and continued emiting, increasing pollution of greenhouse gases into the earth's atmosphere. These greenhouse gases include carbon dioxide (CO_2), methane (CH_4), and nitrogen dioxide (N_2O), and rise in these gas concentrations caused a rise in the amount of the heat from the sun withheld in the Earth's atmosphere, heat that would normally be radiated back into space. This increase in heat has led to the greenhouse effect, resulting in climate change. Warming of the climate change system is now unequivocal. It is now clear that global warming is mostly due to man-made emissions of greenhouses gases (mostly CO_2). Over the last century, atmospheric concentrations of carbon dioxide increased from a pre-industrial value of 278 to 379 parts per million (ppm) in 2005, and average global temperature rose by 0.74 ^{o}C [1]. At the 21st Century, it shows that the global warming will continue and accelerate. The best estimates indicate that the earth could warm by 3° C by 3100. One of climate change impact is seawater flooding, this can occur in the low altitude of coastal area [2].

Seawater flooding causes hazard to coastal region, because the sea level rise can influence social and demographic in the corresponding area. Seawater flooding in the coastal area can make many problems for activities, especially for cities in coastal zone [3]. Recently, the very large cities in this world are located in the coastal zone, for example Jakarta in Indonesia, Chongking in China, and many more. Tens million of people in low altitude coastal areas of South and Southeast Asia can be affected by sea level rise [4]. Indonesia has big cities which are mostly located in the coastal area. One of them is the city of Padang, the capital of West Sumatra Province. Padang City has 11 sub-districts namely Padang Utara Subdistrict, Padang Barat Subdistrict, Padang Timur Subdistrict, Padang Selatan Subdistrict, Nanggalo Subdistrict, Kuranji Subdistrict, Pauh Subdistrict, Bungus Teluk Kabung Subdistrict, Koto Tangah Subdistrict, Lubuk Kilangan Subdistrict, and Lubuk Begalung Subdistrict. BPS Padang city stated that The Padang city has a population around 914,968 people in 2016 and is expected to continue increasing every year. The accelerated increase in the sea level will put 6 vulnerable districts along coastline of Padang city in high risk.

2. Method

2.1. Case Study

Case study in this research is Padang City, the capital of Province West Sumatera. 100°05’05’’ E - 100°34’09” E and 00°44’00’’S - 01°08’35’’ S. Padang City is a coastal region in Indonesia. This region serves as a center of human activity in the West Sumatera, because Padang is a capital of Province. It has 11 subdistricst, they are Koto Tangah, Lubuk Kilangan, Padang Barat, Padang Timur, Padang Selatan, Bungus Teluk Kabung, Pauh, Kuranji, Nanggalo, Padang Utara and Lubuk Begalung. There was an increase in demand of land for human settlement that caused land use change [5].

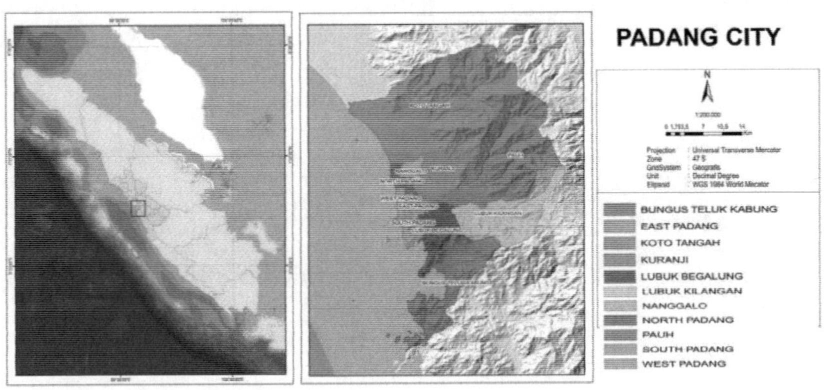

Figure 1. Case study, Padang City, West Sumatera Province, Indonesia

2.2. Material

Spatial data used in this research are taken from various website and governmental sources. Detail of data are listed in table 1.

Table 1. Data for Analysis

Data	Sources	Path and Row
SRTM	USGS EarthExplorer	Path 127 Row 61
Landsat TM5 , 1989	USGS EarthExplorer	Path 127 Row 61
Landsat TM 5, 2001	USGS EarthExplorer	Path 127 Row 61
Landsat OLI 8, 2015	USGS EarthExplorer	Path 127 Row 61
Driving Factor Location	Survey	
Topography	InaGeoportalIndonesia BIG Geospatial Agency	
SPOT 6	LAPAN	
Settlement Sampel Location	Survey	
Administration map	BAPPEDA Padang City	

2.3. Data Analysis

In this research we have three data analysis processes to predict seawater flooding hazard on settlement areas in Padang City as impact of climate changes, such as; increase sweater, land change modeler and seawater flooding on settlement areas. In this research, we predict seawater flooding on settlement areas in Padang City in 2500. We use projection to predict seawater flooding by applying increase value per year as suggested by Intergovernmental Panel on Climate Change (IPCC) [6]. We obtained a high increase in seawater flooding by analysing morphology of Padang City, and to predict settlement areas in Padang City by using Geographical Information System (GIS) and Remote Sensing Technology with Land Change Modeler.

2.3.1. Sea level rise

Sea level rise was predicted using mathematical method and assumption, and digital elevation model (DEM) to get information of altitude and morphology of Padang City. Seawater flooding caused sea level rise. Prediction analysis of sea level rise was carried out by using SRTM imagery to know increase as impact of climate change [7]. The altitude is a basic process as analogic to make spatial data of flooding area. Sea level rise was approximately 0.57 cm/year according to Intergovernment Panel on Climate Change (IPCC). To calculate sea level rise, we used formulae as stated below [8]:

$$Icst = 0,57 \times t \tag{1}$$

where *Icst* is sea level rise prediction, *t* is time. The result are calculation of sea level rise and hazard area of seawater flooding predictions. Geography Information System (GIS) was applied to process digital elevation data in order to extract flooding area by altitude. The flooding area was classified as altitude in number of ranges of seawater flooding.

2.3.2. Land Change Modeler

Urban development resembles behaviour of a cellular automaton in many aspects. The space of an urban area can be regarded as a combination of a number of cells, each cell taking a finite set of possible states representing the extent of its urban development. The state of each cell evolves in discrete time steps according to some local rules [9]. All landscape spatial transition models can be expressed in a simple matrix as in equation (2).

$$LU_{t+1} = LU_t \, x \, P \tag{2}$$

with LU_t is the distribution of land uses among the different types at the beginning of the period and LU_t+1 is showing the distribution of land use types at the end of the projection period, in other words LU_t and LU_{t+1} are vectors composed of the fractions of each landscape type at time *t* and time $t + 1$, respectively, equation (3) illustrates the *P* matrix.

$$P = \begin{bmatrix} P_{11} & P_{12} & \cdots & P_{1n} \\ P_{21} & P_{22} & \cdots & P_{2n} \\ \vdots & \vdots & \ddots & \vdots \\ P_{n1} & P_{n2} & \cdots & P_{nn} \end{bmatrix} \tag{3}$$

A transition areas matrix expresses the total area (in cells) expected to be changed in the next time period. A set of conditional probability images expresses the probability that each pixel will belong to the designated class in the next time period. They are called conditional probability maps since this probability is conditional on their current state [10] as in equation (4).

$$P_m\,(i,j) = \frac{N\,(i,j)}{\sum_{k=1}^{k} n_{ij}} \tag{4}$$

where $N(i, j)$ is the observed data during the transition from state i to j, and n_{ij} is the number of years between time step i and step j, and the total number of years is m; $P(i, j)$ is the yearly transition probability after normalizing the transition probability in multiyear and K is the number of steps [11]. The illustrates a technique of CA-Markov since a Markov chain analysis is performed in order to estimate the transition matrix between the two past dates and to estimate probabilities of change for the third date to be predicted. Then, CA estimates the spatial distribution of land cover at a later date. Equations below illustrate the evaluation of cells from time t to t +1 is determined by a function of state, it neighborhood space and a set of transition rule [12].

$$\mathrm{LU}^n = f(\,LU_{i,j(t)} \cdot S_{i,j(t)} \cdot P_{x,y,i,(t)} \cdot N_{i,j(t)}) \tag{5}$$

$$N_{i,j(t)} = \frac{\sum_t N_{i,j(t)}}{\#\, of\ adjacent\ cells} \tag{6}$$

Where,

$LU_{i,j(t+1)}$	:	The potential of cell i, j to change at time t + 1,
$LU_{i,j(t)}$	:	States of cell i, j at time t,
$S_{i,j(t)}$	:	Suitability indexes of cell i, j at time t,
$P_{x,y,i,(t)}$	:	Probability of cell i, j to change from state x to state y at time t (as shown in equation 4) above, and;
$N_{i,j(t)}$	:	Neighborhood index of cell i, j.

In this research, to take information of the past and existing land use and land cover change (LULC) from Landsat imagery, we interpreted by using remote sensing technology. Land change modeler was used to predict landuse changes in the future. It was applied especially for settlement area in Padang City by processing with GIS [13, 14]. Land use was classified into 8 classes, i.e: 1) primary forest, 2) secondary forest, 3) settlement, 4) mixed garden, 5) weatland, 6) bush and reeds, 7) bare land, and 8) water body. Prediction of settlement area in Padang City in 2500, was selected because Padang City is always a target of development. Development of settlement areas occured because many factors, they namely as driving factors. In this research, driving factors are distance from road, distance from cultivated area, distance from settlement area, land elevation, and distance from river [15]. Landuse change Prediction has three processes, the first step is to interprete Landsat imagery, the second is to analyse driving factors in coastal area and distance from the sources [16], and the last process is to predict settlement area change in the future based on cellular automata and markov chain model [17].

2.4. Seawater flooding on Settlement areas

To analysis impact of sea water flooding on settlement area, we used method of overlay map. It is a simple technique to get information on how the subject can give impact one to each other. Where to complete the data, where to take sample of settlement view location, where to locate hazard probability of seawater flooding in the future, that act with ground check location. Figure 2 displays workflow research of seawater flooding on settlement areas from start to finish step.

3. Result and discussion

3.1 Sea level rise in Padang City

Sea level rise as flooding has predicted to cover many districts. The high increase of sea level at surface in 2500 is 2,8m. The sea level rise can fulfil or cover all the depression area, tidal flat, estuary, and lowland around the coastal. In the city, the character of morphology and its altitude caused the large areas are covered by seawater. The flooding increase started from northern subdistrict of Koto Tangah, and ended in Lubuk Begalung subdistrict. The result of spatial analysis represented an image of altitude, surface and prediction of increase in seawater flooding. The settlement areas affected by seawater flooding in Padang City can be obtained by overlaying maps of sea level rise and settlement areas prediction in 2500 as shown in figure 2.

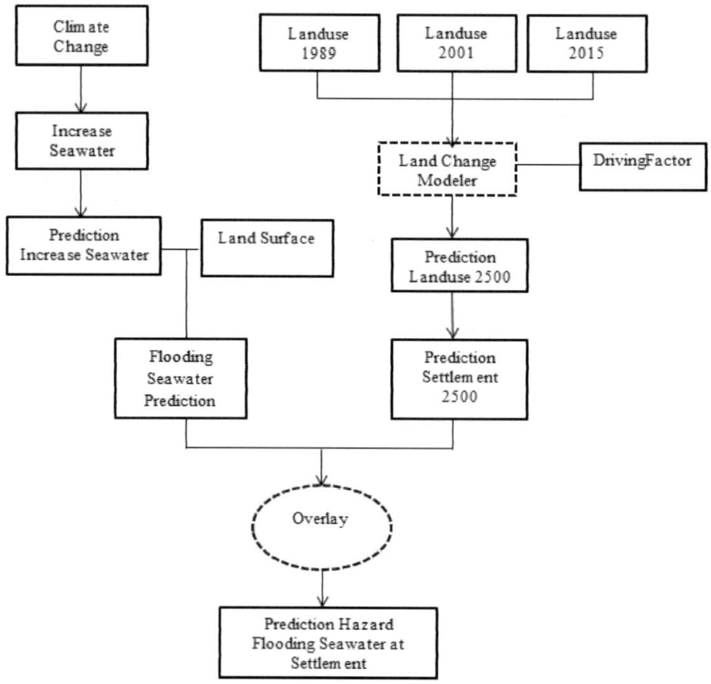

Figure 2. Research Workflow

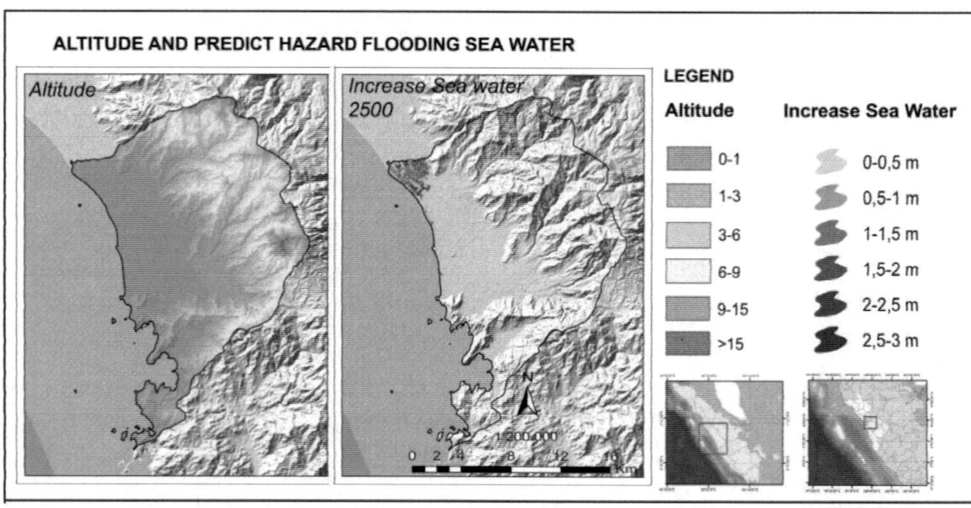

Figure 3. Altitude and prediction of seawater flooding in Padang City

The map shows how sea level rise and flooding in the city. The coastal area is vulnerable to impact of climate changes because of seawater flooding. Overall, in 2500, all of districts located around coastal areas in Padang will experience flooding because of sea level rise. The seawater flooding vulnerable districts were Koto Tangah, Padang Utara, Padang Barat, Padang Selatan, Lubuk Begalung and Bungus Teluk Kabung. The district has the highest vulnerability with the largest area covered by seawater flooding is Koto Tangah. It happens because of the coastal morphology, where land altitude was relatively not much higher than the sea such as lowland, many swam, estuary and land depression. It was so different from other districts, as shown in the surface map, we can observe when we move to southern area around the sea where the land altitude gradually increases such as Bungus Teluk Kabung that has hilly surface. The region that has a low risk of seawater flooding was Lubuk Begalung subdistrict, because it just has a small area in the coast, and the distance from coast was far enough from coastal line.

3.2 Prediction of Settlement areas and Landuse change

We predict the settlement areas and land use change in the year of 2500. Location and landuse change and settlement areas are categorised by years that are 1989, 2001, 2015, and prediction in 2500. The resulted map of landuse change and settlement areas in Padang City and also their prediction can be seen in figure 4.

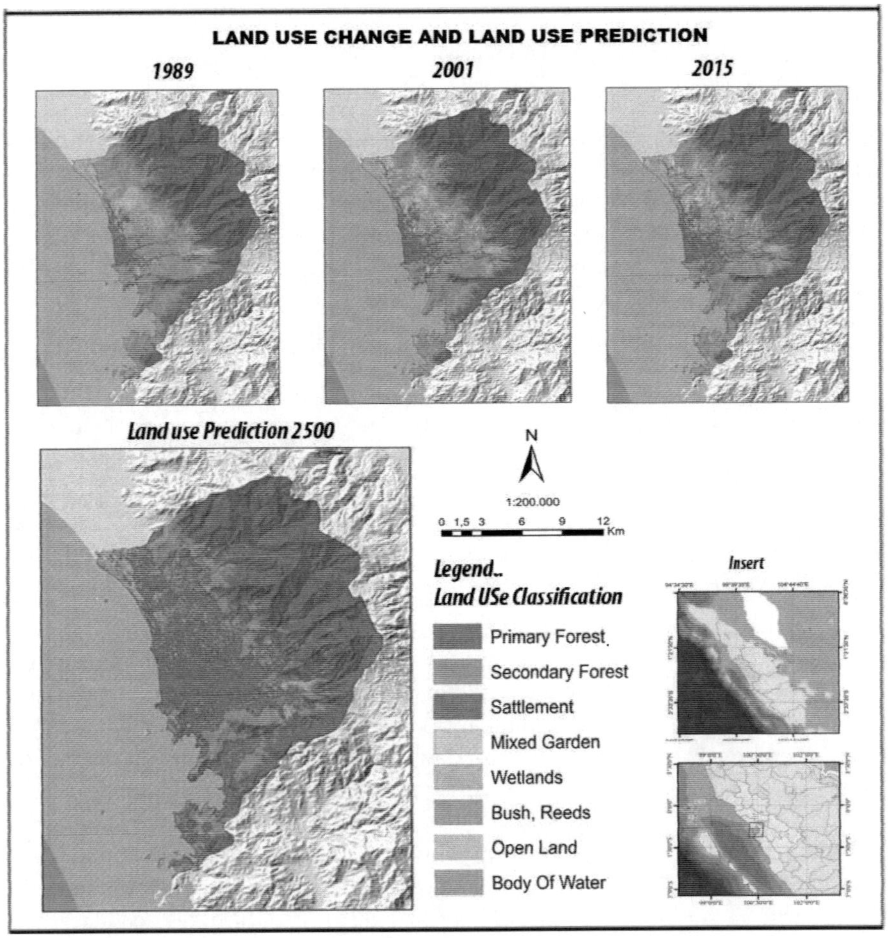

Figure 4. Landuse changes and land change prediction

Landuse change in Padang City resulted from land change modeler value can be seen in figure 4 and figure 5. These figures showed development of settlement areas and landuse change in Padang City from cultivated area, distance from road, distance from river, driver for settlement areas and digital elevation model in Padang City, this prediction use Kappa Index accuracy to get value of prediction.

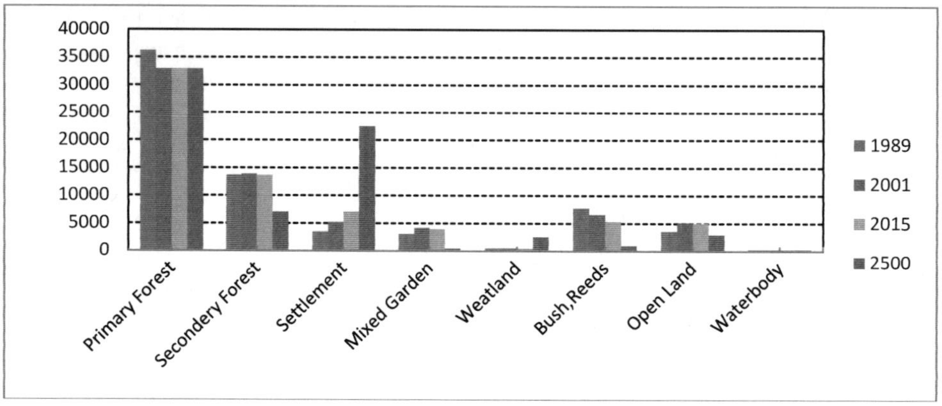

Figure 5. Bar chart of land use changes

The bar chat reveals how the land use changes and land use prediction in 2500. It can be observed that the land use change occurred in the years, some land use type had area increase and some area decreased, but the focus in this case was that the change in settlement area had increased. Generally, landuse change, especially settlement areas, was considerably large. In 1989, settlement areas in Padang City was 3,464 ha. There was an area increase in 2001 to 5,230 ha, and in 2015 to 7,070 ha. The prediction of settlement areas in 2500 could reach as high as 22,522 ha.

From this land use map prediction in figure 4, we can observe that generally, all of the lands with fluctuated topography of land cover are forest and natural land cover. The land was protected because of physiography of the land. The high elevation in rough surface acted as barrier to land use change caused by human activity. Most land cover on the lowland is settlement areas, because driving factor is road, land elevation, and support for human activity to develop the urban, in micro aspect as residential or settlement areas. The future prediction map has showed that around the coastal area or near the coastal line has mainly developed as settlement areas. The assumption that the road as driving factor has led to human accessibility to develop the land and to support human activities. Therefore, land use change was the largest in lowland area.

3.3 Prediction of Seawater Flooding at Settlement areas in Padang City

The sea level rise analysis by using digital elevation model classification can predict sea level rise in 2500. The sea level has increased to 2.8 m from existing coastline. The prediction of settlement areas in 2500 represented the future area. The prediction maps of sea level rise and settlement areas in 2500 have been carried out by using overlay method. We can compare where the settlement areas were impacted by seawater flooding or not, and we can make a matric and then calculate the future affected areas.

Table 2. Seawater flooding at Settlement areas in Padang City in 2500

No	District	Subdistrict Wide (ha)	Hazard Wide (ha)
1	Koto Tangah	23,119	1,062
2	Padang Utara	901	16
3	Padang Barat	624	21
4	Padang Selatan	1,472	62
5	Lubuk Begalung	3,055	6
6	Bungus Teluk Kabung	8,607	154
	Total	37,780	1,322

Table 2 reveals the coverage areas of seawater flooding in many districts in 2500. Overall, the highest area effected by flood because of sea level rise was Koto Tangah, while the smallest was Lubuk Begalung. The Koto Tangah has flooding area of 1,062 Ha, while Lubuk Begalung of 6 Hh. The area with the lowest hazard was represented by Lubuk Begalung. If we look at the administration map, it showed that Lubuk Begalung had further distance area from the coastline than Koto Tangah.

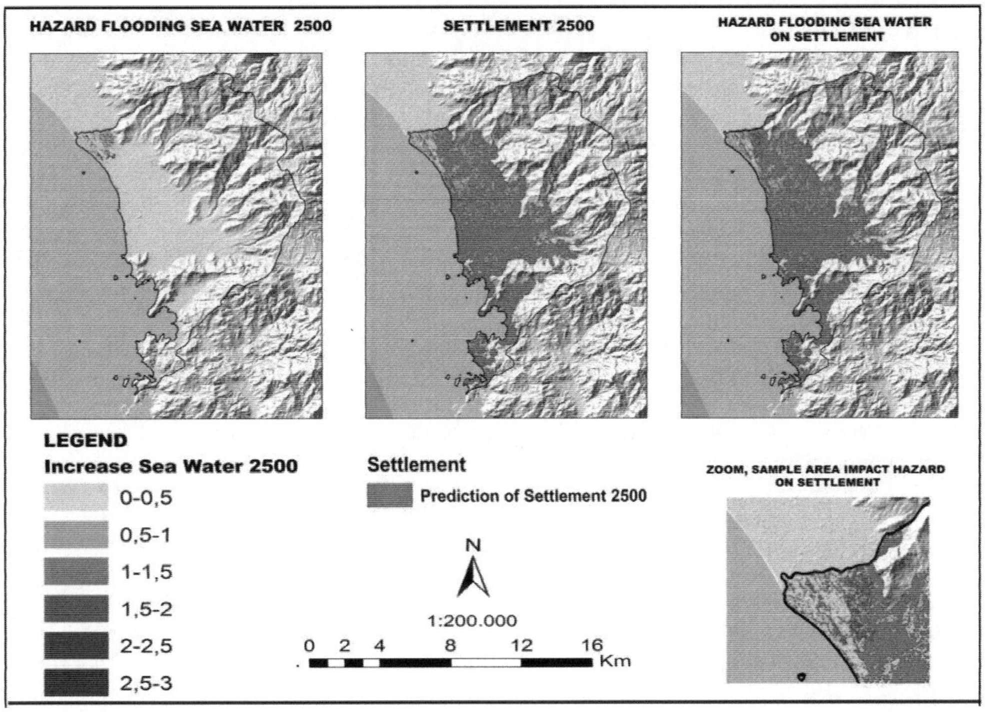

Figure 6. Map of sea level rise and settlement areas predictions in 2500 in Padang City

From figure 6 we can observe how the seawater flooding will affect settlement areas in 2500. It will cause flood in large areas around the coast, and drown many houses. The seawater flooding will also destroy many ecosystem and economic sources, constrain human activities and cause economic loss.

4. Conclusion

Many subdistricts in Padang City have increase in seawater flooding hazard as impact of climate change, especially for settlement areas. The subdistricts near coastline are Koto Tangah, Padang Utara, Padang Barat, Padang Selatan, Lubuk Begalung, and teluk Bayur. It is predicted that sea level rise will cause seawater flooding in 2500 as high as 2.8 m from existing coastline. Development of settlement areas in Padang City is predicted to be as high as 22,522 ha in 2500. Prediction analysis of seawater flooding hazard in Padang City as an impact of climate change showed that 1,322 ha of settlement areas would be affected. The subdistrict has the highest risk was Koto Tangah with predicted affected area of 1,026 ha, while the lowest one was Lubuk Begalung Subdistrict with predicted affected area of 6 ha. It was because Koto Tangah had a long coastline, while Lubuk Begalung has a small area around the coast. This research showed that Remote Sensing and Geography Information System (GIS) can be used to analyse prediction of increase in seawater flooding hazard on settlement areas as an impact of climate change, especially for coastal region, for example Padang City, the capital of West Sumatera Province, Indonesia.

Acknowledgements
I would like to thank to my research team from Community of Geospatial and Technology for their support, and also to Geography Department, Padang State University.

References
[1] Climate Change and Sustainability Committee 2015 Climate Change and Resource Sustainability an overview for Auctaries. *Canadian Institute of Actuaries*.
[2] Heimlich B and Bloetscher F 2011 *Effects of Sea Level Rise and Other Climate Change Impacts on Southeast Florida's Water Resources Florida Water Resources*. Journal September 2011 pp 36-46
[3] Adeyemi O F and Modupeola A G 2014 *The Effect of Sea Water on Compressive Strength of Concrete International*. Journal of Engineering Science Invention. Volume **3** issue 7 pp 23-3
[4] Su T W 2017 *A Study on Coastal Flooding and Risk Assessment under Climate Change in the Mid-Western Coast of Taiwan Water 2017* pp 1-13
[5] Nofrizal A Y 2017 *Normalized Difference Built Up Index (NDBI) Sebagai Parameter Identifikasi Perkembangan Permukiman Kumuh pada kawasan Pesisir di Kelurahan Kalang Kawal Kecamatan Gunung Kijang Kabupaten Bintan*. Jurnal Tunas Geografi Vol **6** No 2. pp 143-150 ISSN 2301-606X.
[6] Hermon D 2017 *Climate Change Mitigation*. (Rajawali Pers Radgagrafindo. Jakarta-Indonesia).
[7] Mardikaningsih et al 2017 *Studi Kerentanan dan Arahan Mitigasi Bencana Banjir di Kecamatan Puring Kabupaten Kebumen tahun 2016*. Jurnal GeoEco Vol **3** No 2 pp 157-163.
[8] IPCC 2001 Summary for Polycemakers in: Climate Change 2001: The Phsycal Science Basis. s.l., *Contribution of working group I to the fourth assesment report of the intergovernmental panel on climate change*. [Houghton J T, Ding Y, Griggs D J, Noguer M, van der Linden P J, Dai X, Maskell K and Johnson CA]. Cambridge University Press, Cambridge, United Kingdom and New York , (New York-USA).
[9] Yan L 2009 *Modelling Urban Development With Geographical Information System and Cellular Automata*. (CRC Press Taylor & Francis Group, Paris-France).
[10] Eastman J R 2012 *IDRISI Selva Manual. IDRISI Tutorial. s.l.* (Clark University, Worcester. www.clarklabs.org Newyork. USA
[11] Tang J, Wang L and Yao Z 2007 *Spatio-Temporal Urban Landscape Change Analysis Using the Markov Chain Model and a Modified Genetic Algorithm*. International Journal of Remote Sensing, **28**, 3255-3271
[12] A Maher E H and Mohammed H O 2015 *Using Cellular Automata-Markov Analysis and Multi Criteria Evaluation for Predicting the Shape of the Dead Sea*. Journal Advances in Remote Sensing Vol **4**, 83-95. Scientific Research Publising.
[13] Daweoud A M and Mulla M M 2012 *Environmental Impacts of Seawater Desalination: Arabian Gulf Case Study International*. Journal of Environtment and Sustainability Vol **1** No 3 pp 22-37.
[14] Rahmstorf 2012 *Comparing Climate Projections to Observations up to 2011*. Environmental Research Letters Lett **7** pp 1-5.
[15] Diva H I, Dewi S, Khairul N, Akina A 2018 *Investigation Volcanic Land Form and Mapping Landslide Hazard Potential at Mount Talang*. Sumatra Journal of Disaster, Geography and Geography Education. Vol **2**, No 1 pp 16-23.
[16] Fiati F and Latubessy A 2015 *Identifikasi Daerah Potensi Banjir Berbasis Expert System*. Prosiding SNATIF ke-**2** tahun 2015 pp 181-190.
[17] Nofrizal A Y et al 2018 *Identifikasi Perubahan Penggunaan Lahan di Kota Solok, Sumatera Barat Berbasis Penginderaan Jauh dan SIG dengan menggunakan Object Base Image Analyst (OBIA)* Prosiding Seminar Nasional GEOTIK 2018 ISSN : 2580-8796 pp 96-104.

IOP Conf. Series: Earth and Environmental Science **303** (2019) 012026 doi:10.1088/1755-1315/303/1/012026

An Application of Markov Chain for Predicting Rainfall Data at West Java using Data Mining Approach

A Azizah[1], R WElastika[1], A Nur Falah[2], B N Ruchjana[2] and A S Abdullah[3]

[1]Magister Student at Department of Mathematics, Universitas Padjadjaran
[2]Department of Mathematics, Universitas Padjadjaran
[3]Department of Computer Science, Universitas Padjadjaran

annisa17056@mail.unpad.ac.id

Abstract. Markov chain model is a stochastic process to determine the transition probability of a state space based on a previous state. We can use a stationary distribution of first order-Markov chain model to determine the long terms probability rainfall phenomena. A rainfall data in West Java area has a big data because we can have a large rainfall data from many cities and regencies both of in spatial and time series observations. Furthermore, in this paper, we demonstrate an application of Markov chain using a Data Mining approach to get the knowledge as a pattern for description and prediction the monthly rainfall data in wet seasons December-January-February (DJF) using Knowledge Discovery in Database (KDD) method through pre-processing, data mining process and post-processing. We simulate the monthly rainfall data from the year 1981-2017 using four-state spaces: low (0), medium (1), high (2), and very high (4). The result of Markov chain shows that the probability of occurrence rainfall phenomena for four state spaces are: low (22.62%), medium (24.86%), high (25.46%), and very high is 27.05%. It means that West Java area over the long term condition will have a very high rainfall probability.

Keywords: Markov chain, Stationary Distribution, Data Mining, Rainfall.

1. Introduction

Java Island is the island with the biggest population in Indonesia and makes various development sectors centers on there. Java is also the largest island affected by climate phenomena compared to other islands in Indonesia, if we predict the rainfall data in unobserved locations, it will be obtained the rainfall data with a pattern similar to observed locations as its neighbor [1]. Rainfall is the height of rainwater that collects in a flat place, not evaporates, not penetrates, and not flow. 1 millimeter of rainfall, which means that in a solid square meter area a flat place is held up to one millimeter of water or can hold one liter or 1000 ml of water [2].

In West Java, rainfall is considered to be one of the major constraints of agriculture plans and policy decisions because of the position of West Java as one of the centers of food based on the production paddy in Karawang Regency. It supports the Sustainable Development Goals to cover social and economic development issues including poverty, hunger, health, global warming, water, etc. Rainfall is the most important part of tropics which influenced the paddy production at West Java. Based on this, it's needed a fairly accurate prediction method especially for West Java Province's rainfall because West Java is one of the regions with the largest agricultural sector [3].

The amount of rainfall that occurs at this time could have been influenced by the amount of rainfall one time before, and the amount of rainfall in the future can be affected by the current rainfall, etc. This phenomenon is a real-life example of the Markov chain event which is a method for modeling in stochastic processes [4]. Markov chain model used to assist in estimating the changes that may occur in the future, where the changes are represented in dynamic variables at certain times. The Markov chain was first coined by Andrey Andreyev Markov (1856-1922)[5]. A stochastic process is said to include the Markov chain if it fulfills the properties of Markov (Markovian property). The properties of Markov stated that the probability of a future event, with known past events and present events, is not dependent on past events and only depends on the present events [4, 6].

The Markov chain is generally classified into two, namely the Markov chain with discrete parameter index and the Markov chain with continuous parameter index. The Markov chain is said to be a discrete parameter index if the shift state occurs with a fixed discrete time interval. Whereas, the Markov chain is said to be a continuous parameter index if the shift state occurs with a continuous time interval [7]. Rainfall data is a time series data that indicates the movement of state in a fixed discrete time interval. Rainfall prediction in the future is needed to anticipate prevention if high rainfall intensity will occur for a long time. Besides that, it indicates that we need to consider the other phenomena can make a significant contribution to increasing the rainfall intensity [8]. Furthermore, in this paper was conducted an analysis of a large database of rainfall data from 27 district/cities using the stationary distribution of the Markov chain and then used to predict rainfall in West Java based on Data Mining approach using KDD method.

2. Method

2.1. Stochastic Processes

A stochastic process $\{X(t), t \in T\}$ is a collection of random variables. That is, for each t in the index set T, $X(t)$ is a random variable. If the time parameter T is a countable set $T = \{0,1,2,...\}$, the process $\{X(n), n = 0,1,2,...\}$ is called a *discrete-time stochastic process*, and if T is a continuum, the process $\{X(t), t \geq 0\}$ is called a *continuous stochastic process*. For a stochastic process $\{X(t), t \in T\}$, a set of all values of $X(t)$ is called a state space [9, 10].

2.2. Basic Concept of Markov Chains

A Russian mathematician, Markov, introduced the concept of a process in which a sequence or chain of discrete states in time for which the probability of transition from one state to any given state in the next step in the chain depends on the condition during the previous step [11]. A first order Markov chain is a stochastic process having the property that probability for future events depends only on the present event, in other words:

$$Pr(X_{n+1} = x | X_1 = x_1, X_2 = x_2, \dots, X_n = x_n) = Pr(X_{n+1} = x | X_n = x_n) \tag{1}$$

For all states $x_1, x_2, \dots, x_n$ and all $n \geq 0$ such a stochastic process is known as a Markov chain [7, 12].

2.3. Discrete Time of Markov Chain

Suppose $\{X(n), n = 0,1,2,...\}$ is stochastic process with discrete parameter index and state space $i = 0,1,2,...$ unless otherwise specified. If

$$\begin{aligned} &P\{X(n+1) = j | X(0) = i_0, X(1) = i_1, \dots, X(n-1) \\ &= i_{n-1}, X(n) = n\} = P\{X(n+1)\} = j | X(n) = 1\} = P_{ij} \end{aligned} \tag{2}$$

for all $i_0, i_1, \dots, i_{n-1}, i, j$ and n, then the process is called a discrete-time Markov chain, and p_{ij} is called a transition probability. The value p_{ij} is called a (stationary) transition probability, it represents the probability that the process will, when in state i, next make a transition into state j. The transition between the states is described by the transition probability matrix, defined as:

$$P = (p_{ij}) = \begin{pmatrix} p_{00} & p_{01} & p_{02} & \cdots \\ p_{10} & p_{11} & p_{12} & \cdots \\ p_{20} & p_{21} & p_{22} & \cdots \\ \vdots & \vdots & \vdots & \cdots \end{pmatrix} \tag{3}$$

Since probabilities are non-negative and since the process must make a transition into some state, we have that $p_{ij} \geq 0, i,j \geq 0$ and $\sum_{j=0}^{\infty} p_{ij} = 1,\ i,j = 0,1,2,\ldots$ [7, 13].

2.4. The n-step Transition Matrix

The one step transition probabilities P_{ij} is already defined. Now define the n-step transition probabilities P_{ij}^{n} to be the probability that a process in state i will be in state j after n additional transitions. Let A be an event. A convenient notation is $P_i(A) = P(A|X_0 = i)$. For example

$$P_i(X_1 = j) = p_{ij}. \tag{4}$$

Given the initial distribution λ, let us treat it as row vector. Then

$$P(X_1 = j) = \textstyle\sum_{i\in I} \lambda_i P_i(X_1 = j) = \sum_{i\in I} \lambda_i P_{ij}. \tag{5}$$

Similarly,

$$P_i(X_2 = j) = \textstyle\sum_k P_i(X_1 = k, X_2 = j) = \sum_k P_{ik}P_{kj} = (P^2)_{ij} \tag{6}$$

$$P(X_2 = j) = \textstyle\sum_{i,k} \lambda_i P_i(X_1 = k, X_2 = j) = \sum_{i,k} \lambda_i P_{ik}P_{kj} = (\lambda P^2)_j. \tag{7}$$

Continuing in this way,

$$P_i(X_n = j) = (\delta_i P^n)_j = (P^n)_{ij} = P_{ij}^{(n)} \tag{8}$$

$$P(X_n = j) = \textstyle\sum_{i_0,\ldots,i_{n-1}} \lambda_{i_0} P_{i_0 i_1} \ldots P_{i_{n-1} j} = (\lambda P^n)_j. \tag{9}$$

Thus, $P^{(n)} = \left(P_{ij}^{(n)}\right)$, the n-step transition matrix, is simply P^n.

Also, for all i,j and $n,m \geq 0$, the (obvious) Chapman-Kolmogorov equations hold:

$$P_{ij}^{(n+m)} = \textstyle\sum_{k\in I} P_{ik}^{(n)} P_{kj}^{(m)} \tag{10}$$

Named for their independent formulation by Chapman, and Kolmogorov (1903-1987) [7].
The Chapman-Kolmogorov equation asserts that

$$p^{(n+m)} = p^{(n)}.p^{(m)} \tag{11}$$

where the dot product represents matrix multiplication. Hence, in particular

$$p^{(2)} = p^{(1+1)} = p^2 \tag{12}$$

and by induction

$$p^{(n)} = p^{(n+1-1)} = p^{n-1}.p = p^n \tag{13}$$

That is, the n-step transition matrix may be obtained by multiplying the matrix p by itself n times [12, 14].

2.5. Stationary Distribution of Markov Chain

There is exist a limiting probability that the process will be in state j after a large number of transitions, and this value is independent of the initial state. In other words, p_{ij}^{n} is converging to some value (as $n \to \infty$) which is the same for all i. Theorem. If an irreducible Markov chain is positive recurrent and aperiodic, there exists the limiting probability

$$\lim_{n\to\infty} p_{ij}^n = \pi_j > 0 \; where \; (i,j = 0,1,2 \ldots) \tag{14}$$

Which is independent of the initial state i, where $(\pi_j, j = 0,1,2,\ldots)$ is a unique and positive solution to

$$\pi_j = \sum_{i=0}^{\infty} \pi_i p_{ij} \text{ where } (j = 0,1,2,\ldots) \; (15)$$
$$\sum_{j=0}^{\infty} \pi_j = 1 \tag{16}$$

and it is called a *stationary distribution* for a Markov chain [10].

2.6. Markov Chain Model for Monthly Rainfall Data

Chain-dependent models treat the occurrence and intensity of daily rainfall events separately [15]. The term "chain dependence" reflects the statistical structure of the occurrence sequence. The monthly rainfall model based on a Markov chain can used to determine rainfall occurrence (i.e., the high or low rainfall conditions) based on transition probabilities [16]. The transition probabilities, estimated from the historic measurements, represent the probabilities of high to high, high to low, low to high, and low to low. If the next day the rainfall is high, then the rainfall intensity is given as a random variable following a probability density function.

The daily rainfall model based on the Markov chain can be explained as follows. First, define X_t as the high and low rainfall condition at the t-h day. That is,

$$X_t = 0, \quad \text{if } t \text{ day rainfall is high}$$
$$X_t = 1, \quad \text{if } t \text{ day rainfall is low}$$

Assuming that the occurrence probability of rainfall at present is dependent on the condition of the previous day, then X_t follows the first-order Markov chain, and then the transition probability of daily rainfall can be divided into the following four cases:

$$P[X_t = 1 \mid X_{t-1} = 0] = p_{01}$$
$$P[X_t = 1 \mid X_{t-1} = 0] = p_{11}$$
$$P[X_t = 0 \mid X_{t-1} = 1] = p_{00}$$
$$P[X_t = 0 \mid X_{t-1} = 1] = p_{10}$$

The above equations express the conditional probabilities of wet or dry on day t depending on the condition of wet or dry on day $t-1$. Therefore $p_{00} = 1 - p_{01}$ and $p_{11} = 1 - p_{10}$. Also, these four probabilities constitute a transition probability matrix:

$$P = \begin{pmatrix} p_{00} & p_{01} \\ p_{10} & p_{11} \end{pmatrix} \tag{17}$$

To derive the number of wet days, we first need to define the n-step transition probabilities:

$$P^n = \begin{pmatrix} p_{00} & p_{01} \\ p_{10} & p_{11} \end{pmatrix}^n \tag{18}$$

The n-step transition probabilities converge to certain probabilities as n increases:

$$\lim_{n\to\infty} P^n = \lim_{n\to\infty} \begin{pmatrix} p_{00} & p_{01} \\ p_{10} & p_{11} \end{pmatrix}^n = \begin{pmatrix} \pi_0 & \pi_1 \\ \pi_0 & \pi_1 \end{pmatrix} \tag{19}$$

These probabilities π_0 and π_1 are represent the mean occurrence probabilities of high and low rainfall conditions. These are called the stationary probabilities [16, 17].

2.7. Data Mining and Knowledge Discovery in Database

Data mining is the process of discovering interesting patterns and knowledge from large amounts of data. The data sources can include databases, data warehouses, the web, other information repositories, or data that are streamed into the system dynamically. As a knowledge discovery process, it typically involves data cleaning, data integration, data selection, data transformation, pattern discovery, pattern evaluation, and knowledge presentation [18]. Data mining also known as knowledge discovery (mining) in databases (KDD), knowledge extraction, data/pattern analysis, data archeology, data dredging, information harvesting, business intelligence, etc.

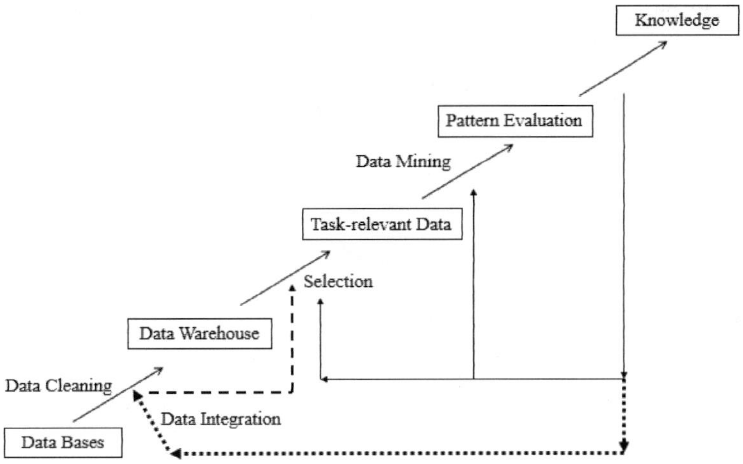

Figure 1. Knowledge Discovery in Database (KDD) Process

The Figure 1. above is a view from typical database systems and data warehousing communities. Data mining plays an essential role in the knowledge discovery process. Data mining has many successful applications, such as business intelligence, Web search, bioinformatics, health informatics, finance, digital libraries, and digital governments [18].

2.8. Data Mining for Rainfall Data in West Java

In this research we use the **Climate Hazards Group InfraRed Precipitation with Station (CHIRPS)** data of rainfall from LAPAN Bandung. CHIRPS is a big data, because it a 30+ year quasi-global rainfall dataset. Spanning 50°S-50°N (and all longitudes), starting in 1981 to near-present. In this research we use a CHIPS data from the year 1981 until 2017. The procedure of Data Mining approach for CHIRPS data is shown as following:

- Data pre-processing is a data mining technique that involves transforming raw data into an understandable format [17]. Raw data is highly susceptible to noise, missing values, and inconsistency. The quality of data affects the data mining results. In order to help improve the quality of data and consequently of the mining results, raw data is pre-processed so as to improve the efficiency and case of the mining process. Data pre-processing methods are divided in to four categories that is data cleaning, data integration, data transformation, and data reduction [19].
- In this research, rainfall data in West Java as a big data is a space-time data that consist of 432 monthly observation, in which 36 years during 1981-2017 in 27 district and city. Data size 432×27 is then carried out using the data cleaning method. Cleaning method is to fill the empty value of the data and to ignore the noisy data and to correct the inconsistencies data. Cleaning method is used by selecting data based on wet months, namely December, January and February (DJF) for each year which began in December 1981 to February 2017 so that the data size become more understandable and easier to process with data size 108×27.

- The data mining process using the Markov chain method starts with defining K state spaces, determine transition frequency matrix with an order of $K \times K$, determine the transition probability matrix, determine the stationary distribution, calculate the long-term proportion of the Markov chain, description and prediction.
- Post-processing component can be categorized into the following groups that is knowledge filtering, interpretation and explanation, evaluation, and knowledge integration [20]. In the case of application of Markov chain for predicting rainfall data in West Java using data mining approach, the knowledge results are to determine the long-term probabilities for rainfall in West Java using the Markov chain method.

3. Results
The process of this research in principle is divided into four parts, namely collecting data, data mining processes, processing data using Markov chains method, and analysis results. The flow of this research is as described in the following flowchart.

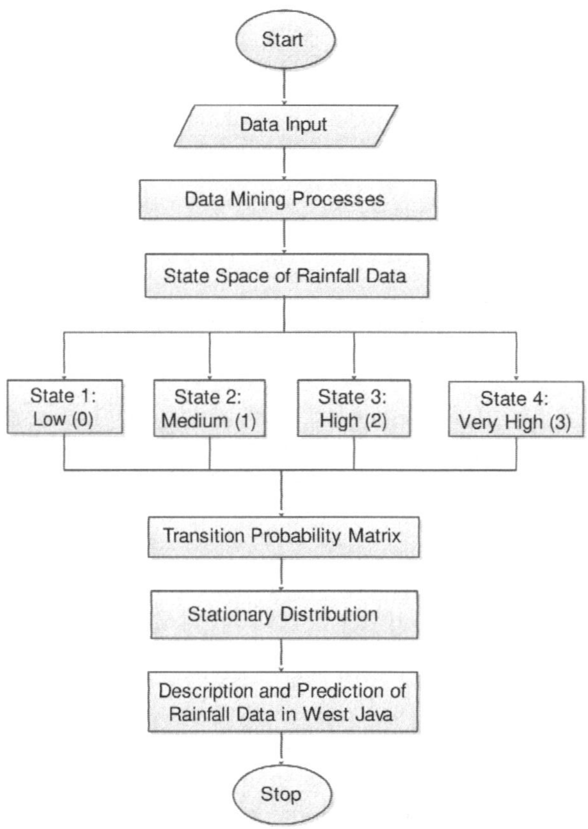

Figure 2. Flowchart of Markov Process using Rainfall Data in 4 State Spaces

Rainfall in a location are generally unpredicted, rainfall data can be daily, weekly, monthly, or yearly. The data that will be used in this research is rainfall data in 27 districts/cities in West Java Province on wet months of December, January, and February calculated from December 1981 to February 2017.

Table 1. Stationary Example of Monthly Rainfall Data at 27 Districts/Cities Year 1981-2017 (source: LAPAN Bandung)

No.	City/District.	Latitude	Longitude	Dec-81	Jan-82	…	Feb-17
1	Dist. Bandung	-7.12	107.57	317	323	…	325

2	Dist. Bandung Barat	-6.87	107.41	305	231	...	229
3	Dist. Bekasi	-6.23	107.15	291	403	...	364
4	Dist. Bogor	-6.58	106.88	408	394	...	445
5	Dist. Ciamis	-7.22	108.39	552	588	...	417
6	Dist. Cianjur	-6.81	107.13	373	315	...	307
7	Dist. Cirebon	-6.77	108.51	336	647	...	388
8	Dist. Garut	-7.39	107.77	361	347	...	264
9	Dist. Indramayu	-6.45	108.16	232	404	...	310
10	Dist. Karawang	-6.29	107.41	298	367	...	325
11	Dist. Kuningan	-6.98	108.50	416	574	...	386
12	Dist. Majalengka	-6.86	108.22	482	535	...	411
13	Dist. Pangandaran	-7.61	108.50	344	361	...	286
14	Dist. Purwakarta	-6.60	107.47	477	318	...	327
15	Dist. Subang	-6.56	107.71	444	390	...	333
16	Dist. Sukabumi	-6.85	106.96	528	326	...	447
17	Dist. Sumedang	-6.82	107.99	351	478	...	322
18	Dist. Tasikmalaya	-7.54	108.14	408	412	...	311
19	Bandung City	-6.92	107.62	171	220	...	167
20	Banjar City	-7.37	108.53	412	612	...	422
21	Bekasi City	-6.27	106.97	309	435	...	387
22	Bogor City	-6.59	106.80	411	357	...	436
23	Cimahi City	-6.88	107.54	227	178	...	144
24	Cirebon City	-6.73	108.56	360	608	...	378
25	Depok City	-6.39	106.80	343	415	...	363
26	Sukabumi City	-6.93	106.93	377	309	...	321
27	Tasikmalaya City	-7.35	108.23	466	474	...	353
28	West Java Average	-7.50	108.48	200	260	...	769

To show that the data is normally distributed, shown on the histogram and QQ plot in the following figure:

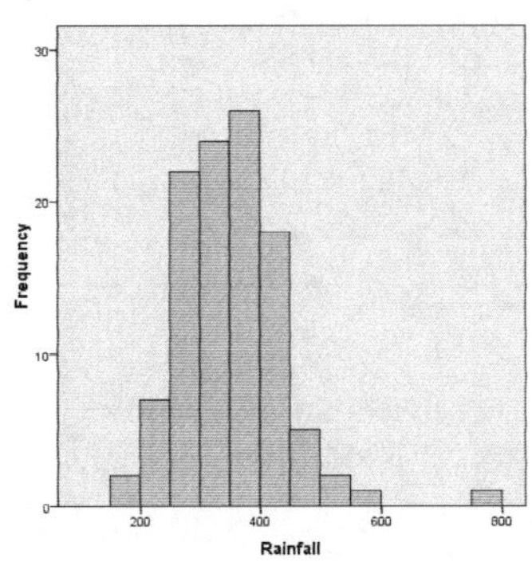

Figure 3. Histogram of Rainfall Data in West Java

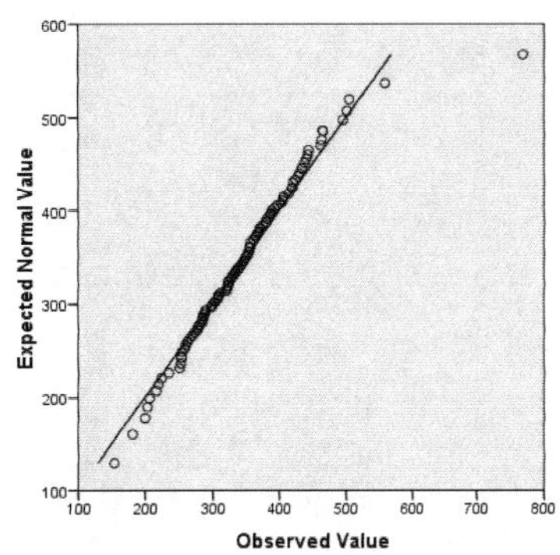

Figure 4. Normal Q-Q Plot of Rainfall Data in West Java

The Figure 3. above shows that the results of data cleaning for the average West Java are normally distributed which is characterized by a bell-shaped normal distribution curve that is quite symmetrical. As for Figure 4. we see the point is match up along a straight diagonal line which shows that the quantiles match and normally distributed [21]. The simulation steps using the Markov chain model are as follows [10]:

- Defining K state spaces.
- Rainfall data is divided in 4 states, i.e. a low rainfall (0), medium (1), high (2), and very high (3). Rainfall is categorized as low (0) if the rainfall value $\leq$ lower quartile, categorized as medium (1) if the lower quartile $<$ rainfall value $<$ middle quartile, categorized as high (2) if

the middle quartile < rainfall value < upper quartile, and categorized very high (3) if the rainfall value ≥ upper quartile.

- Determine a matrix with an order of K × K, with the initial state as a row and the final state as a column, then specify the number of displacements for each state.

Table 2. Frequency of Transition Rainfall 4 State Spaces in West Java

No.	City/District	Transition state i to $j, i = 0,1,2,3$ and $j = 0,1,2,3$															
		00	01	02	03	10	11	12	13	20	21	22	23	30	31	32	33
1	Dist. Bandung	7	5	7	8	4	6	8	7	10	5	7	5	6	9	5	7
2	Dist. Bandung Barat	4	5	8	9	7	5	9	6	9	8	5	5	7	9	4	7
3	Dist. Bekasi	5	6	6	10	10	5	3	9	6	8	8	5	6	8	9	3
4	Dist. Bogor	7	5	5	10	8	7	7	5	6	8	5	8	6	7	9	4
5	Dist. Ciamis	6	5	8	8	8	6	5	7	9	7	7	4	4	9	7	7
6	Dist. Cianjur	4	8	7	8	8	3	8	7	10	8	4	5	5	8	7	7
7	Dist. Cirebon	4	9	6	8	8	3	8	7	10	5	5	7	4	10	8	5
8	Dist. Garut	4	5	11	6	8	7	4	8	7	7	7	6	8	8	4	7
9	Dist. Indramayu	6	8	2	11	5	4	11	7	10	7	5	4	5	8	9	5
10	Dist. Karawang	5	4	10	8	7	6	5	9	7	9	5	6	8	8	6	4
11	Dist. Kuningan	7	6	6	8	6	3	9	8	8	9	7	3	6	8	5	8
12	Dist. Majalengka	8	3	8	8	9	5	4	8	5	9	7	6	5	10	7	5
13	Dist. Pangandaran	5	7	5	10	9	9	5	4	7	6	8	5	6	5	9	7
14	Dist. Purwakarta	5	4	8	10	9	5	7	5	6	9	7	5	7	9	5	6
15	Dist. Subang	6	4	8	9	11	2	7	6	5	9	7	6	5	12	4	6
16	Dist. Sukabumi	6	6	7	8	4	7	8	8	10	7	6	3	7	7	6	7
17	Dist. Sumedang	5	5	7	10	11	2	5	8	5	10	9	3	6	9	6	6
18	Dist. Tasikmalaya	4	4	10	9	11	8	2	5	6	8	7	6	6	7	8	6
19	Bandung City	3	3	12	8	10	4	4	9	11	7	4	5	2	13	7	5
20	Banjar City	5	4	8	10	12	6	4	5	5	10	5	6	5	7	9	6
21	Bekasi City	5	7	7	8	9	4	4	10	9	4	7	7	4	12	8	2
22	Bogor City	5	5	8	9	12	6	4	5	4	5	6	11	6	11	8	2
23	Cimahi City	3	5	5	13	7	6	6	8	10	5	8	4	7	11	7	2
24	Cirebon City	5	8	7	7	7	3	7	9	11	4	8	4	4	11	5	7
25	Depok City	5	8	6	8	6	5	8	8	9	4	6	7	7	10	6	4
26	Sukabumi City	5	1	10	11	8	7	6	5	9	11	3	4	5	8	7	7
27	Tasikmalaya City	6	5	7	9	10	4	6	6	7	11	5	4	4	7	9	7
28	West Java Average	17	5	5	0	5	12	7	3	4	8	8	7	0	2	7	17

- The transition frequency matrix in the previous step is used to determine the transition probability matrix by dividing each transition value with the number in each row (relative frequency). Transition probability matrix 4 space state for rainfall average in West Java

$$P_{\text{West Java Average}} = \begin{matrix} 0 \\ 1 \\ 2 \\ 3 \end{matrix} \begin{bmatrix} 0.62962963 & 0.185185185 & 0.185185185 & 0 \\ 0.185185185 & 0.4 & 0.259259259 & 0.1 \\ 0.148148148 & 0.296296296 & 0.296296296 & 0.259259259 \\ 0 & 0.076923077 & 0.269230769 & 0.653846154 \end{bmatrix}$$

The transition diagram for rainfall average in West Java with 4 state spaces as follow:

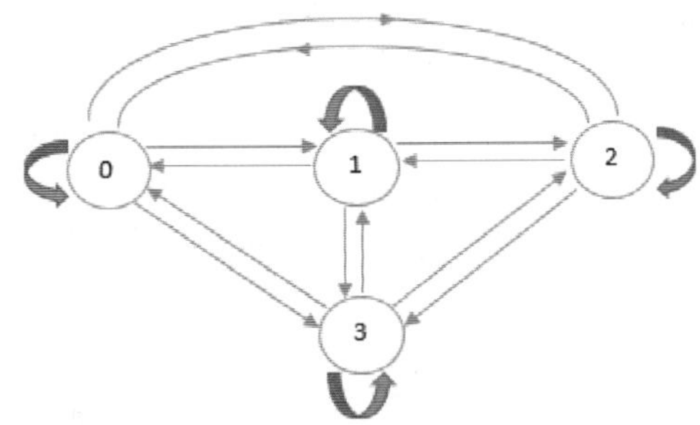

Figure 5. Transition Diagram of Rainfall Average in West Java with 4 State Spaces

- Calculate the long-term proportion of the Markov chain with the K state denoted by π_k using equation (15).
  Next, with $K = 4$, then π_0, π_1, π_2 and π_3 are the long-term proportions for state 0, 1, 2 and 3. The following results of the stationary distribution in percentages are presented in Table 3

Table 3. Stationary Distribution 4 State Spaces of Rainfall in West Java

No.	City/District	Stationary Distribution (%)			
		Low (0)	Quite Low (1)	Medium (2)	High (3)
1	Dist. Bandung	25.76	24.81	25.76	25.76
2	Dist. Bandung Barat	25.09	25.17	24.38	25.37
3	Dist. Bekasi	25.24	25.24	24.34	25.18
4	Dist. Bogor	25.24	25.21	24.43	25.12
5	Dist. Ciamis	25.37	25.13	25.18	24.32
6	Dist. Cianjur	25.18	25.09	24.43	25.3
7	Dist. Cirebon	24.43	25.06	25.29	25.21
8	Dist. Garut	25.14	25.18	24.44	25.24
9	Dist. Indramayu	24.47	25.21	25.29	25.03
10	Dist. Karawang	25.27	25.21	24.35	25.17
11	Dist. Kuningan	25.23	24.3	25.23	25.23
12	Dist. Majalengka	25.38	25.09	24.22	25.31
13	Dist. Pangandaran	25.28	25.29	25.2	24.24
14	Dist. Purwakarta	25.3	25.11	25.31	24.28
15	Dist. Subang	25.41	25.02	24.31	25.26
16	Dist. Sukabumi	25.35	25.24	25.25	24.17
17	Dist. Sumedang	25.23	24.3	25.23	25.23
18	Dist. Tasikmalaya	25.41	25.25	25.05	24.29
19	Bandung City	24.3	25.23	25.23	25.23
20	Banjar City	25.23	25.23	24.3	25.23
21	Bekasi City	25.11	25.47	24.31	25.11
22	Bogor City	25.23	25.23	24.3	25.23
23	Cimahi City	25.02	25.29	24.21	25.48
24	Cirebon City	25.23	24.3	25.23	25.23
25	Depok City	25.23	25.23	24.3	25.23
26	Sukabumi City	25.21	25.12	24.4	25.27
27	Tasikmalaya City	25.43	25.12	25.15	24.3
28	West Java Average	22.62	24.86	25.46	27.05

- Description and prediction.
  Based on the Markov chain stationary distribution, the probability of changes in rainfall from low phenomena is 22.62%, medium is 24.86%, high is 25.46%, and very high is 27.05%. This result indicates that in the long term, the phenomena of monthly rainfall data in West Java area has a tendency to be change is still high with probability 0.27. It is still quite large compared to other state spaces low, medium, and high. The result can be given as a recommendation for a

related institutions such as LAPAN and BMKG. Furthermore, the institutions can announce a special attention to inform the situation of rainfall phenomena to society.

Acknowledgment

The authors thank Rector Universitas Padjadjaran for funding this research through Academic Leadership Grant and RKDU 2018. Also thanks the researchers from LAPAN Bandung for the data and discussion.

References

[1] Abdullah A S, Matoha S, Lubis D A, Falah A N, Jaya I G N M, Hermawan E, Ruchjana B N, 2018 *Implementation of Generalized Space Time Autoregressive (GSTAR)-Kriging Model for Predicting Rainfall Data at Unobserved Locations in West Java*. Applied Mathematics and Information Science, **12** (3), 607-615.

[2] http://bidinagtuns.blogspot.co.id/2010/11/curah-hujan.html, accessed at 13 March 2018

[3] Rustiana S, Ruchjana B N, Abdullah A S, Hermawan E, Sipayung S B, Jaya I G N M and Krismianto. *Rainfall prediction of Cimanuk watershed regions with canonical correlation analysis (CCA)*, Journal of Physics: Conf. Series, 89312021, doi:10.1088/1742-6596/893/1/012021, 2017.

[4] Sandi R 2015 *Simulasi Curah Hujan Harian Menggunakan Stokastik Rantai Markov Orde* 3 × 3 *(Studi Kasus: Daerah Aliran Sungai Kampar)*. Jom FTEKNIK Volume **2** No. 2 October 2015.

[5] Sujatmoko and Bambang 2012 *Analisa Kehandalan Stokastik Rantai Markov untuk Simulasi Data Curah Hujan Harian pada Das Kampar*. Jurnal Sains dan Teknologi **11** (1), ISSN 1412-66257.

[6] Tovler A 2016 *An Introductoin to Markov Chains*. Department of Mathematical Sciences, University of Copenhagen, Denmark.

[7] Ross S M 1996. *Stochastic Processes Second Edition*. University of California, Berkeley, United States of America.

[8] Hermawan E, Ruchjana B N, Abdullah A S, Jaya I G N M, Sipayung S B, Rustiana S *Development of the statistical ARIMA model: an application for predicting the upcoming of MJO index*, Journal of Physics.: Conf. Ser. **893** 012019, doi:10.1088/1742-596/893/1/012019, 2017.

[9] Doubleday K J and Julius N E 2011. *Application of Markov Chains to Stock Trends*. Journal of Mathematics and Statistics, **7** (2): 103-106

[10] Osaki S 1992 *Applied Stochastic System Modelling*. Berlin: Springer-Verlag Berlin Heidelberg.

[11] https://hackernoon.com/what-steps-should-one-take-while-doing-data-preprocessing-502c993e1caa, accessed at 20 January 2019

[12] Abubakar U Y, Lawal A, Muhammed A 2014 *Markov Chain Model to Annual Rainfall Distribution for Crop Production*, American Journal of Theoretical and Applied Statistics, Vol. **3**, No.2, 2014, pp. 39-43.

[13] Firdaniza 2016. *Distribusi Stasioner Rantai Markov untuk Prediksi Curah Hujan di Wilayah Jawa Barat. Proceeding of Seminar Matematika dan Pendidikan Matematika*. ISBN: 978-602-6122-20-9

[14] Munkhammar J and Widen J 2018 *An N-state Markov Chain mixture distribution model of the clear-sky index*. Elsevier Journal. Doi: 10.1016/j.solener.2018.07.056

[15] Chung C H 2007 *Vegetation response to climate change on Jeju Island, South Korea, during the last deglaciation based on pollen record*. Geosciences Journal, vol. **11**, no. 2, pp. 147–155.

[16] Yoo C, Lee J, Ro Y 2016 *Markov Chain Decomposition of Monthly Rainfall into Daily Rainfall: Evaluation of Climate Change Impact*. Research Article, Advances in Meteorology. Doi: 10.1155/2016/7957490

[17] Zohadi B M and Salam A C A 1981. *A Stochastic Model of Daily Rainfall for Universiti Pertanian Malaysia*, Serdang, Pertanika, **4**(1), 1-9

[18] Han J and Kamber M 2006 *Data Mining: Concept and Techniques Second Edition,* Morgan Kaufmann Publishers.

[19] Tamiselvi R, Sivaskhti B and Kavitha R 2015 *An Efficient Preprocessing and Postprocessing Techniques in Data Mining. International Journal of Resesarch in Computer Applications and Robotics*. ISSN: 2320-7345

[20] Diaz J L, Herrera M, Izquierdo J, Garcia R P 2010 *The tasks of pre and post-processing in Data Mining applied to a real world problem*. International Congress on Environmental Modelling and Software. Bringham Young University

[21] Wang Y, Steele T and Zhang E 2016 QQ Plot. https://math.illinois.edu/system/files/inline-files/Proj9AY1516-report2.pdf

IOP Conf. Series: Earth and Environmental Science **303** (2019) 012027 doi:10.1088/1755-1315/303/1/012027

Exploring Standardized Precipitation Index for predicting drought on rice paddies in Indonesia

E Surmaini[1], E Susanti[1], M R Syahputra[2] and T W Hadi[2]

[1] Indonesia Agency for Agro-climate and Hydrology Research Institute, Jl Tentara Pelajar no 1 Bogor 16111, West Java
[2] Faculty of Earth Sciences and Technology, Bandung Institute of Technology

elzasurmaini@gmail.com

Abstract. Droughts have severe consequences for rice crops in Indonesia, as these occur annually and increase during El Niño phenomena. Accordingly, paddy drought assessment is necessary for developing adaptation strategies for successful crop production. We conducted a detailed assessment of paddy drought-climate indices in Indonesia. This was done by looking at the onset and trends of the Standardized Precipitation Index (SPI) in a three-month time scale (SPI-3) and exploring their relationship with paddy drought-affected areas. The Cartesian quadrant was used to illustrate the combination of SPI-3 onset and trend on the paddy drought affected area. This gave four different drought risk levels: low, moderate, high, and very high. The hit rate (HR) as the proportion of drought occurrences were correctly hindcast and percent of correct (PC) as the total number of correct hindcast divided by the total number of hindcast was used to verify the accuracy of drought effect prediction on paddy rice. The results demonstrate that the highest accuracy of paddy drought predictions occurred in the peak of dry season in July, while the accuracy of drought and non-drought occurrences (PC) was higher in the non-peak months, April through September, excluding July.

Keywords : SPI, drought, paddies, rainfall

1. Introduction

Climate change-induced drought intensification has been one of the great challenges faced by the in many countries [1,2, 3], including that in Indonesia [4, 5]. As the extent of drought is expected to worsen as the atmosphere continues to warm [6, 7], plant growth and crop production may be negatively affected in terms of yield and quality [8, 9].

An agricultural drought occurs when a shortage of soil moisture adversely affects the regional agricultural production due to evapotranspiration losses [10], causing dry weight accumulation loss and shortens life cycle [11]. During vegetative growth, drought could reduce photosynthetic rate and depress grain formation during the flowering stage more than any other stages [12].

The Indonesian Ministry of Agriculture reported that during El Niño years, damaged paddy areas due to drought ranged between 350-870 kha [5], where the damage mostly occurred during Dry Season Planting (DSP) within May through October. Furthermore, it is estimated that a 30-day delay in the onset of rainy season would diminish wet season rice production in West/Central Java and East Java/Bali by about 6.5 and 11.0 %, respectively [13]. The recent prolonged drought in 2015 has caused widespread crop failures that resulted in loss of almost 600 kilotons of paddy production. For this reason, the effects of recurring droughts should be considered in the context of crop production, where incoming drought events could be anticipated with some lead time that gives adequate response time for preparing adaptive

IOP Conf. Series: Earth and Environmental Science **303** (2019) 012027 doi:10.1088/1755-1315/303/1/012027

measures. Accordingly, defining the types of drought helps to monitor droughts and to build strategies for the preparation and response to droughts [14, 15, 16].

At present, a drought alert is typically only delivered during the drought period, after it has been detected through monitoring. As drought is a slow onset phenomenon, it is normally only recognized using the monitored indicators once the impacts have started being measured, which is after the onset of the drought itself. This thus leaves little time for the taking of mitigation measures to avoid or alleviate the impacts, consequently, drought management measures should have a strong focus on response [17].

In this study, we used the Standardized Precipitation Index (SPI) to explore paddy drought characteristics. The SPI is a precipitation probability index that was developed to give a representation of abnormal wetness and dryness [18, 19, 20]. SPI is a suitable index to be applied Indonesia, where most of the available data precipitation. However, SPI values by themselves do not completely describes drought characteristics, as drought is not only marked by its intensity, but also by its duration and severity. In attempting to predict drought events, we have defined SPI characteristics using SPI onset and trend. The objective of this study is to explore the correlation between SPI onset/trend of and the paddy drought-affected areas in Indonesia.

2. Data and Method

2.1. Data

The study was conducted using monthly rainfall data from 16 Provinces (Aceh, North Sumatera, South Sumatera, Jambi, Lampung, West Java, Central Java, East Java, DI Yogyakarta, West Nusa Tenggara, East Nusa Tenggara, Central Sulawesi, South Sulawesi, South Kalimantan, Central Kalimantan and East Kalimantan). Monthly rainfall data record from the period of 1989-2015 were collected from the Indonesian Ministry of Public Works, the Indonesian Agency for Meteorology, Climatology and Geophysics, and the Global Precipitation Climatology Centre (GPCC). Monthly data of paddy affected area due to drought from 1989 to 2015 were obtained from the Indonesian Ministry of Agriculture.

2.2. Method

2.2.1. The Standardized Precipitation Index (SPI).

Application of SPI was recommended in the context of meteorological drought monitoring by the World Meteorological Organization (WMO) [21]. The values of SPI are expressed in standard deviations, where positive SPI indicates greater than the median precipitation and negative SPI indicating less than the median precipitation [22] (illustrated in figure 1).

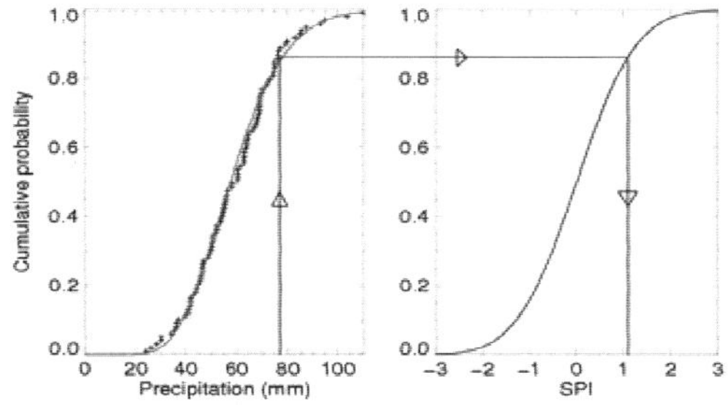

Figure 1. An example of equiprobability transformation from a fitted gamma distribution to the standard normal distribution (After [23]).

2.2.2. The onset and trend of the SPI.

The onset is characterized by the SPI-3 value three months before the drought event. The SPI-3 trend is defined by the linear regression of SPI-3 values after the onset of the drought, up to the actual drought event. For example, paddy drought affected areas in North Aceh District in August 2006 were recorded 13.047 ha (figure 2).

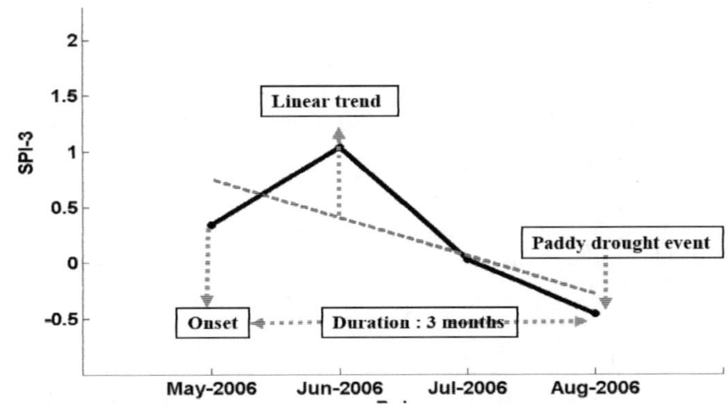

Figure 2. Plot of the onset and SPI trend associated to paddy drought affected areas between May ang August 2006. As illustrated, the onset of SPI-3 was defined three months in advance (May 2006) is 0.45, while the SPI-3 trend from May to August had a gradient of -0.75

2.2.3. Determination of paddy drought risk level.

We used Cartesian plane quadrants to explore the correlations between SPI with paddy drought affected area. The four quadrants represent different combinations of SPI onset and trend values (illustrated in figure 3). In order to be understood easily by user, these indices are simplified in forms of paddy drought risk score. The score was calculated as percentage of drought event x probability of paddy drought affected area using cumulative probability (CP) as follow:

$$CP = P(X > x); f(x) = e^{-\lambda x} \quad (1)$$

$$\lambda = \frac{1}{\bar{x}} \quad (2)$$

Where $\bar{x}$: mean and occurrence rate

We then defined a risk score based on the corresponding probability value, defined as the percentage of drought events multiplied by the probability of paddy drought-affected areas. Based on the risk score, we then classified the paddy drought occurrences into four risk levels: low, medium, high, and very high.

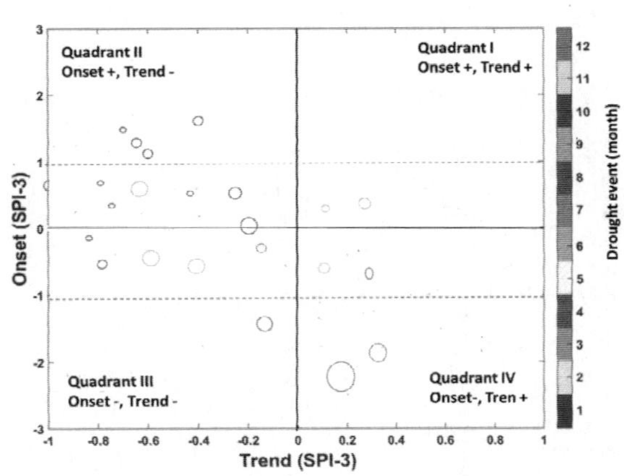

Figure 3. Correlation between the onset of SPI-3 (y-axis) and the trend of SPI-3 (x-axis) and paddy drought affected area. The color of scatter plots indicate time of drought event. The size of circle represent the size of affected area. The red dash lines indicate the range of near-normal SPI-3

2.2.4. Verification of paddy drought prediction.
In order to assess the accuracy of the model's predictive capability on drought occurrence, we used a hindcast on historical SPI onset and trend data as released by IAHRI (http://balitklimat.litbang.pertanian.go.id/) for the period 2011-2015. The accuracy was assessed using a contingency table [5] as illustrated in Table 1.

Table 1. Two-by-two contingency table for verification of a binary forecast system.

Forecast	Observation	
	Yes	No
Yes	Hit (A)	False Alarm (B)
No	Miss (C)	Correct Negative (D)

We then defined the hit rate (HR) as the proportion of occurrences that were correctly forecasted, and the Percent Correct (PC) as the total number of correct forecast (hit + correct negative) divide by total number of forecast. The PC also indicates the accuracy of the forecast. HR and PC are defined as such:

$$\text{Hit Rate} = A/(A+C) \times 100\% \quad (3)$$
$$\text{Percent of Correct} = (A+D)/(A+B+C+D) \times 100\% \quad (4)$$

3. Result and Discussion

3.1. Link between SPI-3 and paddy drought-affected area
Figure 4 shows the relationship betwen SPI-3 and paddy drought-affected area in Deli Serdang District, North Sumatera Province and Cianjur District, West Java Province. The figure shows that paddy drought events mostly occurred following negative SPI events. This is because soil water content is not reduced directly when rainfall is gradually decreased. According to [23] [24], meteorological drought and agricultural drought generally has lagged in time. For that reason, we conclude that SPI-3 can be used to predict the onset of paddy drought events.

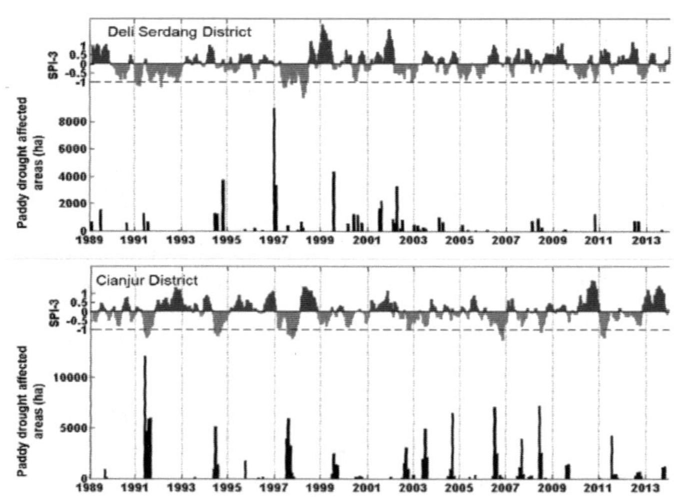

Figure 4. Plots of SPI-3 and paddy drought affected area in Deli Serdang District, North Sumatera Province (top) and Cianjur District, West Java Province (bottom). Bars indicate the paddy drought affected area. The red and blue areas indicates the negative and the positive SPI-3, respectively.

The results of this study also show that the paddy drought can be described properly by SPI within a timescale of three months. This timescale is more effective in showing the condition of groundwater during the growing season. This finding is in agreement with a previous study [25] stating that crops respond to drought after a rainfall deficit of three months. Additionally, it is stated that in general agricultural drought lags behind of meteorological drought by the three months [26].

3.2. Predictive for paddy drought events

Our result in figure 5 illustrates that paddy drought events mostly occur (85% of all occurrences) when the onset of SPI-3 is in the near-normal to wet category (SPI-3 >0), which are then followed by negative SPI trends which indicate drying conditions in the subsequent three months, which leading to high level of drought risk (Quadrant II). Even when the precipitation level and water availability are high enough (SPI-3 > 0) to encourage most farmers to start planting crops at the beginning of the planting season, there is generally no indication of any potential drought occurrence at that point. In particular, in the case of negative SPI-3 trend in the subsequent three months, there is a significant decrease in precipitation levels, causing a severe level of paddy drought occurrence.

Moderate levels of drought risk (SPI-3 values between 0 to -1) indicates near normal condition, is followed by negative trend of SPI results in lower drought risk (Quadrant III). In this case, some farmers would generally avoid planting crops at the beginning of the planting season, or would have already prepared alternative water sources, so that the risk of crop failure is not so dire (medium risk level). Low levels of drought risk (SPI-3 > 1) represent onset conditions that are relatively wet, which are then followed by positive SPI trends, indicating wet conditions in the subsequent three months (Quadrant I). Initial wet conditions, coupled with trends for even wetter conditions, result in low levels of drought risk, given that water is readily available.

Expert judgment was used to define the very high levels of drought risk. For this classification, we chose the SPI-3 value range of -3 < SPI-3 < -1 with subsequent positive or negative SPI-3 trends (Quadrant III and IV). However, lower risk scores were found with higher intensity drought instead, due to the lower number of occurrences of paddy drought (as seen in table 3). This is because for SPI-3 < -1, there are drier conditions at the beginning of the planting season; hence, discouraging most farmers from planting crops in the first place. Drought events found within quadrants III and IV represent the farmers who speculatively planted paddy despite the dry conditions and subsequently experienced paddy drought due to an inadequate water supply via irrigation. Accordingly, for this risk level, the total number of drought occurrences alone could not fully describe the extent of the drought impact, as some farmers would opt to not plant crops, while others would speculate and decide to plant. Expert judgment was used to determine the very high level of drought risk.

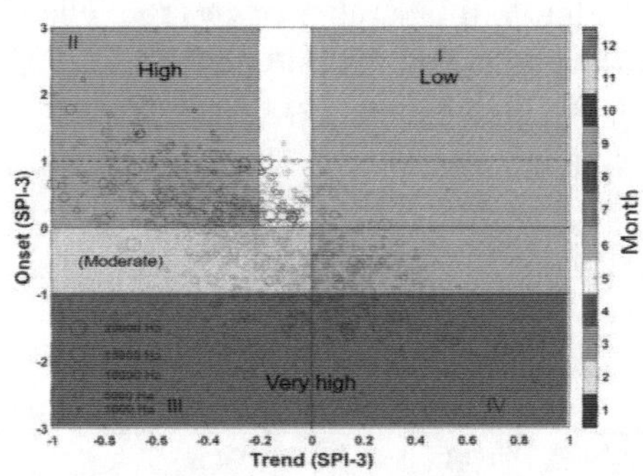

Figure 5. Level of drought risks were classified using quadrant method with the onset of SPI-3 as y-axis and the trend of SPI-3 as x-axis.

Since the paddy drought affected areas are vary greatly in spatial and temporal among districts, thus the minimum drought area was required to be defined. The minimum value was defined using cumulative probability function as described in figure 6. Based on the obtained cumulative probability curve (figure 6), the lower bound paddy drought area value found was approximately 4000 ha.

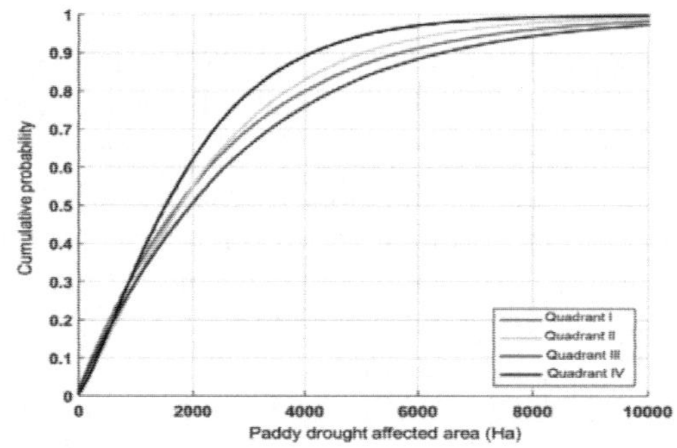

Figure 6. Cumulative probability of paddy drought affected area for quadrant I-IV.

Consequently, there are four level of paddy drought risk, namely low, medium, high, and very high. The higher the score is defined as the higher the risk level as described in table 2. However, for very high level risk we used expert judgement.

Table 2. Criteria used to define drought risk on rice paddies based on the onset and trend of SPI-3.

Level	Criteria	% drought	Probability of paddy affected area > 4000	Score ($x10^2$)	Level
1	Onset SPI-3 > -1, trend > 0	17.16	17.80	3.05	Low
2	Onset Between 0 to 1, trend < 0	24.94	27.96	6.97	Moderate
3	Onset SPI-3 > 0, trend ≤ -0.2	34.65	26.49	9.18	High
4	Onset SPI-3 ≤ -1	14.11	16.95	2.39	Very high

3.3. Accuracy of paddy drought risk prediction

From the total of 5744 hindcast predictions for period 2011-2015, 123 drought events were correctly predicted (hit rate = 28,4%). In addition, the number of drought events that were correctly predicted (hit) and that of non-drought events (correct negative) in total of 4,214 (PC of 75.5%), as shown in table 3.

Table 3. Contingency tables for the paddy drought prediction for period April–September over Indonesia region for the period 2012-2015.

Forecast	Observation		
	Yes	No	Total
Yes	123	954	1077
No	453	4214	4667
Total	576	5168	5744

Examples of paddy drought prediction maps of August 2014 are presented in figure 7. We can see that: i) drought occurrences were correctly predicted in Sumatra, Java and parts of Sulawesi; and ii) drought non-occurrences were correctly predicted in Papua and parts of Sulawesi; but iii) there were a number of inaccurate predictions in Kalimantan and Nusa Tenggara.

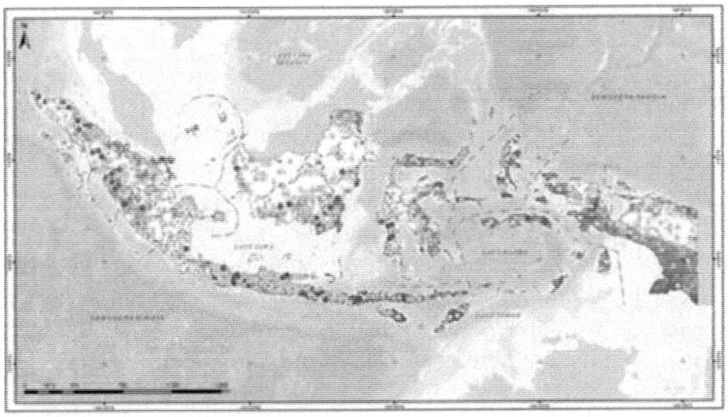

Figure 7. Map of verification of paddy drought events prediction for August 2014. The area with green indicates the area predicted not to experience rice drought. The white circle is an observation data that shows no occurrence of rice drought. The area with pink is an area that is predicted to experience drought, while the red circle is an observation that shows the occurrence of drought.

Verification was conducted for the months April-September, as drought normally occurs during this period. The accuracy of the paddy drought predictions varies significantly from April to September (Table 4). The highest accuracy of drought occurrence prediction (hit rate) is found in the peak dry season (July). However, accuracies during April and September are very low because drought events are rare during this period. Likewise, the accuracy of drought and non-drought occurrences (PC) is higher in other non-peak months (April through September, excluding July), as the number of events that can be predicted correctly is much higher. In fact, the number of drought non-occurrences are higher than that of drought occurrences (correct negative), contributing to higher accuracy. In July, however, the number of correct negatives is low, so that PC is at its lowest due to the smaller number of correct predictions for both drought occurrences and non-occurrences. Overall, the accuracy of the prediction during the peak dry season still needs to be improved.

Table 4. Accuracy of paddy drought prediction using Hit Rate and Percent of Correct method

Target of prediction (Month)	A (Hit)	C (Correct negative)	Hit Rate (%)	Percent of Correct (%)
April	1	1288	3.0	94.2
May	9	990	18.8	85.3
June	31	621	28.2	71.6
July	35	182	56.5	37.1
August	40	386	38.8	61.7
September	7	749	3.2	74.8

Due to climate change severe droughts are expected to become more frequent, increasing the risk of water shortages for agriculture. Future drought outlook may also be possible based on drought indicators with climate change projections. SPI analysis of the rainfall projection data is needed to identify the indication of future drought. According to Pratiwi et al. [27], the SPI analysis results show that the drought period are predicted to occur more often than in the period 2020-2045 in dry season in Cirebon District, West Java. Such climate information may provide evidence of potential risks to climate shocks in the future which may then essential to determine the appropriate adaptation measures in order to cope with paddy yield loss due to drought.

4. Conclusion

We develop a predictive model of paddy drought risk by investigating the correlation between SPI-3 onset and trend in paddy drought-affected. Cartesian plot diagram was used to four different levels of paddy drought risk: low (the onset SPI-3 > -1 and positive trend), moderate (0 < SPI-3 < 1, negative trend)), high (onset SPI-3 > 0, negative trend), and very high (onset SPI-3 ≤ -1). Precise judgment was particularly used to determine the very high level of paddy drought risk due to the low number of events found, given that most farmers would not plant paddy as they anticipated severe droughts.
Based on the verification that was conducted for the months of April-September, when drought normally occurs, the drought events that were correctly predicted (hit rate) is 28.4%. In addition, the number of drought events that were correctly predicted (hit) and that of non-drought events (correct negative) is 75.5%. The highest hit rate was found to occur in July, during peak dry season. The accuracy of drought and non-drought occurrences (PC) is higher in non-peak months, April through September, excluding July.

As the very high level of drought risk could not be properly predicted using SPI-3, we think it is necessary to conduct further studies in order to improve the impact model. One of the ways which the model could be improved is by incorporating paddy crop productivity in the crop simulation mode as this would roughly capture the impact of climate factors on crop growth. Moreover, the projection of SPI and the potential impact on paddy yield due to drought in the future need to be assess.

Acknowledgments

This work is funded by the Indonesian Agency for Agricultural Research and Development under Grant No 83.3/PL.040/I.1/04/2016.K.

References

[1] Coumou C and Rahmstorf S 2012 A decade of weather extremes *Nat. Climate Change* **2** 1-6
[2] Dai A 2001 Drought under global warming *Adv. Rev.* **2** 45-65
[3] Rosenzweig C et al. 2014 Assessing agricultural risks of climate change in the 21st century in a global gridded crop model intercomparison *Proc. Natl. Acad. Sci. USA* **111** 3268-73
[4] Naylor R, Falcon W, Wada N and Rochberg D 2002 Using El Niño – Southern Oscillation Climate Data To Improve Food Policy Planning in Indonesia *Bull. Indones. Econ. Stud* ***38*** *75-91*
[5] Surmaini E Hadi TW, Subagyono K and NT Puspito 2015 Early detection of drought impact on rice paddies in Indonesia by means of Niño 3 . 4 index *Theor. Appl. Climatol.* **121** 669-84
[6] Cook BI, Ault TR and Smerdon JE 2015 Unprecedented 21st century drought risk in the American Southwest and Central Plains *Scie. Adv.* **1** e1400082
[7] Sheffield J and Wood EF 2007 Characteristics of global and regional drought 1950 – 2000 Analysis of soil moisture data from off-line simulation of the terrestrial hydrologic cycle *J. Geophys. Res.* **112** D17115
[8] Boo K, Kwon W, Oh J and Baek H 2004 Response of global warming on regional climate change over Korea : An experiment with the MM5 model *Geophys. Res. Lett.* **31** L21206
[9] Yoo J, Kwon H, Kim T and Ahn J 2012 Drought frequency analysis using cluster analysis and bivariate probability distribution *J. Hydrol* **420–421** 102-11
[10] Peng S, Huang J, Sheehy JE, Laza RC, Visperas RM, Zhong X , Centeno GS Khush GS and Cassman KG 2004 Rice yields decline with higher night temperature from global warming *Proc. Natl. Acad. Sci. USA* **101** 9971-75
[11] Kamoshita A, Babu RC, Boopathi NM and Fukai S 2008 Phenotypic and genotypic analysis of drought-resistance traits for development of rice cultivars adapted to rainfed environments *Field Crop. Res.***109** 1-23
[12] Okami O, Kato Y, Kobayashi N and Yamagishi J 2015 Morphological traits associated with vegetative growth of rice (Oryza sativa L .) during the recovery phase after early-season drought *Eur. J. Agron.* **64** 58-66
[13] Naylor RL, Battisti DS, Vimont DJ, Falcon WP and Burke MB 2007 Assessing risks of climate variability

and climate change for Indonesian rice agriculture *Proc. Natl. Acad. Sci. USA* **104** 7752-57
[14] Ward FA, Hurd BH, Rahmani T and Gollehon N 2006 Economic impacts of federal policy responses to drought in the Rio Grande Basin *Water Resour. Res.* **42** W03420
[15] Harou JJ, Azuara JM, Zhu T, Tanaka SK, Lund JR, Stine S, Olivares M and Jenkins MW 2010 Economic consequences of optimized water management for a prolonged severe drought in California *Water Resour.Res.* 46 W05522
[16] Wilhite DA and Glantz MH 1985 Understanding the Drought Phenomenon The Role of Definitions *Water Int.* **103** 111-20
[17] Werner M, Vermooten S, Iglesias A, Maia R, Vogt J and Naumann G. 2015 *Drought: Research and Science-Policy Interfacing,* ed Andreu et al (London: Taylor & Francis Group).
[18] Mishra AK and Singh VP 2010 A review of drought concepts *J. Hydrol.* **391** 202-16
[19] Rhee J and Im J 2017 Drought forecasting for ungauged areas based on machine learning : Using long-range climate forecast and remote sensing data *Agric. For. Meteorol.* **237–238** 105-22
[20] Hayes M, Svoboda M, Wall N and Widhalmu M 2011 Meteorological and Drought Index, The Lincoln Declaration: Universal Meteorological Drought Index Recommended. *Bull. Amer. Meteor. Soc.* April 485–88
[21] Mckee TB, Doesken NJ and Kleist J 1993. *Proc. Eighth Conf. on Applied Climatology* (California) pp17–22
[22] Edwards DC 1997 *Characteristic of 20TH Century Drought in the United States at Multiple Time Scales* (Colorado:Science Colorado State University)155p
[23] Lee J, Kwon H, Jang H and Kim T 2016 Future Changes in Drought Characteristics under Extreme Climate Change over South Korea *Adv.Meteorol.* **ID** 9164265
[24] Mason SJ and Graham EG 1999 Conditional Probabilities, Relative Operating Characteristics , and Relative Operating Levels *Weather Forecast.* **14** 713-25
[25] Wang H and H. Lin H 2014 Remotely sensed drought index and its responses to meteorological drought in Southwest China *Remote Sens. Lett.* **5** 413-22
[26] Ma'rufah U, Hidayat R and Prasasti I 2017 *IOP Conf. Ser.: Earth Environ. Sci.* **54** 012008
[27] Pratiwi NAH, Karuniasa M and Surosa SA 2018 *1st SRICOENV E3S Web of Conferences* **68** 02007

Temporal and spatial variation of surface heat flux in Makassar Strait

I M Radjawane[1, 2], R B Hatmaja[2], J F Manullang[1] and N J Trilaksono[2, 3]

[1] Oceanography Department, Faculty of Earth Sciences and Technology, Bandung Institute of Technology, Jl. Ganesa no. 10, Bandung – West Java 40132, Indonesia
[2] Graduate Program in Earth Science, Faculty of Earth Sciences and Technology, Bandung Institute of Technology, Jl. Ganesa no. 10, Bandung – West Java 40132, Indonesia
[3] Meteorology Department, Faculty of Earth Sciences and Technology, Bandung Institute of Technology, Jl. Ganesa no. 10, Bandung – West Java 40132, Indonesia

ivonne@fitb.itb.ac.id

Abstract. Indonesian Seas have an important role on the sea-air interaction in Indo-Pacific region. Makassar Strait is located in the center of Indonesian Seas and part of Indonesian Throughflow (ITF), which has been influenced due to the coupled ocean-atmosphere phenomenon in Pacific and Indian Oceans, such as monsoon, El-Niño Southern Oscillation (ENSO), and Indian Ocean Dipole (IOD). The temporal and spatial variation of surface heat flux in Makassar Strait has been evaluated using the multi satellite dataset from 1984-2009. The remote sensing data with resolution of 1° × 1° obtained from Objectively Analyzed air-sea Fluxes (OAFlux) produced by Woods Hole Oceanographic Institution (WHOI). The results show that there is a seasonal fluctuation of surface heat flux related to the wind monsoon. The variation of the shortwave radiation and latent heat flux give a large contribution to the net surface heat flux that associated with ENSO and IOD events. In Makassar Strait, the variation of shortwave radiation and sea surface temperature (SST) is in a good agreement with the El-Niño/La-Niña events. There is a time lag between surface heat fluxes with the ENSO events.

Keywords: Interannual variability, Makassar Strait, monsoon, surface heat flux.

1. Introduction

Indonesian Seas, which is located between Indian and Pacific Oceans, have an important role on the sea-air interaction in Indo-Pacific region. Besides, Indonesian Seas are also influenced by the coupled ocean-atmosphere phenomenon in Pacific and Indian Oceans, such as monsoon, El-Niño Southern Oscillation (ENSO), and Indian Ocean Dipole (IOD). As the part of Indonesian Seas, Makassar Strait has unique oceanographic characteristics and takes part on that important role. Makassar Strait is located in the center of Indonesian Seas, between Kalimantan (Borneo) Island and Sulawesi Island, and also part of Indonesian Through-Flow (ITF). ITF plays a role in transporting the water mass from the Pacific Ocean to the Indian Ocean and also maintaining the heat balance.

The imbalance between the input and output of heat through the ocean surface causes changes in heat stored in the upper layers of the ocean [1,2,3]). The heat exchange through the ocean surface is called surface heat flux. The sum of heat flux into or out of a water column is the heat budget. There are four major terms of surface heat flux, i.e. insolation or shortwave radiation (Q_{SW}), the heat flux of sunlight into the ocean; net infrared radiation loss or long wave radiation (Q_{LW}), the flux of infrared radiation or long wave radiation from the ocean; latent heat flux (Q_L), the heat flux carried by evaporated water; and sensible heat flux (Q_S), the heat flux out the ocean due to conduction process [4].

Heat flux can changes the density of surface water mass and also has role in determining the global water mass distribution and thermohaline circulation [5]. The heat flux in the deeper layers is usually much smaller than the heat flux in the upper layers or surface heat flux. Besides, the total of global surface heat flux which entry and exit the ocean must equals to zero or balance, otherwise the ocean would warm or cool. Therefore, the aim of this research is to find out the process of heat exchange between ocean and atmosphere in Makassar Strait by analyzing the spatial and temporal variation of all terms of surface heat flux there.

2. Data and Methods

Data used in this research are secondary data retrieved from Objectively Analyzed air-sea Fluxes (OAFlux) [4], a research program by Woods Hole Oceanographic Institution (WHOI), including global data of total heat flux, long wave radiation, short wave radiation, latent heat flux, sensible heat flux, wind speed, sea surface temperature and, humidity. OAFlux Program is a sustainable research and development project for global ocean-atmosphere flux. The products of OAFlux are obtained from optimal assimilation of various multi-platform satellite and analysis of three numerical weather predictions (NWP). The OAFlux data are obtained from observation for 26 years, from 1984 to 2009. Moreover, the long wave and short wave radiation data are retrieved from International Satellite Cloud Climatology Project (ISCCP) [6].

Dipole Mode Index (DMI) [7] and Oceanic Niño Index (ONI) [8] are also used in this research to determine the influence of Indian Ocean Dipole (IOD) and El-Niño Southern Oscillation (ENSO) phenomena to the surface heat flux in Makassar Strait. The summary of used data are shown in table 1. The research area location is along the Makassar Strait that lays between Kalimantan Island and Sulawesi Islands. A cross section namely SMR to SMR" transect located along the strait was chosen to analyze the data (see figure 1).

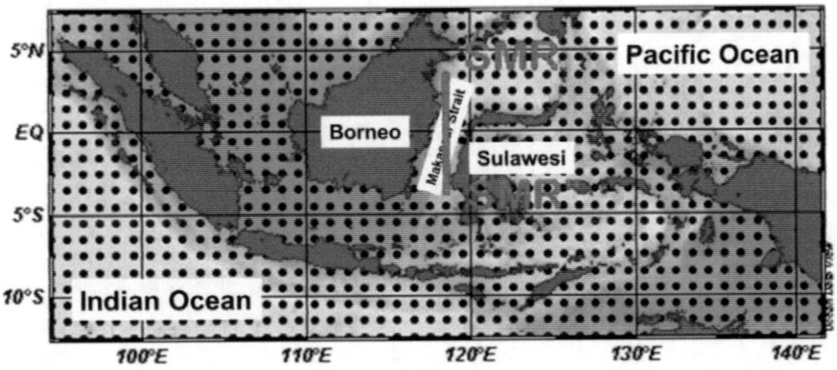

Figure 1. Research area location and a cross section namely SMR to SMR" transect located along the strait (red line)

Conservation of total heat flux is defined as the following equation [9]

$$Q = Q_{SW} - Q_{LW} - Q_S - Q_L \quad (1)$$

Where, Q is the resultant of heat gain or loss. Q and Q_{SW} are positive towards the bottom (sea), whileQ_{LW}, Q_L, andQ_S are positive towardsthe top (atmosphere) and all the parameter unit is W/m^2.

IOP Conf. Series: Earth and Environmental Science **303** (2019) 012028 doi:10.1088/1755-1315/303/1/012028

Table 1. Data description used in this study

Variable	Unit	Resolution	Source
Total heat flux (Q)	W/m^2	1° x 1°	The Objectively Analysed Air-Sea Fluxes (OAFlux)
Insolation or shortwave radiation (Q_{SW})	W/m^2	1° x 1°	International Satellite Cloud Climatology Project (ISCCP)
Net infrared radiation loss or long wave radiation (Q_{LW})	W/m^2	1° x 1°	International Satellite Cloud Climatology Project (ISCCP)
Latent heat flux (Q_L)	W/m^2	1° x 1°	The Objectively Analysed Air-Sea Fluxes (OAFlux)
Sensible heat flux (Q_S)	W/m^2	1° x 1°	The Objectively Analysed Air-Sea Fluxes (OAFlux)
Wind speed (ws)	m/s	1° x 1°	The Objectively Analysed Air-Sea Fluxes (OAFlux)
Sea surface temperature (SST)	°C	1° x 1°	The National Oceanic and Atmospheric Administration (NOAA)
Humidity (HUM)	g/kg	1° x 1°	The Objectively Analysed Air-Sea Fluxes (OAFlux)
Outgoing long wave radiation (OLR)	W/m^2	2.5° x 2.5°	The National Oceanic and Atmospheric Administration (NOAA)
Dipole Mode Index (DMI)	°C	-	Japan Agency for Marine-Earth Science and Technology (JAMSTEC)
Oceanic Niño Index (ONI)	°C	-	The National Oceanic and Atmospheric Administration (NOAA)

3. Results and Discussions

The monthly mean distribution of total surface heat flux (Q) from OAFlux is shown in Figure 2. In January, April, July, and October, the value of total surface heat flux(Q) is positive for almost all Indonesian Seas, where the value in April and October are higher than the value in January and July. This appearance seems to be related to the pseudo-motion of the sun, where in April and October the sun moves to the equator, while in January and July the sun's position is on the mid latitude region.

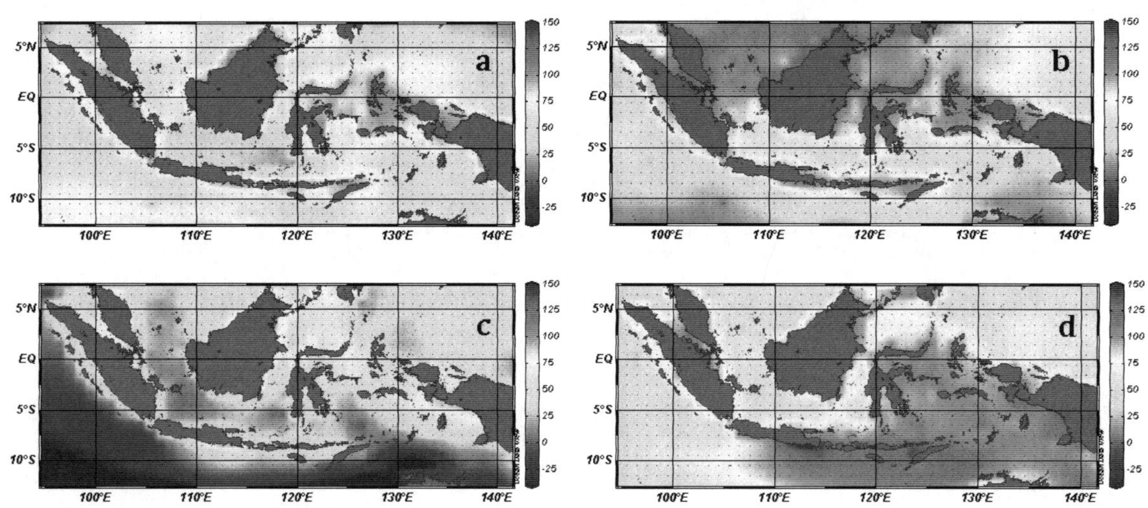

Figure 2. The monthly mean distribution of total surface heat flux (Q) on a) January, b) April, c) July, and d) October. Unit expressed in W/m^2.

However, in July, the total surface heat flux (Q) of southern Indonesian Seas is negative, which means the ocean releases more heat to the atmosphere. It is suggested as the influence of latent heat flux (Q_L) due to high wind speed and low humidity [6]. Based on figure 3, which shows the time series of monthly mean of all surface heat flux terms in Makassar Strait, it is shown that the highest total heat flux (Q) happens in April and the lowest happens in June to July. Moreover, there is also high total heat flux (Q) in November. The high total heat flux (Q) in April and November are caused by high solar insolation at the equator in the boreal spring and autumn seasons. High insolation is also seen from the time series of insolation or short wave radiation (Q_{SW}), which is high in April (spring) and September (autumn) and low in June (summer) and December (winter). The low total heat flux (Q) in June to August is caused by the high latent heat flux (Q_L) due to high wind speed (see figure 4) and low humidity in boreal summer season, especially in August as the peak [6]. The low humidity during boreal summer season is caused by southeastern monsoon wind, which tends to carry dry air masses from Australia.

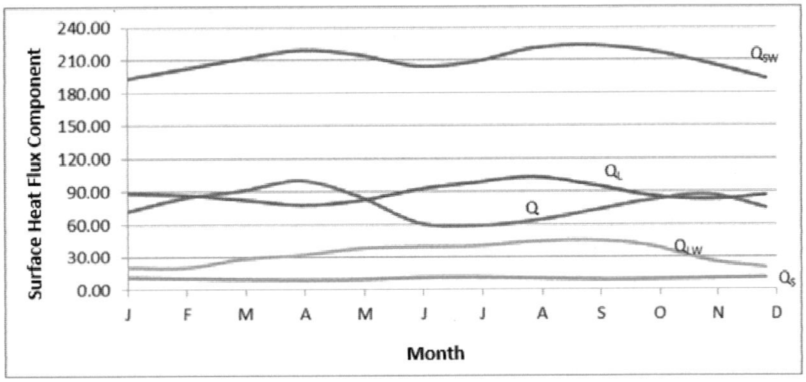

Figure 3. Time series of monthly mean of surface heat flux terms in Makassar Strait. Unit expressed in W/m^2.

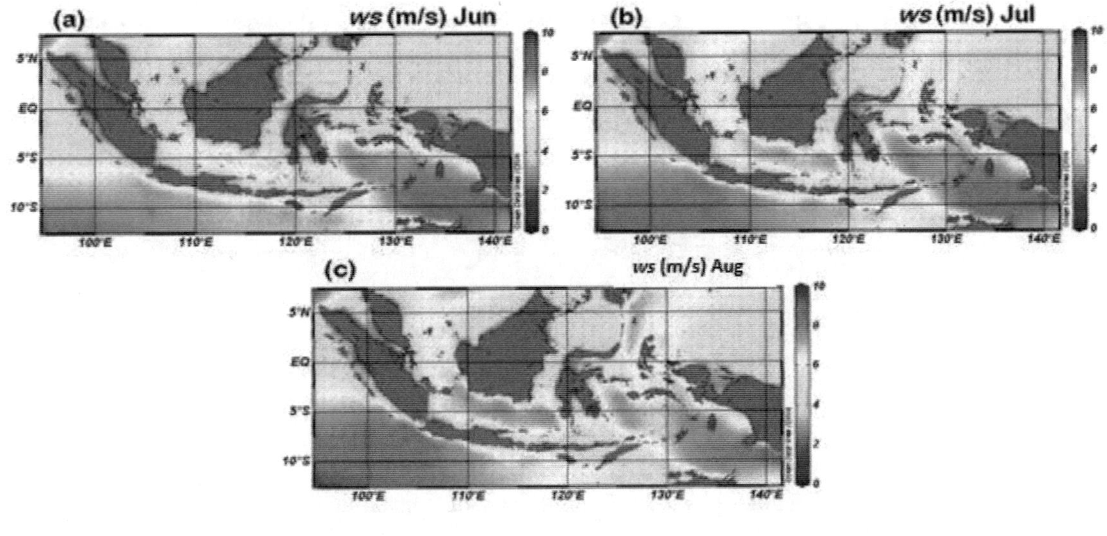

Figure 4. The annual mean distribution of wind speed (m/s) on a) June, b) July, and c) August.

The surface heat flux has a relationship with sea surface temperature (SST). Based on the correlation analysis between all surface heat flux terms and SST in table 2, the SST in the Makassar Strait is significantly related to the sensible heat flux (Q_S) and net infrared radiation loss or long wave radiation

(Q_{LW}). It is shown that the variation of sensible heat flux (Q_S) and net infrared radiation loss or long wave radiation (Q_{LW}) in the Makassar Strait are significantly related to the variation of SST. The higher SST means the higher heat transferred from the ocean to the atmosphere [9]. Previous studies show the phenomena that play roles on the interannual variation of SST in Indonesia, including Makassar Strait, are El-Niño Southern Oscillation (ENSO) [10] and Indian Ocean Dipole (IOD) [11].

Table 2. Correlation of surface heat flux terms to sea surface temperature (SST)

	Q : SST	Q_{SW} : SST	Q_{LW} : SST	Q_L : SST	Q_S: SST
r	**–0.07**[a]	**0.05**[a]	0.16	**–0.04**[a]	0.26
Lag (months)	0	0	0	0	0

[a] No significant relationship between two variables.

Figures 5 and 6 show Hovmöller diagrams of all surface heat flux anomalies along the SMR to SMR" transects from 1987 to 2009 which are compared to ONI and DMI indices. Based on these two figures, it can be seen that there are positive anomalies of total heat flux (Q), insolation or short wave radiation (Q_{SW}), and net infrared radiation loss or long wave radiation (Q_{LW}), and also negative anomalies of sensible heat flux (Q_S) during El-Niño event (ONI > 0.5) and Positive IOD event (DMI > 0.5), as in 1994-1995, 1997-1998, and 2006-2007. Otherwise, during La-Niña event (ONI < –0.5) and Negative IOD event (DMI < –0.5), there are negative anomalies of total heat flux (Q), insolation or short wave radiation (Q_{SW}), and net infrared radiation loss or long wave radiation (Q_{LW}), and also positive anomalies of sensible heat flux (Q_S), as in 1998-1999 and 2008-2009. Besides, there is a lag time between the occurrence of ENSO or IOD with the surface heat flux anomalies, where the anomalies occur several months after.

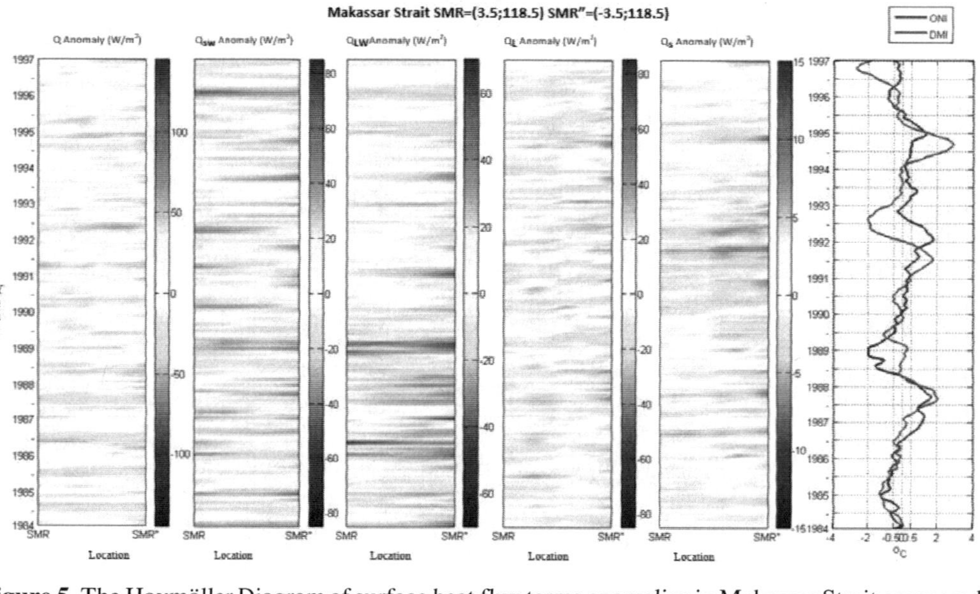

Figure 5. The Hovmöller Diagram of surface heat flux terms anomalies in Makassar Strait compared to ONI and DMI during 1984-1996

Furthermore, for analysing the interannual variability clearly, the anomalies of all surface heat flux were filtered by using low pass filter method with cut off about 13 months and presented in figures 7 and 8. Based on those two figures, it became clear that during El-Niño and Positive IOD events (ONI and DMI > 0.5), such as in 1994-1995 and 1997-1998, there are positive anomalies of total heat flux

(Q), while during La-Niña and Negative IOD events (ONI and DMI <–0.5), such as in 1998-1999 and 2008-2009, there are negative anomalies of total heat flux (Q). During the El-Niño and Positive IOD events, the ocean tends to absorb more heat from the atmosphere, and vice versa.

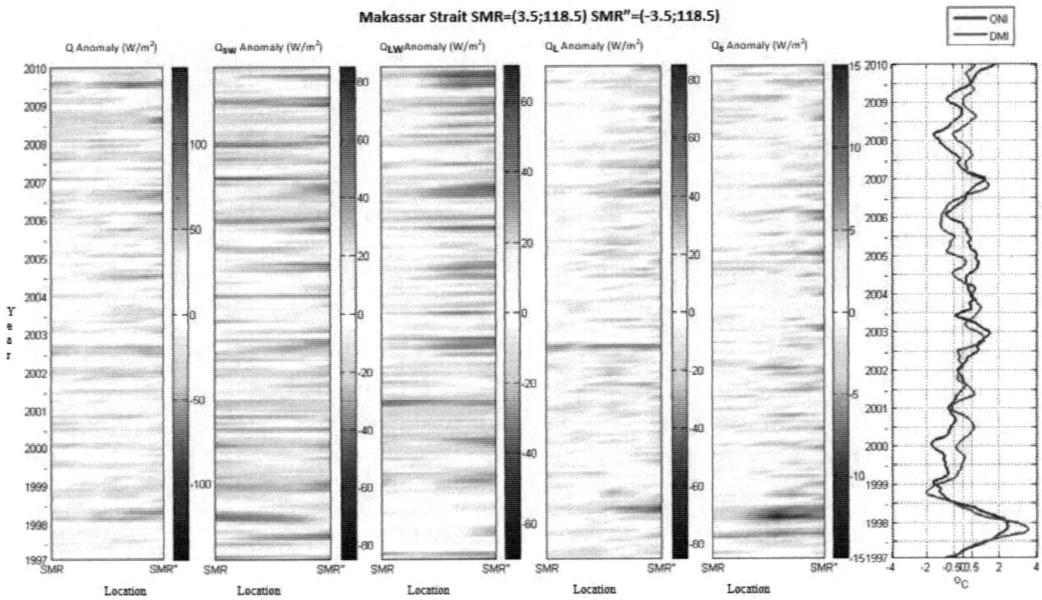

Figure 6. The Hovmöller Diagram of surface heat flux terms anomalies in Makassar Strait compared to ONI and DMI during 1997-2009

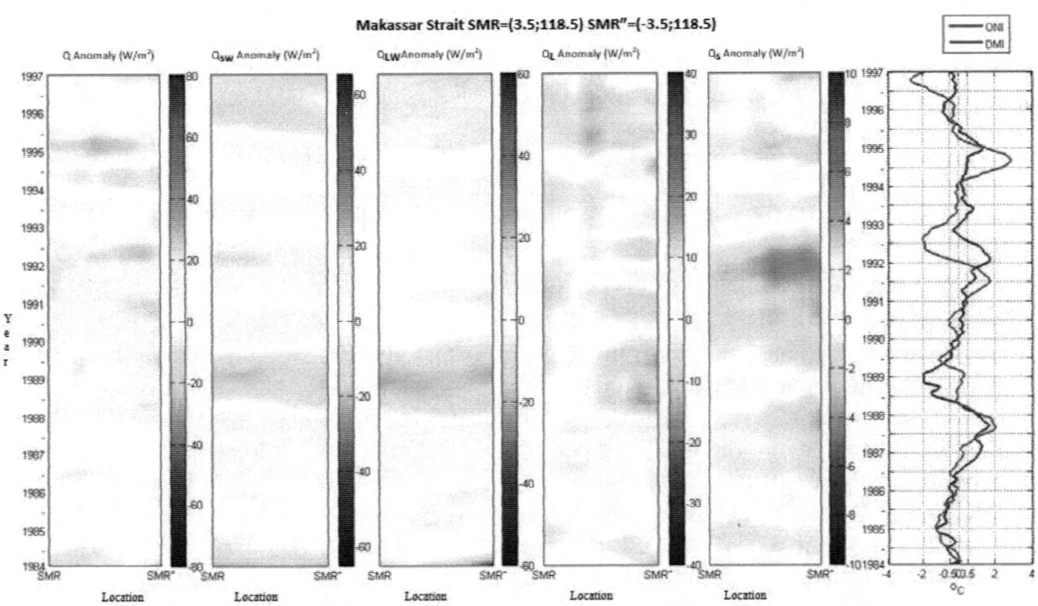

Figure 7. The Hovmöller Diagram of surface heat flux terms anomalies after low pass filter (13 months cut off) in Makassar Strait compared to ONI and DMI during 1984-1996

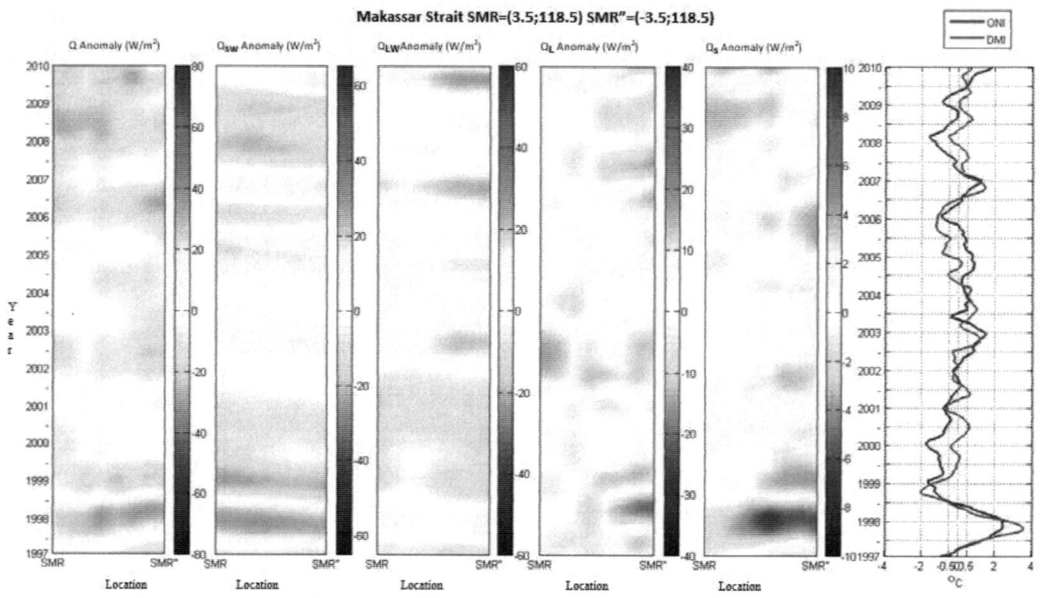

Figure 8. The Hovmöller Diagram of surface heat flux terms anomalies after low pass filter (13 months cut off) in Makassar Strait compared to ONI and DMI during 1997-2009

Based on the correlation analysis in table 3, it can be seen that the correlation between the surface heat flux terms and ONI are higher than the correlation to the DMI. It is shown that greater influence is given by ENSO than IOD. It is also seen in figures 7 and 8, that when the DMI value is more than 0.5 and the ONI value is neutral, there is no significant positive anomaly of total heat flux (Q), but when the ONI value is more than 0.5 and the DMI value is neutral, there is a positive total heat flux (Q) anomaly.

Table 3. Correlation of surface heat flux terms to DMI and ONI in Makassar Strait

	Q	Q_{SW}	Q_{LW}	Q_L	Q_S
DMI	0.16	0.42	0.31	0.13	–0.29
ONI	0.36	0.86	0.62	0.16	–0.55

The most related surface heat flux term to ENSO and IOD phenomena is insolation or short wave radiation (Q_{SW}). In addition, a fairly high and significant correlation was also seen in the net infrared radiation loss or long wave radiation (Q_{LW}) and sensible heat flux (Q_S) to the ENSO and IOD. In contrast to insolation or short wave radiation (Q_{SW}) and net infrared radiation loss or long wave radiation (Q_{LW}), the correlation of sensible heat flux (Q_S) to ONI and DMI are in opposite direction, which means that when the ONI and DMI values more than 0.5, negative sensible heat flux (Q_S) anomaly will occur, and vice versa. In addition, sensible heat flux (Q_S) is associated with lower SST during El-Niño and Positive IOD and higher SST during La-Niña and Negative IOD. Although significantly correlated with ENSO and IOD phenomena, the effect or contribution of sensible heat flux (Q_S) to the total heat flux (Q) tends to be low with values ranging from 2-42 W/m^2 [9]. Therefore, it is suggested that total heat flux (Q) anomalies that occur during ENSO and IOD events are caused by significant insolation or short wave radiation (Q_{SW}) anomalies.

4. Conclusion

The results show that there is a seasonal fluctuation of surface heat flux related to the wind monsoon. The variation of the insolation or shortwave radiation (Q_{SW}) and latent heat flux (Q_L) give a large contribution to the net surface heat flux that associated with El-Niño and Indian Ocean Dipole (IOD) events. In Makassar Strait, the variation of shortwave radiation and sea surface temperature (SST) is in a good agreement with the El-Niño/La-Niña events. There is a time lag between surface heat flux with the ENSO events.

Acknowledgment

This research was supported by the ITB Research Grant. The ocean heat flux products were provided by the WHOI OAFlux Project funded by the NOAA Climate Observations and Monitoring (COM) Program http://oaflux.whoi.edu/. The shortwave and longwave radiation (ISCCP-FD) data were obtained from the International Satellite Cloud Climatology Project (ISCCP) data archives at http://oaflux.whoi.edu/heatflux.html. Advances in understanding the radiative flux from ISCCP *J. Geophys. Res.*, 109, (27pp). We would like to thank to Mr. Giovanni Gunawan who helps editing the figures.

References

[1] Liu H, Lin W and Zhang M 2010 *J. Climate* **23** 1779-92
[2] Trenberth K E and L Smith 2009 *Clim. Dyn.* **32** 1065-79
[3] Katsman C A and G J van Oldenborgh 2011 *Geophys. Res. Lett.* **38** L14610
[4] Yu L and Weller R A 2007 *Bull. Amer. Meteor. Soc.* **88** 527-40
[5] Chiggiato J, Zavatarelli M, Castellari S and Deserti M 2005 *Interannual Variability of Surface Heat Fluxes in the Adriatic Sea in the Period 1998-2001 and Comparison with Observations.* (Bologna: ServizioIdro Meteorologico ARPA Emilia Romgna)
[6] Zhang Y C, Rossow W B, Lacis A A, Oinas V and Mishchenko M I 2004 *J. of Geophys. Res.* **109**
[7] Saji N H, B N Goswami, P N Vinayachandran and T Yamagata 1999 *Nature* **401** 360-3
[8] Trenberth K and National Center for Atmospheric Research Staff (Eds) Last modified 11 January 2019 The Climate Data Guide: Nino SST Indices (Nino 1+2, 3, 3.4, 4; ONI and TNI) Retrieved from https://climatedataguide.ucar.edu/climate-data/nino-sst-indices-nino-12-3-34-4-oni-and-tni
[9] Stewart R H 2008 *Introduction to Physical Oceanography* (Texas: Department of Oceanography of Texas A & M University)
[10] England M and F Huang 2005 *J. Climate* **18** 1435-44
[11] Yuan D and H Liu 2009 *J. Phys. Oceanogr.* **39** 1115-32

IOP Conf. Series: Earth and Environmental Science **303** (2019) 012029 doi:10.1088/1755-1315/303/1/012029

The Vulnerability of Climate Change in Indonesia to Renewable Energy

Mahmud, B Siswanto, E Maryadi, and D R Lubis

Center for Atmospherics Science and Technology-LAPAN, Bandung, Indonesia

mahmud_apl@yahoo.com

Abstract. The need for electrical energy is growing along with economic and population growth. The position of Indonesia is located on the equator line and is a tropical country that has the potential to get maximum solar energy throughout the year. Solar energy is a potential source of energy for Renewable Energy (RE) which is an alternative energy of low-carbon and clean power plants to solve energy problems. Based on the 14-km CCAM model, we have calculated solar radiation indicators at current baseline conditions (1981 – 2010) and 100 years future projections in order to analyze vulnerability and adaptation of the renewable energy sector to climate change in Indonesia. Although the climate change leads to a projected significant increase in mean air temperature 1,72 K across the whole of Indonesia that affect solar radiation indicator value, especially in Kalimantan with an average decrease of 6,5 W/m^2, but the results of the analysis indicate that the solar radiation indicator value of Indonesia region has excellent area and strength for the use of solar energy both now and in the next 100 years.

Keywords: The New and Renewable Energy, CCAM model, solar energy indicators

1. Introduction

The need for energy especially electricity is growing because it is a daily necessity of society along with economic and population growth. Electrical energy demands in Indonesia for industrial, household, public and commercial facilities, predicted by Djojonegoro [1] can reach 8.2 percent on average per year, in 2000 amounting to 122,603 GWh and in 2010 amounting to 258,747 GWh, while forecast of electricity supply from energy sources of coal, gas, oil, diesel, geothermal, water, biomass, solar and wind in 2000 was 37,860 MW and in 2010 was 68,760 MW [1], with the population is around 237.64 million [2], and the geographical condition of Indonesia consists of thousands of islands, the spread and uneven distribution of electricity centers, low levels of electricity demand in several regions, resulting in high marginal costs of supply system development electrical energy.

The position of Indonesia which is located in the equator line and a tropical country obtaining maximum solar energy throughout the year has good solar energy potential, with an average daily radiation approaching 4 kWh/m^2 [3]. The International Energy Agency (IEA) [4] estimates electricity production of 5.3 GW in 2002 and increases to 46 GW by 2030.

Solar energy is a potential energy source for Renewable Energy (RE) which is an alternative to low-carbon power plants. Solar energy is one of the most promising clean and renewable energy sources and has the greatest potential than other resources to solve the world's energy problems [5], At present, solar energy has become more popular as an energy supply in the world [6] . Renewable energy has the role of reducing greenhouse gases, carbon dioxide and methane, in terms of synergy and efficiency in the use of energy associated with climate change [7].

For solar energy, although climate change is expected to affect the distribution and variability of cloud cover, the impact of these changes on the overall technical potential is expected to be small [8]. The Ministry of Energy and Mineral Resources (KESDM) reported that in 2013 and 2014 fossil energy consumption used for energy supply was dominated by coal (69%) followed by natural gas (20%) and petroleum (11%) [9], while Indonesia is rich with the potential of solar and wind powers. Solar and wind powers have been used as a source of energy for Solar Power Plants (SPP) and Bayu Power Plants (BPP). The potential of solar power in Indonesia varies depending on the radiation level of each region, but in general the average solar radiation potential of Indonesia is 4.8 kWh/m^2/day. Wind power potential in Indonesia also varies by region depending on wind speed and surface conditions. In 2013, the potential capacity of BPP that could be built reached 3 - 6 m/s. During the period 2010 - 2014, solar power production experienced a significant increase from 26.4 thousand Barrel of Oil Equivalent (BOE) in 2013 to 32.9 BOE in 2014. This production was calculated by considering solar power electricity production with efficiency of 12.5%. The use of solar power and wind power will increase as the RE mix target is set at a minimum of 23% by 2025 as stipulated in Government Regulation 79/2014 on National Energy Policy [9]. Based on this regulation, RE continues to be developed with a 23% portion in 2025 and greater than 31% in 2050.

In order to anticipate electricity demands, State Electricity Company (PLN) has planned 140 MW of solar photovoltaic (PV) in 2015 and 620 MW in 2020 [10] and according to National Energy Council (DEN) [11] in 2015 electricity consumption in the final sector is 200 tWh and projection resulted on three conditions. First condition is Bussines as Usual (BaU), a basic of energy forecast scenario which is a continuation of historical development without any intervention of government policy that can change historical behavior with the basic assumption of moderate population growth of Gross Domestic Product (GDP) of 5.6% per year. Second condition is alternative condition 1 (Alt 1), the basis of moderate GDP growth of 5.6% and the adoption of renewable energy technology and energy-saving technologies. The third condition is alternative conditions 2 (Alt 2), assuming 7.1% high GDP growth and the application of RE technology and energy-saving technology.

The final energy demand projection in 2025 is 238.8 Mega Ton Oil Equivalent (MTOE) in BaU conditions, 201.5 MTOE in Alt 1 and 244MTOE in Alt 2, while the projection of final energy needs in 2050 is 682.3 MTOE (BaU), 430.3 MTOE (Alt 1) and 621 MTOE (Alt 2). RE final energy demands alone in 2025 amounted to 18 million TOE (BaU), 23 million TOE (Alt 1) and 60 million TOE (Alt 2) while in the year 2050 is 48 million TOE (BaU), 60 million TOE (Alt 1) and 86 million TOE (Alt 2) [11]. The national renewable energy market is projected to grow strongly in the coming decades and beyond, as indicated by current policies and targets, and based on expert projections.

This study aims to determine the indicators of solar radiation on solar radiation events in the Indonesian region to identify potential vulnerabilities of renewable energy to climate change in Indonesia, based on data from the scenario of numerical CCAM models with a focus on solar energy.

2. Data and Methodology

Determination of the value of the radiation indicator on the analysis of climate change vulnerability to renewable energy uses data from the execution of the atmospheric numerical model Conformal-Cubic Atmospheric Model (CCAM), with the climate model Geophysical Fluid Dynamics Laboratory version 3 (GFDL-CM3), in the emission scenario of Representative Concentration Pathways (RCP) 4.5, with 14 km spatial resolution [12], of which CCAM is an atmospheric numerical model developed by the Commonwealth Scientific and Industrial Research Organization (CSIRO) [13], and the GFDL Climate model version 3 (CM3) is a coupled climate model developed by NOAA formulated with the same oceanic and sea ice components as CM2.1 but with extensive development carried out on the atmospheric and land model components. CM2.1 and CM3 show a stable average climate index, such

IOP Conf. Series: Earth and Environmental Science **303** (2019) 012029 doi:10.1088/1755-1315/303/1/012029

as large-scale circulation and sea surface temperature (SSTs). There are several important improvements in CM3 climate simulation relative to CM2.1, including a modified SST bias pattern and a reduced bias in Atlantic sea ice cover, while RCP describes the radiative forcing that will be received by the earth, which is the emission scenario used by IPCC on AR5.

Data processing is carried out by grouping, which is based on a period of 30 years, which is the first 30 years, namely data from 1981 to 2010 and is considered as baseline data which is used as a reference for the next 30 years. Grouping the period of the second 30 years is data from 2011 to 2040, then the data period of the third 30 years is data from 2041 to 2070 and the period of the fourth 30 years is data from 2071 to 2099.

The calculation of the radiation indicator assumes that the use of solar energy depends on the incidence of solar radiation in a particular area, an efficient method of analysis to see the views of this source supply is the radiation balance in a particular region, which can be determined by calculating this radiation and radiation balance is the energy available in Earth's atmospheric system and is defined as the radiation balance of all radiation fluxes leading to and from the surface. The solar radiation indicator (RI) can be determined based on the formulation [14] below:

$$RI = SRI + SER - \alpha\, SRI - SES \qquad (1)$$

Which SRI is short wave radiation incident to the surface (W/m^2); α is albedo; SER is a short wave of emerging radiation at the top of the atmosphere (W/m^2) and SES is a short wave of radiation emerging to the surface (W/m^2). Unit of solar radiation indicator is W/m^2.

3. Results and Discussion

3.1. Temperature

Solar energy equipment can be affected by variations in ambient temperature, therefore it is necessary to analyze the projected increase or decrease in temperature in the future throughout the data processing period.

Based on data in the period of 1981 to 2010 for the first 30 years and considered as a baseline, the land temperature ranged from 286.45 K to 302.86 K, which were the lowest and the highest temperatures occurred on the mainland of Papua, respectively (figure 1a).

The temperature in the next 30 years, which is in the period of 2011 to 2040, has the lowest temperature of 287.76 K and the highest temperature of 303.99 K, meaning that there has been an increase in temperature for the lowest and highest values (figure 1b), which increases the temperature generally in Kalimantan and Sumatera compared to baseline conditions. The temperature in the third 30 years period, which is in the period of 2041 to 2070 (figure 1c) has also increased in temperature, with the lowest and highest temperatures were 288.82 K and 304.62 K, respectively. The temperature in the fourth year period in the span of 2071 to 2099 (figure 1d) has the lowest and highest temperatures of 289.61 K and 305.14 °K, respectively with the temperature increase was all compared to baseline data.

The average temperature rise that occur from the 2nd, 3rd and 4th periods to the 1st period (baseline) for the Kalimantan region, the smallest is 1.72 K, and the largest is 2.28 K

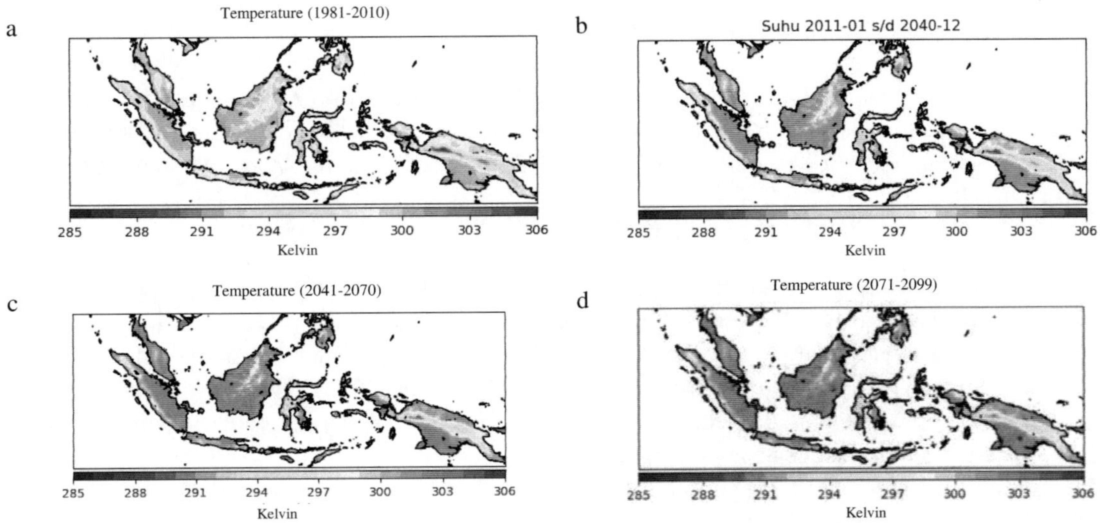

Figure 1. Temperature conditions of the CCAM numerical model output, using the GFDL_CM3 model with a spatial resolution of 14 km, a) period 1981-2010 as baseline, b) period 2011-2040, c) period 2041-2070 and d) period 2071-2099 in unit of Kelvin

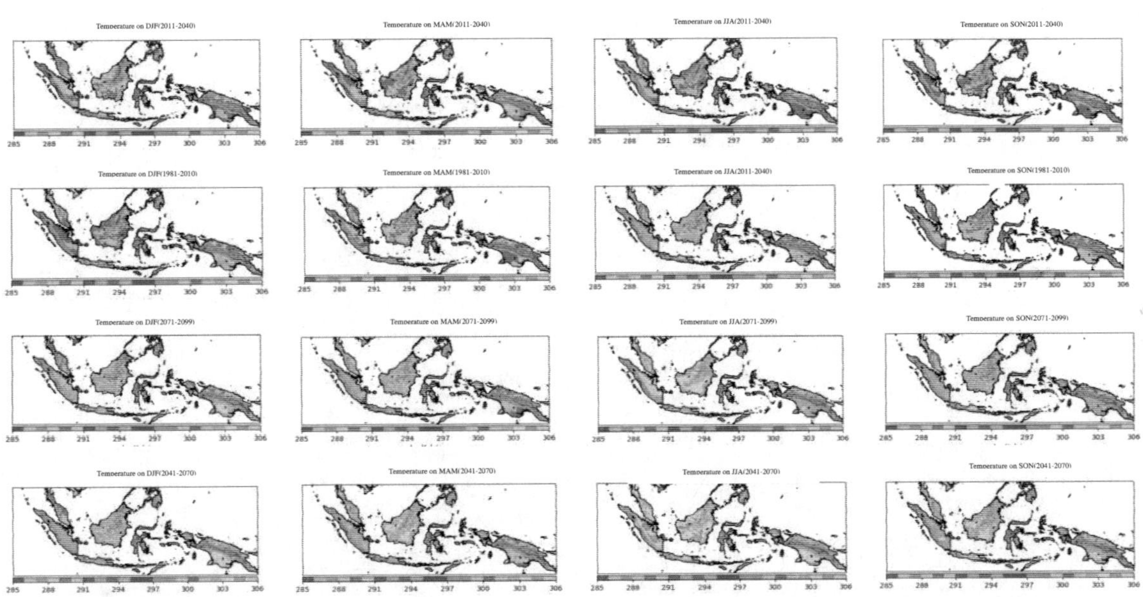

Figure 2. Condition of seasonal mperature released by CCAM numerical models, using the GFDL_CM3 model with a spatial resolution of 14 km, period 1981-2010 as baseline (row 1), period 2011-2040 (row 2), period 2041-2070 (row 3) and period 2071- 2099 (row 4) in units of Kelvin, and the column display shows the condition of the season temperature (DJA, MAM, JJA and SON)

To find out more complete temperature conditions, seasonal temperature analysis will be carried out in each period and figure 2 shows seasonal temperature conditions for December- January - February (DJF), March – April - May (MAM), June - July - August (JJA), September- October - November (SON).

The row display shows the condition of the season temperature, while the column shows the period of temperature. Seasonal temperature conditions based on the same season showed an increase, especially in Kalimantan and Sumatra. A striking increase occurred in the third period, namely the span of 2041 to 2017, both in the same season or in different periods. In the 3rd period, the maximum temperature that occurred was around 302.76 K, except SON season is 301.72 K, while the lowest temperature for all seasons is 299.95 K

3.2. Solar Energy

Results of the calculation of the radiation indicator value for the territory of Indonesia in the 1st period, determined as the baseline, got the smallest and highest radiation indicator values of 401.87 W/m^2 and 600.35 W/m^2, respectively (figure 3.a). The radiation indicator value in the 2nd period has the lowest and highest value of 394.15 W/m^2 (a decrease of 7.72 W/m^2) and 593.32 W/m^2 (a decrease of 7.03 W/m^2), respectively (figure 3.b). The decline generally occurs in all lands, especially in North, East and West Kalimantan, South Sumatra and also in Papua.

The declines still occur in the 3rd period (figure 3.c) and they have the smallest and the highest values of 395.82 W/m^2 (an decrease of 6.50 W/m^2) and 589.86 W/m^2 (a decrease of 10,49 W/m^2). The decline in the second period also generally occurs on all lands, especially in North, East and West Kalimantan, which are increasingly widespread to South Sumatra, East and West Java and also in Papua. While in the 4th period (figure 3.d) has the smallest and the largest values of 396.15 W/m^2 (a decrease of 5,72 W/m^2) and 583.92 W/m^2 (a decrease of 16,43 W/m^2), respectively. The decline in the fourth period still occurs on all lands, especially in North, East and West Kalimantan, South Sumatra, East and West Java and also in Papua. The decrease in the solar radiation indicator is all compared to baseline data. The average reduction in solar radiation indicators that occur from the 2nd, 3rd and 4th periods to the 1st period (baseline) for the Kalimantan region, the smallest is 6.5 W/m2, and the largest is 11.32 W/m^2.

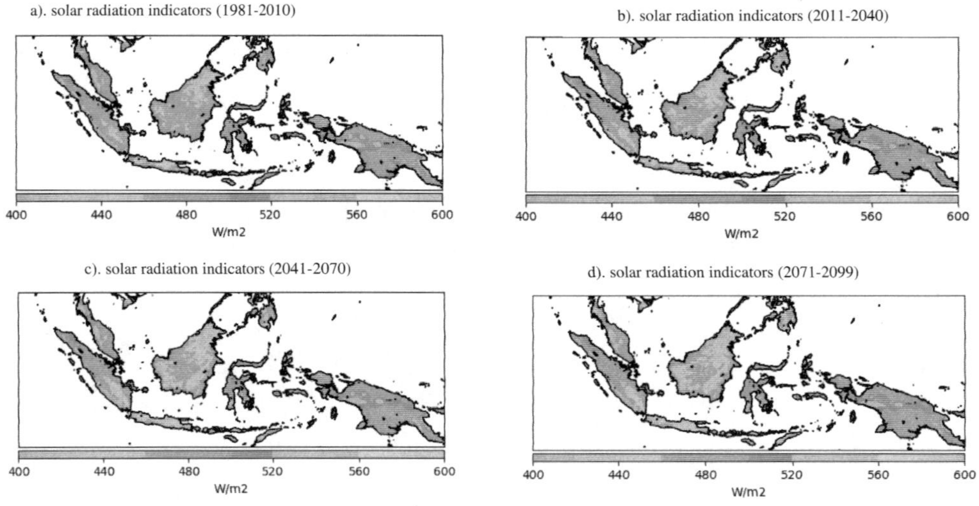

Figure 3 Condition of solar radiation indicator output by CCAM numerical model, using GFDL_CM3 model with 14 km spatial resolution, in a) period 1981-2010 as baseline, b) period 2011-2040, c) period 2041-2070 and d) period 2071-2099 in units of W/m^2

Analysis of the seasonal radiation indicator was carried out for each period to complete information on solar radiation conditions at a location. Figure 4 shows the DJF seasonal radiation indicator shown in the 1st column, MAM in the 2nd column, JJA in the 3rd column, and SON in the 4th column, while the row shows the 1st (as a baseline), the 2nd, the 3rd and the 4th periods.

The baseline period in figure 4 first row shows the highest indicator of seasonal solar radiation occurs in the JJA season, ranged from 451.55 W/m^2 to 567.69 W/m^2. The lowest seasonal solar radiation indicator occurs in the MAM season, ranged from 396.88 W/m^2 to 531.36 W/m^2.

Seasonal solar radiation indicators in the second period across all regions of Indonesia showed decline to the baseline. The highest seasonal solar radiation indicator value of 540.29 W/m^2 (a decrease of 13.27 W/m^2 compared to baseline) occurred in SON season, and the lowest of 1.77 W/m^2 was found in the MAM season.

Likewise, the decline of solar radiation indicator is also observed in the third period, either towards the baseline or the second period. In this 3rd period the highest solar radiation indicator value occurred in the DJF season of 532.13 W/m^2 and the smallest value occurred in the MAM seasons of 396.43 W/m^2.

The 4th period of seasonal solar radiation indicator still shows a decrease compare to the baseline. The biggest value of 536.10 W/m^2 (a decrease of 6.29 W/m^2) occurred in the DJF season, and the smallest value of 397.88 W/m^2 occurred in the MAM season. Comparative results show a decrease in the solar radiation indicator received by a region from a period to period.

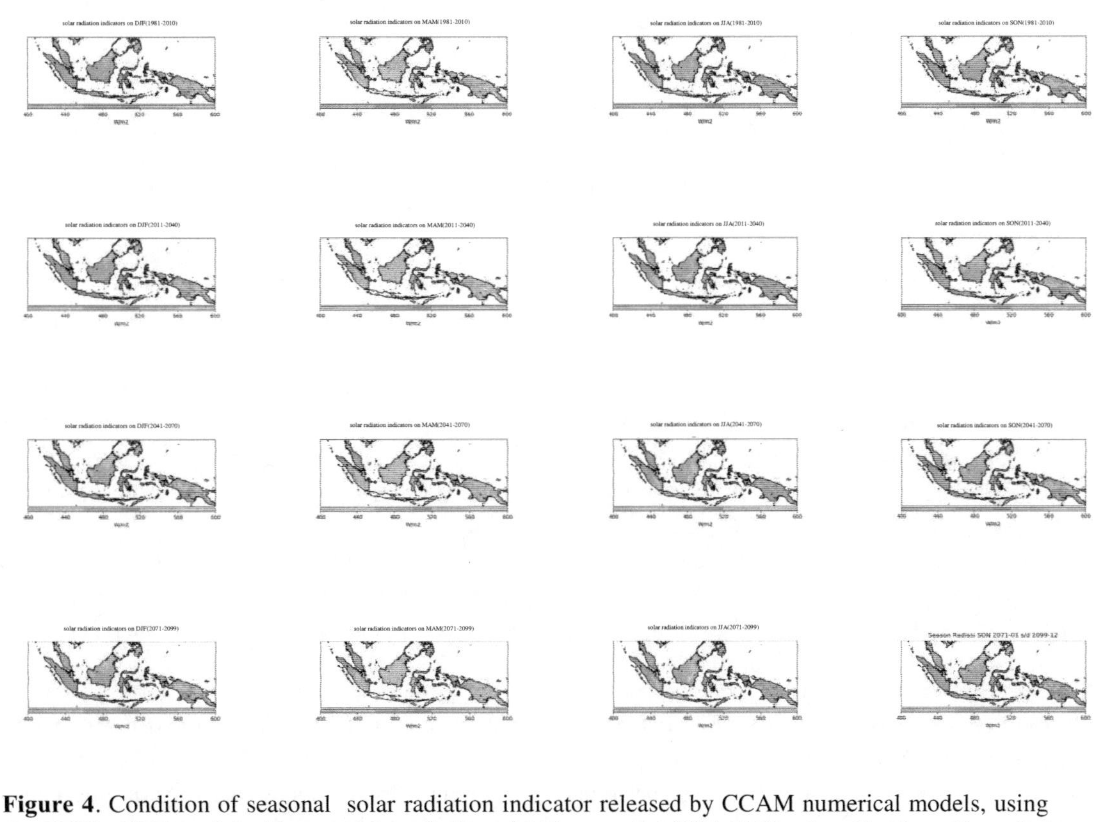

Figure 4. Condition of seasonal solar radiation indicator released by CCAM numerical models, using the GFDL_CM3 model with a spatial resolution of 14 km, period 1981-2010 as baseline (row 1), period 2011-2040 (row 2), period 2041-2070 (row 3) and period 2071-2099 (row 4) in units of W/m^2, and The column display shows the condition of the season DJA, MAM, JJA and SON

Figure 5 showed the percentage reduction of solar radiation indicator. The highest and the lowest differences in the 2nd solar radiation decrease was 2.43% and -2.68%, respectively (figure 5.b). The biggest and the lowest differences for the 3rd period are 2.46 % and -3.98 %, respectively (figure 5.c). The 4th period has the highest and the lowest differences of 2.27% and -4.39%, respectively. These differences are all compared to the baseline (figure 5.a)

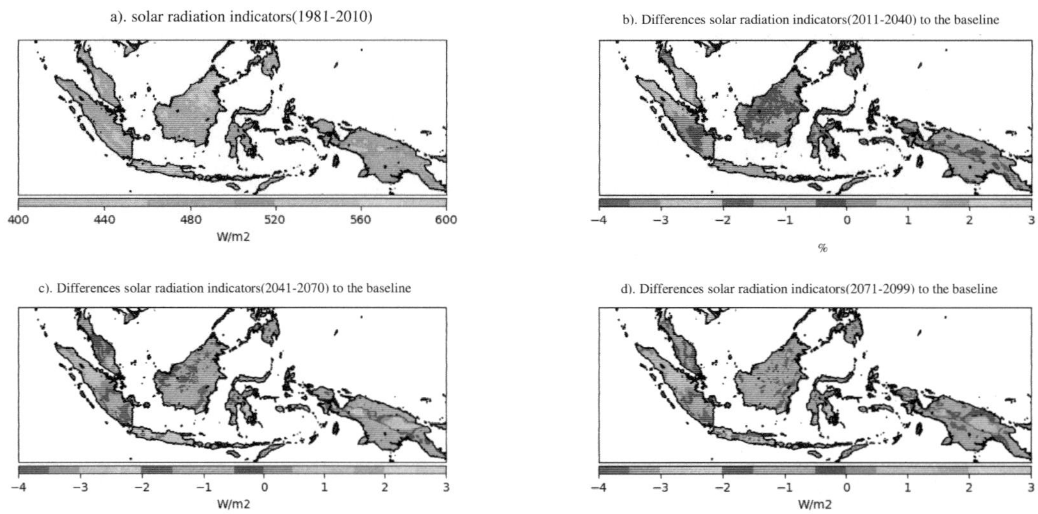

Figure 5. Differences in solar radiation indicators released by CCAM numerical models, using the GFDL_CM3 model with a spatial resolution of 14 km, a) period 1981-2010 as baseline in units of W/ m^2, b) period 20 11-2040 in % , c). period 2041-2070 in % and d.) period 2071-2099 in %

The occurrence of a decrease in solar radiation indicators is not only based on a review of the 30-year period data analysis, but also based on the results of seasonal analysis and the decline generally occurs on all mainland, especially in North, East and West Kalimantan, South Sumatra and also in Papua. Because the regions of North, East and West Kalimantan, South Sumatra and also in Papua has a continuous solar radiation indicator decline, therefore these regions are areas that are vulnerable to climate change. The decrease in the 30-year period solar radiation indicator that occurs in an area is inversely proportional to the increase in surface temperature.

For our future work, when the data of short-wave radiation incident to the surface; albedo; a short wave of emerging radiation at the top of the atmosphere and a short wave of radiation emerging to the surface are available, then we will validate our model input data against the observations.

4. Conclusion

Climate change has increased land surface temperatures and reduced solar radiation received in a particular area. The results of the climate model projections analyzed using a 30-year data period, indicate that the temperature conditions are increasing, while there are declines in the conditions of solar radiation in particular areas of Kalimantan and Sumatera that indicate these regions vulnerability to climate change.

5. Acknowledgments

This paper was prepared in collaboration with the SRIRAMA team at the Center for Atmospheric Science and Technology, LAPAN.

6. References

[1] Djojonegoro,W., 1992, Pengembangan dan penerapan energi baru dan terbarukan, Lokakarya "Bio Mature Unit" (BMU) untuk pengembangan masyarakat pedesaan, BPPT, Jakarta.
[2] Badan Pusat Statistik[BPS] . 2010. Sensus Penduduk 2010. BPS, Jakarta.
[3] Indonesia energy outlook & statistic 2006. Depok, Indonesia: Energy Reviewer, University of Indonesia; 2006
[4] IEA, "World Energy Outlook 2004", International Energy Agency, Paris, IEA/OECD, 2004
[5] Kadir MZAA, Rafeeu YA. Review on factors for maximizing solar friction under wet climate environment in Malaysia. Renewable & Sustainable Energy Reviews 2011, doi:10.1016/ j.rser.2010.04.009
[6] Solangi KH, Islam MR, Saidur R, Rahim NA, Fayaz H. A review on global solar energy policy. Renewable and Sustainable Energy Reviews 2011;15:2149–63.
[7] Intergovernmental Panel on Climate Change (IPCC) (2007). Climate Change 2007: Mitigation. Contribution of Working Group III to the Fourth Assessment Report of the Intergovernmental Panel on Climate Change. Cambridge, United Kingdom and New York, NY, USA, IPCC.
[8] IPCC 2011, Special Report on Renewable Energy Sources and Climate Change Mitigation
[9] KESDM, 2015, Perkembangan Penyediaan dan Pemanfaatan Migas Batubara Energi Baru Terbarukan dan Listrik, ISBN : 978-602-0836-15-7, Desember 2015
[10] ADB, Summary of Indonesia's energy sector assesment, No 09, ADB papers on Indonesia, 2015
[11] Dewan Energi Nasional(DEN), Outlook energi Indonesia 2016, ISSN 2527-300, Jakarta November 2016
[12] SRIRAMA,2017, http://srirama.sains.lapan.go.id/srirama
[13] Mc GREGOR JL and DIX MR (2008) An updated description of the Conformal-Cubic Atmospheric Model. In: Hamilton K and Ohfuchi W(eds.) High Resolution Simulation of the Atmosphere and Ocean. Springer Verlag. 51-76.
[14] Oscar, A. Silva,W. Ruffato, V. (2015) Evaluation of Renewable Energy Vulnerability to Climate Change in Brazil: A Case Study of Biofuels and Solar EnergySmart Grid and Renewable Energy, 2015, 6, 221-232 Published Online August 2015 in SciRes. http://www.scirp.org/journal/sgre http://dx.doi.org/10.4236/sgre.2015.68019

Multi Spectral Satellite Data to Investigate Land Expansion and Related Micro Climate Change as Threats to the Environment

M Hanif, B G Putra, K Nizam H Rahman and A Y Nofrizal

Department of Geography, Padang State University, Padang 25132, Indonesia

hanif12jhenif@gmail.com

Abstract. Technology development of satellite earth observation offers multi-spectral data, variations sensitivity of satellite sensors are very useful to detect existing conditions by using the remote sensing algorithm to extract the land cover and the phenomenon that has occurred. Ecosystems have bound tolerance to nature condition and organisms have adaptation limit to climate change. Method used in this study was the extraction of land cover by applying multi spectral analysis as Maximum Likelihood, object base image analysis, (LST) land surface temperature, and also the formula Threat of Land Expansion (TLE) to the environment. Result showed that multispectral and time series analysis gave the increasing dynamic of land expansion, especially for forest high density, which areas ranged from 29098.73 ha in 2000 to 19216.7 ha in 2018. The surface temperature increase was ± 2.8 C° in eighteen years. The temperature dynamic represented the micro climate changes, has followed stream of land expansion. TLE represented zone threats of land expansion to natural ecosystems. It was found that threat zone because land expansion. The area of high threat vale 0.4 around central human activity was a built up area.

Keywords : satellite data, ecosystem, land expansion, climate change.

1. Introduction

Surface temperatures have an indirect, but significant, influence to air temperatures, especially in the canopy layer, which is closest to the surface. For example, parks and vegetation areas, which typically have cooler surface temperatures, contribute to cooler air temperatures. Dense, built-up areas, on the other hand, typically lead to warmer air temperatures. Because air mixes within the atmosphere, though, the relationship between surface and air temperatures is not constant, and air temperatures typically vary less than surface temperatures across an affected area, therefore urban heat islands and global climate change (global warming) are often similar. For example, some communities may experience longer growing seasons due to either or both phenomenon. Urban heat islands and global climate change can both also increase demand of energy, particularly summertime air conditioning demand, and associated air pollution and greenhouse gas emissions, depending on the electric system power fuel mix [1].

The context study of expanding it is cycle, where land use influence temperatures and climate changes of urban, natural environment and harmonious nature ecosystem like natural forests. Basically, natural land cover is target for expansion, where this area productive in economy context. The various types of nature land cover have wealth, there are in the land that has not involving human activity. The process of expanding land it give impact to supply carbon, emission to air temperatures, as begining for the micro climate changes, and then it will be surplus to global warming. While the land use change also give the impact to biodiversity, changing population biodiversity, and increasing air temperature, and pressing morphology and physiological adaptation of animal and vegetation, it also will make a barrier for animal migration in seasons. Therefore, land expansion will continually increase, if that happens will be basic of disasters. It can damage natural sustainability such as erratic weather, drought and meteorological disasters because the control of the air voltage has changed from the original land covers.

Several publications appeared that use aerial and satellite remote sensing data to study the change of urban environment due to the dynamic increase of population. The assessment was carried out in this study by considering the current study of a significant advancement [2]. Remote sensing and GIS assessments are proved to be useful tools for spatial and temporal patterns mapping of urban surface temperature [3, 4]. The purpose of the research is to investigate the influence of land expansion to micro climate change, and to analysis threat of land expansion to environment, it following the targets (SDGs) sustainable development goals point 13 is climate action, and point 15 is life on land. The case study is Padang city, a capital city of West Sumatra Province, Indonesia. The region is one of the central developments, in order to support the progress of regional development. For maintaining the sustainability of environment, not all of lands could be developed to support human necessary there should be a limit (ecologycal tolerant). If land development is carried out it will contribute to changes and increases in micro climate air temperature and threat to environmental conditions, both biodiversity.

2. Material and Method

2.1. Material

In this study, we used Landsat TM5 and OLI8 images of 2000, 2010, and 2018 data from USGS, at the map scale 1: 100.000. Administration and infrastructure map were obtained from BAPPEDA regional planning, development agency. Terrain data by using DEM was obtained from BIG geospatial information agency Indonesia.

2.2. Method

2.2.1. Multi Spectral Analysis for Land Expansion

In this study we used a multi-spectral image analysis model, namely Maximum Likelihood spectral, based image classification, after which it was combined with image classification based on OBIA objects base image [5], to improve images classification results. Multi spectral image classification Maximum Likelihood is known as a well-established image classification statistical model in grouping pixels for land cover identification [6]. As a parametric classifier, the algorithm relies one each training sample being represented by a Gaussian probability density function, completely described by the mean vector and variance-covariance matrix, using all available spectral bands. These parameters, it possible to compute the statistical probability of a pixel vector being a member of each spectral class [7].

2.2.2. Micro Climate Change

Analysis of micro climate changes is carried out using data from multi spectral satellite, namely thermal band. By using thermal waves, we can identify the condition of the land surface temperature, from the multi spectral satellite data recording information on the surface of the earth. In this analysis, we use the *Land surface temperature* (LST), which represents the surface temperature of the land surface in response to the object's response. There are several steps taken to analyse terrestrial surface temperature, which are spectral values to radians, change radians to degrees of temperature to kelvin, and change the temperature of Kelvin to a degree of Celsius. TOA to Radiance, in the process we must transform top of atmospheric data, known as TOA, which is reflected radiance. At this stage, we will change the digital number (DN) of the TIRS Landsat OLI band image to radiance data, the formula is [8]:

TOA to radiance:

$$L\lambda = ML\ Qcal + AL \tag{1}$$

Radiance to temperature:

IOP Conf. Series: Earth and Environmental Science **303** (2019) 012030 doi:10.1088/1755-1315/303/1/012030

$$T = \frac{\text{K2}}{1n\left\{\frac{K1}{L\lambda} + 1\right\}} \tag{2}$$

Where,

T : Temperature (K)

$L\lambda$: TOA Spectral radiance (Watts/(m^2 *srad*μm))

K1 and K2 : Constant band 10 or band 11

Final process (T – 273) to convert degree from Kelvin to celcius.

2.2.3. TLE Threat of land expansion to environment

The environmental threat in this case, where we used spatial approach with (GIS) Geography Information System, the technology to be implemented from an aspect that it is an area that will be a land expansion threat to the environment. This threat will be calculated from parametric model data that is linked to land expansion, and it will be performed based on the weight and indicator of driving factor. We use an approach with analysis of delivery data that encourage changes in land cover, as a form of land expansion.

3. Results and Discussion

3.1. Land cover and Temperature change

We use data from multi-spectral satellite technology to investigate land expansion. Several experts of the satellite system data explained that the more satellite waves used than it will be able to distinguish more detailed object variations. Results of multi-spectral classification, the Maximum Likelihood is statistical model, showed that variation and distribution of land cover, which is an indication of the land cover expansion. There was category of land expansion on land cover types that are affected by human activities. It was not involving natural land cover. Natural land cover types, such as forests, and some other types will be further processed as a threat to land expansion in natural ecosystems.

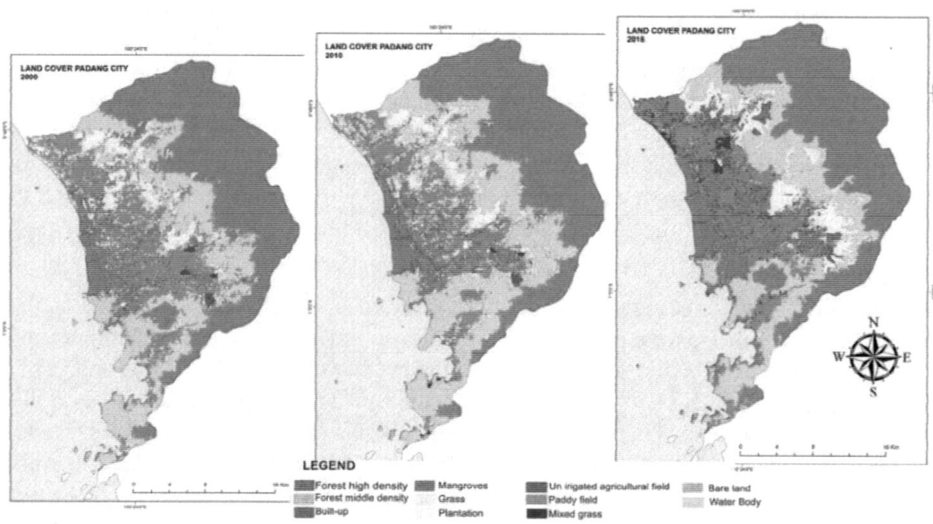

Figure 1. Land cover changes

The results of multi spectral image as land cover map were exported to GIS technology to calculate the land cover area as shown in figure 1. We have compiled chart of the land cover area variations in study time by looking at the largest variant of land expansion. This land expansion tends to move from the central area of the city, and it spreads along the road and some infrastructure. East hills boundary, with pattern of land expansion that follows the physical pattern of the road.

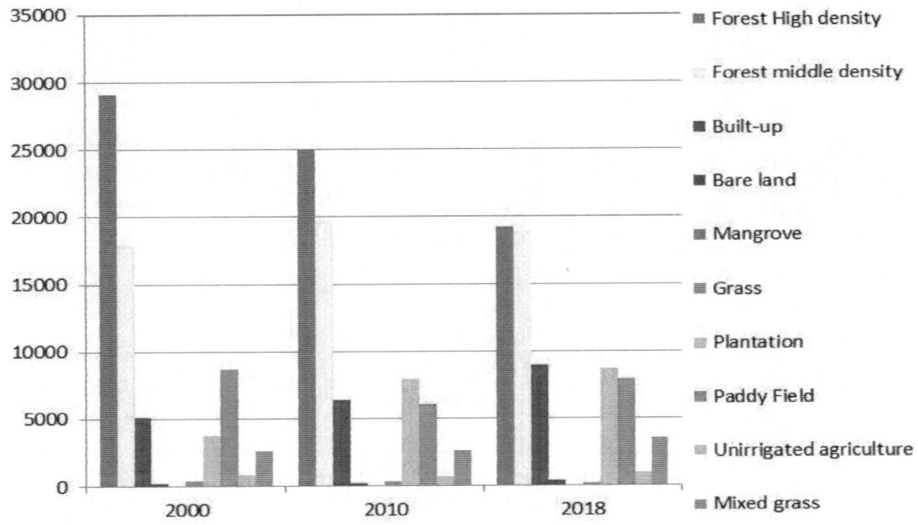

Figure 2. Chart of land cover changes

The bar chart in figure 2 reveals the land cover change because land expansion. Overall, there was a significant difference in the land cover from 2010 to 2018. There was a decrease in land cover changes for forest high density and increase for built up. Forest high density in one decade decreased from 29098.733 ha in 2000 to 25001.7 ha in 2010 and to 19216.7 ha in 2018. The Built-up increased from 5109.78 ha in 2000 to 6394.84 ha in 2010, and to 8956.42 ha in 2018. Plantation area increased dramatically from 3734.79 ha in 2000 to 7931.879 ha in 2010 and to 8636.82 ha in 2018.

The land expansion occurs in some natural land cover and it is threat for ecosystem. They are transformed to serve human necessary. The human activity on the land gives impact as degradation of vegetation spread. When there is loss of vegetation area, then it will disturb air in urban area. The vegetation can influence the micro climate as cooler for air temperature. The impact of land expansion is the most significant because it made loss biodiversity. There were fluctuated changes of some kind of land cover such as forest middle density, grass, and paddy. The land use change brings number of changing in climate pattern. Excessive deforestation leads to decrease in precipitation level and increase in surface temperature on global scale, thus it changes whole global climatic model as forest-covered surfaces are a special form of vegetated surfaces [9].

The band thermal of Landsat it use-full to take information land surface temperature. Result of extraction data has shown in figure 3. Landsat satellite has records of surface object as responses of spectral by pixel value, it is representation temperature object. From this research has found from many years, there was a significant change. The color index of the map represents the area between low to high temperature. From the map has represented the zone of summit maximum temperatures on the land. The zone is namely urban heat island. In one decade, the urban heat island zone was concentrated near coastal, but it is difference in eight years latters. The island spreads around all of lands, especially at every place where the people carry out expansion to develop residential or change the natural area to other function such as plantations, agricultural, etc.

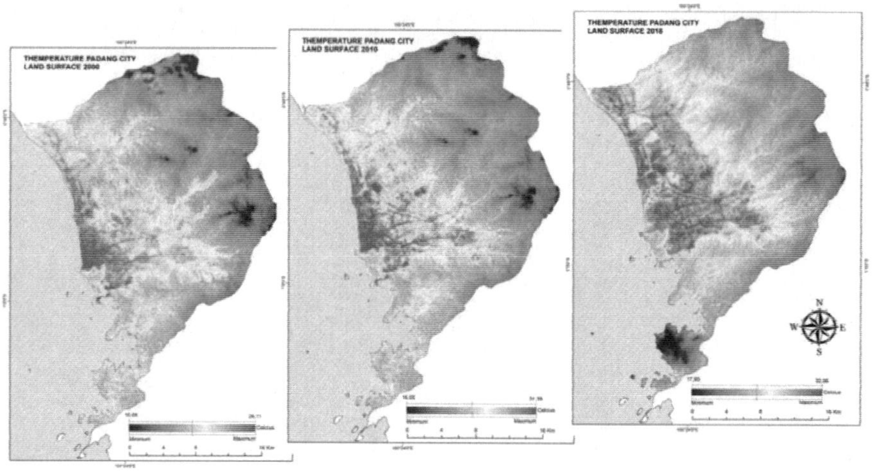

Figure 3. Temperature changes

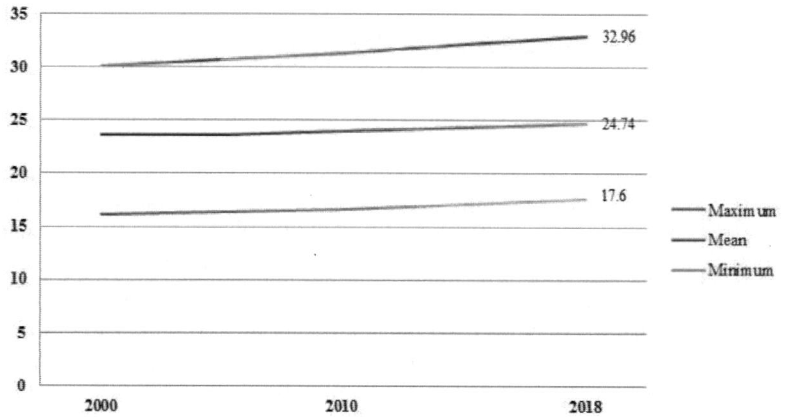

Figure 4. Graphic of surface temperature changes

The graphic in figure 4 reveals how the surface temperature changes in eighteen years. Overall, there was gradual temperature increase of 2.8 C° from 2000 to 2018. We can observe the distribution of surface temperature variants from many years. The maximum surface temperature was made zone around built up area in 2000, and follow the built-up area changes in 2010 and 2018. The maximum surface temperature was widespread around area, where human activities were taken place. This represents the changing conditions of the micro-environment, which is one of the climate aspects of the air temperature, and the surface temperature of the land was very close to the conditions of the land cover variation. The surface temperature increase relatively then spread on area, following land cover changes because impact land expansion.

Urban surfaces are characterized by large roughness elements, wide-spread sealed areas, reduced moisture availability at the surface, and increased possibilities for heat storage. This leads to higher turbulence intensities in the UBL and to stronger sensible heat fluxes from the urban surface [10]. Deforestation and vegetation lost are habitat fragmentation, which occurs when parts of a habitat in a local environment, in which an organism is usually found, become separated from one another because landscape changes, such as road constructions, and the developed land. Our understanding of how climate change will impact natural terrestrial ecosystems and the goods and services is required [11].

In this study at Padang City, the natural habitat was degraded because of deforestation for timber and land expansion for agriculture, paddy field and un-irrigation agriculture. The unproductive areas can be developed to be residential and public facilities as generally we know as built-up. There was a supply of carbon emission and air warming. The disappearance of forest land cover by conversion to plantation, settlement, agriculture and to the needs of other sectors has changed the climate pattern. [12] Human has change ecosystems more rapidly and extensively than in any comparable period of time in human history, to meet rapidly growing demands for food, fresh water, timber, fibre, and fuel. The because of biodiversity loss can be broadly grouped into five categories: habitat loss, fragmentation or change; over exploitation of species; pollution; the spread of invasive species; and climate change, all of which have their origins in human demands placed on the biosphere.

In our research, we found that because of the degradation of vegetation in Padang City, we lost many natural ecosystem areas with the vegetation cover as natural forest to be plantation or agriculture. Some locations with little vegetation have changed into built up areas (urban), resulted in increase of air of temperature. [13] This vegetation loss meant a loss of cooling in urban areas. Indeed, vegetation plays an essential role in preventing the build-up of heat through the process of evapotranspiration and shading of the ground and buildings. During the natural process of evapotranspiration of water vapor, the ambient air is cooled, releasing part of its heat to allow evaporation. Vegetation also contributes to effective control of air quality in cities.

3.2. Threat of Micro Climate change to Environmental,

The fact that smog, which is composed of fine particulate matter and tropospheric ozone, is formed during the reaction between the sun's rays, heat and pollutants nitrogen oxides (NOx) and volatile organic compounds (VOCs) [14, 15]. Measures that have been used to approximate adaptive capacity are the ecosystems diversity, the plasticity of a species, the mobility of individual organisms, including dispersal mechanisms, or the existence of climate gradients [16]. In our research, we found the trend of land surface warming by the years. There was a gradual increase every year. The summit warming of urban heat island was made at the central zone on built-up area. The temperature was gradually dropped away from crowded built-up area and spread out from central. When micro climate change continually, it will be a threat and disturbed circle in environment, such as drying on land surface, depression of soil moisture to physic and chemistry process for vegetation growth. Phenomenon of urban heat island will be a threat to environment such as, the increase of surface temperature, and will finally affect the degradation of air quality by supporting the emission of carbon dioxide (CO_2), dust, and smoke to the atmosphere. In this research the maximum temperature increase was 2.85 °C.

In the longer term, outlying populations of wet forest species in dry areas, and *vice versa*, on edaphically or topographically extreme sites, the impacts of climate change on regional biodiversity can be mitigated only by the protection of large tracts of minimally disturbed forest that are continuous over both altitudinal and rainfall gradients. Existing protected areas are insufficient, so this will require both the rehabilitation of damaged forests and the reforestation of degraded areas [17].

The species, such as vegetation and other biota, have a tolerant and limit temperature for germination. The temperature close to air moist serves as the first media to support vegetation germination to grow. In the future, if the micro climate changes increase every year, and the human do not control and mitigate toward increase of micro temperature, then the climate change will destroy some unique species, or number of species. It is a circle threat of land expansion to micro climate change, and climate to environment.

3.3. Threat of Land Expansion to Environment,

Threat of land expansion to environment (TLE) is approach to measure threat land expansion to general aspect in environment. The environment we mean including vegetation, habitat in ecology landscape, biodiversity, air condition. In this case Padang City is tropical zone and variant geomorphology. We developing and apply the formulation using GIS. We using indicators where adaptation as driving factors that drive land expansion and resistance factor. The considerations we use from several researchers' studies of land cover changes, conditions of distance from roads, distance from residential, land cover excisting, and morphological in region. We not considering TLE can do apply in other region with difference geographical characteristic.

There more available driving factor each other to be threat to environment. The contribution of climate change to future extinctions depends on how quickly species can respond to change [18]. Fragmentation of vegetation makes it difficult for species to move within a habitat, and poses a major challenge for species requiring large tracts of land. The land expansion had press and change the land cover type and causing loss of land cover nature vegetation, where it really had a strong position to control micro climate. When the vegetation decreased, then the micro climate also increased, it made the micro climate change [19]. Vegetation is an important factor affecting the ecological balance, especially in ecologically fragile areas, ecology connectivity to protect biodiversity and human [20].

Table 1. Indicator and weight TLE

Land cover	Weight	Geomorphology	Weight	Distance from residential	Weight	Distance from road	Weight
Forest high density	0.1	Plain	2	100	2	100	2
Forest low density	0.1	Slop slightly	1.8	300	1.2	200	1.8
Built-up	2	Middle aslant	1.1	800	0.8	300	1.4
Bare land	1	Aslant	0.5	1500	0.5	500	1.2
Mangrove	0.1	Steeply, cliff	0.1	< 3000	0.1	700	1
Grass	0.1					1000	0.8
Plantation	1.2					1500	0.6
Paddy field	1.5					2000	0.4
Agricultural	1.5					3000	0.2
Mixed grass	0.1					< 4500	0.1

In this research we using indicator such us distances from road and residential, it have high influence to land expansion. It adopted by principle from many researches using existing land use, distance from road and residential as driving factors, it forward for land use changes. There are also indicators of threats to environment and biodiversity where when land cover changes to residential the vegetation and habitat will disappear and increasing emission to air temperature. In other case when land cover changes to plantation or agricultural area, the vegetation as habitat in environment will change to be fragmentation and corridor in landscape ecology. It meaning the energy of vegetation on the land to control environment and air condition will drop gradually.

Geomorphology is resisting indicator for land expansion, because not all geomorphology characters suitable to develop for human necessary, specific for resident, paddy area, and plantation. It reason we adopted geomorphology as resistance for threat of land expansion to environment, the natural geomorphology can protected environment by the limitation in suitability to development. The threats analysis will obtain from weight sum all indicator and then it must be divided to 4 to get final value analysis. The result of threat land expansion to ecosystem will represented environmental damage, the threat as habitat loss of biodiversity. Where the high value it represent the invulnerable ecosystem from the threat of land expansion and low value it represent the environment in high threat because land expansion.

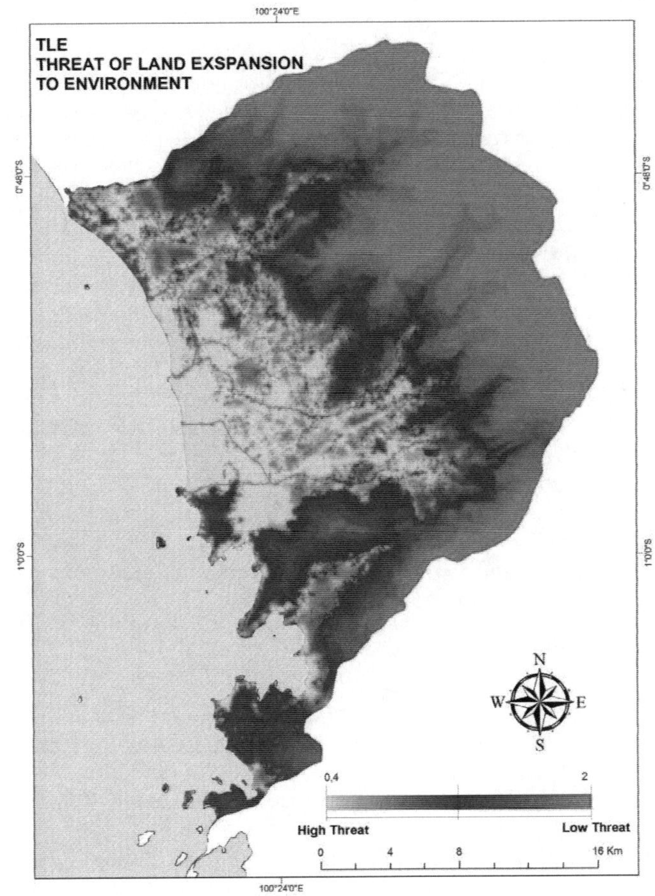

Figure 5. Threat of land expansion to environment (TLE)

From figure 5, the map representing variation and distribution threat land expansion to environment. Colour index as quantitative value to represent distribution of threat. The colour index showed that the area between natural and unnatural ecosystems is the location that receives the highest threat. The increase in purple colour by the value 2 is referred to very high invulnerable ecosystem, that mean threat to environment is low. The colour cyan by the value 0.4 represented the area with a high threat. The factors inhibited the land expansion is geomorphological conditions and accessibility. Geomorphological conditions by fraudulent slope help reduce the threat to ecosystem conditions from land expansion.

4. Conclusion

Increase in the built-up area for years has degraded natural ecosystems. Thermal temperature data was gradually increasing following the land expansion, which shows an increase in microclimate because of the amount of vegetation degradation. There is a strong connection in environmental cycle between land cover changes because of expansion in activities, giving effect to air temperature, and land expansion and climate changes can also be threats to ecosystems and habitats.

Acknowledgements

Special appreciation for research team, and also for members of Community of Geospatial Science and Technology, Padang. Besides of this, authors sincerely honors all referees and appreciates their comments and suggestions that significantly improve this paper.

References

[1] Wong E, Hogan K, Rosenberg J, Denny A 2015 *Reducing Urban Heat Islands: Compendium of Strategies.* Environmental Protection Agency's Office of Atmospheric Programs U.S.

[2] Roth M, Oke T R, Emery W J 1989 *Satellite-derived urban heat islands from three coastal cities and the utilization of such data in urban climatology*. International Journal of Remote Sensing vol **10**, 1699–1720.

[3] Nichol J 2005 *Remote Sensing of Urban Heat Island By Day and Night.* Photogrametric Engineering & Remote Sensing Vol **71**, 613–621. DOI: 10.14358/pers.71.5.613

[4] Orsolya G, Tobak Z, van Leeuwen B 2016 *Satellite Based Analysis Of Surface Urban Heat Island Intensity*. Journal of Environmental Geography Vol **9** (1–2), 23–30.

[5] Hanif M and Nofrizal A Y 2017 *Investigation Natural Forest Ecosystem Use High Resolution and Lidar Data.* Journal Environment and Earth Science IIESTE. Vol **9**.

[6] Hanif M and Adam T 2017 *Application Multi Vegetation Index to Mapping Mangroves Distribution Coastal Environtment Northeast Province of Aceh.* Sumatera Journal Disaster Geography and Geography Education. SJDGGE, Vol **1.** No.1

[7] Weng Q 2011 *Advances in Environmental Remote Sensing, Sensor, Algotihma and Applications*. Indiana State Univeristy. ISBN: 978-1- 4200-9181-6.

[8] http://landsat.usgs.gov/Landsat8_Using_Product.php

[9] Velasco E, Márquez C, Bueno E and Bernabé R M 2007 *Vertical distribution of ozone and VOCs in the low boundary layer of Mexico City*. Atmos Chem Phys Discuss. Vol **7**: 12751–12779

[10] Smith A M S 2014 *Remote sensing the vulnerability of vegetation in natural terrestrial ecosystems.* Journal Remote Sensing of Environment. Vol **154**. 322–337.Esilver

[11] Eimeis S 2011 *Surface Based Remote Sensing of Athmospheric Boundary Layer*. Journal Springer. Dordrecht Heidelberg London New York

[12] WWF 2012 *Ecological Footprint Report Africa Green Infrastructure For Africa's Ecological Security.* The African Development Bank and WWF – the World Wide Fund for Nature. ISBN 978-2-940443-39-0.

[13] Mesev V, Gorte B and Paul L A 2001 *Remote Sensing and Urban Analysis.* First published 2001 by Taylor & Francis 11 New Fetter Lane, London EC4P 4EE.

[14] Akbari H, Berhe A and Levinson R 2006 *Cool Color Roofing Materials.* California Energy Commission, Berkeley, CA,73 p.

[15] Eike L, Catherine M 2013 *Ecosystem Vulnerability to Climate Change, a Literature Review.* World Agroforestry Center.

[16] Salomon T, Aubert C 2004 La fraîcheur sans clim. Terre Vivante, Paris, 160 p.

[17] Corlett R 2010 *Global Climate Change and Biodiversity*. University of East Anglia, Norwich, UK.

[18] Mccarty J P 2001 *Ecology Consequence of Recent Climate Change.* Journal Conservation Biology Page. Vol **15** No. 2 April.

[19] Noonan-Mooney K 2018 *How are people affecting biodiversity*. Ladybird. Julia Kresse

[20] Kang H, Li X and Zang J 2015 GIS *Analysis of Changes in Ecological Vulnerability Using a SPCA Model in the Loess Plateau of Northern Shaanxi, China*. International Journal Environment. Res. Public Health Vol **12**, 4292-4305

An Evaluation Graph of Hourly Rainfall Estimation in Malang

R M Putra [1], A Kurniawan[2], I A Rangga[2], M Ryan[2], Endarwin[1] and A Luthfi[1]

[1]Agency for Meteorology, Climatology, and Geophysics. Angkasa I Street No.2, Kemayoran, Capital Jakarta, 10620 Indonesia
[2]State College of Meteorology Climatology and Geophysics. Perhubungan I Street No.5, Pondok Betung, South Tangerang, 15221 Indonesia

richardmp.rmp@gmail.com

Abstract. Satellite-based rainfall estimation is evolving rapidly. Most studies use data, which is spatially fine, but poorly regarding time. On the other hand, availability of verification data is also quite rare. This study used Hillman Form B report that was corrected by ME-48 from Malang Climatological Station. 2009-2016 IR1 satellite data were used in hourly temporal resolution (only less than 3% data missing). Four estimation methods were compared: Auto Estimator, CST, mCST, and Quantile Analysis Equation. Data processing was carried out using Python and R statistic as a quality control. The analysis was done by creating a graph that combines False Alarm and Miss Information for each rainfall intensity. Binary transformation was done for enabling information to be plotted. All rainfall estimation methods have a high false alarm (more than 74% at 1 mm) but quite low miss (less than 0.03%). By taking into account its error pattern, satellite data can be used in rainfall observation. The Quantile equation is slightly superior to other methods. This study is relatively inexpensive to be duplicated so it can be used as an evaluation tool for rainfall estimation best practice for Meteorological and Climatological Agency's network.

Keywords : rainfall, satellite, CST, Malang.

1. Introduction

One of the most significant weather elements [1] for all human activities is rainfall. Rainfall is the intensity of rain that occurs at a certain duration of time [2]. Rainfall information is expressed in millimeters (mm). Rainfall of 1 (one) millimeter means that the area of 1 square meter has a water level of 1 millimeter or equal to one litter (1 dm^3) assuming that it does not evaporate, does not seep and does not flow [3]. Bulk information is needed primarily for early warning of natural disasters caused by hydro-meteorological factors such as floods, landslides, and droughts as well as in research related to the hydrological cycle [4]. However, the current problem is the uneven distribution of rain gauge [5] so that many locations do not have this information. One alternative solution is to estimate the rainfall intensity using satellite data [6].

Rainfall estimation techniques that have been studied previously produce relatively good values of rainfall estimation based on satellite imagery. Swarinoto and Husein conducted research on satellite data capabilities to estimate rainfall in the Jayapura area and its surrounding area [7] . Other research [8] uses GOES Infrared for making rainfall estimation using regression fit of radar rain rate and GOES – 8 Cloud Top Temperature. Another technique for estimating rainfall using satellite is by clouds classification based on convective and stratiform types [9, 10]. This technique was improved by [11] by using the passive microwave (PMW) to increase performance on rainfall estimation in Indonesia.

This study aims to evaluate the results of rainfall estimation using satellite data for Malang region. In addition, the aim is to identify the best methods for rainfall estimation in Malang and to address problem in Quality Control (QC) before conducting research.

2. Data and Methods

The data used in this research consists of three datasets from January 2009 – December 2017. These data are rainfall observation from Climatological Station in Malang every 3 hours (Form ME.48 Meteorological Archive), rainfall data every 1 hour from Hellman rainfall observation (Form.B Meteorological Archive), and Himawari – 8 Satellite from Kochi University (http://weather.is.kochi-u.ac.jp/archive-e.html) with Portable Gray Map (PGM) data format. Kochi University's satellite data have an hour temporal resolution and have a spatial resolution of about 5.5 km as shown in figure 1.

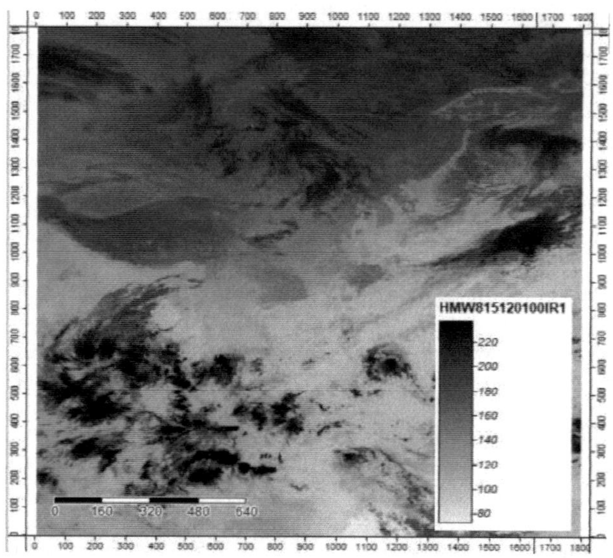

Figure 1. Satellite data from Kochi University in PGM Format

In the PGM format, the Cloud Top Temperature (CCT) cannot be directly obtained from the data. This is because the Portable Gray Map (PGM) format only shows relative CTT based only on the grayscale. The minimum color code is zero, and the maximum is 255. Therefore, in addition to requiring PGM format satellite image data; this study also used CAL data to convert grayscale from PGM into the cloud top temperature (CTT) in Kelvin (K) units.

This research was conducted in the region of Malang Climatological Station located in -7.9° S and 112.6°E. Unfortunately, the pixel position of Malang Climatological Station is at the center of four pixels' satellite from Kochi University. Therefore, in this research, we also tried to find the best pixel for making rainfall estimation in Malang Climatological Station using Himawari – 8 from Kochi University.

Because of that condition of satellite pixel, in this research, we make five schemes for processing the satellite rainfall estimation by using pixel 1, pixel 2, pixel 3, pixel 4 and the average of all pixels around Malang Climatological Station as shown in figure 2. The methods used in this research consist of 4 rainfall estimation methods, i.e. Convective Stratiform Technique (CST), modified Convective Stratiform Technique (mCST), Auto Estimator (AE), and Quantile Analysis formed by regression function of sorted rainfall intensity in Malang for 2009 – 2017 and cloud top temperature (CTT) in that locations using 5 schemes of pixel.

Figure 2. Location of Malang Climatological Station in Pixel's Satellite

2.1. Convective Stratiform Technique (CST) and modified Convective Stratiform Technique (mCST)

The methods used in this research are similar to the procedures proposed by [9, 10]. The technique classifies cloud types into convective or stratiform cloud by using the difference between CTT in the pixel of the study's area with that in the surrounding pixels. [12] Also used brightness temperature difference between the pixel and the eight surrounding neighbour pixels. In this research, we classify convective and stratiform cloud by using eight pixels in two form, referred as "Square" form and "Plus" form as illustrated in figure 3.

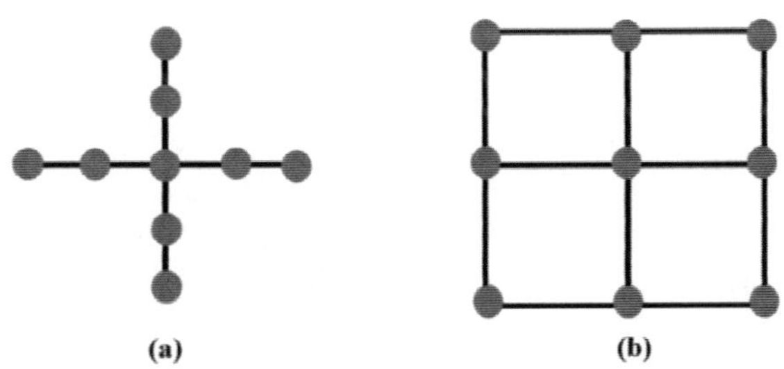

Figure 3. Classification technique for Convective and Stratiform Cloud by "Plus" (a) and "Square" (b) form

The technique of classification whether convective and stratiform cloud is determined by the Slope Parameter (S) using eight pixels around the main location:

$$S = k(T_{i-2,j} + T_{i-1,j} + T_{i+1,j} + T_{i+2,j} + T_{i,j-2,} + T_{i,j-1} + T_{i,j+1} + T_{i,j+2} - 8T_{i,j}) \quad (1)$$

With k = 0.125

Where the values i and j are pixels indices around the center (-7.6 LS and 112.6 BT), while k is a factor that depends on data resolution. Furthermore, the Slope Parameter values that indicate the type of convective cloud must meet the following requirements:

$$S \geq \exp[0.0826\,(T_{min} - 207)] \quad (2)$$

With T_{min} = cloud top temperature (CTT) in Kelvin units

If the slope parameter value does not satisfy the criteria, then the type of is stratiform cloud [11]. The next step to estimate the rainfall is to determine the area of rain. The area of cloud that produce rainfall are determined by the following equation:

IOP Conf. Series: Earth and Environmental Science **303** (2019) 012031 doi:10.1088/1755-1315/303/1/012031

$$A = \exp(15.27 - 0.0492\ x\ T_{min}) \quad (3)$$

The next step is to estimate rainfall from the convective and stratiform cloud by using these equations:

$$Convective\ rainfall\ (mm) = C\left(\frac{A_c}{A}\right)TR_c \quad (4)$$

$$Statiform\ rainfall\ (mm) = S\left(\frac{A_s}{A}\right)TR_s \quad (5)$$

Where c = number of convective cells within a grid, s = number of stratiform cells within a grid, A_c and A_s = rain area from Eq. (3), A= average area covered by each pixel, T = length of period in hour, R_c = convective rain rate in mm h^{-1} (20 mm h^{-1}), R_s = stratiform rain rate in mm h-1 (3.5 mm h^{-1}). The differences between CST and modified CST are the coefficient values of A, R_c, and R_s. [11] has modified this method by the result of best rainfall intensity from the convective cloud (R_c) is 26 mm h^{-1} and stratiform cloud (R_s) is 0.8 mm h^{-1}.

2.2. *Auto Estimator*

This technique uses regression fit of radar rain rate and GOES – 8 Cloud Top Temperature [8]. Rainfall estimation is calculated by using cloud top temperature (CTT) data from a satellite then converted into rainfall intensity by using the following equation :

$$R = 1.183\ x\ 10^{11}.\exp(-3.6382\ x\ 10^{-2} x T^{1.2}) \quad (6)$$

With R = rainfall intensity (mm h^{-1}) and T = cloud top temperature in Kelvin units

2.3. *Quantile Analysis Equation*

In this research, we did not only compare the rainfall estimation from the equations that were studied previously, but also tried to derive equation by regression [13] of sorted rainfall intensity and sorted cloud top temperature called "Quantile Analysis" method. We have eight schemes for this equation which depend on the pixel that we choose from the satellite data. The table below shows all schemes to estimate the rainfall intensity from regression data:

Table 1. Quantile Analysis Equation

Rainfall = B*exp(CTT*A), CTT=Cloud Top Temperature, A and B = Coefficients

Coeff.	Quantile Upper Limit			
	P1	P2	P3	P4
A	-0.15561	-0.15940	-0.15605	-0.15446
B	3.5494E+14	8.0260E+14	4.0251E+14	2.8841E+14
Coeff.	Quantile Lower Limit			
	P1	P2	P3	P4
A	-0.15792	-0.16023	-0.15813	-0.15736
B	5.1820E+14	8.8531E+14	5.8972E+14	5.0726E+14

The coefficient values of A and B are obtained from the regression equation of sorted rainfall intensity and sorted cloud top temperature. However, the condition sometimes has a problem with this pattern from rainfall data and cloud top temperature, which have the same value of rainfall intensity but the cloud top temperature more than one conditions. In this case, we use two approaches namely upper limit and lower limit. Upper limit means we choose the highest data of cloud top temperature for the same-sorted rainfall intensity and lower limit means we choose the lowest data. Therefore, for this condition, we have eight schemes that consist of four pixels and two conditions of approaches.

IOP Conf. Series: Earth and Environmental Science **303** (2019) 012031 doi:10.1088/1755-1315/303/1/012031

2.4. Double Checked Analysis

The rainfall intensity from satellite data is double-checked by both software "R" [14] and "Python"[15]. This step is to ensure that both software agree with each other. The main software for calculating the rainfall intensity from satellite data is "Python" and the result of calculation is compared to result from software "R" for Quality Control. Furthermore, the plotting was also done using QGIS software [16] [17] to ensure that 4 pixels of satellite data are in the right position. In comparison, "R" software has a better configuration script and performance. Therefore, python calculation for rainfall intensity was checked and compared with R calculation. Python have a better processing time than that of R, so we use Python software and checked by R software. Based on this case, it can be concluded that if we use one software, it has to be checked by the other software to ensure that our calculation and script's programming are working well.

2.5. False Alarm and Miss

In this research, verification of rainfall estimation method used 2 x 2 contingency table [18]. In this method, we will calculate the False Alarm and Miss condition for all rainfall estimation's scheme that we used. The previous research from [19, 20, 21], also used FAR for validation.

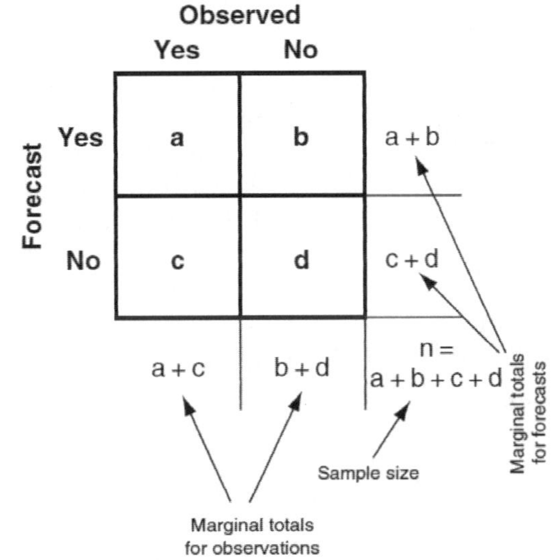

Figure 4. 2 x 2 contingency table (Source: [18])

In term of figure 4, *a, b, c* and *d* are the number of pairs for estimation data and observed data. When the estimation data is same class value with observed data, it is called *hits* (*a*). Similarly, on *b* occasions called *false alarm* (FAR). It is a condition of occurring event in estimation, but did not happen in observed data. There is also *c* that show the number of estimations did not occur, but in observed data is happen. It is called *miss* (MS). The last symbol is *d* means the same condition of estimation and observed show the event does not occur.

In this research, validation using FAR and MS values were done for any rainfall intensity. They are 1 mm, 2 mm, 3 mm, 4 mm, 5 mm, 7 mm, 9 mm, 11 mm, 15 mm, 17 mm, 20 mm, 25 mm, and 30 mm. Hereafter, we plot all of FAR and MS values into one graph~~ie~~ called FAMS picture and analysed it.

3. Result and Discussion

3.1. Data completeness of rainfall observation

The first step before we calculate rainfall intensity using satellite data and compare which one is best method, we want to discuss on completeness of data from 2009 – 2016. The problem that we have to know first is about the different type of report for rainfall intensity before and after 2015. Before 2016, the rainfall intensity of Hellman (Form.B) is reported on the same day as the rainfall occurred. However, starting in 2016, rainfall intensity was reported the day after rainfall occurred. This problem was found by comparing the rainfall intensity every 3 hour from ME.48 and every 1-hour from Hellman observation (Form.B). This figure below is the ratio of corrected data for rainfall intensity caused by that problem:

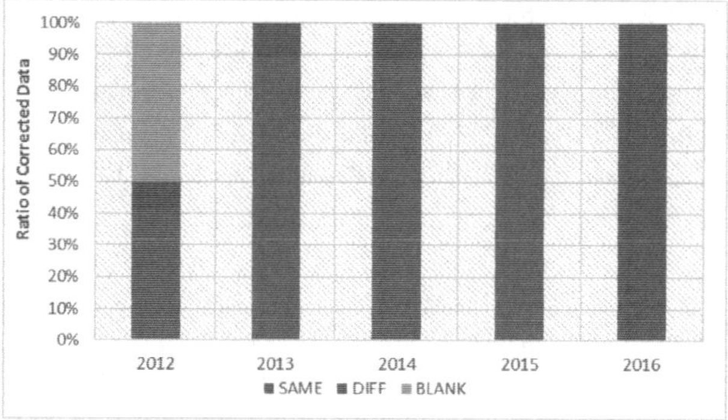

Figure 5. Ratio of corrected data

Based on figure 3, we know that before performing the calculation, we must be performing Quality Control (QC) of our input data. The reason of why data analysis started in 2012 is due to the fact that digitalization of hourly rainfall intensity was only started from this year. Also, in 2012 data is not complete for one year. The green color indicates blank data of rainfall observation. The red color shows different type of report for rainfall observation, every 3 hours and every 1 hour. Also, the blue color shows the same report of rainfall observation every 3-hour and 1 hour based on a different kind of meteorological form. In this discussion, we want to show that rainfall data observation has been verified and has a good quality before we compare them with rainfall estimation method using satellite.

3.2. Data completeness of Satellite data from Kochi University

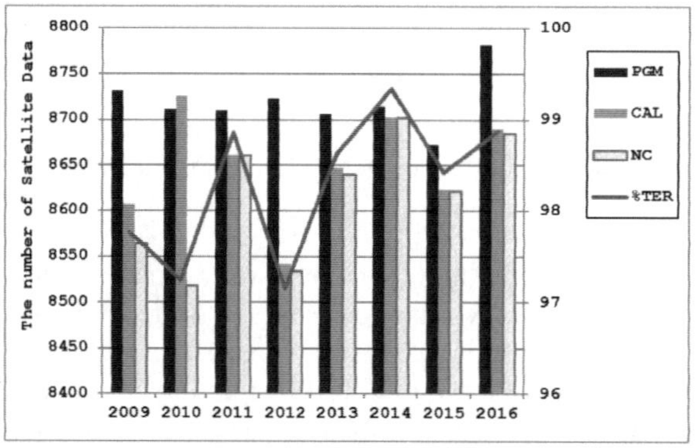

Figure 6. Completeness of satellite data from Kochi University

Satellite data from Kochi University has a different condition with rainfall data observation. Start from 2009 – 2016; the satellite data is existing in PGM format and for Calibration data (CAL). However, since rainfall observation data in digital format was only existing from the middle of 2012, this research is only done when both satellite and observation data exist. Satellite data from Kochi University in PGM format and CAL format was converted into NetCDF (NC) format first for calculating rainfall intensity easily. However, the problem was that we could not convert all of the data to rainfall intensity. So, figure 4 shows the number of data successfully converted to NC format and then rainfall estimation. The highest number of data that could be converted to NC is in 2014; it is almost 100% of data. The lowest number of data that could be changed is 2010, it is about 97.2% of data. However, for all of the data used in this research, successful data conversion are higher than 97% for all of the data.

3.3. Comparison of All Schemes of Rainfall Estimation Using Satellite Data

In figure 7, we plot all of the value for *false alarm* and *miss* for all schemes. Due to its similar result, only best scheme is plotted in black. *False alarm* and *miss* have its own best scheme. Where there is rainfall estimation with high intensity, there will be high *false alarm*. The *false alarm* is about only 0.7 in separating rain and no rain condition. It means that statistically, for every ten times rainfall estimation say there will be a rain, only three times of rains will be true. The best method of rainfall estimation for all schemes based on *false alarm* value is Quantile Analysis in pixel 3 using end time of observation.

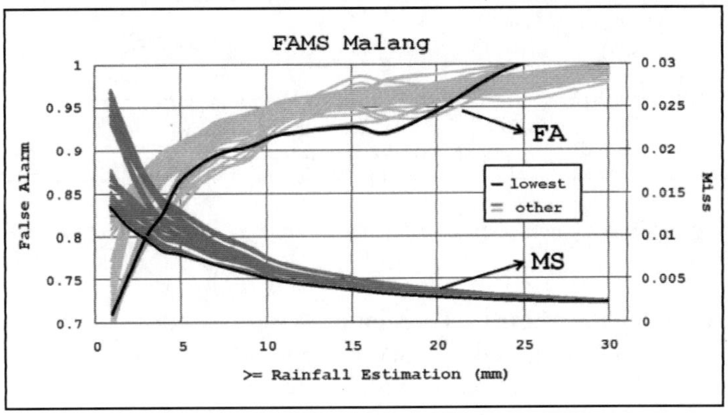

Figure 7. All experiment result of rainfall estimation methods

Miss condition is also shown in figure 7 with axis on the right side. For all methods, miss value is less than 0.03 % and it means that for rainfall estimation, all schemes have a good performance so that if there is condition of rain that declared by someone lie, rainfall estimation is good enough to negate it. Based on the result, the best method for rainfall estimation in Malang based on *miss* value is CST calculated by "square" method for separating cloud type. We can also know that the best rainfall estimation was obtained using pixel that located southwest of Malang coordinate (pixel 3).

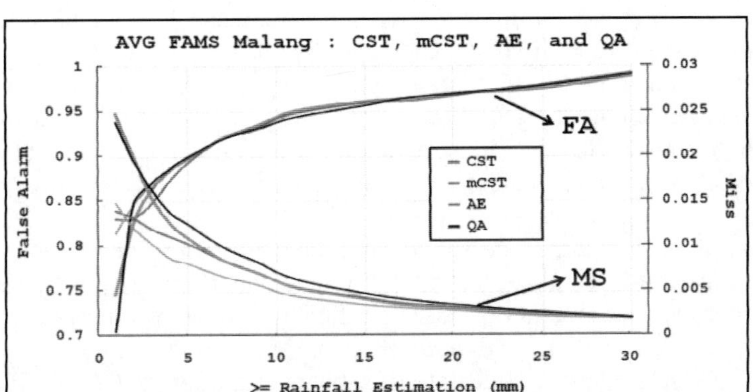

Figure 8. Average result of estimation methods in all method schemes

Unlike figure 7, figure 8 shows average result for all schemes and methods that we used for making rainfall estimation. The red one is CST method, green is mCST method, black is QA method, and blue is AE method. Based on the result, the best average methods for rainfall estimation in all classes is

Quantile Analysis (QA) methods. It has shown from the minimum value of *false alarm* in this method, although *miss* condition is higher than that of others. Miss condition is quite small and not significant compared with *false alarm*.

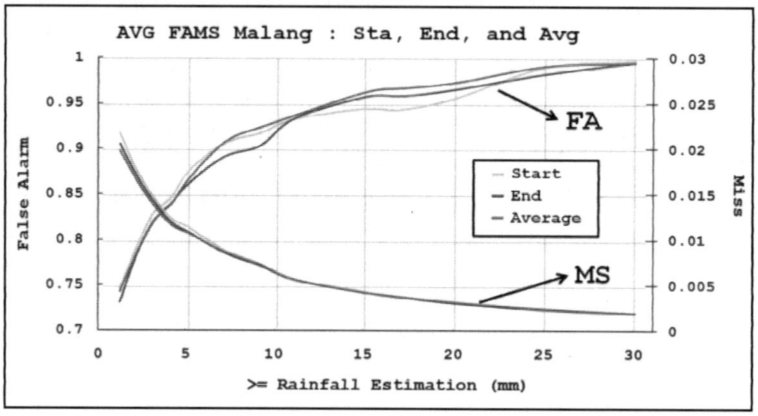

Figure 9. Average result of estimation methods in all time schemes

Figure 9 shows time for making estimation of rainfall. We divide into 3 kind of schemes. First scheme is called "start time". Its means that rainfall estimation is done at first time of satellite data. For example, the satellite data at 08.00 UTC is for making rainfall estimation during 08.00 – 09.00 UTC. Second scheme is called "end time". Rainfall estimation is done at the end of time for satellite data. For example, the satellite data at 08.00 UTC is for making rainfall estimation from 07.00 – 08.00 UTC. The last scheme is called "average time". It means we use both "start time" and "end time" for making rainfall estimation for one hour. In figure 9, the color of orange is the start time. The color of violet is the end time, and brown is an average time. Based on the result, it is shown that the effect of different time is not significant but using start time point should be avoided due to relatively higher FA and MS.

4. Conclusion

Based on the research, we learned that many factors must be considered before proceeding. The most important thing before research is the Quality Control (QC) of the data. In this research, the rainfall observation data from 2009 – 2016 is not existing all in digital format. So, we only use start from 2009 – 2016. Besides, there is difference in reporting for rainfall intensity, so we have to double-check the rainfall data using two types of report (ME.48 and Form B).

The other lesson is performing the calculation by two types of software; they are Python and R. In this research, we must double check calculation system using another software for checking the result of data calculation. In this research, the main software for calculating the rainfall intensity from satellite data is "Python" and the result of the calculation is compared to result from software "R" for Quality Control.

Based on the data analysis, we can understand about the pattern of *false alarm* and *miss* condition in rainfall estimation using satellite by different rainfall intensity and different schemes. *A false alarm* is high when rainfall intensity is high and decreases gradually with decreasing rainfall estimation. For a miss, condition has an opposite pattern with a *false alarm*. Miss value is high when rainfall intensity is slight and decreases gradually until rainfall intensity become moderate – high. The best schemes of rainfall estimation using satellite data are Quantile Analysis using southwest pixel of Malang location and by the end of time observation satellite. The scheme gives the lowest *false alarm* in all intensities, although miss condition is higher than those of others *miss* condition is quite small and not significant compared with the *false alarm*. Quantile analysis is better due to its empirical-based coefficient. It avoids error from generalization process. In the end, this study encourages the idea that local equation is better than the general one so local study for it is urgent to be done.

Acknowledgments
The authors wish to thank Kochi University Weather Home for providing the infrared weather satellite data. We also like to thank to Malang Climatological Station for providing the rainfall intensity in some reported documents.

References

[1] Caraka R E, Supari and Tahmid M 2018 Copula-Based Model for Rainfall and El- Niño in Banyuwangi Indonesia *J. Phys. Conf. Ser*
[2] Caraka R E, Bakar S, Tahmid M, Yasin H and Kurniawan I D 2018 Neurocomputing Fundamental Climate Analysis *Telkomnika*
[3] BMKG 2010 KEP.009 Tahun 2010 Tentang Prosedur Standart Operasional Pelaksanaan Peringatan Dini, Pelaporan dan Diseminasi Informasi Cuaca Ekstrim *Jakarta: BMKG*
[4] Suwarsono P, Kusumaning A D S and Kartasamita M 2009 Penentuan Hubungan Antara Suhu Kecerahan dangan MTSAT dengan Curah Hujan Data QMORPH *J. Penginderaan Jauh* **6** 32–42
[5] Caraka R E and Yasin H 2018 *Spatial Data Panel 1st ed* Wade Group
[6] Phipps M and Rowe S 2010 Seeing satellite data *Public Underst. Sci.* **19** 311–321
[7] Swarinoto Y and Husain 2012 Estimasi Curah Hujan Harian Dengan Metode Auto Estimator (Kasus Jayapura Dan Sekitarnya) *J. Meteorol. dan Geofis*
[8] Vicente G A, Scofield R A and Menzel W P 1998 The Operational GOES Infrared Rainfall Estimation Technique *Bull. Am. Meteorol. Soc.* **79** 1883–1893
[9] Adler R F and Negri A J 1988 A Satellite Infrared Technique to Estimate Tropical Convective and Stratiform Rainfall *Journal of Applied Meteorology* **27** 30–51
[10] Goldenberg S B, Houze J R A and Churchill D D 1990 Convective and stratiform components of winter monsoon cloud cluster determined from geosynchronous *J. Meteorol. Soc. Japan* **68**37 – 63
[11] Endarwin, Tjasyono B, Gunawan D and Siswanto 2014 *Modifikasi Convective Statiform Technique dengan Modifikasi Kombinasi Data Satelit Gelombang Mikro Pasif dan Inframerah Untuk Estimasi Curah Hujan di Indonesia*, ITB
[12] Anagnostou E N and Kummerow C 1997 Stratiform and convective classification of rainfall using SSM/I 85-GHz brightness temperature observations *J. Atmos. Ocean. Technol.* **14** 570 – 575
[13] Jamilatuzzahro, Herliansyah R and Caraka R E 2018 *Aplikasi generalized linear model pada R, 1st ed.* Yogyakarta: INNOSAIN
[14] Sutiksno D U, Gio P U, Caraka R E and Ahmar A S 2018 Brief Overview of STATCAL Statistical Application Program *J. Phys. Conf. Ser.*, **1028**
[15] Payne J 2010 *Beginning Python: Using Python 2.6 and Python 3.1.* Wrox Press Ltd.
[16] Caraka R E and Yasin H 2017 *Geographically Weighted Regression (GWR) Sebuah Pendekatan Regresi Geografis, 1st ed.* MOBIUS GRAHA ILMU
[17] QGIS Development Team 2015 *QGIS Geographic Information System* Open Source Geospatial Foundation Project.
[18] Wilks D S 2006 *Statistical Methods in the Atmospheric Sciences* Elsevier Inc
[19] Dinku T, Chidzambwa S, Ceccato P, Connor S J and Ropelewski C F 2008 Validation of high-resolution satellite rainfall products over complex terrain Validation of high-resolution satellite rainfall products over complex terrain *International Journal of Remote Sensing* **29** 4097 – 4110
[20] Kenabatho P K, Parida B P and Moalafhi D B 2017 Evaluation of satellite and simulated rainfall products for hydrological applications in the Notwane catchment, Botswana, *Phys. Chem. Earth* **100** 19–30
[21] Hossain F and Huffman G J 2008 Investigating Error Metrics for Satellite Rainfall Data at Hydrologically Relevant Scales *Journal of Hydrometeorology* **9** 563–575

Urban Heat Islands analysis towards topographic based land use change and daily commute effect along the Kaliurang Street in Yogyakarta

U Suarma[1], S A Hapsarini[2], N L Isnastuti[2], H R Ikhwani[2] and Durrotunafisah[2]

[1]Department of Environmental Geography, Faculty of Geography, Universitas Gadjah Mada, Yogyakarta, Indonesia
[2] Research assistant, Department of Environmental Geography, Faculty of Geography, Universitas Gadjah Mada, Yogyakarta, Indonesia

utiasuarma@ugm.ac.id

Abstract. This paper's aim is to analyse the correlation between land change pattern and its topography with Urban Heat Islands (UHI) in Kaliurang Street and surrounding area. Secondly, the effects of daily commuting activities from rural or sub urban area towards urban area are well identified. Kaliurang Street is one of the major streets in Sleman that uniquely represented the transition between rural, suburban, and urban area. Rural area is located in the furthest north, suburban located in the center, while the urban area is located in the southern part. Transect method carried out along Kaliurang Street to obtain primary data mainly temperature and vehicles number. Data collected during daytime (between 12-2 PM) and afternoon (between 3 to 5 PM). Furthermore, LANDSAT 8 OLI/TIRS satellite imagery was analyzed to obtain the Land Surface Temperature (LST) Information, while ALOS PALSAR was analyzed to obtain elevation data. One of the causes of temperature differences is the commuting which affects the variation of surface temperature along Kaliurang Street. Therefore, strong correlation between commuting and the UHI phenomenon are expected.

Keywords : urban heat island, Landsat, topography, land surface temperature.

1. Introduction

Road geometry affects the occurrence of UHI phenomena in an area[1] because the road determines the accessibility of a built land. Kaliurang Street is one of the collector roads in Sleman Regency, Yogyakarta Special Province, which stretches from the volcanic cone landform with steep topography area on the north to the volcanic slope landform with flat topography on the south. The topography of the Kaliurang Street, which is flat in the southern part, stretches into steeper area to the north associated with its regional status. Urban area located in flat topography areas, while sub-urban to rural areas are located in sloping to very steep topography areas. This condition underlies that the topography of the Kaliurang Street influences the land change patterns over the past 10 years. From the 10 years land change patterns in Kaliurang Street, the distribution of temperature over the ten years can be found.

Kaliurang Street area is influenced by the diffusion phenomenon of urban areas in Yogyakarta City which has begun to be seen since 2005 [2]. This condition shows that there is a tendency of urbanization in Kaliurang Street since 2005 which started from the south to the north, subsequently occurs along Kaliurang Street topography patterns. The urbanization then followed by the increasing of built-up area and

IOP Conf. Series: Earth and Environmental Science **303** (2019) 012032 doi:10.1088/1755-1315/303/1/012032

the increasing of the rate of transportation which led into surface and land temperature increases. The first objective of this study is to identify the pattern of UHI in the Kaliurang Street area for the past 10 years (2008-2018) through the UHI approach based on its urbanization pattern. The second objective is to identify the correlation between the commuting as a result of urbanization along Kaliurang Street with the increase of air temperature as the indication of UHI phenomenon.

This study examines the urban heat island phenomenon on Kaliurang Street as a road that has unique road geometry. The geometry of Kaliurang Street is subsequently located along with its topographical patterns from south to the north, which is led into urbanization pattern that occurred in Kaliurang Street surrounding area from year to year. Temperature change over years is a phenomenon of climate change that needs temporal analysis [3], therefore the study of land changes, changes in vegetation density, and changes in land surface temperature were studied spatially and temporally using Landsat 7 ETM+ and Landsat 8 OLI / TIRS imagery with LST analysis. Urban Heat Island in Kaliurang Street occured in 2018 then further studied by correlate it with the commuting phenomenon of Kaliurang Street. Commuting in urban area considered as primary activity that also happened as one of the factors of increasing temperature.

2. Method

2.1. Study area

The measurement was carried out along Kaliurang Street as a representative transect from the surrounding area. The location of measurements is shown on Figure 1. The measurement location represents the village administration area located on Kaliurang Street, including Sinduadi, Condongcatur, Minomartani, Sinduharjo, Sardonoharjo, Sukoharjo, Wedomartani, Umbulmartani, Umbulharjo, Harjobinangun, Pakembinangun, Wukirsari, and Hargobinangun Village.

Figure 1. Study Area

2.2. Data Collecting and Analysis

Temperature and number of vehicles that represent the condition of current condition was obtained by direct measurement that was conducted for eight days in two timeframes which are 12:00 AM to 2:00 PM and 3.00 PM to 5.00 PM.

The land cover change of 2008-2010 causes land cover distribution of 2008 and 2018. Land Surface Temperature (LST) of 2008-2018 obtained by land surface temperature extraction from Landsat 8 OLI/TIRS (2018) and Landsat 7 ETM+ (2008). Topography analysis was carried out using landform slope data and ALOS PALSAR DEM (Figure 2).

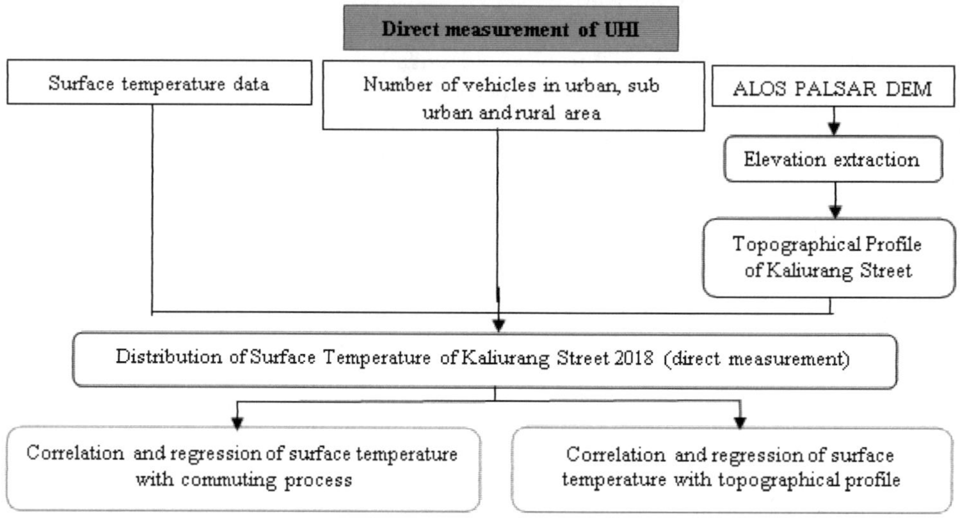

Figure 2. Research methodology

3. Result

3.1 Land use changes along Kaliurang Street and topography function

Urban areas are geographic manifestations which will always experience changes due to population (demographic) and demographic factors [4]. Demographic factors are the most important factor in urban studies because they are related to population growth. Urban areas, over time, are no longer able to provide built-up land for residents which quantity continues to increase, so there is a shift in urban activities in the area around urban areas called function shifting. Shifting the functions of urban areas to the surrounding region is called diffusion of urban areas.

The concept of the urban area of Yogyakarta is known through the Yogyakarta Urban Development Project concept which began in 1990. The concept explains that the area covered by the Yogyakarta Urban Development Project has the potential to carry out urban functions because it includes access to education, housing and services. This concept also explains the development of the urban area of Yogyakarta that leads to the north area of Yogyakarta that is half of some parts of Sleman Regency area and to the south area of Yogyakarta, which is half parts of Bantul Regency area. Diffusion of urban function happened to these areas because in some parts of Sleman Regency and part of Bantul Regency are already have proper access to shelter, education, trades, and services [5].

Kaliurang Street is one of the main roads in Sleman Regency which stretches along Sleman Regency with a length of about 23 km from Mlati Sub District to the Pakem Sub District. The area around Kaliurang Street is one of the areas in the Special Province of Yogyakarta that has experienced diffusion of urban areas. The tendency of diffusion of urban areas in the area around KaliurangStreet is due to accessibility as the main factor [6]. Accessibility conditions for Kaliurang Street services considered accessible until sub urban area in the past 10 years because it is supported by the location of recognized universities located in Kaliurang Street.

The diffusion of urban areas on Kaliurang Street for the last 10 years shows the tendency of spreading into north area of Yogyakarta. Some village areas classified as sub-urban areas, including the area of Sinduharjo Village, Sardonoharjo Village, Sukoharjo Village, and Umbulmartani Village. Sub-urban area happened as the most intensive land conversion area over the past 10 years, as shown in Figures 3a and 3b. The tendency of land conversion in the sub-urban areas is due to several causes, the first condition is due to the increasing number of local residents that the demand for houses with good access also increases. The

IOP Conf. Series: Earth and Environmental Science **303** (2019) 012032 doi:10.1088/1755-1315/303/1/012032

Kaliurang Street area in the sub-urban area is still considered as an area with good and very proper accessibility. The next condition is sub urban areas, more precisely at Umbulmartani Village. There is a university located in Umbulmartani Village, therefore its surrounding considered as educational area and followed by large number of non-permanent migrants, which is university students from other areas in Indonesia. Nevertheless, the presence of university campuses then soonfollowed by business service as supporting facilities to the students who reside around the campus. This condition resulted in the increasing of built up area around the campus.

Land use over the past decade has experienced a significant change in land use, especially from paddy fields to residential land. These changes can be seen in the land use map in 2008 (Figure 3a), where the paddy fields are still on topography that has a flat slope, sloping slope, and rather-steep-slope. After 10 years, the changes in paddy fields continue to be intensive until the settlements dominating sloping topography area, rather steep topography area, and steep topography area. Besides paddy fields, the land cover of forest also decreases its functions into settlements, as shown in Figure 3a and 3b.

The diffusion of urban areas on Kaliurang Road for the last 10 years has occurred to reach rural areas, with intensive land use changes in urban areas and sub-urban. Urban diffusion on Kaliurang Street is associated with topographic paths. Land conversion in Kaliurang Street happened to be intensive in areas with sloping topography and decreases its intensity in areas with steeper topography. Spatial distribution of land use with topography is shown in Figure 3a and 3b.

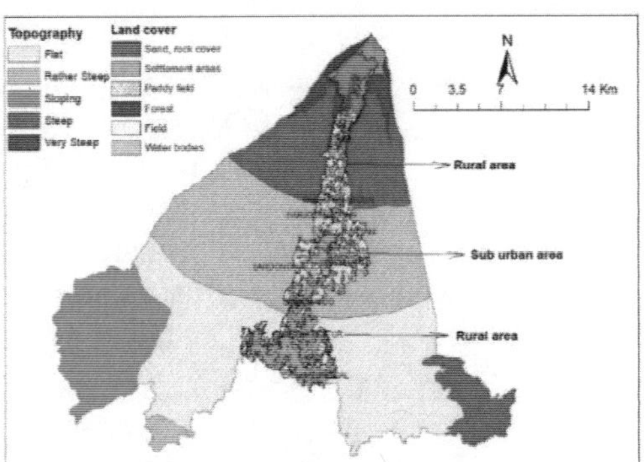

Figure 3a. Distribution of Land use in study area (2008) based on slope aspect in Sleman Regency

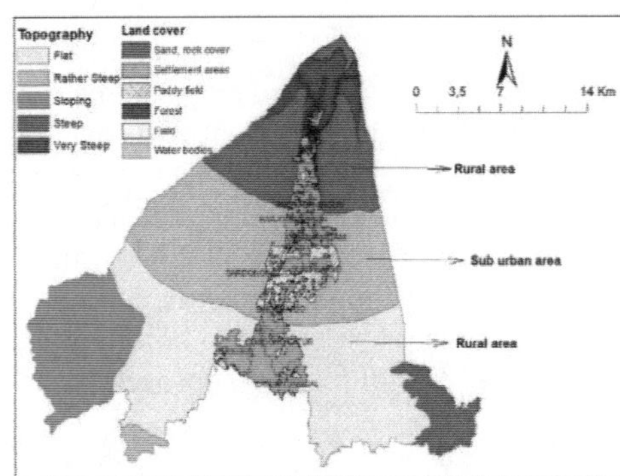

Figure 3b. Distribution of Land use in study area (2018) based on slope aspect in Sleman Regency

3.2. The pattern of temperatures along Kaliurang Street

The characteristic of UHI in Kaliurang Street could be seen periodically and directly. Periodically study of UHI associated with the influence of change of land uses across the years, which is the increase of building area land cover. Periodic study of UHI dealing with the increasing of temperature due to the change of land cover, while in fact the study of UHI dealing with the surface temperature with the environmental conditions. One of environmental condition that has prominent differences at Kaliurang Street is topography, which also describing the pattern of urban diffusion in Kaliurang Street.

Study of periodic UHI of Kaliurang Street obtained from the Land Surface Temperature (LST). The increase of LST showed that the influence of urban diffusion continues to support the process of the increasing of built area that followed by the reduction of vegetation land cover. Reduction of vegetation land cover reduces the effect on reducing the shadow canopy on the ground surface and reducing the supply of moisture in the air [7], resulting the increase of LST over the years.

Calculation of LST 2008 is obtained from the extraction of Landsat 7 ETM that recorded on July 8, 2008. While the calculation of LST 2018 using the extraction of Landsat 8 OLI / TIRS data that recorded on July 20, 2018. Landsat 7 ETM imagery suffered sensor damage in 2003, so the recording results experienced stripping which eliminated imagery data by 24% [8]. Stripping removal on Landsat 7 ETM + imagery is done by filling the gap stripping using image data with the same recording lane but on different recording dates and patterns that recorded in the same year (2008).

Table 1. Average of LST value in sub urban and urban area.

	Year		Status
	2008	2018	
Village	LST average (°C)	LST average (°C)	
Sinduadi	27.44	27.5	Urban
Caturtunggal	28.48	29.11	
Condongcatur	28.08	28.63	
Sinduharjo	26.17	25.75	Sub urban
Sardonoharjo	25.38	24.84	
Sukoharjo	26.22	25.15	
Umbulmartani	25.77	24.30	

The average LST in urban areas has increased in the last 10 years, but there has been a decline in the average temperature in the sub-urban areas (Table 1). LST in sub-urban areas experienced an average decline, but if spatially reviewed, there was an increasing distribution of LST in the past 10 years (Figure 4a and Figure 4b) in sub-urban area. The results of LST extraction in 2008 showed that the surface temperature range is 12.4°C to 34.9°C, while in 2018 it was shown that the surface temperature range is 14.9°C to 32.1°C. A higher temperature range in 2008 can be caused by Landsat 7 ETM + that experiencing the filling gap process beforehand. Filling gap process might use imagery with different weather condition due to its better resolution, so that process resulting higher temperature than it should be.

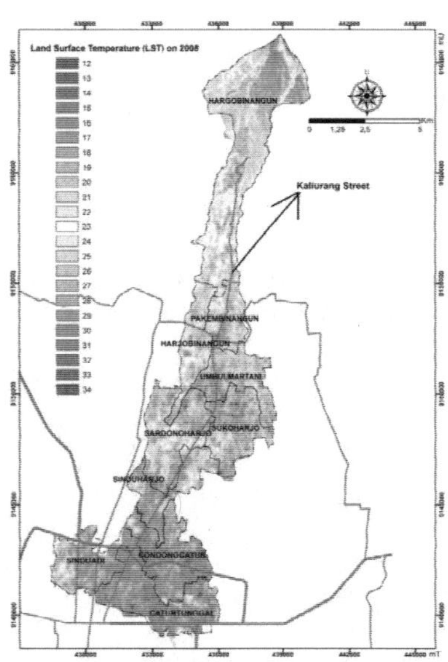

Figure 4a. Spatial distribution of LST (2008).

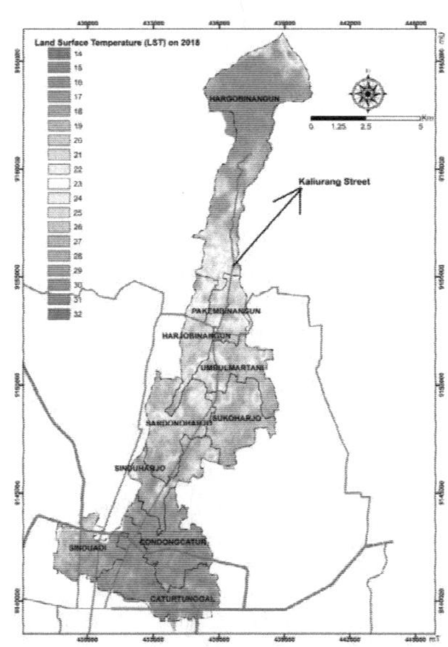

Figure 4b. Spatial distribution of LST (2018).

IOP Conf. Series: Earth and Environmental Science **303** (2019) 012032 doi:10.1088/1755-1315/303/1/012032

Average LST in sub-urban areas which decreased over the last 10 years can be caused by the cloud cover in 2018. Cloud cover is disguised as noise in LST extraction. Extraction of LST affected by cloud cover is resulting in lower LST value than it supposed to be due to the presence of water vapor in the cloud. The presence of cloud disturbances in LST extraction results is shown in areas with low temperatures (14- 17°C) in suburban areas (shown with blue colour) (Figure 4b). The sub urban area is considered as settlement complex area which is impossible to experiencing low LST distribution.

Urban diffusion on Kaliurang Street has reached to the Sub-urban area which includes Sinduharjo Village, Umbulmartani Village, Sardonoharjo Village, and Sukoharjo Village. The 10-year increase of LST that is most spatially visible in sub-urban areas occurs in Sinduharjo Village. The increasing conditions of LST in other suburban areas, such as in Sukoharjo Village, Sardonoharjo Village, and Umbulmartani Village are biased due to cloud cover in the 2018 image.

One of environmental factors that influence the spatial distribution of temperature is topography, because it controls the amount of sun radiation received on the surface [9]. Recent surface temperature and topography condition of Kaliurang Street brought by direct study of UHI. Topographic pattern of Kaliurang Street was extracted from the elevation. Elevation data of Kaliurang Street is obtained from ALOS PALSAR DEM. In order to divide elevation profile of Kaliurang Street into slope segments, the landform data of Sleman Regency obtained from Spatial Planning Agency (*Bappeda*) of Yogyakarta Province was also used. The landform data of Sleman Regency divided the topographical of Kaliurang Street based on volcanic typological landform. The topographical profile of Kaliurang Street is shown in Figure 5. The profile divided into 4 segments based on volcanic landform associated with the break of slope.

Distribution of surface temperature obtained from direct measurements along Kaliurang Street at daytime and evening. Daytime measurement represents maximum temperature due to maximum sun radiation throughout the day, while evening measurement represents the decreasing temperature after maximum condition. The temperature profile shows the mean of surface temperature from 4 segments, while every segment has 3 sub segment points. The sub segment points take role as sampling point that represents surface temperature in different topography along Kaliurang Street. The result of temperature along Kaliurang Street is shown in Figure 6.

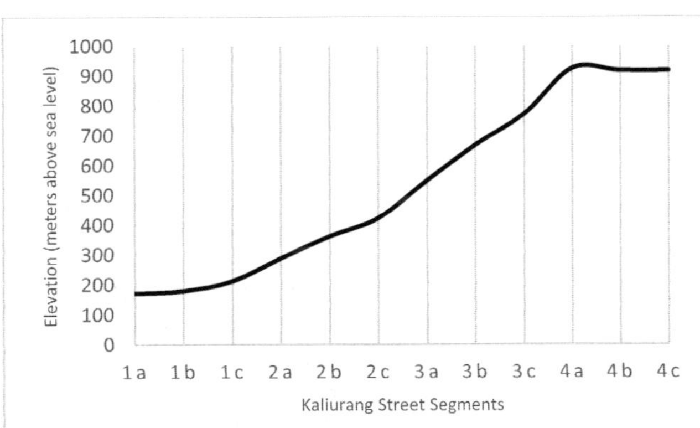

Figure 5. Topographical profile of Kaliurang Street based on elevation and landform.

Based on Figure 6, the highest temperature along Kaliurang Street on daytime and evening located in segment 1. Highest temperature on daytime and evening that occurred at segment 1 mainly caused by the amount of sun radiation in building area with minimum vegetation land cover that also supported by the commuting pattern. The commuting pattern in segment 1 happened as the most crowded traffic out of all these segments. This condition also shows that the most intensive urban diffusion at Kaliurang Street still occurred at urban area, following at the sub urban area. The correlation score (r) of elevation and surface temperature in the daytime is 0.446 and in the evening is 0.4461 which means that the relation between elevation and surface temperature is not strongly related. The UHI pattern of 2018 showed by image and direct measurements basically show the same results. The results concluded that urban area still remains as the centre of the heat island, followed by sub urban area as the second area that dealing with the increasing of temperature.

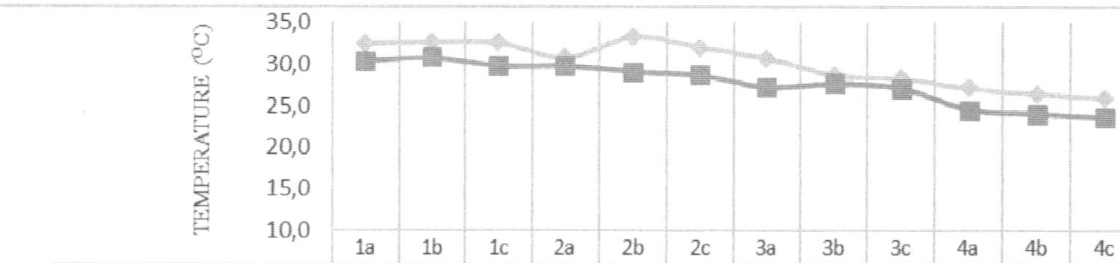

	1a	1b	1c	2a	2b	2c	3a	3b	3c	4a	4b	4c
Daytime Temperature	32,5	32,7	32,6	30,8	33,3	32,1	30,7	28,8	28,4	27,2	26,5	26,1
Evening Temperature	30,4	30,8	29,8	29,8	29,1	28,7	27,3	27,7	27,2	24,6	24,1	23,7

	Segment 1	**Segment 2**	**Segment 3**	**Segment 4**
Elevation (meter)	171-251	252-485	486-817	818-916
Area status	Urban	Suburban	Rural	Rural
Landuse	-Building area	– Building area – Open green area	– Open green area – Building area	– Open green area – forest – Building area
Commuting	Most crowded	Crowded	Crowded-not too crowdeed	Deserted

Figure 6. Distribution of surface temperature of Kaliurang Street.

3.3 Urbanization and correlation between the number of vehicles and atmosphere temperature
Urban Heat Island or UHI is closed isotherms indicating an area of the surface that is relatively warm; most commonly associated areas of human disturbance such as towns and cities. The physiographic analogy derives from the similarity between the pattern of isotherms and height contours of an area on a topographic map. The annual temperature of a city may be 1- 2 ^{o}C warmer than before the development occurs [10]. Figure 4a and 4b can illustrate that some villages like Sinduadi, Condongcatur, and Caturtunggal experiencing an increase in temperature which reached more than 2°C. For example some area in Caturtunggal and Condongcatur which have range of temperatures from 25-26 ^{o}C in 2008, have risen to 29-31 ^{o}C in 2018. Thus, it can be seen that the UHI phenomenon along Kaliurang Street occurs at Sinduadi, Caturtunggal, and Condongcatur.

The large temperature difference between a village and a city that has a high population density and activity indicates the phenomenon of Urban Heat Island. The measurement of UHI in an area cannot only be measured based on the characteristics of the city[11]. Emissions, pollution and greenhouse gases in urban areas encourage the UHI phenomenon [12]. Commuting activities in the study area which included urban, suburban, and rural area were shown in the average number of passing vehicles.

The increased activity of people that use transportation caused the increase of emissions from fossil fuels in the form of carbon emissions [12]. Carbon emissions encourage the UHI phenomenon including the area around Kaliurang Street, which is characterized by the increase of its atmospheric temperature. The results of the measurements show that the highest temperature is in urban areas with maximum temperatures

IOP Conf. Series: Earth and Environmental Science **303** (2019) 012032 doi:10.1088/1755-1315/303/1/012032

recorded reaching more than 33 ^{O}C. The highest temperature in the urban area has a fairly large difference with the highest temperature in the rural area, which is 31.9 ^{O}C.

The correlation between the numbers of passing vehicles at measured temperatures can be found by calculating the correlation value. Positive correlation values indicate that there is a correlation between increased temperatures and an increase in the number of passing vehicles. The correlation value between the temperature and the number of passing vehicles is shown in Table 2 and Table 3.

Table 2. The value of correlation between number of vehicles and atmosphere temperature on the first week of measurement.

	Time of Measurement							
	Monday 16/07/2018		Tuesday 17/07/2018		Saturday 21/07/2018		Sunday 22/07/2018	
	12.00-14.00	15.00-17.00	12.00-14.00	15.00-17.00	12.00-14.00	15.00-17.00	12.00-14.00	15.00-17.00
Urban	0,129	0,775	0,386	Negative value	0,531	0,363	0,431	0,404
Suburban	Negative value	0,847	Negative value	0,607	0,521	Negative value	0,210	0,784
Rural	0,884	0,865	0,789	0,794	0,867	0,683	0,749	0,840

Table 3. The value of correlation between number of vehicles and atmosphere temperature on the second week of measurement.

	Time of Measurement							
	Monday 23/07/2018		Tuesday 24/07/2018		Saturday 28/07/2018		Sunday 29/07/2018	
	12.00-14.00	15.00-17.00	12.00-14.00	15.00-17.00	12.00-14.00	15.00-17.00	12.00-14.00	15.00-17.00
Urban	Negative value	Negative value	0,377	Negative value	Negative value	0,186	0,132	Negative value
Suburban	Negative value	0,745	0,532	0,474	Negative value	0,876	0,664	0,818
Rural	0,384	0,892	0,903	0,849	0,689	0,851	0,728	0,704

The results of the correlative value measurement between the increase in air temperature and the number of passing vehicles in urban, suburban, and rural areas showed that the positive correlations value were dominated. This shows that there is a correlation between the increase in atmospheric temperature and the number of passing vehicles. The positive correlation is in accordance with the carbon emission. The more the number of vehicles passing by will lead more carbon emission emitted to the air.

The weakness of the device causes a bias at the measured temperature. The temperature measurement tool that was used has a resolution of 0.2 with an accuracy of ± 2^{O}C. This level of accuracy affects the accuracy of the temperature measurement results. In addition, temperature gauges have a weakness that is easily heat when it was exposed to the sunlight for too long. Tools that heating due to sunlight can affect the measured temperature result. The results of temperature measurements that are less precise affect the correlation value.

4. Discussion

4.1. Land cover change, topography, and surface Urban Heat Island

Land-surface temperatures are known to be sensitive to changes in latitude, topography, vegetation, land-use, and moisture. The preceding research about the effects of topography on temperature conditions was conducted on Zagreb Greater Area. The UHI development modelling on Zagreb Greater Area was

IOP Conf. Series: Earth and Environmental Science **303** (2019) 012032 doi:10.1088/1755-1315/303/1/012032

implemented on 9-10 June and 17-23 July 2004 with the temperature represented the UHI magnitude was expressed at 150 m height. The study shows that the temperature pattern in the afternoon found in the south-facing slopes of the Medvednica Mountain. The temporal evolution of the UHI halted in the afternoon. The condition caused the spatial distribution of temperature over urban area was homogeneous during the night. The presence of katabatic movements from the southern slopes of Medvenica and from the Pannonian valley caused convergence [13]. The convergence turned the wind flow to the west and removing the warm air masses over urban area and preventing further evolution of the UHI. The adiabatic lapse rate of air is to estimate topography-dependent empirical corrections for known ground-surface temperatures. However, air temperature is only one of the variables influencing ground temperature [14].

This research shows that the topography, specifically slope results in temperature variation. Kaliurang Street has less topography variation than Zagreb Greater Area. Topographic variation of Kaliurang Street is influenced by the landforms of Mount Merapi. The landforms of Mount Merapi that influenced the elevation of Kaliurang Street is Volcanic cone (818 m - 916 m), volcanic slope (486 m - 817 m), volcanic foot (252 m – 485 m), and volcanic foot plain (171 m - 252 m), while the variation of elevation in Zagreb starting from 86 m – 1100 m. The less variation of topography causing the relationship of topography and landform in Kaliurang Street is weaker than in Zagreb Greater Area.

4.2. *Surface Urban Heat Island and commuting patterns*

The correlation between commuting phenomena and air temperature directly shows the influence of urban activity with the UHI phenomenon. This condition based on the fact that the commuting phenomenon produces emissions that also produce heat energy. Vehicles that use fossil fuels are one source of carbon emissions in the world. These carbon emissions have an impact on increasing UHI in an area. The results of this research showed an increase in carbon emissions by 72.97% in Surabaya over the past 20 years[11]. Over the past 20 years, 137 x 106 tons of carbon matrix have been produced by 28 x 106 motorized vehicles. The increase in carbon emissions has positive correlation to the increase in the number of vehicles used in Surabaya. Furthermore, the carbon emission is one of the temperature increasing factors.

The research about emission in Surabaya is not dealing with the condition of temperature, but the research emphasizing that the increasing number of commuting will produce the increase of emission, which brought the increasing of surface temperature later. Urban diffusion is a factor that brought the increasing number of commuting in Kaliurang Street. Commuting process in Kaliurang Street influenced the surface temperature condition in daytime and evening. The correlation value (r) of commuting and temperature on the daytime reached 0.903 in rural area, while on the evening reached 0.876 in the sub urban area.

5. Conclusion

The first conclusion of this research including the change pattern of land use in Kaliurang Street in the past 10 years with its relationship with topography and its connection with the change pattern of spatial distribution of LST. The change pattern of land use due to urbanization in Kaliurang Street is reaching sub-urban area, while the sub-urban area is used to be not considered as settlement complex area in 2008 due its accessibility. Land use changing into building land cover in sub urban area happened as the most intense land use change in the past 10 years that followed by the decreasing of vegetation land cover and resulting the increasing of LST in all over sub urban area. The second conclusion of this research is dealing with commuting as the primary activity in urban area. Positive correlation between commuting and UHI in urban area, sub urban area, and rural area are found in the research. Highest correlation between commuting in afternoon and UHI found in rural area, with $r = 0.903$ while the highest correlation between commuting in the evening and UHI found in sub urban area, with $r = 0.876$.

Acknowledgement
This research funded by Faculty of Geography, Gadjah Mada University as part of the annual grant for research in the faculty of 2018.

References

[1] Oke T R 1982 The energetic basis of the urban heat island *Q. J. R. Meteorol. Soc.* **108** pp. 1–24
[2] Faizah A N and Hendarto M 2013 Analisis Difusi Keruangan di Sekitar Kawasan Perkotaan Yogyakarta **2** pp 1–8
[3] Peng S, Piao S, Ciais P, Friedlingstein P and Ottle C 2012 Surface Urban Heat Island Across 419 Global Big Cities
[4] Yunus H S 1987 *Permasalahan Daerah Urban Fringen dan Alternatif Pemecahannya* Yogyakarta: Fakultas Geografi UGM
[5] Sontosudarmo A 1987 *Studi Tentang Tata Kota di Yogyakarta* Gadjah Mada University Press
[6] Pradoto W 2011 Dynamics of peri-urbanization and socioeconomic transformation : case of metropolitian Yogyakarta, Indonesia *Int. J. Arts 7 Sci.* **4** 19
[7] Dugord P A, Lauf S, Schuster C and Kleinschmit B 2014 Land use patterns, temperature distribution, and potential heat stress risk - the case study Berlin, Germany *Comput. Environ. Urban Syst.* **48** 86–98
[8] Forestian O, Prasetyo L B and Kusmana C 2011 Biomass and density estimation of mangrove vegetation using landsat ETM+: case study on Muara Gembong Protection Forest, Bekasi, West Java *Bonorowo Wetl.* **1** 80–95
[9] Florinsky I V, Kulagina T B and Meshalkina J L 1994 Influence of topography on landscape radiation temperature distribution *International Journal of Remote Sensing* **15** 1–7
[10] American Meteorology Society 2000 American Meteorology Society Glossary of Meteorology, 2nd edition
[11] Kurniati A C, Nitivattananon V and Sulistyarso H 2015 Faktor-faktor yang mempengaruhi urban heat island di Surabaya , Indonesia, *Semin. Nas. Teknol. ITS* 1036–45
[12] Kurniati A C and Nitivattananon V 2016 Factors influencing urban heat island in Surabaya, Indonesia *Sustain. Cities Soc.* **27** 99–105
[13] Nitis T, Moussiopoulos N and Klaic Z B 2005 Effects of topography on the urban heat island *10th International Conference on Harmonisation within Atmospheric Dispersion Modelling for Regulatory Purposes* pp 473–77
[14] Haynes M W, Horowitz F G, Sambridge M, Gerner E J and Beardsmore G R 2018 Australian mean land-surface temperature *Geothermics* **72** 156-62

IOP Conf. Series: Earth and Environmental Science **303** (2019) 012033 doi:10.1088/1755-1315/303/1/012033

Rainfall Scenario of West Nusa Tenggara in 2040 Based on CCAM RCP 4.5

A Nurlatifah, S B Sipayung and B Siswanto

Center of Atmospheric Science and Technology, National Institute of Aeronautics and Space

e-mail: amalianurlatifah92@gmail.com

Abstract. Climate change is one of the global issues with a broad impact to society. Because of its importance, there is a need to build a prediction system that can support mitigation, adaptation or prevention of climate change. West Nusa Tenggara (NTB) is an area where most of the population relies on agriculture for their livelihood so that the availability of rainwater becomes very important for daily survival. In this research, rainfall in NTB was predicted by CCAM (*Conformal Cubic Atmospheric Model*) model based on RCP (*Recent Concentration Pathways*) 4.5 scenario. Based on the verification results, the CCAM model was proven to perform fairly well in estimating rainfall in NTB in 2014-2016. This is evidenced by the high correlation coefficient between CCAM and observed rainfall data in several locations such as Kuripan, Kumbe, and Gunung Sari which reached 0.57, 0.51, and 0.51 respectively. In January 2040 places such as Bima, Sumbawa Besar and parts of Lombok Island are predicted to have a decrease in rainfall when compared to that in January 2018. The decrease ranges from 40-120 mm in one month. While in Kumbe and Kuripan, rainfall is predicted to increase, with an increase of about 0-40 mm in one month. In August 2040 rainfall is predicted to generally rise in NTB with an increase of up to 40 mm in one month.

Keywords : rainfall, climate change, CCAM, RCP 4.5.

1. Introduction

Today climate change is becoming one of the most discussed global issues. This is because climate change can affect many sectors of the global community, for example rising temperatures and changing climatic patterns that lead to changes in water availability, food security, and comfort of residence [1][2][3][4]. Climate change is a global phenomenon triggered by human activities primarily related to the use of fossil fuels and land use change [5] and can cause long-term effects, especially the rise in global temperatures that lead to increased extreme weather phenomenon and hydro meteorological disaster [6][7]. Due to the importance of these issues and widespread impacts, systems are required that can support mitigation, adaptation, or prevention of climate change, one of them is predictions system using scenarios and climate models. Because of this, researchers today have developed many climate change scenarios, one of them is Representative Concentration Pathways (RCPs).

Representative Concentration Pathways (RCPs) is a measure of the fourth greenhouse gasses emissions trajectory used by the IPCC in its 5th Assessment Report in 2014 [8]. RCP itself is a measure of radiative forcing level for 2100 [9]. Radiative forcing is estimated by forcing from greenhouse gases and some other forcing agents against the air that causes global warming [9]. In this case RCP 4.5 is moderate in terms of magnitude of radiative forcing that is above RCP 2.6 and below RCP 6 and RCP 8.5. The number 4.5 here shows the radiative forcing stabilization scale of 4.5 Watt / meter^2 (equivalent to 650 ppm CO2) in 2100 [10][11][12]. The RCP scenario itself is widely aggregated and used in climate modelling using GCM to project–future climatic conditions due to increase in concentrations of greenhouse gases as has been planned in RCP [13]. NTB is one of the areas in Indonesia where climate is much influenced by monsoon and trade wind, so that in this region the season changes every 6 months [14]. NTB is one of the poorest provinces in Indonesia where people rely heavily on agriculture and irrigation for their livelihoods [15]. Therefore, the availability of water becomes one of the most important factors in this region. This study aims to determine the impact of climate change on rainfall in NTB since rainfall is one of the most a crucial source of water for the livelihood of the people of NTB.

2. Data and Methods

The data used in this study include

2.1. Data

2.1.1. Observation data of station rainfall

The rainfall observation data used in this study includes 14 rainfall cultivator stations belonging to PUSAIR (Water Resources Research and Development Center) scattered throughout the Nusa Tenggara Barat region in 2014-2016. The fourteen stations include station rainfall data used to identify rainfall patterns in Nusa Tenggara Barat and to validate CCAM model results.

2.1.2. GFS Data

GFS (Global Forecasting System) is the data used as an input to the model. These data are produced by NCEP (National Center for Environmental Prediction) and contain many atmospheric parameters such as temperature, wind, rainfall, soil moisture and atmospheric ozone concentrations [16].

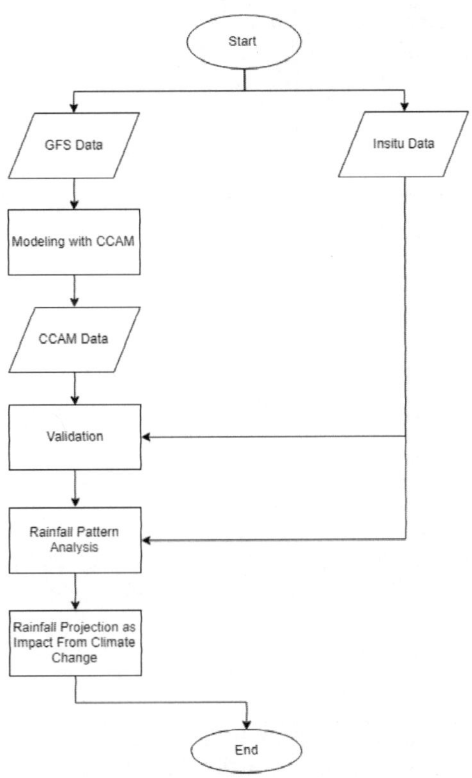

Figure 1. Research Flowchart

2.2. Methods

In this research, climate simulation was conducted using the CCAM model. GFS data was used as an input to the CCAM model, and CCAM output was verified by observation data. After ensuring that the CCAM output data is sufficient, then the rainfall condition was projected in NTB based on the CCAM model with scenario RCP 4.5. In general, this research follows a procedure as shown in figure 1. For this research, the spatial resolution that used from CCAM is monthly resolution and spatial resolution is 14 km (0.125°).

2.2.1. Modelling with CCAM

CCAM (Conformal Cubic Atmospheric Model) is a global atmospheric model developed for climate modelling and regional climate downscaling [17][18]. This model implements a quasi-uniform grid, or principle of projecting the earth's surface into a cube panel [19]. Therefore, this model does not require lateral boundaries in the process of its running. In CCAM, the grid stretch is set by a Schmidt transform (figure 2). This model uses a lagrangian advection system approach with horizontal bicubic interpolation [20][21].

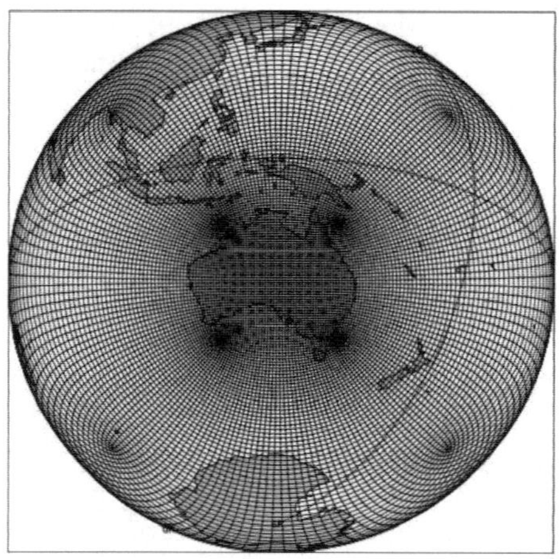

Figure 2. Examples of C48 Global Conformal Cubic Grids that have been stretched according to Schmidt transformation (S=3.3), producing approximate resolution 200 km (average worldwide) and 60 km in around Australia [19].

2.2.2. Validation by Using Correlation and RMSE

Pearson correlation is a measure of correlation or linear relationship between two variables. The value ranges from -1 to +1. The correlation value is calculated by the following formula:

$$r = \frac{n(\sum xy)-(\sum x)(\sum y)}{\sqrt{[n\sum x^2-(\sum x)^2][n\sum y^2-(\sum y^2)]}} \quad (1)$$

Where x is the observed data and y is the model data. Here is the meaning of Pearson correlation value (table 1):

Table 1. The meaning of Pearson correlation

R value	Meaning
r > 0.5	Both variables are strongly correlated positively
0 < r < 0.5	Both variables are weakly correlated positively
r = 0	Both variables are not correlated
-0.5 < r <0	Both variables are weakly correlated negatively
r < -1	Both variables are strongly correlated negatively

RMSE is the average value of the sum of squares error, it can also state the size of the error generated by a forecast model. A low RMSE value indicates that the variation of the value generated by a forecast model approaches the variation in its observed value [22]. The RMSE value is formulated as follows [23]:

$$RMSE = [\frac{\sum_{i=1}^{N}(f_i - O_i)^2}{N}] \tag{2}$$

Where f_i is the forecast generated by the model and O is the observed value.

3. Results and Discussion

Based on monthly rainfall observation data at fourteen rainfall stations owned by PUSAIR, it can be seen that the pattern of rainfall in West Nusa Tenggara generally shows a monsoonal pattern as expressed by Kirono et al. [14]. The peak of the rainy season generally occurs in December and the lowest rainfall occurs in August and September (figure 3). In December to March rainfall is relatively high, whereas in May rainfall begins to decrease.

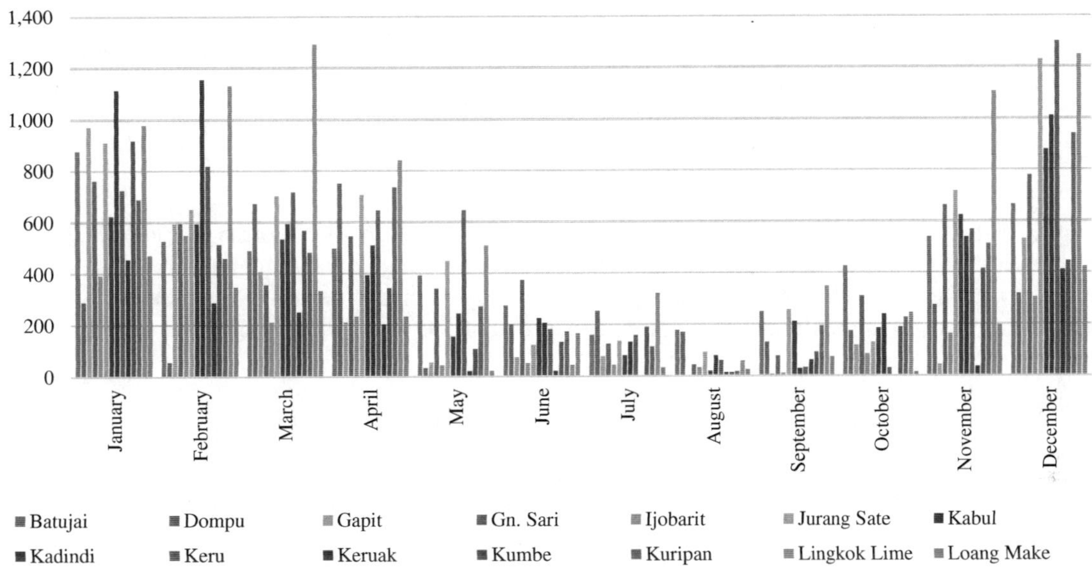

Figure 3. Rainfall Composite in West Nusa Tenggara 2014-2016 Based on Observation Data

Rainfall pattern according to CCAM model also shows similar result. It can be seen that the rainfall pattern in NTB according to CCAM model also follows the monsoonal pattern where the peak rainfall occurs in December and the lowest rainfall occurs in August and September (figure 4). To be able to quantify the difference between the model and the observation, then the correlation and RMSE values of the model and observation was calculated.

Based on the description in figure 3 that shows the rainfall pattern in NTB based on observation data and figure 4 that shows the rainfall pattern in NTB based on CCAM modelling results, it can be seen that there is similar pattern of rainfall between CCAM model and that of observation data. Based on this result, it can be concluded that CCAM model can simulate monsoonal rainfall pattern in West Nusa Tenggara. To reinforce this statement and to demonstrate the performance of the CCAM model in estimating rainfall in West Nusa Tenggara, the correlation and RMSE values between model and observation were calculated. The calculation of the correlation value is intended to calculate the similarity of the patterns between the precipitation described by the CCAM model and the observed rainfall. While RMSE is calculated to describe difference between CCAM model and observation.

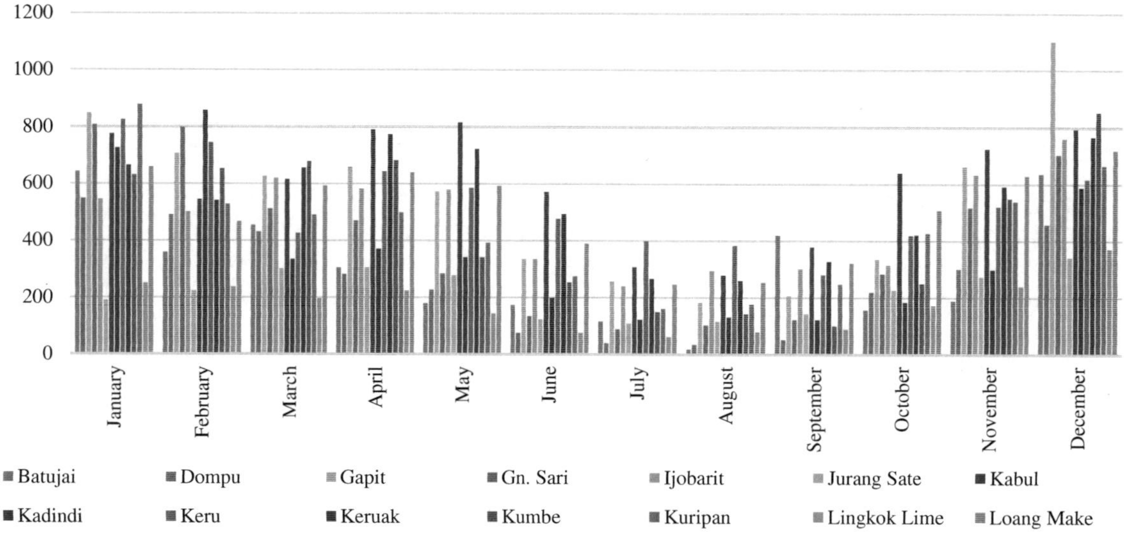

Figure 4. Rainfall Composite in NTB 2014-2016 Based on CCAM Output

Figure 5 shows the correlation result with the highest correlation between observation and CCAM in Kuripan with a correlation value of 0.57. This value is relatively high even though Gulliford categorizes it into the medium to high correlation category. In this case, the value of 0.57 indicates the CCAM climate model can simulate the rainfall pattern in Kuripan quite well. The smallest correlation is found in Dompu with a correlation value of 0.065.

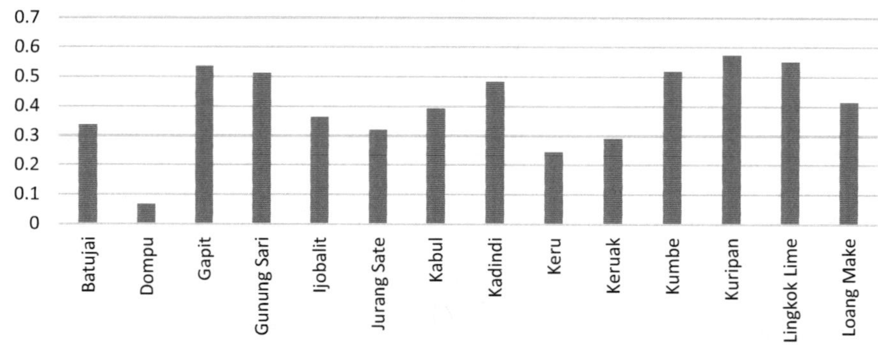

Figure 5. Rainfall Correlation Between Insitu Data and CCAM

For RMSE, the average value is 24.5. When compared with the average rainfall in NTB that ranges up to 200 mm in one month, then this value is relatively small. This means the error bias or the difference between the value of the observation measurement and the CCAM data is not too large. The correlation value between CCAM and observation in Kuripan is also the highest correlation value. Therefore, based on correlation and RMSE values, it can be concluded that the CCAM simulation results in Kuripan is better than in other regions.

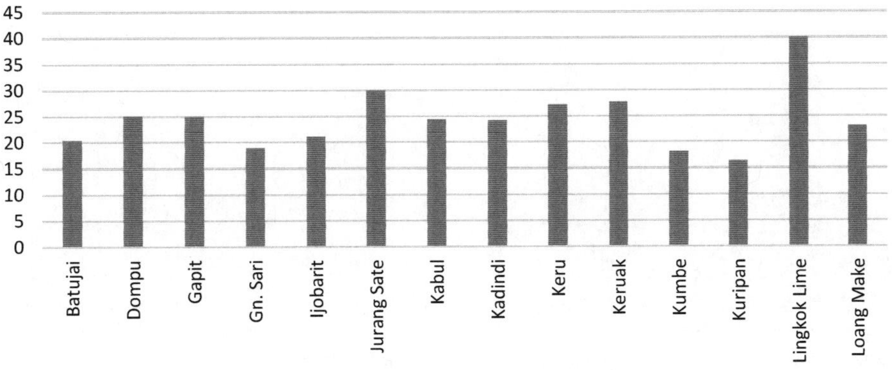

Figure 6. RMSE Between Insitu Data and CCAM

It can be seen from figure 7 that CCAM rainfall trend in Kuripan is quite similar to observation. A large difference only occurs from April 2015 to September 2015. At certain points, there is small difference.

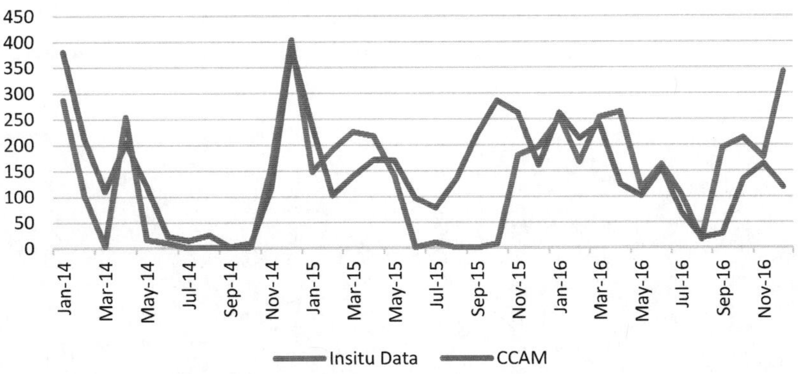

Figure 7. Rainfall trend between insitu data and CCAM in Kuripan

From the description above, it can be seen that CCAM RCP 4.5 performs well in describing the condition of rainfall in Nusa Tenggara Barat. Furthermore, after evaluating the performance of CCAM in simulating rainfall, the model was used to project future rainfall based on climate change scenario RCP 4.5-January 2018 and August 2018 were selected to describe wet and dry months in current conditions in West Nusa Tenggara based on the CCAM model. Projection is done until 2040 each in January 2040 and August 2040.

Figure 8a shows that in January 2018, rainfall in the Indian Ocean to the south of NTB is predicted to reach 400 mm. However, the rainfall in the mainland of Nusa Tenggara Barat on the island of Lombok is about 80-240 mm. While on the island of Sumbawa, rainfall is about 120-240 mm, and at the point around the District of Bima reaches 80 mm.

Figure 8b shows that in January 2040, the rainfall generally tends to decrease. Rainfall in Central and North Lombok is around 40-200 mm. On the island of Sumbawa rainfall decreased to about 40-120 mm in Bima, Sumbawa Besar, and part of Lombok Island. Indications of declining rainfall occurred in the Indian Ocean to the south of NTB where rainfall decreased to about 320 mm. Thus, from both figures it can be seen that there is an indication of rainfall decrease in some areas in Nusa Tenggara on January 2040 compared to that in January 2018.

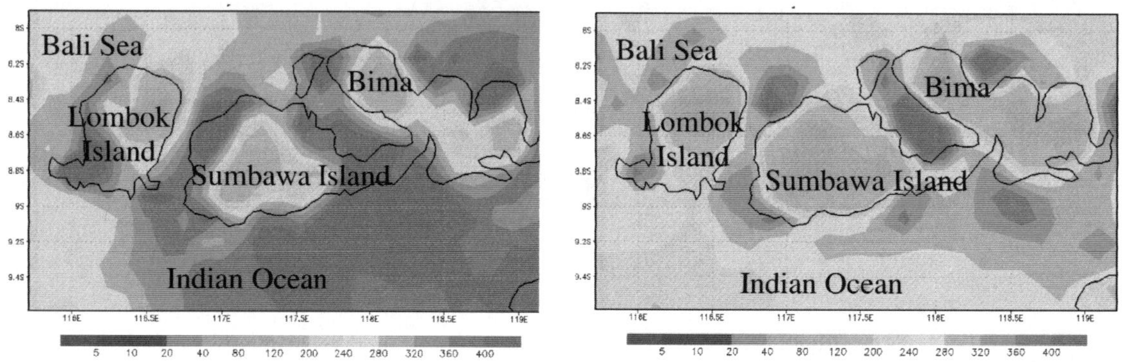

(a) (b)

Figure 8. Rainfall in NTB (on mm) based on CCAM output in January (a) 2018, (b) 2040

Figure 9a shows that generally the rainfall in most of West Nusa Tenggara in August 2018 is around 5-20 mm. Relatively high rainfall in this month only occurred in Bima that reached 40 mm. Figure 9b shows that rainfall in West Nusa Tenggara has increased in some areas such as South Lombok, Sumbawa, Bima and Bali Sea is about 40 mm. Thus, from both figures, it can be seen that there is an indication of the increase of rainfall in August 2040 compared with that in August 2018.

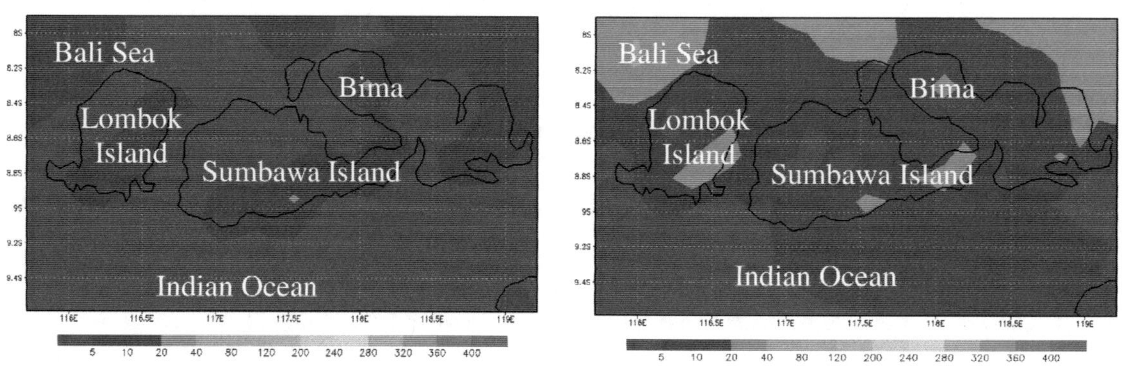

(a) (b)

Figure 9. Rainfall in NTB (on mm) based on CCAM output in August (a) 2018, (b) 2040

For further indication whether there is a change in rainfall in NTB, an analysis of trends in rainfall scenarios was projected by CCAM with RCP 4.5. This analysis is performed on CCAM projection at 3 points which has the highest rainfall correlation with observation data and the lowest RMSE. In other words, 3 points that has the best CCAM performance were selected. This trend analysis was conducted in Kuripan, Kumbe, and Gunung Sari.

In January, it is estimated that in some areas the rainfall will decrease. It can also be seen in the linear regression of rainfall trend line in January from 2018-2040 where the slope at Gunung Sari is negative (table 2). This means that the rainfall in Gunung Sari will decrease by -0.0001 mm per month from 2018 to 2040. In contrast to Gunung Sari, rainfall in Kumbe and Kuripan is expected to increase. It is characterized by slope of rainfall trend line regression in January in 2018 to 2040 which is positive in these locations (table 2).

Table 2. The value of slope and linear regression intercept of rainfall January 2018-2040

No	Location	Slope	Intercept
1	Gunung Sari	-0.0001	306.88
2	Kumbe	0.0071	-42.526
3	Kuripan	0.0004	257.44

Based on spatial analysis, it is estimated that rainfall in NTB will generally increase in August 2040 compared to August 2018. In trend line analysis, the slope trend line for August rainfall is positive in two points (table 3). This means that in all these points rainfall in August is expected to continue to rise until August 2040.

Table 3. The value of slope and linear regression intercept of rainfall January 2018-2040

No	Location	Slope	Intercept
1	Gunung Sari	0.003	-122.7
2	Kumbe	0.0039	-154.67

4. Conclusions

The pattern of rainfall in West Nusa Tenggara based on observation and CCAM data shows monsoonal pattern. This means that rainfall in NTB generally reaches a maximum in December and reaches a minimum in August and September. Both the observation data and CCAM show a similar pattern, the Kuripan rainfall correlation reached 0.64. This indicates that CCAM can well simulate rainfall in NTB especially in the Kuripan region. In August 2040 rainfall is projected to increase based on the output of RCP 4.5 scenario by CCAM when compared to that in August 2018. This is shown by the results of the CCAM spatial plot that shows rainfall will increase in the area of South Lombok, Sumbawa, Bima, and Bali Sea with an increase of about 40 mm in August 2040 compared with that in August 2018. This result is also supported by the plot trend line which shows its positive slope in Gunung Sari, Kumbe and Kuripan. While in January 2040, spatially it is estimated that there are some areas where rainfall will decrease compared to that in January 2018, for example in Kabupaten Bima, Sumbawa Besar, Gunung Sari, and part of Lombok Island. Climate change indicated can change the hydrological cycle in the world so that the rainfall in some areas can decrease and vice versa, in other area the impact of climate change is increasing the rainfall.

Acknowledgments

Author wish to thank to the Centre of Atmospheric Science and Technology who has provided material support in the form of facilities so that the writer can complete this research.

References

[1] Rosenzweig C, Iglesias A, Yang X B, Epstein P R, Chivian E 2001 *Climate Change and Extreme Weather Events; Implications for Food Production, Plant Diseases, and Pests. Global Change and Human Health*, December 2001, Volume **2**, Issue 2, pp 90–104.

[2] Piao S, Ciais P, Huang Y, Shen Z, Peng, S., Li, J., Zhou, L., Liu, H., Ma, Y., Ding, Y., Friedlingstein, P., Liu, C., Tan, K., Yu, Y., Zhang, T. & Fang, J. 2010. *The impacts of climate change on water resources and agriculture in China*. Nature volume **467**, pages 43–51 (02 September 2010). doi:10.1038/nature09364

[3] Turral H, Burke J, Faurès J M 2011 *Climate change, water and food security*. FAO water reports. ISBN : 9789251067956.

[4] Charles J, Vörösmarty, Green P, Salisbury J, Lammers R B 2000 *Global Water Resources: Vulnerability from Climate Change and Population Growth.* Science 16 March 2018 Vol **359**, Issue 6381.

[5] Harmoni A 2005 *DAMPAK SOSIAL EKONOMI PERUBAHAN IKLIM*. Proceeding, Seminar Nasional PESAT 2005. ISSN 18582559.

[6] Schipper L and Pelling M 2006 *Disaster risk, climate change and international development: scope for, and challenges to, integration*. Disasters Volume **30**, Issue 1, March 2006, Pages 19–38.

[7] O'Brien G, O'Keefe P, Rose J, Wisn B 2006 *Climate change and disaster management.* Disasters Volume **30**, Issue 1, March 2006, Pages 64–80.

[8] Moss R, Babiker M, Brinkman S, Calvo E, Carter T, Edmonds J, Elgizouli I, Emori S, Erda L, Hibbard KA 2008 *Towards new scenarios for analysis of emissions, climate change, impacts, and response strategies.* IPCC Expert Meeting Report on New Scenarios. Intergovernmental Panel on Climate Change,

[9] van Vuuren D P, Edmonds J, Kainuma M, Riahi K, Thomson A, Hibbard K, Hurtt G C, Kram T, Krey V, Lamarque J F, Masui T, Meinshausen M, Nakicenovic N, Smith S J, Rose S K 2011 *The representative concentration pathways: an overview*. Climatic Change November 2011, **109**:5.

[10] Clarke L, Edmonds J, Jacoby H, Pitcher H, Reilly J, Richels R 2007 *Scenarios of the greenhouse gas emission and atmospheric concentrations*. Sub-report 2.1A of Synthesis and Assessment Product 2.1 by the U.S. Climate Change Science Program and the Subcommittee on Global Change Research. Washington, USA, Department of Energy, Office of Biological & Environmental Research, p 154.

[11] Smith S J, Wigley T M L 2006 *Multi-gas Forcing stabilisation with the MiniCAM*. Energ J (Special Issue **3**): 373–391.

[12] Wise M A, Calvin K V, Thomson A M, Clarke L E, Bond-Lamberty B, Sands R D, Smith S J, Janetos AC, Edmonds JA 2009 *Implications of limiting CO_2 concentrations for land use and energy*. Science **324**:1183–1186

[13] Knutti R and Sedláček J 2013 *Robustness and uncertainties in the new CMIP5 climate model projections*. Nature Climate Change volume **3**, pages 369–373 (2013).

[14] Kirono D G C, Butler J R A, McGregor J L, Ripaldi A, Katzfeya J, Nguyen K 2016 *Historical and future seasonal rainfall variability in Nusa Tenggara Barat Province, Indonesia: Implications for the agriculture and water sectors*. Climate Risk Management Volume **12**, 2016, Pages 45-58.

[15] Butler J R A, Suadnya W, Puspadi K, Sutaryono Y, Wise R M, Skewes T D, Kirono D, Bohensky E L, Handayani T, Habibi P, Kisman M, Suharto I, Supartarningsih S, Ripaldi A, Fachry A, Yanuartati Y, Abbas G, Duggan K, Ash A 2014 *Framing the application of adaptation pathways for rural livelihoods and global change in Eastern Indonesian islands*. Global Environ. Change **28**, 368–382

[16] https://www.ncdc.noaa.gov/data-access/model-data/model-datasets/global-forcast-system-gfs

[17] McGregor J L, Dix M R 2008 *An Updated Description of the Conformal-Cubic Atmospheric Model*. In: Hamilton K., Ohfuchi W. (eds) High Resolution Numerical Modelling of the Atmosphere and Ocean. Springer, New York, NY

[18] McGregor J L, Dix M R 2005 *CCAM: Geometric aspects and dynamical formulation*. CSIRO Marine and Atmospheric Research Technical Report **70**.

[19] Thatcher M and McGregor J L 2008 *Using a Scale-Selective Filter for Dynamical Downscaling with the Conformal Cubic Atmospheric Model*. Ametsoc. https://doi.org/10.1175/2008MWR2599.1

[20] McGregor J L 1996 *Semi-Lagrangian advection on conformal-cubic grids*. Mon Weather Rev **124**: 1311-1322.

[21] McGregor J L 1993 *Economical determination of departure points for semi-Lagrangian models*. Mon Weather Rev., **121**, 221-230.

[22] Makridakis S 1982 *The Accuracy of Extrapolative (Time Series Methods): Results of a Forecasting Competition*, Journal of Forecasting, Vol. **1**, No. 2, pp. 111-153

[23] Barnston A G 1992 *Correspondence among the Correlation, RMSE, and Heidke Forecast Verification Measures; Refinement of the Heidke Score*. Ametsoc. Dec 1992. https://doi.org/10.1175/1520-0434(1992)007<0699:CATCRA>2.0.CO;2

The influence of environmental conditions (vegetation, temperature, equator, and elevation) on tropospheric nitrogen dioxide in urban areas in Indonesia

A.D Syafei[1], T. D Irawandani[1], R. Boedisantoso[1], A. F Assomadi[1], A. Slamet[1], and J. Hermana[1]

[1]Department of Environmental Engineering, Faculty of Civil, Environmental and Geo Engineering, Institut Teknologi Sepuluh Nopember, Surabaya, Indonesia

Email: dipareza@enviro.its.ac.id

Abstract. Nitrogen dioxide (NO_2) is one important species in the tropospheric layer that has a contribution towards air quality. In urban areas in Indonesia, urbanization and growing industrial activities have increased rapidly, which may have affected tropospheric NO_2. This study investigates the effect of green spaces and other environmental variables (e.g., temperature, city, distance from the equator) had on tropospheric NO_2 in over thirty (30) urban areas in Indonesia between 2007 to 2016. Data on land cover were obtained from Google Earth's satellite imagery data, tropospheric NO_2 readings came from Metop-A, and information on temperature was obtained from the Terra MODIS satellite. Our results show an annual decrease in green space over 10 years in all urban areas except Surabaya. Multiple linear regression analysis also reveals that a decline in vegetation decreases tropospheric NO_2 significantly. Other influential environmental factors are surface temperature and distance to the equator. Despite a promising model being obtained from this study, adding variables is important to improve the forecasting.

Keywords: Green space, Metop-A, Tropospheric NO_2, Urban area.

1. Introduction

NO_2 is one among many gas pollutants that are emitted largely through combustion processes, such as those in the transportation sector [1]. A large concentration of NO_2 will have a negative impact on health, namely decreased lung function and visibility, and it also inhibits plant growth [2].

The increase in NO_2 concentration has occurred mainly in urban areas. In Indonesia, almost all of the provincial capitals have increased air pollutant concentrations, which is in line with increased urbanization activities. Based on United Nation's data from 2015, urbanization in Indonesia is predicted to increase by 64% by 2050 and cause dramatic changes in land cover and use [3]. Many facts support vegetated land becoming tighter. This condition indicates that vegetated land as an absorber for various air pollutants is getting smaller, whereas, on the other hand, the number of buildings, industries, and vehicles as sources of emissions keeps increasing [4].

Vegetation has an important role in controlling temperature and air quality, including the amount of NO_2 in the atmosphere. Vegetation has an effect on the climate through two processes: providing canopy shade and evapotranspiration. Canopy shade in the form of leaves and twigs reduces the amount of solar radiation that reaches the area under trees. The amount of sunlight transmitted through the canopy varies by plant species. Evapotranspiration cools the air by using heat from the air to evaporate water [5]. The mechanism of NO_2 processes in plants is divided into three major processes: assimilation into amino acids, accumulation in vacuoles, and re-emission as NOx [6]. Specific types of vegetation have high NO_2 absorption capabilities since plant morphology affects NO_2 absorption. One influencing factor is stomata size. The larger the size of stomata, the higher the absorption rate [7].

Almost all urban area in Indonesia are losing green space every year. For example, between 2001 and 2014, Jakarta lost as many as 5.1% of its trees while growing by approximately 13% [8]. Further, vegetation in Yogyakarta decreased, on average, by 1% during 2003-2014 [9]. Finally, the decrease in vegetation in Semarang was approximately 0.26% and in Bandung 1.4% in the period 1994-2002 [10]. The dominant factors that dictate urban land cover changes in Indonesia are agricultural activities, industry, and land ripe for development [11].

Studies related to the control of NO_2 are not limited to a micro scale, but can also be done on a macro scale because the dispersion of the gas pollutant will be on a large and predominant scale in the atmospheric layer. Therefore, this study aims to investigate how the decreasing trend in green space over thirty (30) large cities in Indonesia has affected tropospheric NO_2 concentrations using satellite images.

2. Method

This quantitative research is based on remote sensing data. Satellites used are Google Earth Pro to identify land cover, GOME MetOP-A for tropospheric NO_2 concentrations, and Terra MODIS for temperature readings. Data used as variables in this research are proportion of vegetation area (%), tropospheric NO_2 concentration (molecules/cm^2), urban surface temperature (°C), elevation, and perpendicular distance of an urban center from the equatorial line (km). We observed thirty (30) capital cities of provinces, which may represent the urban conditions in Indonesia. All data were processed using ArcGIS 10.0, and statistical analysis was conducted using Minitab 16.

2.1. Data Generation for Vegetation Area

Google Earth imagery data is classified based on two classes: non-vegetation and vegetation. The time span of imagery data used is from 2007 to 2016. Magnification of the map was done up to one thousand meters, which is represented by one centimeter on the map. Identification of vegetation was done by digitizing (creating a polygon following the shape of the visual green area), then the area of each polygon was calculated. Once these were summed up, the percentage of green area was obtained.

2.2. Data Generation for Tropospheric NO_2

GOME-2 MetOP-A satellite datasets were used in this research, and the images were converted into an ArcGIS format file. Once a visual map appeared in the program for each city, the pixel colors in the legend were adjusted. The value of troposphere NO_2 was obtained by averaging values from the pixels that cover each city.

2.3. Data Generation of Urban Surface Temperature

The Terra-MODIS data used was the monthly MOD11C3 product. Based on the Terra MODIS data processing guide, the MOD11C3 product has a 0.02 factor scale used in surface temperature identification algorithms. A unit conversion process by map calculation uses the following formulation:

$$(°\ \text{Celsius}) = ([\text{LST\_Day}] * 0.02) - 273.15 \quad (1)$$

Urban surface temperature (°C) was then calculated by averaging all pixel values until all of the data in the range 2007-2016 was incorporated.

2.4. Data for Distance from the Equator and Elevation

Elevation categories are either low or high hill based on secondary data published by each municipal government agency. In this categorization, an area with an elevation of more than 200 meters above sea level is referred to as a high hill [8][12]. The distance from the city to the equator line was measured perpendicularly.

Statistical analysis was used to obtain the relationship between the above independent variables and tropospheric NO_2. Regression was done using Minitab 16. It was found that the correlations between variables are small, except for the correlation between temperature and vegetation. As expected, more vegetation reduces the surface temperature (Table 1).

Table 1. Correlations among variables (p-values inside brackets).

	Vegetation (%)	NO_2 (molec/cm^2)	Temperature (°C)	High Hill
NO_2 (molec/cm^2)	-0.036			
	(0.000)			
Temperature (°C)	-0.577	0438		
	(0.000)	(0.000)		
High Hill	0.066	-0.027	-0.116	
	(0.000)	(0.106)	(0.000)	
Distance from equator (km)	-0.494	0.261	0.404	-0.013
	(0.000)	(0.000)	(0.000)	(0.443)

3. Results and Discussion

In this study, the green cover classification does not include vegetation type or density. Over the ten year period (2007 - 2016), the change in each city is different. Overall, green land cover decreased in area. Figure 1 explains that almost all of the 30 provincial capitals in Indonesia experienced a decrease in the area of green land in each year of the study period. The average decrease in green land area in the 30 cities over the given time period was 14.56%. The cities with the highest percentages of green area decline were Mataram (35.7%), Banda Aceh (32.8%), Kupang (26.6%), Pangkal Pinang (25.2%), and Jambi (24.2%).

The decrease in green areas is due to the conversion of land into land for building. This is supported by the increasing number of residents and various activities making the developmental changes in these cities dynamic. The five cities with the lowest percentage of green areas during the last ten years were Jakarta (4%), Yogyakarta (7.7%), Jambi (16.7%), Bandung (21.3%), and Denpasar (34.9%). Cities with the highest percentages of green areas were the capital cities in eastern Indonesia and Kalimantan; the five cities with the highest green areas were Ambon (95.5%), Jayapura (94.5%), Palangkaraya (up to 97%), Samarinda (75.5%), and Palu (84.4%). It should be mentioned that Surabaya is an exception. It increased its percentage of green space between 2012 and 2016 by 4.7%. The overall trend in urban green spaces areas in Indonesia can be seen in Figure 1. For ease visualization of thirty cities, we create three graphs showing overall conditions of all cities.

IOP Conf. Series: Earth and Environmental Science **303** (2019) 012034 doi:10.1088/1755-1315/303/1/012034

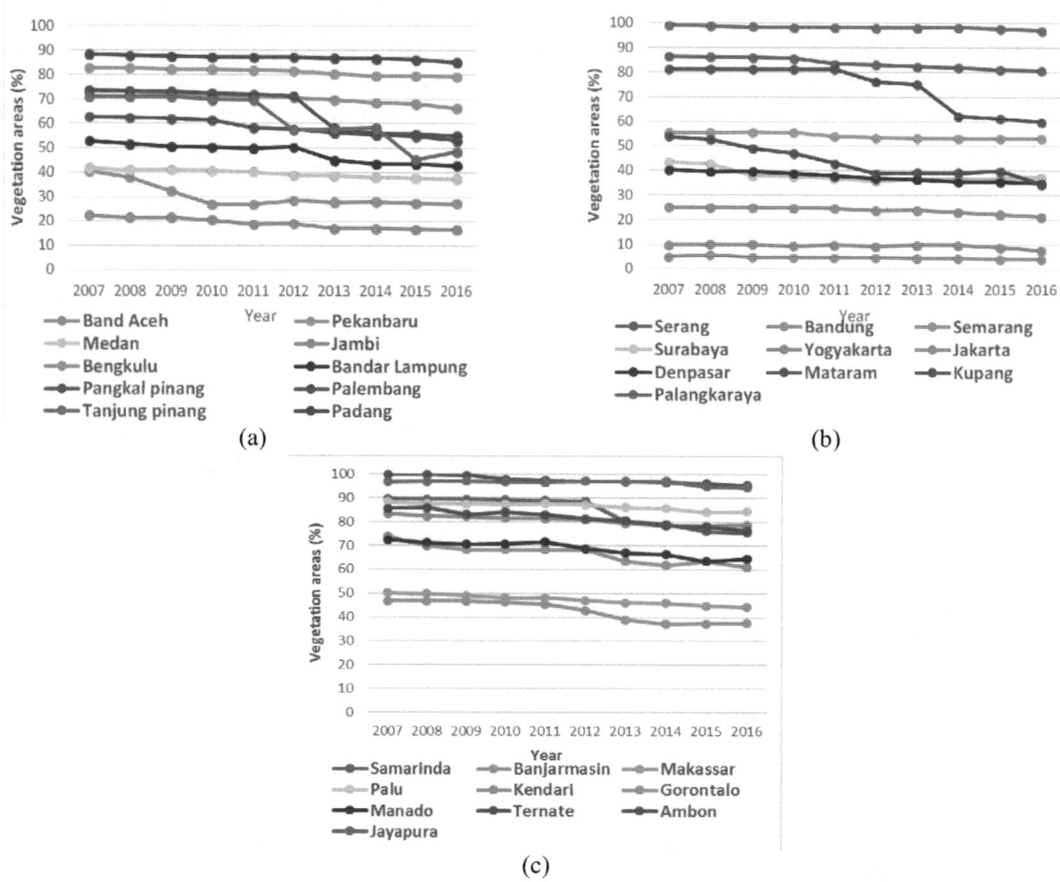

Figure 1. Trend in Urban Vegetation Areas (%) of thirty cities (displayed in three graphs for clear visual) in Indonesia (a) Banda Aceh, Medan, Bengkulu, Pangkal Pinang, Tanjung Pinang, Pekanbaru, Jambi, Bandar Lampung, Palembang, Padang, (b) Serang, Surabaya, Denpasar, Palangkaraya, Bandung, Yogyakarta, Mataram, Semarang, Jakarta, Kupang, (c) Samarinda, Palu, Manado, Jayapura, Banjarmasin, Kendari, Ternate, Makassar, Gorontalo, Ambon.

The percentage of urban green land area and its fluctuations are influenced by many factors. The density of a city; as controlled by the area, type, and shape of the topography of the urban land surface; population; and development of the city; is one such factor. There are also socio-economic factors that come into play which influence government development policies, such as urban sprawl (uncontrolled urban sprawl), the need for land for agricultural activities, and the growth of new hierarchical centers [11]. Some cities in Java, Bali, Sumatra, and Nusa Tenggara have a tendency to change the area given over to green land through expanded city development, while some cities located in Kalimantan, Sulawesi, and Papua have a tendency to change the extent of green land through urban development and mining activities.

The results of the satellite Metop-A present the tropospheric NO_2 concentration on a macro scale well. Over the past ten years, global NO_2 gas concentrations in thirty cities in Indonesia have been increasing. Some of the cities with the highest NO_2 gas concentrations are Sumatera and Java, while, in the eastern part of Indonesia, concentrations tend to be lower. The five cities with the highest average of NO_2 concentrations were Jakarta (826.3), Serang (570.1), Surabaya (384.0), Bandung (299.7), and Tanjung Pinang (298.8). The five cities with the smallest average concentrations of the NO_2 gas were

Ternate (32.2), Ambon (32.9), Gorontalo (38.8), Manado (39.0), and Jayapura (39.6). The trends in tropospheric NO_2 concentrations are seen clearly in Figure 2.

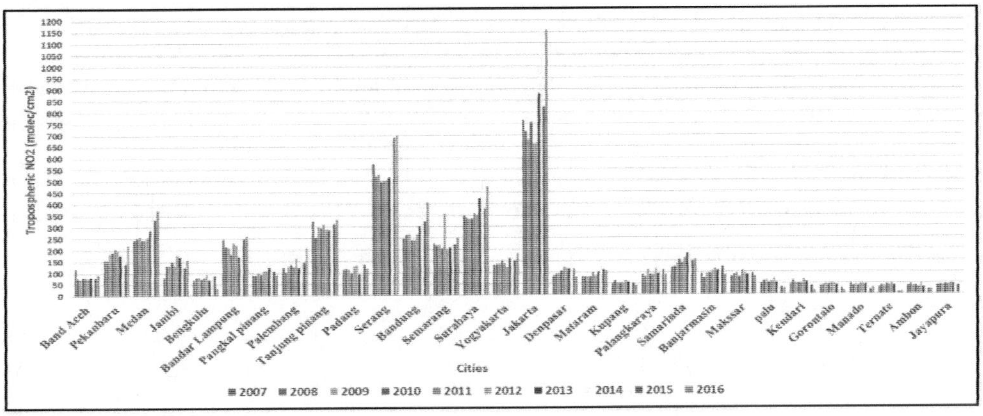

Figure 2. Trends in Tropospheric NO_2 Concentrations in Urban Area of Indonesia, 2006 till 2017.

In 2014, NO_2 emissions spiked in almost all the cities. The process of NO_2 dispersion in the atmosphere is influenced by meteorological conditions (solar radiation, temperature, wind distribution, and air humidity). However, pollutants from anthropogenic activities are the main cause of increasing NO_2 production, especially the transportation sector. Overall conditions for troposphere NO_2 concentrations in Indonesia, based on the average value of the 30 cities, point to an increase of 18.49% between 2006 and 2017. Air pollution prevention documents state that nitrogen dioxide in urban NO_2 gas loads varies according to time, season, and meteorological conditions. Commonly, NO_2 concentrations in urban areas increase in the morning and afternoon due to peak traffic in the morning and photochemical reaction at noon. In addition to being peak times for anthropogenic activity, especially transportation and other fuel usage activities, these are also times of higher solar intensity, leading to increases in NO_2 concentrations during these periods [13].

The cities with the highest surface temperature averages were Jakarta (38°C), Yogyakarta (34.9°C), Surabaya (34.4°C), and Makassar (34°C). Some of the cities with the lowest surface temperature averages were Ambon (26.3°C), Ternate (26.9°C), Jayapura (28°C), Manado (28.8°C), and Padang (28.9°C). The trend of surface temperatures from 2006 until 2017 in urban area of Indonesia can be seen in Figure 3.

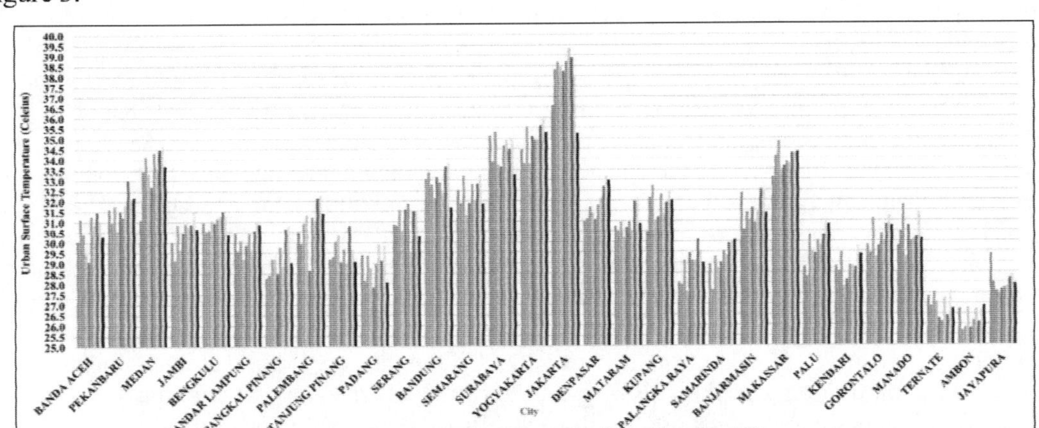

Figure 3. Trend in Urban Surface Temperatures in Indonesia 2006 till 2017.

IOP Conf. Series: Earth and Environmental Science **303** (2019) 012034 doi:10.1088/1755-1315/303/1/012034

The urban surface temperatures in the 30 cities during the study period fell in an interval from 25-40 °C. During those ten years, most cities experienced the highest temperatures in 2014 and 2015. The years 2010 and 2011 were most likely to show decreases in temperature. Increasing the surface temperature of the earth impacts many aspects of life. These impacts include increased energy consumption for air conditioning and fans, need for use of motorized vehicles, and hunger due to decreased crop production or crop failure, evaporation which limits the availability of water, and pest and disease attacks.

The magnitude of the effect of vegetation area, urban surface temperature, distance to the equator, and hill category on tropospheric NO_2 were analyzed statistically. The results of the analysis can be seen in Table 2.

Table 2. Multiple Regression for Tropospheric Nitrogen Dioxide Gas.

Variable	Coef.	F Stat. (P-Value)	R_{square}	VIF
Constant	-441.80	0.000		
Vegetation (%)	-1.0649	0.000	22%	1.709
Temperature (°C)	21.434	0.000		1.559
Distance from equator (km)	0.03651	0.001		1.364
High Hill	13.273	0.136		1.015

Standard error, indicated significance 95%

Based on the VIF values, there is no multicollinearity among the independent variables. Hence, the multiple regression equation can continue on to influence analysis. Table 2 shows that each variable has a different coefficient of influence. The result has a coefficient of determination (R*square*) of 22%. Based on the R*square* value, 22% of all variation in tropospheric NO_2 concentrations can be explained by these four variables, while the remaining 78% of the variability may explained by other factors not taken into account in this model.

Due to its nature, NO_2 reacts easily with other components. However, ozone can bind subsequently with NO in the troposphere to form NO_2 once again. These conditions, along with the micro NO_2 conditions on the surface, result in the rapid accumulation of tropospheric NO_2 [14]. The very rapid fluctuations in and reactions of NO_2 changing its stability are supported by the increasing sources of NO_2 on earth, especially industrial and transportation sources.

The regression coefficient of area vegetation is -1.0649. The negative value means that the relationship between vegetation areas and tropospheric NO_2 is inversely proportional, i.e., as vegetation increases, tropospheric NO_2 concentration decreases. These results agree with previous research on both laboratory and macro scales using satellite image databases. An experiment on the reduction of NO_2 gas by various species of plant species yielded the information that *Erythrina cristagali* and *Caliandra sp* have the highest absorption capabilities of 68.3 and 41.01 μg/g, respectively [15]. Another review showed that stomata play an important role in air control in the troposphere layer, especially in terms of ozone and NOx. In these processes, the surface temperature factor also has an effect, mainly on the dispersion and chemical reactions that occur [16]. In an examination of the absorption of NO_2 using eight types of plants under different environmental conditions, it was found that NO_2 absorption rates by plants are higher in higher temperatures (approximately 30°C) than in lower temperatures (approximately 20°C) [7].

The regression coefficient for surface temperature is 21.343, which means that urban surface temperature has a positive effect on tropospheric NO_2. The higher the urban surface temperature, the higher the concentration of tropospheric NO_2 gas. The results of this analysis contradict, or are not in accordance with, the basic theoretical relationship established between temperature and concentrations of NO_2 or various other types of gas pollutants. Results from previous research, as mentioned above, indicate that the lowest concentration of NO_2 occurs when the air temperature is high, and the highest

IOP Conf. Series: Earth and Environmental Science **303** (2019) 012034 doi:10.1088/1755-1315/303/1/012034

concentration of NO_2 occurs when the air temperature is low. Although contrary to most theories, this study's result is possible. Jakarta, for instance, has produced results that show that a high temperature indicates high concentrations of pollutants in the air. In urban areas, the disparities between NO_2 conditions during the day and afternoon are larger than they are in rural areas. The main factors driving NO_2 concentrations are, as noted previously, anthropogenic activity on the part of the transportation sector and fuel consumption in industry. The next most influential factor is the difference in air temperature that occurs [17]. It is also known that high ozone concentrations indicate high concentrations of NO_2 gas.

Another study has also explained that, overall, starting from the surface of the earth up to a height of about 17 km, ozone concentration and temperature have a positive correlation such that the lower the concentration of ozone, the lower the temperature. The height of 17 km includes the troposphere layer. At a height of 17-27 km, the positive correlation between the concentration of ozone and temperature persists. Conversely, at a height of 27-40 km, the correlation turns negative, meaning that the higher the temperature, the lower the ozone concentration [18].

Based on elevation, cities are classified as being high or low hill. In this analyze, being at 'high hill' in elevation has a positive regression coefficient of 13.273. Hence, the higher elevation of a city means that it is likely that the tropospheric NO_2 is higher too. This condition has a meteorological explanation. A high hill or plain has a low temperature. High temperatures cause the air to release water and become less dense, making the convection of various pollutant gases, including NO_2, flow upward. This means that NO_2 concentration will be low. On the other hand, low air temperatures cause the surrounding air density to be almost equal to the density of the upper air layer. Therefore, the flow of air convection moves more slowly and causes NO_2 concentrations to be high as it accumulates on the surface [19].

The position of a city, as based on its latitude, which determines its distance to the equator, also affects tropospheric NO_2 concentration. Cities in Indonesia generally have a fairly small distance from the equator in accordance with the position of Indonesia, which is a tropical country. The result of the regression analysis indicates that the distance of a city from the equator positively affects troposphere NO_2 gas concentration with a regression coefficient of 0.03651. According to a 2016 publication which shows that ozone conditions in developing countries, Southeast Asia, in particular, increased its ozone concentrations rapidly during the period 1980-2010 [20]. The mapping of dispersions based on the equator shows that pollutants, especially ozone, have high concentrations in areas with low latitudes (close to equator). If the ozone concentration is high in areas with low latitudes, then NO_2 is low in these areas. A city close to the equator will have a relatively high surface temperature. In high surface temperature conditions, the pollutant concentration, including that of NO_2 gas, tends to be low [21].

4. Conclusion

Urban growth in developing countries, especially in Indonesia, is marked by substantial growth in the number of urban buildings, which is usually followed by a decrease in green space over time. However, this need not be the case if local leaders are aware of the importance of vegetation and increase public green space by adding more vegetation and trees in the city. Among the 30 cities observed in Indonesia, this was the case only in Surabaya, where the green space grew as much as 0.94 % annually between 2012 and 2016.

However, a growth in green space will fail to outweigh massive additions of anthropogenic sources that emit nitrogen dioxide, leading to an increasing trend in tropospheric NO_2 concentration, despite a decreasing trend in surface temperature. The statistical analysis also shows that temperature and elevation affect tropospheric NO_2 concentrations. This result implies that a green space expansion policy should receive attention from local governments and other stakeholders in order to improve air quality not only in the troposphere but also on the surface. Despite the model having quite low accuracy, we have been able to show that using data from satellites and other secondary sources may further improve the model. Therefore, this may be a promising method with which to derive conclusion concerning air quality as well as suggest policy.

IOP Conf. Series: Earth and Environmental Science **303** (2019) 012034 doi:10.1088/1755-1315/303/1/012034

References

[1] Vallero D 2008 *Fundamentals of Air Pollution 4th Ed* United Kingdom: Elsevier

[2] Pannullo F, Lee D, Waclawski E and Leyland A 2015 Improving Spatial NO_2 Predict Using Diffusion Tubes: A Case Study in West Central Scotland 01 Atm. Env. J. **118** pp. 227-235

[3] UNESCAP 2015 *Statistical Yearbook Asia and Pacific 2015* [Online] December 2016 21 http://www.unescap.org/resources/statistical-yearbook-asia-and-pasific-2015

[4] Wijaya N 2015 Detection of Land Use Change With Landsat Image J Geo and Planning **02** No.02 pp. 82-89

[5] EPA 2014 www.epa.gov Online accessed 22 October 2017 https://www.epa.gov/sites/production/files/201406/documents/treesandvegcompendium.pdf

[6] Hu Y 2011 *NO_{2-} Drived NO_{3-}* Metabolism in Leaves *Insceince* J **01** No.02 pp 90-101

[7] Abdurahim D 2010 Effect of Vegetation and Environmental Temperature on Gas Pollutant Absorption NO_2, Thesis:Institut Pertanian Bogor

[8] Ramdhoni S, Rusyahati S B dan Prasetyo L B 2016 Open green space development priority based on distribution of air temperature change in capital city of Indonesia Jakarta Procedia Envi. Scie. **33** pp 204-213

[9] Wardana I K 2015 Analysis of Urban Surface Temperature For Green Spaces Planning In Bandung Indonesia Thesis University of Twente

[10] Tursilowati L 2015 Urban Heat Island And Its Contribution In The Climate Change And Its Relationship To Land Change *Proceeding National Conference Global Warming And Mitigation Fact* Indonesia

[11] Pribadi D O, Shiddiq D and Ermiyanila M 2006 *The Model Of Land Change And The Factors That Influence It* J. Teknik Lingkungan P3TL BPPT **7** No.1 pp 35-51

[12] Widyatmanti N and Dini W 2008 *Geogrph: Atm. and Geo Conditions* (Jakarta: Grasindo)

[13] World-Bank 1998 *Nitrogen Oxides Pollution Prevention and Abatement Handbook* World Bank Group Word Bank

[14] Lövblad G, Palmgren F, Aalst R Van, Allegrini, Bauman R, Derouane A, dwards L, Fiala I, Hauer A, Lahtinen T, Nilsson M, Pohjola, Saeger E D and Santis F D 1997 *Position paper on Air Quality: NO_2* (Europe: ECDG XI)

[15] Nasrullah N, Gandanegara S, Suharsono H, Wungkar M and Gunawan A 2000 Measurement of NO2 Gas Poll. Absorp. On Tree Type Plant, Bush and Closing Land by Using NO_2 Marked Mark 15_N *Minutes of Penelilian* Scientific Meeting and Development of Isotope and Radiation Tech

[16] Monks S 2015 Tropospher O_3 and Its Percusor From Urban to Global **15** pp 8889-8973

[17] Risalah N 2011 *Linkage of Air Pollutant and Mainland Surface Temperature and Its Distribution In DKI Jakarta* (Thesis: Universitas Indonesia Depok)

[18] Risdianto D 2013 Analysis of the Relationship between Ozone With Temperature (Case Study of Lapan Watukosek 2000-2010) *Proceeding National Conference of Research, Edu, And Scie App* Yogyakarta

[19] Antika Riadi and Sugiyanto S 2013 *Effect of Temperature, Humidity and Wind Speed Against Accumulation of Nitrogen Monoxide and Nitrogen Dioxide* (Thesis: Universitas Riau Pekanbaru)

[20] Zhang Y, Cooper A, Gaudel T, Nédélec P, Ogino S-Y and West J 2016 Tropospheric Ozone Change From 1980 To 2010 Dominated by Equatorward Redistribution of Emissions Nature Geoscience J Vol **12** NOAA Scientific Publications Report November 21 2016

[21] NOAA 2012 *Air-Pollution-Composition-Of-Air-Structure-Atmosphere-Concise-Environmental-Engineering.* Online accessed 11 January 2018 http://www.zainbooks.com/read.php?i=air-pollution-composition-of-air-structure-atmosphere-concise-environmental-engineering&b=133&c=9

Chemical Composition of Fine Particulate Matter from Peat Forest Fires at Palangka Raya and Its Dispersion using HYSPLIT

D D Lestiani[1], M Santoso[1], S Kurniawati[1], D K Sari[1], I Kusmartini[1], A Manurung[2] and A Riadi[2]

[1] Center of Nuclear Technology for Materials and Radiometry, National Nuclear Energy Agency (BATAN), Jln. Tamansari 71, Bandung 40132, Indonesia
[2] Environmental Protection Agency of Palangkaraya, Tjilik Riwut Km 5.5, Palangka Raya, Indonesia

diahdwi@batan.go.id

Abstract. An air quality monitoring was conducted to investigate the chemical characteristics of particulate emissions in peat forest fires at Palangka Raya, Indonesia during 2012 where a peat forest fire episode occurred. Aerosol samples were collected at Palangka Raya where represents a sub urban site downwind of the peat forest fires. The samples were collected using Gent stacked filter unit sampler. The sampler provides two size fractions: coarse (2.5-10 μm equivalent aerodynamic diameters EAD) and fine (<2.5 μm EAD). Black carbon (BC) was measured using Smoke Stain Reflectometer, and elemental compositions were analyzed using X-ray Fluorescence (XRF). The concentrations of $PM_{2.5}$ during a haze episode were significantly different about seven to eight times higher when compared to the concentrations in non-haze conditions. The major components of $PM_{2.5}$ in peat smoke haze consist of sulphur, biomass burning component (BC, K), and crustal components (Al, Fe, Si). The sulphur mainly comes from the peat as it is the earliest stage of transition from compressed plant growth to the formation of coal. The dispersion pattern to show the effect of wind direction using the Hybrid Single Particle Lagrangian Integrated Trajectory Model (HYSPLIT) simulation was also applied.

Keywords : particulate, emissions, peat, forest fires.

1. Introduction

In Indonesia, forest fires from uncontrolled burning of large areas of lands are major problem for air quality. Tropical peat/forest fires in Indonesia are generally caused by illegal human activities including land clearing, converting to agricultural lands through heavy logging and slash and burn techniques, use of fire as weapon in land tenure/disputes, fire for resource extraction and accidental fire [1]. Usually the fires take place in the dry season in September-November. Most of these fires particularly occurr in Kalimantan and Sumatra islands, which have the world largest areas of tropical peat lands. Indonesia has approximately 83% (225,000 km^2) of peat land in Southeast Asia of which 26% and 13% are in Kalimantan and Central Kalimantan respectively. A large Mega Rice Project (MRP) was carried out near Palangka Raya, Central Kalimantan. The MRP area consists of 9191 km^2 peat swamps. This kind of large-scale land development resulted in extensive deforestation. MRP peat with roughly 4500 km^2 is more than 3 m depth and is the main source of fire-related air pollution [2].

During the haze episode, several studies of the chemical composition of the fire aerosol have been performed at Palangka Raya, Central Kalimantan. Most of them focus on TSP and PM_{10}, organics components and some on inorganic material. Study related to peat fires in 1997 reported during peat fires that total suspended particulate (TSP) were detected more than 4000 ug/m3 exceeding the Indonesian national air quality standard by a factor of 15 [3]. In particular, peat fires could become an even worse air pollution source due to its underground fires and also its impact to other regions nearby as trans-boundary haze pollution. Neighbouring country such as Malaysia and Singapore experienced severe air pollution exceeding the typical air quality standards due to the peat fires in Kalimantan [4], and the impact of it also appeared in Brunei and even Southern Thailand [5]. In order to reduce trans-boundary haze pollution from vegetation fires, information on pollution sources and their impacts need a thorough investigation. For urban air pollution sources, there are several reports in Indonesia, but only a few for peat fires [6, 7]. Studies about fine particulate ($PM_{2.5}$) and its chemical composition are still lack.

Several epidemiological studies conducted in different parts of the world indicate that there are strong correlations between exposure to airborne particulate matter and increased morbidity and mortality [8]. Moreover, $PM_{2.5}$ emitted from biomass burning is a major contributor for bronchial-related problems [9]. The content of some metals in aerosols, when deposited in the lower airways, can lead to acute and chronic effects on the lung [10]. Therefore, the monitoring and assessment of the impact of forest fires on air quality is required especially on its content of certain elemental concentration.

In this study, the observation of air quality during peat fires episode in Palangka Raya is discussed which focused on the fine particulate to investigate the chemical characteristics of particulate emissions in peat fires and its dispersion to show the effect of the wind that travel through transboundary and influence to the neighbouring countries.

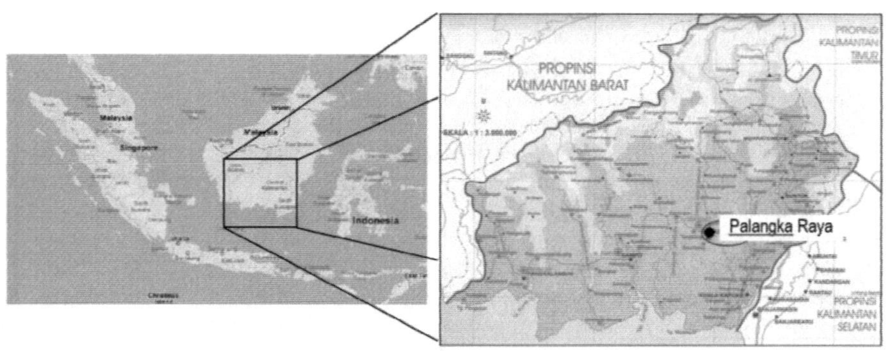

Figure 1. Sampling site in Palangka Raya, Kalimantan

2. Materials and methods

2.1. Sampling

Samplings were carried out in Palangka Raya, Kalimantan (Borneo). Palangka Raya is the capital city of Central Kalimantan Province, has the area width around 2678.51 km^2. Palangkaraya is located at 75 metres above sea level. In general, Palangka Raya has two seasons, dry season (June to October) and wet season (November to March). Palangka Raya population in the year of 2011 are recorded 224663 based on population census 2010 (Palangka Raya City in Figures, 2011). The sampling site is located on 113° 52'49" E and 02°10'30" S. Samples were collected in January – December 2012 using a Gent stacked filter unit sampler provided by International Atomic Energy Agency that is capable for collecting particulate matter in the $PM_{2.5-10}$ (coarse) fraction and $PM_{2.5}$ size (fine) fraction [11]. The coarse fraction $PM_{2.5-10}$ samples were collected on an 8.0 μm nuclepore filters, whereas the fine fraction $PM_{2.5}$ samples were collected on 0.4 μm nuclepore filters. Sampling was performed at a flow rate of 15-18 $L.min^{-1}$ for 24 hours with a frequency of once a week. Sampler was placed on the rooftop of a container 2-3 m above ground level and the intake nozzle of the sampler is located 1.8 m above the roof. The total of 51 pairs of $PM_{2.5}$ and $PM_{2.5-10}$ samples were collected from this site.

2.2. Mass concentrations, black carbon and elemental analysis
The particulate matter mass of each fine and coarse fractions were determined by gravimetric using a six digits microbalance. The filters were stored for 24 hours prior to weighing in controlled environmental condition room. Mass concentrations of $PM_{2.5}$ and $PM_{2.5-10}$ ($\mu g.m^{-3}$) were obtained by dividing the gravimetric mass by the volume of air that passed through the filter. The concentration of black carbon (BC) in the samples was determined by reflectance measurement using a Smoke Stain Reflectometer model 43D. Secondary standard of known reflectance was used to calibrate the reflectometer. The elemental composition analysis was conducted using Epsilon 5 EDXRF spectrometer from Panalytical, which has a 100kV excitation source and a polarising optical path give detection limits in the low ng/cm^2 range for most elements. The calibration was set up using macromatter standard, reference material and a blank standard for each element, Mg, Al, Si, S, K, Ca, Ti, Mn, Fe, Ni, Zn and Pb. The method validation of this application was applied by analyzing the standard reference material air dust on air filter SRM NIST 2783.

3. Results and Discussions

3.1. Particulate matter concentrations
Fine and coarse fraction concentrations were resumed in Table 1. Twenty four hours fine fraction $PM_{2.5}$ concentration ranged from 1.64 to 55.73 $\mu g\ m^{-3}$ with average value of 12.54 $\mu g\ m^{-3}$, and for coarse fraction mass concentrations ranged from 3.69 to 80.67 $\mu g\ m^{-3}$ with average value of 14.88 $\mu g\ m^{-3}$, while the PM_{10} concentration ranged from 7.94 to 136.4 $\mu g\ m^{-3}$. Time series of $PM_{2.5}$ and PM_{10} concentrations is shown in Figure 2.

Table 1. Summary of the coarse and fine fraction mass concentrations for period January – December 2012

Parameter	Sampling site, Palangka Raya	
	Fine Fraction ($\mu g\ m^{-3}$)	Coarse Fraction ($\mu g\ m^{-3}$)
Average	12.54	14.88
Median	7.75	10.96
Standar deviation	12.54	14.08
Minimum	1.64	3.69
Maximum	55.73	80.67

Even none of the annual average and the twenty four hour of $PM_{2.5}$ and PM_{10} concentrations exceed the Indonesian National Ambient Air Quality Standard NAAQS (15, 65 and 150 $\mu g\ m^{-3}$ respectively), but there were periods in September – October 2012 that high concentrations were 7-8 times higher than concentrations in other periods observed. These periods have been identified that the peat forest fires occurred. LAPAN National Institute of Aeronautics and Space reported that more than hundreds hotspot were observed in Kalimantan on 15 and 19 September 2012.

3.2. Black carbon and elemental concentrations
Black carbon concentrations in $PM_{2.5}$ time series is shown in Figure 3. The BC concentrations ranged from 0.19 to 3.08 $\mu g\ m^{-3}$, with average 1.37 $\mu g\ m^{-3}$. The average percentage ratio of BC in $PM_{2.5}$ was 16%. The BC concentration did not show high peak and any fluctuation as shown in $PM_{2.5}$ and PM_{10} concentration on September-October 2012, when the period of peat forest fires occurred. The BC concentrations in this period are 2-3 times higher than other periods, even though the magnitude is not high as the increasing of the mass concentration, it seems that the forest fires episodes contribute to the increasing of BC concentrations in this period. The peat forest fires are substantial sources of BC, as well as biomass burning since the peat forest fires burn through thick layers of peat (dead plant

debris/biomass) that blanket the ground in ecosystems [12, 13]. Besides BC, the increasing of amount of several elemental concentrations also contributes to the increasing of the mass, as well as other organic compounds [13].

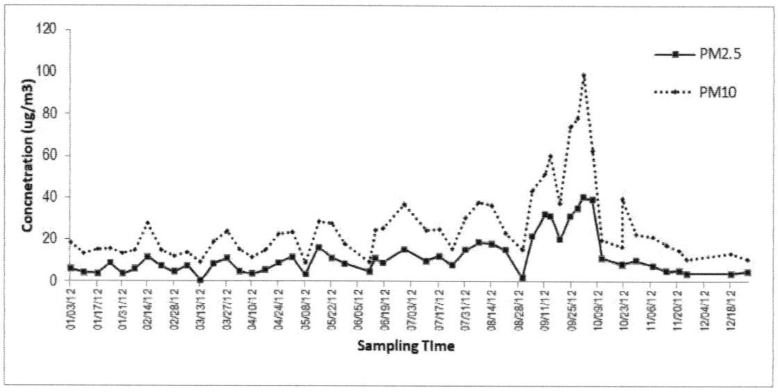

Figure 2. Time series of $PM_{2.5}$ and PM_{10} concentrations

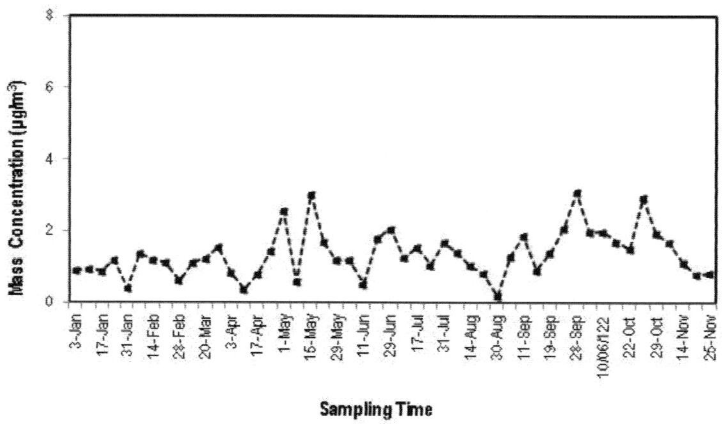

Figure 3. Time series of BC concentrations in $PM_{2.5}$

Table 2. Elemental concentrations in fine fraction (ng/m3) in 2012 period at Palangka Raya

Element	Mean	Range
Mg	32.7	1.2 - 79.3
Al	147	110 - 252
Si	72.5	19.4 - 175
S	400	40.3 - 1757
K	107.7	4.17 - 357
Ca	24.1	5.76 - 52.5
Ti	5.84	2.20 - 15.1
Mn	2.94	1.15 - 6.01
Fe	207	168 - 381
Ni	1.35	0.51 - 3.44
Zn	6.79	3.45 - 18.34
Pb	2.28	0.15 - 6.04

Elements such as Mg, Al, Si, S, K, Ca, Ti, Mn, Fe, Ni, and Zn were detected in the most of samples. The concentrations of elements were resumed in Table 2. The elemental concentrations in $PM_{2.5}$ in Palangka Raya sampling site are shown as box and whisker plot in Figure 4. The box represents 25% to 75% of the distributions of the concentration. The box represents 25% to 75% of the value. The horizontal bar in the box indicates that the median and the plus symbol (+) denotes the mean data. The points lying outside the range defined by the whiskers are plotted as outlier dots. It can be seen that high concentrations of S, K, and crustal elements Al, Si, Fe were identified. The time spatial distribution of S and K showed that the high concentrations of S and K were observed in the period of September – October 2012 when peat forest fires were occurred (Figure 5 and Figure 6). Peat is the earliest stage of transition from compressed plant growth to the formation of coal [14], therefore a lot of sulphur contained in the peat and it released some amounts of sulphur emission due to the burning [15]. Although the types of plants and combustion conditions may influence the composition of particles emitted to the atmosphere, but Ikegami has reported on forest fires at Kalimantan in 1997 that high accumulation of sulphur in particles in the atmosphere after their emission was not in the form of sulfuric acid since ammonia would be generated from biomass burning and would neutralize sulfuric acid [16]. Okada et al (2001) found that the haze particles over Kalimantan were identified as ammonium sulphate [17]. Beside sulphur, other key elements associated with biomass burning was found in high concentration was potassium. Potassium is one of the key elements of biomass burning [12, 18].

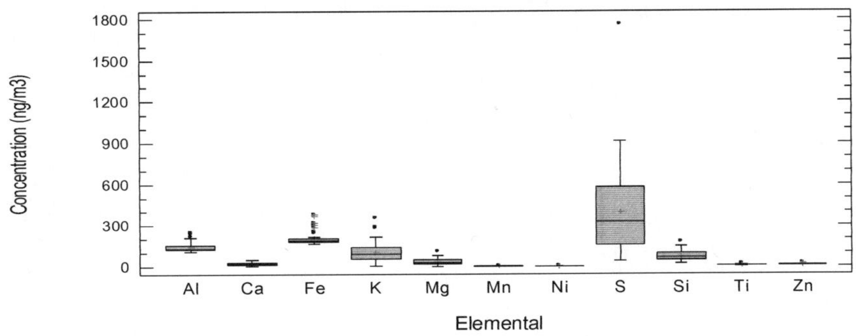

Figure 4. Elemental concentrations in $PM_{2.5}$

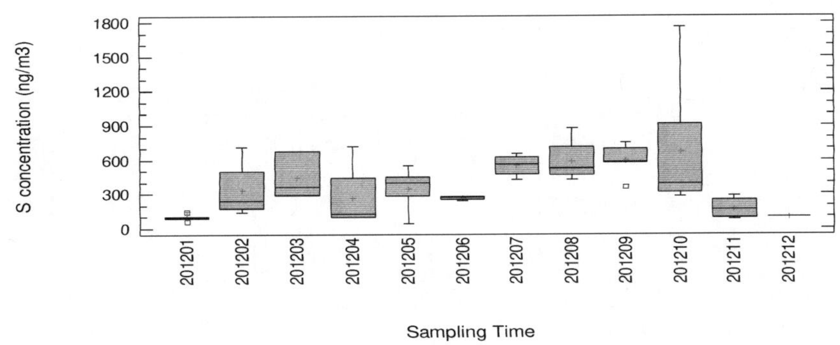

Figure 5. Box and whisker plot of S concentrations in $PM_{2.5}$

3.3. Particulate Dispersion

In 2012 peat forest fire period, the dispersion pattern to show the effect of wind direction using the Hybrid Single Particle Lagrangian Integrated Trajectory Model (HYSPLIT) simulation were applied. The result is shown in Fig. 7. The dispersion distribution showed that the particulate matter from this area was not ableto reachto the other regions, which within 48 hours the concentration of particulate was found to reduce as it dispersed in the surrounding area. It indicated that distribution of particulate matter in the surrounding areas was not influenced by the wind direction and can be concluded that the peat forest period occurred in 2012 at Palangka Raya did not influence the other countries nearby since it has been dispersed before traveling trans-boundary. This fact could present information about public concern since Indonesian has been accused for several peat forest fires due to the haze.

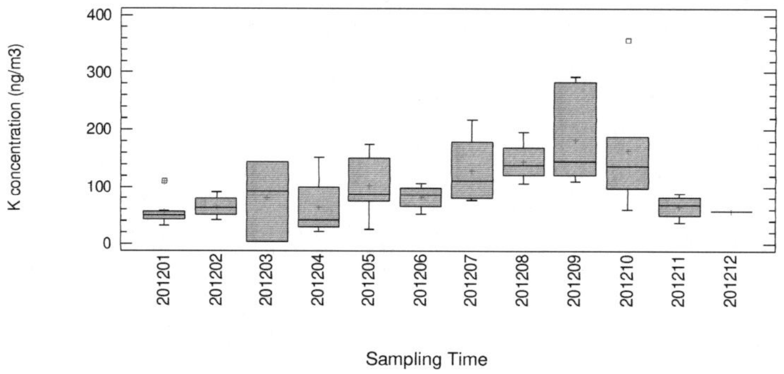

Figure 6. Box and Whisker plot of K concentrations in $PM_{2.5}$

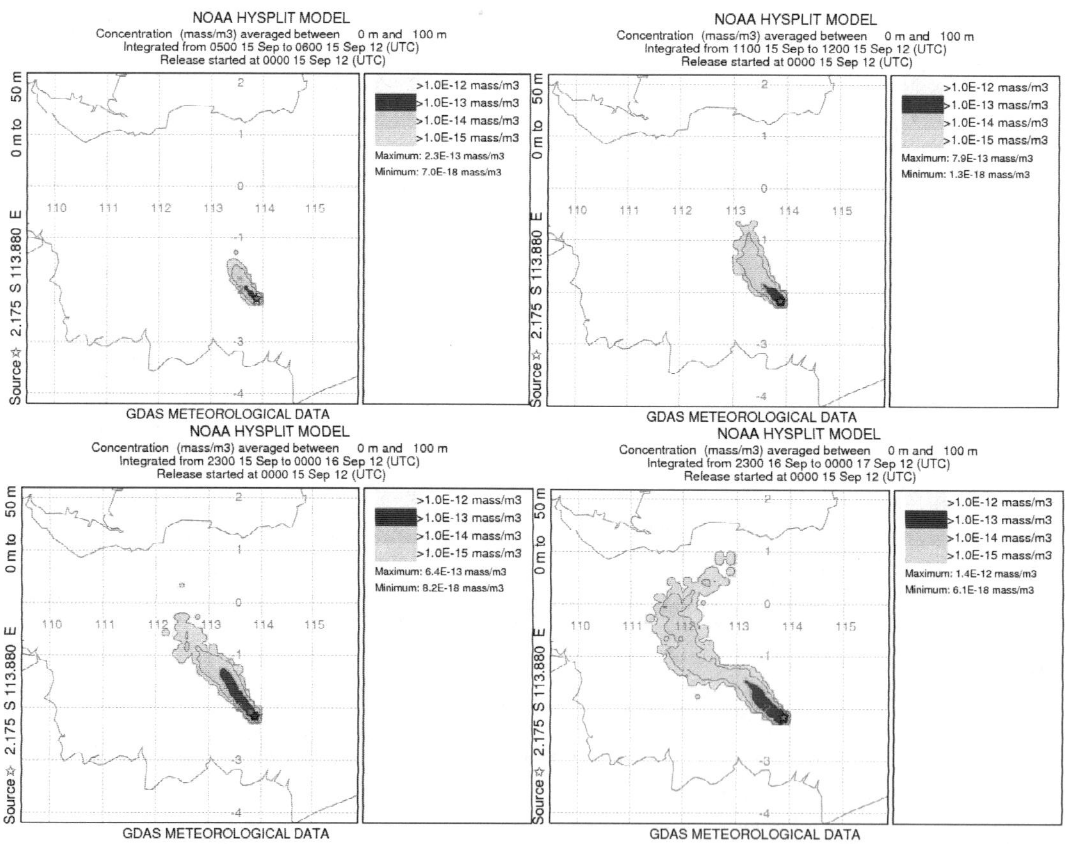

Figure 7. The dispersion distribution of particulate matter within 48 hours on Sept 15, 2012

Haze is problematic for several reasons as it indicates significant forest fire could release big amount of emissions, affects economic activities, tourism, and threatens diplomatic relation between neighbouring countries [19]. Some facts were also revealed that haze was identified as major problem in 1997 and was exacerbated by unusually long droughts connected to El Nino, and it has returned in years unaffected by El Nino, most seriously in 2005, 2006 and 2013. In 2013, it was found from several local newspapers that the haze has been affected Singapore badly [19]. Further study is needed in order to clarify the case of the peat forest fires in 2013 which affected Singapore, through applying the similar approach in this study.

4. Conclusions

The chemical characteristic of particulate emission of peat forest fire in 2012 at Palangka Raya on Central Kalimantan, Indonesia was investigated. The mass concentration during the peat forest fires was 7-8 times higher than normal condition. The high concentration of S, BC and K which associated with the biomass burning was identified. Peat is the earliest stage of transition from compressed plant growth to the formation of coal which contains large amount of sulphur. Further analysis of organic compound should be conducted in the samples to ensure the consistency the form of sulphur and carcinogen elements in the samples. It was also found that the peat fires happened in the period did not affect to other neighbouring countries like happened in 1997, 2005, and 2006.

Acknowledgments

Authors wishing to acknowledge assistance or encouragement from Environmental Management Center – Minister of Environmental of Indonesia and International Atomic Energy Agency (TC IAEA INS7007) for the financial support and cooperation throughout this work. The authors also gratefully acknowledge all colleagues at the Radiometry Analytical group and Palangka Raya Environmental Protection Agency for their technical assistance.

References

[1] Harrison ME, Page SE, Limin SH 2009 *Biologist* **56**(3) 156-163
[2] Hiroshi H, Noguchi I, Erianto IP, Yulianti N, Vadrevu K 2014 *Environ. Poll.* **195** 257–266
[3] Heil A, Goldammer A 2001 *Reg. Environ. Change.* **2**(1) 24-37
[4] Heil A, Langmann B, Aldrian E 2007 *Mitig. Adapt. Strateg. Glob. Change.* **12** 113-133
[5] Wooster MJ and Strub N 2002 *Global Biogeochem. C* **16**(1)
[6] Raghu B, Pradani M, Lestari P, Umid Man Joshi, Jeffrey SR, Rajasekhar B 2013 *Atmos. Res.* **122** 571-578.
[7] Pradani M, Lestari P 2001 Correlation between hotspots and ambient air quality as Impact of Forest Fire (Case study: Central Borneo). In Indonesian, available at http://www.ftsl.itb.ac.id/kk/air_waste/wp-content/uploads/2010/10/PI-AP4-MAHARANI-PRADANI-15305022.pdf.
[8] Dockery DW, Pope CA, Xu X, Spengler JD, Ware JH, Fay ME, Ferris BG Speizer FE 1993 N. Engl. J. Med **329** 1753-1759
[9] Karthikeyan S, Balasubramanian R, Iouri K 2006 *J. Toxicol. Environ. Health A.* 69 1895-1908
[10] Sobaszek A, Boulenguez C, Frimat P, Robin H, Haguenoer, JM, Edme JL 2000 *J. Occup. Environ. Med.* **42** 923-931
[11] Hopke PK, Xie Y, Raunemaa T, Bieglski S, Landsberger S, Maenhaut W, Artoxo P, Cohen DD 1997 *Aerosol Sci.Technol.* **27** 726-735
[12] Begum BA, Kim E, Jeong CH, Lee DW, Hopke PK 2005 *Atmos. Environ.* **39** 3719–3724.
[13] Benscoter B, Page S, Rein G, Van der Werf G, Watts A 2014 *Nat. Geo.* **8** 11-14.
[14] Shankar RK 2003 Proc. Indian Natn. Sci. Acad, **69A** 759-774
[15] Hamada Y, Darung U, Limin SH, Hatano R 2013 *Atmos. Environ.* **74** 177-181.
[16] Ikegami M, Okada K, Zaizen Y, Makino Y, Jensen JB, Gras JL, Harjanto H 2001 *Atmos. Environ.* **35** 4237-4243.
[17] [Okada K, Ikegami M, Zaizen Y, Makino Y, Jensen JB, Gras JL 2001 *J. Aerosol Sci.* **32**(11) 1269-1279
[18] Santoso M, Hopke PK, Hidayat A, Dwiana DL 2008 *Sci. Total Environ.* **397** 229-238.
[19] Forsyth T 2014 *Glob. Environ. Change* **25** 76-86.

Elemental Composition of Particulate Matter Air Pollution Collected Around Industrial Area in East Java

M Santoso[1], D D Lestiani[1], S Kurniawati[1], E Damastuti [1], I Kusmartini[1], D Prakoso[1], Dimas[2], Triyambodo[2] and Rita[3]

[1]Center for Applied Nuclear Science and Technology, National Nuclear Energy Agency (BATAN), Jl. Tamansari 71 Bandung, Indonesia
[2]The Environmental Protection Agency of East Java Province, Jl. Wisata Menanggal 38 Surabaya, Indonesia
[3]Environmental Management Center, Ministry of Environment and Forestry, Puspiptek, Serpong, Indonesia

hayat@batan.go.id

Abstract. Air pollution is one of the key environmental and public health concerns in urban areas in Indonesia. The environment impacts are particularly severe in Java region where the cities have a combination of intense industrial activities, large populations, and increase of motor vehicle usage. This paper describes results from a detailed elemental characterization study conducted in Surabaya, East Java. Samples of airborne particulate matter, $PM_{2.5}$ and $PM_{2.5-10}$ were collected once a week using a Gent stacked filter sampler during 2012-2015. The samples were analyzed for their elemental composition and black carbon (BC) content by Energy Dispersive X-ray Fluorescence (EDXRF) and light absorption, respectively. The mass concentration of $PM_{2.5}$ and PM_{10} ranged from 4.8 to 30.8 μg/m^3 and 9.4 to 114.9 μg/m^3, respectively. The annual average $PM_{2.5}$ obtained during over the period of 2012 to 2015 would be in violation of the Indonesian annual average standard concentrations of 15 μg/m^3. The distributions of the ratio of $PM_{2.5}$ to PM_{10} is less than 0.5. The average ratio of BC concentrations compared to the $PM_{2.5}$ concentrations was 22.6 %. Several composition of elemental concentration were also determined and the results showed that lead level in $PM_{2.5}$ and $PM_{2.5-10}$ was significantly higher with the maximum concentrations were 2779 and 2258 ng/m^3 respectively. The Pb concentrations in Surabaya were tens to thousand times higher than in other cities. These results could be used to provide an overview of the measurements of the concentrations and composition of particulate air pollution in Surabaya and as a key in identification and estimation of pollutant sources to design the appropriate and right policies.

Keywords : pollution, $PM_{2.5}$, $PM_{2.5-10}$, Energy Dispersive X-ray Fluorescence, black carbon

1. Introduction

In Indonesia, air pollution is one of many serious environmental problems faced by major urban cities. There are several factors that cause this pollution and the main factors were population growth and rapid urbanization. In 2015, the population of Indonesia was estimated to reach 255.4 million and about 58 % live on the island of Java [1]. Therefore, increased economic development in Indonesia, especially in Java, has led to rapid and unplanned urbanization. The change of urban economic structure will increase the transportation, energy consumption and also caused severe environmental pollution, particulary air pollution.

Air pollution in an urban area is a dynamic mixture of pollutants emitted from numerous sources including motor vehicles and other anthropogenic activities. The impact of air pollution can be one of the reasons for civilization diseases [2]. Health effects are mostly associated with the particles in the size range of about 10 um and 2.5 um (PM_{10} and $PM_{2.5}$). High concentration of particulate matter, particularly $PM_{2.5}$ have raised great attention. Santoso et al. in 2008 reported the composition and source apportionment of fine and coarse particle samples collected in Bandung and Lembang [3], while the $PM_{2.5}$ concentration in Serpong Tangerang, especially with respect to high concentration of Pb were reported in 2011 [4]. However, monitoring $PM_{2.5}$ in Indonesia were rare and limited.

Surabaya as second largest city in Indonesia was also face the environmental problem especially air pollution. Syafei et al. reported the spatial and temporal factors of air quality in Surabaya and the trend of air pollution indicates a strong increase in daily concentration proportionally to the growth of Surabaya's vehicle volumes [5]. Airborne particulate matter characterization and preliminary study of the determination of pollutant sources in Surabaya have been carried out for the October 2012 - February 2014 monitoring period [6]. To monitor air quality, the environmental protection agency of Surabaya has introduced Air Quality Management System (AQMS), continuous automated monitoring for CO, SO_2, NOx, O_3, and PM_{10}. This system itself had limitation and could not monitor particulate matter smaller than 2.5 micrometers. Therefore it could not detect a range of potentially harmful air pollutants. The main objective of this research was to provide an overview of particulate air pollution especially $PM_{2.5}$ in this city from the results of 4 years monitoring in 2012-2015. This report describes the distribution of mass and some major species concentration across this region.

2. Materials and methods

2.1. Sampling

Air particulate samples were collected using a Gent stack filter unit particle sampler capable of collecting $PM_{2.5-10}$ and $PM_{2.5}$ size fractions. The samples were collected on an 8 µm and 0.4 µm pore nucleopore filter for the coarse fraction ($PM_{2.5-10}$) and the fine fraction ($PM_{2.5}$) sample respectively. The samples were collected for a 24 h period, once per week in Surabaya at an area of local EPA of East Java province (07°20.770' S, 112°44.024' E) between March 2012 to November 2015 (Figure 1). Surabaya is a capital of East Java and as second largest city in Indonesia covers the area 374.78 km^2 with over 3 million in population. Surabaya industrial area located in the Surabaya city territory with an area of 245 hectares accommodates more than 300 companies and about 70 % of the total land is used for industry while the rest is for public use. Surabaya features a tropical wet and dry climate, with distinct wet and dry seasons. The wet season is usually established from November to June, while the dry season covers the remaining five months. The average temperatures are very consistent throughout the year from 26 °C to 31 °C [5,6].

2.2. Mass and Black Carbon measurement

PM mass of each samples were determined by weighing the filter using a microbalance (Mettler Toledo XP6) in the Center for Applied Nuclear Science and Technology Laboratory, National Nuclear Energy Agency (BATAN), Bandung Indonesia. The aerosol masses of both $PM_{2.5-10}$ and $PM_{2.5}$ fractions were determined by weighing the filters before and after exposure. The difference in weights for each filter was calculated and the mass concentrations for each samples were determined, then divided by the volume of air passing through the filter to obtain the concentration of $PM_{2.5}$ and $PM_{2.5-10}$ ($\mu g.m^{-3}$). The PM_{10} concentration was obtained by summing these two values. Black carbon concentration on a filter was measured by an EEL-type Smoke Stain Reflectometer (Diffusion Systems, Ltd). Secondary standards of known black carbon concentrations were used to calibrate the reflectometer [7]. The concentrations are defined based on the amount of reflected light that is absorbed by the filter sample and an assumed mass absorption coefficient. It is related to the concentration of light absorbing carbon through standards of carbon with known areal density [8].

Figure 1. Location of sampling site at Surabaya.

2.3. *Elemental Analysis*

The elemental analysis were all carried out in BATAN Bandung that has implemented ISO/IEC 17025:2005 accredited by National Accreditation Body. The analysis was conducted using Epsilon 5 EDXRF spectrometer from Panalytical, which has a 100kV excitation source and a polarising optical path give detection limits in the low ng/cm^2 range for most elements. EDXRF Epsilon 5 with its polarization method produces an indirect excitation/secondary target excitation gives significantly lower background. A special sample insert enables air filters to be fitted directly into the sample cup. No additional sample preparation is needed and the sample changer makes continuously analysis possible. The EDXRF Epsilon 5 installed in BATAN Bandung has 9 secondary targets i.e: Al_2O_3, Fe, CaF_2, Ge, Zr, CeO_2, Mo, Ag and Al. The calibration was set up using multi-standard reference material, standard Micromatter® and a blank standard for each element: Mg, Al, Si, S, K, Ca, Ti, V, Cr, Mn, Fe, Co, Ni, Cu, Zn, As and Pb. The validation of XRF measurements in this application was applied using standard reference material SRM NIST 2783 Air Particulate on Filter Media [9].

3. Results and Discussion

The annual average of $PM_{2.5}$ and $PM_{2.5\text{-}10}$ at Surabaya over the period of 2012 to 2015 were 20.71, 14.55, 16.95, 14.86 and 30.59, 14.80, 21.93, 21.91 respectively. Adding these two values together estimates of PM_{10} and the annual average of PM_{10} were 51.29, 29.35, 38.88 and 36.77. The detail on the number of samples collected during period of study and the range of twenty-four hour mass concentrations are provided in Table 1. The Indonesian National Ambient Air Quality Standards (NAAQS) for annual average $PM_{2.5}$ is 15 µg/m^3, while the 24 h standard is 65 µg/m^3. It can be seen that the data provided by 2012 and 2014 would be in violation of the Indonesian annual average standard for $PM_{2.5}$, while the values for 2013 and 2015 almost reach the threshold. It should be noted that Gent sampler will underestimate $PM_{2.5}$ since the 50% collection point is closer to 2.2 µm [10]. Therefore, there is a strong indication that in this area there was problem with air quality.

The mass concentration for 2012 was found higher than the values for 2013 to 2015. The annual average of $PM_{2.5}$ at Surabaya in 2012 was also higher than the value in several urban areas in Indonesia over the period of 2011 as reported by Santoso [7]. The average $PM_{2.5}$ concentrations for Bandung, Jakarta, Serpong, and Yogyakarta were 18.35, 16.50, 16.68, and 8,78 µg/m^3 respectively. While for PM_{10} at these areas were 35.81, 41.58, 32.05, and 20.13 µg/m^3. The higher mass concentration during 2012 could be due to a longer period of dry season and less rainfall during the period of 2012.

Table 1. Mass concentrations of $PM_{2.5}$ and $PM_{2.5\text{-}10}$ in Surabaya during 2012-2015.

Year	n	$PM_{2.5}$ (ng/m^3) Mean	$PM_{2.5}$ (ng/m^3) Range	$PM_{2.5\text{-}10}$ (ng/m^3) Mean	$PM_{2.5\text{-}10}$ (ng/m^3) Range
2012	17	20.71 ± 3.86	16.30 - 30.85	30.59 ± 14.66	12.71 - 71.03
2013	42	14.55 ± 4.76	4.92 - 22.45	14.80 ± 5.98	4.47 - 30.82
2014	38	16.95 ± 4.19	8.11 - 24.55	21.93 ± 15.09	6.15 - 99.33
2015	22	14.86 ± 3.86	6.17 - 20.95	21.91 ± 8.81	5.66 - 34.57

The average ratio of $PM_{2.5}$ to PM_{10} at Surabaya site were 0.46 and in general, the ratio is less than 0.5. This values indicating there are higher coarse particle ($PM_{2.5\text{-}10}$) masses. The urban site in Australia, Bangladesh, Malaysia, India, Pakistan, Sri Lanka, Thailand, and Vietnam have ratio of $PM_{2.5}$ to PM_{10} values lower than 0.5 [11], while for Jakarta and Yogyakarta were 0.4 and 0.48 [7]. These values were similar with the results obtained for Surabaya site.

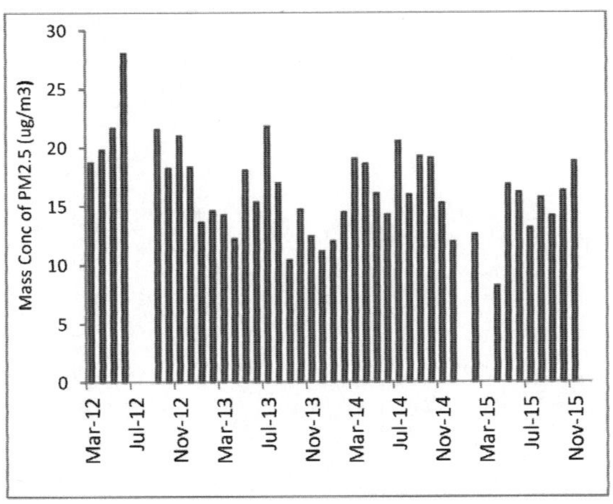

Figure 2. The average monthly data for the $PM_{2.5}$ mass in Surabaya during 2012-2015.

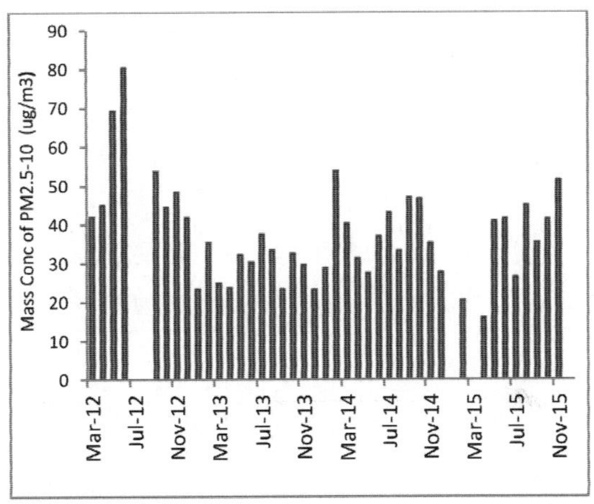

Figure 3. The average monthly data for the $PM_{2.5\text{-}10}$ mass in Surabaya during 2012-2015.

The time variations of monthly $PM_{2.5}$ and $PM_{2.5\text{-}10}$ concentrations over the period of 2012 to 2015 are shown in Figure 2 and Figure 3. Due to technical issues, the samples were not collected during the period of January to February 2012, July to August 2012, January 2015, March 2015 and December 2015. Eventhough the annual average of $PM_{2.5}$ was slightly fall the Indonesian NAAQS, however, none of daily concentrations exceeded the 24 h Indonesian NAAQS for $PM_{2.5}$ (65 µg/m^3). In Figure 2 and Figure 3, the seasonal variations are also clearly shown, where the average monthly data for the total $PM_{2.5}$ and $PM_{2.5\text{-}10}$ concentrations are plotted as a function of time for the Surabaya site. Dry seasons peaks (June to August) in the $PM_{2.5}$ and $PM_{2.5\text{-}10}$ concentrations are clearly visible with wet variations being more than a factor of two in some years. The seasonal variation obtained in this research were also similar with the results for Bandung sites [12]

There are several high concentration of $PM_{2.5}$ and $PM_{2.5\text{-}10}$ levels in Surabaya were also observed from Figure 2 and Figure 3. The high mass concentration at November 2013, February 2014, July 2015 and November 2015 were resulted from the samples collected during the volcano eruption at Mount Merapi, Kelud volcano, Raung volcano, and Rinjani volcano, respectively. Mount Merapi is the most active volcano in Indonesia located on the border between Central Java and Yogyakarta. On November 18, 2013, mount Merapi ejecting smoke up to a height of 2000 meters. While, Kelud volcano erupted on February 13, 2014, destroying the lava dome and ejecting boulders, stones, and ashes up to west Java about 500 kilometers from mount Kelud. The Kelud eruption has also significantly influenced the air quality in Java area especially at Yogyakarta in central Java, due to the volcano spit ash up to 17 kilometers into the sky and the wind was strongly blown to the west direction [13]. Raung is one of the most active volcanoes on the Java island and located in the province of East Java. Raung standing almost 3332 meters above sea level. The volcano eject material causing a dust cloud during June to July 2015 that influences the air quality to the residents surrounding the volcano. Several airports throughout Indonesia were also closed due to the ash being produced by the Raung volcano eruptions. Mount Rinjani is an active volcano on the island of Lombok Indonesia, started erupting on 31 October 2015. After that a lot of major activities had occurred in November 2015 and this eruption has forced the closure of Ngurah Rai airport in Bali and disrupted international flights in and out of Bali. During the periods of study, several volcanoes eruptions in Java and surrounding areas were significantly influence mass concentration of $PM_{2.5}$ and $PM_{2.5-10}$, such as high level coarse particle (100 ug/m^3) was observed during Kelud volcano eruption at February 14, 2014.

Table 2. Black carbon and elemental concentrations of fine and coarse in Surabya during 2012-2015

	$PM_{2.5}$ (ng/m^3)					**$PM_{2.5-10}$ (ng/m^3)**				
		Mean		Range			Mean		Range	
Na	534	±	412	17	2237	524	±	405	44	2724
Mg	42	±	39	0.4	259	113	±	68	9	517
Al	59	±	46	0.3	341	358	±	379	3	3891
Si	132	±	103	16	1026	954	±	990	190	10569
S	711	±	279	167	1487	299	±	164	45	699
K	231	±	111	4.3	632	161	±	110	37	905
Ca	125	±	59	16	298	628	±	343	132	2394
Ti	4.37	±	2.97	0.2	20	32	±	21	6	195
Mn	9.29	±	7.40	0.1	47	18	±	12	3	82
Fe	124	±	84	6.8	568	486	±	320	56	2620
Zn	285	±	205	8	1091	290	±	236	8	1596
Pb	420	±	518	2.4	2779	234	±	399	1	2258
BC	3472	±	1036	673	7427					

Table 2 shows the mean and range of values black carbon (BC) concentration for $PM_{2.5}$ and the elemental concentration for $PM_{2.5}$ and $PM_{2.5-10}$ at Surabaya site over the period of 2012-2015. The BC content of the samples were measured using a reflected light instrument and BC was measured only for the fine fraction filters ($PM_{2.5}$). The average and range concentration of BC were 3472 and 673 to 7427 ng/m^3. While the average ratio of BC concentrations compared to the $PM_{2.5}$ concentrations was 22.6 %. The highest BC concentration was found in June 2012 and July 2015, it could be due to forest fires from surrounding areas of Surabaya. It can also be shown from Table 2 that there are several samples have high concentration of crustal elements especially Si. The high concentration of crustal elements was observed from the samples collected during the eruption episode, especially from mount Kelud at February 2014. The high amount of silica should be considered due to its impact on human health. These are free crystalline silica known to cause silicosis, a disabling and potentially fatal lung disease typically found in miners and quarry workers exposed to high concentrations of siliceous dust over long periods of time [14].

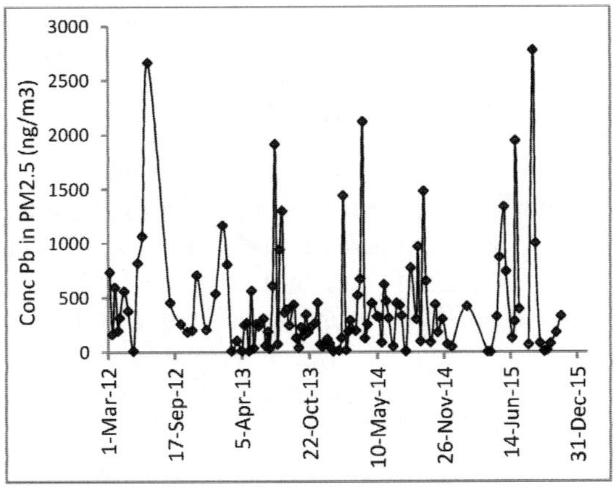

Figure 4. The distributions of Pb in $PM_{2.5}$ at Surabaya site during 2012-2015.

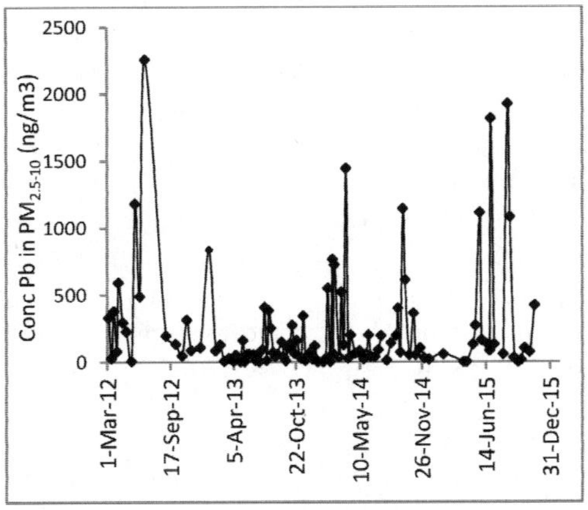

Figure 5. The distributions of Pb in $PM_{2.5\text{-}10}$ at Surabaya site during 2012-2015.

Figure 4 and Figure 5 shows the measured distributions of lead for the fine and coarse particle fractions. The result obtained from this study showed that the average level of lead in $PM_{2.5}$ and $PM_{2.5\text{-}10}$ collected during 2012-2015 in Surabaya site were 420 and 234 ng/m^3, respectively. While the maximum concentration observed were 2779 ng/m^3 in $PM_{2.5}$ and 2258 ng/m^3 in $PM_{2.5\text{-}10}$. In Figure 5, It could be shown that lead concentration was also high in the coarse fraction and the distributions have the same pattern with the fine fraction. The lead in coarse particle could arise from road dust that has been re-suspended by the motion of tires over the road surface [4]. The annual average of lead in $PM_{2.5}$ and $PM_{2.5\text{-}10}$ during 2012 to 2015 were 560, 323, 417, 525 and 422, 124, 244, and 448 ng/m^3 respectively, while the maximum concentrations were 2662, 1911, 2119, 2779 and 2258, 835, 1446, 1924 ng/m^3 respectively. These annual average values were significantly higher and several data fall to the 24 h National Ambient Air Quality Standard of Indonesia for lead in total suspended particulate matter (2000 ng/m^3). These results revealed that lead concentration in Surabaya were much higher compared to other cities in Asia countries [12] and even higher than Tangerang that already polluted with lead [4]. The results of the preliminary study of source apportionment in Surabaya showed that biomass, vehicle, soil and industrial activities are the main sources [6]. To improve health protection for at-risk group, especially children, In 2008, the USEPA has revised NAAQS for Lead in total suspended particulate matter 10 times tighter than previous standards (from 1500 to 150 ng/m^3 as a quarterly average value). The results obtained from this study showed that the annual average lead in Surabaya site during 2012 to 2015 have exceeded the US standard (150 ng/m^3). Much of lead in fine fraction was due to emissions from the emissions of vehicles burning fuel containing tetraethyl lead. However, Indonesia has eliminated leaded gasoline, the high lead concentrations could be from other lead sources such as smelter industry and battery recycling plant, as reported in the previous studies related the high Pb concentration in Tangerang [4]. This finding gave very important information as early warning and scientific research based references to avoid greater financial disadvantage and human health impact. A better solution and proper action should be taken into this matter. The local government institution and Ministry of Environment have to take the follow up of this fact.

4. Conclusions

The annual mean concentrations of $PM_{2.5}$ collected at Surabaya sites during March 2012 to November 2015 showed that majority exceeded the annual air quality standards (15 μg/m^3). In general, the ratio of $PM_{2.5}$ to PM_{10} were less than 0.5, while the average ratio of BC concentrations compared to the $PM_{2.5}$ concentrations was 22.6 %. The elemental concentrations of Na, Mg, Al, Si, S, K, Ca, Ti, Mn, Fe, Zn and Pb in $PM_{2.5}$ and $PM_{2.5\text{-}10}$ were determined and the results showed that high concentration of crustal element, especially Si were observed in samples collected during volcanic eruption episode. While the lead level in $PM_{2.5}$ and $PM_{2.5\text{-}10}$ at Surabaya site showed significantly higher than in other sites and should be further followed by identification and quantitative apportionment of the particle sources. Through these results, important finding related to heavy metal pollution in Surabaya were achieved and these results could significantly support national program of air quality improvement.

Acknowledgments
The authors acknowledge the National Nuclear Energy Agency for the financial support throughout this work. The authors also gratefully acknowledge to all the staff members in Radiometry Analytical Techniques group in BATAN Bandung, and P3KLL Ministry of Environment and Forestry for their technical assistance. The authors also gratefully acknowledge to the International Nuclear Energy Agency (RAS7023 and TCINS7007).

5. References

[1] "Population of Indonesia by Province 1971, 1980, 1990, 1995, 2000 and 2010". Central Bureau of Statistics. Retrieved 18 May 2015.
[2] S.S. Lim, et al. A comparative risk assessment of burden of disease and injury attributable to 67 risk factors and risk factor clusters in 21 regions, 1990–2010: a systematic analysis for the Global Burden of Disease Study 2010 Lancet 380, 2224-2260, 2012.
[3] Santoso M, Hopke PK, Hidayat A, Lestiani DD 2008 Science of The Total Environment 397 229-237
[4] Santoso M, Lestiani DD, Mukhtar R, Hamonangan E, Syafrul H, Markwitz A and Hopke PK 2011 Atmospheric Pollution Research 2 190-196.
[5] Syafei AD, Fujiwara A and Zhang J 2014 Procedia – Sosial and behavioral sciences 138 612-622.
[6] Ahmad EF and Santoso M 2016 Jurnal Kimia Valensi 2(2) 97-103.
[7] Santoso M, Lestiani DD, Hopke PK 2013 Journal of the Air & Waste, Management Association 63 (9), 1022-1025, 2013.
[8] Islam JB, Lutfor AKM, Rahman M. Sarkar, Ahmed KS and Begum BA 2014 Jagannath University Journal of Science 3-1 87-96.
[9] Santoso M and Lestiani DD, Application of energy Dispersive X-Ray Fluorescence (EDXRF) In Supporting National Program of Air Quality Program of air Quality Improvement In Indonesia XRF Newsletter of the IAEA Laboratories, Seibersdorf, Issue No 26, April 2014.
[10] Hopke PK, Xie Y, Raunemaa T, Biegalski S, Landsberger S, Maenhaut W, Artaxo P, Cohen DD 1997 Aerosol Sci Technol 27 726–35.
[11] Hopke PK, Cohen DD, Begum BA, Biswas SK, Bangfa N, Pandit GG, Santoso M et al 2008 Science of the Total Environment 404 103-112.
[12] Santoso M, Lestiani DD, Kurniawati S, Markwitz A, Trompetter WJ, Barry B and Davy P K 2014 International Journal of PIXE 24-3&4 157–165, DOI: 10.1142/S0129083514400087.
[13] Kristiansen NI, Prata AJ, Stohl A, Carn SA 2015 Geophysical Research Letters 42-2 588-596.
[14] Peter JB, Claire JH 2015 Chapter 60 – Impacts of Eruptions on Human Health. The Encyclopedia of Volcanoes (Second Edition), Pages 1035–1047.

The Distribution of Rainwater Acidity in Bandung

A Indrawati, D Aries Tanti, T Budiwati and Sumaryati

Indonesian Nasional Institute of Aeronatutic and Space (LAPAN), Jl dr djundjunan no 133 Bandung 40173, Indonesia

asri.indrawati@lapan.go.id

Abstract. Acidity value (pH value) of rainwater, are influenced by the concentration of anions and cations dissolved in rainwater and it was also influenced by several sources of pollutants emitted into the atmosphere. This research was conducted to determine the sources of pollutants that affect the pH value in Bandung. Sampling locations were conducted in 10 locations, both in Bandung City and Regency. We used data monthly averages of pH value from 2006 – 2015, to see the distribution of pH values in Bandung. As for knowing the effect of pollutant sources that affect the pH value, Cipedes area was chosen to represent urban areas from the sampling location. We used data anion and cation concentrations of SO_4^{2-}, NO_3^-, Cl^-, NH_4^+, Na^+ and Ca^{2+} in 2006 - 2015. The statistical method used for data process was PCA method. Result shows that pH value in Bandung fluctuated from 2006-2015, with the lowest average pH value in 2010 and 2014 was 5.2. From result of data processing with PCA gained 3 factors influencing pH value Bandung, which were Factor 1: SO_4^{2-}, NO_3^-, K^+ and Mg^{2+}, Factor 2: Ca^{2+}, and NH_4^+, Factor 3: Cl^- and Na^+.

Keywords : acidity, SO_4^{2-}, NO_3^-, Cl^-, NH_4^+, Na^+ and Ca^{2+}.

1. Introduction

Research on rainwater chemistry has been widely carried out by researchers around the world, because it is related to acid rain and the impact on the environment, especially human health [1]. Rain is one of the most effective ways to remove pollutants in the atmosphere. Rainwater chemistry is the result of a series of complex chemical reactions that involve the interaction between the microphysic process and cloud dynamics [2].

Rainwater composition is different for each location, because it is very difficult to control the source of pollutants derived from both anthropogenic and natural sources [3]. The level of the acidity parameter of rainwater can be used as an indicator of the pollution level of a region [4]. For example industrial areas, where industrial processes emit large amounts of sulfur and nitrogen oxides into the atmosphere, and these gases are converted in a series of chemical reactions in the atmosphere to form strong acidic compounds. When this strong acid compound is dissolved in rain water will result low acidity (pH) of rainwater [2].

According to several studies that have been carried out regarding the trend and characteristics of the pH value of rainwater, in Beijing the lowest pH value was obtained during high rainfall, i.e. between July - September [5]. In India, the pH value is strongly influenced by the presence of suspended particulate matter, which will have a neutral effect on the acidity of rainwater [6]. In the Brazilian region, the lowest pH value occurs during the summer, i.e. December, January and February, while the highest pH values occur during spring and autumn [7].

Bandung is an urban area that has a unique topography like a basin area. Bandung is famous tourism city in Indonesia with its rapid development as an urban and industrial area. Previous research on Bandung air quality related to acid deposition, shows in 2013 acid rain has been occurred in Bandung, because the pH value of rainwater was below 5.6. [8]. Based on this, the objective of the research was conducted using rainwater acidity data monitoring in a longer period, from 2006 - 2015 to see the distribution of acidity in Bandung, both the city and regency, and also to see the influence of pollutant sources on the acidity level in Bandung, especially for urban areas. This study will provide useful information for identification of potential source in rainwater and improvement of air quality in Bandung.

2. Data and Method

2.1. Sampling location and sample collection
The location of the study was conducted in 10 locations for the city and regency of Bandung. Its location were Cipedes, Dago, Kebon Kalapa, Kopo, Jl. Riau, Cikadut, Ciparay, Cililin, Soreang and Padalarang (figure 1). Cipedes, Dago, Kebon Kalapa, Jl. Riau and Cikadut were include in city area, meanwhile Kopo, Ciparay, Cililin, Soreang and Padalarang, are regency area. Bandung is a Basin area surrounded by mountains and hills, where transport of pollutants from the center of Bandung city will be carried by dominant wind then turn around due to collisions with hillsides [8]. Rainwater sampling is done every rainy day from January 2006 - December 2015 by using automatic rain sampler for Cipedes and raingauge for other locations.

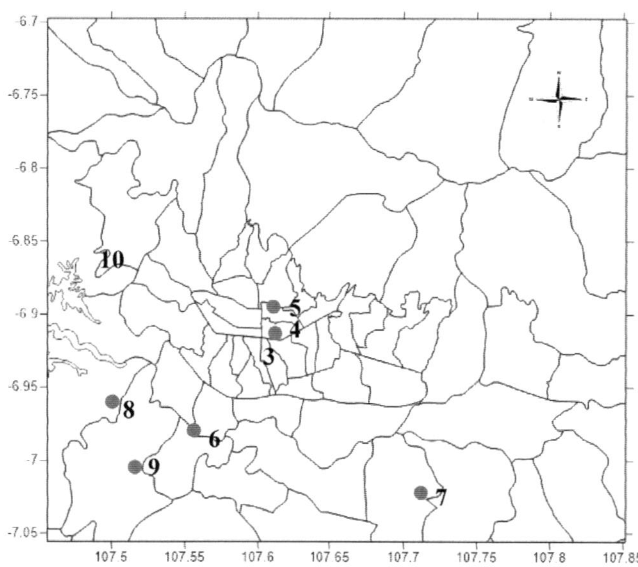

City area :
1. Cipedes
2. Dago
3. Kebon Kalapa
4. Jl. Riau
5. Cikadut

Regency area :
6. Kopo
7. Ciparay
8. Cililin
9. Soreang
10. Padalarang

Figure 1. Sampling location

2.2. Sampling analysis and quality control
Samples of rainwater are then measured in volume and placed in polyethylene bottles. Measurement of rainwater pH was done by using the pH meter Horiba F 51 with measurement control at 25°C. After measurement of pH value, the sample of rainwater filtered by using Millipore filter paper with poor size of 0.45 μm. Anion and cation concentrations in rainwater were determined using Dionex ICS 1500 ion chromatography for anions and ICS 1600 for cations. The quality of analytical data was also checked

IOP Conf. Series: Earth and Environmental Science **303** (2019) 012037 doi:10.1088/1755-1315/303/1/012037

by a cation-anion balance and by comparison of the measured conductivity with the conductivity calculated from the concentration of all measured ions and their specific conductivity refer to EANET guidelines.

2.3. Data analysis

We used data monthly average of pH value to see the distribution of rainwater acidity in Bandung from 2006 - 2015 for 10 monitoring locations. To see the effect of pollutant sources on rainwater acidity values in Bandung, we choose Cipedes data anion and cation concentration (representing Bandung city area) which is also in the form of monthly averages from 2006 - 2015. Data processing is done using PCA (Component principle Analysis) processed with SPSS.23 software

3. Result and Discuss

3.1. Distribution of acidity value rainwater

The acidity level of rainwater is determined by the relative amount of acids and bases dissolved in rainwater. The dominant strong acid in influencing the level acidity of rainwater, globally is sulfur, nitrate and hydrochloride and the dominant weak acids such as formic acid, acetic acid and carboxylic acids. Organic acids play an important role in the acidity of rainfall when the concentration of strong acids is low, but the deposition of organic acids is considered not important for the ecosystem because organic acids are the consumption of microorganisms on the surface of the earth.

The distribution of the acidity value rainwater for Bandung area varies from year to year. By using interpolation data from 10 sampling points both city and regency areas, contour distribution of rainwater acidity was obtained for Bandung. Based on Figure 2, in 2006 the acidity of rainwater ranged from 5.93 – 6.23 for the regency. But for the city itself, the acidity of rainwater has ranged between 5.63 - 5.73. The acidity value of rainwater is still above the reference value for acid rain, which is 5.6.

From 2007 until 2010, the distribution of acidity in rainwater showed a lower pH value, especially for city area, with a range of pH values was 4.43 - 5.32. Distribution of low pH values from the city to the southwest part of Bandung, Soreang and Cililin and suggesting acid rain events in this area were serious. Because acidity value of rainwater shows below the reference value for acid rain. This condition assumed that anthropogenic source contributed to decrease the acidity of rainwater value during this sampling period, such as fossil fuel buring, industrial and vehicular emission [9].

In 2011until 2014 it was seen that the spread of low pH values was still to the west of the city area with a range of pH values between 4.43 - 5.33. The distribution of rainwater acidity in 2015 showed that the most locations in Bandung regency had a condition low pH values. But it's different from some locations in the city, where the pH value ranges from 6.03 to 6.33. According to the research that has been carried out the existence of Na^+, Ca^{2+} and NH_4^+ which is dissolved in rainwater has a great influence on neutralizing the acidity of rainwater for the area of Bandung [10]. Acidic pH reveals the presence of strong acids while neutral or alkaline pH indicates neutralization of acids by carbonates, mineral dust or by ammonium. This may be due to the reaction of sulphuric and nitric acid absorbed in the aerosols with alkaline carbonates in the particulate matter [2].

An average pH value has been carried out for 10 locations, showing that since 2008 acid rain has been occurred in Bandung. Based on Figure 3, from 2006 until 2015, there were two valleys which stated the lowest value of rainwater acidity in Bandung, it was in 2010 and 2014, with pH value around 5.2. Fluctuations in pH values correlate with fluctuations in the content of ions dissolved in rainwater. When concentration of SO_2 and NO_2 in air is high, and then they are dissolved in rain water it forms SO_4 and NO_3, then the pH value of rainwater will be low [11].

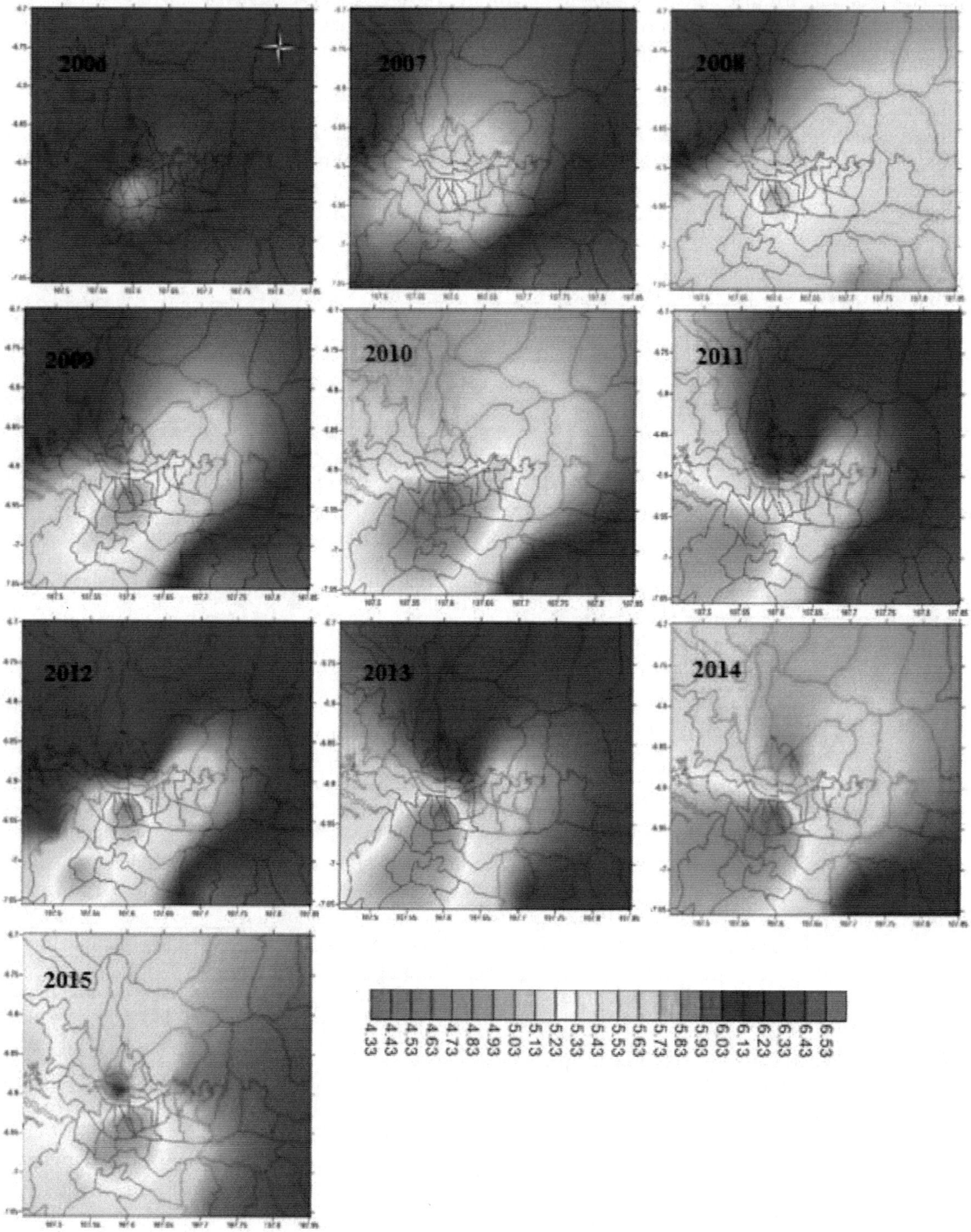

Figure 2. Distribution of pH value Bandung 2006 – 2015

Year	*pH*
2006	6.06
2007	5.72
2008	5.52
2009	5.51
2010	5.20
2011	5.48
2012	5.65
2013	5.46
2014	5.24
2015	5.38

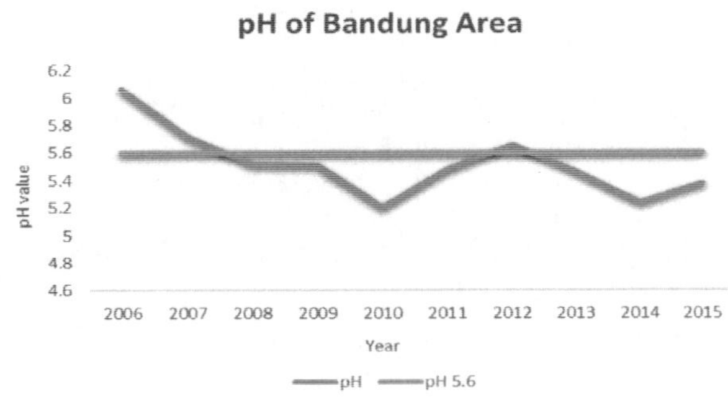

Figure 3. pH value Bandung 2006 – 2015

3.2. pH and rainfall

Rainfall volume is related to the process of washing pollutants in the atmosphere which will ultimately affect the chemical content of rainwater, especially the acidity of rainwater. During the dry season, the tendency to wash pollutants in the atmosphere does not run optimally, because of the large amount of pollutants in the atmosphere and the least rainfall. Several studies have been carried out related to the relationship between pH value and rainfall, it was found that if rainfall is high, especially in the rainy season, the pH value will be low, whereas when rainfall is low, the pH value will be high [5]

However, based on research conducted [4] there was no correlation between rainfall and pH value. In addition, [12] in his research was found that rainfall does not directly affect the acidity of rainwater, but there is a tendency of pollutant dispersion during heavy rain the previous day or the same day, the pH value will be low (acid).

In Figure 4, it can be seen that in each sampling location, the relationship between rainfall and the acidity of rainwater does not provide a significant correlation. In certain months it shows when the rainfall volume is low, the pH value will be high, and vice versa. From the calculation of correlation for several locations shows the significance of the relationship between rainfall and pH value with a negative relationship, while other locations do not provide a significant relationship. Some researches were needed regarding the effectiveness of washing the atmosphere carried out by rainwater against air pollution, especially in Bandung, which is correlated with the value of dissolved ion concentration also the acidity of rainwater.

3.3. Analysis of PCA

Data processing was performed using PCA statistical analysis method to identify the influence of anthropogenic sources and natural sources on the value of rainwater acidity in the Bandung basin and for the location we choose Cipedes represents for urban location. The variables used are the values of the concentration of rainwater anions and cations such as SO_4^{2-}, NO_3^-, Cl^-, NH_4^+, Na^+, K^+, Ca^{2+} and Mg^{2+}.

The result from PCA analysis shows the Barlett's test of sphericity for the overall significance of all correlations in a correlation matrix, which is characterized by significance (p value <0.05). There is a significance level of 0.00 (p value <0.05). To measure the adequacy of sampling, by comparing the magnitude of the correlation coefficient observed with a partial correlation coefficient, the KMO (Kaiser-Meyer-Olkin) value shows a value of 0.218 (KMO value <0.05), this means that the correlation between pairs of variables can be explained by other variables .

From the total variance explain result (figure 5) there are 3 factors that will explain the sources that affect the pH value in Cipedes. Of the three variables can explain the variance of 8 items by 87.99%. This value is quite good, because it is proven to explain more than 50% of the variance of the variable.

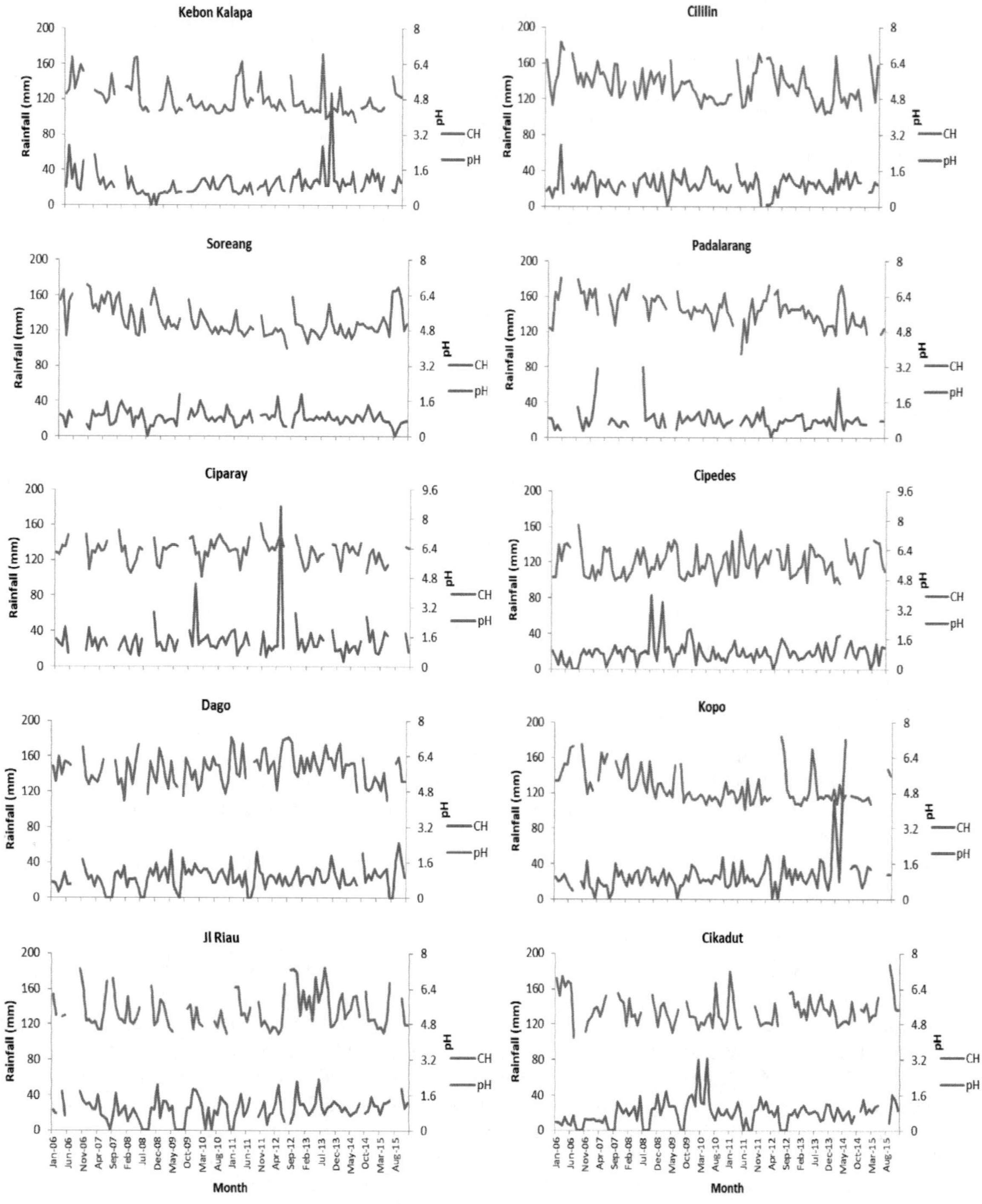

Figure 4. Rainfall and pH value

IOP Conf. Series: Earth and Environmental Science **303** (2019) 012037 doi:10.1088/1755-1315/303/1/012037

Component Matrix[a]

	Component		
	1	2	3
so4	.578	.632	-.250
no3	.850	.256	-.257
cl	.217	.263	.911
nh4	-.450	.842	-.182
na	.746	.131	.573
k	.890	.137	-.078
ca	.720	-.527	-.097
mg	.956	-.105	-.215

Extraction Method: Principal Component Analysis.

a. 3 components extracted.

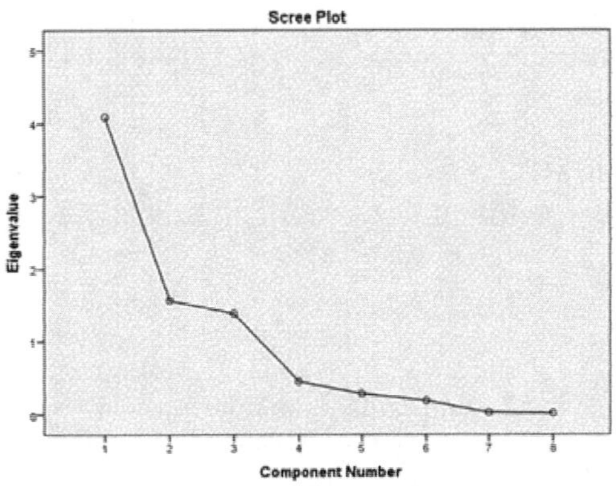

Figure 5. The results of component matrix and screen plot

Rotation analysis is then carried out to explain clearly the distribution of variables, where the results show 3 rotations of the matrix component, according to the number of factors obtained. There are 3 variables that are highly correlated. Factor 1 explained about 51.11% of the total variance with strong positive loading for SO_4^{2-}, NO_3^-, K^+ and Mg^{2+} (0.849; 0.902; 0.817 and 0.813). Showing the pH value is influenced by transportation factors, where the source of pollutants comes from motorized vehicles which contribute to the concentrations of SO_2 and NOx in the air, while K^+ and Mg^{2+}, indicate the source of soil (dust) from the road and the flying land into the air. The loadings of all ions in factor I give some result that industrial activity gives way to vehicular activity which causes stirring of soil dust raised by the convective activity coupled with the dust raised by vehicles themselves [12].

Based on data from the Central Bureau of Statistics in Bandung Municipality in Figure 2007 and 2012, the number of motorized vehicles was 699,320 units in 2006, and increased by 1,252,230 units in 2011. Based on this, it is seen that transportation is indeed the main factor that contributes to the acidity of rainwater Bandung. Besides that, since the opening of Cipularang Toll Road, which made it easy for tourists to visit Bandung, they also contributed to the source of vehicle emissions, especially during holidays. Recorded in 2009 the number of vehicles entering Bandung via Pasteur Toll Gate was 9,999,213 vehicles and increased by 11,035,023 vehicles in 2011.

Factor 2 which explained 19.60% of the total variance has strong positive correlated with Ca^{2+}, and negative correlation with NH_4^+. NH_4^+ has a negative correlation of 0.964, while Ca^{2+} is 0.809. This result showed pH value more influenced by limestone which contributes Ca^{2+} to the air and NH_4^+ in urban area produce from industrial processes, vehicular emissions and volatilization from soil [13] and commonly associated to biomass decomposition [7]. Negative loading for NH_4^+ means concentration of NH_4^+ greatly affects the acidity value of rainwater in Bandung in the opposite direction of correlation. Factor 3 with strong positive loading 17.28% for Cl^- and Na^+ (0.970 and 0.778), both of them are very strongly influenced by sea resources (sea salt breeze) [7].

Based on research conducted by Xiao, 2016, using PCA to identify sources of pollutants that affect air quality in Xi'an, China. The results showed there are two factors that affect air quality in Xi'an, China, the first was affected by mixed source such as crustal, anthropogenic and marine origins, possibly transported over a long distance and the second factor were from neutralization process which is also from anthropogenic source that relase fluoride contents. In another research, done by Cerqueira et al, 2014 was found 4 factors that affect the concentration value of anion and cation in rainwater. There are soil and road dust, anthropogenic source, sea salt and from agricultural and cattle farming.

4. Conclusion

Acid rain results from human activities that affect the environment. The distribution of rainwater acidity value for the Bandung area has shown the distribution of acidic pH values to the city and regency of Bandung in 2015. However, there are conditions where the city area has a pH value> 6.0 which is thought to be due to neutralization by the presence of Ca^{2+}, Na^{+} and NH_4^{+} ions . By using PCA, obtained variables that affect the high and low pH values for Cipedes , were vehicles, street dusts and also the influence of marine sources.

Acknowledgement
The authors are grateful for Center for Atmospheric Science and Technology, LAPAN to funding this research.

References

[1] Bharti P. K, V. Singh, and P. K. Tyagi 2017 Assessment of rainwater quality in indutrial area of rural Panipat (haryana), India Archive of Agricultural and Environment Science 2(3) p 219-223
[2] Chungtai M, S. Mustafa, and M. Mumtaz 2014 Study of physicochemical parameter of rainwater: a case study of Karachi, Pakistan American Journal of Analytical Chemistry vol 5 p 235-242
[3] Venkateshwarlu P, and J. Satyanarayana 2014 Chemical composition of rainwater at an urban area during southwest monsoon Interbational Journal of Modern Chemistry and Applied Science 1(1) p 1-4
[4] Gaddamwar A. G 2011 Analytical study of rainwater for determination of polluted or unpolluted zone International Journal of Environmental Science vol 1 p 1317-1322
[5] Yang F, J. Tan, Z. B. Shi, Y. Cai, K. He, Y. Ma, F. Duan, T. Okuda, S. Tanaka, and G. Chem 2012 Five-year record of atmospheric precipitation chemistry in urban Beijing, China Atmospheric Chemistry and Physics vol 12 p 2025-2035
[6] Al Obaidy A. H. M. J, and H. Joshi 2006 Chemical composition of rainwater in a tropical area of nothern India Atmospheric Environment 40(2006) p 6886-6891
[7] Cerqueira M. R. F, M. F. Pinto, I. N. Derossi, W. T. Esteves, M. D. R. Santos, M. A. C. Matos, D. Lowinsohn and R. C. Matos 2014 Chemical characteristics of rainwater at a southestern site of Brazil Atmospheric Pollution Research 5(2014) p 253-261
[8] Indrawati, A and D. A. Tanti 2014 Chemical composition and seasonal variation of wet and dry deposition in Bandung basin area Proceedings of National Seminar of Atmospheric Science, LAPAN
[9] Gupta, A., and B. Chakraboty 2014 Rainwater quality analysis in selected areas of eastern and notheastern India Proceeding of the International Conference on Advances in Civil, Structural, Environmental & Bio-Technology – CSEB 2014 p 23 - 27
[10] Indrawati, A and D. A. Tanti 2013 Analysis of chemical composition from wet and dry deposition in Ciater and Padalarang Proceedings of National Seminar of Atmospheric Science, LAPAN
[11] Budiwati T, D. A. Tanti and A. Indrawati 2012 Influence of transporrtation (NO2 and SO2) to wet deposition in Bandung Journal of Tropical Environment vol 6 p 81-89
[12] Xiao, J 2016 Chemical composition and source identification of rainwater constituents at an urban site in Xi'an
[13] Behera, S. N, M. Sharma, V. P. Aneja and R. Balasubramanian 2013 Ammonia in the atmosphere a review on emission source, atmospheric chemistry and deposition on terrestrial bodies Environmental Science and Pollution Research DOI 10.1007/s1 1356-013-2051-9

Greenhouse gases and air pollutants monitoring project around Jakarta megacity

M Nishihashi[1], H Mukai[1], Y Terao[1], S Hashimoto[1], Y Osonoi[1], R Boer[2], M Ardiansyah[2], B Budianto[2], G S Immanuel[2], A Rakhman[2], R Nugroho[3], N Suwedi[3], A Rifai[3], I M Ihsan[3], A Sulaiman[3], D Gunawan[4], E Suharguniyawan[4], M S Nugraha[4], R C Wattimena[4] and A F Ilahi[4]

[1] National Institute for Environmental Studies (NIES), 16-2 Onogawa, Tsukuba, Ibaraki 305-8506, Japan
[2] Bogor Agricultural University (IPB), Gedung Fisik dan Botani Lantai 2, Kampus IPB Baranangsiang, Jl. Pajajaran No.1, Bogor, Jawa Barat 16143, Indonesia
[3] Agency for the Assessment and Application of Technology (BPPT), Laboratorium Geostech 820, Kawasan Puspiptek Serpong, Tangerang Selatan, Banten 15314, Indonesia
[4] Meteorological, Climatological, and Geophysical Agency (BMKG), Jl. Angkasa I No.2, Kemayoran, Jakarta Pusat, DKI Jakarta 10610, Indonesia

nishihashi.masahide@nies.go.jp

Abstract. National Institute for Environmental Studies (NIES) has been implementing a joint monitoring project of greenhouse gases (GHGs) and air pollutants in Indonesia with Bogor Agricultural University (IPB), Agency for the Assessment and Application of Technology (BPPT), and Meteorological, Climatological, and Geophysical Agency (BMKG). To estimate the amount of anthropogenic emissions from Jakarta megacity (Jabodetabek) and compare with city activities, we developed a ground-based comprehensive monitoring system of GHGs and air pollutants and installed it at Bogor (center of Bogor city) in March 2016, Serpong (Jakarta suburb) in August 2016, and Cibeureum (mountainous area, background-like site) in March 2017. The monitoring system consists of data acquisition/control units and the instruments for continuous measurements of CO_2, CH_4, CO, NO_x, SO_2, O_3, aerosol concentrations ($PM_{2.5}$, PM_{10}, BC) and the chemical components, and meteorological parameters. Flask sampling of air is also done to analyze N_2O, SF_6, and carbon isotopes (^{13}C, ^{14}C) in CO_2 and to validate the continuous measurement data. The result shows that CO_2 mole fractions observed at three sites have clear diurnal variations representing the minimum values from 12 to 15 local time while the values at Bogor and Serpong are 6.8 and 7.1 ppm higher than Cibeureum, respectively.

Keywords : CO_2, CH_4, CO, NO_x, SO_2, O_3, aerosol.

1. Introduction

Megacities, being hotspots of human activity, have emitted large quantities of greenhouse gases (GHGs) and air pollutants into the atmosphere with distinct climate effects [1]. Jakarta, the capital and largest city of the Republic of Indonesia, including suburban cities (Bogor, Depok, Tangerang, South Tangerang, and Bekasi; locally known as "Jabodetabek"), has a population of 32 million people and has been listed as the second largest megacity in the world and the largest megacity in Southeast Asia [2]. The increasing GHGs and air pollutants emissions from the megacity in developing country such as

Jakarta due to rapid economic growth in recent years have been recognized as one of the important issues to be solved [3]. Since reduction strategies of GHGs and air pollutants emissions in developing countries have been relatively insufficient compared to developed countries, further urbanization will lead to further increase of GHGs and deterioration of air quality. Urban climate change-related risks, such as heat stress, extreme precipitation, inland and coastal flooding, and air pollution, are increasing with widespread negative impacts on people (and their health, livelihoods, and assets) and on local and national economies and ecosystems [4]. Hence new interdisciplinary research studies have been needed to increase our current understanding of the interactions between emissions, air quality, and regional/global climates taking place in megacities [1].

Several urban air quality monitoring studies in Indonesia were conducted in Jakarta [5, 6, 7, 8], Serpong [8, 9], Bandung [8, 10, 11], and Makassar [12]. However, highly accurate and continuous monitoring has been insufficient due to the limitations of environmental monitoring budgets and experts [3, 6]. Such situations may not promote better understanding of GHGs and air pollutants variability in Indonesia, which is considerably important for making quick actions for global warming mitigation.

National Institute for Environmental Studies (NIES) has been implementing a joint monitoring project of GHGs and air pollutants in Indonesia with Bogor Agricultural University (IPB), Agency for the Assessment and Application of Technology (BPPT), and Meteorological, Climatological, and Geophysical Agency (BMKG). To estimate the amount of anthropogenic emissions from Jakarta megacity (Jabodetabek) and compare with city activities, we developed a ground-based comprehensive monitoring system of GHGs and air pollutants and installed it at three sites around Jakarta. In this paper, we introduce the monitoring system and the initial results obtained at three monitoring sites.

2. Monitoring system

We developed a ground-based comprehensive monitoring system of GHGs and air pollutants. By monitoring both GHGs and air pollutants, it will be possible to gain a greater understanding of the complex relationships between these species during different seasons and times of day as a result of emissions from anthropogenic and natural sources as well as secondary atmospheric chemical reactions [13]. After developing the monitoring system in NIES, we transferred it from Japan to Indonesia and then installed it at Bogor (center of Bogor city) in March 2016, Serpong (Jakarta suburb) in August 2016, and Cibeureum (mountainous area, background-like site) in March 2017 (Figure 1, Table 1).

The monitoring system consists of data acquisition and system control units (GSCT-14, Kimoto) and the instruments for continuous measurements of carbon dioxide (CO_2), methane (CH_4) (G2301, Picarro), carbon monoxide (CO) (CO-30r, Los Gatos Research), nitrogen oxides (NO_x) (Model 42i-TL, Thermo), sulfur dioxide (SO_2) (Model 43i-TLE, Thermo), ozone (O_3) (OA-787, Kimoto), aerosol concentrations (particulate matter ($PM_{2.5}$, PM_{10}), black carbon (BC)) and the chemical components of $PM_{2.5}$ and PM_{10} (nitrate ion (NO_3^-), sulfate ion (SO_4^{2-})) (ACSA-14, Kimoto), and meteorological parameters (WXT520, VAISALA) (Table 2). Flask sampling of air has also been done automatically at 1:30 PM in local time once a week to analyze nitrous oxide (N_2O), sulfur hexafluoride (SF_6), and carbon isotopes (^{13}C, ^{14}C) in CO_2 in NIES and to validate the continuous measurement data of CO_2, CH_4, and CO (these flask sampling data are not shown in this paper). Figures 2 and 3 show the framework and schematic diagram of the monitoring system, respectively. Also Figures 4 and 5 express the appearance of the system installed at IPB in Bogor. This system allows us to control remotely not only all instruments but also peripheral devices (e.g., pumps, valves) through the IPsec virtual private network constructed between three monitoring sites and NIES. The system also has an automatic operation function for power failures because electric power supply is sometimes unstable due to high lightning activity in Indonesia. Moreover, the system is corresponding to redundant operation for any device failures (i.e., there are dual air sampling lines, dual dust filters, dual sampling pumps, etc.; Figure 3). For instance, if any problems occur in one sampling pump, we can change the pump to alternative one remotely. To minimize missing data by any instrument and/or program errors, the system has an e-mail alert function to send alert mails on errors and also has a status mail function to send a status mail every morning. The monitoring data obtained from each measuring instrument and sensor, such as flow meters, pressure sensors, and temperature sensors, are visualized on the monitoring display (Figure 6) in real-time and recorded every

1 minute in unified data format with GPS time stamps. These data files are transferred to a cloud server at regular intervals.

For highly accurate measurements, the observed values of CO_2, CH_4, CO, NO_x, and SO_2 mole fractions have been automatically calibrated with standard gases periodically. We have utilized 9 standard gas cylinders in Bogor and Cibeureum and 6 standard gas cylinders in Serpong (4 concentrations of CO_2/CH_4, 3 concentrations of CO, 1 concentration of NO, and 1 concentration of SO_2). These standard gases of CO_2, CH_4, and CO were calibrated with NIES-09, NIES-94, and NIES-09 scales, respectively [14], in advance of the shipment from NIES. As for calibration for NO_x and SO_2, we used commercially available NO and SO_2 standard gases which had higher concentrations (e.g., ppm level). Such gases were diluted by zero gas (i.e., NO_x and SO_2 free air) to make span gases for the analyzers.

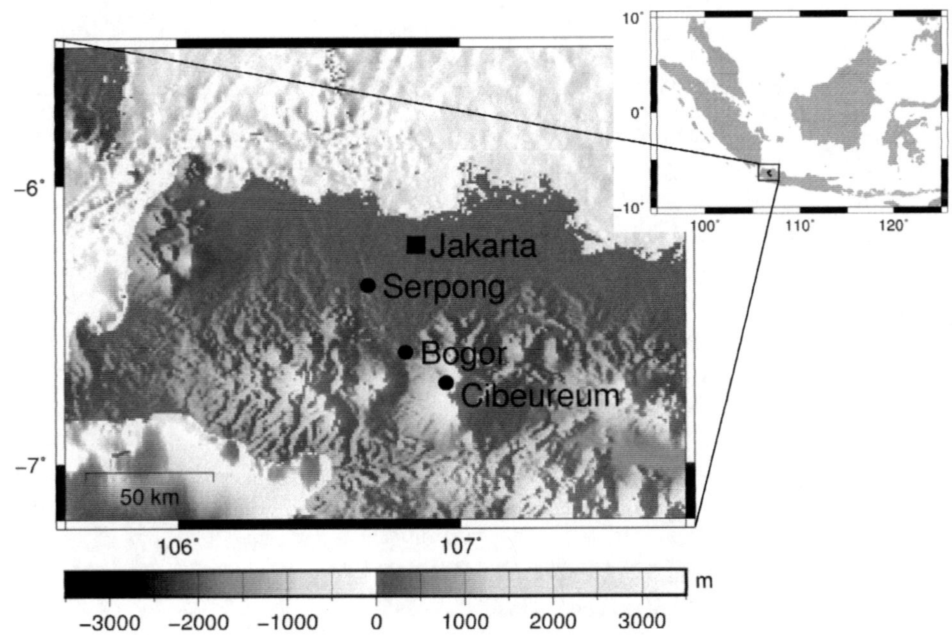

Figure 1. Map of the monitoring sites.

Table 1. Location and affiliation of the monitoring sites and monitoring starting date.

Location	Affiliation	Latitude	Longitude	Altitude	Starting date
Serpong *(Jakarta suburb)*	Geostech Laboratory, Agency for the Assessment and Application of Technology (BPPT)	6.36°S	106.67°E	63 m	26 Aug 2016
Bogor *(center of Bogor city)*	Center for Climate Risk and Opportunity Management in Southeast Asia Pacific, Bogor Agricultural University (IPB CCROM-SEAP)	6.60°S	106.81°E	266 m	1 Jun 2016
Cibeureum *(mountainous area)*	Meteorology Station Cibeureum, Meteorological, Climatological, and Geophysical Agency (BMKG)	6.71°S	106.95°E	1160 m	10 Mar 2017 (O_3, PM) 6 Jun 2017 (all)

IOP Conf. Series: Earth and Environmental Science **303** (2019) 012038 doi:10.1088/1755-1315/303/1/012038

Table 2. Monitoring species and instruments.

Species	Characteristics	Monitoring instrument	Serpong	Bogor	Cibeureum
CO_2 CH_4	GHG	$CO_2/CH_4/H_2O$ analyzer (G2301, Picarro)	√	√	√
CO	Air pollutant	CO analyzer (CO-30r, Los Gatos Research)		√	√
O_3	GHG, air pollutant	O_3 analyzer (OA-787, Kimoto)	√	√	√
$PM_{2.5}$, PM_{10} NO_3^- SO_4^{2-} Black carbon	Air pollutant, radiation related material	Continuous dichotomous aerosol chemical speciation analyzer (ACSA-14, Kimoto)		√	√
NO_x	Air pollutant, converted to NO_3^-	NO_x analyzer (Model 42i-TL, Thermo)	√	√	√
SO_2	Air pollutant, converted to SO_4^{2-}	SO_2 analyzer (Model 43i-TLE, Thermo)	√	√	√
CO_2, CH_4, N_2O, SF_6	GHG	Automatic flask sampler (S-KM-IS, Koshin-RS) and analysis in NIES	√	√	√
CO	Air pollutant				
CO_2 stable isotope (^{13}C and ^{18}O) CO_2 radiocarbon (^{14}C)	CO_2 indicator				
Temperature Relative humidity Wind direction Wind speed Rain, Pressure	Meteorology	Weather sensor (WXT520, VAISALA)	√	√	√

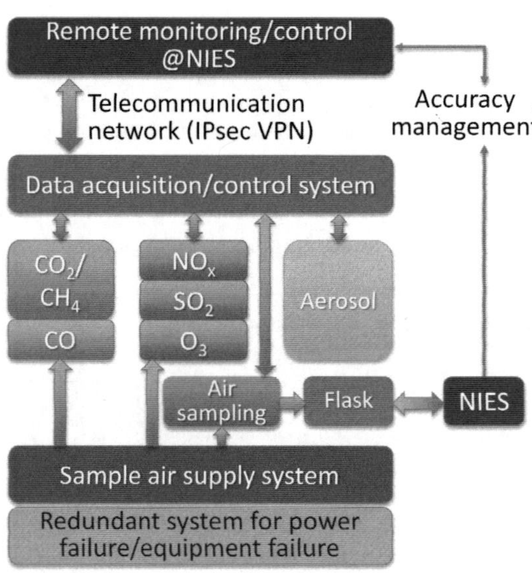

Figure 2. Framework of the monitoring system.

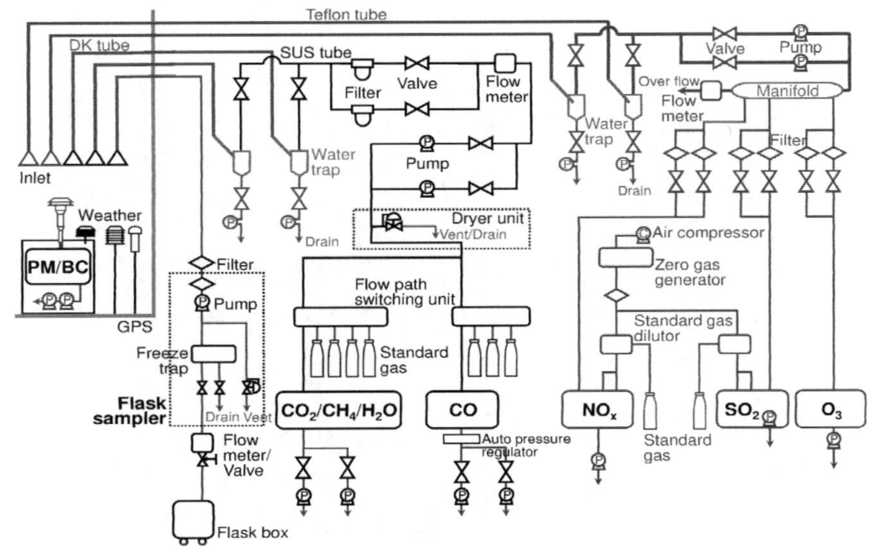

Figure 3. Schematic diagram of the monitoring system.

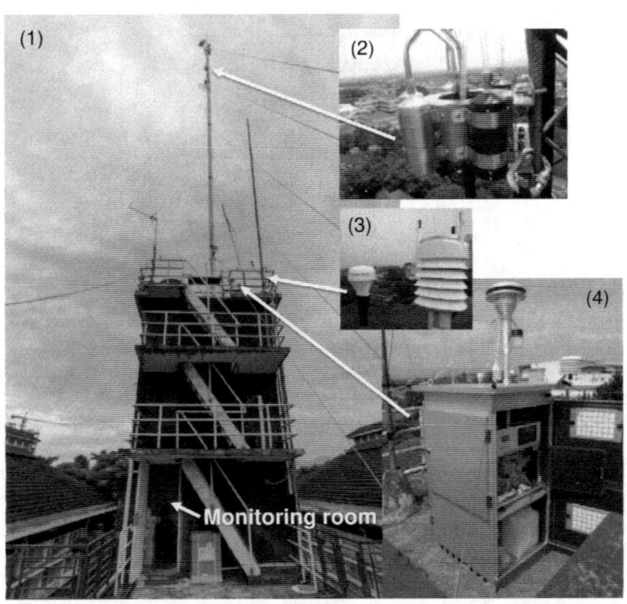

Figure 4. Appearance of the monitoring site at Bogor. (1) Monitoring room and tower, (2) inlets on the tower (35 m above the ground), (3) GPS antenna and weather sensor (WXT520, VAISALA), (4) continuous dichotomous aerosol chemical speciation analyzer (ACSA-14, Kimoto).

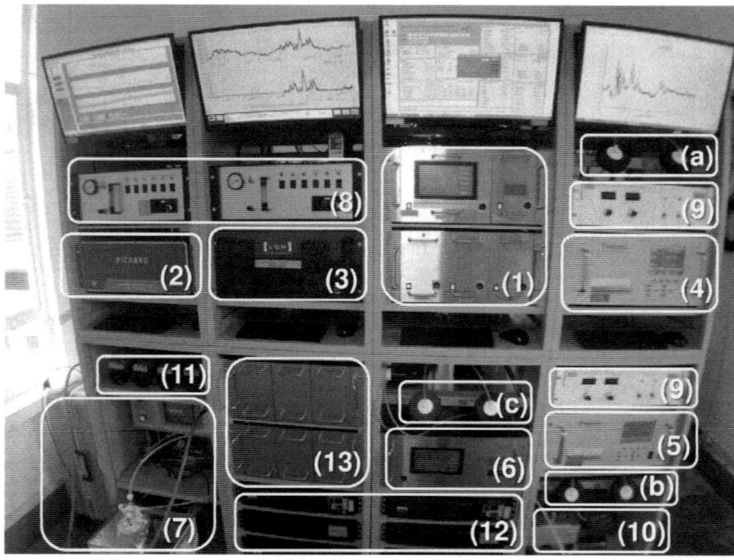

Figure 5. Monitoring system in the monitoring room at Bogor site. (1) Data acquisition and system control units (GSCT-14, Kimoto), (2) $CO_2/CH_4/H_2O$ analyzer (G2301, Picarro), (3) CO analyzer (CO-30r, Los Gatos Research), (4) NO_x analyzer (Model 42i-TL, Thermo), (5) SO_2 analyzer (Model 43i-TLE, Thermo), (6) O_3 analyzer (OA-787, Kimoto), (7) automatic flask sampler (S-KM-IS, Koshin-RS), freeze trap, and flask box, (8) flow path switching units for $CO_2/CH_4/H_2O$ and CO analyzers (GBJT-2001-01, Glovebox Japan), (9) standard gas dilutors for NO_x and SO_2 analyzers (103S, Nippon thermo), (10) zero gas generator for NO_x and SO_2 analyzers and exhaust pump for NO_x analyzer, (11) exhaust pumps for $CO_2/CH_4/H_2O$ and CO analyzers, (12) uninterruptible power supplies, (13) power supply units for GSCT-14. There are three sets of dust filters for NO_x (a), SO_2 (b), and O_3 (c) analyzers.

(a) (b)

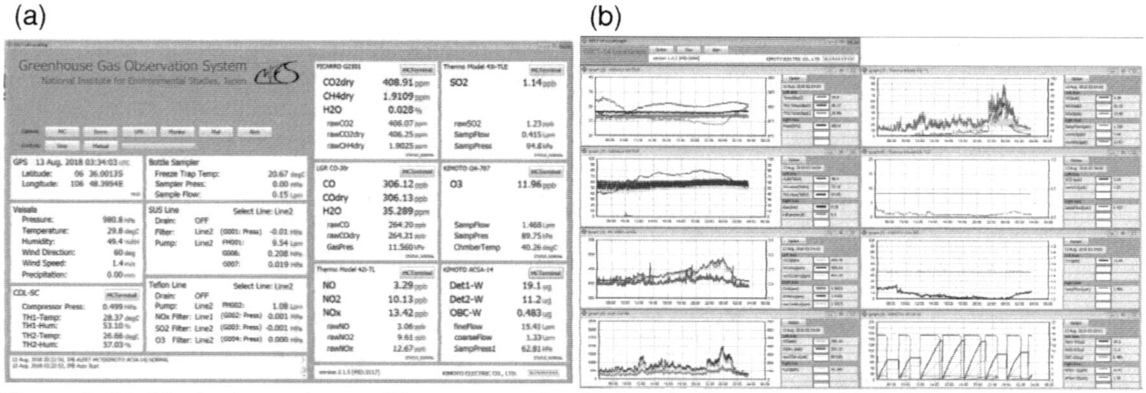

Figure 6. (a) Monitoring and controlling program for the monitoring system, (b) real-time monitoring data visualization program.

3. Monitoring data

We report the initial results obtained at Serpong, Bogor, and Cibeureum.

3.1. CO_2, CH_4, and CO

Figure 7 shows the temporal variation of hourly averaged CO_2, CH_4, and CO mole fractions observed at Serpong, Bogor, and Cibeureum from 1 June 2016 to 11 March 2018. The mole fractions of CO_2 and CH_4 at Serpong and Bogor and of CO at Bogor are relatively high value compared with those at Cibeureum. In particular, very high mole fraction (3–5 ppm) of CH_4 were observed at Serpong from evening to early morning as well as CO_2 at Serpong (Figure 7, 8). Figure 9 shows the diurnal variations of CO_2 mole fractions (hourly median) observed at three sites from 1 June 2017 to 28 February 2018. They have clear diurnal variations representing the minimum values at 3 sites during early afternoon (12–15 local time) and the maximum values at Serpong and Bogor during early morning and at Cibeureum during midnight. Hence we computed daytime values of CO_2 mole fractions at each site using three hours averages from 12 to 15 local time (Figure 10). The daytime values at Bogor and Serpong are 6.8 and 7.1 ppm higher than Cibeureum in average between June 2017 and March 2018, respectively. These results indicate that the mole fractions of CO_2 at Serpong and Bogor and of CO at Bogor have been affected by anthropogenic emission originating from fossil fuel combustion and industrial processes in the urban area. Meanwhile, those values observed at Cibeureum indicate that they have background-like characteristics. Furthermore, we need to survey the reason of the high concentration of CH_4 and CO_2 at Serpong and also any effects of vegetation and anthropogenic emissions around three sites using carbon isotopes obtained from flask samplings.

The daytime values at Bogor and Serpong in December and January are 5–10 ppm lower than the other months. We also observed no frequent night-time enhancements of CO_2 and CH_4 mole fractions at Serpong in December and January. The wind rose diagrams of the three sites in April, July, and October 2017 and January 2018 in Figure 11 show that most of winds in January (mid-rainy-season around Jakarta) blow from westerly direction obviously, which is relatively strong wind compared with the other months. There are no large source in the west of Serpong and Bogor. Thus we estimate that the lower values of CO_2 mole fractions in December and January are caused by such distinctive westerly and relatively strong wind.

3.2. NO, NO_2, SO_2, and O_3

Figure 12 shows the temporal variations of hourly NO, NO_2, SO_2, and O_3 mole fractions observed at Serpong, Bogor, and Cibeureum from 1 June 2016 to 11 March 2018. The mole fractions of NO, NO_2, and SO_2 at Serpong and Bogor tend to be higher than those at Cibeureum. While the NO mole fractions at Cibeureum almost stayed at low level (1–2 ppb), the NO mole fractions at Serpong and Bogor sometimes exceeded 40 ppb. This result indicates that there are many anthropogenic emission sources around Serpong and Bogor while there are few sources around Cibeureum. The clear diurnal variations of O_3 mole fractions are recognized at three sites. Although the peak values of the O_3 mole fraction in the afternoon observed at three sites were almost same level, the values of O_3 mole fraction at Cibeureum during night-time were 10–20 ppb higher than Serpong and Bogor. Moreover, the mole fractions of NO_2, SO_2, and O_3 at three sites decreased in December and January as well as CO_2 and CH_4.

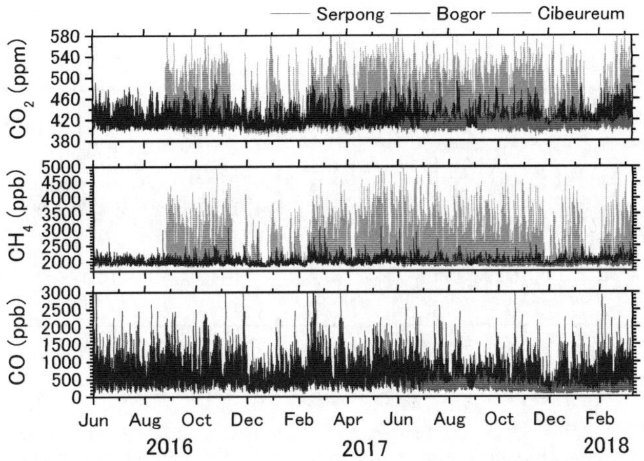

Figure 7. Temporal variation of hourly CO_2, CH_4, and CO mole fractions observed at Serpong, Bogor, and Cibeureum from 1 June 2016 to 11 March 2018.

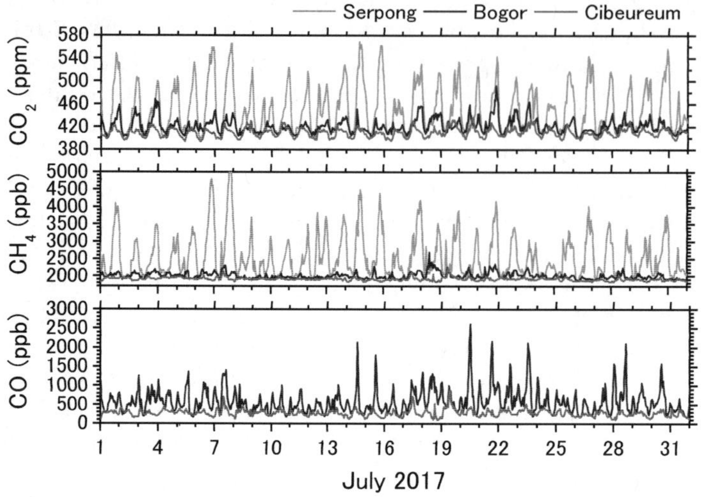

Figure 8. Temporal variation of hourly CO_2, CH_4, and CO mole fractions observed at Serpong, Bogor, and Cibeureum in July 2017.

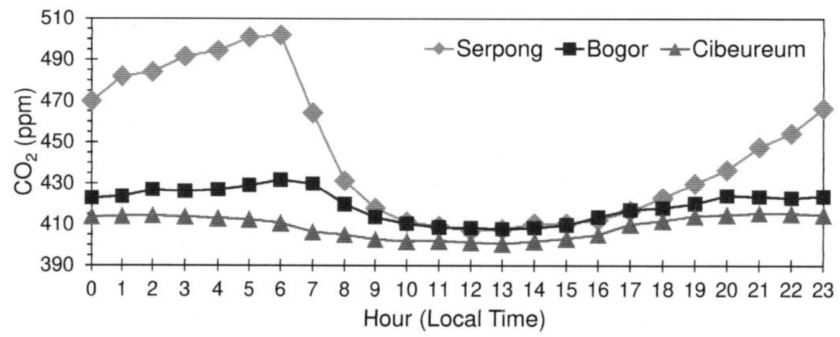

Figure 9. Diurnal variation of hourly CO_2 mole fractions observed at Serpong, Bogor, and Cibeureum from 1 June 2017 to 28 February 2018.

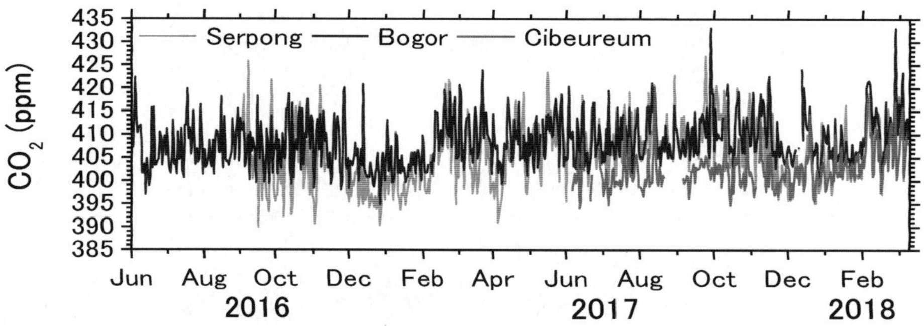

Figure 10. Daytime value (average from noon to 3 PM) of CO_2 mole fraction observed at Serpong, Bogor, and Cibeureum from 1 June 2016 to 11 March 2018.

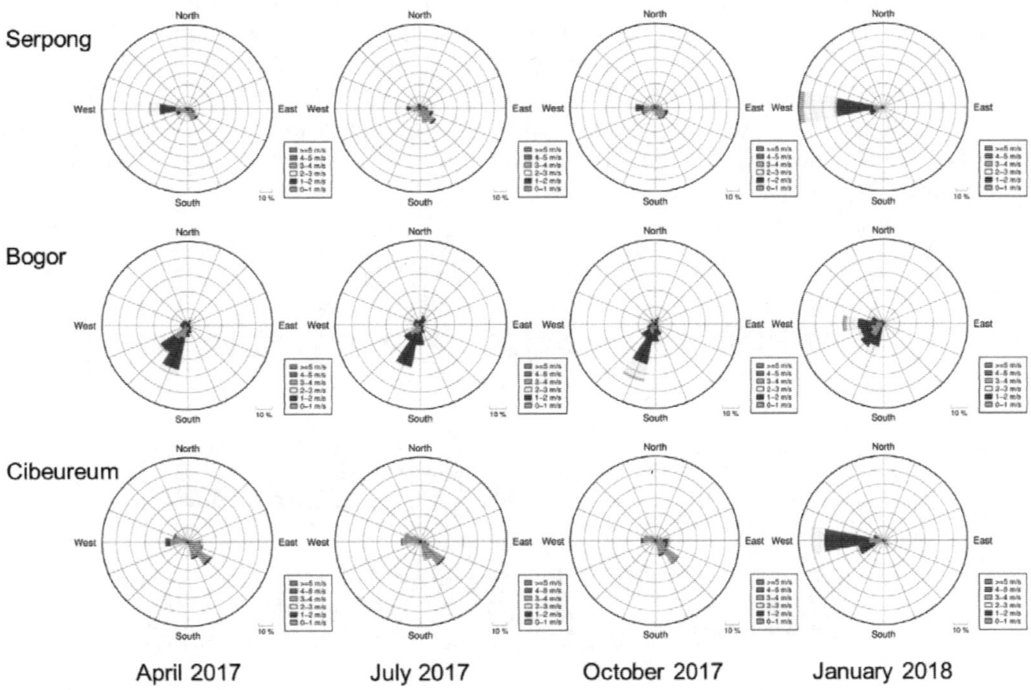

Figure 11. Wind rose diagrams of Serpong, Bogor, and Cibeureum in April, July, and October 2017 and January 2018.

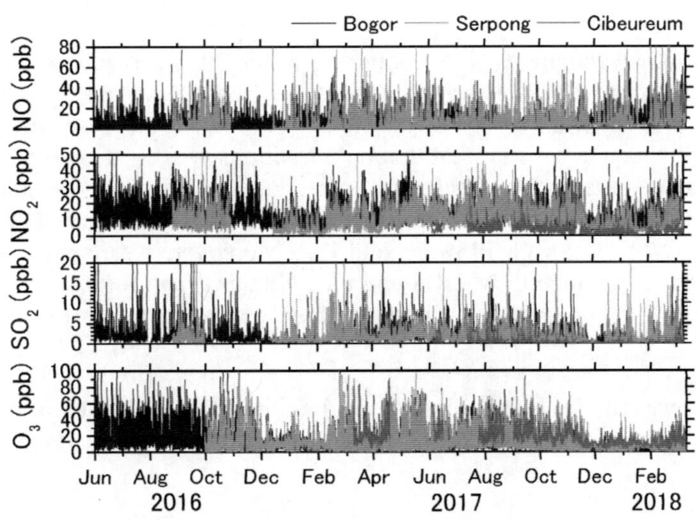

Figure 12. Temporal variations of hourly NO, NO_2, SO_2, and O_3 mole fractions observed at Serpong, Bogor, and Cibeureum from 1 June 2016 to 11 March 2018.

3.3. $PM_{2.5}$, $PM_{10-2.5}$, and BC

Figure 13 shows the temporal variation of $PM_{2.5}$, $PM_{10-2.5}$, and BC concentrations observed at Bogor and Cibeureum from 1 June 2016 to 11 March 2018. The measurement interval is 3 hours to extend the replacement interval of filter tape and chemical reagents for the chemical component analysis of $PM_{2.5}$ and $PM_{10-2.5}$. The average values of $PM_{2.5}$ observed at Bogor and Cibeureum during dry season, approximately from May to October, are 2.0 and 2.7 times higher than those of wet season, from November to April, respectively. While the seasonal trend of $PM_{2.5}$ at Bogor is similar with Cibeureum, the average $PM_{2.5}$ at Bogor is 1.8 times larger than Cibeureum. The difference of $PM_{10-2.5}$ and BC between Bogor and Cibeureum is large, compared with $PM_{2.5}$, and the average values of $PM_{10-2.5}$ and BC observed at Bogor is 2.8 and 4.2 times larger than Cibeureum, respectively. These results suggest that the $PM_{2.5}$, $PM_{10-2.5}$, and BC at Bogor is affected by peripheral anthropogenic emissions compared with Cibeureum.

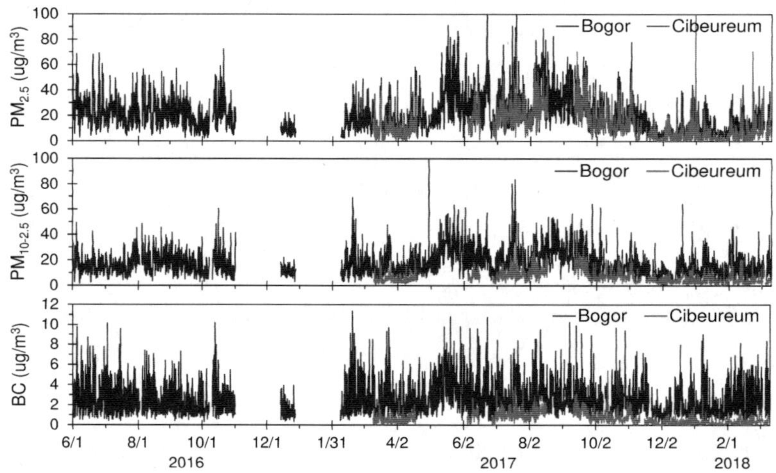

Figure 13. Temporal variation of $PM_{2.5}$, $PM_{10-2.5}$, and BC concentrations observed at Bogor and Cibeureum from 1 June 2016 to 11 March 2018. The measurement interval is 3 hours.

4. Summary

We developed a ground-based comprehensive monitoring system of GHGs and air pollutants specialized for remote automatic operation in Indonesia and installed it at Bogor, Serpong, and Cibeureum in 2016–2017 to estimate the amount of anthropogenic emissions from Jakarta megacity and compare with city activities. The monitoring results indicate that the mole fractions of CO_2 at Serpong and Bogor and of CO at Bogor have urban characteristics affected by anthropogenic emission from the urban area while those values at Cibeureum have background-like characteristics. This characteristics are also recognized in the other species, NO_x, SO_2, $PM_{2.5}$, $PM_{10-2.5}$, and BC. Moreover, we found diurnal and seasonal characteristics. We need to continue the GHGs and air pollutants monitoring in Indonesia for a long time to observe the variability and to predict future local and global environment.

Acknowledgement

This monitoring project was funded by the Ministry of the Environment, Japan. We would like to thank Keiichi Katsumata for his support of installation and maintenance of instruments.

References

[1] Baklanov A, Molina L T and Gauss M 2016 *Megacities, air quality and climate Atmos.* Environ. **126** 235–49
[2] Demographia 2018 *Demographia world urban areas* 14th annual edition: 201804 (Belleville, Demographia)
[3] Santosa S J, Okuda T and Tanaka S 2008 *Air pollution and urban air quality management in Indonesia Clean Soil Air Water* **36** 466–75
[4] IPCC 2014 Climate change 2014*: Impacts, adaptation, and vulnerability. Part A: Global and sectoral aspects. Contribution of working group II to the fifth assessment report of the intergovernmental panel on climate change* (Cambridge: Cambridge University Press)
[5] Hamonangan E, Kondo A, Kaga A and Inoue Y 2002 *Simulation and monitoring of sulphur dioxide and nitrogen oxides in the Jakarta Metropolitan Area* Asian J. Energy Environ **3** 159–83

[6] Kondo A, Hamonangan E, Soda S, Kaga A, Inoue Y, Eguchi M and Yasaka Y 2007 *Impacts of converting from leaded to unleaded gasoline on ambient lead concentrations in Jakarta metropolitan area* J. Environ. Sci. **19** 709–13

[7] Permadi D A and Kim Oanh N T 2008 *Episodic ozone air quality in Jakarta in relation to meteorological conditions Atmos*. Environ. **42** 6806–15

[8] Santoso M, Lestiani D D and Hopke P K 2013 *Atmospheric black carbon in $PM_{2.5}$ in Indonesian cities* J. Air Waste Manag. Assoc. **63** 1022–5

[9] Santoso M, Lestiani D D, Mukhtar R, Hamonangan E, Syafrul H, Markwitz A and Hopke P K 2011 *Preliminary study of the sources of ambient air pollution in Serpong, Indonesia Atmos. Pollut*. Res. **2** 190–6

[10] Santoso M, Hopke P K, Hidayat A and Diah Dwiana L 2008 *Sources identification of the atmospheric aerosol at urban and suburban sites in Indonesia by positive matrix factorization Sci*. Total Environ. **397** 229–37

[11] Lestari P and Mauliadi Y D 2009 *Source apportionment of particulate matter at urban mixed site in Indonesia using PMF Atmos*. Environ. **43** 1760–70

[12] Rashid M, Yunus S, Mat R, Baharun S and Lestari P 2014 *PM_{10} black carbon and ionic species concentration of urban atmosphere in Makassar of South Sulawesi Province, Indonesia Atmos*. Pollut. Res. **5** 610–5

[13] Mitchell L E, Crosman E T, Jacques A A, Fasoli B, Leclair-Marzolf L, Horel J, Bowling D R, Ehleringer J R and Lin J C 2018 *Monitoring of greenhouse gases and pollutants across an urban area using a light-rail public transit platform Atmos*. Environ. **187** 9–23

[14] Machida T, Tohjima Y, Katsumata K and Mukai H 2011 *A new CO_2 calibration scale based on gravimetric one-step dilution cylinders in National Institute for Environmental Studies-NIES 09 CO_2* scale Proc. 15th WMO/IAEA Meeting of Experts on Carbon Dioxide, Other Greenhouse Gases and Related Tracers Measurement Techniques (Jena) (Geneva: WMO) pp 114–8

Linear Regression Model for Predicting Daily $PM_{2.5}$ Using VIIRS-SNPP and MODIS-Aqua AOT

N Cholianawati[1], W E Cahyono[1], A Indrawati[1] and A Indrajad[1]

[1]Indonesia National Institute of Aeronautics and Space (LAPAN), Bandung, Indonesia

nani.cholianawati@lapan.go.id

Abstract. Small-sized particulates ($PM_{2.5}$) are very dangerous if inhaled by humans. The measurement of the particle concentration is carried out by the highly variable AQMS. Several studies have shown a positive relationship between aerosol optical thickness to surface $PM_{2.5}$. The study aims to predict $PM_{2.5}$ from Aerosol Optical Thickness using Linear Regression Model in Indonesia. The aerosol optical thickness is obtained from the VIIRS sensor on the SNPP satellite and from the MODIS sensor on the Aqua satellite. Linear regression model for predicting daily $PM_{2.5}$ using AOT VIIRS SNPP is $PM_{2.5}$=56.683*AOT-3.807 with r=0.7. Linear regression model for predicting daily $PM_{2.5}$ using AOT MODIS-Aqua is $PM_{2.5}$=104.3*AOT+1.0118 with r=0.7.

Keywords : AOT, VIIRS, MODIS, SNPP.

1. Introduction

Particulate matter (PM) or, more appropriately, the atmospheric aerosol is a multi-phase system of solid and liquid particles suspended in the air with aerodynamic diameters ranging from 10^{-3} μm–10 μm and includes both the particles and the air. Typically, the small solid and liquid particles are regarded as aerosols [1]. Aerosols have an important influence on atmospheric visibility, the radiation budget of the Earth-atmosphere system, the cloud-forming process, precipitation and human health.

Aerosol Optical Thickness is the degree to which aerosols prevent the transmission of light in the atmosphere. An aerosol optical thickness of less than 0.1 indicates a crystal-clear sky with maximum visibility, whereas a value of 4 indicates the presence of aerosols very dense so that people would have difficulty seeing the Sun, even at mid-day [2]. From an environmental standpoint, aerosols also constitute an important policy issue in air quality and climate sciences. In fact, PM pollution is probably the most pressing issue in air quality regulation worldwide, but at the same time it represents one of the biggest sources of uncertainty in current climate simulations. [3]

A reduction in long-term exposure to PM_{10} and $PM_{2.5}$ would have postponed 8% to 9% of all-cause mortality or about 37,000 deaths. One-third of them were associated with cardiopulmonary mortality and one-ninth of them were associated with lung cancer mortality. Current air pollution levels in Southeast and East Asian countries (including Indonesia) have a non-negligible public health impact.[4]

Many studies discuss the differences between aerosol properties as measured locally by surface stations and the large-scale ones measured by satellites. [5] Numerous researchers have attempted to estimate ground $PM_{2.5}$ levels using satellite-derived atmospheric aerosol optical depth. There are several approaches to estimate $PM_{2.5}$ from AOD, namely statistical approach, model scaling, and data assimilation. [6] Multiple linear regression (MLR) [7], mixed-effect model (MEM) [8], chemistry

IOP Conf. Series: Earth and Environmental Science **303** (2019) 012039 doi:10.1088/1755-1315/303/1/012039

transport model (CTM) [9], and geographically weighted regression (GWR) [10] were the models most commonly used to predict $PM_{2.5}$. [10]

Remote sensing of $PM_{2.5}$ in Indonesia has not been studied yet. Our research aims to predict $PM_{2.5}$ concentration from aerosol optical thickness observed by satellite and $PM_{2.5}$ in situ relationship with the linear regression model in Indonesia. The $PM_{2.5}$ concentration spatial map will provide pollutant information in monitor sparse regions.

2. Data and Methods

2.1. Data

This study used three data sources. First is $PM_{2.5}$ concentration data recorded by Air Quality Monitoring System in three stations (figure 1): Pekanbaru1 (0°31'21" N 101°28'34" E), Pekanbaru2 (0°31'18" N 101°26'31" E), and Pekanbaru3 (0°27'54" N 101°23'15" E). Temporal resolution is half hourly. Pekanbaru City is a city located in Riau Province. The Riau region is very vulnerable to land fires. Pekanbaru City people experience health problems due to the highest smoke haze. But compared to other Riau regions, if seen from its geographical location, it is not possible to burn forests in pekanbaru City which can cause haze of forest fires. [11]

The second data source is Aerosol Optical Thickness from VIIRS which can be downloaded from ftp://ftp.star.nesdis.noaa.gov/pub/smcd/jhuang/npp.viirs.aerosol.data/edraot550 with spatial resolution 0.1° and temporal resolution is daily. Aerosol Optical Thickness (AOT), providing a measure of the aerosol content of the atmospheric column. The NASA/NOAA Visible Infrared Imaging Radiometer Suite (VIIRS) sensor was launched 28 October 2011 on the Suomi National Polar-orbiting Partnership (SNPP) satellite in a near-circular, sun-synchronous, near-polar orbit. VIIRS was having an 840 km mean altitude, 98.7° inclination, an orbital period of 101.5 min, and a nominal ascending equator local crossing time of 13:30 ±10 min. [12].

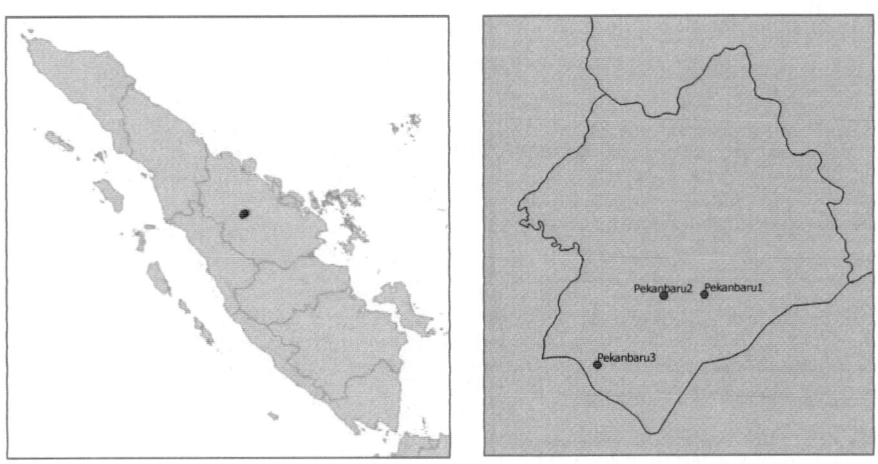

Figure 1 Three AQMS in situ Stations: Pekanbaru1, Pekanbaru2, Pekanbaru3.

The third data source is Aerosol Optical Depth or Thickness from MODIS which can be downloaded from https://ladsweb.modaps.eosdis.nasa.gov/. The MODIS instrument is a highly sensitive radiometer operating onboard Terra and Aqua. Aqua has a 1:30 am/pm equator crossing time while Terra has a 10:30 am/pm equator crossing time, meaning that MODIS data is typically available on a daily basis. Daily Level 2 (MOD 04) data are produced at the spatial resolution of a 10 x 10 1-km (at nadir) pixel array. This study used Aqua product only (MYD04_L2). MODIS bias that occurs in the aerosol optical thickness data set can be found in [13]. Validated against ground-based measurements, some studies have suggested that MODIS can retrieve aerosol optical depth with good accuracy over the ocean and has some skill over land. [14] Note that aerosol optical thickness and aerosol optical depth (τ) are used interchangeably to describe this data set.

2.2 Methods
This study used Linear Regression methods to produce mathematics model to predict $PM_{2.5}$ concentration using Aerosol Optical Thickness data (figure 2). To get the prediction model, AOT must be collocated with $PM_{2.5}$ in situ data averaging at 13:00, 13:30, and 14:00 adjusted satellite overpass. Regression analysis provides the first approximation of surface $PM_{2.5}$ mass concentration and air quality and it works best when the boundary layer is well mixed, there is no significant aerosol aloft, and in small particle dominated regions.

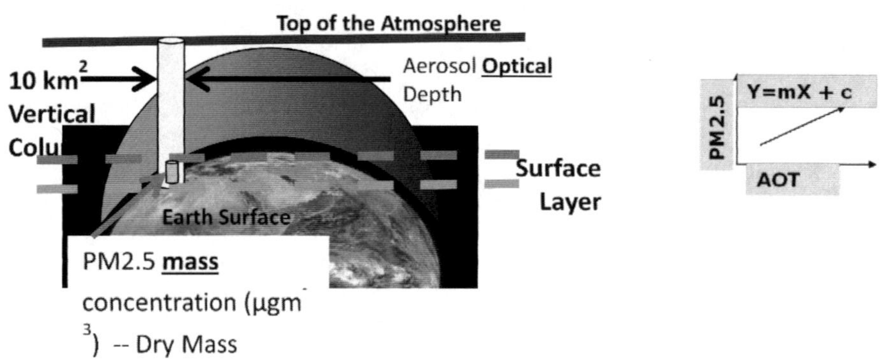

Figure 2 Illustration of predicting daily $PM_{2.5}$ using AOT data.

To plot, map, and collocate all data, we used Python version 3.6 and GrADS. To get 'closest distance' in collocating data, we used the Haversine Formula. [15, 16]

$$a = \sin^2(\Delta\varphi/2) + \cos\varphi_1 \cdot \cos\varphi_2 \cdot \sin^2(\Delta\lambda/2) \quad (1)$$

$$c = 2 \cdot \text{atan2}\left(\sqrt{a}, \sqrt{(1-a)}\right) \quad (2)$$

$$d = R \cdot c \quad (3)$$

where φ is latitude, λ is longitude, R is earth's radius (mean radius = 6,371 km).
In addition, the effect of different AOD sampling box sizes was studied. The box size was 3 by 3 and 5 by 5 MODIS pixels, whose size is 10 km x 10 km.

3. Results and Discussions

3.1. $PM_{2.5}$ and Aerosol Optical Thickness
$PM_{2.5}$ concentration half hourly time series recorded in three stations from 1 January 2015 00:00 until 31 December 2015 23:30 were shown in figure 3. Both three stations showed very high $PM_{2.5}$ concentration in September – October as biomass burning result.

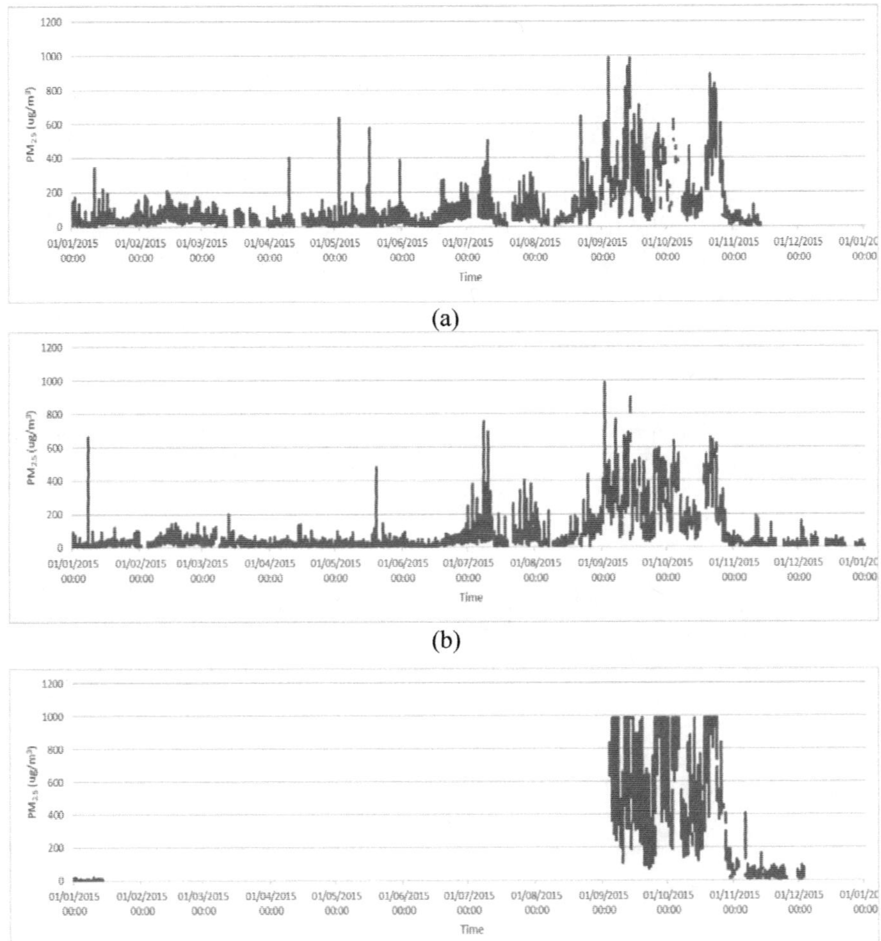

(c)

Figure 3 $PM_{2.5}$ concentration ($\mu g/m^3$) in situ half-hourly time series (1 January 2015 00:00 to 31 December 2015 23:30) in three stations: Pekanbaru1 (a), Pekanbaru2 (b), and Pekanbaru3 (c).

Aerosol Optical Thickness time series from 1 January until 31 December 2015 was shown in Figure 4. On September – October 2015, SNPP cannot detect AOT may be because of the existence of very thick smoke then satellite had difficulty in differentiating between cloud and smoke. The highest AOT recorded by VIIRS on September 22, 2015, amounted to 1.53 and was the peak of forest fires that hit Sumatra. Subsequent high-value AOT occurs on 20, 24 and 30 August 2015, which is around 1.4 which is the initial period of forest fires. AOT at Pekanbaru 1 Station were highest recorded by VIIRS on September 22, 2015 at 1.53 and was the peak of forest fires that hit Sumatra. Subsequent high-value AOT occur on 20, 24 and 30 August 2015, which is around 1.4 which is the initial period of forest fires. While at Pekanbaru 2 Station, the highest aerosol was found on August 31, 2015 at 1.18. Then at Pekanbaru 3 Station, the maximum AOT occurred on August 20, 2015 at 1.62.

During 1 January to 31 December 2015 (365 days), VIIRS-SNPP only detected AOT for 68 days in Pekanbaru1 (18.6%); 53 days in Pekanbaru2 (14.5%); and 66 days in Pekanbaru3 (shown in Figure 4). Compared with MODIS-Aqua (not shown), VIIRS-SNPP detects more AOT than MODIS-Aqua. It may be because of VIIRS scan wider and fewer gaps than MODIS. MODIS-Aqua only detected AOT for 15 days in Pekanbaru1 (4.12%); 11 days in Pekanbaru2 (3%); and 12 days in Pekanbaru3 (3.3%).

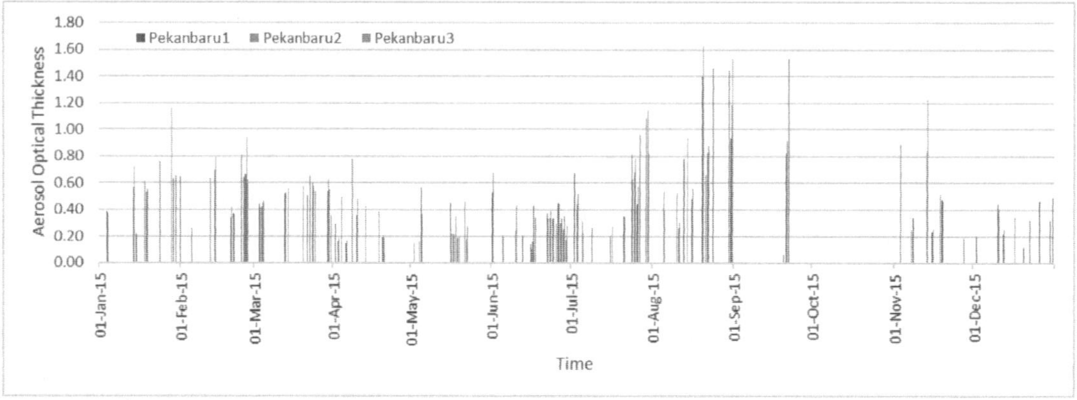

Figure 4. Aerosol Optical Thickness daily time series observed by VIIRS-SNPP (1 January 2015 to 31 December 2015) in three locations: Pekanbaru1 (a), Pekanbaru2 (b), and Pekanbaru3 (c).

3.2. Linear Regression Model

Scatter plots between $PM_{2.5}$ concentration (AQMS) and Aerosol Optical Thickness (VIIRS-SNPP) are shown in Figure 5. This figure was produced by collocating $PM_{2.5}$ and AOT between in situ locations with VIIRS SNPP regularly spaced data (gridded) coordinates. The correlation coefficient between $PM_{2.5}$ and AOT for three sites: 0.8 (Pekanbaru1) with 47 data pairs, 0.75 (Pekanbaru2) with 45 data pairs, and 0.6 (Pekanbaru3) with 4 data pairs. For all data pairs by combining Pekanbaru1, Pekanbaru2, and Pekanbaru3 data, the correlation coefficient between $PM_{2.5}$ and AOT is 0.7. Linear regression model for predicting daily $PM_{2.5}$ using AOT VIIRS SNPP is $PM_{2.5}$=56.683*AOT-3.807.

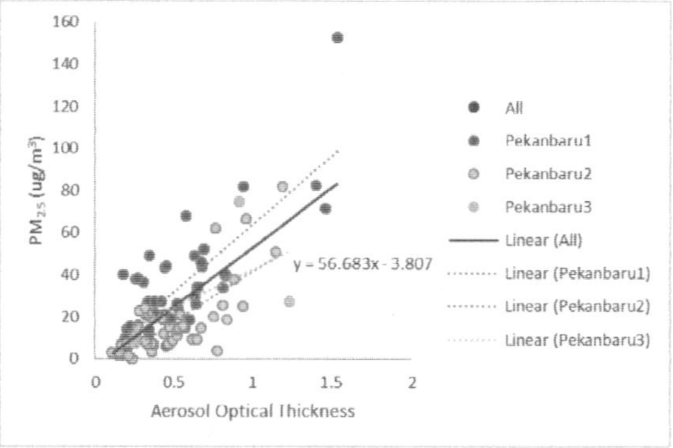

Figure 5. Scatter plot $PM_{2.5}$ concentration in situ collocated with AOT VIIRS SNPP in three sites: Pekanbaru1, Pekanbaru2, and Pekanbaru3 from 1 January to 31 December 2015.

Scatter plots between $PM_{2.5}$ concentration (AQMS) and Aerosol Optical Thickness (MODIS-Aqua) are shown in Figure 6. Figure 6a was produced using "closest distance" to collocate between in situ locations with MODIS-Aqua coordinates in radius 10 km. The correlation coefficient between $PM_{2.5}$ and AOT for two sites: 0.85 (Pekanbaru1) with 10 data pairs and 0.9 (Pekanbaru2) with 10 data pairs. Another site (Pekanbaru3) only had 2 data pairs then correlation coefficient was not computed. Figure 6b was produced using average values in "radius 3x10 km". The correlation coefficient between $PM_{2.5}$ and AOT for two sites: 0.6 (Pekanbaru1) with 24 data pairs and 0.58 (Pekanbaru2) with 28 data pairs. Another site (Pekanbaru3) only had 3 data pairs then the correlation coefficient was not computed. Figure 6c using average values in "radius 5x10 km". The correlation coefficient between $PM_{2.5}$ and AOT for three sites: 0.3 (Pekanbaru1) with 53 data pairs, 0.5 (Pekanbaru2) with 52 data pairs, and 0.5 (Pekanbaru3) with 9 data pairs. Wider radius (30 km and 50 km) will add the number of $PM_{2.5}$ and AOT pairs but not increasing correlation coefficient.

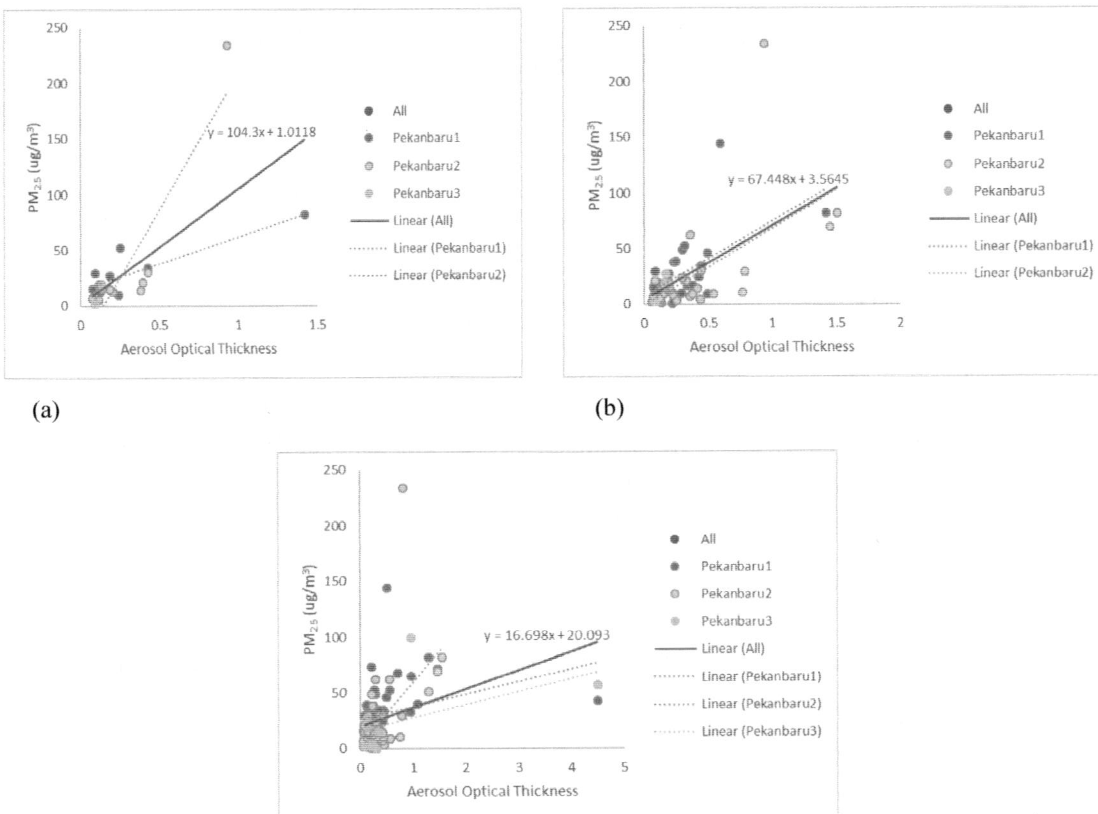

(c)

Figure 6. Scatter plot $PM_{2.5}$ concentration in situ collocated with AOT MODIS Aqua in three sites: Pekanbaru1, Pekanbaru2, and Pekanbaru3 during 1 January to 31 December 2015 using three different methods: **a)** closest distance (radius 10 km), **b)** radius 3x10 km, and **c)** radius 5x10 km.

For all data pairs by combining Pekanbaru1, Pekanbaru2, and Pekanbaru3 data, the correlation coefficient for three methods: 0.7 (closest distance with 22 data pairs), 0.6 (radius 3x10 km with 55 data pairs), and 0.3 (radius 5x10 km with 114 data pairs). These results are different from the results of previous studies which showed that AOD does not change remarkably as a function of box size around their study area. [17] Then we suggest to use closest distance methods and results linear regression model for predicting daily $PM_{2.5}$ using AOT MODIS-Aqua is $PM_{2.5}=104.3*AOT+1.0118$.

4. Conclusion

Both aerosol optical thickness from MODIS-Aqua and VIIRS-SNPP can be used to predict daily $PM_{2.5}$ concentration with correlation coefficient 0.7. The formula to predict $PM_{2.5}$ from VIIRS-SNPP AOT is $PM_{2.5}=56.683*AOT-3.807$. The formula to predict $PM_{2.5}$ from MODIS-Aqua is $PM_{2.5}=104.3*AOT+1.0118$. To improve the accuracy of predictions, we should add the number of $PM_{2.5}$ and AOT pairs.

Acknowledgments
This work was funded by Center for Atmospheric Science and Technology – Indonesia National Institutes of Aeronautics and Space (LAPAN). We say thank you for all who contribute to this paper.

References

[1] Fuzzi S, U Baltensperger, K Carslaw, S Decesari, H Denier van der Gon, M C Facchini, D Fowler, I Koren, B Langford, U Lohmann, E Nemitz, S Pandis, I Riipinen, Y Rudich, M Schaap, J G Slowik, D V Spracklen, E Vignati, M Wild, M Williams, and S Gilardoni 2015 Particulate matter, air quality and climate: lessons learned and future needs Atmos. Chem. Phys., 15, 8217–8299, 2015.doi:10.5194/acp-15-8217-2015
[2] Exploring aerosol optical thickness using MODIS satellite imagery in NEO https://serc.carleton.edu/usingdata/datasheets/aerosolopticalthick.html
[3] Shao P, J Xin, J An, L Kong, B Wang, J Wang, Y Wang, D Wud 2017 The empirical relationship between $PM_{2.5}$ and AOD in Nanjing of the Yangtze River Delta Atmospheric Pollution Research https://doi.org/10.1016/j.apr.2016.09.001.
[4] Yorifuji T, B Sanghyuk, K Saori, T Toshihide, D Hiroyuki, H Yasushi, K Ho, H Yun-Chul 2015 Health impact assessment of PM_{10} and $PM_{2.5}$ in 27 Southeast and East Asian cities Journal of Occupational and Environmental Medicine: July 2015 - Volume 57 - Issue 7 - p 751–756, doi: 10.1097/JOM.0000000000000485
[5] van Donkelaar A, Martin R, Brauer M, Kahn R, Levy R, Verduzco C, Villeneuve P 2010 Global estimates of ambient fine particulate matter concentrations from satellite-based aerosol optical depth: development and application, Environ. Health Perspect., 118, 847–855, doi:10.1289/ehp.0901623, 2010
[6] Zhang Y and Z Li 2015 Remote sensing of atmospheric fine particulate matter (PM2.5) mass concentration near the ground from satellite observation. Remote Sens. Environ., 160, 252–262.
[7] Zheng Y, Zhang Q, Liu Y, Geng G and K He 2016 Estimating ground-level PM2.5 concentrations over three megalopolises in china using satellite-derived aerosol optical depth measurements. Atmos. Environ. 2016, 124, 232–242.
[8] Zou B 2016 High-resolution satellite mapping of fine particulates based on geographically weighted regression IEEE Geosci. Remote Sens. Lett. 2016, 13, 495–499.
[9] Wang B and Z Chen 2016 High-resolution satellite-based analysis of ground level PM2.5 for the city of Montreal. Sci. Total Environ. 2016, 541, 1059–1069.
[10] Chu Y, Liu Y, Li X, Liu Z, Lu H, Lu Y, Mao Z, Chen X, Li N, Ren M, Liu F, Tian L, Zhu Z, Xiang H 2016 A review on predicting ground $PM_{2.5}$ concentration using satellite aerosol optical depth. Atmosphere 2016, 7, 129; doi:10.3390/atmos7100129
[11] Yuliarti V and Irdayanti 2016 Peran Dinas Kota Pekanbaru dalam menanggulangi dampak kabut asap kebakaran hutan di Kota Pekanbaru. Jurnal Penelitian sosial keagamaan, Vol.19, No.1 Januari-Juni 2016
[12] Wolfe RE, G Lin, M Nishihama, KP Tewari, JC Tilton, AR Isaacman 2013 Suomi NPP VIIRS prelaunch and on-orbit geometric calibration and characterization https://doi.org/10.1002/jgrd.50873
[13] Zhang J and JS Reid 2006 MODIS aerosol product analysis for data assimilation: Assessment of over-ocean level 2 aerosol optical thickness retrievals. Journal Of Geophysical Research, Vol. 111, D22207, doi:10.1029/2005JD006898, 2006
[14] NASA ARSET Applied Remote Sensing Training Program 2016 Converting AOD to PM2.5: a statistical approach https://arset.gsfc.nasa.gov/sites/default/files/airquality/workshops/BAQ16/P9.pdf
[15] Anisya and GY Swara 2017 Implementation of Haversine formula and Best First Search method in searching of Tsunami evacuation route, IOP Conf. Ser.: Earth Environ. Sci. 97 012004, doi: 10.1088/1755-1315/97/1/012004
[16] Farid and Y Yunus 2017 Analisa algoritma Haversine formula untuk pencarian lokasi terdekat Rumah Sakit dan Puskesmas Provinsi Gorontalo ILKOM Jurnal Ilmiah Volume 9 Nomor 3 Desember 2017, ISSN online 2548-7779
[17] Natunen A, A Arola, T Mielonen, J Huttunen, M Komppula, and KEJ Lehtinen 2010 A multi-year comparison of PM2.5 and AOD for the Helsinki region Boreal Environment Research 15: 544-552 ISSN 1797-249 (online)

The influence of aerosol on the sunlight divergence in the atmospheric Indonesia

Rosida

Center for Atmospheric Science and Technology
National Institute of Aeronautics and Space

rosida2009@gmail.com

Abstract. Solar radiation is the source of energy for the earth. The amount of solar radiation that reaching the earth's surface, is strongly influenced by the interaction of solar radiation with the atmospheric that is passed through. This paper analyzes the effect of aerosols on solar radiation balance in the Indonesian atmosphere using radiation data from the Clouds and the Earth's Radiant Energy System (CERES) and aerosol optical depth that is obtained from the Moderate Resolution Imaging Spectroradiometer (MODIS) - Terra, from March 2000 to February 2016. Aerosol optical depth data (AOD) and shortwave (SW) radiation that reach the top of atmosphere and the earth's surface were analyzed spatially and temporally with the observation area of the 15^0S – 15^0N and 90^0W – 150^0E. Moreover, the sample of several locations was analyzed such as Palangkaraya, Pontianak, Banjarmasin and Samarinda, to look at the cases of forest fires which generate aerosols that quite high. Results of data analysis showed a significant influence of aerosols on the divergence of short-wave solar radiation in the atmosphere of Indonesia. The divergence of short-wave radiation shows how much solar radiation is absorbed in the atmosphere. Some cities such as Palangkaraya and Pontianak showed a strong correlation (0.7 - 0.95) between AOD and the value of shortwave radiation divergence. Temporal pattern from time series analysis showed that the increase in the value of AOD due to the fires incident, give the biggest influence on the decrease in solar radiation that reaches the earth's surface.

Keywords : aerosol, divergence, CERES, MODIS, radiation.

1. Introduction

Radiation is fundamental in climate. Almost all of the energy entering the climate system comes from the sun in the form of electromagnetic waves. Although there are other additional sources such as geothermal heating, but the contribution is very small so that the effect can be neglected [1]. In the IPCC [2] report explained about the solar radiation absorbed, infrared radiation emitted, and the balance that occurs between these things which include the process of absorption, transfer and emission from radiation that occurs throughout the globe for many years. McGuffie and Henderson-Sellers [3] explain that in the process of radiative transfer there is heating on the surface through absorption of short waves, and heating or cooling in the atmosphere that occurs through absorption and emission of infrared radiation. Whereas the controller of the interaction is cloud albedo and absorbtivity to short waves.

Net radiation flux is one of the important parameters that can show energy loss or stored energy in one layer. This concept is important for defining radiative heating or cooling from each element of the atmosphere or surface. Heating or cooling that occurs due to energy stored or lost, which will affect the energy balance. Therefore this energy balance is important, and the simplest concept of radiation balance has become the basis in climate models by considerating how radiation is absorbed, transferred and emitted back by the atmosphere. Sellers [4] explained that in the radiative transfer process, there are several factors that control the interaction, namely albedo and absorptivity to shortwaves, which absorptivity to radiation is strongly influenced by the composition of the atmosphere. Thus, fluctuations in absorption of radiation in the atmosphere can be one of the benchmarks for changes in the components in the atmosphere. Based on the statements of Palmen and Newton [5] and also from the results of Alexander and Schubert's research [6] elucidate that the storage and transfer of energy in the ocean-atmosphere system haves an important role in determining the average climate and evolution of climate anomalies. The energy balance model has proven to be very useful in determining processes that affect climate by providing a balance in storage, flux divergence, energy sources and sinks in the related region.

Divergence of radiation flux is a value that can show how much stored or lost energy is obtained from the difference between net radiation flux in the top of atmosphere (TOA) and on the surface [7][8]. Observation of the divergence of radiation flux was first carried out by Funk [9] and Fuggle and Oke [10]. They observed it on grasslands at altitudes of 0.5 and 1.5 m and recorded the divergence of radiation flux at night which tends to show cooling speeds at -120 K day^{-1}. Lieske and Stroschein [11] reported the results of their first measurements on snow at two air levels of 1 m and 5 m above Alaska, they found an average radiation heating effect at 120 K day^{-1}. Rascke et al [8] explained that the absorption of sunlight in all conditions is positive, and for the average annual value in the tropics reaches a value of up to 90 watts/m^2. Bohren et al [16] found negative flux divergences related to heating and almost all occur in the atmospheric layers at altitudes below 30 km. Under 15 km, heating occurs by the absorption of solar radiation which rapidly increases like a cooling event that occurs due to emissions of terrestrial radiation. Total flux divergence between 8 km and 14 km is positive, while between 8 km to ground is negative.

The role of aerosols in the climate system and radiation balance has been widely discussed by researchers, including Crutzen and Andreae [17], Ichoku et al [18], Keil and Haywood [19], Kaufman, et al [20] and Breon et al [21]. They explained that during the period when the aerosol load was very high, the balance of the solar radiation budget would be disrupted. Kaufman et al [20] asserted that the accumulation of aerosol particles that occurred in the Indian Ocean in 1998-1999 came from the transportation of particulate pollutants which is emitted from the Indian sub-continent. The accumulation of aerosol particles has caused a reduction in the amount of solar radiation that reaches the surface by 15%.

Fouquart et al [22] observed aerosol potential in influencing global radiation budgets from Earth-atmosphere systems in desert regions, the results showed that aerosols on a small scale were able to have a significant impact on the balance of radiation energy in the atmosphere and on the surface. In the case of aerosol dust, Fourquart et al [22] found a drastic reduction in the budget for short-wave radiation caused by the accumulation of aerosols particle dust originating from the desert. From research of Benas et al [23] in the Mediterranian region where environmental conditions were heavily polluted by dust, showed that the accumulation of aerosol particles covering the Mediterranian region had caused cooling on the surface as a result of reduced radiation reaching 215 watts/m^2, and reduction of radiation at the top of atmosphere reaches 46 watts/m^2.

Rosida and Susanti [24] conducted a spatial analysis of the effects of aerosols on short-wave radiation (SW) which was then linked to forest fires that occurred in Kalimantan and Sumatra. The results of the analysis show that the forest fires that occurred in 2002 and 2006 in Kalimantan and Sumatra, have caused an increase in AOD values that reach more than 300% of the average value, and showed a strong linkage to changes in short-wave radiation fluxes that reached the surface ($r > 0.8$) [24]. In 2014, Rosida et al [25] also analyzed spatially seasonal influences and the results showed that there was a strong link between net radiation flux, short-wave radiation flux (SW) and long-wave radiation flux (LW) to the average high of AOD value which is occurred in the transitional season of September-October-November. In that cases anthropogenic factors was estimated to provide a significant role in influencing the balance of radiation budget [26].

The purpose of this study is to determine the magnitude of sunlight divergence in the atmosphere of the Indonesian region at the earth's surface level, namely at the lower troposphere level at an altitude of below 1 km and at the top of atmosphere (TOA) level, which is around 100 km above the surface of the Earth. Data of net radiation flux, short-wave radiation flux, and long-wave radiation flux were used in this study to obtain the magnitude of the radiation flux divergence value. The magnitude of divergence value of the radiation flux which were obtained is then used to gain the amount of radiation energy stored or released. In addition, the aims of this study also to analyze the influence of aerosols on the divergence of sunlight in the Indonesian atmosphere. The importance of the role of aerosols in radiation balance has been widely discussed which generally states that when the aerosol load is very high, the balance of the solar radiation budget will be disrupted. If this is the case, then the amount of aerosol will show a high correlation to the amount of energy stored or released from radiation based on the divergence value of the radiation flux.

2. Research methods

Radiation data used in this study was obtained from the NASA Langley Research Center Atmospheric Science Data Center. The Clouds and the Earth's Radiant Energy System (CERES) experiment is one of the highest priority scientific satellite instruments developed for NASA's Earth Observing System. CERES products include both solar-reflected and Earth-emitted radiation from the top of the atmosphere to the Earth's surface [28]. The radiation data products available on this instrument were short-wave (SW), and long-wave (LW) radiation on the top of atmosphere (TOA), and on the Earth's surface. From one's own the Earth's surface radiation products was consist of components of downward radiation, and upward radiation.

The aerosol data which was used in this study is aerosol optical depth (AOD) that is obtained from the Moderate Resolution Imaging Spectroradiometer (MODIS) level 3 (MOD08). This parameter can be downloaded from ftp: // ladsftp. nascom.nasa.gov/allData/6/MOD08_M3/. [29] The spatial resolution of aerosol data and radiation data used in this study is 1 degree, and the temporal resolution of the data in the month averages with the data period from March 2000 to June 2017. Study area boundaries are 15^0 LU - 15^0 LS and 90^0BT - 150^0BT and Format data in NetCDF.

Radiation flux divergence is obtained through a series of stages of determining the radiation net SW and LW involving the radiation components downward and upward respectively for the radiation component on the surface and in the TOA [7][8][27]. The value of divergence of the radiation flux which was gaine is used to indicate the magnitude of radiation energy stored or released.

Referring to the explanation of Raschke et al. [8] that the terminology of divergence used here has the opposite meaning to the classical mathematical definition. So if the flux divergence value is positive (+) then it means that there is a release of radiation energy which is usually related to the radiation cooling rate, and vice versa for divergences that are negative (-) then it means that there has been absorption of radiation energy associated with the rate of radiation heating.

3. Result and discussion

3.1. Characteristics of Indonesia's regional divergence

From the observation's results, it can be seen that divergences in the Indonesian region run into fluctuations which were influenced by natural and anthropogenic factors. The results of Rosida's study, et al. [27] related to the radiation divergence of three components of longwave radiation, shortwave radiation and net radiation showed a trend of specific temporal variations and varied fluctuations. The results of data processing for a period of 17 years from March 2000 to June 2017, different characteristics of the climatological patterns of SW divergence and LW divergence were obtained.

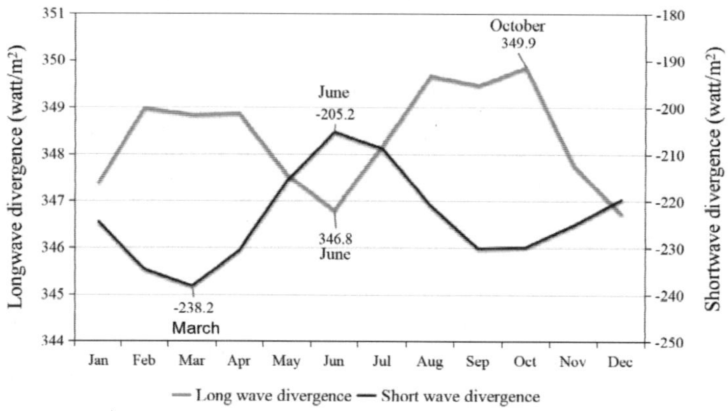

Figure 1. Climatological characteristics pattern of longwave (LW) and short wave (SW) radiation divergence in Indonesian region.

Based on the analysis, the average characteristics for the SW divergence and LW divergence patterns present a seasonal cycle that is influenced by the sun's position factor. The minimum value of the average SW divergence is around -238.2 watts/m^2 which is usually happening in every March. The maximum SW divergence average value occurs in June with a magnitude of approximately -205.2 watts/m^2. LW divergence indicates a reverse trend of the SW divergence pattern by reaching a minimum in June of 346.8 watts/m^2 and a maximum of 349.9 watts/m^2 which is usually occurred in October (Figure 1). Rosida, et al [27] estimated that LW divergence increases by 0.001 watts/m^2 per month, or 0.012 watts/m^2 per year.

The spatial distribution of each radiative divergence flux shows very different spatial variations and illustrated the differences in the flux of radiation divergence between land and sea. In this study the divergence flux values calculated are the average land and sea. Hoch [7] and Raschke [8] explain that absorption of solar radiation tends to be largely controlled by the accumulation of material emitted by anthropogenic factors such as aerosol particles, dust, trace gases or other factors such as desert dust, water vapor, clouds and others. Forest fires that often occur in the territory of Indonesia are thought to be the causes that contribute significantly to the fluctuations in radiation divergence in the territory of Indonesia.

Based on the information, forest fires that occurred in 2002, 2006 and 2015 in the western part of Indonesia, exactly in Kalimantan and Sumatera, were events that adversely affected environmental conditions in the region and beyond. The emission of aerosol particles in the form of smoke and dust is very disturbing to public health conditions, especially to respiratory health and interfere with transportation activities both land and air because it affects atmospheric visibility. Several incidents of forest fires, that occurred in 2015 were the largest that had ever occurred in Indonesia.

The trend of SW divergence anomalies during the period January 2001 to June 2017 showed that the fluctuations in the radiation energy absorbed and released were quite low, which the averaged values were no more than -5 watts/m^2 and 5 watts/m^2. The data indicated that in October 2006 there was an increase in the radiation energy released by 7.9 watts/m^2. The highest peak of SW divergence occurred in October 2015, which showed an increase in radiation energy released by 18.1 watts/m^2 (Figure 2). The trend of net divergence anomalies and LW divergence did not indicate a trend of increasing or decreasing striking or extreme in the magnitude of radiation energy. Fluctuations in radiation energy absorbed and released were in the range of -5 watts/m^2 to 5 watts/m^2. Results of the data's analysis can be explained that the emission of aerosol particles from forest fires in Kalimantan and Sumatra is more likely to affect SW radiation fluctuations.

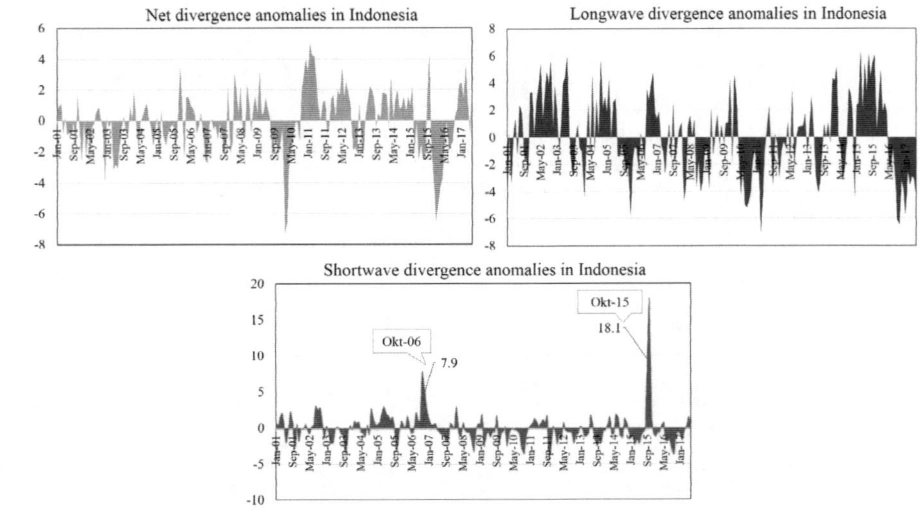

Figure 2. Trends in net divergence anomalies, long wave divergences and shortwave divergences for Indonesian regional averages in the period January 2001 to June 2017.

The spatial analysis presented in Figure 3 follows respectively the spatial patterns of the net radiation divergence anomalies, the anomalies of LW radiation divergence and the SW radiation divergence anomalies that occurred in 2015 in the Indonesian region. By using the same period, the anomalous trend of SW divergence and LW divergence was analyzed for its effect on the distribution of aerosol's (AOD) particles spatially. The analysis focused on August, September, October, and November, because of all the observed data, the trend of changes in spatial patterns that indicate a change to increase or decrease is more often found in those periods.

Assumed that the change in SW divergence anomaly that occurred in October 2015 was due to the accumulation of aerosol particles emitted from the forest fire at that time. Changes that occur, are then identified based on changes in the areal extent of the red spatial pattern, both for the spatial pattern of the radiation and the spatial pattern of the aerosols. Results of the data analysis it was concluded that the effect of aerosols on radiation flux would appear to be quite strong in the event of a very large forest fire. For example in the case of forest fires that occurred in 2002, 2006 and 2015, the concentration of aerosols drastically experienced a very sharp increase which was presented by an increase in the AOD value > 2.5. The incidence of increasing the AOD value is very high, generally found almost always occurs in the period September-October-November (SON). The event of forest fires always recurs in almost every year in the same area, namely between Kalimantan and Sumatera with a quantity that is quite varied.

Figure 4 presents the relationship between the increase in the concentration of aerosols emitted from forest fires and the fluctuations in the radiation energy absorbed and lost/released based on the divergence flux value is presented. The accumulation of aerosol particles in the form of smoke and dust emitted from average forest fires almost always occurs in the SON season in the western part of Indonesia. The indication of the extent of burning land, in Figure 4, is estimated from the spatial red changes that indicate the level of aerosol concentration. The accumulation of aerosol particles emitted during the event of forest fires, has enough to have an impact on changes in radiation energy.

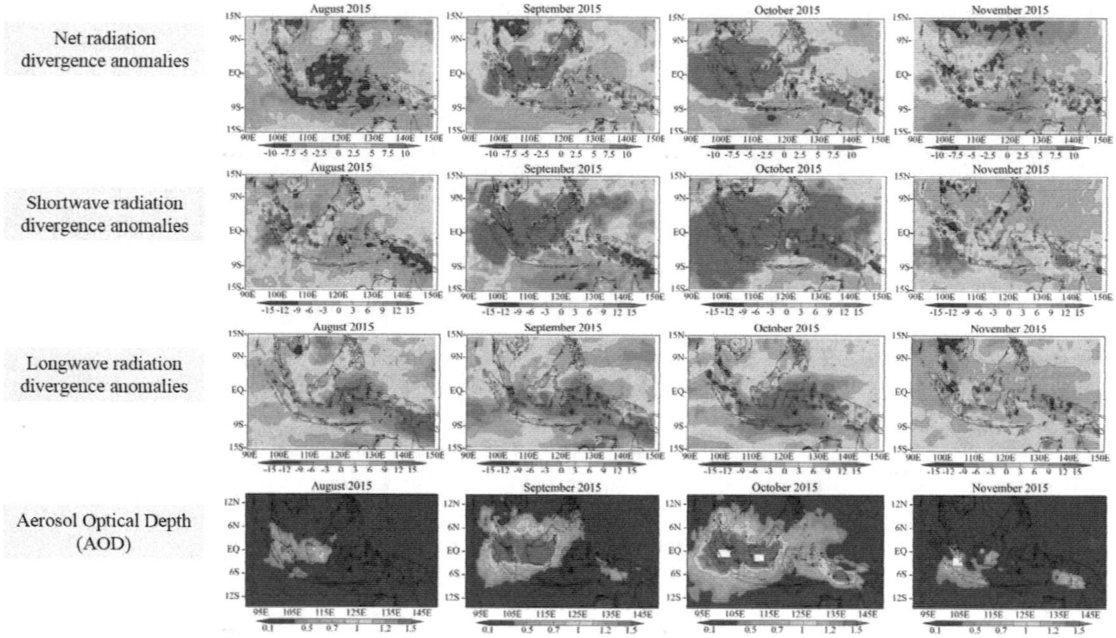

Figure 3. Spatial pattern of flux anomalies of net radiation divergence, longwave divergence flux (LW) and shortwave divergence (SW) flux, and aerosol optical depth (AOD) distribution patterns in 2015.

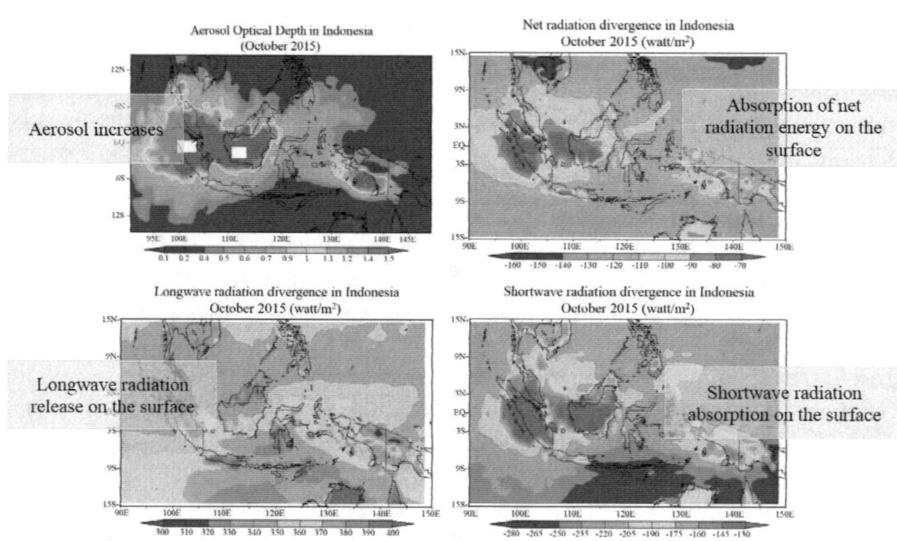

Figure 4. Patterns of linkage of optical depth aerosol enhancement to net radiation divergence flux, long wave radiation divergence (LW) flux and seasonal shortwave divergence (SW) flux.

In this study, an increase in aerosol concentrations from extreme forest fires that occurred in Kalimantan in the 2015 SON season showed an AOD value of > 1.5. The increase in aerosol accumulation at that time has resulted in a decrease in net radiation with the magnitude of the flux of the divergence of the net radiation released is > -80 watts/m^2. Slightly different from the results of research by Rascke et al [8], the accumulation of aerosol dust particles observed by Rascke et al [8] came from highly polluted environmental areas. In extreme conditions, Rascke et al [8] found an increase in the accumulation of aerosol dust particles that showed an AOD value of ~ 2 and resulted in a decrease in radiation flux with a divergence flux value of solar radiation released was > 190 watts/m^2.

At the same time, namely, in the event of very extreme forest fires in Kalimantan in 2015 above, there was also a decrease in SW radiation energy with a large flux of lost SW radiation divergence of > -130 watts/m^2. In addition, the accumulation of aerosol particles also results in absorption of LW radiation energy with a flux amount of LW radiation divergence of approximately > 370 watts/m^2.

3.2. Urban divergence

To find out the characteristics within the local scale which showing a high concentration of aerosol's particles, in this study was carried out cropping data area that is analyzed. In this study the area in Kalimantan was determined within the area boundary of (108,29^{0}W -119.6^{0}E; -4.6^{0}S - 7.07^{0}N) that is considered to meet the expected criteria. Figure 5 shows the differences in the spatial pattern of the SW radiation divergence, the divergence of net radiation and the divergence of LW radiation which were analyzed with the accumulation of aerosol particles in the period September-October-November 2015.

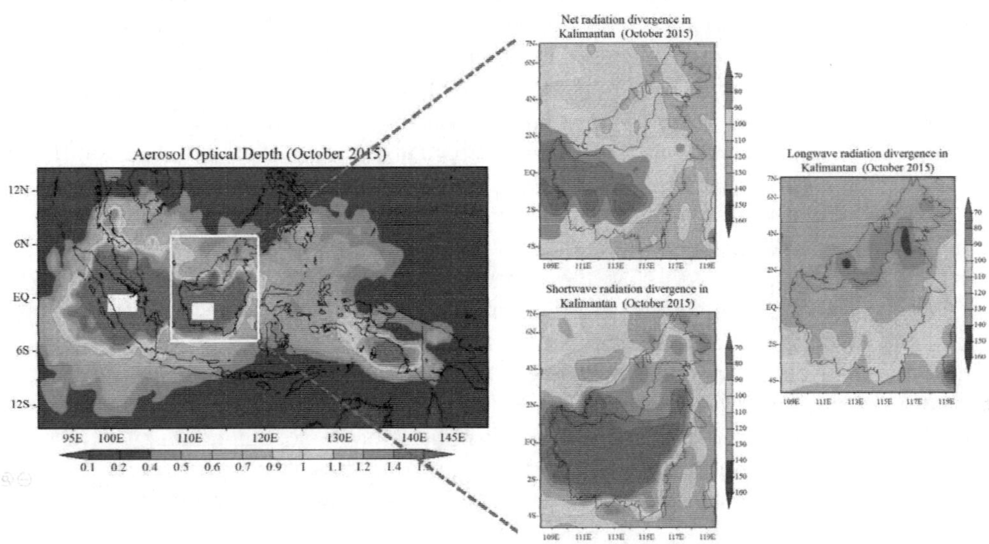

Figure 5. Differences in spatial patterns of shortwave radiation divergence, net radiation divergence and longwave divergences in Kalimantan.

The result explains that the accumulation of aerosol particles in Kalimantan tends to have more influence on the pattern of SW radiation compared to LW radiation or net radiation. The concentration of aerosol particles with an average value of AOD = 1.4 is found in this region which is estimated to have caused the absorption of radiation energy of approximately -180 watts/m^2. Spatially, changes in the area of red spatial patterns also show a stronger influence on SW radiation.

Furthermore, as a comparison, the author analyzes the smaller scope of the urban sphere. Four major cities in Kalimantan are Pontianak in the boundary (109.17^{0}W-109.23^{0}E; -0.045^{0}S - -0.016^{0}N), Palangkaraya with a limit (113,50^{0}W - 113,59^{0}E; -2.16^{0}S -2.093^{0}N), Banjarmasin in the limit (114,30^{0}W- 114,40^{0}E; -3.221^{0}S - -3.154^{0}N) and Samarinda with limits (117,01^{0}W-117,20^{0}E; -0.44^{0}S--

0.17^0N). For these cities, the divergence of the radiation is determined, then it is analyzed the possibility of its correlation to the accumulation of aerosol loads that occur at the same time.

The results of section 3.1 analysis, it can be concluded that the effect of the accumulation of aerosol particles is more dominant on the flux of the SW radiation divergence compared to the effect on long-wave radiation flux or on net radiation flux. Based on these results, the urban analysis will only examine the flux of the SW radiation divergence. The results of radiation data analysis during the period of January 2001 to June 2017, for each of the urban boundaries mentioned above, the magnitude of the radiation divergence of SW is highly variable. The average amount of radiation energy absorbed is -204.4 watts/m^2 for Pontianak city, -167.8 watts/m^2 for Palangkaraya city, -186.1 watts/m^2 for Banjarmasin city and -181.8 watts/m^2 for the city of Samarinda.

From divergence analysis in the urban area, the accumulation of aerosol particles that affect the divergence flux of radiation varies with the AOD value of 1.5 - > 3. Figure 6 shows how the increase in aerosol concentration affects the short-wave radiation changes in these cities. From the events of forest fires in the Kalimantan region, extreme conditions were found in 2002, 2006 and 2015. The fires that occurred in those years were indicated by a reduction in the amount of shortwave radiation (SW) above these cities caused by the accumulation of aerosol particles.

As happened in 2002, the accumulation of aerosol particles emitted from forest fires at that time, caused the absorption of SW radiation to only reach -95.6 watts/m^2 in Palangkaraya, where the AOD value reached 3.15. The relationship between the accumulation of aerosol particles and the amount of radiation absorption that occurs is quite high (R^2 = 0.70). While the smaller AOD value of 2.99 was found in Pontianak and showed the effect of radiation absorption of SW of -125.5 watts/m^2, a greater value compared to what happened in Palangkaraya. Statistically, the correlation of relation shows a very high correlation with R^2 = 0.91. In Banjarmasin and Samarinda, the events of the 2002 forest fire also showed an increase in the AOD value, only the accumulation of aerosol particles that occurred at that time, did not have a significant effect on the flux of the radiation divergence.

Similar to what happened in 2002, in the event of a forest fire in 2006, an increase in aerosol concentrations occurred in Palangkaraya and Pontianak, each of which was indicated by an increase in AOD values of 2.53 and 3.15, respectively. But the response to the increase in AOD values in the two cities was quite different from what happened in 2002. Compared to the increase in aerosol concentrations in Palangkaraya (AOD = 2.53), the concentration of aerosols in Pontianak experienced a higher increase (AOD = 3.15). But in influencing radiation, the accumulation of aerosol particles covering the city of Palangkaraya shows a stronger influence (shortwave divergence values in Palangkaraya -48.1 watts/m^2), compared to Pontianak which shows the amount of radiation energy absorption reaches -128.3 watts/m^2.

The extreme conditions in the event of forest fires in Kalimantan were discovered in 2015. The accumulation of aerosol particles that occurred in 2015 has also led to increased concentrations of aerosol particles in Palangkaraya, Pontianak, Banjarmasin, and Samarinda. However, as happened in 2002 and 2006, a significant increase in the concentration of aerosol particles occurred only in Palangkaraya (AOD = 3.32 ; R^2 = 0,95) and Pontianak (AOD = 2.65 ; R^2 = 87). While the other two cities, that is Banjarmasin and Samarinda, only showed an increase in AOD value <1.5, and the effect on radiation flux was not significant. Based on the explanation of Rascke et al [8] that the different radiation responses to the AOD magnitude are very dependent on the chemical composition, the diameter size distribution and the type of aerosol particles that greatly affect the reflection power and absorption power.

Kalimantan is known as an area that has peatland content. According to Stockwell et al [30] that the chemical content of peatlands consists of the same chemical composition as carbon, CO, CO_2, CH_4, SO_2, and some heavy metals. The difference that occurs between Palangkaraya and Pontianak both from its AOD value and from the amount of absorption of short-wave radiation that occurs above the two cities, the authors suspect that it is influenced by many factors. Referring to Stockwell et al's explanation [30] the differences that occur in Palangkaraya and Pontianak are probably caused by differences in the concentration of each chemical element contained in peat, the type of peat burning itself, or possibly other factors. This is very interesting and needs to be studied more deeply to be able to prove it.

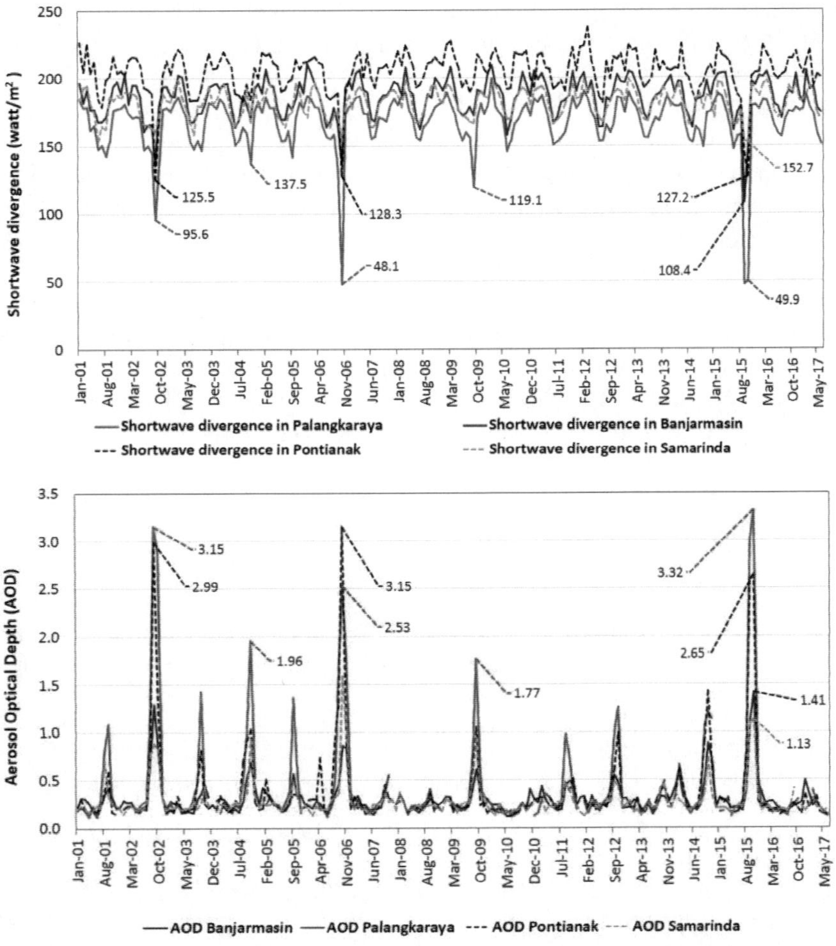

Figure 6. (above) Trends in short-wave radiation divergence in the cities of Pontianak, Palangkaraya, Banjarmasin and Pekanbaru, and (below) trends in the size of aerosol particle emissions (in AOD quantities) for Pontianak, Palangkaraya, Banjarmasin and Pekanbaru in the same time period.

4. Conclusion

The characteristics of short-wave divergence (SW) and long-wave divergence (LW) for the average Indonesian region show seasonal cycles that depend on the position factor for the sun. The minimum SW divergence on average occurs in March, which is around the value of -238.2 watts/m^2 and the maximum SW divergence on average occurs in June, which amounts to approximately -205.2 watts/m^2. LW divergence shows a reverse trend of the SW divergence pattern by reaching a minimum in June of 346.8 watts/m^2 and a maximum of 349.9 watts/m^2 in October. Anomalous trends for the three components of SW divergence flux, LW divergence flux and net radiation divergence flux during the period January 2001 to June 2017 showed that the energy fluctuations absorbed and released were quite low, which averaged no more than -5 watts/m^2 and 5 watts/m^2.

The emission of aerosol particles originating from very large / extreme forest fires will increase the attenuation value of aerosols (AOD) which will affect the radiation flux of the energy. The accumulation of high concentration aerosol particles (AOD > 2.5) will have a significant effect especially on SW radiation energy flux. As an example of extreme forest fires that occurred in the period September-October-November 2015 in Palangkaraya. In this period the high concentration of aerosol particles shown by the value of AOD = 3.32 has affected the absorption of radiation energy that reaches the surface of the earth.

From the analysis of temporal variations on a smaller scale (urban area) the effect of aerosols on SW radiation flux can be very clearly seen in the case of very large/extreme forest fires. In the urban area, the accumulation of very high concentration of aerosol particles (presented with AOD values > 2.5), shows a very high correlation with changes in the flux of the SW radiation divergence (R^2 0.70 - 0.95).

Acknowledgments
This research is the part of research activities that is carried out and funded by the LAPAN Atmospheric Science and Technology Center Bandung. For this reason the authors would like to thank the LAPAN Bandung Atmospheric Science and Technology Center for its support. These data were obtained from the NASA Langley Research Center's Atmospheric Science Data Center, for which the authors would like to thank to NASA (National Aeronautics and Space Administration), which has provided very useful data. Thank you also to the authors for telling colleagues who have provided much support.

Reference

[1] Gray L J, Beer J, Geller M, Haigh J D, Lockwood M, Matthes K, Cubasch U, Fleitmann D, Harrison G, Hood L, Luterbacher J, Meehl G A, Shindell D, van Geel B, White W 2010 *Solar Influence On Climate* Rev Geophys 48 RG4001 doi:10.1029/2009RG000282

[2] IPCC 2007 *Climate Change* The Physical Science Basis. Contribution of Working Group I to the Fourth Assessment Report of the Intergovernmental Panel on Climate Change [Solomon S, Qin D, Manning M, Chen Z, Marquis M, Tignor K B M and Miller H L (eds.)] Cambridge University Press Cambridge United Kingdom and New York NY USA

[3] McGuffie K and Sellers A H 2005 *A Climate Modelling Primer* John Wiley & Sons Ltd Third Edition ISBN: 0-470-85750-1

[4] Sellers W D 1965 *Physical Climatology* The University of Chicago Press 272

[5] Palmen E and Newton C W 1969 *Atmospheric circulation systems* Academic Press 603

[6] Alexander M A and Schubert S D 1990 *Regional Earth-Atmpsphere Energy Balance Estimates Based on Assimilations with a GCM* Journal of Climate vol 3 pp 15-3

[7] Hoch S W 2005 *Radiative flux divergence in the surface boundary layer–A study based on observations at Summit Greenland.* Ph.D. thesis ETH Zu¨rich Switzerland 181

[8] Raschke E S King and Zhang Y C 2012 *Vertical Radiative Flux Divergence in the Atmosphere, in GEWEX Radiative Flux Assessment* (RFA) World Climate Research Programme (WCRP) Report vol 1 pp 189-208

[9] Funk J P 1960 *Measured radiative flux divergence near the ground at night* Q.J. Royal Meteorology Soc 86 382-389 doi: 10.1002/qj.49708636910

[10] Fuggle R F and Oke T R 1976 *Long-wave Radiative Flux Divergence and Nocturnal Cooling of the Urban Atmosphere* Bound. Layer Meteorology vol 10 pp 113-120 doi:10.1007/BF00229279

[11] Lieske B J and Stroschein L A 1967 *Measurements and radiative flux divergence in the Arctic* Arch. Meteorol Geophys Bioklimatol Ser B 15 pp 67– 81 doi:10.1007/BF02319112

[12] Nunez M and Oke T R 1976 *Long-wave radiative flux divergence and nocturnal cooling of the urban atmosphere* Boundary Layer Meteorology vol 10 pp 121–135 doi:10.1007/BF00229280.

[13] Garratt J R and Brost R A 1981 *Radiative cooling effects within and above the nocturnal boundary layer* Journal Atmospheric Science vol 38 pp 2730–2746.

[14] Andre J C and Mahrt L 1982 *The nocturnal surface inversion and influence of clear-air radiative cooling* Journal Atmospheric Science vol 39 pp 864–878

[15] Raisanen P 1996 *The effect of vertical resolution on clear-sky radiation calculations: Test with twoschemes* Tellus Ser. A 48A pp 403–423 doi:10.1034/j.1600-0870.1996.t01-2-00004.x.

[16] Bohren C F and Clothiaux E E 2006 *Fundamentals of Atmospheric radiation* Wiley-VCH pp 472 (Adidactic jewel: discusses the fundamental concepts, laws and equations of the emission and transmission of radiation through the atmosphere, focusing on physical interpretation and avoiding the use of complex mathematics).

[17] Crutzen P J and Andreae M O 1990 *Biomass Burning in the Tropics: Impact on Atmospheric Chemistry and Biogeochemical Cycles* Science vol 250 pp 1669–1678.

[18] Ichoku C, Remer L A, Kaufman Y J, Levy R, Chu D A, Tanr´e D and Holben B N 2003 *MODIS Observations of Aerosols and Estimation of Aerosol Radiative Forcing Over Southern Africa During SAFARI 2000* Journal of Geophysical Research vol 108 p 8499 doi:10.1029/2002JD002366.

[19] Keil A and Haywood J M 2003 *Solar Radiative Forcing by Biomass Burning Aerosol Particles During SAFARI 2000 A Case Study Based on Measured Aerosol and Cloud Properties* Journal of Geophysical Research vol 108 p 8467 doi:10.1029/2002JD 002315.

[20] Kaufman Y J, Tanr´e D and Boucher O 2002 *A Satellite View of Aerosols in the Climate System* Nature vol 419 p 215–223.

[21] Breon F M, Tanr´e D, and Generoso S 2002 *Aerosols Effect on the Cloud Droplet Size Monitored from Satellite* Science vol 295 p 834-838.

[22] Fouquart Y, Bonnel B, Brogniez G, Buriez J C, Smith L, Morcrette J J, Cerf A 1987 *Observations of Saharan Aerosols: Results of ECLATS Field Experiment Part II: Broadband Radiative Characteristics of the Aerosols and Vertical Radiative Flux Divergence* Journal of Climate and Applied Meteorology vol 26 p 38-52.

[23] Benas N, Hatzianastassiou N, Matsoukas C, Fotiadi A, Mihalopoulos N, and Vardavas I 2011 *Aerosol Shortwave Direct Radiative Effect and Forcing Based on MODIS Level 2 Data in the Eastern Mediterranean (Crete)* Atmospheric Chemistry and Physics vol 11 p 12647–12662 doi:10.5194/acp-11-12647- 2011.

[24] Rosida dan Susanti I 2011 *Spatial Analysis of Atmospheric Aerosol Impact on Shortwave Radiation over Indonesia* International Symposium on the 10th Anniversary of the Equatorial Atmospheric Radar (EAR) ISBN: 978979-1458-52-8, 231-239.

[25] Rosida, Susanti I dan Cahyono W E 2014 *Analisis Dampak Kebakaran Hutan Terhadap Kesetimbangan Radiasi di Permukaan Menggunakan Data CERES* Prosiding Seminar Sains Atmosfer 2014 Pusat Sains dan TeknologiAtmosfer LAPAN ISBN: 978-979-145884-9, hal 418-431.

[26] Rosida dan Susanti I 2014 *Analisis Spasial Perubahan Radiasi di Top of Atmosphere (TOA)* Kualitas Udara dan Komposisi Atmosfer Indonesia Buku2 – PSTA 2014 Penerbit Adira ISBN: 978-9791458-82-5 hal 89-99.

[27] Rosida, Susanti I, Cholianawati N 2016 *Variabilitas Divergensi Fluks Radiasi Di Indonesia* Kontribusi Sains dan Teknologi Atmosfer serta Aplikasinya untuk Meningkatkan Daya Saing Nasional Prosiding Seminar Nasional Sains Atmosfer 2016 PSTA-LAPAN Bandung ISBN : 976-602-6465-05-4 p 205-216.

[28] Doelling D 2014 Atmospheric radiation NASA Langley Research Center (LaRC) Atmospheric Science Data Center(ASDC) https://ceres.larc.nasa.gov/

[29] Platnick S et al 2015 MODIS Atmosphere L3 Monthly Product. NASA MODIS Adaptive Processing System Goddard Space Flight Center USA http://dx.doi.org/10.5067/MODIS/MOD08_M3.061

[30] Stockwell C E, Jayarathne T, Cochrane M A, Ryan K C, Saharjo B H, Nurhayati A D, Albar I, Blake D R, Putra E I, Simpson I J, Yokelson R J, Stone E A 2016 *Field measurements of trace gases and aerosols emitted by peat fires in Central Kalimantan, Indonesia, during the 2015 El Niño* Atmos Chem Phys Vol 16 p 11711–11732 doi:10.5194/acp-16-11711-2016

Diurnal Pattern of PM10 and CO in Riau as a Forest Fire Risk Area

Sumaryati[1], T Budiwati[1], A Indrawati[1] and Syahrial[2]

[1]Centre for Atmospheric Science and Technology, National Institute of Aeronautic and Space, LAPAN. Jl.Junjunan 133 Bandung, 40173 Indonesia
[2]Environmental and hygiene services of Pekanbaru, Jalan Datuk Setia Maharaja No. 04 Kelurahan Simpang Tiga, Kecamatan Bukit Raya Pekanbaru, Riau.

Email: sumaryati@lapan.go.id

Abstract. PM10 and CO are air pollutants emitted by fossil fuel combustion and biomass burning have the bad impact on human being. Riau as a forest fire risk area, it is supposed that air quality related to PM10 and CO affected by the fire. This paper studies diurnal pattern of PM10 and CO in Riau during 2015 when a strong El Nino phenomena occurs. The diurnal pattern of pollutant is useful for hazard mitigation of the pollutant especially when its concentration reaches maximum and potent to exceed the threshold value of ambient air quality. The result of the study shows plot scatter of PM10 and CO has got the correlation coefficient of 0.72 when no fire period and 0.91 when fire period. The strong correlation in fire period is caused by the similarity of major sources of biomass burning that ignored other sources. Both diurnal patterns, especially CO shows a bimodal shape when no fire period, and the pattern changes to being flat all of the day when a great forest fire occurs in Riau and its surrounding. The bimodal shape was characterized by two concentration peaks in the morning at 08.00 and in the evening at 20.00 local time that was related on rush hour, atmospheric stability, chemical reaction rate of CO. During one year of el Nino phenomena in 2015, CO concentration is still below national air quality standard although in a great fire period and the diurnal peak concentration, while PM10 concentration in some days in fire period exceeds the national air quality standard.

Keywords : PM10, CO, forest fire, pollution

1. Introduction

The forest fire is one of the environmental problems in Indonesia, especially in Sumatera and Kalimantan islands. Although various efforts of forest fire mitigation and its risks reducing have been carried out, forest fire occurs repeatedly. This event is caused many factors such as economic crisis, the price of log and wood product, CPO prices, the Ministry of Forestry's budget, and supported by natural factors, namely El Nino [1]. Some compounds including air pollutants and greenhouse gases are released to the atmosphere during the forest fire.

The important air pollutants released from the forest fire are PM10 and CO. Carbon elements which have almost half of the dry matter of burning biomass is a source of PM10 and CO formation [2,3]. PM10 is a particulate which has an aerodynamic diameter size of less than 10 μm. It consists of various elements especially carbon and mixed with water vapor appeared as smoke haze. The presence of particulates in the atmosphere reduces air quality because of bad impact on health and climate change [7]. Particulates are poisonous when they enter through the respiratory system and cause death [8]. Particulates reduce the intensity of solar radiation reaching the earth's surface and play a role in cloud formation which is both climate parameters. Another negative impact is on the transportation sector because of reduced visibility

Carbon monoxide (CO) is an odorless and colorless gas which has an important role in studying atmospheric chemistry. It is formed by oxidation of carbon compound when fuel content carbon burned incompletely. CO is classified as a gas pollutant because it has the bad impact on human health, photochemical smog contributor, and surface ozone pollutant precursor [4,5,6]. Its character is very reactive to hydroxyl radical (OH), so it has the short lifetime, and finally CO is oxidized to CO_2 that is greenhouse gas with warming effect on the atmosphere.

Both PM10 and CO are pollutants which have the same main source of biomass burning in rural areas and fossil fuels combustion in urban areas. Therefore, research on PM10 and CO particulates is often carried out simultaneously [9,10]. Riau is a high-risk forest fire area and its urban area is a developing region that has busy transportation. The two main sources have contributed to the air quality in Riau, especially for CO and PM10 parameters.

There have been many publications indexed journals studying in forest fire impact on economy and ecology [11]. There have been many indexed journal publications studying the impact of forest fires on economics and ecology [11]. In this study, diurnal patterns of PM10 and CO were assessed in fire period and no fire period indicated by the number of hotspots observed in the Riau area and its surroundings. Observations were carried out in 2015 when a strong El Nino phenomenon occurred which produced the largest fires in Sumatra and Kalimantan in the last decade [12]. The understanding of air pollutant diurnal pattern is useful for air pollutant management to minimize the bad impact of air pollutant

2. Data and Method

This study analyzed PM10 and CO data yield of monitoring air quality using air quality monitoring system (AQMS). The monitoring was located at three places in Pekanbaru city, Riau province, namely in Tampam (0°28.0`N, 101°23.0`E), Kulim (0°27.5`N, 101°32`E), and Sukajadi (0°31.5`N, 101°26.5`E). Sukajadi was an urban area with a dense population and high transportation activities. Whereas Tampan and Kulim were suburban areas which at the first time intended to monitor the impact of forest fires, but finally Tampan area developing to be crowded traffic, so traffic emission affected the observation of PM10 and CO. The monitoring system was also supported by observing meteorological parameters, namely relative humidity (RH) and temperature (T). The type of data its location could be seen in Figure 1. The observation was carried out for one year, January-December 2015 coincided with El Nino Event, and the data was collected every 30 minutes.

Figure 1. Location of Monitoring of PM10 and CO

Forest fire intensity was identified by the hotspot counts observed by satellite. The hotspot was provided by National Institute of Aeronautics and Space (LAPAN) and could be downloaded at http://modis-catalog.lapan.go.id/monitoring/hotspot/index#. The hotspot was detected using Terra and Aqua satellites with MODIS sensors (moderate resolution imaging spectroradiometer), and SNPP satellites (Suomi National Polar-orbiting Partnership) with VIIRS (Visible/Infrared Imaging Radiometer Suite) sensors. Hotspot positions were considered to study for the effects of distance at a distance of 100 km, 200 km and 500 km from the monitoring location. Hotspot criteria was selected with a confidence level of more than 80%, which in forest fire management must be taken action to control forest fires. Wind direction was not considered in this study assuming the hotspot was distributed evenly around observation location. Temporal analysis of correlation between the PM10 and CO concentrations and the number of hotspots was grouped every month.

As an important product of biomass burning and fossil fuels, it was necessary to investigate relationship between PM10 and CO. The correlation was distinguished in fire period and no fire period. The relationship analysis was based on Pearson correlation. Diurnal pattern was also distinguished in fire period and no fire period. Fire period was the month when more hotspot detected. No fire period was not only no fire period but also included when just a few hotspots detected in the month. The diurnal pattern of CO and PM10 concentration was calculated by averaging at the same hour each day. Diurnal pattern analysis was associated with emission sources and meteorological parameters including temperature and relative humidity which affect PM10.

Analysis of CO and PM10 quality criteria was based on the national ambient air quality standards in the annex of PP No. 41 of 1999 concerning air pollution control, which was 30 mg / Nm^3 for CO for an hour sampling or 10 mg / Nm^3 for one day sampling and 150 µg / Nm^3 for PM10 for one day sampling. Because the diurnal PM10 and CO pattern was not flat, the daily average was calculated if only the data was more than 80% of the complete data in one day, the daily average both of CO and PM10 were calculated if only the amount of data more than 80% of the complete data in one day, while the monthly both of CO and PM10 were calculated if only the amount of daily average more than 60% of the days in one month.

3. Result and Discussion

3.1. Relationship of PM10 and CO concentrations and hotspot

Hotspot is an area with a relatively hot temperature in comparison to its surroundings where this difference in certain threshold values can be detected by satellites. Forest fires which have high temperature relatively can be detected by satellites as hotspots. However, not all hotspots are forest fires, as well as forest fires if they only occur in a narrow area, they are not detected by the satellite as hotspots. Hotspot is suspected as fire dependence on confidence level. Higher confidence level is more potential as fire and the more hotspot is the greater the potential for burning, so there is a positive correlation between the number of hotspots and the extent of forest fires [13].

The amount of PM10 and CO emissions from forest fire are determined by the amount of biomass burned and emission factors of each compound. The amount biomass burned is multiplication of the mass density of forest biomass, forest area burned, and the fraction of biomass burned. Emission factors of CO and PM10 that consist most carbon compound depend on the combustion process. The combustion process is influenced by oxygen supply in the fire. More oxygen supply is more carbon compound formed as CO_2 and the consequently less of CO and PM10, and vice versa. Whether of calm wind, wet biomass and higher density biomass conditions like a tropical forest or peat land are difficult to supply oxygen into combustion process of biomass.

Between hotspot count and area burned is correlated positively, as well as the area burned and the amount of emission [14], consequently, the hotspot count and emission is correlated too. Sukitpaneenit and Kim Oanh NT study of correlation analysis shows that CO and PM10 daily concentrations associated with the hotspot count especially in the rural area has a maximum correlation coefficient (r) of 0.59 for CO and 0.65 for PM10 [15]. In this study the relationship between the hotspot count observations with PM10 and CO concentrations is presented in Figure 2. In this figure shows that the hotspot observed by Terra sensor has the best correlation with PM10 and CO concentrations compared to Aqua and SNPP sensors. The monthly hotspot count within an area in radius about 100 to 500 m from location of observation is almost same, especially in radius of 100-200 km from the location. At a distance of less than 200 km from the location observation which affects the concentration of PM10 and CO, the more hotspots the higher the concentration of PM10 and CO.

PM10 and CO concentrations are observed to be high in August and October, although the hotspot count in these months was detected slightly. The higher concentrations in these months are affected by forest fires in the previous month where many hotspots detected in July and September. This means that the impact of severe forest fires on CO and PM10 concentrations in ambient air continues until the following month around in the forest fire area.

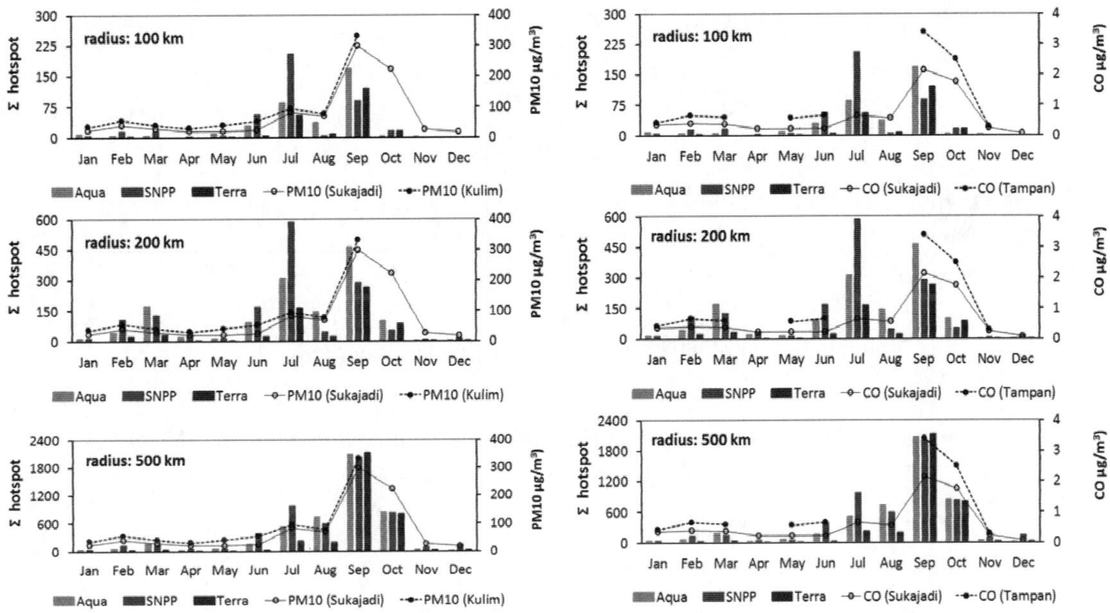

Figure 2. Related hotspot counts and PM10 and CO concentrations

Great forest fires generally occur in Indonesia at the end of the dry season period where weather and biomass are in dry conditions. According to Aldrian and Susanto [16], in normal weather condition, most of Sumatra island, especially Riau and its surrounding areas have two peaks rainfall in March and October. However, the peak of rainfall in March is not as high as October and even rainfall in March is less than rainfall in November and December. July until September is a dry season period when rainfall is very low. Many hotspots were observed during the dry season, i.e July to September. Most hotspots occur in September which is the end of the dry season. It has started increasing rainfall in October cause decreasing hotspot intensity. Based on observing hotspot, the period of fire is from July to September, and the other months are classified no fire period.

The concentration of PM10 in urban area of Sukajadi is lower than in sub-urban area of Kulim, as well as the concentration of CO in the urban area of Sukajadi is lower than in sub-urban area of Tampan. It shows that in Riau as a forest fire risk area, sub-urban areas are more polluted by PM10 and CO pollutants than in urban areas. Local sources of PM10 and CO in urban areas are dominated by burning fossil fuels, while in urban areas other than burning fossil fuels, an important source is burning biomass. The source of biomass burning is borne from cooking traditional and especially during forest fires.

Scatter plot between daily concentrations of PM10 and CO in fire period and no fire shows the strength of correlation both of them presents in Figure 3. The scatter plot has a correlation value ($r = (R^2)^{0.5}$) of 0.73 for no fire period and 0.91 for the fire period. According to Taylor [17] if ($r > 0.68$) then both variables are classified to have a strong or high correlation, whereas according to Ratner [18] the correlation is strong if ($r > 0.7$) or ($r < - 0.7$). Based on these two criteria, the correlation between PM10 and CO in Riau has a strong relationship for all conditions. The strong relationship is caused by the same of the main source emission, both fossil fuels when there are no forest fires and from burning biomass when forest fires occur.

Despite the relationship between PM10 and CO is strongly correlated in all conditions, but it is stronger in the fire period. The stronger correlation is caused the emissions from forest fire process are so great that ignore the other sources of CO and PM10. The strong correlation between PM10 and CO during the fire period in Riau as a forest fire risk area can be applied to estimate if one of the data is known and the other data unknown.

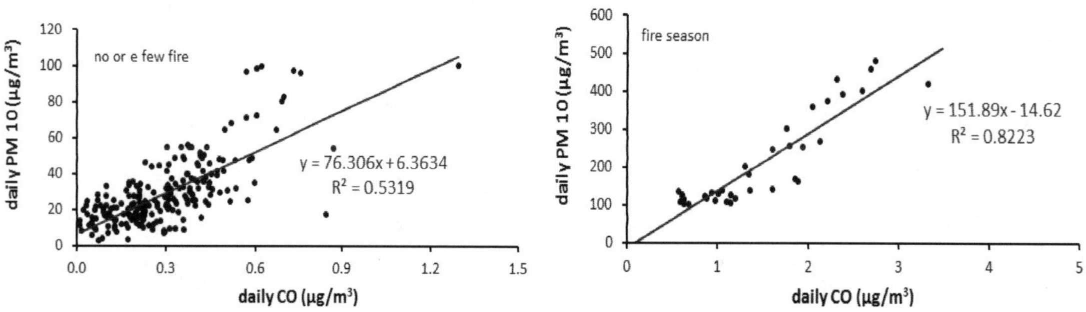

Figure 3. Scatter plot between PM10 and CO daily concentrations

Another difference besides the correlation coefficient of the relationship between PM and CO in the two conditions is a gradient of the correlation equation. The gradient of the correlation equation in the fire period is almost twice the value of the gradient of correlation equation in no fire. It means that there is an increase in emission of PM10 higher than CO when a fire period or the other hand, the emission factor of particulate from forest fire is very high compare from fossil fuel combustion.

3.2. Quality and diurnal pattern of PM10 and CO

PM10 and CO have a bad impact on human health due to their toxic if they inhaled. The smaller size inhaled particulate penetrates the deeper respiratory system, so that only fine particulate be able to reach penetration on blood circulation. CO is toxic for life because it is very reactive with hemoglobin forming carboxyhemoglobin (HbCO). Hemoglobin binds CO of 200 times faster than with oxygen. If the environment is rich in CO, then the body will lack oxygen, while oxygen is a vital matter for life. Because both PM10 and CO are dangerous pollutants, both of pollutants regulated in Government Regulation in PP No. 41 of 1999 which regulates ambient air quality standards. In the annex of the PP No 41 of 1999 is listed national air quality standards (NAQS) for some pollutants. According to the annex, the ambient air quality standard of CO is 30 mg/ Nm^3 for an hour sampling and PM10 is 150 µg / Nm^3 for one-day sampling.

The hourly average of CO and the daily average of PM10 are shown in Figure 4. Air quality in Riau when no fire is very clean against PM10 and CO and this value can be used as the background value. When the forest fire period indicated by more hotspot observed, the concentration of PM10 and CO rise very high from the background value. Despite CO concentrations very high during great forest fires, the concentration is still below NAQS. Unfortunately, PM10 concentration is very high when forest fires occur. During July-October, there is 28% of days in PM10 conditions above NAQS.

As explained above, that forest fire has a high emission factor. The consequencesare air quality of PM10 exceeds national air quality standard. Although people can use masks to reduce the impact of PM10, particles in very small sizes can escape in the filter mask, while the particulate hazard level is higher for smaller sizes. Hence, the particulate is an important concern in mitigating the effects of the forest fire in Riau and its surrounding.

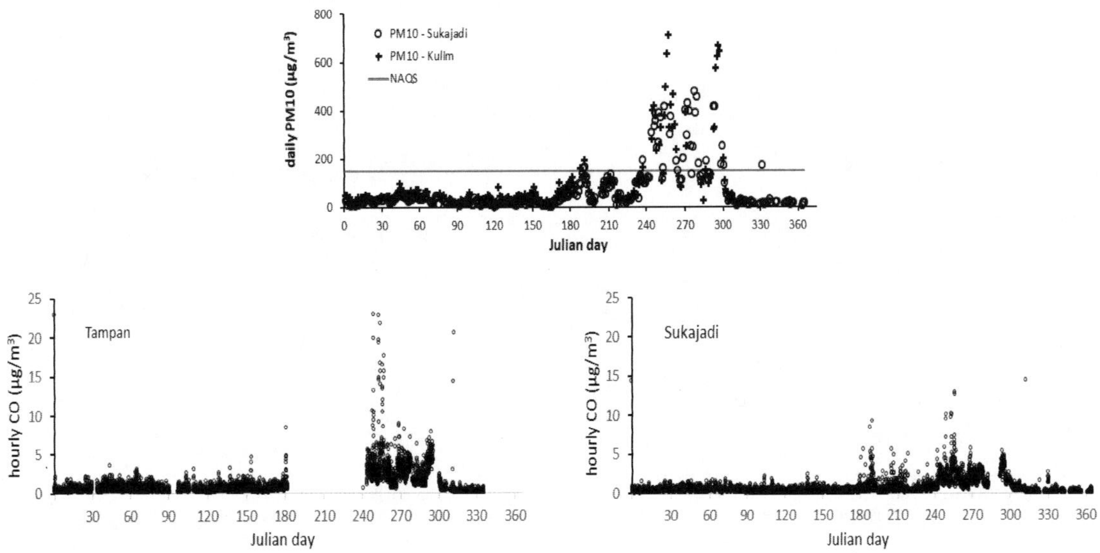

Figure 4. Daily concentration of PM10 and hourly concentration of CO

Diurnal pattern of PM10 and CO concentrations in different fire are shown in Figure 5. The concentration of a pollutant in the atmosphere is determined by the number of emissions, and it is influenced by weather conditions that causes pollutant removed from the atmosphere through diffusion, dilution, dispersion, deposition, and chemical reactions. The number of emissions combined with meteorological conditions that removal pollutant generates ambient air concentrations, which it compiles time series data and creates temporal, such as the diurnal, pattern.

Diurnal CO concentration configurates a bimodal pattern clearly, while PM10 concentration configurations bimodal pattern slightly when the emission from biomass burning can be neglected because of no fire period. Two peak concentrations occur on the morning at 08.00 and at night at 20.00 local time. The bimodal CO pattern of research conducted by [19] in semi-urban areas in India is caused by transportation activities and planetary boundary layer.

There is different in peak shape concentration in the morning and the evening. The concentration peak in the morning is more sharply than in the evening. The sharp concentration peak in the morning is caused CO is oxidized to form CO_2 by solar energy immediately after sunrise [20], through the process:

$$O_3 + h\nu \rightarrow O(^1D) + O_2 \quad (\nu \text{ is solar energy}) \quad (R1)$$
$$O(^1D) + H_2O \rightarrow 2\,OH \quad (R2)$$
$$2\,[CO + OH \rightarrow CO_2 + H] \quad (R3)$$

$$\text{Net: } O_3 + h\nu + H_2O + 2\,CO \rightarrow O_2 + CO_2 + H$$

In the day, atmospheric stability condition tends to unstable and the mixing layer is higher than the night. This weather condition causes the pollutant to be mixed vertically well so that the concentration on the surface becomes low. Therefore the concentration in the morning quickly drops and its concentration is low during the day even though traffic activity is still crowded.

IOP Conf. Series: Earth and Environmental Science **303** (2019) 012041 doi:10.1088/1755-1315/303/1/012041

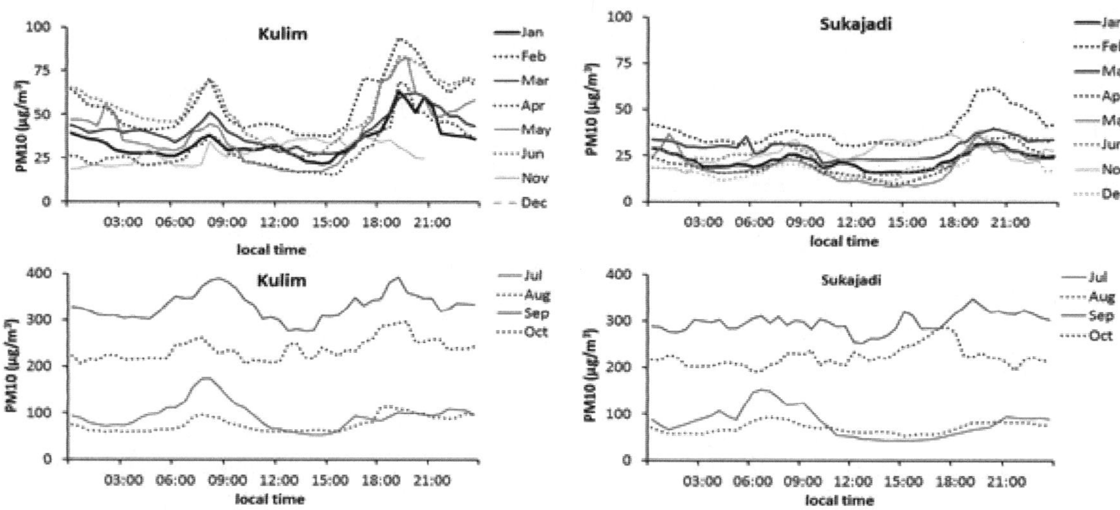

Figure 5. Diurnal patern of concentrations of PM10 and CO

Based on the R1-R3 reactions, the net reaction shows that CO removal from the atmosphere need OH hydroxyl produced using solar radiation energy. In the night, there is no solar radiation to support producing OH and according to Voulgarakis [21] that lifetime of OH is very short only on the order of a few of seconds, so the removal of CO from the atmosphere through chemical reactions is ineffective. The lower mixing layer in the night and no significant CO removal through oxidation by OH process in the night makes CO emitted by rush hour in the evening accumulated around the way where observation located. Reduction of CO concentration at night at the location of the observation occurs because it is transported to a large area and decrease traffic activity as a source of CO. Because of the slow process of decreasing CO concentration at the night, the peak concentration in the evening is widened not as sharp as in the morning.

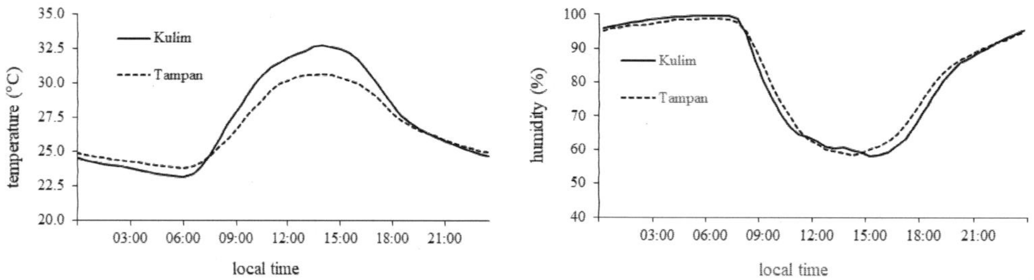

Figure 6. Diurnal pattern of temperature and relative humidity

Particulate is not reactive unlike CO, but its concentration can be affected by meteorological factors namely temperature and pressure. Particulate concentrations are positively correlated with temperature and negatively correlated with pressure (Hasheminassab, 2013; Jayamurugan 2013) because of its hygroscopic. Concentration peak of PM10 in the morning is not as high as in the evening, because in the morning when rush hour occurs, the temperature is still low and pressure is high (Figure 6). While in the day when temperature rising, pressure decreasing, and supported the higher mixing layer condition creates low PM10 concentration. Although the mixing layer is low at night, the source of emission is not active, finally PM10 concentration at night becomes as low as during the daytime.

During the fire period, the diurnal bimodal pattern of CO and PM10 turn out to be the almost flat pattern all of the day. The great emission from forest fire can change diurnal pattern under normal condition and neglects the influence of meteorological factors that should reduce the presence of these two compounds. It shows high PM10 and CO emissions by forest fires that affect air quality in Riau not only in the quantity but also in diurnal pattern.

4. Conclusion

The diurnal pattern of CO and PM10 in Riau as a forest fire risk area for one year in the 2015 El Niño event can be distinguished when a fire period and no fire period. During no fire period, the diurnal pattern forms a bimodal pattern, with peak concentrations in the morning around 8:00 and evening at 20:00 local time due to the influence of source activities and weather conditions. In fire period, a great emission during severe fire changes the diurnal pattern to be flat regardless of meteorological factors. The great fire not only changes the diurnal pattern, but also changes the correlation coefficient between PM10 and CO, where the correlation coefficient between CO and PM10 is 0.72 when no fire, and it rises to 0.91 during great fire.

Acknowledgement

We would like to thank LAPAN and Environmental Agency of Pekanbaru which have supported this research, both in providing data and other facilities.

References

[1] Cahyono A A, Warsito S P, Andayani A, Darwanto D H 2015 *JSL.* Vol. **3** No. 1 (103-112)
[2] Samsonov Y N, Ivanov V A, McRae D J, and Baker S P 2012 *International Journal of Wildland* vol **21** no 7, 779-801
[3] Singh P and Lodhiyal L S 2009 *New York Science Journal* **2**(6):49-53
[4] Gosh R E, Close R, McCann L J, Crabble H, Garwood K, Hansell A L and Leonardi G 2016 *Jornal of Public Health*, Vol **38**, 1:76-83
[5] Haiduc I and Beldean-Galea M S 2011 *Air Quality - Models and Applications.* publisher InTech, Rijeka
[6] Xu J, Zhang X, Xu X, Zhao X, Meng W and Pu W 2011 *Procedia: Earth and Planetary Science* **2** 255-261
[7] Fuzzi S, Baltensperger U, Carslaw K, Decesari S, van der Gon H D, Facchini M C, Fowler D, Koren I, Langford B, Lohmann U, Nemitz E, Pandis S, Riipinen I, Rudich Y, Schaap M, Slowik J G, Spracklen D V, Vignati E, Wild M, Williams M and Gilardoni S 2015 *Atmos. Chem. Phys.*, **15**, 8217–8299
[8] Yorifuji T, Sanghyuk B, Kashima S, Tsuda T, Doi H, Honda Y, Kim H and Hong Y 2015 *JOEM*, **7**: 751 – 756
[9] Bartington S E, Bakolis I, Devakur D, Kurmi O P, Gulliver J, Chaube G, Manandhar D S, Saville N M, Costello A, Osrin D, Hansell A L and Ayres J G 2017 *Environ Pollut*; **220** 38–45
[10] Yadav R, Sahu L K, Beig G, Tripathi N and Jaaffrey S N A 2017 *Elsevier: Environmental Pollution* **225** 291 – 303
[11] Juárez-Orozco S M, Siebe C and Fernández D 2017 *Tropical Conservation Science*, Vol. **10**:1-14
[12] Yananto A and Dewi S 2016 *Jurnal Sains and Teknologi Modifikasi Cuaca*, Vol, **17** No. 21, 11-19
[13] Tansey K, Beston J, Hoscilo A, Page S E, and Paredez Hemández C U 2008 *JGR*, Vol. **113**, D23112
[14] Saharjo B H and Yungan A 2014 *Jurnal Silvikultur Tropika* Vol. **05** No. 2, 124-130
[15] Sukitpaneenit M and Kim O N T 2014 *Environmental Monitoring and Asesment* Vol. **186** isuuse 4 2249 - 2504
[16] Aldrian E and Susanto R D 2003 *Int. J.Climatol*, **23**: 1435 -1452
[17] Taylor C 1990 *JDMS* **1**:35-39
[18] Ratner B 2009 Journal of Targeting, Measurement and analysis for Marketing **17**, 139-142
[19] Vaughan S, Ingham T, Whalley L K, Stone D, Evans M J, Read K A, Lee J D, Moller S J, Carpenter L J, Lewis A C, Fleming Z L and Heard D E 2012 *Atmos. Chem. Phys.*, **12**, 2149–2172
[20] Voulgarakis A, Naik V, Lmarque J F 2013 *J. Atmos. Chem. Phys.*, **13**. 2563-3587
[21] Jayamurugan R, Kumaravel B, Palanivelraja S and Chockalingam MP 2013 *International Journal of Atmospheric Sciences* Volume **2013** 1 – 7

Identification of Potential Source Location of Sulfur at Urban Area Bandung using Conditional Probability Function (CPF)

S Kurniawati, M Santoso, D D Lestiani, D P D Atmodjo, D K Sari and I Kusmartini

Center for Applied Nuclear Science and Technology, Jl. Tamansari no.71 Bandung

syukria@batan.go.id

Abstract. Bandung area is known as the center of economic growth in West Java. Bandung is also surrounding by mountains, therefore the morphological form of its territory is like a giant bowl. The previous study showed that the concentration of Sulfur in air particulate matter from Bandung is high. The objective of this study is to examine the use of conditional probability function (CPF) to identify directions of Sulfur sources using data collected from Bandung. CPF method is an advanced tool that analyzes pollutant source contributions in relation to the winds at the site and has been applied to find source directions. Data collection were done by air particulate sampling using GENT Sampler for 24-hours, once a week in Bandung during 2012-2014. EDXRF technique was applied for Sulfur analysis. The CPF values for Sulfur presented in polar plots, for fine and coarse particles. From this polar plot, the contributions of Sulfur in fine particles were coming from all sectors in Bandung whereas, in coarse particles, Sulfur dominated from the northern sector of Bandung.

Keywords : conditional probability function, sulfur, wind, particles.

1. Introduction

Sulfur (S) is the tenth most abundant element in the universe. The atmosphere plays a complex and critical role in the global cycle of Sulfur. The presence of sulfur in the atmosphere is a few days with wide variations dependent on meteorological and other factors [1]. The atmosphere is also the recipient of the majority of anthropogenic sulfur, which considered as one of the air pollution sources due to its possibility to oxidized and form sulfur dioxide (SO_2). The compound can be transformed into sulfuric acid and mixed with water vapor in the atmosphere to form acid rain, which is one of the serious environmental problems of transboundary nature [2]. Furthermore, SO_2 emissions that lead to its high concentrations in the air could also form other sulfur oxides (SO_x) that can react with other compounds in the atmosphere to form small particles. These particles contribute to particulate matter pollution. Airborne particulate matter (APM) consists mostly of organic matter, metals, water-soluble ions and mineral oxides [3]. This particle may penetrate deeply into sensitive parts of the lungs and cause additional health problems.

Air pollution studies in Indonesia have been conducted for health effect examination as well as source apportionment using multivariate receptor models such as positive matrix factorization (PMF). The technique was developed to provide a flexible modeling approach that can effectively use the information contained in the data [4,5]. The process of identification and apportionment of pollutants to their sources is an important step in air quality management. Source apportionment studies in several urban areas Indonesia have been reported. Santoso et al. (2008) presented the composition and source apportionment of fine and coarse particle samples collected in Bandung and Lembang, Indonesia between 2002 and 2004 [6]. This study also showed that the concentration of Sulfur in air particulate matter from Bandung during that period was high. Beside sulfur, Lestari and Muliadi (2009) also reported that concentration of sulfate (SO_4^{2-}) in fine and coarse particle from Bandung during 2001 – 2007 also high [7].

However, understanding the nature and sources of airborne particulate matter where there are limited or few impacts from a local source is useful. The conditional probability function (CPF) method is an advanced tool that analyzes pollutant source contributions in relation to the winds at the site and has been applied to find source directions [8]. The conditional probability function (CPF) has been successfully used to analyze point source impacts and to identify source direction in multiple studies, and most of it used in conjunction with PMF [9-12]. CPF plots constructed from the PMF modeling results by Lee and Hopke (2006) clearly showed factor specific wind directions consistent with the locations of corresponding industrial point sources identified from emission inventories [11]. In Indonesia, only limited publications related to CPF utilization to identify the local source direction of air pollution.

In this study, the Conditional Probability Function (CPF) method was utilized to identify the local source directions of Sulfur in Bandung area, which were known to be high from the previous study by Santoso et al. (2008). Air particulate matter (APM) sampling in Bandung area was conducted during 2012-2014, and the Sulfur concentration in APM during the time will be determined and compared with the previous study. Bandung is located at latitude 6°91′ S, longitude 107°60′ E and 630 m above sea level. It is the provincial capital of West Java and is categorized as an industrial city with a population which is more than 2.6 million inhabitants in an area of approximately 167.67 km^2 [6]. Bandung area is known as the center of economic growth in West Java. Many small-scale factories are also located around the city. Bandung is also surrounded by mountains, therefore the morphological form of its territory is like a giant bowl. Hourly measurements of APM mass and Sulfur along with CPF are used to identify the directionality of sources contributing to observed pollutant concentrations in urban area Bandung from 2012-2014.

2. Experimental methods and data analysis

2.1. Air particulate sampling

The sampling was located at the rooftop of Environmental Agency of West Java, Naripan Street, Bandung City (latitude: 6°92′ S and longitude: 107°61′ E) as shown in Figure 1. The sampling site represents an urban mixed site (commercial, offices and traffic areas). The samples were collected with a Gent stacked filter unit sampler capable of collecting particulate matter in the $PM_{2.5-10}$ and $PM_{2.5}$ size fractions [13]. The fine and coarse fraction samples are collected on a 0.4 μm pore size filter and coated 8 μm pore size filter, respectively. Samples were collected for 24 h once a week during 2014, at a flow rate of 15–18 L/min.

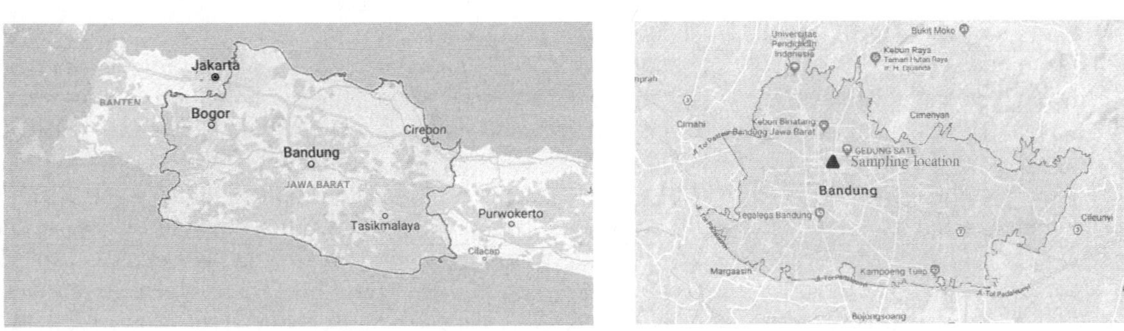

Figure 1. Maps of Bandung, West Java and sampling location.

2.2. Sample Analysis

After sampling, all exposed filters were stored in plastic petri dishes with diameter 47mm for 24 h prior to weighing in controlled environmental condition at 18–23°C and 40–60 % humidity. The particulate matter mass of each of the coarse and fine fractions was determined by gravimetric method. Mass concentrations were obtained by dividing the gravimetric mass by the volume of air that passed through the filter to obtain the concentration of $PM_{2.5}$ and $PM_{2.5-10}$ (μg/m^3). Determination of Sulfur was carried out by X-Ray Fluorescence Spectrometry (PANalytical Epsilon 5 XRF Spectrometer). The methode of the validation was performed by measuring Standard Reference Material (SRM) NIST 2783 Air Particulate on Filter Media.

2.3. Wind rose diagram

The wind rose is graphically presenting the wind conditions, direction and speed, over a period of time at a specific location. Wind speed (WS) and wind direction (WD) data are needed for creating wind rose diagram. Data values for WS and WD were obtained from the nearby Airport. Software WRPLOT was utilized for generating the wind rose diagram. The collected wind data is then sorted by wind direction so that the percentage of time that the wind was blowing from each direction can be determined. The wind direction data is sorted into twelve equal arc segments, 30° each segment, in preparation for plotting a circular graph in which the radius of each of the twelve segments represents the percentage of time that the wind blew from each of the twelve 30° direction segments.

2.4. Conditional Probability Function (CPF)

The CPF method was developed to assess the impacts of sources from various wind directions using the source contributions coupled with the surface wind direction data [8]. The CPF predicts the probability that a given source contribution from a given wind direction will exceed a predetermined threshold criterion [14]. The same daily contribution was assigned to each hour of a given day to match to the hourly wind data. CPF is a simple model, but it provides a powerful approach for identifying the directionality of point sources and/or source areas whose emissions significantly impact the receptor site [15]. The CPF is defined with the following equation:

$$CPF_{\Delta\theta} = \frac{m_{\Delta\theta}}{n_{\Delta\theta}} \tag{1}$$

where $m_{\Delta\theta}$ is the number of occurrence from wind sector Δθ that exceeded the threshold criterion and $n_{\Delta\theta}$ is the total number of data from the same wind sector. In this study, 12 sectors (Δθ = 30°) were used. Calm winds (<1ms^{-1}) periods were excluded from the calculation due to the isotropic behavior of the wind vane for such conditions. The threshold criterion was set at the upper 25 percent quartile of the fractional source contributions for each source. In urban areas, the sources, which are close to the receptor site, can be clearly identified by CPF [11].

3. Results and Discussion

Sampling site at Naripan, Bandung located in the center of Bandung City. The area represents an urban mixed site (industrial, commercial, tourism and traffic areas). Total of 210 samples which consist of fine and coarse particulate matter were collected on weekdays, once a week during 2012-2014. Determination of Sulfur in the fine particulate matter (FPM) and coarse particulate matter (CPM) was carried out using X-Ray Fluorescence (XRF) Spectrometer at Center for Applied Nuclear Science and Technology Laboratory Bandung. Quality control (QC) of the method was also applied by measuring SRM NIST NIST 2783 Air Particulate on Filter Media in the same conditions as the samples. Table 1 presents the analytical accuracy and precision, described as %recovery and % coefficient of variance (%CV). The value was acceptable, compared to AOAC International guidelines [16]. Figure 2 shows the periodic measurement of SRM for method validation.

Table 1. Method validation of Sulfur.

Conc. (mg/kg)	Cert.Value (mg/kg)	% Recovery	% CV
107.3 ± 15.7	105 ± 26	102.2	7.3

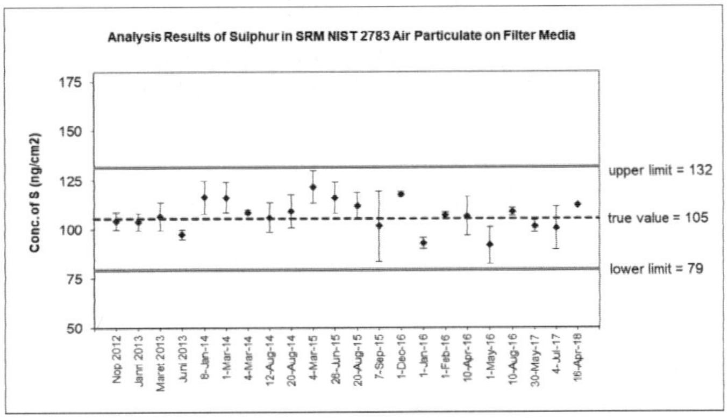

Figure 2. Periodic measurement of SRM for method QC.

The analysis results of mass concentration $PM_{2.5}$ and PM_{10} and Sulfur concentration in FPM and CPM from Bandung during this study period were shown in Table 2. Most of the filters contain Sulfur. The sulfur in FPM and CPM are associated with the sources of pollution in Bandung. Natural sources of pollution that have significant effects on the city are the secondary aerosol, soil dust, road dust, lime dust, and volcanic dust. Anthropogenic sources consist of industry, biomass burning, and vehicular emissions [7]. Time variation of Sulfur concentration in FPM and CPM during 2012-2014 is shown in Figure 3. The significant increasing average occurred for FPM in 2014. That might be correlated with volcanic activity from the eruption of several mountains in Indonesia. Mount Sinabung in Karo land, North Sumatra, was erupted by spraying hot lava and pyroclastic into the atmosphere on January 21st and released a hot cloud on 3rd February 2014. Mount Kelud near the town of Kediri in East Java erupted on February 13. Bursts of volcanic dust clouds and other materials reach a height of several kilometers [18]. The volcanic ash from Kelud spreading through Java and Madura Island [19]. Fine particles can remain airborne for long periods and travel hundreds of miles. It can be seen from Table 2 that mass concentration of fine particles in 2014 is highest compare to 2012 and 2013 due to the traveling volcanic dust. The variation of PM and Sulfur concentration from 2012 – 2014 in Table 2 shows the same pattern, therefore the high sulfur concentration is related to the high PM concentration. The concentration of Sulfur in 2012 is higher than 2013, due to the increasing activity of Tangkuban Perahu Mountain. The status of the mountain was also increased to an alert level since August 23, 2012 [28].

Table 2. The average of $PM_{2.5}$, PM_{10} and Sulfur concentration in FPM and CPM Bandung during the period of study

Year	$PM_{2.5}$ ($\mu g/m^3$)	PM_{10} ($\mu g/m^3$)	S in FPM (ng/m^3)	S in CPM (ng/m^3)
2012	17.28±7.02	44.33±14.79	745.69 ± 341.95	501.17 ± 254.58
2013	14.39±5.30	38.62±10.97	624.86 ± 270.33	427.73 ± 229.01
2014	20.48±7.83	45.50±14.42	1341.60 ± 789.58	743.68 ± 305.51

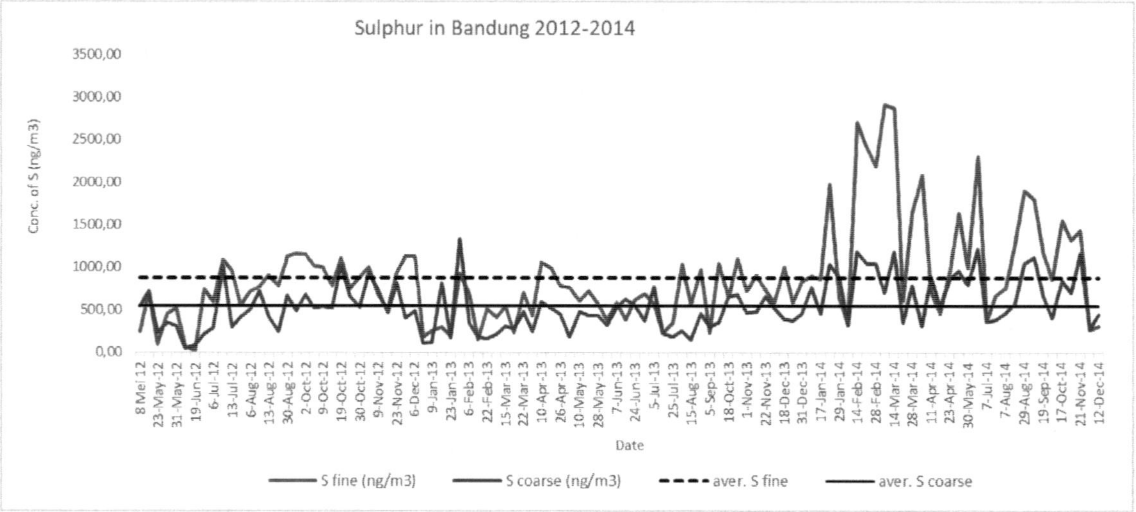

Figure 3. Daily concentration of Sulfur in FPM and CPM Bandung 2012-2014.

Comparison of Sulfur in FPM and CPM Bandung with previous study in Bandung and with other countries in Asia was also presented in Table 3. The average Sulfur in FPM Bandung in 2012 and 2013 are lower than previous study (Bandung 2002-2004), but higher in 2014 due to the volcanic activity from other cities. The range of Sulfur in FPM from Serpong is generally higher than Bandung as a result of the high anthropogenic activities whereas in CPM is lower than Bandung. The source apportionment of FPM in Serpong showed that the contribution of Sulfur was from the formation of lead sulfate in old batteries. The emissions of diesel vehicles were also make contribution since Indonesia still has a high sulfur content in diesel fuel (~5 000 ppm) [6]. High values of Sulfur probably arises from multiple fuelled power plants that burn coal as their main fuel and residual oil. There are two power plants within 50 km of the sampling site in Serpong [20]. Sulfur in FPM and CPM in Ulanbataar, Mongolia was higher than Bandung due to the coal combustion. The emissions from coal combustion sources consist of two distinct source types, the first identified as a high–temperature combustion source originating from power station emissions in the western part of Ulaanbaatar and the others from coal combustion used for domestic heating during winter [21].

Sulfur in FPM from two sites (AQM and AEA) in Srilanka almost similar with this study. Sulfur mostly comes from motor vehicle particularly that is using diesel fuels [22]. In Thailand, both average of Sulfur in FPM and CPM were higher than Bandung. Traffic and biomass burning were contributed since the sampling area was in Bangkok. Industrial or coal–fired power plant as well as diesel vehicles burning high Sulfur fuel possibly also made contribution to this factor [23]. The concentration of Sulfur in FPM from Kuala Lumpur, Malaysia is higher than Bandung whereas Sulfur in CPM almost similar with CPM Bandung. High concentration of Sulfur in FPM from Kuala Lumpur are correlated with anthropogenic activity such as biomass burning, industry as well as exhaust from motor vehicles. It was reported that the city of Kuala Lumpur was among the Asian cities that were heavily affected by vehicle emissions [24]. The average Sulfur in FPM from Dhaka, Bangladesh is higher than Bandung 2012-2013 but lower than Bandung 2014 whereas in CPM is higher. Study by Begum et al (2010) was found that the pollution contributions from the brick kiln increased over the years. Because of the increasing economic activities mainly around the Dhaka city, there has been a significant population growth, which created extra demand on transportation and infrastructures. The building boom has led to establishment of additional brick kilns. Currently, over 1,000 brick kilns are operating around Dhaka, and primarily use coal as fuel [25].

Table 3. The average and range of Sulfur in FPM and CPM in Bandung sites 2012-2014, comparison with previous studies and comparison with other studies.

Year	Fine Fraction (ng/m^3)		Coarse fraction (ng/m^3)	
	Sulfur (average)	Sulfur (range)	Sulfur (average)	Sulfur (range)
Bandung 2012[a]	745.69 ± 341.95	25.68 – 1168.64	501.17 ± 254.58	49.57 – 1040.66
Bandung 2013[a]	624.86 ± 270.33	145.97 – 1100.41	427.73 ± 229.01	115.53 – 1329.38
Bandung 2014[a]	1341.60 ± 789.58	264.38 – 2918.90	743.68 ± 305.51	278.82 – 1212.15
Bandung 2002-2004[6]	1049.20 ± 684.30	-	-	-
Serpong (Setu)[20]		681–1409		456–746
Serpong (EMC)[20]		892–1533		146–697
Serpong (BSD)[20]		687–1362		113–674
Serpong (Batan Indah)[20]		776–1721		311–847
Ulaanbaatar, Mongolia[21]	1969±3978	125 – 40079	1458±1659	129 – 13831
Colombo,Sri Lanka (AQM)[22]	709.3 ± 511.4	585.8 – 3583.7		
Colombo,Sri Lanka (AEA)[22]	484.7 ± 304.8	436.4 – 1990.9		
Thailand (Bangkok)[23]	1546 ± 1218		946 ± 1002	
Malaysia[24]	1935 ± 949		479 ± 288	
Dhaka, Bangladesh[25]	1200 ± 793		961 ± 839	

[a]This study

Bandung has a tropical weather with average temperature of 23°C and rainfall intensity of 120 mm h^{-1} [7]. Identification of the local source directions of Sulfur in Bandung area using CPF needs metrological data particularly wind speed and wind direction. In this study, the metrological data during 2012-2014 were obtained from nearby airport. Wind rose and wind class frequency distribution were generated using WRPLOT software. Figure 4, 6 and 8 presents wind rose that shows characteristics of wind speed and direction per year in Bandung during the study. Wind rose shows that the frequencies of wind directions and are a simple and popular way to represent local transport information [17]. The major wind blowing in 2012 was from east with majority speed, average about 2.10 – 5.70 m/s and from west with majority speed, ranged from 3.60 – 5.70 m/s. Wind direction in 2013-2014 mostly comes from west and north west direction, with speed average about 3-5 m/s. Wind direction frequency from east increases in 2014 than 2013, although it was not dominant compared to 2012. In general, Indonesia has two seasons which are dry season (occurs during April–October) and wet season (occurs during October–April). Major wind blows during dry season from east direction while during wet season major wind blows from west direction [7].

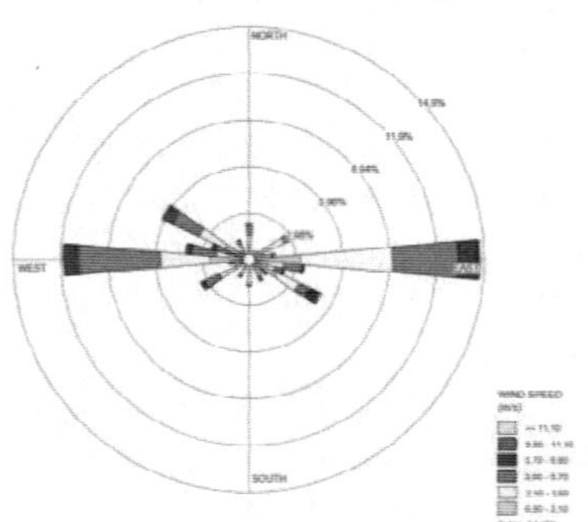

Figure 4. Wind rose of 2012.

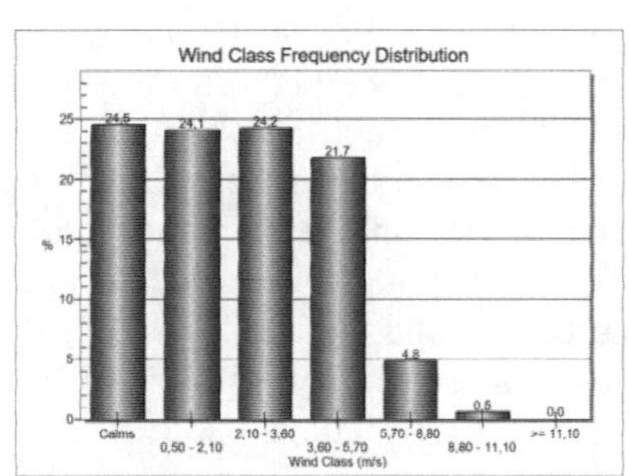

Figure 5. Wind class frequency distribution.

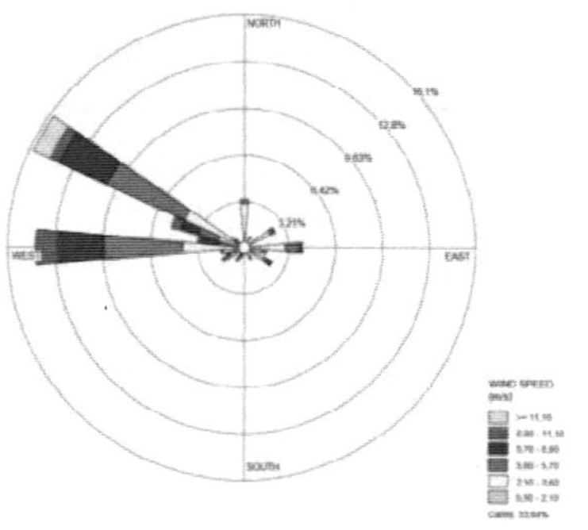

Figure 6. Wind rose of 2013.

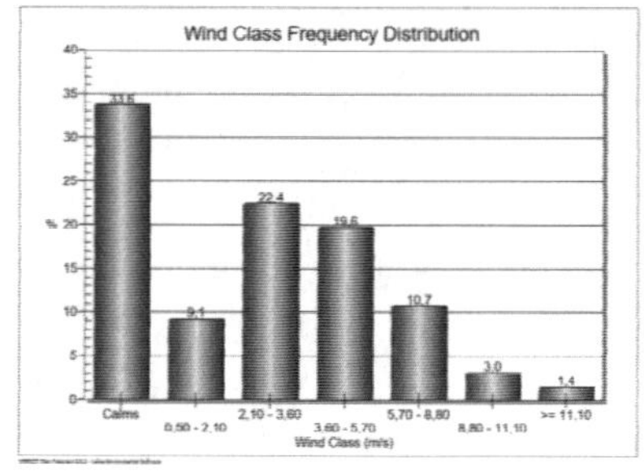

Figure 7. Wind class frequency distribution.

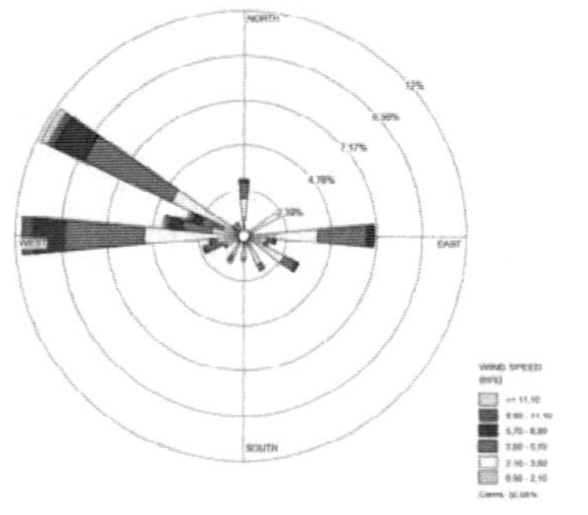

Figure 8. Wind rose of 2014.

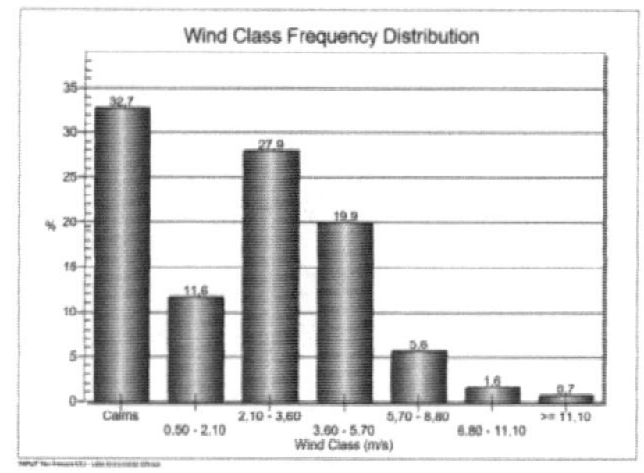

Figure 9. Wind class frequency distribution.

A CPF method is needed to find the directions of high values of estimated source contributions that are likely to be related to the directions of sources. Sources are likely to be located in directions that have high CPF values. CPF was applied to fine and coarse particulate matter. The CPF values for Sulfur presented in polar plots, for fine and coarse particles per year (2012 – 2014) as presented in Figure 10 – Figure 15. The contributions of Sulfur in fine particles for each year shows the different pattern. In 2012, the major source directions of S were coming from west, south-west and north-west from the sampling sites whereas in 2013 were coming from east, north, and west direction. CPF polar plot of FPM in 2014 shows that the Sulfur sources dominantly comes north, south and south-east. The sampling site at local EPA of West Java is located at the centre of Bandung city and surrounded by the commercials, offices, and roads that near the busy intersection with traffic coming from multiple directions that could be one of the emission sources of Sulfur. The gas station in the east side of the sampling site could also become one of the sulfur source that is emitted during the refueling activity. In the west side of sampling site, there are several industries such as pharmacy, textiles, etc. that could be release sulfur during the production process. The distinguishable directionalities for Sulfur in coarse particles were detected where Sulfur is dominated from the northern sector of Bandung. In the northern sector of Bandung, there is Tangkuban Perahu Mountain that located approximately 30 km from the city of Bandung, that become tourist attractions and famous for its Sulfur.

Fine particulate matter usually dominantly comes from anthropogenic sources whereas coarse particulate matter dominantly comes from natural sources. Previous study by Santoso et al. (2008) related to source apportionment of fine and coarse particulate matter in Bandung using PMF suggest that Sulfur as major constituent identified as secondary sulfate. This factor partly contributed from refuse burning or transport of biomass burning from the limestone kilns that are located in Padalarang district, approximately 25 km north-west side away from the sampling site. The Large amount of wood is burnt in those kilns. Other industry in Bandung such as textiles industries using coal as the main energy supply

IOP Conf. Series: Earth and Environmental Science **303** (2019) 012042 doi:10.1088/1755-1315/303/1/012042

for the dying process [26]. The high concentration of sulfur is likely the result of high sulfur concentration in motor vehicle fuels. Generally, the motor vehicle fuels consist of hydrocarbon and contain sulfur since the fuels are produced from underground extraction [27]. Thus, the locally formed sulfate is condensing onto the surface of the existing particles [6].

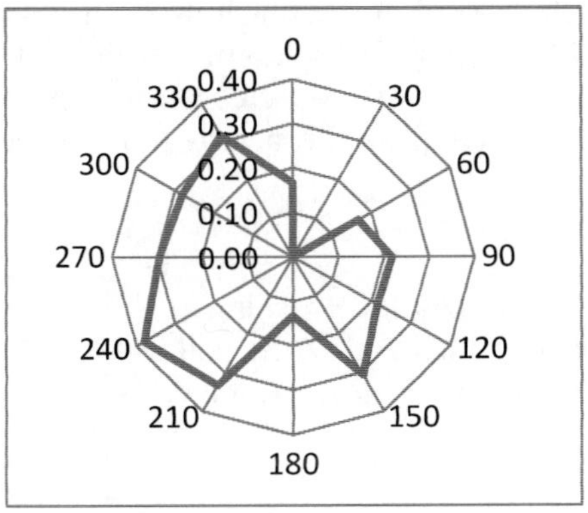

Figure 10. CPF for FPM Bandung 2012.

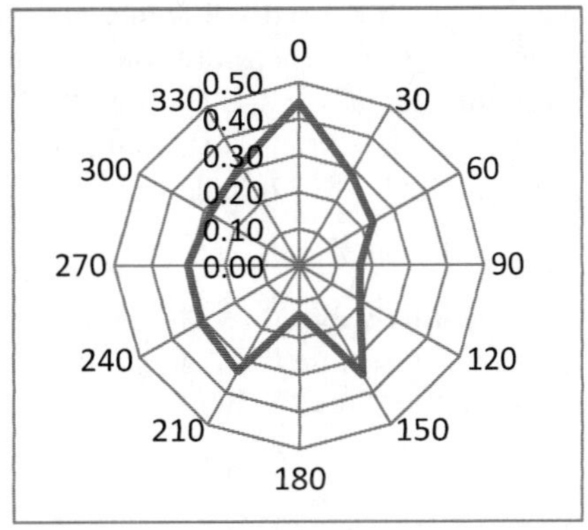

Figure 11. CPF for CPM Bandung 2012.

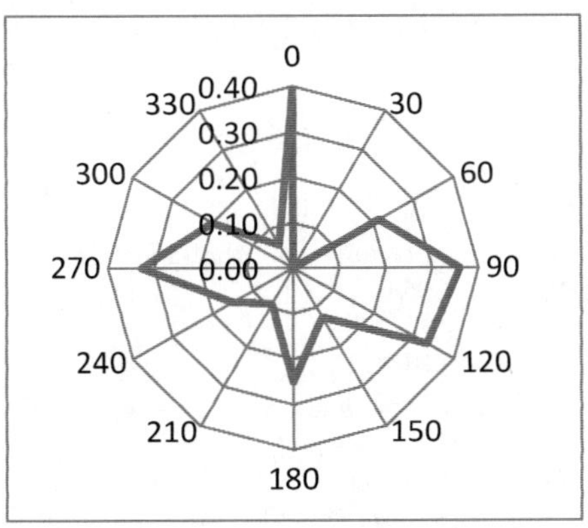

Figure 12. CPF for FPM Bandung 2013.

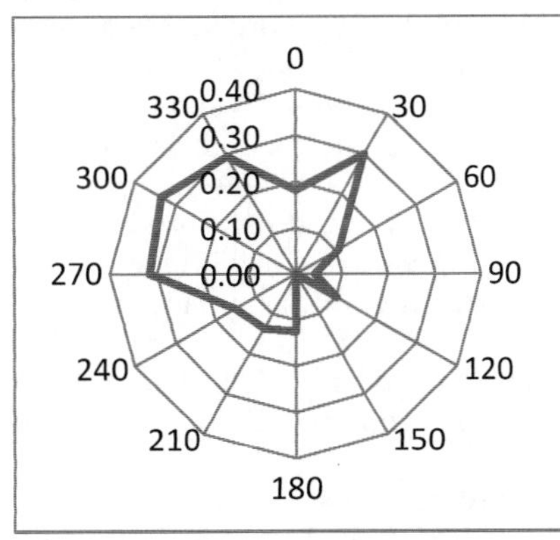

Figure 13. CPF for CPM Bandung 2013.

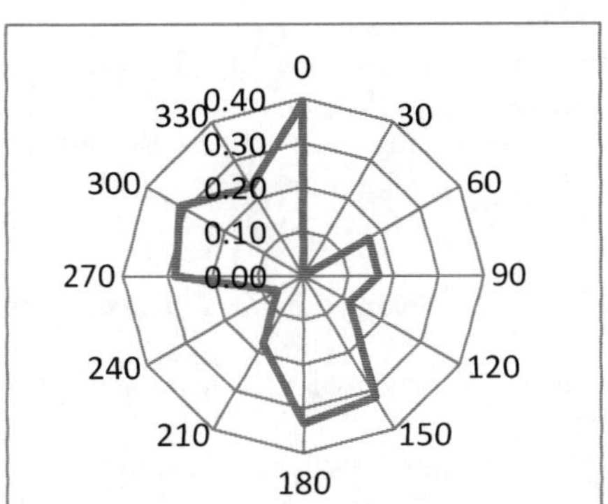

Figure 14. CPF for FPM Bandung 2014.

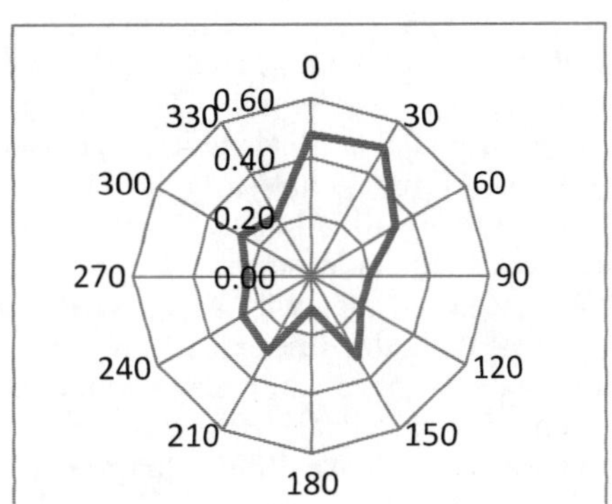

Figure 15. CPF for CPM Bandung 2014.

4. Conclusions

Fine particulate matter usually dominantly comes from anthropogenic sources whereas coarse particulate matter dominantly comes from natural sources. The CPF polar plot of Sulfur for fine and coarse particles showed that the contributions of Sulfur in fine particles shows different pattern for each year. From the previous study, the sulfur in fine particle mostly comes from industrial activity and motor vehicle that spread in all sectors of Bandung. Significant increase was found for Sulfur concentration in FPM that probably correlated with volcanic activity from several mountains in Indonesia during 2014 such as Kelud and Sinabung, therefore the CPF in FPM Bandung 2014 shows that the source of S in FPM was from south east, north and northwest of Bandung. In coarse particles, Sulfur dominated from the northern sector of Bandung. This source probably comes from the nearby volcano, Tangkuban Perahu that is 30 km north of Bandung.

Acknowledgments

This study was financially support by National Energy Agency. Authors wishing to acknowledge special work by staffs from Environmental Agency of West Java during the samples collection and Dian from School of Nuclear Technology for her assistance in summarizing metrological data.

References

[1] Krouse HR and McCready RGL 1979 Studies in Environmental Science, Biogeochemical Cycling of Sulfur, Volume 3 Chapter 6.4 pp 401-430, https://doi.org/10.1016/S0166-1116(08)71065-3
[2] Prihatin J, Corebima AD, Ariffin and Gofur A 2015 Procedia Environmental Sciences 23 186 –191
[3] Hand JL, Schichtel BA, Pitchford M, Malm WC and Frank NH 2012 J. Geophys. Res. 117
[4] Paatero P 1997 Atmos Environ. 37 23-35
[5] Paatero P and Tapper U 1994 Envirometrics 5 111-126
[6] Santoso M, Hopke PK, Hidayat A and Lestiani DD 2008 Science of Total Environment 397 229-237
[7] Lestari P and Mauliadi YD 2009 Atmos Environ. 43 1760-1770
[8] Kim E, Hopke PK and Edgerton E 2003 J. Air Waste Manage Assoc. 53 731-739
[9] Begum B, Kim E, Biswas SK and Hopke PK 2004 Atmospheric Environment 38 3025-3038
[10] Kim E and Hopke PK 2008 Atmospheric Environment 42 6047-6056
[11] Lee JH and Hopke PK 2006 Atmospheric Environment 40 360-377
[12] Heo JB, Hopke PK and Yi SM 2009 Atmos. Chem. Phys. 9 4957-4971
[13] Hopke PK, Xie Y, Raunemaa T, Bieglski S, Landsberger S and Maenhaut W 1997 Aerosol Sci. Tech. 27 726-735
[14] Kim E and Hopke PK 2004 Atmospheric Environment 38 4667-4673
[15] Bae M-S, Schwab JJ, Chen W-N, Lin C-Y, Rattigan OV and Demerjian KL 2011 Atmospheric Environment 45 2531-2540
[16] Gonza´lez AG and Herrador MA 2007 Trends in Analytical Chemistry 26-3 227-238
[17] Xie Y and Bercowitz CM 2007 Atmospheric Environment 41 5831-5847
[18] Deutsche Welle, 2014 Indonesian Volcano Eruption (in Indonesian), accessed at 18 September 2018 from https://www.dw.com/id/letusan-gunung-berapi-indonesia-2014/g-18172846
[19] BBC Indonesia, Jakarta, Indonesia, February 14, 2014, accessed at 18 September 2018 at https://www.bbc.com/indonesia/berita_indonesia/2014/02/140214_gunung_kelud_meletus
[20] Santoso M, Lestiani DD, Mukhtar R, Hamonangan E, Syafrul H, Markwitz A and Hopke PK 2011 Atmospheric Pollution Research 2 190-196
[21] Davy PK, Gunchin G, Markwitz A, Trompetter WJ, Barry BJ, Shagjjamba D and Lodoysamba S 2011 Atmospheric Pollution Research 2 126-137
[22] Seneviratne MCS, Waduge VA, Hadagiripathira L, Sanjeewani S, Attanayake T, Jayaratne N and Hopke PK 2011 Atmospheric Pollution Research 2 207-212
[23] Wimolwattanapun W, Hopke PK and Pongkiatkul P 2011 Atmospheric Pollution Research 2 172-181
[24] Rahman SA, Hamzah MS, Wood AK, Elias MS, Salim NAA and Sanuri E 2011 Atmospheric Pollution Research 2 197-206
[25] Begum BA, Biswas SK, Markwitz A and Hopke PK 2010 Aerosol and Air Quality Research 10 345–353
[26] Kompas Newspaper, Jakarta, Indonesia, July 15, 2005
[27] Tempo Newspaper, Jakarta, Indonesia, July 27, 2010
[28] Liputan 6, Jakarta, 3 September 2012. Available at https:// www.liputan6.com/news/read/434466/kandungan-belerang-di-tangkuban-perahu-di-atas-normal

IOP Conf. Series: Earth and Environmental Science **303** (2019) 012043 doi:10.1088/1755-1315/303/1/012043

Fine and Coarse Particle Concentration and Composition Measured In Urban And Non Urban Area Bandung, West Java - Indonesia

P Lestari and Y Hendra

Faculty of Civil And Environmental Engineering, Institute of Technology Bandung
JL. Ganesha No.10, Bandung - Indonesia. Fax No: +62-22-2534189

pujilest@indo.net.id

Abstract. Atmospheric particulate concentrations for fine (particulate matter less than 2.5 micron) and coarse particles (particle with diameter 2.5-10 micron) were measured in 2 locations which are Tegalega and Dago Pakar Bandung. Tegalega represents an urban mixed site while Dago Pakar represents rural site. Samples on each location were collected over one or two days-time periods during dry and wet season using Dichotomous sampler and Low Volume Sampler (LVS). Meteorological conditions were also measured simultaneously in the sampling locations. Samples were further analyzed to determine concentrations of particulate mass, elements, sulfate, nitrate and ammonium. The results from this study show that the concentrations of the particulate mass were higher during the dry season than in the wet season for both locations, mixed and background sites. The average fine particles concentrations at mixed site were 37 and 49 $\mu g/m^3$ for the wet and dry seasons respectively. The average coarse particles concentrations were 27 and 33 $\mu g/m^3$ for wet and dry seasons respectively. The average mass concentrations of fine particles at the rural site were 30 and 45 $\mu g/m^3$ for wet and dry seasons respectively, while for the coarse particles the concentrations were 13 for wet season and 17 $\mu g/m^3$ in dry season. Concentration of Black Carbon, sulfate, nitrate, and ammonium mostly existed in the fine fraction while crustal elements dominated the coarse fraction. Black Carbon, Sulfate, nitrate and ammonium contributed about 55 % to the total mass of the fine particles at Mixed site (Tegalega) and about 52 % at rural site (Dago Pakar) in dry season. Fine particulate contribute about 57 % to the PM10 in Tegalega and about 73% in Dago Pakar (rural site). This indicates that fine particles may come from other places (transported) in the rural area.

Keywords : total suspended particle (TSP), PM10 , PM2.5, Black Carbon (BC), nitrate, sulfate, ammonium, elements, fine particle, coarse particle

1. Introduction

Atmospheric particulate matter is one of the major pollutants concerned in the ambient air due to the effect of material to the environment and human health. Particulate with size particles less than 10 μm diameter could go through the human respiration system, and particulate with size less than 2.5 μm could penetrate deep into the lung and cause lung damage. Particle larger than 10 μm diameter generally do not penetrate deep into the lung. However, these size particles cause the major dry deposition to the earth surface which cause the problem to the environment [2] and [3]. Because the effect of the particulate matter to human health and environment is very large therefore chemical compositions such as sulfate, nitrate, metals contained in the particles (PM10, PM2.5) are very important and need to be investigated.

Bandung is the capital city of West Java province with unique climate and topography. The pollution levels in Bandung are incredibly exceeded the Indonesian Ambient Air Quality standard for particulate matter. Results from earlier study in Bandung [1] showed that the particulate concentration from 14 monitoring locations out of 16 monitoring locations exceeded the Indonesian ambient air quality standard for 24 hour average (260 µg/m3). The study conducted in Braga Street [4] still show that the concentration of atmospheric particulate in Jalan Braga exceed the ambient air quality standard and the PM10 contributes from 75 to 95 % of the total suspended particulate matter. While study conducted in Ganesha and Merdeka streets also indicated a big portion of fine particles to the total PM10 [8]. Since the contribution of the PM10 as well as PM2.5 to the total suspended particulate is very significant, therefore further or intensive study on PM10, PM2.5 were conducted in Bandung

The purpose of this study is to measure the fine and coarse particle concentrations in the ambient air and to determine the chemical compositions of such particles. The study was conducted during the wet and dry season in Bandung during the period of 2001-2003.

2. Methodology

Fine and coarse particulate samples were collected at Tegalega (mixed site) and Dago Pakar (rural site) during wet and dry season 2001- 2003. Tegalega is located on the roof of two story buildings in mixed area (commercial and traffic), while Dago Pakar is located in the north side of Bandung with rural activities. Description of sampling location is presented in the figure 1 and table 1. Atmospheric fine and coarse particulate concentrations were measured over 24-36 hours period using a dichotomous sampler and Low Volume Sampler (LVS). Dichotomous sampler has two cut size diameters, which are PM smaller than 2.5 µm (fine) and PM with size 2.5-10 µm (coarse). This sampler has a low-flow rate of 17.68 L/min. while the LVS were operated at 3.5 L/min and 20 L/min for PM2.5 and PM10.respectively. Meteorological conditions were also measured simultaneously during the sampling period.

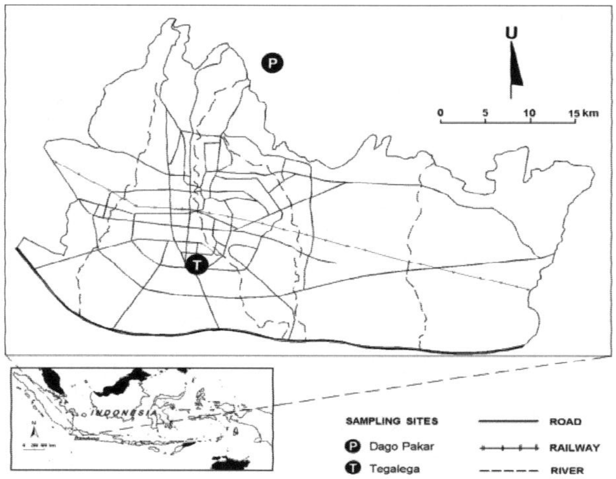

Figure 1. Map of Sampling Locations in Bandung

Particles collected on the filters from dichotomous sampler were analyzed using gravimetric method by weighing the filters before and after sampling to determine ambient particulate concentration. Element concentrations were measured using Atomic Absorption Spectrophotometer with flame ionization and graphite furnace detectors. Prior to AAS analysis, the samples were extracted using Microwave Digestion. The water soluble ions were analyzed by ion chromatography (DX-100). In analyzing anions the eluent of 1.7 mM $NaHCO_3$/1.8 mM Na_2CO_3 was used at flowrate

IOP Conf. Series: Earth and Environmental Science **303** (2019) 012043 doi:10.1088/1755-1315/303/1/012043

2.5 L/min. The concentration of black carbon (BC) in the fine fraction of the samples was determined by reflectance measurement using an EEL type Smoke Stain Reflectometer. The BC values were measured from coarse fraction filters and then converted to fine fraction values. Secondary standards of known black carbon concentrations are used to calibrate the reflectometer.

Background contamination was routinely monitored by using operational blanks (unexposed filters) which were processed simultaneously with field samples. The concentrations would then be blank corrected.

Table 1. Geoghrapical location of the sampling sites

Sites	Latitude	Longitude	Elevation (m asl)
Tegalega	6°56'32"S	107°37'28"E	708
Dago.Pakar	6°51'41"S	107°37'42"E	825

3. Results and Discussion

3.1. Fine and Coarse Particle Concentrations

The mass concentrations of fine and coarse particles during wet and dry seasons from urban mixed site (Tegalega) and non-urban (Dago Pakar) site are presented in the Figure 2 and 3. Basically, the concentrations of both fine and coarse particles were higher during the dry seasons. In Tegalega, concentrations of coarse particles increased during dry season and the contribution between fine and coarse particles was almost balance. This can be caused by increasing the windblown dust and road dust during the dry season. Different from Tegalega site, at Dago Pakar site, the fine particle concentrations were consistently high compared to the coarse particles during wet or dry season as presented in Figure 3.

The PM concentrations at Dago Pakar site (non-urban site) during the dry season were mostly lower than at Tegalega site. The average fine ($PM_{2.5}$) concentration was very similar to the Tegalega site during the dry season, however, the coarse particle concentrations were much lower at Dago Pakar. The average coarse particle concentration in Tegalega (Urban) was almost two times higher than the coarse concentration in Dago Pakar (non-urban). This indicates that there is possible transport of fine particles from urban to non-urban areas, and there was no significant contribution of coarse particles from the local source at non-urban, Dago Pakar site. It can also be seen that the concentrations were much higher during the transition periods (transition between wet and dry season). This may be caused by the various wind directions and speeds occurred during this time period.

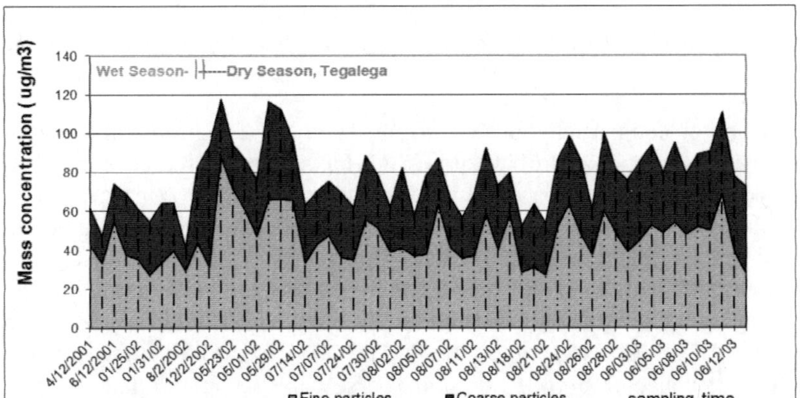

Figure 2. Concentrations of fine and coarse particles at Tegalega during wet and dry season

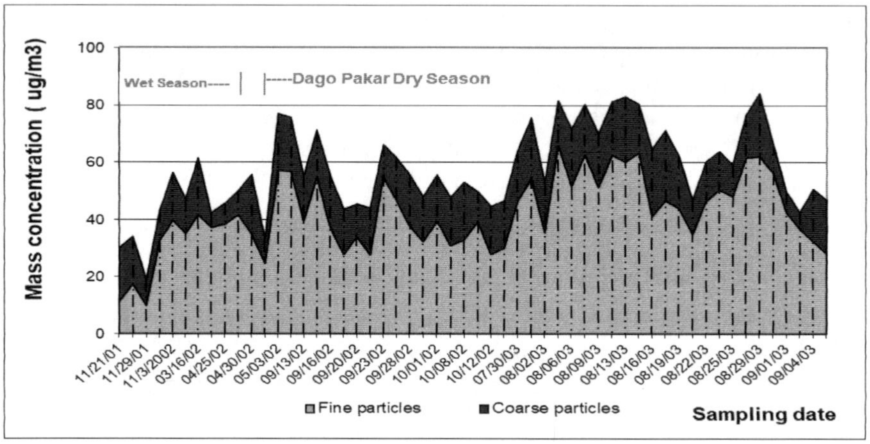

Figure 3. Concentrations of fine and coarse particles at Dago Pakar in Wet and Dry season

The fine particles contributions to the PM_{10} were very high for both locations. Figure 4 shows the contribution of fine particles to PM_{10} at both locations. It can be seen that fine particles contribute about 67% to the PM_{10} at Tegalega and about 80% at Dago Pakar site during wet season. While in the dry season, $PM_{2.5}$ (fine particles) contributions to PM_{10} were 57 % and 73 % at Tegalega and Dago Pakar sites respectively. Again, this indicates that there is no local source of the coarse particles at background site and may only small amount that could be transported from other places.

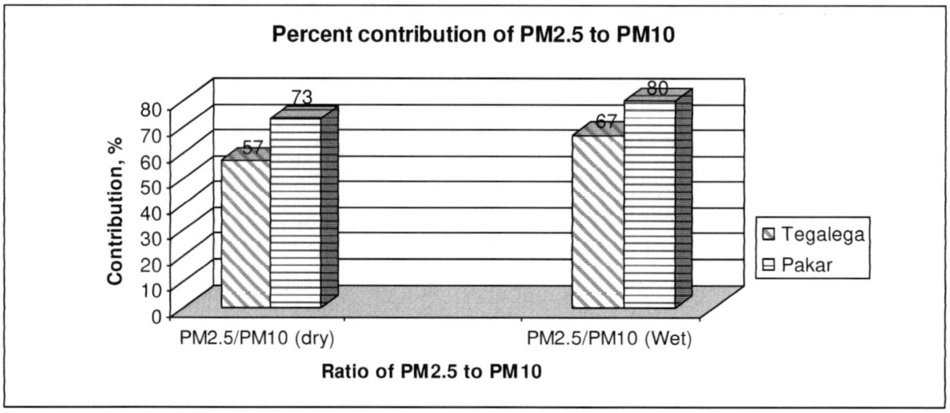

Figure 4. Average mass contribution of $PM_{2.5}$ to PM_{10} at Tegalega and Pakar Sites

The average fine and coarse particle concentrations, standard deviation and range of concentrations at both mixed site (Tegalega) and non-urban site (Dago Pakar) are presented in the Table 2. The tables show the average concentrations, standard deviations, minimum and maximum concentrations as well as slope and R^2 of the $PM_{2.5}$ and PM_{10} correlation. The concentrations in the mixed site area Tegalega were consistently higher compared to the rural site Dago Pakar. The concentration of fine particle and PM10 in Tegalega ranged from 27 to 88 and 52 to 117 $\mu g/m^3$ respectively. This results are higher compared to the study in the semi industrial area in Bandung which was range from 4 to 22.2 $\mu g/m^3$ for fine particle and 24.5 to 77.1 $\mu g/m^3$ for the PM10 [10]. In addition, the study by Santosa et.al showed that the average concentration of fine and coarse particles in Bandung were 14.03 and 11.88 $\mu g/m^3$ respectively [11]. The average coarse particle concentration were similar to that of in Dago Pakar.

Table 2. Summary of Dichotomous particulate matter measurements (Avg±std and range) in μg/m3 in Bandung in the years 2001-2003.

Stations	Fine		Coarse		PM10		PM2.5/PM10	
	Av±std	range	Av±std	range	Av±std	range	Slope	R2
Dry season								
Tegalega (Mixed)	49±14	27-88	33±8	15-50	81±16	52-117	1.01	0.80
Dago Pakar (rural)	45±12	28-66	17±4	6-25	61±13	43-84	1.06	0.91
Wet season								
Tegalega	37±8	27-55	27±14	12-62	55±30	0-93	1.1	0.53
Dago Pakar	30±12	10-42	13±5	5-21	40±17	0-56	0.95	0.81

The correlation between the fine ($PM_{2.5}$) and the PM_{10} concentrations is presented in the figure 5 and 6. The high correlations between the $PM_{2.5}$ and PM_{10} indicated that there is possible similar sources for $PM_{2.5}$ and PM_{10}. For Tegalega site, lower correlation was obtained for wet season, since the number of sample was small.

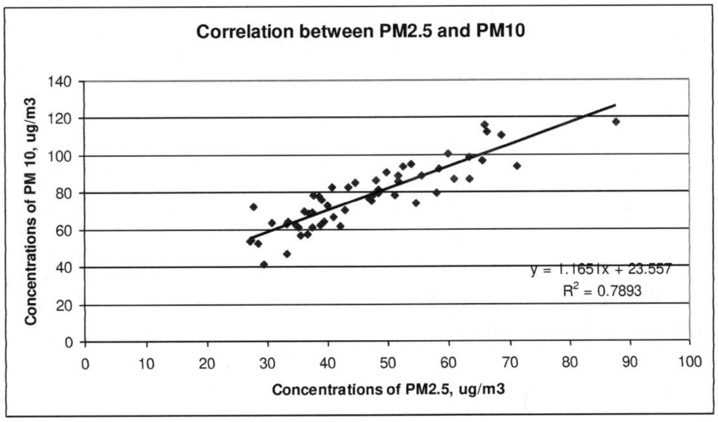

Figure 5. Correlation between $PM_{2.5}$ and PM_{10} for Tegalega Site during wet and dry season

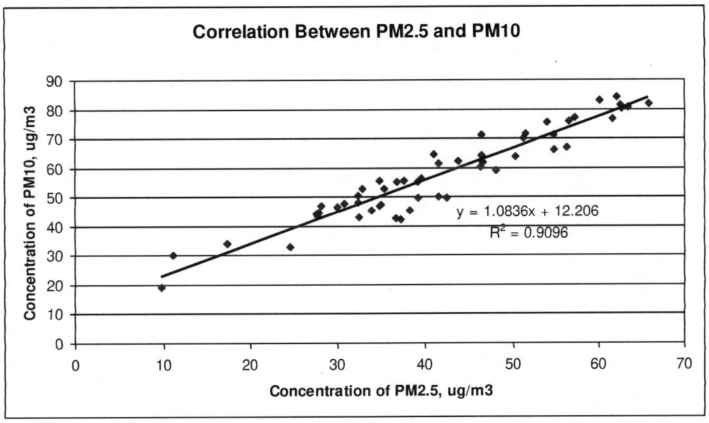

Figure 6. Correlation between $PM_{2.5}$ and PM_{10} for Dago Pakar Site during wet and dry season

IOP Conf. Series: Earth and Environmental Science **303** (2019) 012043 doi:10.1088/1755-1315/303/1/012043

3.2. Chemical Composition

The compounds analyzed in this study include ions (NH4, SO4, NO3, Cl, F), Black carbon (BC) and elements. The elements were grouped into two categories which are crustal elements (K, Mg, Ca, Si, Al, Na, Mn, and Ti), and anthropogenic elements (Pb, Cd, Co, Cu, Sb, As, Fe, Cr, Zn, V, Ni, La, Sc, Th, I).

The average Concentrations of Elements, Ions (Sulfate, Nitrate, Chloride, Fluoride, and Ammonium) and Black Carbon (BC) from both locations during wet and dry seasons are presented in Figure 7 and 8. Figure 7 shows the average concentrations of compounds (NH4, SO4, NO3, Cl, F), BC, crustal and anthropogenic elements in the fine particles from both locations during the wet and dry season. While average concentrations of compounds in the coarse particles are presented in Figure 8.

During the Wet season (for Tegalega and Pakar sites) Black Carbon (BC) was not analyzed due to the equipment problem. Most of the fine particles contained NH4, SO4, NO3, elemental carbon (Black Carbon) and metals. Element carbon, Sulfate and Nitrate take a big portion of the fine particles for both seasons. This agrees with the current work on PM2.5 composition under Spartan Net- Work, which shows that most of the urban PM2.5 compose of Ammonium sulfate, BC, and Ammonium Nitrate as well as crustal material [5]. As a comparison, the study in Beijing also found that Sulfate, Nitrate and Ammonium contributed a big portion of PM2.5. [7]. The major component of PM2.5 in most of the cities in Asian cities (Bangkok, Beijing, Manila, Chennai, and Hanoi) were OM, EC/BC, SO42-, NH4+, and NO3-. [7]. Meanwhile, crustal elements and nitrate took a big portion of coarse particles. Coarse particle are usually came from materials of geological origin through mechanical activities such as grinding, and wind blowing [9]

Concentration of Black Carbon, Sulfate, Nitrate, and Ammonium mostly existed in the Fine fraction while Crustal elements dominated the coarse fraction. Based on the figure 7, Black Carbon, Sulfate, Nitrate, and Ammonium contributed about 56.4 % to the total mass of the Fine particles at Urban Mixed site (Tegalega) and about 51 % at rural site (Dago Pakar) during Dry season. Black Carbon itself contributed about 20.8% to the total fine particles in Tegalega. The study in Serpong area also found that BC contributed about 13-25% to total fine particles [12]. In the wet season, Sulfate, Nitrate, and Ammonium Contributed about 32 % to the total mass of Fine Particle in Tegalega and 29% in Dago Pakar. The high concentration of Sulfate in Dago Pakar may be caused by the active volcano which located about 10 km in northwest side of the Dago Pakar site.

The coarse fractions were mostly dominated by Crustal and Anthropogenic elements. The average contributions in Dry season were about 26 % and 17 % for Tegalega and Dago Pakar sites respectively. In the Wet season, Crustal and Anthropogenic elements contributed about 44% and 43% for Tegalega and Dago Pakar sites respectively. The high Crustal elements concentrations in the coarse particles may be caused by the windblown dust as the high wind speeds were found during the Wet season. During the Wet season, the wind was mostly blown from west to east direction, where there are many limestone productions in the west side of Bandung city that could potentially contribute generating coarse particles containing of Ca and Mg. Figure 8 also shows that the concentrations of anthropogenic elements were much higher in the urban site Tegalega than in the non-urban area. Steel smelters near the urban mixed site may cause the high concentrations of anthropogenic elements such as Fe and Zn.

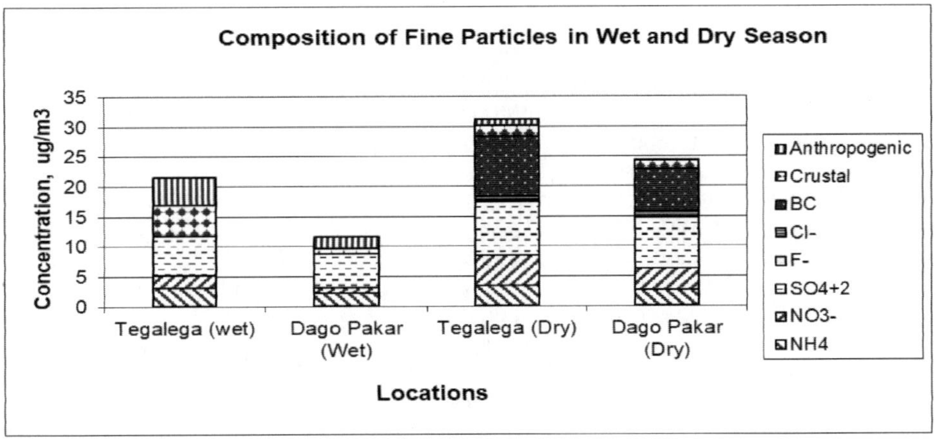

Figure7. Concentrations of Ions, Elements and Black Carbon in the Fine Particles during Wet and Dry season

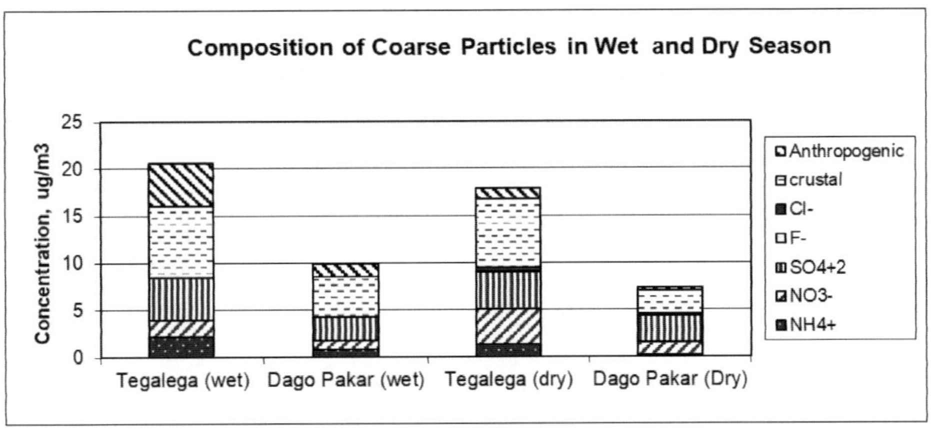

Figure 8. Concentrations of Ions, Elements and Black Carbon in the Coarse Particles during Wet and Dry season

4. Conclusions

The concentrations of fine and coarse particles were mostly higher during dry season than in the wet season for both locations. Coarse particles concentrations were two times higher in Tegalega site than in Dago Pakar in both wet and dry season. Fine particles in Bandung take a big portion of PM10 and there is no difference between concentrations of fine particles in the mixed site and rural site. Fine particles mostly consisted of the element carbon or BC, Sulfate, Nitrate, Ammonium and small portion of metals. The coarse particles mostly consisted of big portion of crustal material. Crustal element concentrations in the coarse particles were higher in the wet season than in the dry season.

Acknowledgement:
This research project was funded by Swedish International Development Cooperation Agency (SIDA) through Asian Regional Research Program on Environmental Technology coordinated by Asian Institute of Technology (AIT). This project would also like to acknowledge to students (Yulia, Yenny, Icha, Ira and Lilik, Haryono) who have been involved in the research project.

References

[1] Lembaga Penelitian(LP) ITB, 1993. Laporan penelitian, Institut Teknologi Bandung, Indonesia.
[2] Noll K.E. and FangK.Y.P.(1989a).Atmospheric Environment 23, 585-594.
[3] Noll, K.E, Puji Lestari, and Holsen,T.M (1997) "Dry deposition of sulfate and nitrate in Chicago" conference proceeding at the Air and Waste Management Association's90th annual meeting & exhibition,Toronto Ontario Canada.
[4] P Lestari, and Savitri,(2003) "Atmospheric Particulate Concentration Measured in an Urban Area Bandung" J. PAGEOPH, 160 (2003) p.107-116
[5] Graydon Snider et.al, 2016. Atmos. Chem. Phys., 16, 9629–9653, 2016
[6] Crystal L. Weagle et.al, 2018. Environ. Sci. Technol. 2018, 52, 11670−11681
[7] N.T Kim Oanh. et.al, 2006. Atmospheric Environment. 40 (2006), 3367-3380
[8] P. Lestari et al.(2002) "Characterization of atmospheric PM2.5 and PM10 in Bandung" presented on the IAC in Taiwan on September 8-13, 2002
[9] Chow.J.C, 1995, Journal of Air and Waste Management Association, 45, 320-382
[10] Lestiani DD, Santosa M, Kurniawati S, and Markwitz A 2013, Indonesian Journal of Chemistry 13. 3, 271-277
[11] Santoso M, Hopke PK, Hidayat A and Lestiani DD 2008 *Science of Total Environment* **397** 229-237
[12] Santoso M, Lestiani DD, Mukhtar R, Hamonangan E, Syafrul H, Markwitz A and Hopke PK 2011 *Atmospheric Pollution Research* **2** 190-196

Forecast of Tropical Cyclone Occurrences based on Fuzzy Logic Algorithm

A M Irawan [1*], H A Nugroho [2], P P Simanjuntak [1] and S I Sugiarto [1]
[1] Department of Climatology, School of Meteorology Climatology and Geophysics (STMKG), Jakarta, 15221, Indonesia
[2] Instrumentation Laboratory, Department of Instrumentation, School of Meteorology Climatology and Geophysics (STMKG), Jakarta, 15221, Indonesia

amir.irawan@stmkg.ac.id

Abstract. Most studies of the tropical cyclone (TC) forecasting are focused on the track and the wind radius forecasting, even though a tropical cyclone formation alerts also developed. This paper takes a different approach and explores and forecast the TC occurrences. This study presents the development of fuzzy logic (FL) models for predicting the occurrence of TC from five primary TC genesis parameters. These parameters are low-level relative vorticity (θ), horizontal wind of upper troposphere (u), sea surface temperature (SST), equivalent potential temperature (θe), and specific humidity (q). The FL model was developed by employing the trapezoidal and triangular fuzzy membership functions for the input and output variables. The fuzzy rules were inferred from the TC genesis parameters data, with a daily calculation period from 1989 to 2018. The amount of TC genesis parameters when the cyclone occurred of the lowest, middle, and upper tercile reconstruction was used as a threshold to build FL model. The model satisfactorily simulated the occurrence of TC with comparable error measures. The result exhibits the accuracy at 0.75 (range: 0 to 1, perfect score: 1). The evidence shows that the result provides insights into the adequacy of FL methods for forecasting the TC occurrences.

Keywords : tropical cyclone, fuzzy logic, relative vorticity.

1. Introduction

TC is the generic term for a non-frontal synoptic scale low-pressure system originating over tropical or subtropical waters with organized convection and definite cyclonic surface wind circulation. Tropical cyclones with maximum sustained surface winds of less than 17 ms^{-1} are generally called "tropical depressions (TD)." Once a tropical cyclone achieves surface wind strengths of at least 17 m/s, it is typically called a "tropical storm" or "tropical cyclone" and assigned a name [1]. In the northern hemisphere, TC occurs between June and November peaking in September; moreover, in the southern hemisphere, the season lasts from November to April. More than one TC can occur in the same ocean and region at once. Among cyclones that occur in the Southern Hemisphere, almost half are formed above the Northern coast of Australia, a quarter appear above the South Pacific Ocean, and others occur above the ocean of South Indonesia [2]. Tropical cyclones are perhaps the most devastating of natural disasters both because of the loss of human life they cause and the large economic losses they induce [3-7].

One of the most challenging subjects in term of ocean-atmosphere interaction is that of understanding the tropical cyclone (TC) occurrences. Many opinions stated that the environmental conditions and the physical mechanisms which bring about tropical cyclone formation is important to understand why and how TC form. Among scientists who have studied this phenomenon there lies a wide variety of opinion. For instance, Neiburger, Edinger and Bonner [2] observed that TC possibly form when sea surface temperatures (SST) depicts the magnitude more than 27°C and do not form between the 4°N and 4°S (tropical areas) of the equator. Based on this TC form requirements, TC will be less likely to pass through Indonesia due to a tropical characteristic of Indonesia region. However, the effects of TC that occur around Indonesia could affect bad weather and meteorological disaster in various places in Indonesia. Vulnerability to tropical cyclones is becoming more pronounced because the fastest population growth is in tropical coastal regions. In Indonesia, for example, TC Cempaka that occurred on 28 October 2017 generated extreme weather has caused floods, landslides, and tornadoes in 21 regencies/cities in Java and Bali. Temporary data collected by National Agency for Disaster and Management (BNPB) post, the disaster occurred in the large area of Java such as Semarang, Kulon, Progo, Ponogoro, Bantul, and Kudus [8]

In this research, fuzzy logic (FL) algorithm was used to forecast both TC and TD occurrences. This method has some advantage, i.e. (i) it mimics the human thinking and reasoning, (ii) it can detect the uncertainties of dynamic system behaviors, (iii) it is based on training and experiences rather than of the theory. This new approach is necessary since it could have some implications for the TC and TD forecast.

2. Data and Method

2.1 Study Site

To assess TC forecasting related to the main parameters in the formation of TCs such as wind—pressure relationship, finding a suitable study site is important [9]. Tropical cyclones generally occur in the tropics with low latitudes between 10° and 20° N / S from the equator [10]. In addition, the most active TC growth area for the Southern Hemisphere is Australia's northern coast which experienced 124 incidents within 100 years [11]. As explained by the Australian Bureau of Meteorology (BOM), this region reaches an average of 10 times of TC occurrences per year. From this literature, Australia's northern coast is an area that has higher TC occurrences more than many places in the world. In the Southern Hemisphere, almost TC occurrences are formed above the northern Australia ocean, South Indonesia and South Pacific Ocean [2]. Therefore, in this study, the Southern Ocean of East Nusa Tenggara (15°-10°S and 115°-130° E), Indonesia, is employed as the study site. This area is an example of a populated TC occurrence that is highly vulnerable to the storm which is mainly caused by TC.

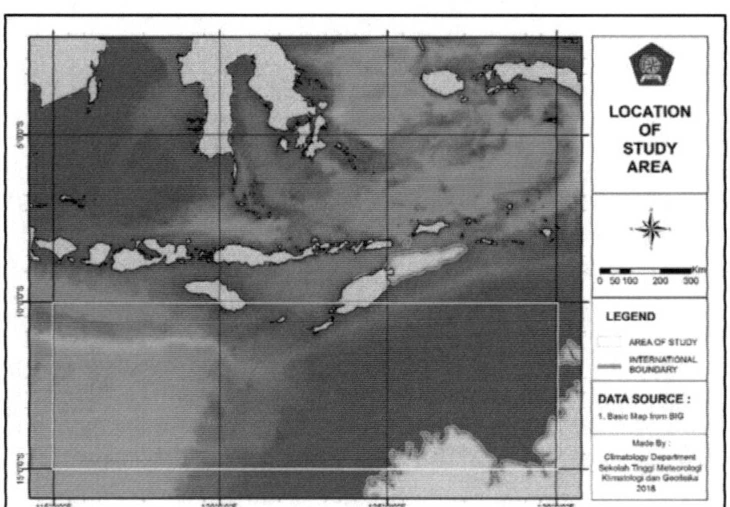

Figure 1 Domain of the study

As illustrated in figure 1, the layer of the study domain shows an ocean between East Nusa Tenggara and Australia North coast. Based on the archive of tropical cyclones [12], there were 56 occurrences of tropical cyclones in the research grid during the period 1989-2018 in winter and spring periods.

2.2 Input Selection

Gray [13] explained that the frequency of tropical cyclones could be directly related to a combination of several physical parameters called the main parameters in the formation of tropical cyclones. These parameters are:

1. The amount of vorticity in the lower troposphere. Tropical cyclones only form in large areas of vorticity in the lower troposphere.
2. The zonal wind vertical shear role. According to Gray [13], the requirements of tropical cyclones are a weak zonal wind shear approximately at 0-1 m/s.
3. The role of ocean thermal energy. Tropical cyclones have a large influence on sea surface temperature, when the sea surface temperature changes, it changes the cyclone characteristics. Tropical cyclones can have a profound influence on the temperature of the ocean over which they travel. The altered ocean temperature, in turn, can feedback and alter the character of the cyclone.
4. The influence of the surface to the equivalent potential temperature in the middle troposphere. Cyclone formation is related to moist buoyancy potential or the extent of the layer boundary of the atmosphere to the middle troposphere. Representative Buoyancy to study TC characteristics is the difference between the surface (1000 mb) and 600 mb layer.
5. The role of moisture in the middle troposphere. Moisture parameters vary from 0 to 1. The development of cyclones does not occur if the humidity in the layer 500-700 mb is less than 40%. This factor increases linearly by 1 in the humidity between layers 700-500 mb reaching 100%, or humidity parameter is between 40% and 70%. So tropical cyclones only form in regions with relatively high humidity (RH) or specific humidity (q) in the middle troposphere.

Therefore, the input parameters used in this study are the main parameters in the formation of tropical cyclones, i.e., low-level relative vorticity (ϑ), the horizontal wind of upper troposphere (u), sea surface temperature (SST), equivalent potential temperature (θe), and specific humidity (q). The input variables are selected during the rainy season in Indonesia on December, January, February (DJF) and March, April, May (MAM). These periods were selected due to the highest frequency of TC occurrences in the south of Indonesia started from October to May with a peak in December to March [14].

All of the input parameters are obtained from the Japan Meteorological Agency (JMA), Tokyo, Japan [15]. Furthermore, this study employed the input parameter data with a daily calculation period from 1989 to 2010 as a training data and 2011 to 2018 as a validation data.

2.3 Fuzzy Logic (FL) Reconstruction

Bellman and Zadeh [16] suggested that fuzzy is a different aspect of uncertainty with randomness. Then they proposed a mathematical form to see how obscurity can be expressed in a human language whose approach is called "fuzzy logic (FL)". In fuzzy it is known that membership degrees have intervals [0,1]. With fuzzy logic, we can express the same information (less specific), then manipulate it, and conclude from that information. FL is used in this study because it can handle the determination of event predictions based on input data.

Many scientists have employed the FL algorithm to meteorology and hydrology problems [17-20]. The method of FL algorithm is to consider the system depicts in the form of subsets or fuzzy approaches, each of which is stated with words such as “low”, “medium”, “high”, etc. An FL algorithm consists of parameters that have varying membership degrees in the set. This idea is in contrast with classical algorithm theory because for crisp sets an element either is a member of that set or is not. FL rule-based on systems can be used as useful representations of either simple or complex physical problems. A small number of fuzzy sets leads to unrepresentative forecasts whereas a large number leads to many calculations. In this research, we set 3 number of fuzzy and labeled with, low, medium, and high. This study depicts the following steps to construct an FL algorithm. These are:

(i) Fuzzification of the input and output parameters by considering convenient linguistic subsets, i.e., low, medium, high
(ii) Fuzzy rules construction based on IF-THEN rules to simulate the model. The rules relate the combined linguistic subsets of input parameters to the output fuzzy sets. The input subsets within

IOP Conf. Series: Earth and Environmental Science **303** (2019) 012044 doi:10.1088/1755-1315/303/1/012044

the premise part are combined interchangeably with the logical "and" or "or" conjunction whereas the rules of this study used a logical "and" conjunction based on the formation of tropical cyclones,

(iii) The implication part of a fuzzy system is defined as the shaping of the consequent based on the premise (antecedent) part, and finally

(iv) The result is a fuzzy set, and therefore, requires defuzzification to arrive at a crisp value, which is required to simulate the model.

In the applications of the FL algorithm in control and forecasting, there are mainly two approaches: the first one is the Mamdani method, and the other is Takagi-Sugeno. For the Mamdani method, they use clear procedures, i.e., fuzzification, logic decision, and defuzzification procedure. For Takagi-Sugeno algorithm [21], however, does not have a clear defuzzification method. For the Mamdani approach, the outcome of each IF-THEN rule will be a fuzzy set for the output variable so that the step of defuzzification is indispensable to obtain the crisp value of the output variable. Therefore, this study utilising Mamdani procedure to construct FL rules.

2.4 Tropical Cyclone (TC) Forecasting Model

Before determining the threshold of the fuzzy membership function, this paper aims at examining the lagged input data to predict TC and TD occurrences. We used the cross-correlation method to evaluate the relationship between the lag time of each input variable and both TC and TD occurrences. The lag time with best biserial correlation with TC and TD occurrences were chosen as predictors.

Furthermore, we categorized the predictor threshold based on 33th percentile of TC and TD occurrences for 1991-2010 periods and divided into three classes (low, moderate & high). After the fuzzy rule base is identified and a defuzzification algorithm is selected, forecast accuracy is tested using a different set of the historical data set (training data) from the one used to obtain the rule base. If it is unsatisfactory, then the number of fuzzy membership functions and/or shape of the fuzzy membership functions can be changed, and a new FL rule base is obtained. The iterative process of designing the rule base, choosing a defuzzification algorithm, and testing the system performance may be repeated several times with a different number of fuzzy membership functions and/or different shapes of fuzzy memberships. The FL rule base that provides the minimum error measure for the test set is selected for real-time forecasting.

This study used trapezoidal and triangular membership function in term of the main characteristics of five TC and TD input parameters. The results of the model will be verified using 'categorical statistics' from the 'contingency table'. Contingency table or also called cross-tabulation table or crosstab is a table arranged based on data tabulation according to 2 or more categories that are displayed because an element with other elements is in conformity or relationship. Each column of the matrix is an example in the prediction class, while each row represents an example in the actual class. Dimensions or sizes of various contingency tables, the simplest are dimension 2, like the example below:

Table 1. Contingency Table Form

Forecast	Observed		Total
	Yes	No	
Yes	Hits	False alarms	Forecast yes
No	Misses	Correct negatives	Forecast no
Total	Observed yes	Observed no	Total

The four combinations of forecasts (yes or no) and observations (yes or no), called the joint distribution, are hits (event forecast to occur, and did occur), misses (event forecast not to occur, but did occur), false alarm (event forecast to occur, but did not occur) and correct negative (event forecast not to occur, and did not occur).

$$Accuracy\ (AC) = \frac{Hits + Correct\ negatives}{Total} \quad (1)$$

According to Provost and Fawcett [22], performance measurement algorithms that can be measured include Accuracy (AC) which is the correctness of the whole model and is calculated as the sum of the correct classifications divided by the total number of classifications. This skill was calculated based on the proportion of the correct number of predictions and is determined using the equation of the contingency table above.

3. Results and Discussion

3.1 Time Lag Analysis and Fuzzy Rules Construction

Figure 2 shows the relationship between the lag time of each model input with the occurrence of TC and TD. All of the inputs exhibit the highest correlation in lag 1, and those relationships decrease with an increase of lag time.

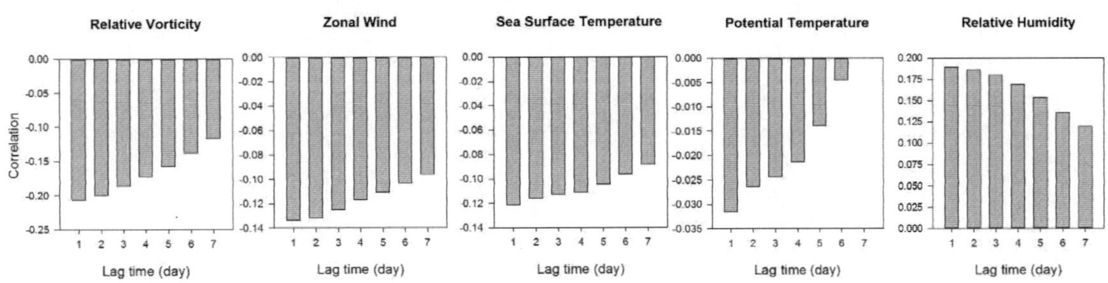

Figure 2. Biserial correlation between input parameters and TC and TD occurrences

Among input parameters, negative relationships were found in low-level relative vorticity (ϑ), the horizontal wind of upper troposphere (u), sea surface temperature (SST), and equivalent potential temperature (θe).

In contrast, specific humidity (q) employed positive correlation with TC and TD occurrences in the south of East Nusa Tenggara. Based on the above time lag analysis, we used input variables based on the highest correlation with TC and TD occurrences. Furthermore, we construct FL rules regarding the input parameters that were chosen. The class of FL rules was a construct based on 33th percentile of TC occurrences data. Table 2 exhibits the threshold value of each input parameter in each class.

Table 2. The threshold values of input parameters were divided into three classes based on the lowest, medium, and highest tercile of TC occurrences from 1991 to 2010.

	Θ (10^{-6}, S^{-1})	U (ms^{-1})	SST (°C)	θe (°C)	Q (10^{-4}, gkg^{-1})
Low	-31 to -8	-18.75 to -8.64	26.78 to 28.18	7.39 to 12.20	8.57 to 41.56
Moderate	-8.1 to 0	-8.65 to -3.55	28.19 to 29.09	12.21 to 14.10	41.57 to 49.55
High	>0	-3.56 to 14.56	>29.09	>14.10	>49.55

After the FL rule base is identified and a defuzzification algorithm is selected, forecast accuracy is tested using a different set of the historical data set (training data) from the one used to obtain the rule base. FL rules were obtained and changed based on the best accuracy of training data. From Table 2, we build 243 rules for the occurrence of TC and TD. We set the rules regarding the highest accuracy of training data from 1991 to 2010. It should be noted that the FL approach will work reasonably as the degree of fuzziness is suitable with the variables and the natural behavior of the system. Furthermore, FL rules are constructed between input parameters in IF-THEN format regarding the variables and expert decision. Lastly, results are defuzzified to a specific number as an output thought to the best represent fuzzy (figure 3).

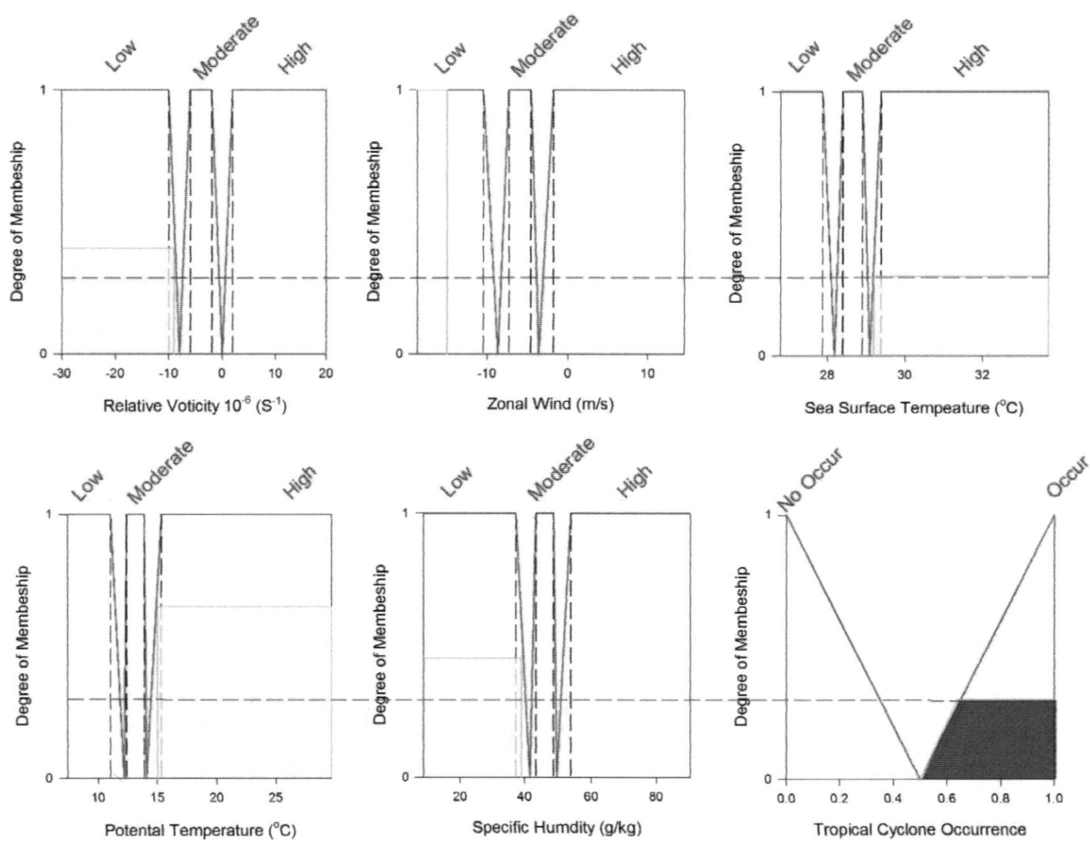

Figure 3. The instance of FL model reconstruction (rule 1) by using centroid method for ϑ= -9.5 S^{-1}, u= -16 m/s, SST= 29.5°C, θe=15°C, and q= 39.5 g/kg. In this study, we build and simulate 243 FL rules that depict the complexity of the model.

Many scientists used centroid as defuzzification method to model FL algorithm [23]. Therefore, in this study, we continue to use the centroid method to compute FL algorithm of TC and TD occurrences based on the expert decision. Figure 3 was made to facilitate understanding of the relationship between the input parameter and both TC and TD occurrences. It depicts the rule relationships between all input parameters and both TC and TD occurrences. TC and TD will occur when they employ value from 0.51 to 1. For instance, Figure 3 exhibits FL computation by using centroid method for ϑ= -9.5 S^{-1}, u= -16 m/s, SST= 29.5°C, θe=15°C, and q= 39.5 g/kg. With those values, the rules depict the probability of either TC or TD occurs with a degree of membership approximately at 0.3. The rule illustrated in Figure 3 exhibits the condition of TC and TD to appear is fulfilled. However, this study builds fuzzy rule up to 243 rules that make the computation run complexly.

3.2 The System of Tropical Cyclone (TC) Forecast Design

The Matlab® commercial FL Toolbox was used to design the Graphical User Interface (GUI) of TC occurrences forecast. This structure is stored inside a GUI tool. All the information for a given fuzzy inference system is contained in the GUI design, including variable names, input parameter definitions, and so on. This system can itself be thought of as supervised learning to predict the occurrences of both TC and TD.

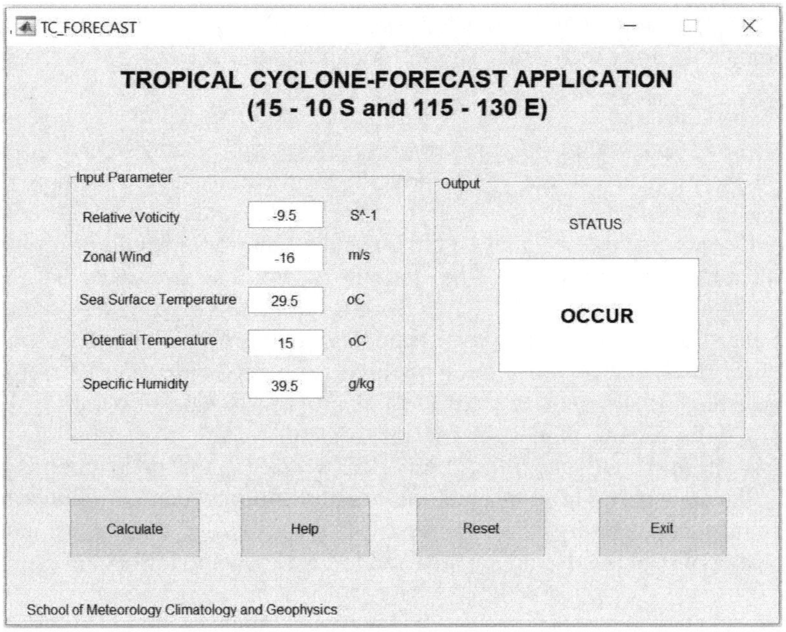

Figure 4. The GUI main screen of the TC occurrences forecast.

The implementation of the GUI of the TC occurrences forecast is analogous to the GUI used for Type-1 FL in the Matlab® Fuzzy Logic Toolbox, thus permitting the experienced user to adapt easily to the use of this GUI. This GUI was built through the process of an algorithm learning from the training dataset that can be thought of as a teacher supervising the learning process based on the FL rule that we constructed. The supervised learning problem in this study was classified into two groups, i.e., 'OCCUR, and NO OCCUR'. OCCUR is indicate that all the input variables are possible to make TC appeared, while NO OCCUR is the contrast condition. Figure 4 exhibited the example of all input variables that enable to make TC occurred in the study area.

3.3 The accuracy of the Forecast

The accuracy of the forecast was calculated from the total numbers of observed and forecast occurrences and non-occurrences which are given on the lower and right sides of the contingency table. The validation of the forecast was calculated based on daily data from 2011 to 2017 in DJF and MAM periods.

Table 3. Forecast validation from contingency table of 2011-2017 periods

Forecast	Observed		Total
	Yes	No	
Yes	10	329	339
No	4	1022	1026
Total	14	1351	1365

In table 3, a yes/no forecast of TC occurrences for 21 months are presented, and the high values of hit and correct negatives indicate that the model may be considered as practically accurate. Interestingly, the results reveal that the model predictions are in good agreement with the observed TC and TD events. Table 3 depicts that this system can predict 1032 event that matched with TC observation and only 333 events that miss predicted during 2011 to 2017 periods. During these periods, 10 TC occurrences can be predicted accurately from 14 number of TC events. From table 3, the accuracy of the forecast can be calculated based on the correct number of predictions. The model satisfactorily simulated the occurrence of TC with comparable error measures. The result exhibits the accuracy at 0.75 (range: 0 to 1, perfect score: 1).

3.4 Comparing of FL model with previous studies in term of TC and TD forecasting

Most studies of the tropical cyclone (TC) forecasting are focused on the track and the wind radius forecasting, even though a tropical cyclone formation alerts also developed. For instance, Vitart, Leroy and Wheeler [24] compared the accuracy of the European Centre for Medium-Range Weather Forecasts (ECMWF) forecast system to predict the genesis of tropical cyclones (TCs) over the Southern Hemisphere during intraseasonal (10-60 days) periods have been investigated and compared to the skill of a statistical approach. Other studies have been developed to investigate the potential link between TCs and its impacts both at a regional and global scale in worldwide [25-29]. For instance, in terms of extremes, Khouakhi, Villarini and Vecchi [29] employing annual maximum (AM) and peak-over-threshold (POT) methods, the highest impact of tropical cyclone to rainfall are found in East Asia, followed by Australia and North and Central America, with fractional contributions generally decreasing farther inland from the coast. He found that the relationship between extreme rainfall induced by TC and ENSO exhibits that extreme rainfall caused by TC tends to occur more frequently in Australia and along the East Coast of the U.S. during La Niña and in East Asia and the northwest Pacific islands during El Niño.

For the forecast of TC formation, Li, Fu, Ge, Wang and Peng [30] used QuikSCAT satellite products and the Tropical Rainfall Measurement Mission Microwave Image data to document temporal and spatial structures of Rossby wave to investigate cyclogenesis formation in the western North Pacific. However, there is still a lack of studies to predict the genesis of TC occurrences. Therefore, this paper takes a different approach and explores and forecast the TC occurrences based on unsupervised machine learning (self-organized) system. This study presents the development of fuzzy logic (FL) models as an unsupervised machine learning approach [31] for predicting the occurrence of TC from five primary TC genesis parameters. This machine learning approach is a family of statistical methods with origins in the area of artificial intelligence. This method is developed as holding great breakthrough for the improvement of understanding and prediction about the earth-science system. This modeling approach is quite flexible to resolve complex problems with various interacting parameters and typically outcompete traditional methods (e.g., generalized linear models), performing them ideal for modeling atmospheric environment [32]. In this study, for the results, the skill of the model exhibits good agreement with the observed TC and TD events.

3.5 Limitations

There are some limitations for this study to develop FL model. A small number of fuzzy sets leads to unrepresentative forecasts whereas a large number leads to many calculations. In previous studies, many numbers of fuzzy sets are selected initially from 3 to 6 [33, 34]. Therefore, it is difficult to determine the length of interval for FL approaches. In this study, we set 3 number of fuzzy and labeled with, low, medium, and high based on the highest accuracy of the training dataset. In addition, we used 3 number of fuzzy sets to save computational resources. In FL systems, membership functions play a very important role in representing problems [35]. The membership function that is used in this study is a trapezoidal and triangular membership function. We used the trapezoidal and triangular membership function regarding the behavior of TC that visually matched with its pattern criteria. This TC behavior can come from expert knowledge and training data. In this study, we construct TC behavior based on the training data in a daily calculation period from 1989 to 2010. The longer periods that potentially be used in future research may result in different membership function.

Furthermore, the selection of training data can also be a difficulty, because the data used must represent the actual data. The selection of training data will determine the knowledge and accuracy of the generated fuzzy logic. Ultimately, in this study, the accuracy of fuzzy logic depends on the expert or training data. In this method, it is also possible to combined FL approach with another method such as adaptive neuro-fuzzy inference system (ANFIS) to get the better skill of TCs prediction.

From the results, the forecasting of TC occurrences using FL algorithm is generally accurate based on the accuracy score at 0.75. However, this model only can be used in this study area (15°S-10° S and 115°E-130°E). It is a need either to do a cluster analysis or to explore other TC occurrences at potential areas such as North Atlantic, West Pacific or the South China Sea before making a forecast. Different locations are possible to produce different rules of FL algorithm. For further research, it also possible to employ another input variable to increase the forecast accuracy of TC occurrences.

4. Conclusion

This contribution explored the new forecasting of TC occurrences method utilizing FL algorithm as an unsupervised machine learning approach. We built-up FL rules and identified a defuzzification algorithm based on 33th percentile of TC and TD historical data set from 1991-2010. Forecast accuracy is tested using a different set of training data from the one used to obtain the rule base. The model satisfactorily simulated the occurrence of TC with comparable error measures in the study area. The result exhibits the accuracy at 0.75 (range: 0 to 1, perfect score: 1) showing a good agreement with the observed TC and TD events. In addition, we employed The Matlab® commercial FL Toolbox to design the Graphical User Interface (GUI) of TC genesis (occurrences) forecast. The evidence shows that the result provides insights into the adequacy of FL methods for forecasting the TC occurrences.

5. References

[1] Henderson-Sellers A, Zhang H, Berz G, Emanuel K, Gray W, Landsea C, Holland G, Lighthill J, Shieh S-L and Webster P 1998 Tropical cyclones and global climate change: A post-IPCC assessment *Bulletin of the American Meteorological Society* **79** 19-38
[2] Neiburger M, Edinger J G and Bonner W D 1995 Memahami Lingkungan Atmosfer Kita *Edisi kedua, Penerbit ITB, Bandung*
[3] Anthes R 1982 Tropical Cyclones: Their Evolution, Structure and Effects. American Meteorological Society, 208 pp., and DP Baumhefner, 1984: A diagram depicting forecast skill and predictability *Bull. Amer. Meteor. Soc* 65,701-3
[4] Gray W M and Landsea C W 1992 African rainfall as a precursor of hurricane-related destruction on the US East Coast *Bulletin of the American Meteorological Society* **73** 1352-64
[5] Gray W M, Landsea C W, Mielke Jr P W and Berry K J 1993 Predicting Atlantic basin seasonal tropical cyclone activity by 1 August *Weather and Forecasting* **8** 73-86
[6] Tonkin H 1996 Investigating thermodynamic models of tropical cyclone intensity. MS thesis, Climatic Impacts Centre, Macquarie University, Sydney, Australia)
[7] Diaz H F and Pulwarty R S 2012 *Hurricanes: climate and socioeconomic impacts*: Springer Science & Business Media)
[8] Management N A f D 2017 Tropical Cyclone Cempaka Hit Indonesia Causing Floods, Landslides and Tornadoes.
[9] Knaff J A and Zehr R M 2007 Reexamination of tropical cyclone wind–pressure relationships *Weather and Forecasting* **22** 71-88
[10] Dyahwathi N, Effendy S and Adiningsih E S 2007 Tropical Cyclone Characteristic and Its Impact on Rainfall Anomaly in Indonesia *J. Agromet Indonesia* 61-72
[11] Khotimah M 2008 Klimatologi Siklon Tropis di Indonesia *Bul. Met Klim Dan Geo* **4** 251-64
[12] Underground W 2018 Hurricane and Tropical Cyclones.
[13] Gray W M 1975 Tropical cyclone genesis *Atmospheric science paper; no. 234*
[14] Sinclair M R 1994 An objective cyclone climatology for the Southern Hemisphere *Monthly Weather Review* **122** 2239-56
[15] Reanalysis J 2018 iTacs (Interactive Tool for Analysis of the Climate System).
[16] Bellman R E and Zadeh L A 1970 Decision-making in a fuzzy environment *Management science* **17** B-141-B-64
[17] Tayfur G and Singh V P 2006 ANN and fuzzy logic models for simulating event-based rainfall-runoff *Journal of hydraulic engineering* **132** 1321-30
[18] Alvisi S, Mascellani G, Franchini M and Bardossy A 2006 Water level forecasting through fuzzy logic and artificial neural network approaches *Hydrology and Earth System Sciences Discussions* **10** 1-17

[19] Xiong L, Shamseldin A Y and O'connor K M 2001 A non-linear combination of the forecasts of rainfall-runoff models by the first-order Takagi–Sugeno fuzzy system *Journal of hydrology* **245** 196-217
[20] Chang F-J and Chang Y-T 2006 Adaptive neuro-fuzzy inference system for prediction of water level in reservoir *Advances in water resources* **29** 1-10
[21] Tagaki T and Sugeno M 1985 Fuzzy identification of systems and its application to modelling and control *IEEE Trans. Syst. Man and Cybernetics* **15** 116-32
[22] Provost F J and Fawcett T 1997 Analysis and visualization of classifier performance: Comparison under imprecise class and cost distributions. In: *KDD,* pp 43-8
[23] Ibrahim A 2003 *Fuzzy logic for embedded systems applications*: Elsevier)
[24] Vitart F, Leroy A and Wheeler M C 2010 A comparison of dynamical and statistical predictions of weekly tropical cyclone activity in the Southern Hemisphere *Monthly Weather Review* **138** 3671-82
[25] Knight D B and Davis R E 2009 Contribution of tropical cyclones to extreme rainfall events in the southeastern United States *Journal of Geophysical Research: Atmospheres* **114**
[26] Chang C P, Lei Y, Sui C H, Lin X and Ren F 2012 Tropical cyclone and extreme rainfall trends in East Asian summer monsoon since mid-20th century *Geophysical Research Letters* **39**
[27] Lavender S L and Abbs D J 2013 Trends in Australian rainfall: contribution of tropical cyclones and closed lows *Climate dynamics* **40** 317-26
[28] Villarini G and Denniston R F 2016 Contribution of tropical cyclones to extreme rainfall in Australia *International Journal of Climatology* **36** 1019-25
[29] Khouakhi A, Villarini G and Vecchi G A 2017 Contribution of tropical cyclones to rainfall at the global scale *Journal of Climate* **30** 359-72
[30] Li T, Fu B, Ge X, Wang B and Peng M 2003 Satellite data analysis and numerical simulation of tropical cyclone formation *Geophysical research letters* **30**
[31] Lin C-T and Lee C S G 1991 Neural-network-based fuzzy logic control and decision system *IEEE Transactions on computers* **40** 1320-36
[32] Olden J D, Lawler J J and Poff N L 2008 Machine learning methods without tears: a primer for ecologists *The Quarterly review of biology* **83** 171-93
[33] Stuber M, Gemmar P and Greving M 2000 Machine supported development of fuzzy-flood forecast systems. In: *European Conference on Advances in Flood Research*: Citeseer) pp 504-15
[34] Kucukali S and Baris K 2010 Turkey's short-term gross annual electricity demand forecast by fuzzy logic approach *Energy policy* **38** 2438-45
[35] Zadeh L A 1996 *Fuzzy Sets, Fuzzy Logic, And Fuzzy Systems: Selected Papers by Lotfi A Zadeh*: World Scientific) pp 775-82

Simulation of Sea Breeze Events in Gulf of Jakarta under Different Synoptic Condition: An Application of WRF Model

D M Pera Rosita Dewi, D Fatmasari and I Gustari
Meteorological, Climatological, and Geophysical Agency (BMKG), Jakarta, Indonesia

desak.dewi@stmkg.ac.id

Abstract. The behaviour of the sea breeze developing along the Gulf of Jakarta during days with different synoptic condition is examined using observational data and simulations from the Weather Research and Forecasting (WRF). An objective and systematic selection method is used to detect sea breeze days from observational data. The simulation used three nested grids at 27, 9 and 3 km resolution covering Gulf of Jakarta and its joining region for sea breeze event at November 22th, 2016 which formed under offshore large-scale ambient wind, January 20th, 2017 which formed under westerly large-scale ambient wind, and August 10th, 2017 which formed under easterly large-scale ambient wind. Results show that the development of sea breeze circulation under offshore ambient wind produced lower 10-m wind speed simulation, while sea breeze conditions under westerly large-scale ambient wind are associated with highest surface wind speeds reaching 9.5 ms^{-1} in Gulf of Jakarta. Sea breeze simulation also indicates the formation of calm zone which is 22 - 56 km away from the Jakarta Gulf coastline. Model also shows the formation of thermal internal boundary layer that are about 300-700 m near the coast gradually merged with the mixing layer inland. Performance of WRF on the simulated circulation is studied by conducting value of correlation, RMSE, and MBE. Overall, model simulation agrees fairly well with the observations taken at a near-surface meteorological station.

Keywords : sea breeze, WRF, offshore ambient wind, boundary layer.

1. Introduction

Meteorological conditions in coastal areas are very different from inland areas. The air flow pattern in coastal areas is determined by differences in land-sea temperature, topography, coastline shape, and heterogeneity of land use [1]. Sea breeze is one of the meteorological phenomena caused by land-sea temperature differences. Sea breeze emerges as mesoscale atmospheric circulation which occurs when air masses on land heat up more intensely and faster than air masses in the sea. This thermal contrast will create a pressure gradient force that leads from the sea to the land which causes the onset of air flow to land [2]. Jakarta is a densely populated Metropolitan City located on the north coast of the western part of Java Island, one of the 5 largest islands in Indonesia [3]. Jakarta is located at 6 ° – 7 ° S and 106 ° - 118 ° E, bordering with Bay of Jakarta in the north, that the sea breeze circulation has the potential to influence the weather system in Jakarta.

Research on sea breeze using balloon pilots has long been carried out in Batavia (now Jakarta) by van Bemmeln in 1912 [3]. However, evidence of the existence of sea breeze circulation that can penetrate on the land had only been reported by Hadi [4] based on observations of L-band radar boundary layer (BLR) in Serpong. Climatological conditions of sea breeze based on BLR observations by Hadi [3] also found that sea breeze fronts could well developed along the northern coastal plains of West Java and spread to the land until a distance of about 60-80 km from the coastline. However, despite the many

IOP Conf. Series: Earth and Environmental Science **303** (2019) 012045 doi:10.1088/1755-1315/303/1/012045

studies on sea breeze in Jakarta, there is a lack of attention about the influence of large-scale ambient wind on sea breeze characteristic.

Azorin-Molina and Chen stated that large-scale ambient wind is a meteorological factor that mostly influences the characteristic and evolution of sea breeze. Previous research has shown that offshore large-scale ambient wind will strengthen the frontogenesis convergence in core of sea breeze convergence zone, and lead to the reduction in sea breeze intrusion over land [2]. Steele's research found the formation an offshore calm zone when the large-scale ambient wind was offshore [6]. Calm zone is an area with a wind speed of 10 m less than 1 ms^{-1} which formed as the large-scale ambient wind pressure gradient are in equilibrium with the local thermal gradient force between land and ocean [2]. Steele found that when large-scale ambient wind is paralleled the coast with land to the left of the gradient wind (northern hemisphere), sea breeze circulation will induce the formation of coastal jets. Coastal jet is areas that have local maximum wind speeds and are formed about 1 km from the coast. In contrast, the interaction between gradient winds parallel to the coast, where the land is in the right of the gradient wind, sea breeze associated with lower surface wind speeds ranging from 2-3 ms^{-1} [6].

Numerical modelling is a method for simulating sea breeze events which is carried out to improve accuracy of the analysis of atmospheric conditions. Weather Research and Forecasting weather modelling (WRF) is the future generation of weather models that can be used to predict medium-scale atmospheres and assimilate these prediction results with local observation data. The influence of large-scale ambient wind is carried out using WRF-ARW to understand sea breeze interactions with synoptic scale flow in Gulf of Jakarta, and its joining area.

2. Data and Methods

2.1 Research area

The research location is Jakarta and Gulf of Jakarta according to Figure 1. Jakarta's topographic conditions are relatively homogeneous and flat. The city of Jakarta is directly adjacent to the Gulf of Jakarta in the north, and the Bogor highlands in the south which is about 50 km from the capital of Jakarta with the highest peak reaching more than 1200 meters. The shape of the Jakarta Gulf coastline is concave. Shoreline orientation of Jakarta Gulf is 270° - 90°. The axis of the sea breeze center is at 360° and the axis of the land breeze is at180°. Sea breeze defines as wind direction that deviates ± 90° from the axis of the sea breeze center [7].

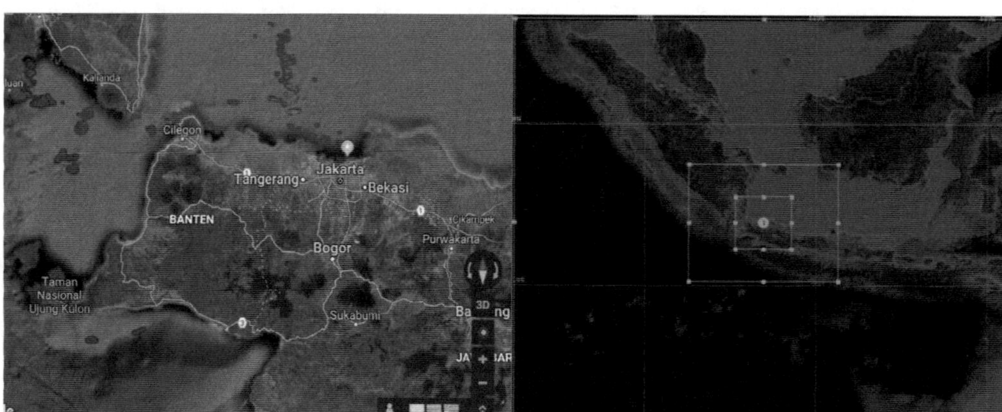

Figure 1. Map showing (a) Jakarta and its joining areas (b) three model domains

2.2 *Research data*

Data used in this study are described as follows:

1. Meteorological observation data. This study uses the meteorological observation data from August 2016 to July 2017 obtained from the Tanjung Priok Meteorological Station (6.108 °S, 106.88 °E). These data are used for sea breeze identification and model verification.

2. Zonal and meridional wind component data. The zonal and meridional data at the mid latitude (700 hPa) is obtained from the ECMWF ERA-Interim analysis fields which have 0.125°x0.125° spatial resolution and 6 hours temporal resolution. These data are accessed from http://apps.ecmwf.int/datasets/ and used for sea breeze identification.
3. Sea surface temperature data. The sea surface temperature data is obtained from the ECMWF ERA-Interim analysis fields which have 0.125°x0.125° spatial resolution and 6 hours temporal resolution. Used for sea breeze identification.
4. Sunrise and sunset data. This study uses sunrise and sunset data accessed from https://www.timeanddate.com, for which data are used for sea breeze identification.
5. FNL (Final Analysis) data. This study uses the FNL data accessed from http://rda.ucar.edu which have 1°x1° spatial resolution and 6 hours temporal resolution. These data are used for sea breeze simulation.

2.3 Data processing

2.3.1 Sea breeze day identification

Sea breezes identification is carried out by using Borne method. This method identifies sea breeze events when a noticeable wind shift and an immediate period of constant wind-direction occurs during daytime, not due to synoptically forcing, and a threshold sea-land temperature gradient is exceeded [9]. This method is consisted by 6 different filters, to be considered as sea breeze day all requirement must be met.

The first filter eliminates the day with a large change in wind direction of the 70 hPa layer. No change in wind direction is greater than 90°, and is calculated as a change between 13:00 local solar time (LST) with LST as UTC + 7, the previous day and the actual day at 13:00 LST. The second filter also relates to the upper layer wind, and eliminates days with gradient wind speed change higher than 6 ms^{-1} for 12 hours (from 1:00 to 13:00 LST) at a layer of 700 hPa. The third filter aims to exclude days with synoptic wind speeds that are too high more than 11 ms^{-1}. These three filters aim to ensure a stable synoptic condition that is favourable for the formation of sea breezes circulation.

The last 3 filters are related to the surface conditions. The fourth filter determinates that minimum temperature difference between the sea and the land needed for developing sea breezes is 3 degrees. The fifth filter shows at a time interval between 1 hour after sunrise and 5 hours before sunset there is a change in surface wind direction >30°. Quick changes in surface wind direction are characteristics of sea breezes. The last filter rejects days in which the hourly variation mean of the wind direction during 5 h subsequent to the observed wind shift is not at least 6 times smaller than the shift in question. With this last filter, the wind direction constancy is assured [9].

2.3.2 Running process of the WRF ARW model.

Table 1. WRF-ARW configuration used in the research.

	Domain 1	Domain 2	Domain 3
Resolution	27 km	9 km	3 km
Geographic resolution	5 m	2 m	30 s
Microphysics scheme	WSM-3	WSM-3	WSM-3
Long wave radiation scheme	RRTM	RRTM	RRTM
Short wave radiation scheme	Dudhia	Dudhia	Dudhia
Cumulus scheme	Kain-Fritsch	Kain-Fritsch	Kain-Fritsch
Planetary boundary layer scheme	YSU	YSU	YSU
Surface scheme	NOAH Land Surface	NOAH Land Surface	0

The mesoscale model used in this study is the WRF ARW Version 3.8.1 model. The WRF ARW model was ran by using the Mercator map projection and divided into three domains as shown in Figure 1 b.

The inner domain covers the entire Gulf of Jakarta and Jakarta regions with horizontal grid resolution of 3 km. The remaining 2 domains have horizontal resolution 9 and 27 km. For each simulation the model is run for 30 h, the first 12 h are considered as spin-up time. The simulated time never exceeds 24 h to avoid the possible influence of simulated conditions precedent [6]. The parameterization that used in this simulation is default as shown in Table 1.

The synoptic wind condition is determined using 925 hPa wind speed and direction from the output of WRF simulation. Large-scale ambient flow is divided into three types of flow based on its direction, namely (a) offshore flow is a synoptic scale flow from land to sea and ranges between 112.5 °- 247.5 °, westerly flow is a synoptic scale flow that blows from west to east and ranges between 247.5° - 292.5°, and easterly flow is a synoptic scale that blows from east to west and ranges from 67.5° - 112.5°.

2.3.3 *Verification*

The accuracy of the model simulating the observed variables is evaluated using the coefficient correlation (r), the Mean Bias Error (MBE) and the Root Mean Square Error (RMSE). These statistical indexes are computed as follows:

$$\text{Correlation (r)} = \frac{\sum mo - \frac{1}{n}\sum m \sum o}{\left[\sum m^2 - \frac{1}{n}(\sum m)^2\right] * \left[\sum o^2 - \frac{1}{2}(\sum o)^2\right]} \quad (1)$$

$$\text{RMSE} = \sqrt{\frac{\sum_1^n (M-O)^2}{n}} \quad (2)$$

$$MBE = \frac{1}{n}\sum_1^n (M-O) \quad (3)$$

where *m* and *o* represent the modelled and observed values of the variables respectively, and n is number of data.

3. Result and Discussion

3.1. Sea breeze under different synoptic condition

After applying the sea breeze selection method to the observational data, 17 sea breeze days are obtained in Tanjung Priok. A summary of selected sea breeze day by each filter is shown in Table 2. The partial acceptance percentage is the ratio of the output days and the input days in each filter, while the absolute acceptance percentage is a ratio of the output days in each filter and the number of days of total input. Filters 1-3 are used to ensure a synoptic stable atmosphere, where in Filter 2 receives all input day. This indicates that throughout the year, the synoptic conditions are favorable for the formation of sea breezes. The Filter 4 is responsible for the most rejection of input days, which rejects 68% of the days that pass through the Filter 4. The high number of rejected days is due to differences in surface temperature of the sea-land that do not meet the requirements for sea breeze formation.

Table 2. Summary of the selected sea breeze days by each filter of the selection method for Tanjung Priok

Filter	Input days	Output days	Partial acceptance percentage (%)	Absolute acceptance percentage (%)
1	365.0	312.0	85.5	85.5
2	312.0	312.0	100.0	85.5
3	312.0	284.0	91.0	77.8
4	284.0	91.0	32.0	24.9
5	91.0	45.0	49.5	12.3
6	45.0	17.0	37.8	4.7

IOP Conf. Series: Earth and Environmental Science **303** (2019) 012045 doi:10.1088/1755-1315/303/1/012045

In order to analyze an appropriate sea breeze condition during different synoptic condition, 3 days of the selected sea breeze days have been chosen: November 22th, 2016 which formed under light offshore large-scale ambient wind ranged 1-2 m s^{-1}, January 20th, 2017 which formed under strong westerly large-scale ambient wind ranged 7 – 9 m s^{-1}, and August 10th, 2016 which formed under moderate easterly large-scale ambient wind ranged 5-7 m s^{-1}.

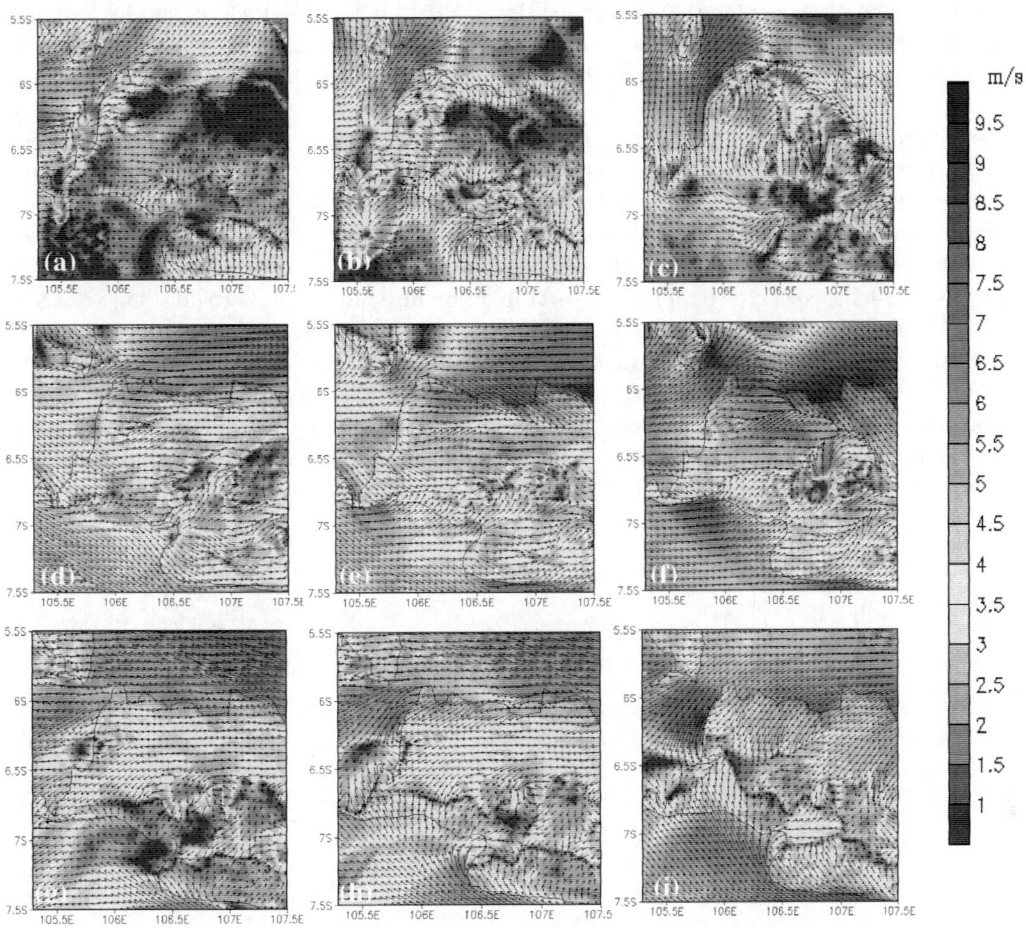

Figure 2. Plot of 10m wind speed (shading, ms^{-1}), and direction (arrows) of sea breeze events identified in Jakarta by the filter method at 10:00 UTC, 12:00 UTC, and 16:00UTC respectedly for 22 November 2016 (a-c), 20 January 2017 (d-f), and 10 August 2016 (g-i)

From the Figure 2 a,d,g, it is shown that sea breeze onsets occurred between 1000-1100 LST, which these results are consistent with Hadi who found that sea breeze onsets in the Jakarta area generally occurs between 11:00 - 14:00 LST [3]. Sea breeze condition under large-scare offshore ambient wind indicates the formation of regions with minimum wind speed or calm zone which is about 22-56 km offshore from the Gulf of Jakarta as shown on Figure 2 a-c. In the literature, the calm zone phrase in the context of sea breeze is the minimum wind speed caused by the orientation of the pressure gradient ambient wind which is opposite to the orientation of the land-ocean thermal pressure gradient. According to Arrit, the existence of the calm zone shows the limit of sea breeze circulation offshore. This is shown by the pattern of surface wind divergence in the calm zone region. The thermal pressure gradient of land-sea causes the wind to blow from the sea to the land, in the opposite direction to the

IOP Conf. Series: Earth and Environmental Science **303** (2019) 012045 doi:10.1088/1755-1315/303/1/012045

offshore gradient wind that leads from the land to the sea (south to north). The calm zone first appears near the coastline and moves towards the sea as shown on Figure 2 a-c, indicating that as the heating in the mainland reaches its maximum, the sea breeze circulation on offshore region or the water body area affected by the sea breeze is increasing. Sea breezes under the large-scale offshore ambient wind are associated with lower surface wind speeds than the other ambient wind patterns. This is due to the gradient wind being in the opposite direction to the sea breeze that the resultant sea wind speed is low.

Sea breeze conditions under westerly large-scale ambient wind are associated with highest surface wind speeds than the other sea breeze events reaching 9.5 ms^{-1} in Gulf of Jakarta, and lower wind speed in land as shown by Figure 2 d - f. Sea breeze direction blows from northwest due the influence of westerly large-scale ambient wind. When the ambient wind is parallel to the coastline, the thermal pressure gradient and the synoptic scale flow pressure gradient are not completely opposite on the same dimension, so the thermal pressure gradient does not have to overcome the synoptic scale flow pressure gradients for the sea breeze to enter the land, this causes the calm zone does not form when the sea breeze blows under the westerly or easterly large-scale ambient wind. Sea breeze conditions under easterly large-scale ambient wind are associated with higher surface wind speeds than sea breezes under offshore flows, but are relatively lower with sea breeze under westerly flow as shown by Figure 2 g - i. It is caused by the wind speed of easterly large-scale ambient wind which is lower than westerly large-scale ambient wind, therefore the resultant sea breeze under easterly flow is lower than westerly flow. Due of the influence of easterly large-scale ambient wind, sea breeze also blows from northeast.

Horizontal intrusion of sea breeze can be seen from how far the sea breeze convergence zone (SBCZ) entered the land. SBCZ is defined as an area where two or more thermally induced air masses from the Java Sea, Sunda Strait, Indian Ocean south of Java Island, and Gulf of Jakarta combined to form areas with low horizontal speeds and relatively high vertical wind speeds.

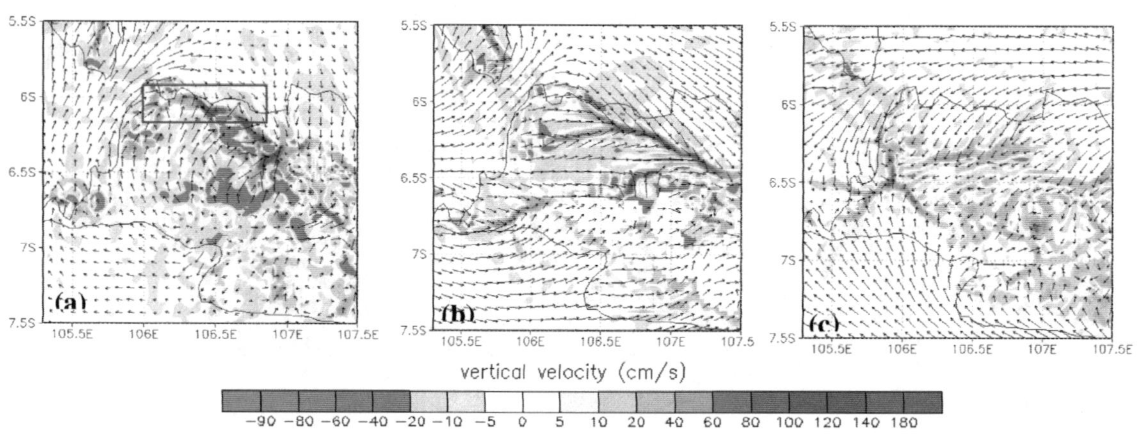

Figure 3. Plot of simulated vertical velocity (shaded, cms^{-1}), and simulated surface wind direction (arrows) at 15:00 UTC for sea breeze events at (a) 22 November 2016, (b) 20 January 2017 (d), and (c) 10 August 2016.

The SBCZ conditions under offshore large-scale ambient wind indicates the presence of sea breeze that is held near the northern coastline of Java island as shown on Figure 3 a. This is due to the offshore ambient wind, that makes sea breezes originating from the southern side of the island of Java can intrude further on land, and obstruct sea breeze intrusion from the north side of the island of Java. This is in accordance with the findings of Miller explaining that offshore ambient winds can reduce the maximum intrusion of sea breeze on land [2]. In contrast, the SBCZ conditions under westerly large-scale ambient wind show that strong westerly ambient wind pushes the SBCZ towards the other direction and reduces the intrusion of SBCZ as shown on Figure 3 b. And during the easterly large-scale ambient wind, due to the orientation of the easterly ambient wind which has the opposite direction to the sea breeze which originates from the west side of Java Island, it causes sea breeze originating from the Sunda Strait,

IOP Conf. Series: Earth and Environmental Science **303** (2019) 012045 doi:10.1088/1755-1315/303/1/012045

unable to intrusion further on land, and stuck near the coastline as shown on Figure 3 c. These all implies, the sea breeze from Gulf of Jakarta can intrude further inland during easterly large-scale ambient wind, and its position is pushed to southwest.

The vertical velocity analysis of SBCZ for the 3 sea breeze events show that under offshore ambient wind, the vertical velocity value at the core of SBCZ reaches more than 1.8 m s^{-1}, relatively higher than the vertical velocity value under westerly/easterly ambient wind. This indicates that sea breezes under offshore ambient wind conditions experience a strengthening of convergence in the SBCZ region or a strengthening of the frontogenesis convergence. This finding is in accordance with Arritt's research which states that the interaction of weak-medium offshore ambient wind with sea breeze results in a reinforcement of frontogenesis convergence [10]. Behind SBCZ, an area of updraft with a lower positive vertical velocity value is formed. Secondary updraft behind this SBCZ can be formed due to low level convergence, friction with the surface, thermal instability that causes the rising of air masses [2]. Around the SBCZ, a downdraft area with a negative vertical velocity value is also seen. This downdraft area is part of the repeated updraft and downdraft pattern associated with convective activities on land [11].

To study the vertical extent of the sea breeze circulation, the north-south vertical cross section of meridional wind component, and water vapor mixing ratio is examined. The Figure 4 and 5 show the vertical circulation at 1500 UTC, which is representative of the hour in which sea breeze reached its maximum development on surface. From Figure 4, the vertical extent of sea breeze under offshore ambient wind reaches 820 hPa, relatively higher than westerly ambient wind which reaches 910 hPa and easterly ambient wind which reaches 870 hPa. The vertical intrusion of sea breeze is found at the very front of the sea breeze, and can be identified as a sea breeze head (SBH). Figure 5 shows that behind the SBH, the air mass with a high-water vapor mixing ratio value flowing on land, following the spread of SBF on land and is indicated as sea breeze gravity wave (SBG). SBG is a spreading of moist, and cold air masses carried by the sea breeze. At the top of the sea breeze circulation, a return flow that blows from the south towards the north is detected on the 3 sea breeze events, this indicates that the sea breezes are able to maintain their closed circulations.

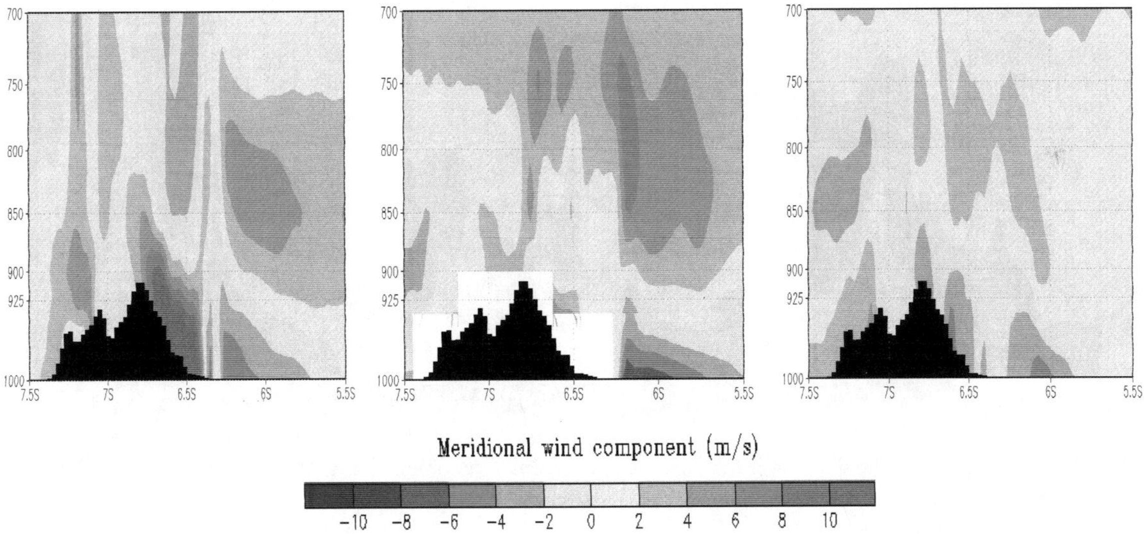

Figure 4. Cross section N-S of simulated Meridional wind component at 15:00 UTC for sea breeze events at (a) 22 November 2016, (b) 20 January 2017 (d), and (c) 10 August 2016.

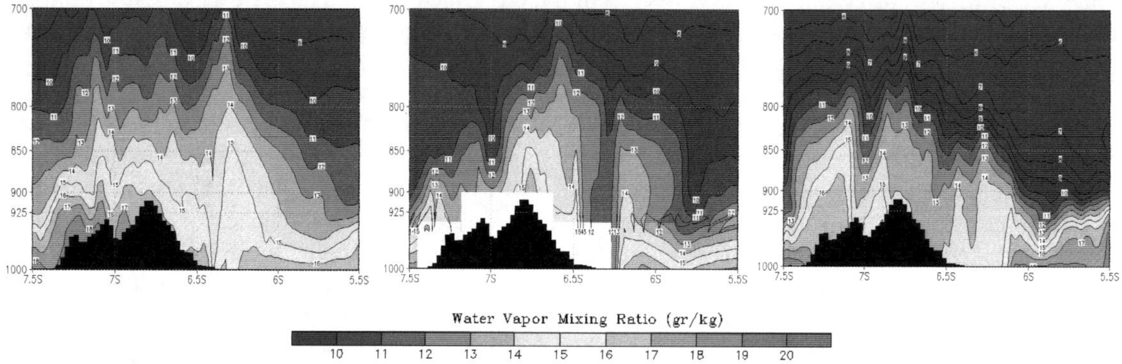

Figure 5. Cross section N-S of simulated water vapor mixing ratio at 15:00 UTC for sea breeze events at (a) 22 November 2016, (b) 20 January 2017 (d), and (c) 10 August 2016

Planetary boundary layer height (PBLH) is an important parameter as it represents limiting of the boundary layer heating, convection, pollutant mixing, and low-level circulating, in which PBLH is influenced by sea breeze advection leading the formation of thermal internal boundary layer (TBIL) [12]. Figure 6 shows deep boundary layer over land reaches more than 1600 meters, while on ocean and coastal areas, a shallower boundary layer is seen. The region of shallower boundary layer on coastal areas can be identified as TIBL, and the value ranges between 300-700 meters. From Figure 6, we can see the difference in characteristic between TIBL under westerly ambient wind, and easterly ambient wind. Under westerly ambient wind, the formation of TIBL over land leads to the southeast, following the direction of the sea breeze that blows from northwest (Fig. 6b) to the southeast. In contrast, under easterly ambient wind, the formation of TIBL over land leads to the southwest following the direction of the sea breeze (Fig. 6c). Here also can be seen that TIBL gets merged with the generic inland PBL much away from in land.

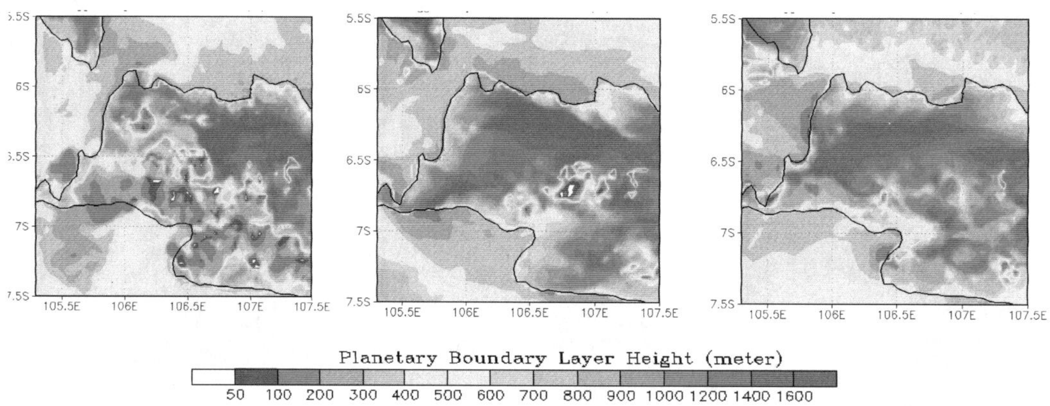

Figure 6. Simulated boundary layer height at 15:00 UTC for sea breeze events at (a) 22 November 2016, (b) 20 January 2017 (d), and (c) 10 August 2016

3.2 Model Evaluation

Evaluation of the WRF model is done by comparing the simulation values with the observational values from the Tanjung Priok Meteorological Station. Wind direction and wind speed are chosen as the main meteorological variables to compare, because they are used to study wind fields which are the main objectives of this study. Surface temperature (*T)* and relative humidity (*RH)* were also compared as well because they can be used to study the main features of sea breezes such as: daily evolution, sea wind front (SBF), and land sea wind intrusion.

Table 3. Statistical skill scores for different parameters from sea breeze simulation

Simulation Time	Parameter	R	RMSE	MBE
22 November 2016	T	0.82	1.93	1.04
	RH	0.80	10.91	-7.93
20 January 2017	T	0.75	1.42	-0.31
	RH	0.15	11.46	2.63
10 Augustus 2016	T	0.88	1.82	-1.40
	RH	0.88	7.67	3.29

In general, the verification results indicate that the WRF model is able to simulate the daily pattern and observed trend of surface temperature in Tanjung Priok. This is shown through the high positive correlation coefficient, and low RMSE. The verification results of relative humidity also indicate that the WRF model is able to simulate the daily pattern and observed trend of relative humidity in Tanjung Priok, except for January 20th 2017 where the correlation value is relatively low.

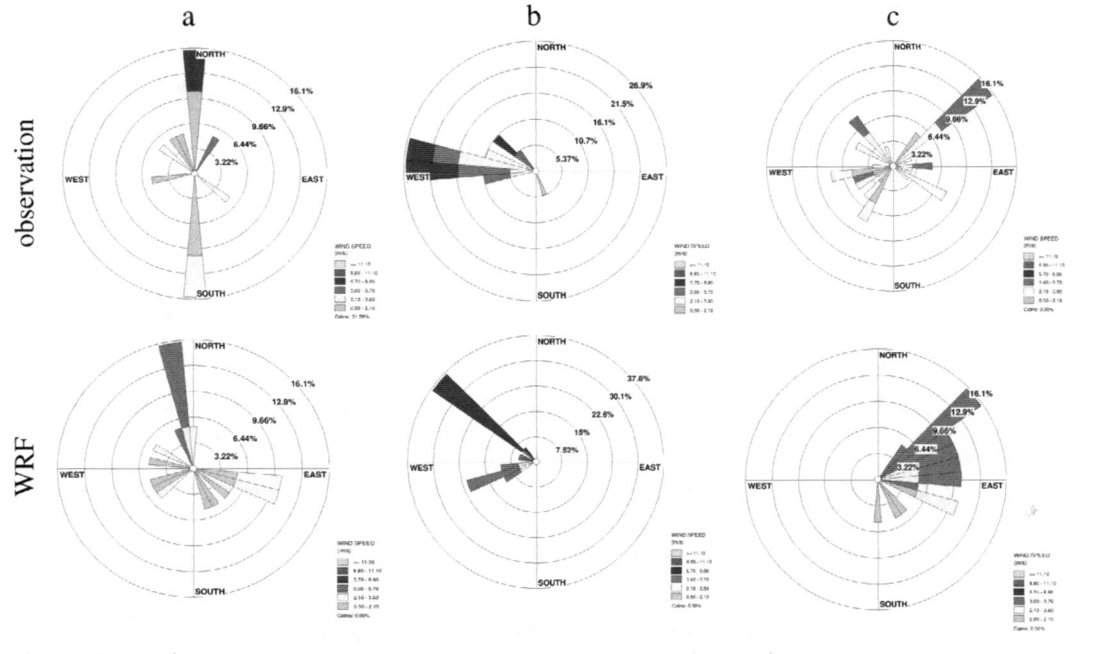

Figure 7. Wind speed and direction comparison of observed and simulated sea breeze days in Tanjung Priok on (a) 22 November 2016, (b) 20 January 2017, and (c) 10 Augustus 2016

Figure 7 shows wind speed and direction comparisons of WRF model simulations and observations in Tanjung Priok. Overall, it can be seen that the WRF model is able to simulate the wind direction of sea breeze, although there are differences in simulating the direction of the land breeze. Under offshore large-scale ambient wind, sea breeze direction is from north as shown by Figure 7 a, indicating sea breeze is able to penetrate inland. Sea breeze blows from west-northwest during westerly large-scale ambient wind and from northeast during easterly large-scale ambient wind, as shown by Figure 7 a and Figure 7 b respectively. This indicates the direction of sea breeze is highly influenced by direction of large-scale ambient wind. The wind speed and direction comparison of observed and simulated also show that the sea breeze intensity of the WRF model is overestimate to the observed sea breeze intensity.

IOP Conf. Series: Earth and Environmental Science **303** (2019) 012045 doi:10.1088/1755-1315/303/1/012045

4. Conclusion

The aim of this study was to investigate the influence of synoptic condition to sea breeze characteristic by applying WRF model to Jakarta and Gulf of Jakarta. An objective and systematic selection method was used to detect sea breeze days from observational data. The model simulations refer to the 22 November 2016, 20 January 2016, and 10 August 2016 which was chosen to analyse an appropriate sea breeze condition during different synoptic condition, therefore the drawn conclusions could be generalised: The development of sea breeze circulation under offshore ambient wind produced lower 10-m wind speed simulation, while sea breeze conditions under westerly large-scale ambient wind are associated with highest surface wind speeds reaching 9.5 ms^{-1} in Gulf of Jakarta. Sea breeze simulation indicates the calm zone formation which is 22 - 56 km away from the Jakarta Bay coastline. SBCZ intrusion under offshore ambient wind showed a reduction on its spatial intrusion over land, while SBCZ under westerly/easterly ambient wind showed that SBCZ position pushed towards the other direction. Model also shows the formation of thermal internal boundary layer that are about 300-700 m near the coast gradually merged with the mixing layer inland. Performance of WRF on the simulated circulation is studied by conducting value of correlation, RMSE, and MBE. Overall, model simulation agrees fairly well with the observations taken at a near-surface meteorological station.

References

[1] Pielke, R A 1984 *Mesoscale Meteorological Modeling* (New York: Academic Press)

[2] Miller S T K, Keim B D, Talbot R W, and Mao H 2003 Sea Breeze: Structure, Forecasting, and Impacts *Rev. of Geophysics* pp 1011

[3] Hadi T W, Tsuda H H, and Fukao S 2000 Tropical Sea-Breeze Circulation and Related Atmospheric Phenomena Observed With L-Band Boundary Layer Radar in Indonesia *Jour. of Meteorology Society of Japan* Vol. 78 123–140.

[4] Hadi T W, Horinouchi T, Tsuda T, Hashiguchi H, and Fukao S 2002 Sea Breeze Circulation over Jakarta, Indonesia: A Climatology Based on Boundary Layer Radar Observations *Mont. Weth. Rev* Vol. 130 pp 2153– 2166.

[5] Azorin-Molina C, and Chen D 2008 A Climatological Study of The Influence of Synoptic-Scale Flows on Sea Breeze Evolution in the Bay of Alicante (Spain), *Theoretical and Applied Climatology* Vol. 96 249–260.

[6] Steele C J, Dorling S R, Glasow1 R V, and Bacon J 2014 Modelling Sea-Breeze Climatology and Interactions on Coast in the Southern North Sea: Implication for Offshore Wind Energy *Quarterly Journal of Royal Meteorology Society* Vol.141 pp 1821–1835.

[7] Prtenjak M T, and Grisogono B 2007 Sea/Land Breeze Climatological Characteristic along the Northern Croatian Adriantic Coast *Theory of Applied Meteorology* Vol. 91 pp 1-15.

[8] Borne K, Chen D, and Nunez M 1998 A Method for Finding Sea Breeze Days Under Stable Synoptic Conditions and Its Application to The Swedish West Coast *International Journal of Climatolology* Vol. 18 pp 901–914.

[9] Arrigala J A, Carlos Y, Mariano S and Carlos R 2016 A characterisation of sea-breeze events in the eastern Cantabrian coast (Spain) from observational data and WRF *simulations Atmospheric Research* Vol.81 pp 265-280.

[10] Arritt RW 1993 Effects of the large-scale flow on characteristics features of the sea breeze *Journal of Applied Meteorologi* Vol.32 pp 116–125.

[11] Stephan K, Kraus H, Ewenz C M, and Hacker J M 1999 Sea-breeze front variations in space and time *Meteorol. Atmos. Phys.* Vol.*70* pp 81–95.

[12] Challa V S, Indracanti J, Rabariso M K, Patrick C, Baham J M, Young J, Hughes R, Hardy M G, Swanier S J, Yerramilli A 2008 A simulation study of mesoscale coastal circulation in Mississippi Gulf coast *Atm. Research* Vol 91 pp 9-25.

IOP Conf. Series: Earth and Environmental Science **303** (2019) 012046 doi:10.1088/1755-1315/303/1/012046

Bias correction of daily precipitation from downscaled CMIP5 climate projections over the Indonesian region

F Amsal, H Harsa, A Sopaheluwakan, U A Linarka, R P Pradana and R Satyaningsih

Center for Research and Development, Agency for Meteorology Climatology and Geophysics (BMKG), Jakarta 10720, Indonesia

ferdikaamsal@gmail.com

Abstract. Global Climate Models (GCMs) have been the primary source of information for constructing climate scenarios, and they provide the basis for climate change impacts assessments of climate change for a range of scales, from global down to regional scale. Due to the coarse spatial resolution, the GCM outputs have to be downscaled to resolve the scale discrepancy between the resolutions required for impact assessments and the model's resolution. However, it is important to bias-correct (BC) the raw climate projection outputs which ideally correct the discrepancy between a model's climate and the observed historical climate. In this study the results of bias correction of daily precipitation over the Indonesian region from downscaled CMIP5 GCM climate simulations using an optimized configuration of the Regional Climate Model (RegCM) for a baseline period of 16 years (1990–2005) with respect to observation is discussed in detail. The statistical bias correction method validated in this study is based on the initial assumption that both observed and simulated intensity distributions are well approximated by the Gamma distribution and the correction is made by matching the quantiles of the Gamma cumulative distribution functions. Overall, the results suggest that when the bias-correction is applied on dynamically downscaled model, it improved the skill in simulating the precipitation over Indonesia and this is a useful tool for further regional downscaling studies.

Keywords : Global Climate Models, CMIP5, Regional Climate Model, Gamma distribution.

1. Introduction

Precipitation is one of important climate variables for several sectors such as hydrology, health, agriculture, food supply. Those sectors eventually might affect social and economic sectors. Precipitation also the most sensitive to model formulation because of its dependence on parameterization schemes and their interaction with the model dynamics [1]. The International Panel on Climate Change (IPCC) have stated that warmer climate in the future are likely to increase the frequency and intensity of the extreme precipitation events [2]. Therefore, it is crucial to have reliable model output particularly in impact studies of extreme precipitation in the context of the climate change.

Global Climate Models (GCMs) have been used as the primary source of information to understand and quantify such extreme events for both the present-day climate and possible future climates. However, GCMs are generally run at coarse resolution and therefore not all processes, especially those on mesoscales, are reasonably simulated for regional scale [3]. This may lead to misrepresentations of extreme precipitation event [4,5]. To bridge this resolution gap, dynamical and/or statistical downscaling techniques were applied in order to obtain plausible regional climate change projections [1,6].

The dynamical downscaling approach is based on regional climate models (RCMs) running over limited geographical domains with boundary conditions given by the GCM to be downscaled [3]. These RCMs explicitly solve mesoscale atmospheric processes and provide spatially and physically consistent outputs. However, they still have considerable biases [7,8,9], which are typically adjusted in practical applications using a variety of bias correction (BC) methods. Formally, in the BC downscaling approach, the target variable (e.g., precipitation) simulated by the RCM is directly corrected against the available local-scale observations using appropriate statistical techniques [1,10,11].

This study aims to applying statistical bias correction to regional climate model output and evaluate the performance of the bias correction. In Section 2, we describe the data and methodology used in this study. The results are discussed in Section 3, followed by conclusions in Section 4.

2. Data and methodology

2.1. Datasets

In this study we use the Regional Climate Model system RegCM4 [12] to downscale CSIRO Mk3.6. RegCM4 is a hydrostatic and compressible model that uses sigma-p vertical coordinates, run on an Arakawa B-grid. RegCM4 has the dynamical core of the mesoscale model version 5 (MM5) developed by the National Center for Atmospheric Research (NCAR) and Pennsylvania State University [14]. It incorporates the radiation scheme of the NCAR Community Climate System Model version 3 (CCSM3) [15,16] for radiation and infrared spectra, following Collins et al. [17]. The Biosphere-Atmosphere Transfer Scheme (BATS) for land-surface processes follows Dickinson et al. [18] and Gao et al. [19]. The CLM [20] is also available as the land-surface scheme in RegCM4. To represent urban and suburban environments, two new land-use types have been added to BATS in RegCM4. Urban development modifies the surface albedo and alters the surface energy balance; it also heavily affects runoff and evapotranspiration. To describe the land–atmosphere exchanges of energy, momentum, water and carbon, RegCM4 uses Community Land Model version CLM3.5 [21], which applies a series of biogeophysically-based parameterizations. Moreover, RegCM4 uses other physical processes, including those of Holtslag et al. [22] for the Planetary Boundary Layer (PBL) and Zeng et al. [23] for ocean flux parameterization approaches. Multiple cumulus convection schemes are available in RegCM4. The new scheme allows the user to select either Grell or Emanuel in the function of the ocean-land mask. The Arakawa–Schubert type closure or the Fritsch and Chappell [24,25] type closure are available for use in the Grell convective parameterization scheme.

A set of sensitivity experiment in RegCM4 physical parameterizations for Southeast Asia region has been conducted, combining six cumulus parameterization schemes and three ocean surface flux schemes [26,27,28]. Following this experiment, in this study RegCM4 was run with the main configuration as shown in Table 1.

Table 1. The main configuration of RegCM4 run in this study.

Model physics	Scheme selected in this study
Planetary Boundary Layer (PBL) scheme	Holtslag PBL [22]
Radiation scheme	CCSM
Large scale moisture	SUBEX
Land-surface treatment	BATS1e
Cumulus parameterization	MIT Emanuel
Ocean flux scheme	Zeng et al. [23]

As reference data for statistically correcting the aforementioned climate model output we use the Tropical Rainfall Measuring Mission (TRMM) 3B42 product [30]. The precipitation estimates are available for spatial resolution of $0.25° \times 0.25°$ and temporal resolution of 3 hourly. The bias correction in this study covers the period of 1990-2005.

IOP Conf. Series: Earth and Environmental Science **303** (2019) 012046 doi:10.1088/1755-1315/303/1/012046

2.2. Methodology

The quantile mapping (QM) has been widely used in RCMs, e.g. by Piani et al. [30] and Ngai et al. [31]. QM is an empirical statistical technique that matches the quantile of simulated model values to the observed values at the same quantile. The technique is applied by first constructing the cumulative distribution function (CDF) of each datasets, the model and the reference. The quantile of each value in the model is extracted using model's CDF. This process yields a new dataset consisting quantile value of each model datum. Using this dataset, an inversion procedure is carried out to every quantile datum by looking up a value in reference data that lies on the same quantile with the datum being observed. To illustrate the methodology, the QM is applied to a dataset generated randomly (Figure 1). The black line denotes the synthetic observed data while the red line represents the synthetic model data.

The method was then applied to the daily precipitation of RegCM4 output. We evaluate the performance of the bias correction in three ways. First, we compare the seasonal bias of uncorrected and corrected precipitation of RegCM4 compared to TRMM data. Second, we analyse the performance of the method by comparing the number of wet days for each month over all the period of the study. Third, we examine extreme precipitation in terms of RX1DAYS and 95th percentile of seasonal mean.

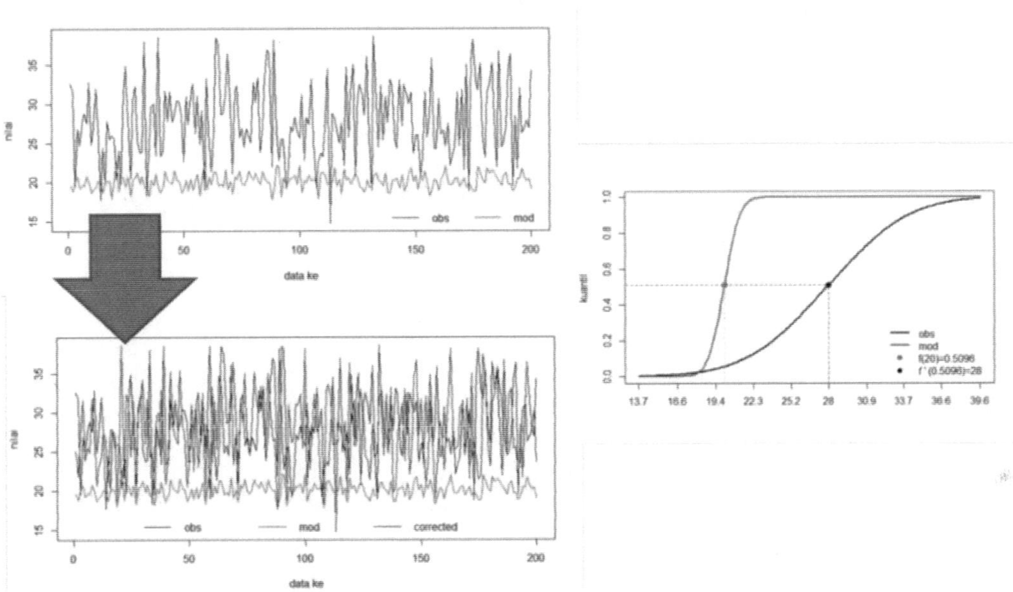

Figure 1. QM simulations using 1000 set (observation and model) of random sampling

3. Result and discussion

3.1. Bias corrected monthly and seasonal precipitation

Figure 2(a) shows quantile-quantile (Q-Q) plots of the uncorrected and corrected RegCM4 precipitation againts the TRMM estimated precipitation for each month and in all grid boxes of the domain. The uncorrected RegCM4 output tends to underestimate low precipitation but overestimate high precipitation in all months. After bias correction, the corrected quantiles are close to the observation. RegCM4 also shows a large monthly mean bias on each grid and these occur almost in every month, even though the standard deviation is relatively constant (Figure 2(b)). The standard deviations of the bias of corrected model output are more varied from month to month compared to the standard deviations of the bias of the uncorrected one.

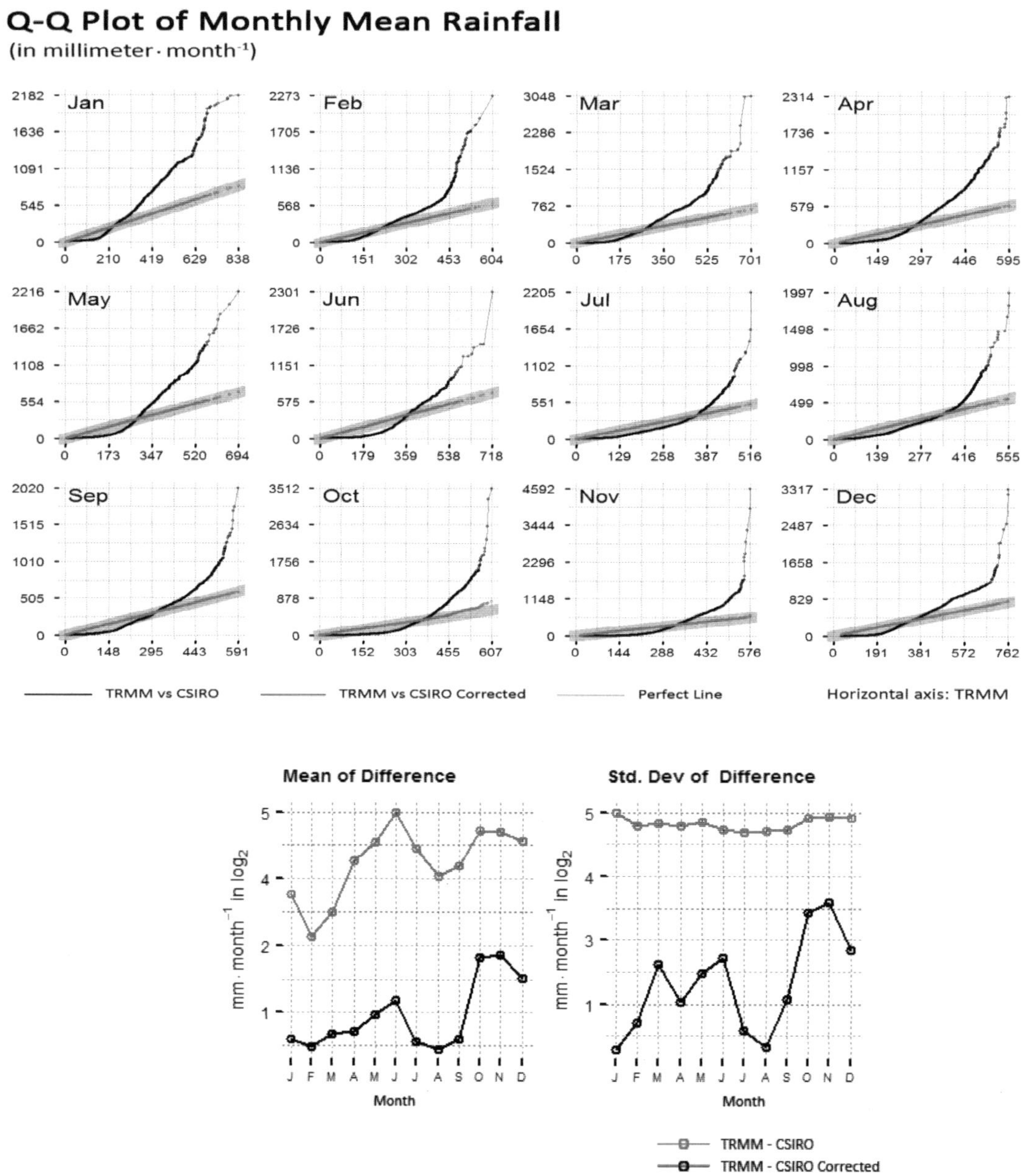

Figure 2. (a) Q-Q Plot of monthly mean precipitation (above) and the mean bias of uncorrected and (b) corrected ReGCM4 monthly mean precipitation in each grid (below).

The spatial distribution of mean biases in monthly total precipitation for each season are shown in Figure 3. The largest systematic bias over land area occurred in the centre part of Sumatera island, north of Sulawesi island, and the centre of Papua island, where the large biases are observed almost in all month. All these regions are characterized by high topography and therefore physically affect most of the heavy precipitation in the model, which result in profound differences between the observation and the model. Big negative ofthe relative bias occurs in mountain range from December to May. We also can see in almost all other seasonal and regional regions cases, observational estimates are within the range of the model simulation (bias ± 500 mm/month). After applying bias correction, precipitation

IOP Conf. Series: Earth and Environmental Science **303** (2019) 012046 doi:10.1088/1755-1315/303/1/012046

biases generally decrease in all seasons. This is consistent with the previous study conducted by Ngai et al. [31].

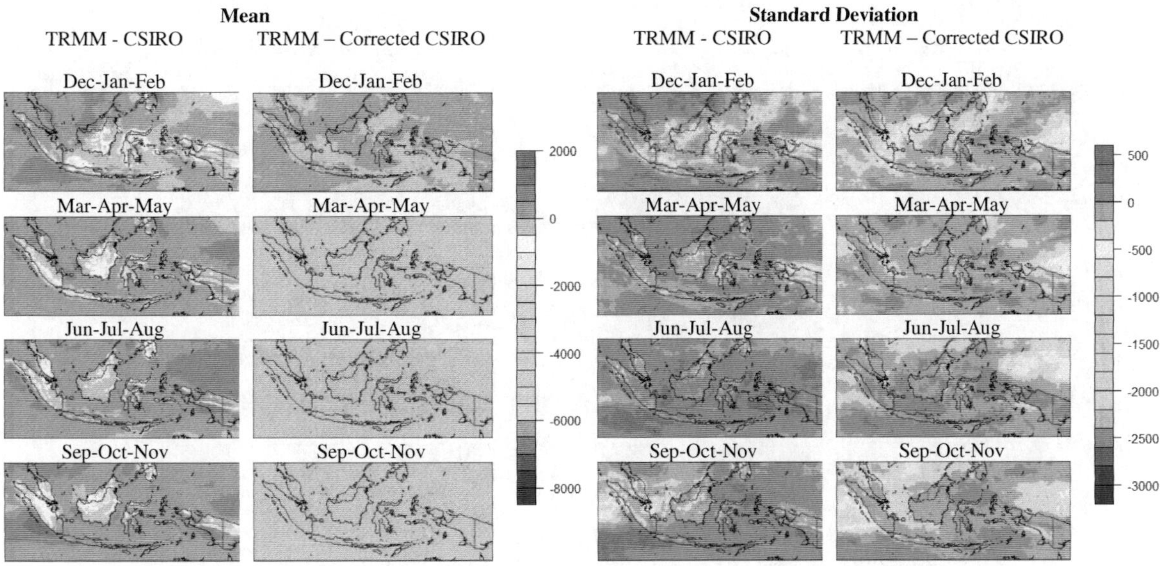

Figure 3. The difference of mean and standard deviation of seasonal precipitation

3.2. The number of wet-days characteristics

Figure 4 shows a distinctive wet and dry seasonal pattern as observed by TRMM due to the Asian-Australian monsoon, which affects Southeast Asia and Australia through the winds from the southeast and northwest during the cooler and warmer months, this is respectively to the Indonesian monsoon type based on the rainfall patterns and characteristics. Therefore, we can see the effect of ITCZ over the middle of Indonesia region. We also can see that the observation can capture the effect of warm pool over western Pacific Ocean and eastern Indian ocean that is related to the surface evaporation exposure time, marked by increasing the likelihood of cumulus formation due to greater convection over warmer seas.

RegCM4 was able to simulate the north–south rainfall gradient realistically over Indonesia, as shown in Figure 5. However, the model tends to shift the number of days with precipitation so that in boreal autumn (September-October-November; SON) we can see longer wet-days over Sumatera. The model underestimates the number of wet-days over the sea and overestimate them over the land. The model was also able to capture the orographic precipitation and mostly drive the longer period of rainfall over the high topography area such as the middle of Sumatera island, northern Sulawesi island, and southern Papua island.

QM bias correction was realistically able to improve the model simulation so that we can see most of the pixel on the corrected model have a similarity with the observation (Figure 6). We can see the same pattern of wet-days period over Indonesia region.

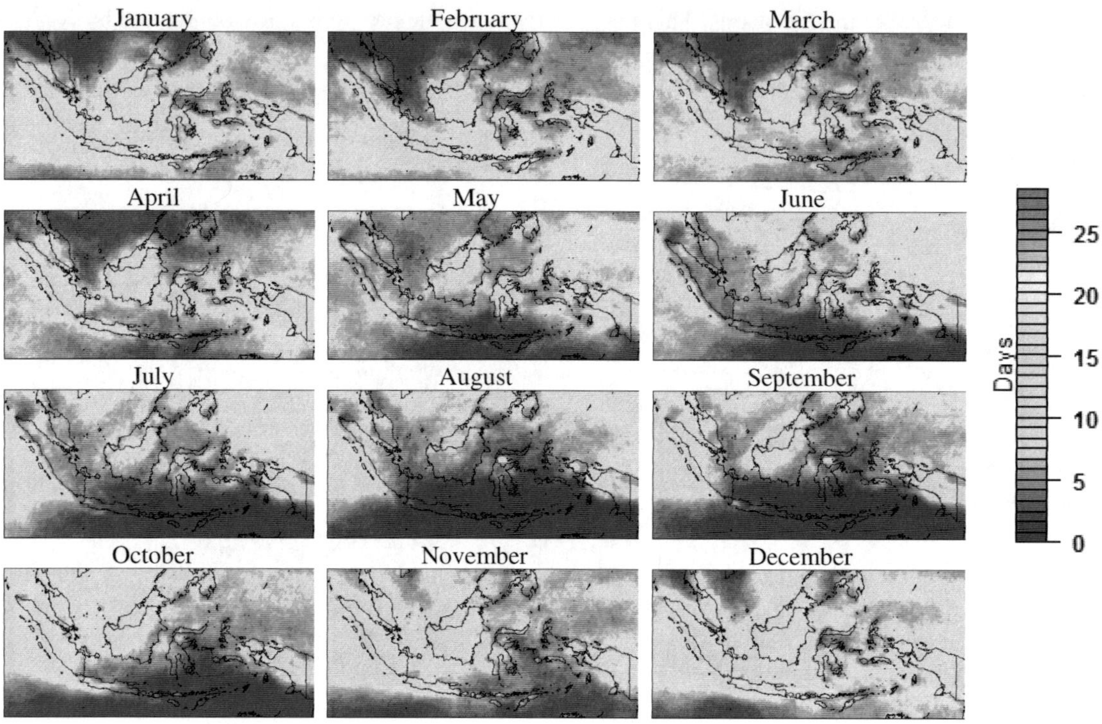

Figure 4. TRMM (observation) mean monthly number of wet-days

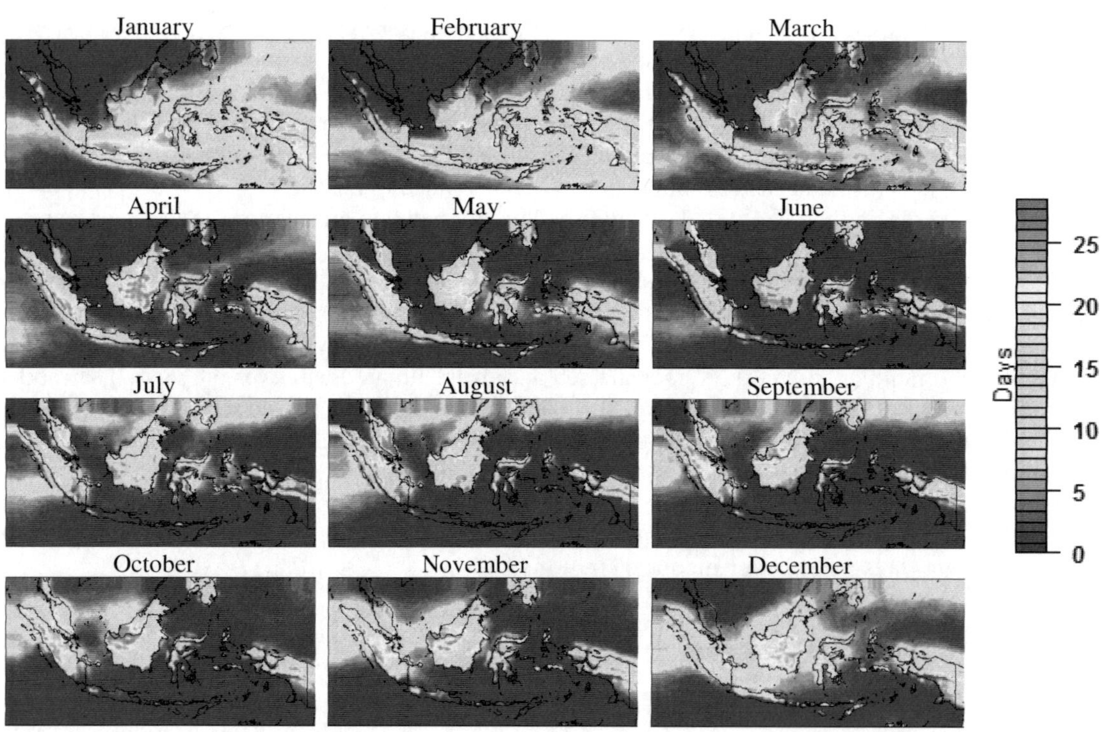

Figure 5. CSIRO (model) mean monthly number of wet-days

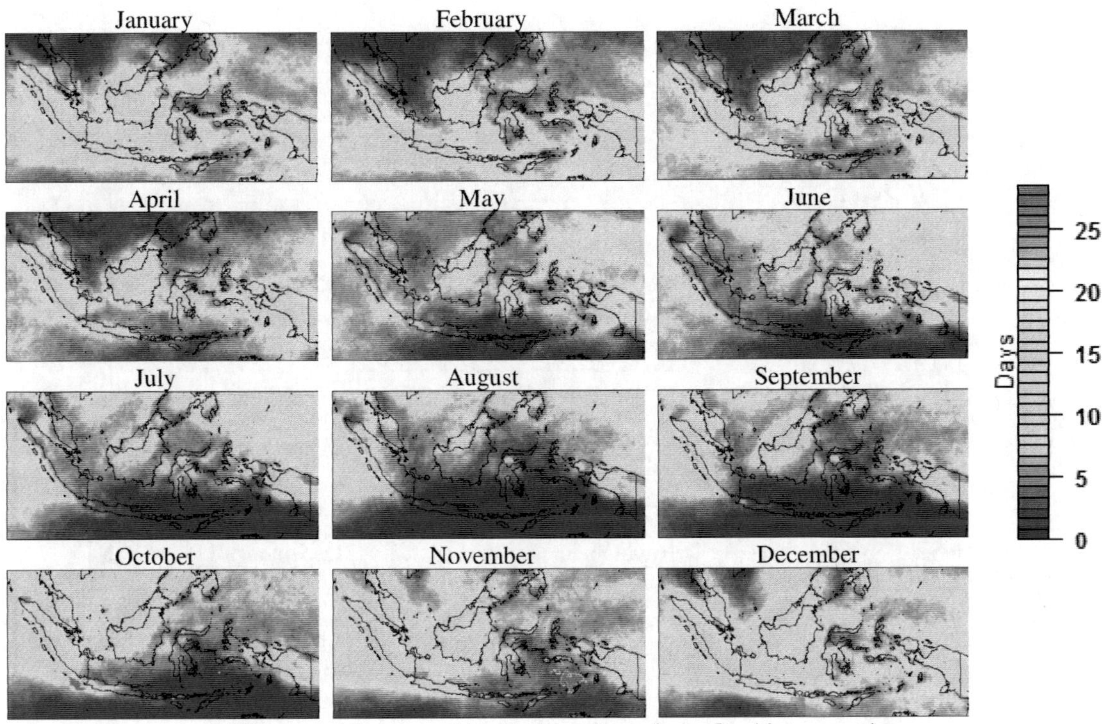

Figure 6. Mean monthly number of wet-days after bias correction

3.3 Extreme precipitation

Extreme value distributions are fitted to these data in order to evaluate rare precipitation events, such as one-day amounts that are exceeded on average only every 20 years.

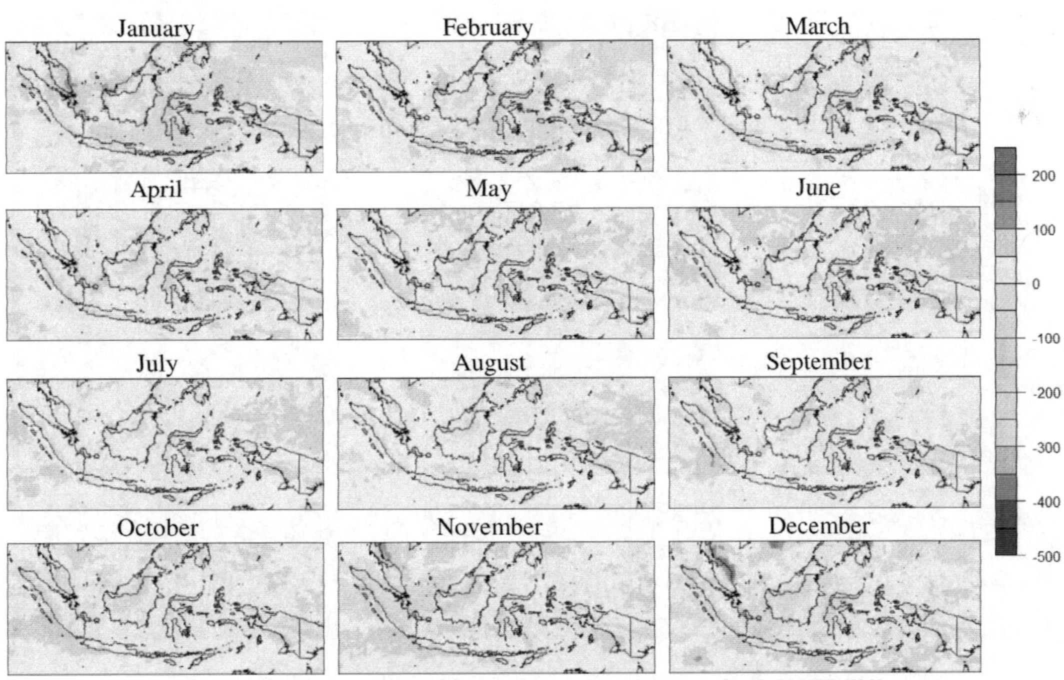

Figure 7. Mean monthly of daily maximum precipitation (RX1DAY) different TRMM – CSIRO (observation – uncorrected model)

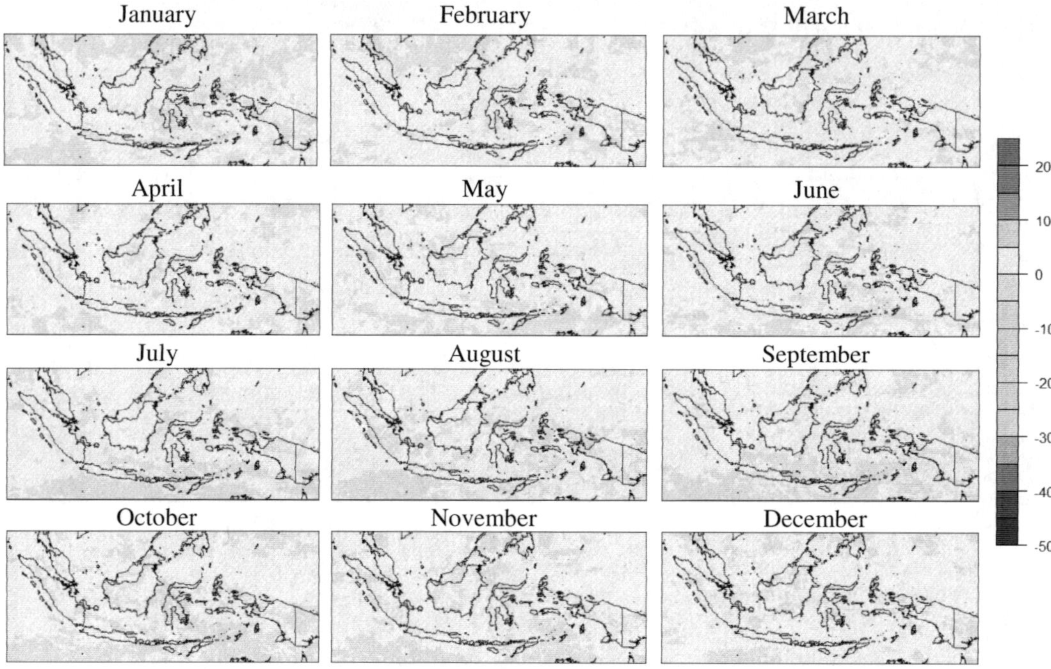

Figure 8. Mean monthly of daily maximum precipitation (RX1DAY) different TRMM – Corrected CSIRO (observation – corrected model)

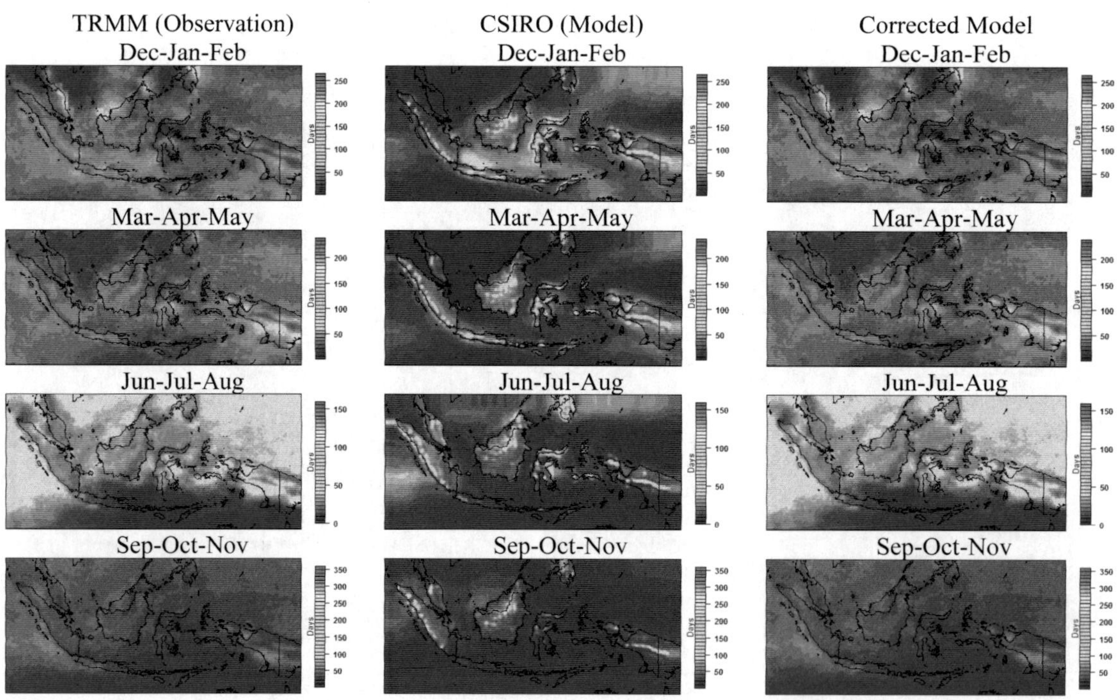

Figure 9. Percentile 95 of seasonal mean of daily precipitation

4. Conclusion

From this study we have investigated how well the quantile mapping bias correction method improve the performance of the RCM for a set of precipitation indices. The analysis focuses on precipitation over Indonesia region. The results from this study would serve as a useful reference for an extended evaluation of regional climate models over the study area. We assess that RegCM without bias correction have shown large uncertainty in climate but the bias correction tool helped to simulate the reliable climate output with reference to observed climate data. Therefore for framing of better management practices, adaptation programme and planning and policy-making based on climate model output, we must ensure to develop the more reliable and validated regional climate scenarios before using that. Our future work also aims to characterize spatial variability over a larger region of Southeast Asia using gridded precipitation products.

References

[1] Maraun D, Wetterhall F, Ireson AM, Chandler RE., Kendon EJ, Widmann M, Brienen S, Rust H W, Sauter T, Themeßl M, Venema VKC, Chun KP, Goodess CM, Jones RG, Onof C, Vrac M, and Thiele-Eich I 2010 Precipitation downscaling under climate change: Recent developments to bridge the gap between dynamical models and the end user *Rev. Geophys.* **48** RG3003

[2] Intergovernmental Panel on Climate Change (IPCC) 2007 *Summary for policymakers, in climate change 2007: The Physical science basis. Contribution of working group I to the fourth assessment report of the Intergovernmental Panel on Climate Change*, ed Solomon S et al. (Cambridge: Cambridge University Press) pp1–18

[3] Giorgi F and Mearns L O 1991 Approaches to regional climate change simulation: A review *Rev. Geophys.* **29** 191-216

[4] Willems P, Arnbjerg-Nielsen K, Olsson J, and Nguyen VTV 2012 Climate change impact assessment on urban rainfall extremes and urban drainage: methods and shortcomings *Atmos. Res.* **103** 106-118

[5] Tabari H., Taye MT, and Willems P 2015 Water availability change in central Belgium for the late 21st century *Global Planet. Change* **131** 115–123

[6] Fowler HJ, Blenkinsop S, and Tebaldi C 2007 Linking climate change modelling to impacts studies: recent advances in downscaling techniques for hydrological modelling *Int. J. Climatol.* **27** 1547–78

[7] Christensen JH, Boberg F, Christensen OB, and Lucas-Picher P 2008 On the need for bias correction of regional climate change projections of temperature and precipitation *Geophys. Res. Lett.* **35** L20709

[8] Herrera S, Fita L, Fernandez J, and Gutierrez JM 2010 Evaluation of the mean and extreme precipitation regimes from the ENSEMBLES regional climate multimodel simulations over Spain *J. Geophys. Res.* **115** D21117

[9] Turco M, Sanna A, Herrera S, Llasat M-C, and Gutiérrez JM 2013 Large biases and inconsistent climate change signals in ENSEMBLES regional projections *Clim. Change* **120(4)** 859–869

[10] Marzban C, Sandgathe S, and Kalnay E 2006 MOS, perfect prog, and reanalysis *Mon. Weather Rev.* **134(2)** 657–663

[11] Ruiz-Ramos M, Rodríguez A, Dosio A, Goodess C, Harpham C, Mínguez M, and Sánchez E 2016 Comparing correction methods of RCM outputs for improving crop impact projections in the Iberian Peninsula for 21st century *Clim. Change* **134(1–2)** 283–297

[12] Giorgi F et al 2012 RegCM4: model description and preliminary tests over multiple CORDEX domains *Clim. Res.* **52** 7–29

[13] Collier M, Jeffrey S, Rotstayn L, Wong K, Dravitzki S, Moeseneder C, et al. 2011 The CSIRO-Mk3-6-0 Atmosphere-Ocean GCM: participation in CMIP5 and data publication *MODSIM 2011 - 19th International Congress on Modelling and Simulation* (Perth) (Modelling and Simulation Society of Australia and New Zealand Inc) pp 2691-97

[14] Grell GA, Dudhia J, and Stauffer DR 1994 A description of the fifth-generation Penn State/NCAR Mesoscale Model (MM5) *NCAR Tech. Note, NCAR/TN-398+STR* pp 122

[15] Kiehl JT, Hack J, Bonan G, Boville B, Briegleb B, Williamson D, and Rasch P 1996 Description of the NCAR Community Climate Model (CCM3) *Technical Report 30 NCAR/TN-420+STR*, National Center for Atmospheric Research, Boulder, Colorado, pp 152

[16] Collins WD and others 2006 The Community Climate System Model Version 3 (CCSM3) *J. Climate* **16** 2122-43

[17] Dickinson RE, Henderson-Sellers A, and Kennedy PJ 1993 Biosphere-atmosphere transfer scheme (BATS) version 1e as coupled to the NCAR Community Climate Model *NCAR Technical Note NCAR/TN-387+STR* p 72
[18] Gao X et al. 2006 Projected changes in mean and extreme precipitation over the Mediterranean region from high resolution double nested RCM simulations *Geophys. Res. Lett.* **33** L03706
[19] Oleson KW, Niu Gy, Yang ZL, Lawrence DM and others 2008 Improvements to the Community Land Model and their impact on the hydrologic cycle *J. Geophys. Res.* **113** G01021
[20] Tawfik AB, Steiner AL 2011 The role of soil ice in land–atmosphere coupling over the United States: a soil moisture precipitation winter feedback mechanism *J. Geophys. Res.* **116** D02113
[21] Holtslag A, de Bruijn E, Pan HL 1990 A high resolution air mass transformation model for short-range weather forecasting *Mon. Weather Rev.* **118** 1561–75
[22] Zeng X, Zhao M, Dickinson RE 1998 Intercomparison of bulk aerodynamic algorithms for the computation of sea surface fluxes using TOGA COARE and TAO data *J. Clim.* **11** 2628–44
[23] Fritsch JM and Chappell CF 1980 Numerical prediction of convectively driven mesoscale pressure systems. Part I: Convective parameterization *J. Atmos. Sci.* **37** 1722–33
[24] Juneng L, Tangang F, Chung J, Ngai S, Tay T, Narisma G, et al. 2016 Sensitivity of Southeast Asia rainfall simulations to cumulus and air-sea flux parameterizations in RegCM4 *Climate Research* **69(1)** 59–77
[25] Ngo-Duc T, Tangang FT, Santisirisomboon J, Cruz F, Trinh-Tuan L, Nguyen-Xuan T, et al. 2016 Performance evaluation of RegCM4 in simulating extreme rainfall and temperature indices over the CORDEX-Southeast Asia region *International Journal of Climatology* **37** 1634-47
[27] Cruz FT, Narisma GT, Dado JB, Singhruck P, Tangang F, Linarka UA, et al. 2017 Sensitivity of temperature to physical parameterization schemes of RegCM4 over the CORDEX-Southeast Asia region. *International Journal of Climatology* **37** 5139-53
[28] Huffman GJ, Adler RF, Bolvin DT, Gu G, Nelkin EJ, Bowman KP, et al. 2007 The TRMM Multisatellite Precipitation Analysis (TMPA): Quasi-Global, Multiyear, Combined-Sensor Precipitation Estimates at Fine Scales *Journal of Hydrometeorology* **8** 38–55
[29] Piani C., Haerter J, and Coppola E 2010 Statistical bias correction for daily precipitation in regional climate models over Europe *Theor. Appl. Climatol.* **99** 187–192
[30] Ngai ST, Tangang F, and Juneng L 2017 Bias correction of global and regional simulated daily precipitation and surface mean temperature over Southeast Asia using quantile mapping method *Global and Planetary Change* **149** 79–90

IOP Conf. Series: Earth and Environmental Science **303** (2019) 012047 doi:10.1088/1755-1315/303/1/012047

Assimilation of Weather Radar Data Using WRF 3DVar Modelling for Rainfall Prediction

M I Hastuti[1,2] and J I A Paski[2]

[1] School of Meteorology Climatology and Geophysics (STMKG)
[2] Indonesian Agency for Meteorology and Geophysics (BMKG)

mynameisindrie@gmail.com

Abstract. Data assimilation is one method of improving the initial condition of the atmosphere in numerical weather predictions. The weather-based radar data assimilation, which generates high-resolution data of about 250 meters with coverage of hundreds of kilometers, is expected to improve the quality of weather forecasts and analysis, especially in determining the precipitation forecasting or Quantitative Precipitation Forecasting (QPF). This study aims to investigate the role of assimilation of Doppler weather radar data in mesoscale models for the forecast of heavy rain events occurring in the Greater Jakarta area on 20 February 2017 and 13 June 2017. For this purpose, the reflectivity data and radial radar speeds obtained from weather radar Tangerang will be assimilated with the 3DVar system into the Weather Research Forecasting (WRF) numerical model version 3.9.1 and Tropical Physics Suite parameterized configuration recommended for the tropics. The output of assimilation model of radar data and without assimilation of the numerical model of WRF is verified by point and spatial with some observation data of Automatic Weather System (AWS) and GSMAP rainfall in Jabodetabek region. In general, the WRF model assimilation of radar data provides a better simulation of spatial rain events and better points compared to the WRF model without assimilation.

Keywords : Quantitative Precipitation Forecasting, WRF, Doppler, AWS.

1. Introduction

One of the weather modeling systems is the Numerical Weather Prediction (NWP). The NWP responds to the challenge with the scientific method required in predicting in complex environments [1]. In recent years, NWP has always been the Weather Research and Forecasting (WRF) meso scale model. The advantages of WRF are efficient and flexible because it can be used in laptops for various interactions with meter scales up to thousands of kilometers [2]. In addition, WRF includes open source, model configurations that can be customized and tailored to the research needs. In the WRF model, there are several ways to increase the weather prediction value. One of them is by assimilating data. Assimilation of data is done by improving the data as input model by calculating the data into the grid system model [2].

WRF Data Assimilation (WRFDA) is a special WRF program to assimilate data. WRFDA has several assimilation techniques, including Three Variable Dimensions (3DVar), Four Dimensional Variational (4DVar), and Ensemble Kalman Filter (EnKF). From various studies, 3DVar is considered to have the best performance of all types of techniques with good efficiency [3]. Data that can be assimilated in WRFDA are surface air, satellite and radar data. One of the most frequent researches is the assimilation of radar data. Excess radar data compared to other data is having high resolution (higher resolution than meso scale resolution model) with wide coverage. By this, assimilation of radar data is expected to be used.

Outside Indonesia, Xiao et al. assimilated 3DVar radar data using radial velocity data [4]. The study said that the assimilation of radial velocity data can improve the ability of predicting quantitative rainfall or Quantitative Precipitation Forecasting (QPF) in the short term. Furthermore, research on the assimilation of 3DVar radar data uses two data, i.e. radial velocity and radar reflectivity data [5, 6]. Based on these studies, the assimilation of radial velocity data and radar reflectivity has a positive effect on QPF capability by using one of the two radar data.

In Indonesia, Satrya (2012) conducted research on the assimilation of 3DVar radar data (reflectivity and velocity) in Bandung area. In conclusion, the largest sampling technique on radar data reflectivity as a WRFDA input is the best in virtualizing heavy rain [6]. The same research has conducted by Paski and Gustari (2017) in Lampung area [7]. In conclusion, the largest sampling technique on radar data reflectivity as a WRFDA input is the best in virtualizing heavy rain. In addition, Paski et al. (2017) also conducted a recent engineering study on the assimilation of 3DVar radar data in Jakarta using products from the reflectivity radar data, i.e. Constant Altitude Plan Position Indicator (CAPPI) as the assimilation data of WRFDA which was better information compared to the WRF model without assimilation [7].

The use of CAPPI in radar assimilation is something new in Indonesia. Therefore, it is interesting to conduct research on the assimilation of 3DVar radar data technique using CAPPI data. The difference with previous research is the comparison test between the CAPPI data assimilation performance model, radial velocity, reflectivity, and the combination of reflectivity and radial velocity in terms of data that successfully initiated WRF model for heavy rain events in Jabodetabek. The most possible model improvement is to increase the predicted value of heavy rainfall and temporality.

2. Data and Methodology

In this study some primary data are used, Global Forecast System (GFS) is used as initial data model and C-Band Doppler Radar (CDR) data is used as observation data. The data used in model verification in this research are Global Satellite Mapping of Precipitation (GSMaP) data for spatial analysis and observational data from the meteorological station owned BMKG around Jakarta region including Cengkareng, Pondok Betung, Kemayoran, Tanjung Periuk, and Citeko for point analysis.

WRF model for this research uses a three domain configuration. The third domain or the last domain has a resolution of three kilometers and covering Jabodetabek area. Domain is not created too large to decrease model run time. Model parameterization configuration uses tropical configuration in WRF 3.9.1 for tropical Indonesia. Weather assimilation model is obtained from the package of WRFDA with 3DVar technique using radar data as observation.

Radar volumetric (.vol) volumetric data is converted into two types of formats, the polar coordinates netCDF and csv using EDGE applications because of the EEC doppler weather radar of the EEC type. Then, both types of radar data formats will each go through different processes. The polar coordinates netCDF radar data will be processed with the wradlib-python application to convert the format into netCDF cartesian coordinates, as well as process the data into a netCDF format CAPPI product. The CAPPI calculation specifications have been designed in accordance with research from Permana et al. (2016) [8]. Furthermore, CAPPI netCDF format data is processed by taking the maximum sampling value in one grid model of 3 km x 3 km (according to the domain of previous *.ctl) using the R Studio application [7]. The output data of the R application will result in the output of radar data in ASCII format or according to the WRFDA application input format.

In csv format data, separate data is based on elevation or sweep. In each sweep, the data is also separated based on reflectivity (Z), radial (V), spectral width (W) and filter unfilled (U) reflectivity. This research uses Z and V data as assimilation data. Csv radar data formats will be processed in numerical data analysis and computing applications to combine Z and V data in one .txt format file. The combined data Z and V (.txt) will be processed by the largest sampling technique on a 3 km x 3 km grid (according to the domain of previous *.ctl) using numerical data analysis and computation applications [6,9,10]. The output data of the R application will result in the output of radar data in ASCII format or according to the WRFDA application input format.

IOP Conf. Series: Earth and Environmental Science **303** (2019) 012047 doi:10.1088/1755-1315/303/1/012047

3. Result and Discussion

3.1 The Effect of Radar Data Assimilation on WRF Parameter

Effect of radar data assimilation on WRF is to improve the initial WRF (WRFinput) data more accurately. Here is a table that specifies what WRF parameters are improved on each model of radar data assimilation.

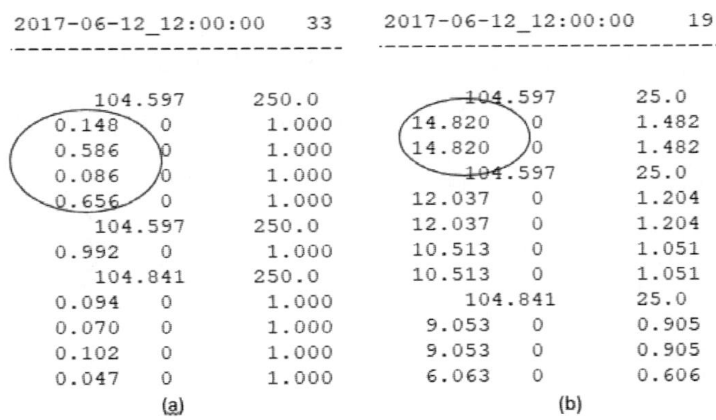

Figure 1. Differences in Z value on reflectivity data (a) and CAPPI data (b).

Table 1. The effect of assimilating radar data on WRF initial data

WRF Data Assimilation	WRF Parameter			
	Temperature	Humidity	*Mixing Ratio*	Wind
CAPPI	√	√	√	√
Reflectivity (Z)	√	√	√	-
Radial Velocity (V)	-	-	√	√
Combination of *Reflectivity* (Z) & Radial Velocity (V)	√	√	√	√

WRF with assimilation CAPPI and Z have the same data type, i.e. data reflectivity, only in CAPPI data is reflectivity data at certain fixed height, while Z data is data reflectivity of all volume scanning. In 3DVAR's WRF system, CAPPI assimilation has a larger value than the Z assimilation (figure 1) therefore we change the wind parameters through negative temperature change scheme, since Z assimilation is not able to change the wind parameter.

3.1.1 Temperature Parameter

In the 3DVAR WRF assimilation, initial air temperature data improvements are calculated from warm-rain schemes using reflectivity data or CAPPI. Air temperature parameters play a role in the convection process where the higher the temperature, the greater the convection process that is likely to the occurrence of heavy rain.

Figure 2 shows the difference of WRF initial data for surface air temperature parameters between WRF without assimilation and WRF with assimilation on February 20, 2017 (rainy season). In the southeast-to-west study area, the CAPPI assimilated WRF air temperature (Figure 2b) gets warmer than WRF without assimilation (Figure 2a), ranging from 24-22 °C to 26-27 °C, as did study areas such as Kemayoran, Tanjung Periuk, Pondok Betung, and Cengkareng, where the area of warmer temperatures is widespread. In Figure 4.10c, the WRF Z assimilation also changes the air temperature to warmer in the southeast to the west of the study area. However, the temperature of study areas such as Kemayoran, Tanjung Periuk, Pondok Betung, and Cengkareng are getting colder. Meanwhile, in Figure 4.1e, the WRF assimilation of the Z & V combination (figure 2d) is no different than the WRF Z assimilation. In addition, the WRF V assimilation (figure 2d) does not change at all the air temperature parameters.

Figure 3 shows the difference of WRF initial data for surface air temperature parameters between WRF without assimilation and WRF assimilation on June 13, 2017 (dry season). When viewing Figure 3b, it can be seen clearly that the warmer surface air temperature expansion in WRF CAPPI assimilation such as Kemayoran, Tanjung Periuk, Pondok Betung and Cengkareng areas near the radar, ranges from 26 - 26,5°C to 27 - 27,5°C. However, the special Citeko area does not change temperature. In the WRF Z assimilation (figure 4.12c) and Z assimilation & V combinations (figure 3d) also gets warmer the greater the temperature, while only in the southwest region of study and the special Citeko region away from the radar location display the cooler air temperature. Meanwhile, the WRF V assimilation (figure 3d) does not change the air temperature parameters at all.

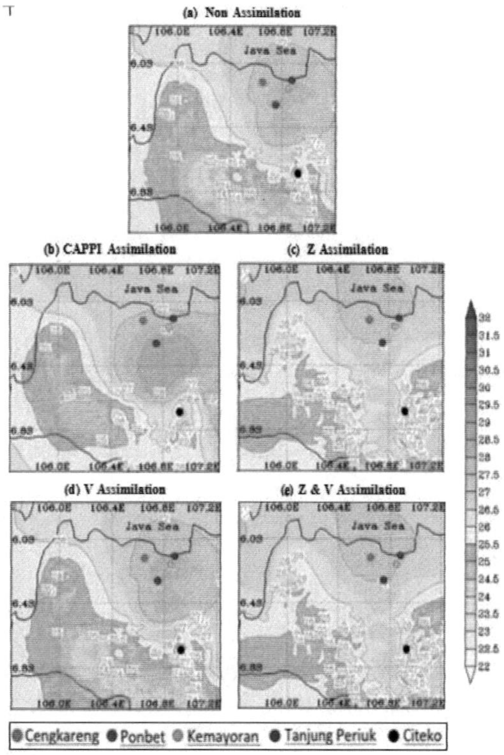

Figure 2. Initial data of surface air temperature parameter in case of heavy rains on February 20, 2017.

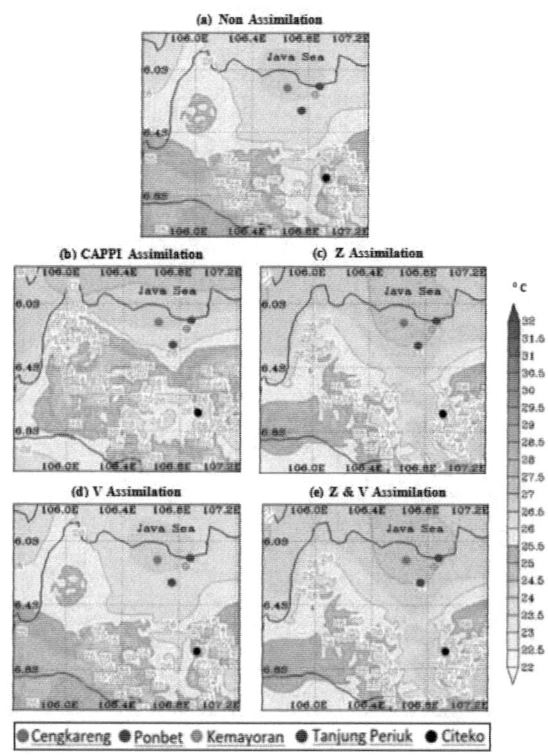

Figure 3. Initial data of surface air temperature parameter in case of heavy rains on June 13, 2017

3.1.2 Humidity Parameter

The humidity parameter is the concentration of water vapor in the atmosphere. In this study, we will examine the air humidity in the 850 mb layer because if the value is high, then it indicates the moisture content is large enough for convective cloud growth process that has the potential to cause moderate to heavy rain.

Figure 4 shows the initial data difference on the 850 mb layer moisture parameters between WRF without assimilation and WRF with assimilation on February 20, 2017 (rainy season). Compared to WRF without assimilation (Figure 4a), WRF CAPPI assimilation (Figure 4b) and Z assimilation (figure 4c) do not change significantly. However, generally the air gets damper so it can be ascertained that rainfall will be high as well. Location of value changes occurring in areas near radar such as Kemayoran, Tanjung Periuk, Pondok Betung, and Cengkareng by 84% to 86%. Similarly, areas far from radar like Citeko, the humidity value is getting higher, i.e. 96% to 98%. Meanwhile, air humidity improvements in WRF Z assimilation & V combination (figure 4e) are similar to WRF Z assimilation, while WRF V assimilation (figure 4d) shows no change at all or equal to WRF without assimilation.

Figure 5 shows the initial data difference for the 850 mb layer air humidity parameter between WRF without assimilation and WRF assimilation on June 13, 2017 (dry season). In the northern area of study including Kemayoran, Tanjung Periuk, Pondok Betung, and Cengkareng areas, the CAPPI assimilation WRF air humidity (Fig. 5b) changed more humidly than WRF without assimilation (figure 5a). However, the southern region including Citeko region did not change significantly. In WRF Z assimilation (figure 5d), there is no significant difference, as did with WRF Z assimilation & V combination (figure 5e) which is similar to WRF assimilation Z. Meanwhile, WRF V assimilation (figure 5d) has not changed at all.

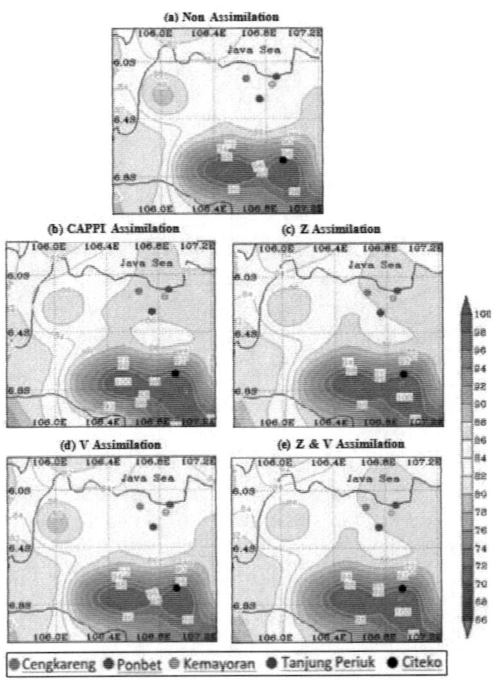

Figure 4. Initial data of 850 mb air humidity parameter in the case of heavy rains on February 20, 2017

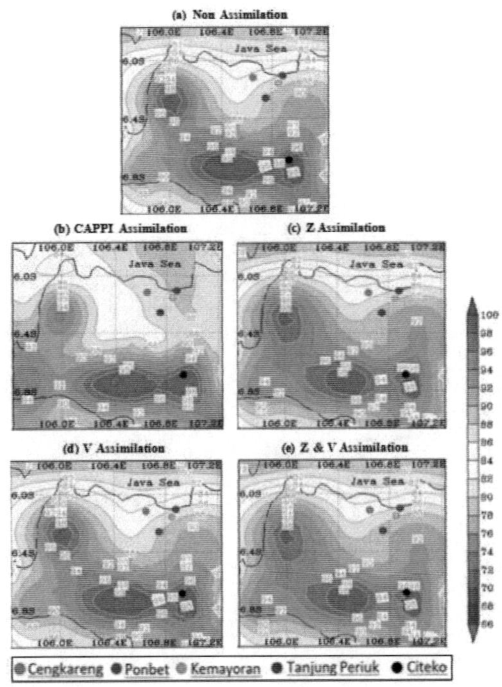

Figure 5. Initial data of 850 mb air humidity parameter in the case of heavy rains on June 13, 2017

3.1.3 Water Vapor Mixing Ratio Parameter

The parameter of the water vapor mixing ratio is the ratio of moisture mass to the dry air mass present in one particular volume [11]. The value of this mixing ratio also represents the moisture content of the moisture present in the air that the higher the mixing ratio, the greater the chance of rain.

Figure 6 shows the initial data difference for the water vapor mixing ratio between WRF without assimilation and WRF assimilation on February 20, 2017 (rainy season). Compared to WRF without assimilation (Figure 6a), significant changes in water vapor mixing in WRF assimilation of CAPPI (figure 6b), Z assimilation (figure 6c), and Z assimilation & V combination (figure 6d) are shown in the western region study. In areas close to radar sites such as Kemayoran, Tanjung Periuk, Pondok Betung, and Cengkareng, there is an increase in mixing ratio from 0.0132 to 0.0136 in WRF Z assimilation and Z assimilation & V combination, whereas 0.014 in WRF CAPPI assimilation). This increase indicates the higher moisture content of the water or wetter conditions that there is more potential for heavy rains. In contrast, areas far from radar such as Citeko display decrease in the value of mixing ratios, from 0.0152 to 0.014 in WRF assimilation of CAPPI, Z assimilation, and Z assimilation & V combination. Furthermore, WRF V assimilation (figure 6d) no change is detected at all in mixing ratio.

Figure 7 shows the initial data difference for the water vapor mixing ratio between WRF without assimilation and WRF assimilation on June 13, 2017 (dry season). Compared to WRF without assimilation (Figure 7a), significant changes in water vapor mixing in WRF assimilation of CAPPI (figure 7b), Z assimilation (figure 7c), and Z assimilation & V combination (figure 7e) in the northern region such as Kemayoran, Tanjung Periuk, Pondok Betung, and Cengkareng, there was an increase in mixing ratio from 0.0128 - 0.0136 to 0.0132 - 0.014 in the WRF Z assimilation and Z assimilation & V combination, whereas 0.0136-0.0144 on WRF CAPPI assimilation (tend to be larger). Likewise, Citeko increases the value of mixing ratio, from 0.014 – 0.0148 to 0.0148 – 0.0152 at WRF assimilation CAPPI, Z assimilation, and assimilation of combination of Z & V). Furthermore, in WRF V assimilation (figure 7d), there is no change in the mixing ratio at all.

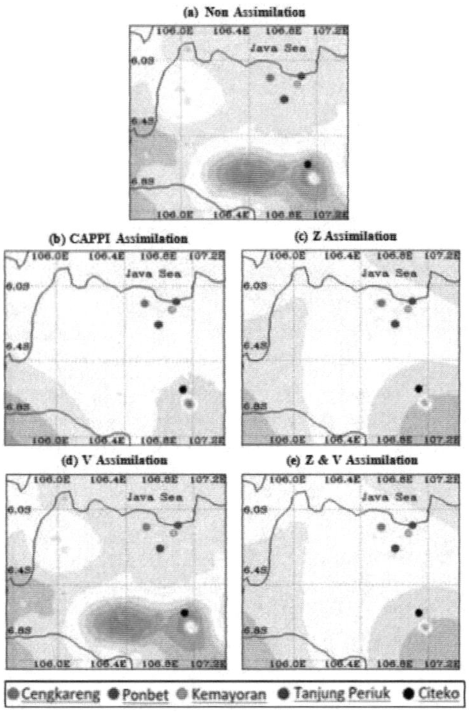

Figure 6. Initial data of 850 mb water vapor mixing ratio parameter in the case of heavy rains on February 20, 2017

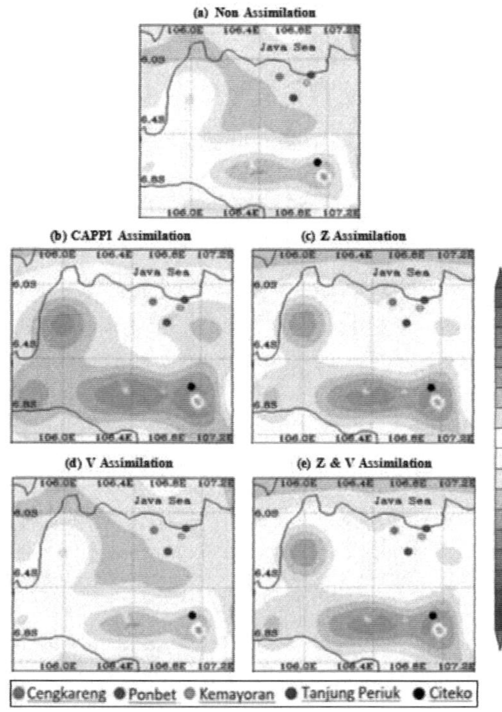

Figure 7. Initial data of 850 mb water vapor mixing ratio parameter in the case of heavy rains on June 13, 2017

3.1.4 Wind Parameter

The wind parameter acts as vapor mass supply from the outside or as humidity transport. In WRF 3DVAR assimilation, the initial wind data improvement is calculated from CAPPI data and radial velocity. CAPPI data assimilated through negative temperature change, while assimilation of radial velocity data directly from wind parameter element u, v, w.

Figure 8 shows the initial data difference on 850 mb layer wind parameters on February 20, 2017 (rainy season). Spatially, the CAPPI assimilation WRF (Figure 8b), V assimilation (Figure 8d) and Z assimilation & V combinations (Fig. 8e) significantly alter the wind parameters, shows that calm wind speed is widespread in most Southwest Jabodetabek and middle area. This indicates that convective cloud growth is not hindered by the wind. Kemayoran area, Tanjung Periuk, Pondok Betung, and Cengkareng (near radar location) wind speed also weakened, namely Cengkareng and Pondok Betung from 10 knots to 8 knots and Kemayoran and Tanjung Periuk areas from 14 knots to 10 knots. Meanwhile Citeko, away from the radar location did not change significantly, i.e. calm wind speed on WRF without assimilation and WRF with other assimilation. Meanwhile, wind direction parameters did not change significantly in each WRF assimilation. Furthermore, the WRF Z assimilation (figure 8b) does not change wind parameters in both direction and speed.

Figure 9 shows the initial data difference on 850 mb layer wind parameters between WRF without assimilation and WRF assimilation on June 13, 2017 (dry season). Compared to WRF without assimilation (Figure 9a), the CAPPI assimilation WRF wind parameter (Figure 9b), V assimilation (Figure 9d), and Z assimilation & V combinations (Figure 9e) change significantly. Generally, wind speeds of CAPPI assimilation WRF, V assimilation and Z assimilation & V combination are weakened, except in CAPPI assimilation WRF, wind speed of Tanjung Periuk, Kemayoran, Citeko region are unchanged, while in WRF V assimilation and Z assimilation & V combination only Citeko that does not change. Meanwhile, wind direction parameters did not change significantly in each WRF assimilation. Furthermore, the WRF Z assimilation (figure 9b) does not change the wind.

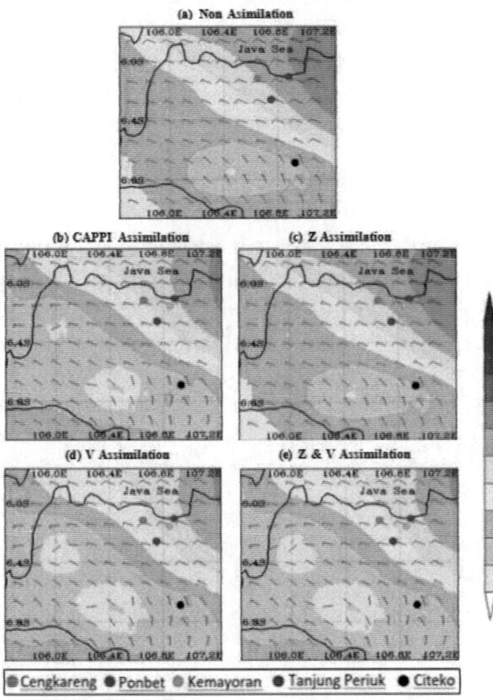

Figure 8. Initial data of 850 mb wind parameter in the case of heavy rains on February 20, 2017

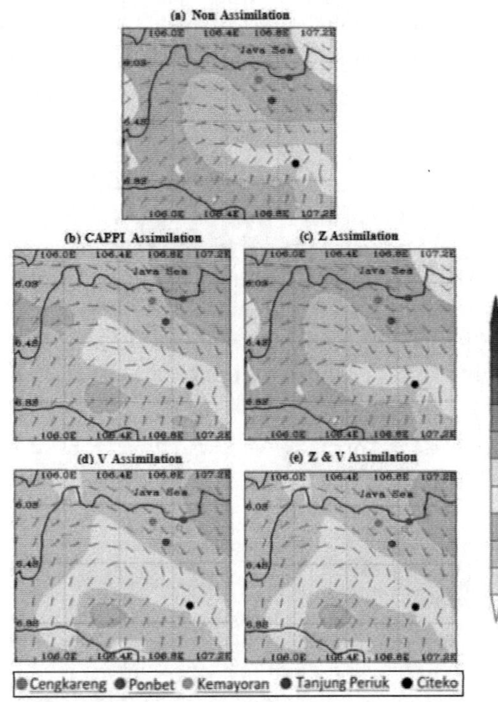

Figure 9. Initial data of 850 mb wind parameter in the case of heavy rains on June 13, 2017

3.2 Analysis and Verification of Spatial Rain Distribution

3.2.1 The case of heavy rains on 20 February 2017

Based on Figure 10, it shows the rainfall simulation results and the amount of rainfall between GSMAP observation data and the simulation results using WRF without assimilation, WRF assimilation CAPPI, WRF Z assimilation, WRF V assimilation, and WRF assimilation combination Z & V on 20 February 2017. GSMAP data (figure a) shows the rainfall of Jabodetabek area dominated by moderate rain (20 - 50 mm). However, the southern part of Bogor area displays light rain (<20 mm). In addition, the western and eastern parts of Jabodetabek show heavy rains (50 to 100 mm), namely southwestern Tangerang, northern Bogor, eastern Depok, eastern Bekasi, southwestern Bekasi and most of Java Sea.

Compared to GSMAP observation data, the moderate WRF rainfall area without assimilation (Figure 10b) is narrowed. As a result, the area of light rain extends and turns to be in the northwest and northeast of Jabodetabek, which is in most of Tangerang, West Jakarta, South Jakarta, western Depok, and the southeast and northeast of Bekasi. Likewise, the area of heavy rain also change places and slightly narrowed in the southern region of Greater Jakarta, which is in Bogor.

IOP Conf. Series: Earth and Environmental Science **303** (2019) 012047 doi:10.1088/1755-1315/303/1/012047

Compared to WRF without assimilation, WRF spatial rainfall assimilation does not change significantly, but generally the broad-range rainfall area in WRF with assimilation is wider than WRF without assimilation, as well higher rainfall. The area of moderate - heavy rain remain narrower than GSMAP data. The highest rainfall is in WRF CAPPI assimilation (figure 10c) and assimilation of Z & V combination (figure 10f) occurring in the southeastern part of Jabodetabek, the northeastern Bogor and southeastern Bekasi. In addition, the WRF's light-weight CAPPI assimilation area and the Z & V combination assimilation in northwest Jabodetabek also appear to be narrowing.

In the WRF Z assimilation (figure 10d), the light rain area also narrows in the northwest part of Jabodetabek and the heavy rain area is seen in the southeastern part of Jabodetabek, Bogor and surrounding areas. Meanwhile, the light rain area in WRF V assimilation (figure 10e) extends to the south of Jabodetabek, i.e. in most areas of Bogor. However, the area of heavy rain on WRF V assimilation is almost absent in Jabodetabek area. In general, the simulated GSMAP rain-simulation results are similar to WRF assimilation, differ only in rain intensity with the most superior models are WRF assimilation of CAPPI and Z assimilation & V combination because it is more susceptible to heavy rainfall than other WRF assimilation.

3.2.2 The case of heavy rains on June 13, 2017

Based on Figure 11, it shows the rainfall simulation results and the amount of rainfall between the GSMAP observation data and the simulation results using WRF without assimilation, WRF assimilation CAPPI, WRF Z assimilation, WRF V assimilation, and WRF assimilation combination Z & V on June 13, 2017. GSMAP observation data (Figure 11a)shows that almost all Jabodetabek areas are dominated by light rain (<20 mm), except for moderate rainfall (20 - 50 mm) in most of western and western part of Bogor. Then, there is no rain in a small part of northern Bekasi. In addition, there is no heavy rain (50 - 100 mm) in the Greater Jakarta area and surrounding areas.

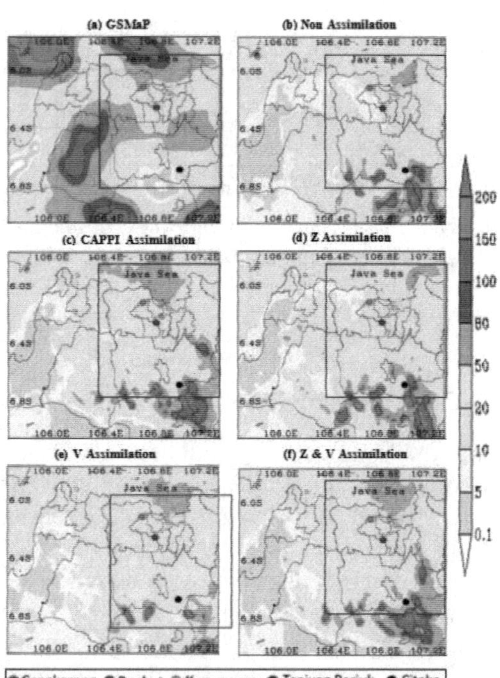

Figure 10. Spatial rain distribution in the case of heavy rains on February 20, 2017

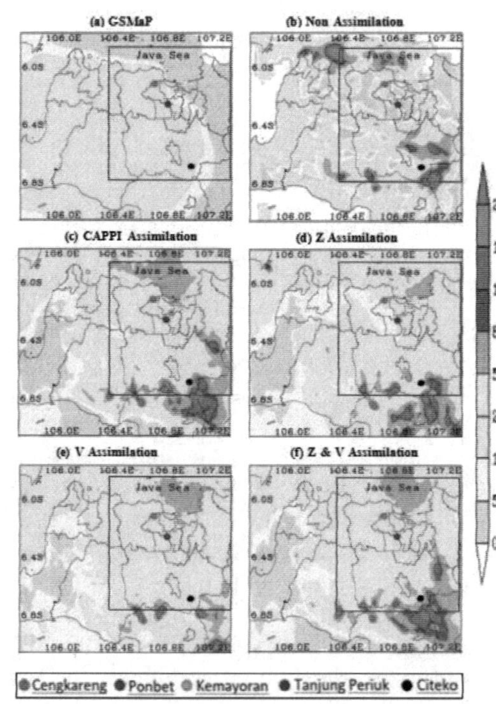

Figure 11. Spatial rain distribution in the case of heavy rains on June 13, 2017

IOP Conf. Series: Earth and Environmental Science **303** (2019) 012047 doi:10.1088/1755-1315/303/1/012047

Compared to the GSMAP observation data, the distribution of WRF rainfall without assimilation (Figure 11b) is significantly different. WRF rain distribution without assimilation looks uneven and there is heavy rain in some areas of eastern Bogor and northern Java Sea and parts of southern Bogor. In contrast, the distribution of WRF CAPPI assimilation area (Figure 11c), Z assimilation (figure 11d), V assimilation (figure 11e), and Z assimilation & V combination (Figure 11f) look similar to GSMAP observation data but bulk intensity of the rain is different that it is quite different compared to WRF without assimilation. In the four models of WRF assimilation, rainfall is dominated by moderate rain, while light rain occurs in most of western part of Tangerang and part of western Bogor where according to GSMAP observation data of the area occurs moderate rain. Then, the WRF specific V assimilation area of light rain extends to the southern part of Bogor. However, in general the widest area of light rain occurred in WRF Z assimilation. In addition, in the four WRF assimilation, there was heavy rain in Jabodetabek area, namely in the southern part of Bogor. While in fact, according to GSMAP data there is no heavy rain at all in the Greater Jakarta area. The largest density of heavy rainfall occurs in WRF CAPPI assimilation that it has the greatest amount of rainfall than any other.

3.3 Rain Dichotomy Verification (yes / no)

In this discussion, we will analyze and verify skill prediction based on the 5-point observation of the rainfall from the nearest to the farthest from the radar. This is done because the further distance the radar observation, the greater the beam radar is formed and ultimately impact on the sensitivity and accuracy of radar parameters used for data assimilation. Based on a summary of the results of verification of prediction skills on dichotomy yes / no on the observation area (table 2), the best PC, POD, and FAR skills also occur in WRF assimilation CAPPI and assimilation of Z and V combination data. In addition, WRF assimilation radar data is generally good for station areas closer to radar sources such as Cengkareng within 5.31 km, Pondok Betung within 15.28 km, Kemayoran within 21.65 km, and Tanjung Periuk within 26.85 km compared to areas farther from radar sources such as Citeko which is 66.6 km from the radar source and also the topographic area is hilly (mountainous). This can be seen from the skills of PC, POD, and FAR in the region which has always been the worst with an average of the lowest PC value of around 0.52, the average POD value is at least 0.76, and the average FAR value is the lowest height of 0.5.

Table 2. Forecast Verification

Skill	Region	WRF model best order					Average
		1	2	3	4	5	
PC	Cengkareng	S1	S4	S3	S2	S0	0,85
	Pondok Betung	S1	S4	S3	S2	S0	0,79
	Kemayoran	S4	S1	S2	S3	S0	0,67
	Tanjung Periuk	S1	S4	S2	S3	S0	0,64
	Citeko	S1	S4	S2	S3	S0	0,52
POD	Cengkareng	S1	S4	S3	S2	S0	0,85
	Pondok Betung	S1	S4	S3	S2	S0	0,9
	Kemayoran	S4	S1	S2	S3	S0	0,96
	Tanjung Periuk	S1	S4	S3	S2	S0	0,86
	Citeko	S4	S1	S2	S3	S0	0,76
FAR	Cengkareng	S1	S4	S3	S2	S0	0,4
	Pondok Betung	S1	S4	S2	S3	S0	0,36
	Kemayoran	S1	S4	S3	S2	S0	0,39
	Tanjung Periuk	S1	S4	S3	S2	S0	0,42
	Citeko	S1	S4	S3	S2	S0	0,5

Information :

S0 : WRF without assimilation
S1 : WRF CAPPI data assimilation
S2 : WRF Z data assimilation
S3 : WRF V data assimilation
S4 : WRF Z and V combination data assimilation

4. Conclusion

Based on the results of the research described in the previous chapter, some conclusions can be drawn as follows; In general, the WRF model assimilates radar data given to spatial rainfall event simulations and better points compared to the WRF model without assimilation where the rain prediction performance will be more visible in areas close to the radar and not obstructed from fixed objects such as mountains or hills, also more visible during the rainy season. From the four WRF model assimilation radar data conducted in this study, it can be concluded that the best sequence of WRF improvements is WRF assimilation of CAPPI data, Z & V data assimilation, V data assimilation, and Z data assimilation.

Acknowledgments

Authors wishing to acknowledge assistance from Donaldi Sukma Permana, Ph.D. for the CAPPI data assimilation. Acknowledgments for weather radar data, working facilities and financial support from Indonesian Agency for Meteorology Climatology and Geophysics (BMKG).

Reference

[1] Sagita N 2017 Asimilasi Model Weather Research and Forecasting (WRF) dengan Data Observasi Untuk Prediksi Curah Hujan di Wilayah Jawa (IPB: Bogor)

[2] Skamarock, William C, Klemp Joseph B, Dudhia J, Gill D O, Barker D M, Duda M G, HuangX. Y, Wang W and Powers J G 2008 A Description of the Advanced Research WRF Version 3 NCAR Technical Note Mesoscale and Microscale Meteorology Division (*National Center for Atmospheric Research* : Boulder Colorado USA)

[3] Liu J, M Bray and D Han 2013 A Study on WRF Radar Data Assimilation for HydrologicalRainfall Prediction *Hydrological Earth System* vol **17** 3095-3110

[4] Xiao Q and Sun J 2007 Multiple Radar Data Assimilation and Short-Range QuantitativePrecipitation Forecasting of a Squall Line Observed during IHOP_2002 *American Meteorological Society* vol **135** 3381-3404

[5] Sugimoto S, N.A Crook, Sun J and Xiao Q 2009 An Examination of WRF 3DVar Radar DataAssimilation on Its Capabilityin Retrieving Unobserved Variables and Forecasting Precipitation through Observing System Simulation Experiments *American Meteorological Society* vol **137** 4011-4029

[6] Satrya L I 2012 Asimilasi Data Radar dalam Penerapan Prediksi Cuaca Numerik di Indonesia(Studi Kasus di Jawa Barat) (ITB: Bandung)

[7] Paski, J A I, Permana D S and Hastuti M I 2017 Pemodelan Cuaca Data Asimilasi Radar CBand Doppler Produk Constant Altitude PPI (CAPPI) *Jurnal Meteorologi dan Geofisika (in Review)*

[8] Permana D S, Hutapea T D F, Praja A S, Fatkhuroyan and Muzayanah L F 2006 PengolahanMulti Data Format Radar Cuaca Menggunakan Wradlib Berbasis Python *Jurnal Meteorologi dan Geofisika* vol **17** No.3 157-164

[9] Paski J A I 2018 Pengaruh Asimilasi Data Radar C-Band dalam Prediksi Cuaca Numerik (Studi Kasus di Lampung) *Jurnal Meteorologi dan Geofisika* vol **18** 2

[10] Paski J A I 2018 Pengaruh Asimilasi Data Penginderaan Jauh (Radar Dan Satelit) Pada Prediksi Cuaca Numerik Untuk Estimasi Curah Hujan (Impact of Remote Sensing Data Assimilation (Radar And Satellite) On Numerical Weather Prediction For Rainfall Estimation) *Jurnal Penginderaan Jauh dan Pengolahan Data Citra Digital* vol **14** 2

[11] Rogers R R 1979 *A Short Course in Cloud Physics 2nd Edition* (Pergamon Press: Salby)

Indonesian Climate under 2°C and 4°C Global Warming: Precipitation Extremes

Supari[1,2], A Sopaheluwakan[3], U A Linarka[3], J Rizal[3], R Satyaningsih[3] and J X Chung[2]

[1]Center for Climate Change Information, Indonesia Agency for Meteorology, Climatology and Geophysics-BMKG, Jakarta, Indonesia
[2]School of Environmental and Natural Resource Sciences, National University of Malaysia – UKM, Selangor, Malaysia
[3]Center for Research and Development, Indonesia Agency for Meteorology, Climatology and Geophysics-BMKG, Jakarta, Indonesia

supari.bmg@gmail.com

Abstract. The Paris Agreement, signed in 2015, is intended to limit global warming well below 2°C. This paper is aimed to assess the potential key impact of 2°C and 4°C global warming on the characteristics of precipitation extremes over Indonesia. For this purpose, the CSIRO-Mk3.6.0 global projections forced by the representative concentration pathway (RCP8.5) scenario is dynamically downscaled using the RegCM modelling system. The results show that under these two global warming level, total annual precipitation (PRCPTOT) will decrease in most regions. Consistently, the dry spell duration (CDD, consecutive dry days) is projected to increase. On the other hand, the frequency and the intensity of precipitation extremes (R50mm and RX1day) are projected with mix increase and decrease tendency. Seasonally, the contrast changing of PRCPTOT is projected. PRCPTOT tends to decrease during dry season (June-July-August, JJA) and tends to increase during wet season (December-January-February, DJF). The similar pattern is found for other indices e.g. CDD, R50mm and RX1day. In general, changes at 4°C global warming are statistically more significant and more intensified compared to that at 2°C. Our findings suggest the benefit of limiting global warming at a lower level.

Keywords : RCP8.5, dry spell duration, total annual precipitation, global warming.

1. Introduction

The Paris Agreement which was signed in December 2015 by world leaders bounds the participated countries to limit global warming well below 2.0°C and to pursue efforts to limit the warming even further to 1.5°C above the pre-industrial level [1]. A global warming level (GWL) of 2.0°C (above the pre-industrial level) has been considered as a threshold where society has to try to keep, in order to limit the unanticipated effects of anthropogenic climate change. This global ambition is partly pursued by all parties by reducing their national greenhouse emissions where the result should be reported regularly.

In line with the global response to the Paris Agreement, in the last recent years, scientists reported their studies on the potential consequences of different global warming levels at regional and local scale [2–6]. While there are many studies on the topic of global warming level in the literature, this is not a case for Indonesia, a country with highly vulnerable to the climate change impact. Moreover, for the research quantifying differences in precipitation extremes between the 1.5◦C and 2.0◦C global warming levels. Thus, it is an urgent need for more studies on the potential impacts of a 2 ◦C warmer world.

Hence, this paper aims to examine the potential key impacts of climate change on precipitation extremes over Indonesia under 2.0◦C global warming levels. As a comparison, a GWL of 4.0◦C is also investigated to give insight on the response of precipitation extremes if the warming continues with double magnitude. By doing so, we hope to provide a quantitative evaluation of what may occur when the global surface temperature warms reaching 2.0◦C and 4.0◦C which is relevant for policymakers.

2. Data and Methods

In this study, we analysed the downscaling product involving a single global climate model (GCM) and a single regional climate model (RCM) run by the centre for research and development, BMKG. This experiment is intended to participate in the CORDEX-SEA project (Coordinated Downscaling Experiment – Southeast Asia) [7]. In this collaborative experiment, BMKG contributes to the project by downscaling the CSIRO-Mk3.6.0 global model using a regional model of RegCM4. The control simulations driven by observed natural forcing and observed GHG (greenhouse gases) were run for period of 1950–2005 while climate projections forced by two RCPs (Representative Concentration Pathways [8] were run for period of 2006-2100. Even though RCP4.5 is also available, the RCP8.5 is selected in this paper because the projection under RCP4.5 never reaches global warming level of 4°C. Also, RCP8.5 is considered as a realistic business-as-usual scenario referring to the current trajectory of greenhouse gases emissions. The satellite-based precipitation product i.e. TRMM 4B42v7 [9] is also used to validate the model.

To characterize the precipitation extremes, four indices of ETCCDI (Expert Team for Climate Change Detection and Indices) [10] were selected i.e. PRCPTOT (annual/seasonal total of wet days), CDD (consecutive dry days), R50mm (frequency of days with rainfall at least 50mm) and RX1day (annual/seasonal maxima of daily rainfall). For more detailed characterization of possible changes, the analysis was also performed based on sub-country level (Figure 1) partly based on Kang et al., [11], particularly over Sumatra and Java. This sub-country boundary is also an extension of regionalization in Amsari et al., [12] who used 8 sub-countries in their El Nino impact analysis.

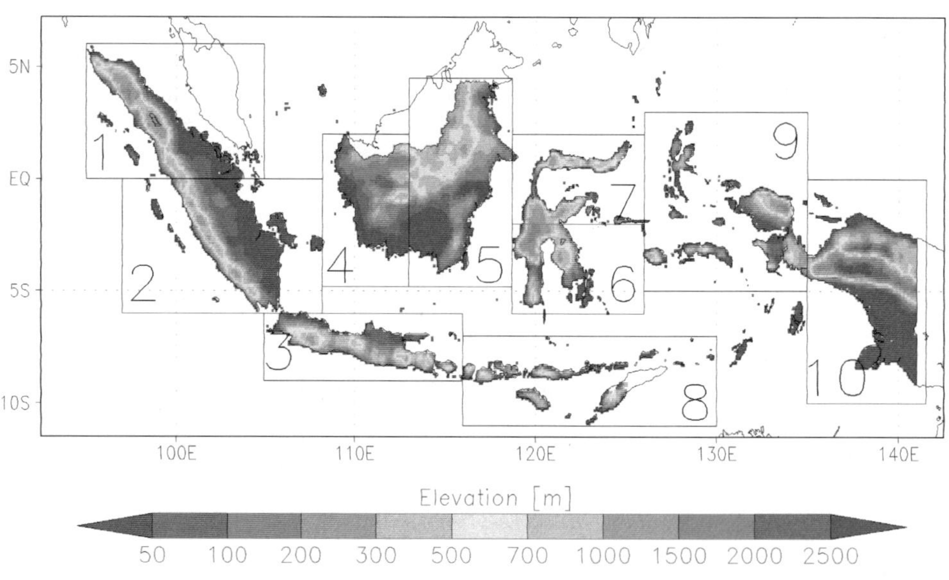

Figure 1. Indonesia and its topography. Boxes indicate the ten sub-countries level used for more detailed analysis. 1 for northern Sumatra, 2 for southern Sumatra, 3 for Java-Bali, 4 for western Kalimantan, 5 for eastern Kalimantan, 6 for southern Sulawesi, 7 for northern Sulawesi, 8 for Nusa Tenggara, 9 for Maluku and western Papua, 10 for eastern Papua.

The timing year of GWL (global warming level) of 2°C and 4°C was defined based on study by Karmalkar & Bradley [13]. They analysed 32 GCMs from CMIP5 and found that the ensemble mean of those GCMs is projected to cross the threshold of 2°C and 4°C by 2041 and 2083, respectively (Figure 2). However, it is reported also that there is large spread among ensemble member suggesting the uncertainty in the exact timing of thresholds. Hence, to reduce the uncertainty, the climate over 20-year period around the timing year is chosen for the analysis. To be more specific, for GWL 2°C (and 4°C), the changes are calculated by subtracting indices over the 2031–2050 (2073-2092) period with indices over reference period i.e. 1986-2005. Additionally, besides assessing the magnitude of possible changes, we also evaluate the significance of changes using Monte-Carlo test [14].

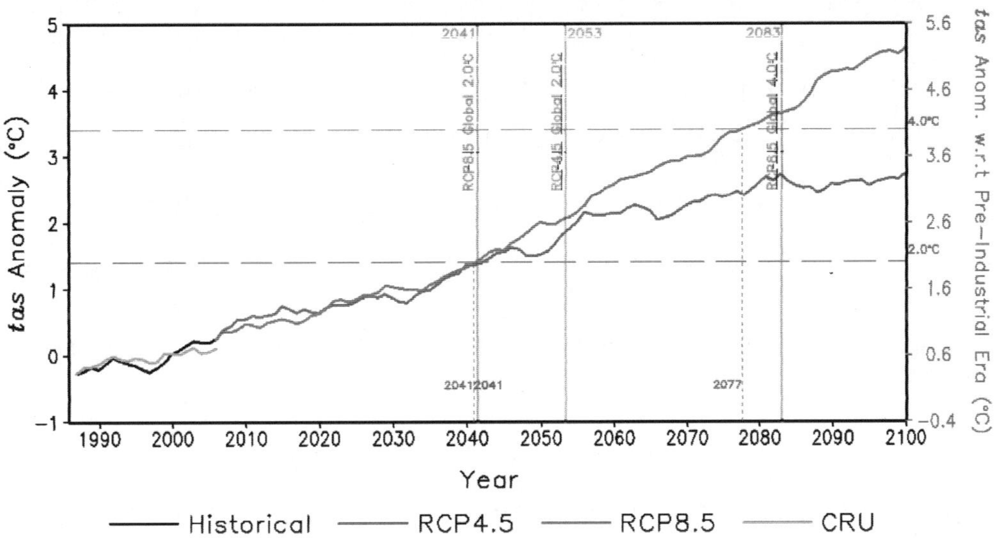

Figure 2. Annual mean of surface temperature anomalies in the CSIRO(RegCM) simulation relative to reference period of 1986–2005 (left Y-axis) and relative to Pre-industrial period (right Y-axis) under RCP4.5 (blue line) and RCP8.5 (red line) averaged over CORDEX-SEA domain. Also plotted in the figure is the timing of global warming level (GWL) based on GCMs analysed in Karmalkar & Bradley [13] (solid vertical line) and based on CSIRO(RegCM) (dashed vertical line).

3. Results and Discussion

3.1. Model Validation

Figure 3 presents the spatial pattern of four indices during 1998-2005, a common period for those two datasets i.e. TRMM and simulation data. The model simulates realistically the spatial pattern of observed PRCPTOT but with slightly higher magnitude indicating the wet bias of model. In TRMM data, PRCPTOT of more than 4000 mm is only found over mountainous area in Papua Island. On the other hand, it is observed over all islands in model data. The wet bias of model is also shown by CDD index even though its spatial feature in TRMM data is captured well in model data. The observed TRMM data estimates the largest CDD value (over Nusa Tenggara Islands) is more than 50 days while in the simulation data, it is around 30-50 days. Over Kalimantan and Sumatra, the difference in CDD between TRMM and model data is also clear where the shortest dry spell (lowest CDD) is around 5-10 days in TRMM and less than 5 days in model data. Similarly, the pattern of observed R50mm and RX1day are spatially replicate in model data with wet bias in magnitude. Overall, the model data is able to reproduce the spatial pattern of observed indices. However, the model tends to produce more rainfall (higher PRCPTOT), more rainy days (shorter CDD), more frequent heavy rainfall (higher R50mm) and more intensified extreme rainfall (higher RX1day). The wet bias in the output of RegCM downscaling of CORDEX-SEA is known related to the selection of parameterization [15].

IOP Conf. Series: Earth and Environmental Science **303** (2019) 012048 doi:10.1088/1755-1315/303/1/012048

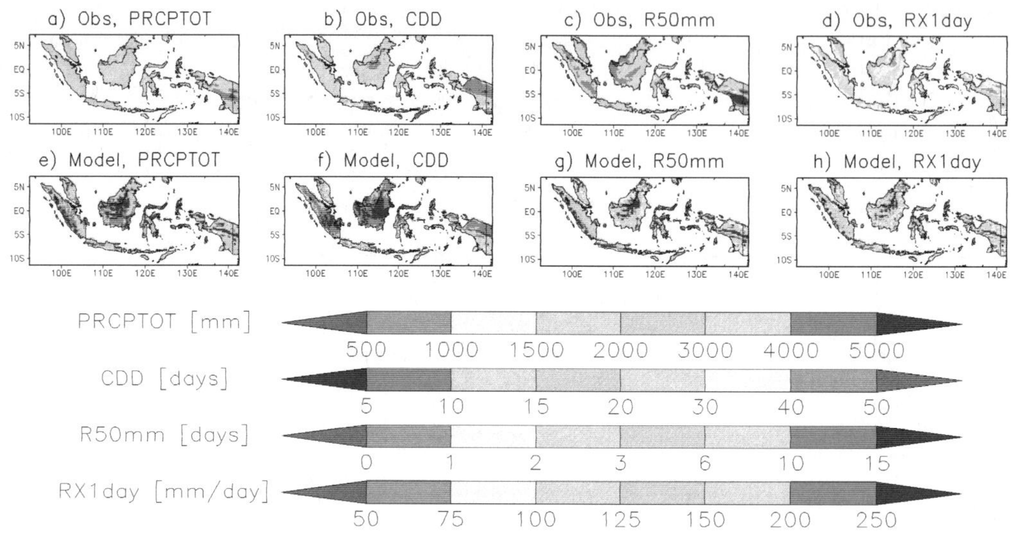

Figure 3. The spatial pattern of indices based on TRMM 4B32v7 (top row) and CSIRO(RegCM4) (bottom row) during 1998-2005.

3.2. Projection of Annual Indices

The future projection of PRCPTOT under GWL 2°C and 4°C is illustrated in Figure 4. Under 2°C global warming, PRCPTOT over Indonesia is projected to decrease with magnitude up to -20%, particularly over south of equator line. In contrast, for the area north of equator line, PRCPTOT is projected to increase including northern Sumatra, western and eastern Kalimantan, northern Sulawesi and northern Papua. The decrease of PRCPTOT is significant over South Kalimantan, East Java, Bali, Nusa Tenggara and part of Papua. The increase in PRCPTOT over northern Kalimantan and northern Papua is also significant.

Under a warmer environment, GWL 4°C, the decrease of PRCPTOT becomes more obvious covering wider area compared to that under GWL 2°C, in particular over Kalimantan and Sulawesi. We found that the magnitude of decrease is also larger in this warmer temperature. Over Java, Bali and Nusa Tenggara the PRCPTOT may decrease up to -50% while over Sumatra, Kalimantan, Sulawesi and Papua the decrease in PRCPTOT may up to -30%. On the other hand, the increase of this index over northern Sumatra and northern Papua is found to have a larger magnitude as well. Besides, the negative change is statistically significant over almost all areas including a large part of Sumatra, Kalimantan, Sulawesi and Papua where the decrease is not significant in the GWL 2°C. Overall, under GWL 2°C, PRCPTOT decrease up to -20% and the decrease becomes more intensified and more significant when global temperature warming reaches 4°C.

In figure 5, the projection of CDD when the global mean temperature reaches 2°C and 4°C above pre-industrial level is presented. Consistent with PRCPTOT, the CDD index shows a drying tendency both under GWL 2°C and 4°C. The CDD is projected to lengthen up to 20% over Java, Bali, Nusa Tenggara, southern Sumatra, Kalimantan, Sulawesi and southern Papua. Over certain spots such as South Sumatra and Central Kalimantan, even the increase of CDD reaches 30%. Also seen from the figure 5, the magnitude of increase of CDD under 4°C is higher than that under 2°C.

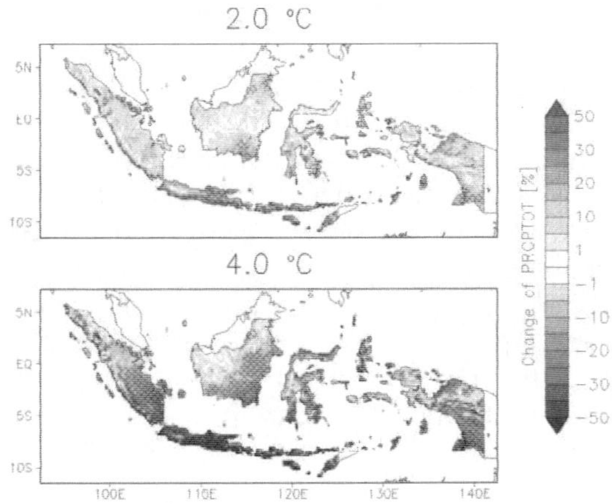

Figure 4. Projected changes of PRCPTOT under GWL 2°C (top) and 4°C (bottom). Areas with statistically significant changes are marked by hatching.

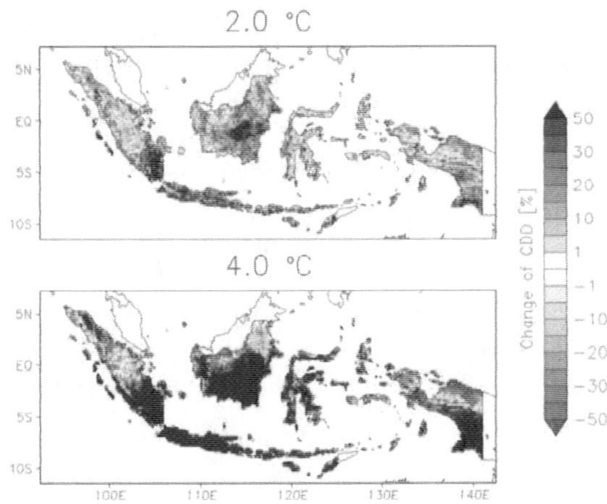

Figure 5. Projected changes of CDD under GWL 2°C (top) and 4°C (bottom). Areas with statistically significant changes are marked by hatching.

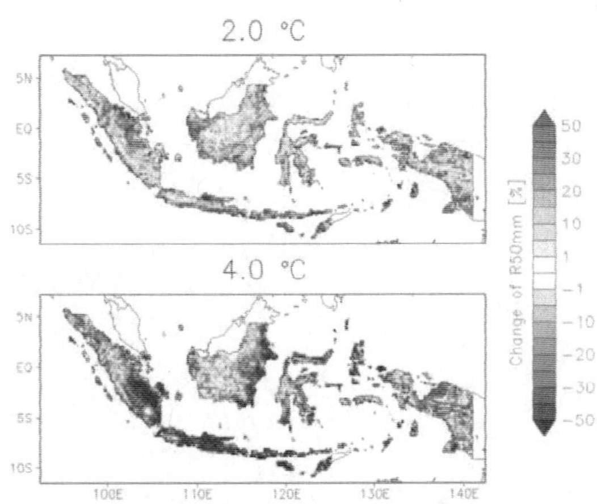

Figure 6. Projected changes of R50mm under GWL 2°C (top) and 4°C (bottom). Areas with statistically significant changes are marked by hatching.

Figure 6 displays the possible changes of R50mm under 2°C and 4°C global warming. Under 2°C global warming, R50mmm is projected to increase over northern Sumatra, western and eastern Kalimantan and over Papua. In contrast, it may decrease over Central part of Sumatra, Java, Bali-Nusa Tenggara and most of Sulawesi. Under 4°C global warming, while the other parts show similar characteristics with that under 2°C, southern Sumatra and eastern Kalimantan depict contrast pattern. In the southern Sumatra and eastern Kalimantan where R50mm are projected to increase under 2°C, it is projected to decrease under 4°C global warming with high magnitude (> 30%) and are statistically significant. In general, the magnitude of both increase (e.g. over northern Sumatra) and decrease (e.g. over Java and Bali-Nusa Tenggara) is higher compared to that under 2°C. Differently, the changes of RX1day is less spatially coherent (Figure not shown). Mixed positive and negative changes are projected over all islands, except Java, Bali and Nusa Tenggara Islands under 4°C global warming that may indicate a more independent characteristic of extreme intensity. Also, we note that the changes of RX1day are not significant.

3.3. Projection of Seasonal Indices

To get more detailed characteristics, the seasonal changes are also assessed. As shown in Figure 7, PRCPTOT is projected to decrease almost over whole country during JJA under those two GWLs with magnitude of 5-50%. Under 4°C global warming, the negative change is found to be significant everywhere. During DJF, the contrast tendency is found. The PRCPTOT is projected to increase during this season except over Java, Bali, Nusa Tenggara, Maluku and small part of Papua. Over Kalimantan, the increase in PRCPTOT is significant under both 2°C and 4°C global warming.

For CDD, the changes during JJA is consistent with that for PRCPTOT (Figure 7). The lengthening of CDD (an indication of drying tendency) is projected by 10-50% particularly over Sumatra, Kalimantan, Java, Bali and Nusa Tenggara under 2°C and including Sulawesi, Maluku and Papua under 4°C global warming. Moreover, the drying tendency is statistically significant for almost all areas under 4°C global warming. During DJF, the CDD shows a wetting tendency particularly over Sumatera, Kalimantan and Papua. Over southern part of the country (e.g. Java, Bali, Nusa Tenggara) the drying tendency is found, similar to that during JJA. The significant changes are seen only under 4°C global warming, in particular over Java (significant drying) and over Papua (significant wetting).

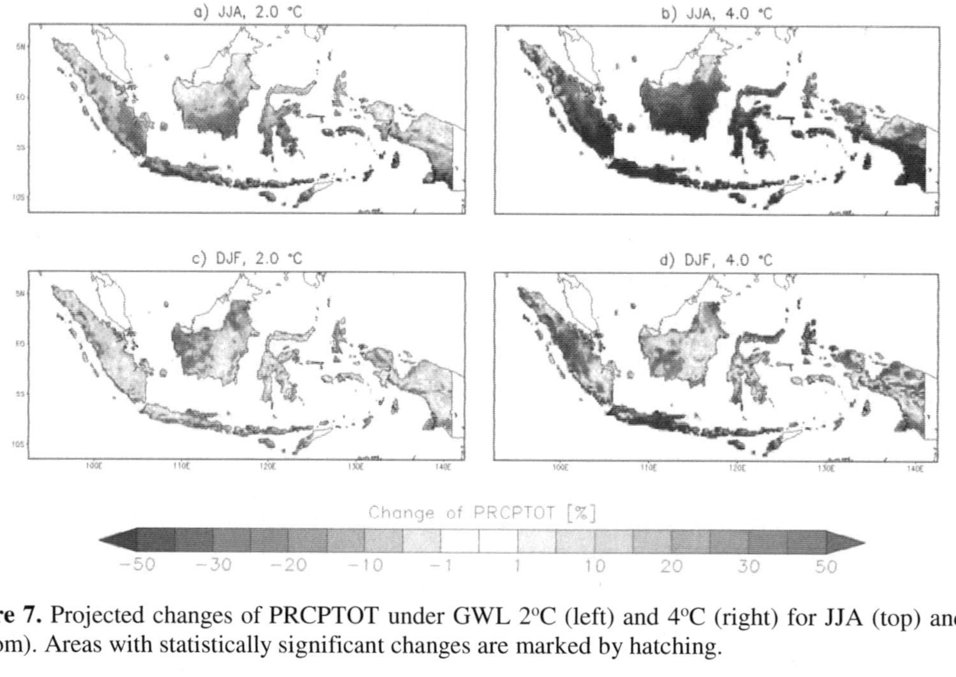

Figure 7. Projected changes of PRCPTOT under GWL 2°C (left) and 4°C (right) for JJA (top) and DJF (bottom). Areas with statistically significant changes are marked by hatching.

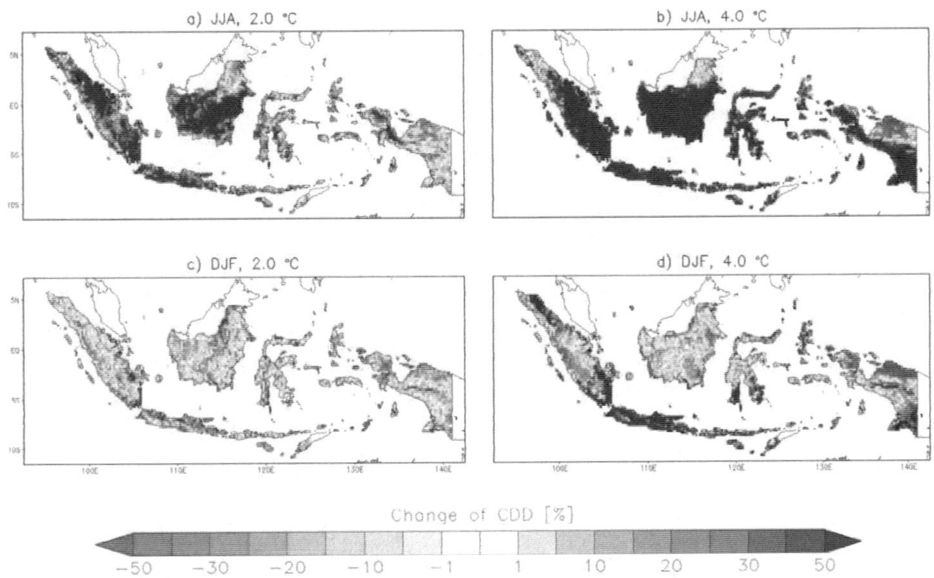

Figure 8. Projected changes of CDD under GWL 2°C (left) and 4°C (right) for JJA (top) and DJF (bottom). Areas with statistically significant changes are marked by hatching.

A similar drying tendency is also projected for R50mm and RX1day during JJA both under those two GWLs. For R50mm, the changes are generally less spatially coherent (Figure not shown) while for RX1day, the drying tendency is clearly observed over most of the country. The changes of R50mm and RX1day are generally not significant under GWL 2°C and 4°C except for RX1day where the drying tendency is significant and in high magnitude under 4°C GWL (Figure not shown). During DJF, those two indices show a wetting tendency over Sumatera, Kalimantan and small part of Papua and a drying tendency over eastern Java, Bali, Nusa Tenggara and Maluku. The wetting tendency over Sumatera and drying tendency over Java during this wet season are significant under 4°C global warming both for those two indices.

3.4. Sub-Country Analysis

Figure 9 presents the projected changes of all studied indices for each sub-country. In general, the changes of indices under 4°C global warming is in similar direction but with higher magnitude compared to that under 2°C global warming. The changes of all indices under 2°C global warming are less than 30 % for all sub-country while those under 4°C global warming may reach more than 100% (e.g. CDD over Java-Bali, sub-country 3). From this sub-country analysis, some remarkable findings are found. First, we note that Java-Bali (sub-country 3), northern Sulawesi (sub-country 7) and Nusa Tenggara (sub-country 8) show a consistent drying tendency in all indices. Over those regions, the decrease in PRCPTOT (decrease in annual rainfall) is followed by increase in CDD (longer dry spell), decrease in R50mm (lower frequency of very heavy rainfall) and decrease in RX1day (lower magnitude of annual maxima of daily rainfall). This consistent drying tendency is in agreement with seasonal analysis showing a negative changes both during dry season (JJA) and wet season (DJF). Second, there are also regions (e.g. sub-country 4, western Kalimantan) where, under 2°C global warming, the increase in CDD is accompanied by increase in RX1day. CDD, in general, represents a condition of dry season while RX1day represents a condition of wet season. Thus, this combination of changes indicates a drier dry season followed by a wetter wet season.

Overall, we found that the country may experience drying tendency in particular over southern part when global warming reaches 2°C, relative to pre-industrial level. This drying tendency may occur with higher magnitude and cover wider area when the global surface temperature warms with double magnitude. With different emphasize, study by Kang et al., [11] found a similar drying tendency in the future climate of western Maritime Continent (WMC). They proposed that the main mechanism behind the simulated decrease in rainfall over the WMC is the enhancement and expansion of sinking motion of Hadley circulation under global climate change, particularly in RCP8.5 scenario. Their explanation is in agreement with that stated by Fu [16].

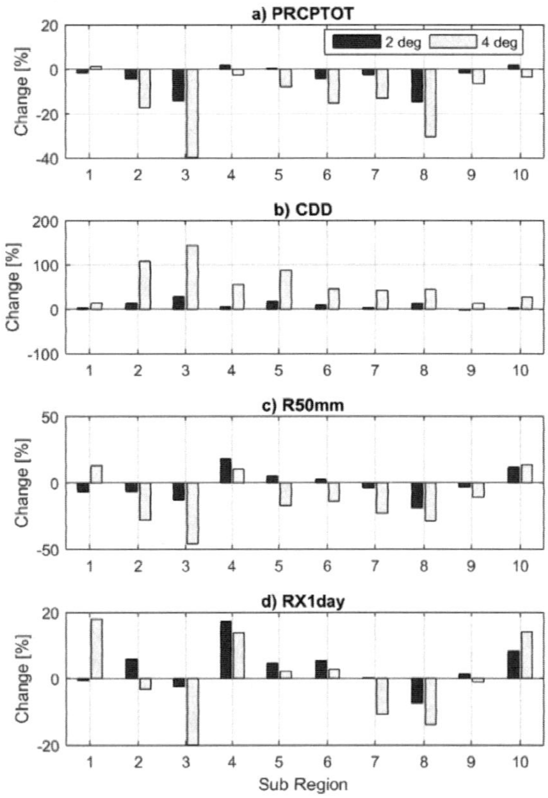

Figure 9. Projected changes (in %) in annual indices under GWL 2°C (blue) and 4°C (yellow) relative to the reference period (1986–2005) for RCP8.5. The changes are shown in term of areal mean for each of the 10 sub-country (Figure 1).

4. Conclusion

The possible changes of annual and seasonal precipitation extremes over Indonesia are assessed in this study for the case of global mean temperature reaching 2°C and 4°C above pre-industrial level under RCP8.5 as the unmitigated climate change scenario. The assessment is based on the downscaling product involving the CSIRO-Mk3.6.0 as GCM (global climate model) and RegCM as the regional modelling system. We find that the decrease in PRCPTOT and the increase in CDD over most of the country is the most notable changes under those two global warming levels. On the other, mixed negative and positive changes are found for R50mm and RX1day. Seasonally, the southern part of the country may experience consistent drying tendency both during dry and wet season. These findings suggest that there will be some implications for water resources and food security if we are not able to mitigate well the global warming so that it continues above 2°C.

References

[1] UNFCCC Conference of the Parties (COP) 2015 *Paris Climate Change Conference-November 2015, COP 21* **21932**

[2] Osima S E, Indasi V, Zaroug M A H, Edris H, Gudoshava M, Misiani H, Nimusiima A, Anyah R, Otieno G, Ogwang B, Jain S, Kondowe A, Mwangi E, Lennard C, Nikulin G and Dosio A 2018 *Environ. Res. Lett.* **13** 065004

[3] Klutse N A B, Ajayi V, Gbobaniyi E O, Egbebiyi T S, Kouadio K, Nkrumah F, Quagraine K A, Olusegun C, Diasso U J, Abiodun B J, Lawal K A A, Nikulin G, Lennard C and Dosio A 2018 *Environ. Res. Lett.* **13** 055013

[4] Kumi N and Abiodun B J 2018 *Environ. Res. Lett.* **13** 89502

[5] Mba W P, Longandjo G-N, Moufouma-Okia W, Bell J P, James R, Vondou D A D, Haensler A, Nguemo T C F, Guenang G M, Tchotchou A L D, Kamsu-Tamo P H, Takong R R, Nikulin G, Lennard C and Dosio A 2018 *Environ. Res. Lett.* **13** 055011

[6] Vautard R, Gobiet A, Sobolowski S, Kjellström E, Stegehuis A, Watkiss P, Mendlik T, Landgren O, Nikulin G, Teichmann C and Jacob D 2014 *Environ. Res. Lett.* **9** 034006

[7] Kang H S, Tangang F and Krishnan R 2016 *APJAS* **52** 77

[8] Moss R H R H, Edmonds J A J A, Hibbard K A K A, Manning M R, Rose S K S K, van Vuuren D P D P, Carter T R T R, Emori S, Kainuma M, Kram T, Meehl G A G A, Mitchell J F B J F B, Nakice-novic N, Riahi K, Smith S J S J, Stouffer R J R J, Thomson A M A M, Weyant J P J P, Wilbanks T J T J and Nakicenovic N 2010 *Nature* **463** 747–56

[9] Huffman G J and Bolvin D T 2015 *TRMM and Other Data Precipitation Data Set Documentation* p1-44

[10] Zhang X, Alexander L, Hegerl G C, Jones P, Tank A K, Peterson T C, Trewin B and Zwiers F W 2011 *Wires: Climate Change* **2** 851–70

[11] Kang S, Im E S and Eltahir E A B 2018 *Clim. Dyn.* 1–18

[12] Setiawan A M, Lee W S and Rhee J 2017 *Int J Climatol* **37** 4700–19

[13] Karmalkar A V. and Bradley R S 2017 *PLoS ONE* **12** 1–17

[14] Dufour J M and Khalaf L 2007 *Companion to Theoretical Econometrics* **1** 494–519

[15] Chung J X, Juneng L, Tangang F and Jamaluddin A F 2018 *Int J Climatol* **38** 794–810

[16] Fu R 2015 *PNAS* **112** 201503231

Sensitivity of Planetary Boundary Layer Scheme in WRF-Chem Model for Predicting PM_{10} Concentration (Case study: Jakarta)

M Musthafa[1], A Turyanti[1] and D E Nuryanto[1,2]

[1] Department of Geophysics and Meteorology, FMIPA Bogor Agricultural University, Kampus IPB Darmaga, Bogor 16680, West Java, Indonesia.
[2] Research and Development Center, The Indonesian Agency for Meteorology Climatology and Geophysics, Jl Angkasa I No 2 Kemayoran Jakarta Pusat 10720, Indonesia.

danang.eko@bmkg.go.id

Abstract. Weather Forecasting-Chemistry (WRF-Chem) is an air pollution model that combines meteorological factors and atmospheric chemistry simultaneously (online coupled). Many schemes selected on WRF-Chem according to boundary layer conditions of the study area. The objectives of this study were to evaluate WRF-Chem model and analyze the sensitivity of Yonsei University (YSU) scheme and Mellor Yamada Janjic (MYJ) scheme as physical boundary layer parameters, especially for Jakarta region. Particulate matter 10 micrometers or less in diameter (PM10) selected as the air pollutant parameter. The model outputs compared to observational data in five locations in Jakarta. The regression analysis between air temperature data shows that the values of R2 are 0.5 - 0.6 in February and 0.5 - 0.9 in August. Another result, the wind velocity from WRF-Chem overestimates but the hourly averages have a similar diurnal pattern with the observation. The PM10 concentration model output is underestimated which is caused by local emissions that are not accommodated in the model. The sensitivity analysis shows a higher sensitivity index for MYJ than YSU scheme, especially in August, so the model outcome of the MYJ scheme has a good response to the observations in Jakarta. It is seen that the extreme concentration of PM10 produced through the WRF-Chem with both schemes in February and August have the same time in morning. The model output of WRF-Chem through both schemes also demonstrate the minimum concentration of PM10 that follows during afternoon. The Planetary Boundary Layer (PBL) height analysis shows that in morning until noon, the PBL height with MYJ scheme is lower than YSU. It is different for other hours, that from the afternoon through the night until next morning, the height of PBL with MYJ scheme is thicker than YSU. At the point when the PBL is high, the pollutants assorted in a more volume so the estimation of the pollutant concentration estimated on the surface will be reduced.

Keywords : planetary boundary layer, WRF, PM10, wind

1. Introduction

Air pollution and its impact have become a concern in several countries including Indonesia. In general, air pollution occurs in large cities due to high transportation and industrial activities. However, pollutants transported to other areas in the vicinity, thus affecting the air quality of the area.

Air quality data becomes important for use in indicating public health around the measurement site. However, continuous and even measurable air quality measurements throughout the region require considerable cost. Limitations of air quality data become one of the problems that often encountered in conducting research. Modelling on air pollution can be an alternative to predict air quality and pollutant distribution in different regions. Air pollution models have built from simple to complex. Due to the large influence of meteorological factors on the dispersion of pollutants, a combined model involving meteorological factors with atmospheric chemistry expected to be closer to the actual conditions on the surface. The current model of air pollution sufficiently developed to model numerical weather predictions combined with atmospheric chemistry models, one of which is Weather Research Forecasting-Chemistry (WRF-Chem) model.

WRF-Chem is a model that combines atmospheric meteorological and chemical factors simultaneously (online coupled). Based on the research that has been done, WRF-Chem has been widely used to estimate the concentration of pollutants in sub-tropical regions [1,2,3,4]. While research on WRF-Chem for the tropics is still little done. The WRF-Chem model uses several parameterization schemes that can selected according to the location conditions that modelled or analyzed for the quality of the air. Selection of the parameterization scheme will affect the model output, so the choice of the parameterization scheme is essential to obtain the closest scheme in a model. The use of parameterization schemes in sub-tropical regions will be different from tropical regions. Therefore, it is necessary to understand the most appropriate scheme for use in a particular area, and this can be done through the comparison of the outputs of different schemes, so the sensitivity can be known.

Misenis and Zhang [5] in Texas and Cuchiara *et al.* [6] in Texas and Zhong *et al.* [3] in East Asia have done research on the sensitivity of the WRF-Chem model among others. Because there is not many uses the WRF-Chem model in Indonesia, therefore this research is the first step to know the parameter scheme that is good enough for the Indonesia region by taking case study of Jakarta and its surroundings. Jakarta is chosen to be a study area because it has a continuous ambient air quality monitoring station (SPKU) that makes it easy to validate or compare the outcome of the model.

In this research, the parameterization scheme analyzed by the sensitivity is the Planetary Boundary Layer (PBL) physical parameter with Yonsei University (YSU) and Mellor Yamada Janjic (MYJ) scheme. Types of pollutants to study are the Particulate matter 10 micrometers or less in diameter (PM_{10}). This study aims to evaluate the use of WRF-Chem model as well as to test the sensitivity of the physical parameterization scheme of PBL for PM_{10} in Jakarta area.

2. Data and Methods

The following is the data used in this study. The model initial and boundary condition are driven from the Global Data Forecasting System (GFS) with a spatial resolution of 0.5° x 0.5°. The global EDGAR-HTAP emissions data for 2010 are used for initial pollutants of WRF-Chem. The ambient air quality data is obtained from SPKU observation (Fig. 1) in Jakarta area 2010 (DKI 1), 2011 (DKI 2) and 2013 (DKI 3, DKI 4 and DKI 5).

Simulations were conducted on February 9, 2010, and August 7, 2010, taken at 00:00 UTC. The data collection on the land represents the highest rainfall in the rainy season and the lowest rainfall in the dry season. The grid used during the simulation of 4 km x 4 km with longitude and latitude points is located at coordinates 105.2 ° east to 107.8 ° E -5.4 ° LS to -7.6 ° LS. The study uses only one domain with focus Jakarta area (Fig. 1). The simulation was for 7 days and ends at 12.00 UTC.

Several studies have used the YSU and MYJ schemes with the same pollutant or region. Kumar *et al.* [7], Ge *et al.* [8], Nuryanto and Nuraini [9], Heriyanto and Nuraini [10] have used the YSU scheme. While the MYJ scheme has been used by Turyanti [11], Beck *et al.* [12], Amnuaylojaroen *et al.* [13]. The difference between the YSU and MYJ schemes is not only visible in terms of their turbulence flux, but also in the surface clay parameter (sf_sfclay_physics) for the YSU scheme is one (1), while the MYJ scheme is two (2). The choice of the schemes used in the research shows in Table 1.

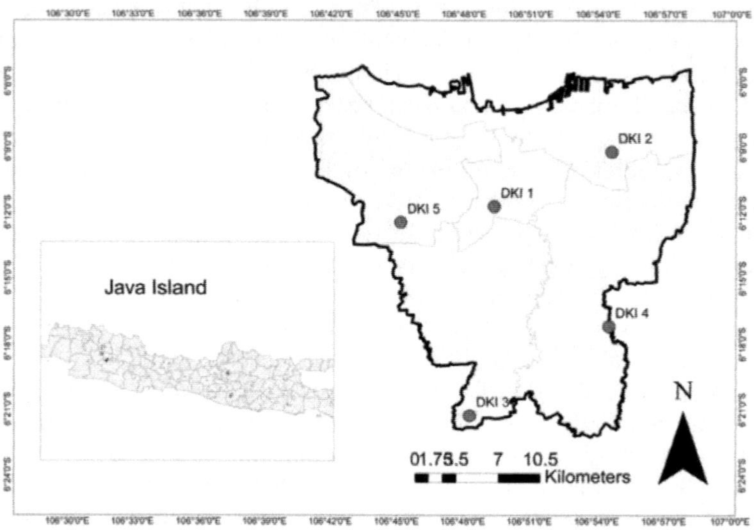

Figure 1. The study area.

Table 1. The schemes used in the study.

Parameters	Scheme	Remarks
Microphysics	Lin et al.	The single moment scheme includes several modifications such as saturation adjustment and is associated with minor crystal ice sedimentation [1]
Cumulus	Grell 3D	An increase of the Grell D scheme that can be used at high resolution [15]
PBL	YSU	Nonlocal schemes with explicit entertainment layers and parabolic K-profiles in unstable mix layers [16]
	MYJ	Local scheme with vertical mixing and 1.5 TKE prognostic [16]
Long wave radiation	RRTMG	A scheme that can handle the direct and indirect effects of aerosols in WRF simulations [17]
Short wave radiation	RRTMG	A scheme that can handle the direct and indirect effects of aerosols in WRF simulations [17]
Chem option	RADM2 Chemistry and GOCART aerosols	There are 158 chemical reactions among 26 species [17]

The sensitivity test of the PBL schemes of the model results refers to Pianosi et al. [14]. Sensitivity analysis performed using the Standardized Regression Coefficients (SRC) method [14]. SRC is one of the easiest methods used in analyzing sensitivity by using linear regression [14]. The results obtained from SRC are values between -1 and 1. Sensitivity values that are close to 1 indicate that the model has a good response with observation, whereas the sensitivity value approaching -1 indicates that the model does not respond well to observation. Sensitivity is calculated using equation (1).

$$Si = b_i \frac{SD(x_i)}{SD(y)} \quad (1)$$

Si = sensitivity index
bi = slope
SD(xi) = standard deviation of model i
SD(y) = standard deviation of observation

Sensitivity analysis of parameter scheme were tested based on surface layer thickness (level 1). The thickness of this layer was calculated from the value of ph and phb output of the model, which is still in the form of geopotential perturbation data. The *pbl* value obtained through equation (2).

$$pbl = \frac{ph+phb}{gravity} \quad (2)$$

pbl = layer thickness on each of the outer layers of the model in this case the surface layer (m)
ph = *Perturbation geopotential* (m^2/s^2)
phb = *Base-state geopotential* (m^2/s^2)
gravity = 9.8 (m/ s^2)

The stages of research can be seen in Figure 2.

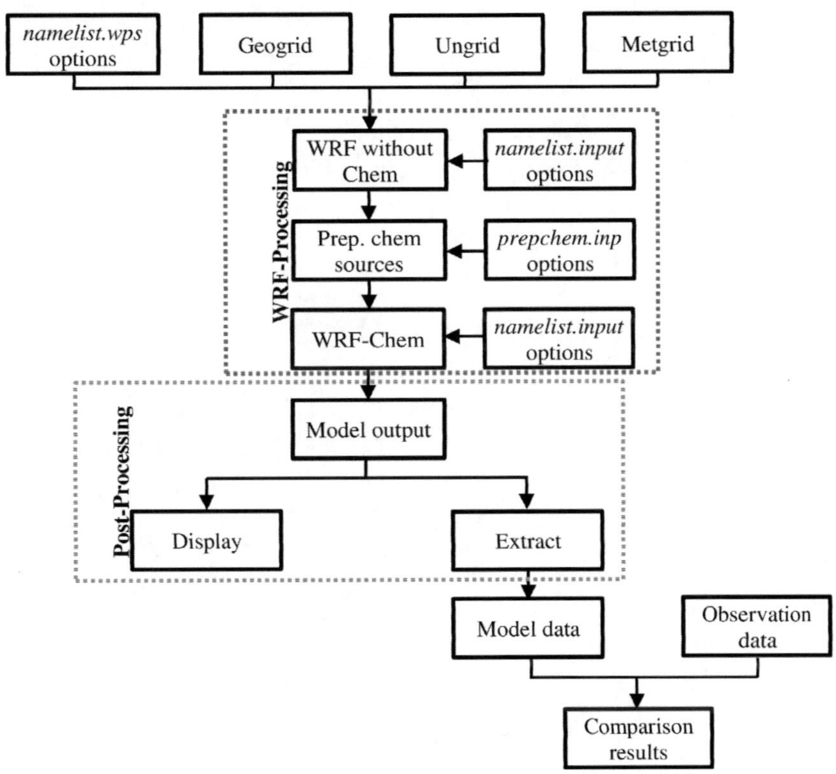

Figure 2. Process flow diagram.

3. Results and discussions

3.1. Spatial Distribution Pattern Concentration PM_{10} Model

Model simulations conducted in February and August, where February represents the rainy season and August represents the dry season. In February, the maximum concentration of PM_{10} using the YSU scheme at 07:00 LT occurred in Pandeglang District at 9,599 µg / m^3. In contrast to night time at 7 pm, maximum concentration of PM_{10} is in Lebak Regency with concentration value of 2,991 µg / m^3. PM_{10} concentration using YSU scheme in August at 07:00 LT, result maximum concentration around Tangerang 17,023 µg / m^3. Serang Regency is also a region with high PM_{10} concentration. At 19:00 LT, maximum concentration occurred in Bogor and Tangerang areas with a maximum concentration of 3,819 µg / m^3.

Different patterns of PM_{10} concentration distribution in the morning and evening also shown in Fig. 3. It might be caused by the influence of atmospheric stability. In the morning until noon, atmospheric conditions tend to be unstable resulting in the removal of air masses and pollutants located on the surface. While at night, atmospheric conditions tend to be stable and there is a decrease in air mass along with pollutants to the surface so that pollutants (PM_{10}) at night are more diffuse on the surface than in the morning and afternoon.

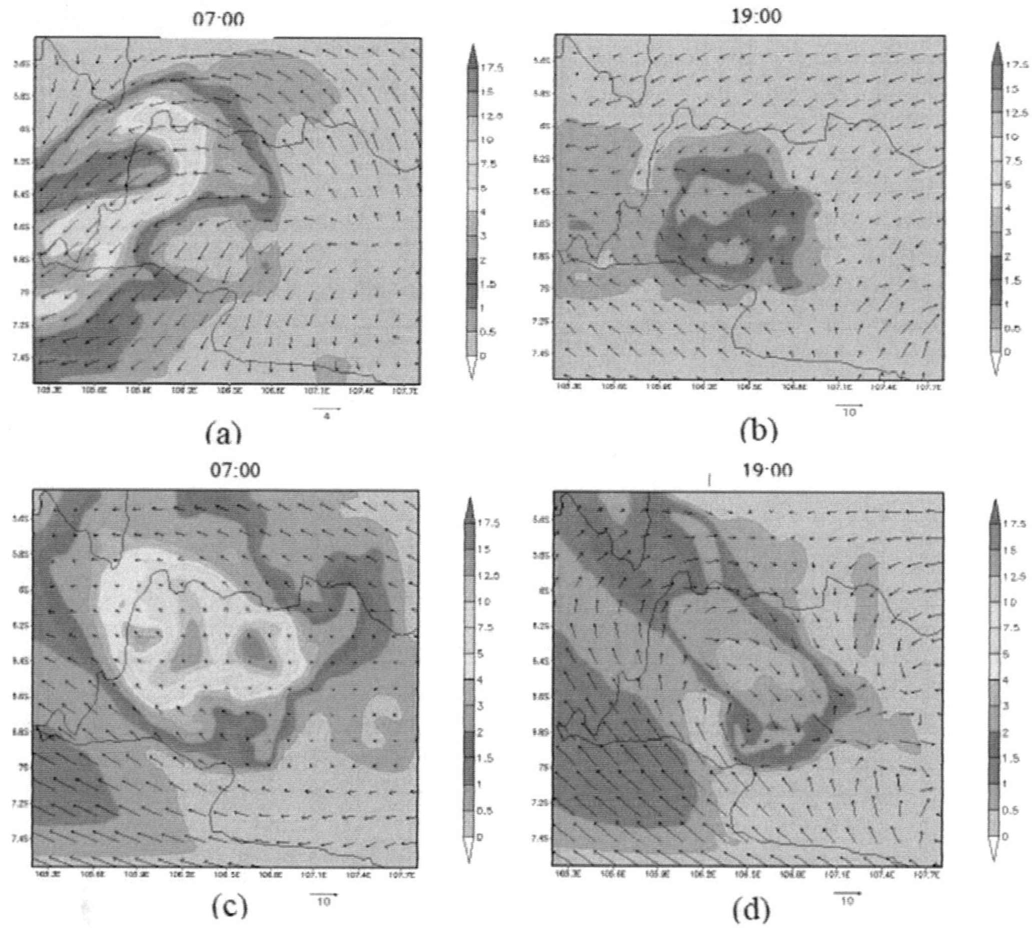

Figure 3. Concentration of PM10 on an average hourly basis using the YSU scheme in February (a) at 07:00 LT (b) at 19:00 LT (b) and August (c) at 07:00 LT (d) at 19:00 LT.

3.2. Comparison of PM_{10} Spread Concentration between YSU and MYJ Schemes

The concentration of PM_{10} in February in the Jakarta area is relatively low compared to August. The PM_{10} concentration values generated from the two different moons also look different; it can be due in February to be a wet season so pollutants will be carried away by rain. The concentration of PM_{10} using the MYJ scheme shows more concentration higher than the YSU scheme, for the same location (Fig. 4). The YSU scheme with a non-local approach causes mixing to involve a wider scale of both vertical and horizontal than MYJ.

The maximum concentration of PM_{10} in February is not in the study area but spread to the Tangerang region while the maximum value is PM_{10}. In August, using the maximum concentration the YSU scheme occurred around Tangerang and Depok areas, whereas for the MYJ scheme the maximum concentration of PM_{10} occurred in Jakarta area.

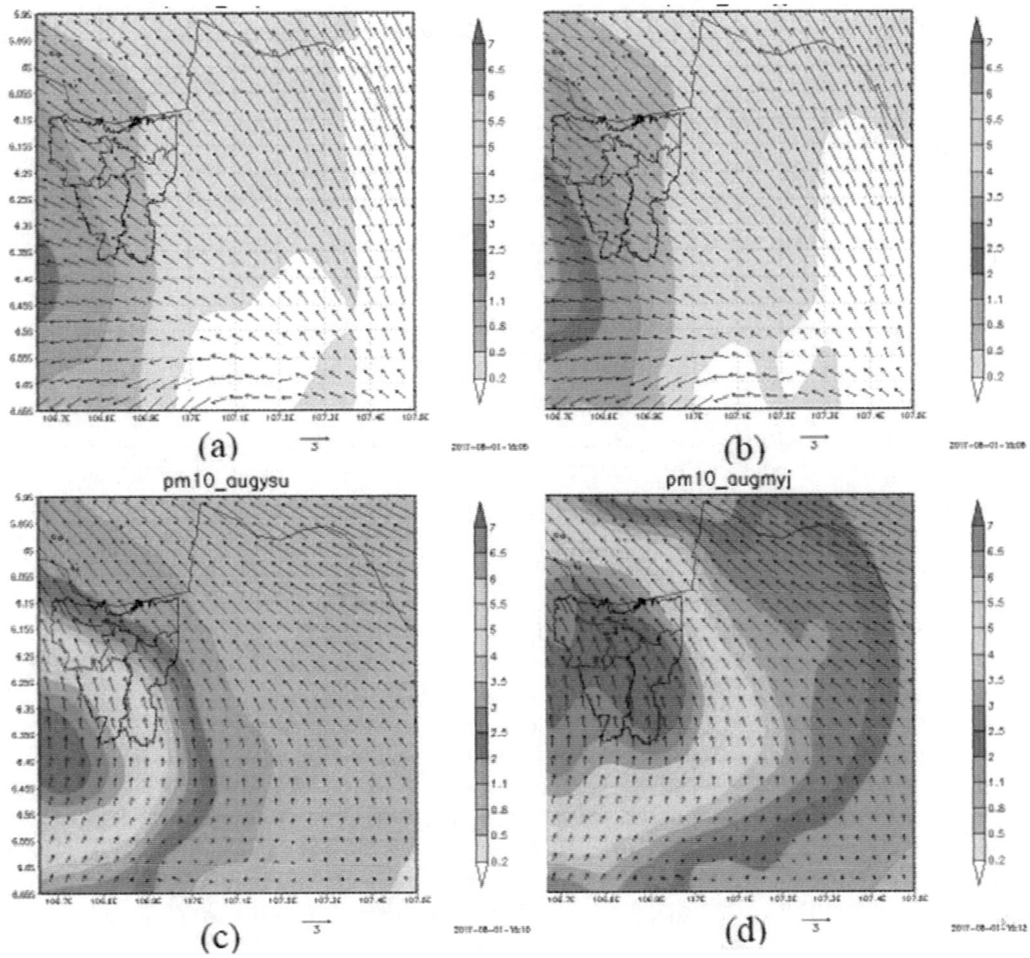

Figure 4. The February PM10 concentration distribution uses (a) the YSU scheme and (b) the MYJ scheme and in August it uses (c) the YSU scheme and (d) the MYJ scheme in Jakarta and surrounding areas

Figure 5 shows the average PM_{10} concentration fluctuations per hour in DKI1. The model simulation results show that the average PM_{10} concentration value of the MYJ scheme is greater than the YSU scheme, except at night there is a tendency of MYJ concentration value near or below YSU. This is expected from the influence of wind speed. At night, the wind speed of the MYJ scheme is higher than the YSU scheme, so the PM_{10} concentration will be lower. The difference in means values of PM_{10} concentration using the YSU and MYJ schemes obtained through the difference of the two schemes (YSU-MYJ). In August the difference from YSU with MYJ is negative, the PM_{10} concentration using the MYJ scheme is always greater than the YSU scheme. In contrast to February, the difference between the two schemes is a positive value indicating that the average PM_{10} concentration using the YSU scheme is larger than the MYJ scheme.

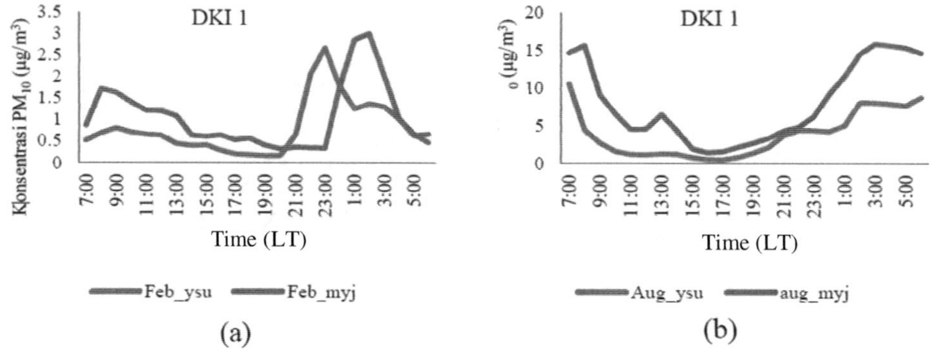

Figure 5. Hourly mean of PM_{10} concentration during (a) February and (b) August.

3.3. Comparison for Meteorological Factor of Model and Observations

Based on the results of data processing, both using the YSU and MYJ schemes in different months (February and August), showed that the largest wind speed occurred in the afternoon at around 15:00 LT (YSU) and at 16:00 - 17:00 LT (MYJ) which can be seen in Fig. 6. Diurnal fluctuations in observation data are closer to the output of the YSU scheme. Observation data in DKI 1 around Bundaran Hotel Indonesia showed that the maximum wind speed based on an hourly average occurs at around 15:00 LT. This could be caused by solar radiation that reaches a maximum during the day and there is a considerable pressure difference so that wind speed increases [18].

Model wind speed is higher than observation data. Similar results also occur in the study of Li *et al.* [19] that the output wind speed of the model with the scheme (YSU, MYJ) is greater than the observation results. This can be caused by the height of the model layer is different from the height of the observation that the model is at an altitude of 10 m while the observation is about 2 m, this is in accordance with the research of Turyanti [11] which uses the MYJ scheme. Surface wind speed in the February period has a lower range than August, according to observation data. The opposite is true for the wind speed at the upper layer (10 m) which has a larger range in February in the both scheme options. The wind speed of the YSU and MYJ schemes began to look different in size towards the afternoon until the middle of the night, that is, the value of the wind speed of the MYJ scheme was greater than YSU. This shows the influence of different approaches to turbulence kinetic energy used in both PBL schemes.

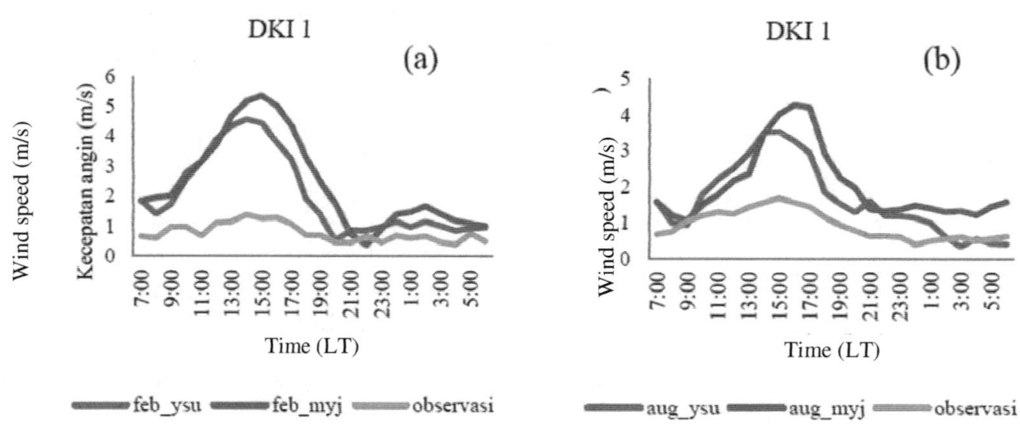

Figure 6. Hourly mean of wind speed during (a) February and (b) August.

IOP Conf. Series: Earth and Environmental Science **303** (2019) 012049 doi:10.1088/1755-1315/303/1/012049

Based on the WRF-Chem model simulation results for air temperature values at five points (5 SPKU) tend to approach the observed value. The results of the regression analysis show that the coefficient of determination (R^2) of the outputs of the two schemes has a value of 0.55-0.92 (Table 2). Model simulations in February resulted in a coefficient of determination between the models with observations ranging from 0.5 to 0.6. The output of the model in August for temperature has a coefficient of determination that varies from the two schemes, namely 0.5 to 0.9. Variations of the determination coefficient caused by the use of observation data used. Regarding the limitations of observation data, temperature data in 2010 only existed in Jakarta, while in 2011 for DKI 2, DKI 3, DKI 4 and 2013 for DKI 5.

Table 2. The value of coefficient of determination between the models with observations

Areas	Coefficient of determination			
	February		August	
	YSU	MYJ	YSU	MYJ
DKI1	0.5468	0.5818	0.7166	0.7406
DKI2	0.5934	0.6293	0.8962	0.9155
DKI3	0.5821	0.5695	0.8716	0.8894
DKI4	0.5895	0.6010	0.9111	0.9083
DKI5	0.6996	0.8996	0.6625	0.5687

3.4. PM_{10} Concentration of Model and Observation

The PM_{10} concentration of the WRF-Chem model shows underestimated values using both the YSU scheme and the MYJ scheme for both months of modeling simulation (Fig. 7). The value issued by the model is very different from observations that is caused by local emissions that are not accommodated in the model. There are differences in PM_{10} peak points from February and August. In February, the maximum PM_{10} value using the YSU scheme occurs at night, which is around 23:00 LT, while the MYJ scheme shows the maximum PM_{10} value occurs at around 00:00 LT. Observation data in February shows the maximum value of PM_{10} occurs at night which is around 00:00 LT.

The maximum PM_{10} value on the model using the YSU scheme in August occurs at 07:00 LT while the observation results show the maximum PM_{10} value occurs in the morning around 10:00 LT. This is happening because of the observation data where an effect of local emissions from motor vehicles that is accumulated in. Likewise, with the MYJ scheme, the maximum concentration of PM_{10} from the model output occurs in the morning around 8:00 LT. So that it has been seen that, the maximum concentration of PM_{10} generated through the models of the two schemes has the same time. The model output also shows the minimum concentration of PM_{10} that occurs during the daytime, using both the YSU scheme and the MYJ scheme. This is caused by the influence of vertical diffusion that change every hour during the day [19]. Vertical diffusion is affected by the presence of solar radiation that bounces to the surface, so that solar radiation is one of the factors of vertical diffusion.

The results of the correlation analysis of the output model and observation of pollutant concentration showed a small value. Therefore, a correlation analysis performed on the average hourly concentration values is shown in Table 3. Significant tests on the correlation value between concentrations the average model output and observation pollutants shown in Table 3 with a t table value of 1.711. The results of the YSU scheme show that DKI 1, DKI 3, DKI 4 and DKI 5 have insignificant correlation coefficients, while for the MYJ scheme; only DKI 4 has insignificant correlation. Overall, the correlation coefficient value in DKI 4 is not significant, so the possibility of the model has not been able to predict PM_{10} concentration in the region.

IOP Conf. Series: Earth and Environmental Science **303** (2019) 012049 doi:10.1088/1755-1315/303/1/012049

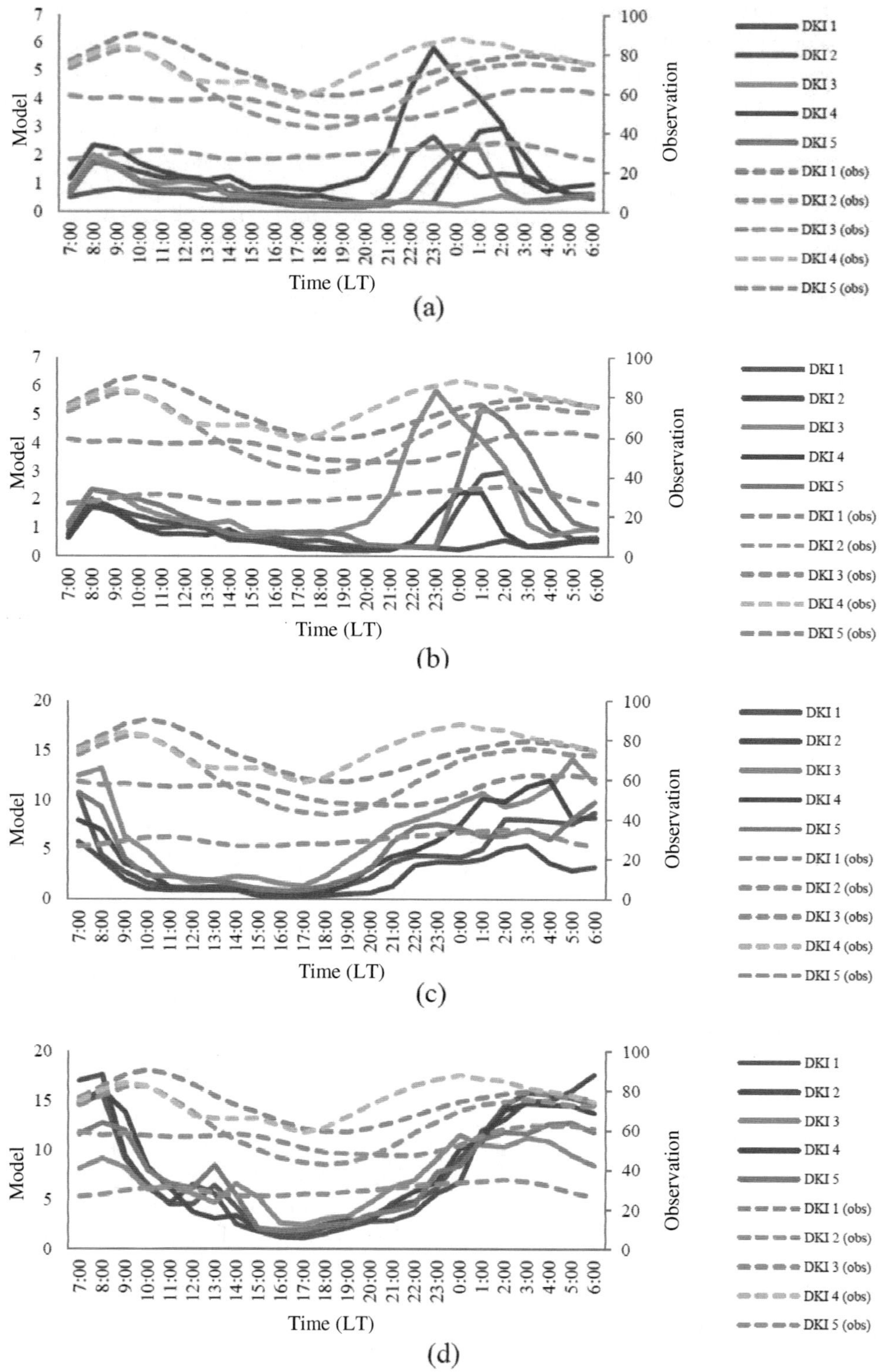

Figure 7. Diurnal patterns of PM_{10} ($\mu g / m^3$) from observation and WRF-Chem model output in February using (a) YSU and (b) MYJ Schemes and in August using (c) YSU and (d) MYJ Schemes.

Table 3. The value of coefficient of correlations between the models with observations

Areas	Coefficient of determination			
	February	August		
	YSU	MYJ	YSU	MYJ
DKI 1	0.21	0.58	0.26	0.53*
DKI 2	0.63*	0.54*	0.63*	0.76*
DKI 3	0.42*	0.63*	0.30	0.60*
DKI 4	-0.32	0.02	-0.27	0.14
DKI 5	0.15	0.43*	0.47*	0.86*

* Significance values

3.5. The Sensitivity of the Parameterization Scheme to Observation

The model output of each scheme or parameters are different. The difference in the output of the model shows the response of the parameter changes (sensitivity). Sensitivity analysis can determine the output variation of the model to changes in one or more parameters [20]. Figure 8 shows the sensitivity value in DKI 1 - DKI 5. Based on the results of the sensitivity analysis, in August the MYJ scheme consistently had a higher sensitivity value than the YSU scheme. In contrast to August, the YSU scheme has a higher sensitivity than the MYJ scheme for DKI 2 and DKI 3. In February, it is suspected by the influence of rain and the regional characteristics of DKI 2 and DKI 3.

3.6. Comparison of PBL Height at Surface Using the YSU and MYJ Schemes

PBL height is one of the most important variables for air quality modelling. PBL height will have an impact on the intensity of vertical pollutants mixing so that it will affect the characteristics of pollutant dispersal [21]. The WRF-Chem model simulation results in a higher level of PBL with the MYJ scheme larger than the YSU scheme although the difference between the two schemes is not too large in both February and August. This result is the same as the research conducted by Banks *et al.* [16] that the PBL height of the YSU scheme is smaller than the MYJ scheme. However, in certain hours (morning to noon) the height of PBL YSU is higher than MYJ (Fig. 9). The PBL differences are seemed very small. According to Cuchiara *et al.* [6], the height of PBL using the YSU scheme produces greater value during the day and a smaller value at night. Based on PBL height data from the five SPKU points, it shows that in February the height of PBL with MYJ scheme was higher than YSU around 12:00 to 07:00 LT, while for August it was around at 14:00 to 21:00 LT at night.

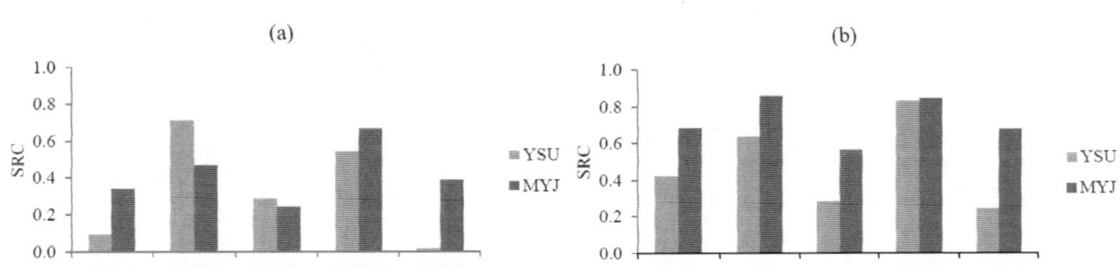

Figure 8. Sensitivity results from model output and observation of PM10 in (a) February and (b) August.

IOP Conf. Series: Earth and Environmental Science **303** (2019) 012049 doi:10.1088/1755-1315/303/1/012049

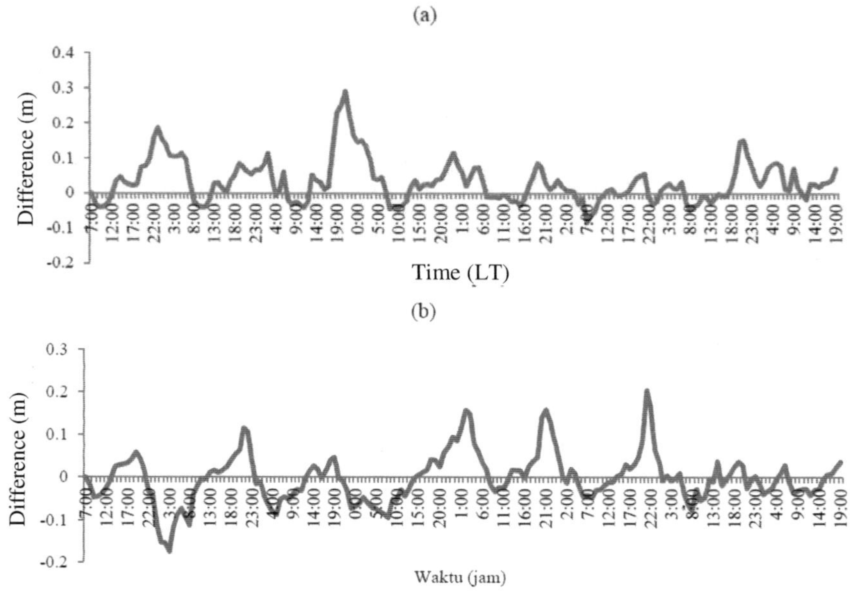

Figure 9. Difference of PBL height from MY Time (LT) schemes in (a) February and (b) August.

For the explanations delineated above, the concentration of simulated PM10 during daytime was lower than at night. In summary, in the model, the effects of vertical dispersion of the hourly change tendency of PM10 during daytime are much more essential compared with the effects of daily averaged PM10. The difference of PM10 concentration in YSU and MYJ schemes is still little in separate stations; meanwhile, the modelling results in different location contain certain differences.

Figure 10 shows that there is a relationship that is inversely proportional between the height of PBL and PM_{10} concentration in DKI 1. The greater the height of PBL, the smaller the PM_{10} concentration produced. The statement applies to the results of the model using the YSU scheme and MYJ scheme in February and August in the five SPKU points. The process of dispersing pollutants can cause inversely proportional relationships. When the height of the PBL is high, the pollutants mixed in a larger volume so the value of the pollutant concentration measured on the surface will be smaller.

It should be noted that the different representations of vertical dispersion in these PBL schemes might have different influences on PM_{10} simulation under different conditions of atmospheric stability in different regions. It seems to be that the little difference of diffusivity calculation between haze and clean days by the PBL schemes calculation might lead to this interesting phenomenon, which is probably the main way to improve PM_{10} forecasting in coastal city.

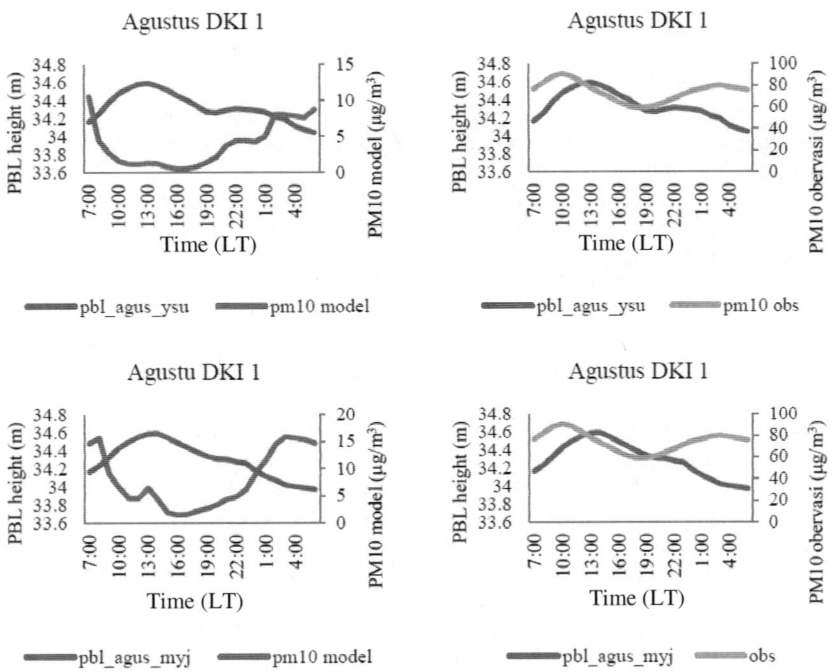

Figure 10. Comparison of PBL height and mean of PM10 at DKI1

4. Conclusions

The WRF-Chem model can produce a model output that is good for temperature because it has a coefficient of determination of 0.5-0.6 for February and 0.5-0.9 for August. The model output for the wind is still to overestimate and PM_{10} is still underestimate that are affected by resident emissions that are not put up in the model. In this study, the PBL scheme has a small impact compared to difference in observation. The results of the PM_{10} concentration model output with the MYJ scheme produce a greater concentration value compared to the YSU scheme. Sensitivity analysis of the two schemes against observation shows that the MYJ scheme has greater sensitivity than the YSU scheme. The height ratio of PBLs from the two schemes shows that in February the PBL height with the MYJ scheme was higher than YSU around 12:00 to 07:00 LT. It has been understood that the highest concentration of PM_{10} created by the WRF-Chem with both schemes in the February and August has the equivalent time in the morning. The model output of WRF-Chem over both schemes also displays the smallest concentration of PM_{10} that monitors during the afternoon. Whereas for August, it is around 14:00 to 21:00 LT. PBL height analysis also shows a relationship that is inversely proportional between the height of PBL and PM_{10} concentration. Exactly when the PBL is high, the pollutants arranged in a more volume so the estimation of the pollutant concentration assessed on the surface will be decreased. The reasons for the reduced image of diurnal difference in the PBL schemes, resulting in PM_{10} miscalculations in numerical models, need to be studied in detail and then modifications need to be done to improve results for different locations.

References

[1] Baro R, Guerreo P J, Balzarini A, Curci G, Forkel R, Grell G, Hirtl M, Honzak L, Langer M, Perez J L, Pirovano G, Jose RS, Tucella P, Werhahn J and Zabkar R 2015 Sensitivity analysis of the microphysics scheme in WRF-Chem contributions to AAQMEII phase 2 *Atmos. Environ.* **30** 1-10

[2] Werner M, Kryza M, Geels C, Ellermann T and Skjoth C A 2015 Spatial, temporal and vertical distribution of ammonia concentrations over Europe comparing a statistic and dynamic approach with WRF-Chem *Atmos. Chem. Phys.* **15** 22935-22973
[3] Zhong M, Saikawa E, Liu Y, Naik V, Horowitz LW, Takigawa M, Zhao Y, Lin N H and Stone EA 2016 Air quality modeling with WRF-Chem v3.5 in East Asia: sensitivity to emissions and evaluation of simulated air quality *Geosci. Model Dev.* **9** 1201-1218
[4] Vela A V, Andrade M F, Kumar P, Ynoue R Y and Munoz A G 2016 Impact of vehicular emissions on the formation of fine particles in the Sao Paulo Metropolitan Area: a numerical study with the WRF-Chem model *Atmos. Chem. Phys.* **16** 777-797
[5] Misenis C and Zhang Y 2010 An examination of sensitivity of WRF/Chem predictions to physical parameterizations, horizontal grid spacing and nesting options *Atmos. Res.* **97** 315-334
[6] Cuchiara GC, Li X, Carvalho J and Rappengluck B 2014 Intercomparison of planetary boundary layer parameterization and its impacts on surface ozone concentration in the WRF/Chem model for a case study Houston Texas *Atmos. Environ.* **96** 175-185
[7] Kumar A, Jimenez R, Belalcazar L C and Rojas N Y 2016 Application of WRF-Chem model to simulate PM10 concentration over Bogota Aerosol Air Qual. Res. **16** 1206-1221
[8] Ge C, Wang J and Reid J S 2014 Mesoscale modeling of smoke transport over the Southeast Asian Maritime Continent: coupling of smoke direct radiative effect below and above the low-level clouds *Atmos. Chem. Phys.* **14** 159-174
[9] Nuryanto D E and Nuaraini T A 2015 Pengaruh skema cumulus yang berbeda pada polutan dalam model WRF-Chem *Prosiding SNSA* **39**
[10] Heriyanto E and Nuraini T A 2015 Perbandingan konsentrasi NO2, SO2 dan partikel debu luaran WRF-Chem dengan data pengukuran studi kasus DKI Jakarta 2009 *Prosiding SNSA* **61**
[11] Turyanti A 2016 Pemodelan dispersi PM10 dan SO2 dengan pendekatan dinamika stabilitas atmosfer di lapisan perbatas pada kawasan industri Disertasi Bogor Institut Pertanian Bogor
[12] Beck V, Gerbig C, Koch T, Bela M M, Longo KM, Freitas S R, Kaplan JO, Prigent C, Bergamaschi P and Heimann M 2013 WRF-Chem simulations in the Amazon region during wet and dry season transitions: evaluation of methane models and wetland inundation maps *Atmos. Chem. Phys.* **13** 7961-7982
[13] Amnuaylojaroen T, Barth M C, Emmons L K, Carmichael G R, Kreasuwun J, Prasitwattanaseree S and Chantara S 2014 Effect of different emission inventories on modeled ozone and carbon monoxide in Shoutheast Asia *Atmos. Chem. Phys.* **14** 12983-13012
[14] Pianosi F, Beven K, Freer J, Hall J W, Rougier J, Stephenson D B and Wagener T 2016 Sensitivity analysis of environmental models: a systematic review with partical workflow Environ. Modell. *Softw.* **79** 214-232
[15] Skamarock W C, Klemp J B, Dudhia J, Gill D O, Barker D M, Duda M G, Huang X Y, Wang W and Powers J G 2008 A Description of The Advanced Research WRF Version 3 Colorado US National Center for Atmospheric Research
[16] Banks R F, Alsina J T, Baldasano J M, Rocadenbosch F, Papayanis A, Solomos S and Tzanis C G 2016 Sensitivity of boundary-layer variables to PBL schemes in the WRF model based on surface meteorological observations, lidar, and radiosondes during the HygrA-CD campaign *Atmos. Res.* **176-177** 185-201
[17] Chen D, Xie X, Zhou Y, Lang J, Xu T, Yang N, Zhao Y and Liu X 2017 Performance evaluation of the WRF-Chem model with different physical parameterization schemes during an extremely high PM2.5 pollution episode in Beijing Aerosol *Air Qual. Res.* **17** 262-277
[18] Ullva D S 2010 Analisis karakteristik angin wilayah pantai menggunakan model meteorologi skala meso (MM5) sebagai dasar pendugaan potensi penyebaran pencemar udara (studi kasus: DKI Jakarta) Skripsi Bogor Institut Pertanian Bogor
[19] Li T, Wang H, Zhao T, Xue M, Wang Y, Che H and Jiang C 2016 The impacts of different PBL schemes on the simulation of PM2.5 during severe haze episodes in the Jing-Jin-Ji region and its surroundings in China *Adv. Meteorol.* 1-15
[20] Poosarala V V, Kumar A and Kadiyala A 2010 Estimation of uncertainty in predicting ground level concentrations from direct source releases in an urban area using the USEPA's AERMOD model equations *Air Quality* 169-200
[21] Xie B, Fung JCH, Chan A and Lau A 2013 Evaluation of nonlocal and local planetary boundary layer schemes in the WRF model *J. Geophys. Res.* 117: 1-26.

Comparison of seasonal prediction outputs based on dynamic atmosphere model and observations (case study: seasonal prediction in 2016 - 2017)

H Satyawardhana and Gammamerdianti

Center for Atmospheric Science and Technology, National Institute of Aeronautics and Space, Bandung, Indonesia

hariessatha@gmail.com

Abstract. Seasonal predictions are considered as an important factor for many sectors in Indonesia, especially for the agriculture sector. In Indonesia, the variability of seasonal characteristics is influenced by various regional atmospheric phenomena, such as monsoon, ENSO, and IOD. With such urgency, this research is conducted using dynamic models to generate accurate medium-term predictions. The dynamical atmospheric model, Cubical Conformal Atmospheric Model (CCAM) is used to produce daily predictions up to 8 months. CCAM result has been tested on major islands in Indonesia such as Java, Sumatra, Sulawesi, Borneo and Papua. It shows that during the rainy season, the value of rainfall predictions is lower than observations. Meanwhile, the rainfall predictions are overestimate when compared to observations during dry season. Nevertheless, the patterns of rainfall predictions are following satellite rainfall observations patterns, especially in Java.

Keywords : seasonal, prediction, dynamic atmosphere model, rainfall, CCAM

1. Introduction

Seasonal prediction is an attempt to provide information about atmospheric conditions, especially temperature and rainfall, on a time scale of one or several seasons ahead. The ability of numerical and dynamical models in predicting the season has increased sharply since the start of the development of data assimilation systems in the 1990s. The system of dynamical season prediction began operating in the United States around 1995 with the National Meteorological Centre's Medium Range Forecast (NMC). The first seasonal prediction that operated is CFS version 1 in August 2004, providing predictions for the world's climate [1] and showed reasonable skills in predicting the major features of large-scale climate such as ENSO [2] and Asian monsoon [3]. CFSv1 was replaced to NCEP CFSv2 (National Version for Environmental Prediction Climate Forecast System version 2) in 2011 [4].

In Europe, the ECMWF (European Centre for Medium-Range Weather Forecasts) season prediction system began in 1996 as ECMWF Ensemble Prediction System (EPS) [5]. The seasonal prediction system in Australia has developed since 2002 and known as the POAMA (Predictive Ocean Atmosphere Model for Australia). The first operational version of POAMA called POAMA-1 that ran from 2002 to 2007, and was subsequently replaced by version POAMA-1.5b in 2007. At the end of 2011, further updates resulted in the multi-model version POAMA-2.4 (POAMA-2 herein) [6]. Development of numerical season prediction systems that are operational in these countries through years of intensive research. While the performance of CFSv2 for the Indonesian region has been studied by Zhang et al. (2016) which investigating CFSv2 output in Maritime Continent (MC) in wet and dry seasons [4].

Nowadays, Indonesia's climate was difficult to predict. The main reason is Indonesia as the maritime continent is like a boiler box [7] which has a large amount of latent heat due to the rain process, and moved by Hadley Cell and Walker Circulation. In addition, Indonesia has a unique and complex geographical location, among others the distribution of land, oceans and terrains which contribute to rainfall variations in both spatial and temporal [8]. Indonesia is located between the Indo-Pacific heat pools, where evaporation increases atmospheric moisture, and the circulation transmits water vapor from the ocean toward Indonesia [9]. This is a new challenge for atmospheric researchers in terms of predicting, especially predictions until the next few seasons. Conformal-Cubic Atmospheric Model (CCAM) is an atmospheric numerical model developed by CSIRO, Australia which previously developed the Division of Atmospheric Research Limited Area Model (DARLAM). CCAM is one of the global atmospheric models developed effectively starting in 1994 by the Commonwealth Scientific and Industrial Research Organization (CSIRO), Australia [10]. CCAM not only has been adapted and developed with the purpose to increase the weather prediction capability in producing more reliable forecasts but also can be integrated with high resolution on a specific domain [12]. A lot of researchers using CCAM in predicting weather conditions with different data. For example, the prediction of temperature and rainfall of CCAM which validated using observation data at 128 stations in Vietnam shows that forecast temperature tends to be lower than observation data in most climatic regions across the country, but generally it has low value of error. The correlation coefficient between CCAM and observation is very high [13]. Yulihastin et al. (2015) has been using CCAM to show the correlation between rainfall anomalies over Indonesia when Madden Julian Oscillation (MJO) activated. The result shows that CCAM can identify MJO activities over Indonesia quite well [13]. Therefore, this study has a purpose to compare the predicted rainfall of the season using CCAM models and GSMaP rainfall for the Indonesian region. Also, this research is aimed to improve Indonesia's prediction by the season and the first step to make the seasonal predictions in Indonesia using dynamic models, especially using CCAM.

2. Data and Methods

2.1. Data

The data that used in this study are the input for CCAM which includes 2 data, namely initial condition data from the Global Forecasting System (GFS) and input data on sea surface temperature (SST) from the predicted POAMA results.

The season prediction simulation using CCAM is carried out for the period March 2016 to March 2018. The initial condition data that is used in this study is the first GFS data of each month with a resolution of 1°, and simulations are conducted monthly, starting March 2016 to March 2018. CCAM also can be used to simulate atmospheric by using the forcing of SST with initial condition from GFS. CCAM can be run as Atmosphere General Circulation Model (AGCM) with only SST forcing. Therefore, this research using SST prediction data from POAMA which has 2.5° spatial resolution, where this data is actually a surface temperature data that has been converted to sea surface temperature by eliminating data on the mainland. The SST data from POAMA prediction has 3 classes, namely m24a, m24b and m24c where in each class consists of 11 ensembles (e00, e01, ..., e10) and 1 mean ensemble (emn). But in this study, only one predicted surface temperature data of POAMA is used i.e. m24b_emn. The spatial resolution of CCAM prediction results is 0.28° or equivalent to 31.8km. GSMaP satellite data provided by JAXA (http://sharaku.eorc.jaxa.jp/GSMaP/) is used for the spatial and temporal comparison of rainfall parameters. The spatial distribution of rainfall from GSMAP has spatial resolutions $0.1^{o} \times 0.1^{o}$, daily and near real time observation.

The locus of this research is the entire territory of Indonesia with a limit of 95° - 145°E and 6°N - 12°S. This is the result of the downscaling from domain 1 (global - 1° resolution) to domain 2 (Indonesia – 0.28°), as shown in Figure 1a.

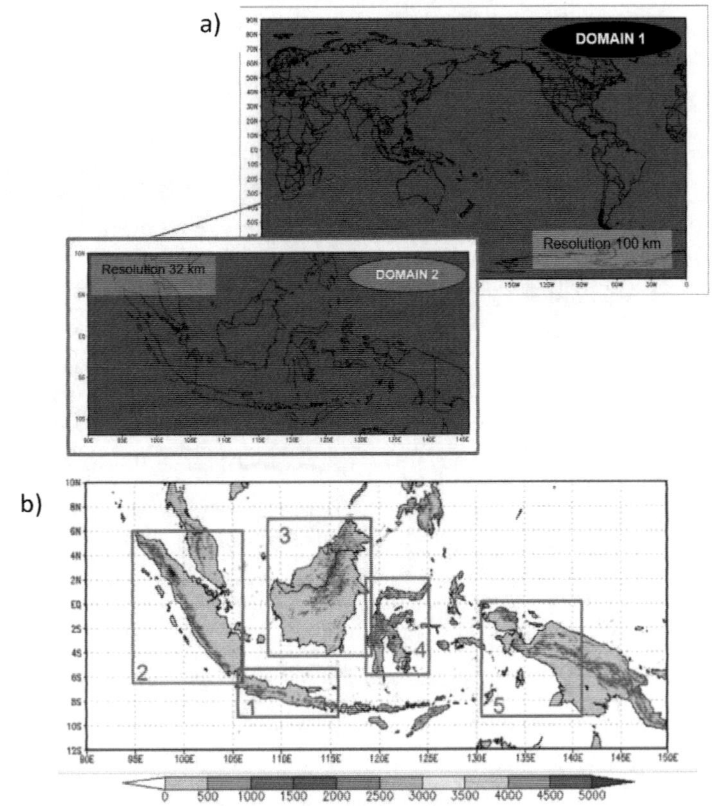

Figure 1. a) Domain of the CCAM simulation and b) topography of the Indonesian Region using USGS DEM data and research study area. The box in Figure 1b is the area boundary for the analysis of rainfall per large island in Indonesia (1. Java-Bali, 2. Sumatra, 3. Borneo, 4. Sulawesi, and 5. Papua).

2.2. Methods
CCAM uses conformal cubic grid which each panel has 48x48 grid points (for C48 grid format), 6 cube sides, and 18 vertical levels, so the number of grid points is 248,832 [11]. But in this study using the C96 format, which has 96x96 grid points in each panel. This aims to be able to produce a broader domain with a denser resolution. The value of 32 km of the prediction output in this study is the optimal consideration from the use of the C96 format. Another advantage of CCAM is that it has the ability to provide high flexibility for dynamical downscaling results from GCM data by using sea surface temperature data and some nudging from the host model [11]. But in this study, CCAM was used to make seasonal predictions with GFS data input and SST predictions from POAMA. Before running the CCAM model, the input data have to be set into cubical so it can be processed in CCAM. Furthermore, CCAM can be run after adjusting land settings to obtain the results of the topographic and vegetation description of a domain area using executable file called *runtopo*. The domain area used in the CCAM in this study is the territory of Indonesia with an output resolution of 0.28° or equivalent to 31.8km. After the prediction output is obtained, the monthly rainfall calculation and analysis plot are carried out. Prediction simulations are carried out at the beginning of each month (March 2016 to March 2018) and produce predictions for the next 8 months from the initial condition, resulting in daily temporal resolution. The research workflow can be seen in Figure 2a and simulation scheme of the seasonal prediction in this study is shown in Figure 2b.

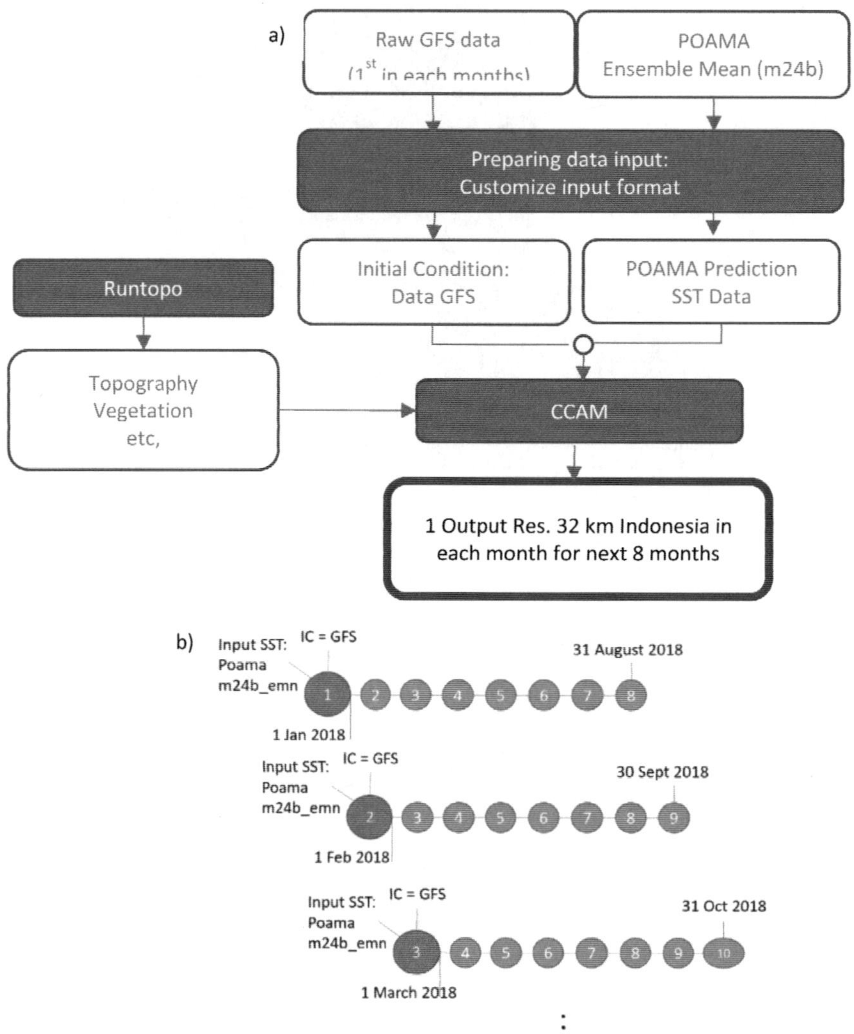

Figure 2. a) Research flowchart and b) scheme simulation of the seasonal prediction using CCAM (the date for both the initial and end date of simulation are just example)

Calculation of the correlation coefficient between the results of the season prediction simulation and GSMaP data is used to analyze the validation of the prediction. In addition, an area average method is used from several large islands in Indonesia to find out the accuracy of rainfall simulation and which island has a pattern and value of rainfall close to GSMaP data.

3. Results and Discussions

3.1. Temporal comparison

Comparison of monthly rainfall values between GSMaP satellite data and CCAM output on 5 island in Indonesia is shown in Figure 1. Rainfall values are the average area of rainfall in Java (105-116E, 5.5S-9.5S), Sumatra (95-108E, -6N-7S), Borneo (108-120E, 8N-5S), Sulawesi (105-116E, 9.5S-5.5S) and Papua (130-141E, 1N-10S), where the rainfall used is rainfall on land (rainfall in the ocean is eliminated).

IOP Conf. Series: Earth and Environmental Science **303** (2019) 012050 doi:10.1088/1755-1315/303/1/012050

From Figure 3, it can be seen that predicted precipitation has a tendency to be wetter during the dry season and drier during the rainy season. Of the several major islands in Indonesia, the similarity of predictive rainfall patterns that approach GSMaP's monthly rainfall patterns are in Java and Bali. Predictive rainfall in Java and Bali has a pattern that resembles monthly rainfall from GSMaP data, although rainfall from the dry season and rainy season shifts slightly from 1 to 2 months compared to GSMaP. While the monthly rainfall pattern on other islands is still not good / does not resemble the GSMaP rainfall pattern.

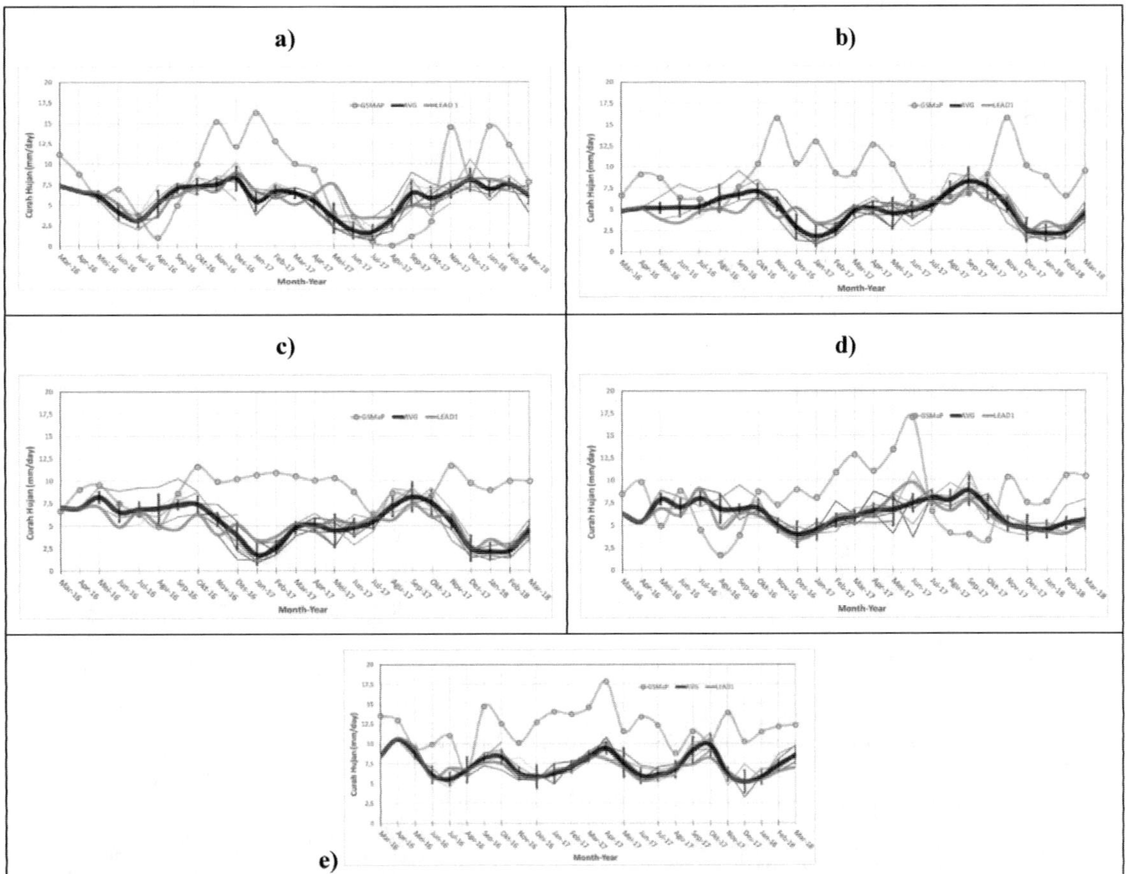

Figure 3. Comparison of the average rainfall area predicted by CCAM in various initial conditions and GSMaP for: a) Java Island, b) Sumatra, c) Borneo, d) Sulawesi and e) Papua. Small lines are the monthly rainfall of CCAM ensemble results in different initial conditions, thick black lines are the average rainfall and standard deviation of CCAM ensembles, thick grey lines are rainfall every 1st month of each CCAM output (lead month 1) and the blue line is GSMaP rainfall.

Linear correlation analysis between monthly rainfall prediction and GSMaP can be seen from Figure 4, both on average and lead 1-month ensemble. Figure 4 shows more clearly that in general the predicted rainfall in Java and Bali (both the average and the predicted 1-month lead) has not been able to show high rainfall from GSMaP. Even so, the prediction of ensemble average rainfall has a correlation of 0.64 and rainfall prediction of 1-month lead has a correlation of 0.69 when compared to GSMaP rainfall.

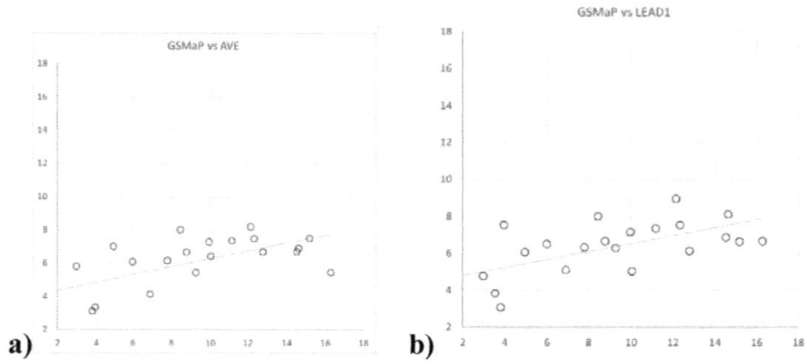

Figure 4. Scatter-plot of the average rainfall area predicted by CCAM a) average and b) lead1 in various initial conditions and GSMaP for Java

3.2. Spatial Comparison

Spatial comparisons of monthly rainfall between prediction results and GSMaP are shown in Figures 5 and 6. In the dry months, namely JJA (as shown in Figure 5), CCAM prediction rainfall tends to show a lower value compared to GSMAP in almost all region. Rainfall falls gradually from June to August. The peak of the dry season for the southern region (Java, Bali and Nusa Tenggara) is in August. The decrease in rainfall is also seen in the CCAM prediction results even with a value below the GSMaP rainfall value. Rainfall in northern Sumatra such as Aceh and North Sumatra which despite entering the JJA period still shows high rainfall, cannot be shown by CCAM.

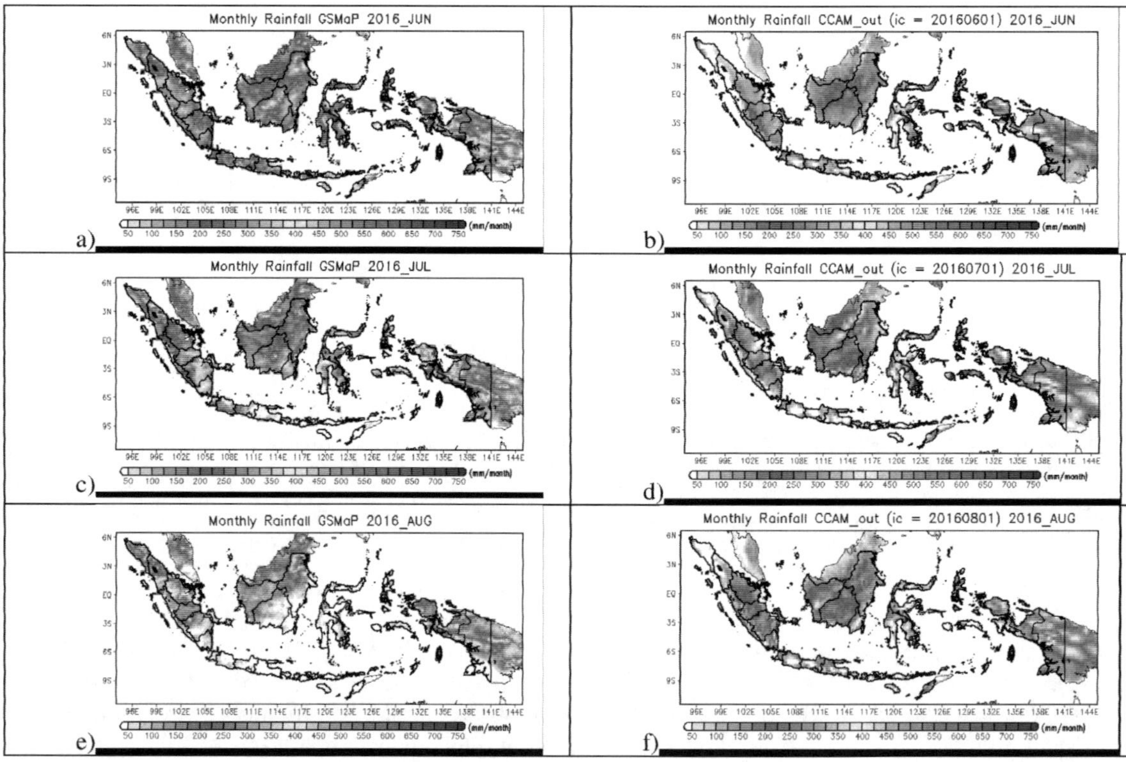

Figure 5. Spatial pattern comparison of monthly rainfall predicted by CCAM on 1-month lead and GSMaP during dry months 2016 (JJA), which are respectively a) GSMaP for June, b) CCAM for June, c) GSMaP for July, d) CCAM for July, e) GSMaP for August and f) CCAM for August.

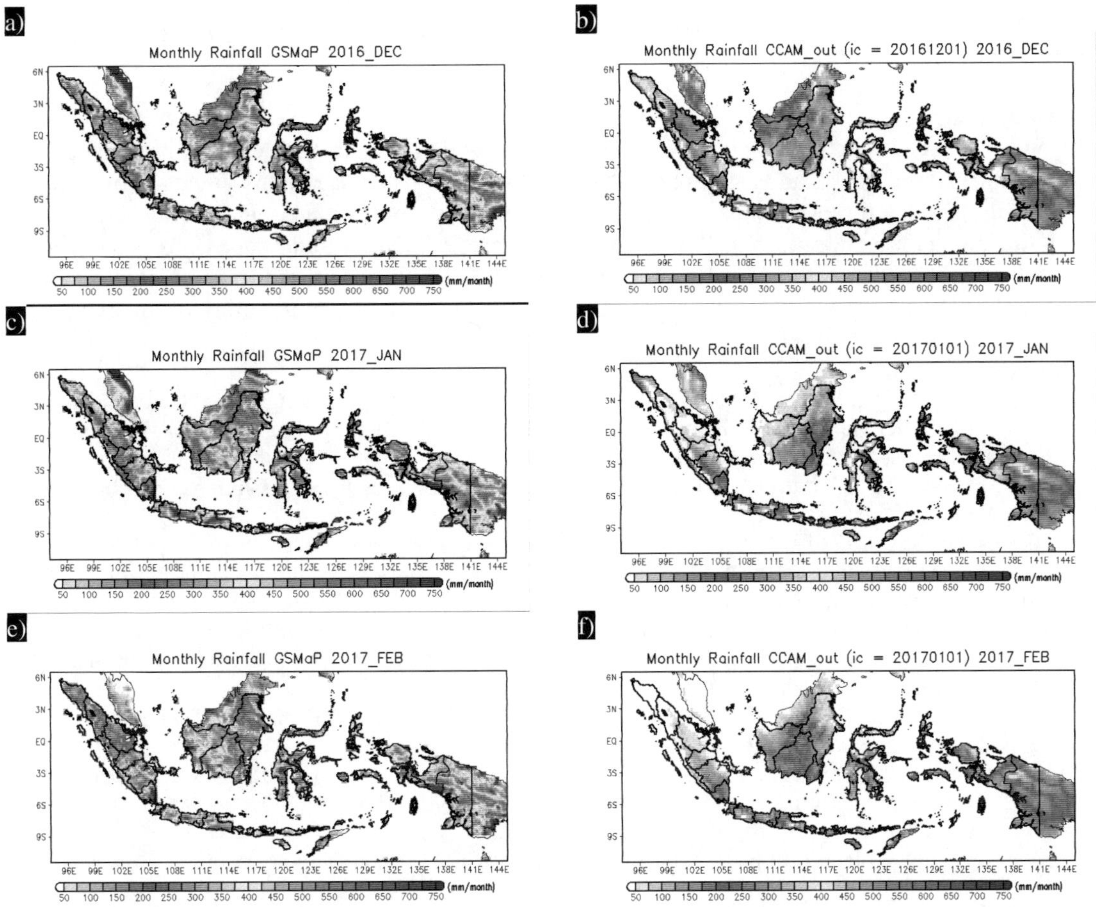

Figure 6. Same as Figure 5, but during wet months 2016/2017(DJF), which are respectively a) GSMaP for December 2016, b) CCAM for December 2016, c) GSMaP for January 2017, d) CCAM for January 2017, e) GSMaP for February 2017 and f) CCAM for February 2017

While the prediction of CCAM for rainfall in the DJF 2016/2017 (wet months) (Figure 6), has a high difference with the underestimate value compared to GSMaP. High rainfall that occurs in these months, still cannot be shown by the results of the prediction, although there is an increase in rainfall predicted results when compared with the predicted rainfall in the dry months. But the results are almost the same as the prediction in the dry months, namely the northern region (northern Sumatra, northern Borneo). This will be the focus for future research, the other way is to multiply experiments using different SST ensemble predictions, different initial data or different parameterize models.

4. Conclusion

The season prediction conducted in this study resulted in a prediction of spatial rainfall with a spatial resolution of 32 km for the Indonesian region. CCAM's predicted monthly rainfall for Java and Bali Island has a high correlation when compared to GSMaP rainfall. Even so, in general the predicted precipitation tends to be below the GSMaP rainfall value. Whereas for the northern Sumatra and northern Borneo regions, the CCAM prediction results are still unable to provide good results, both predictions of wet and dry months. And the high rainfall that occurs during the wet months is still not well demonstrated by the CCAM prediction results, although the decrease of rainfall during the dry months and the increase of rainfall in the wet months in the southern part of Indonesia already well predicted by CCAM.

Acknowledgments
Thank you to CSIRO whose developing CCAM and JAXA which provides GSMAP rainfall data, so the Indonesia's rainfall prediction and satellite could be analyzed in this study. Thank you also for all author who works on the related fields which is cited in the references, so this paper can be finished. Last but not least, the authors are grateful to the National Institute of Aeronautics and Space (LAPAN), which is the institution where the author works and also as the organizer committee to this international conference (ICTMAS) especially to reviewer who give good input in the completion of this paper.

References

[1] Saha S, Moorthi S, Wu X, Wang J, Nadiga S, Tripp P P, Behringer D, Hou Y, Chuang H, Iredell M, Ek M, Meng J, Yang R, Mendez M P, Dool H V, Zhang Q, Wang W, Chen M, and Becker E 2006 The NCEP Climate Forecast System *J. Climate* **19** 3483-517

[2] Wang B, Ding Q, Fu X, Kang I S, Jin K, Shukla J and Doblas-Reyes F 2005 Fundamental challenges in simulation and prediction of summer monsoon rainfall Geophys *Res. Lett.,* 32, L15711 doi: 10.1029/2005GL022734

[3] Yang S, Wen M and Higgins W 2008 Subseasonal features of the Asian summer monsoon in the NCEP climate forecast system Acta Oceanol. Sin **27** 88-103

[4] Zhang T, Yang S, Jiang X and Zhao P 2016 Seasonal-Interannual Variation and Prediction of Wet and Dry Season Rainfall over the Maritime Continent: Roles of ENSO and Monsoon Circulation *J. Clim.* **29(10)** 3675–95 https://doi.org/10.1175/JCLI-D-15-0222.1

[5] Molteni F, Buizza R, Palmer T N and Petroliagis T 1996 The ECMWF ensemble prediction system: methodology and validation *Q. J. R. Meteorol. Soc.* **122**: 73–119

[6] Cottrill A, Hendon H, Lim E-P, Langford S, Shelton K, Charles A, McClymont D, Jones D and Kuleshov Y 2012 Seasonal Forecasting in the Pacific Using the Coupled Model POAMA-2 AMS Weather and Forecasting **28**: 668-80 doi: 10.1175/WAF-D-12-00072.1

[7] Ramage C S, 1968 Role of a tropical "Maritime Continent" in the atmospheric circulation *Mon. Wea. Rev.* **96** 365-9.

[8] Hamada J-I, Yamanaka M D, Matsumoto J, Fukao S, Winarso P A and Sribimawati T 2002 Spatial and temporal variations of the rainy season over Indonesia and their link to *ENSO J. Meteor. Soc.* **80** 285-310.

[9] Qian J H 2008 Why rainfall is mostly concentrated over islands in the Maritime Continent *J. Atmos. Sci.* **65** 1428-41

[10] McGregor J L 2005 CCAM: Geometric aspects and dynamical formulation CSIRO Atmospheric Research Technical Paper No.**70,** CSIRO, Australia

[11] Thatcher M 2007 Training Module "CCAM General Introduction for New Users" Training Weather and Climate Modelling based on CCAM Jakarta 15-25 January 2008.

[12] Park R 2010 Development of The New Conformal-Cubic Atmospheric Model (CCAM) in Capturing the Past Season's Major Rain Events CSIR Natural Resources and the Environment, South Africa

[13] Khiem M V, Minh H T and Nam P Q 2018 Study on Sensitivity of CCAM Model to The Sea Surface Temperature Boundary Conditions Vietnam *Journal of Science Technology and Engineering* Vol.**60** No 1: 83–9

[14] Yulihastin, E, Adikusumah N and Eddy H 2015 Identifikasi Fase Aktif MJO di Benua Maritim Indonesia dalam Model CCAM Fisika, Kimia dan Dinamika Atmosfer di Indonesia CV. Media Akselerasi Bandung 43-58

The Indonesia In-House Radar Integration System (InaRAISE) of Indonesian Agency for Meteorology Climatology and Geophysics (BMKG): Development, Constraint, and Progress

D S Permana[1*], T D Hutapea[1], A S Praja[1*], J A I Paski[1], E E S Makmur[1], U Haryoko[2], I H Umam[3], M Saepudin[4] and R Adriyanto[5]

[1]Meteorology Research Division, Center for Research and Development, BMKG, Kemayoran, Jakarta 10720, Indonesia
[2]Director, Center for Research and Development, BMKG, Kemayoran, Jakarta 10720, Indonesia
[3]Remote Sensing Management Division, Center for Public Meteorology, BMKG, Kemayoran, Jakarta 10720, Indonesia
[4]Weather Prediction and Early Warning Division, Center for Public Meteorology, BMKG, Kemayoran, Jakarta 10720, Indonesia
[5]Marine Meteorology Management Division, Center for Public Meteorology, BMKG, Kemayoran, Jakarta 10720, Indonesia

donaldi.permana@bmkg.go.id; alfan.sukmana@gmail.com

Abstract. In 2017, BMKG has 41 weather radars covering most of Indonesia region and most of its data are automatically sent to the BMKG headquarter every 10 minutes. There are four different weather radar brands with its specific data format and software analysis. In recent years, the weather radar community has developed open-source software to handle several radar data formats. Based on this, BMKG has developed the Indonesia In-House Radar Integration System (InaRAISE) of BMKG using the open-source weather radar software. InaRAISE has been developed using Python-based libraries Wradlib and Py-ART for processing weather radar data. BMKG radar data have been successfully extracted and transformed into Cartesian coordinates for post-processing. The multiple radars have been successfully composited by comparing column-maximum reflectivity. Web-based near real-time radar images has been experimentally operated, but not officially launched. The main constraint is the susceptible communication network between radar sites and BMKG headquarter causing real-time data transfer problems. InaRAISE serves feasible data radar extraction for data assimilation in the numerical weather prediction model. InaRAISE could serve as a supporting of the existing radar integration system or possibly as a replacement.

Keywords : InaRAISE, radar, reflectivity, BMKG

1. Introduction

As the national meteorological weather service agency in Indonesia, BMKG is responsible to provide the weather forecast and early warning information to the public. BMKG uses various numerical weather prediction (NWP) models and considers the real-time satellite and weather radar data to provide the short-term and nowcasting (1 to 6 hours) forecasts. The weather radar data is important in providing a high spatial resolution early warning and weather forecasts information. In 2017, BMKG has 41 weather radars across Indonesia archipelago consisting of four different brands namely Gematronik, EEC, Baron, and Vaisala (Figure 1) [1]. The recent weather radar was installed in 2017 at Sintang, West Kalimantan. Most of the radars are C-band Doppler single polarization with optimum radius coverage of 150 - 240 km. Most of its data is available at 10 minutes interval and is automatically sent to the BMKG headquarter. Each radar brand has a different data format that can only be processed using the software of each radar manufacturer. Thus, analysis and processing of radar data from different brands can be limited.

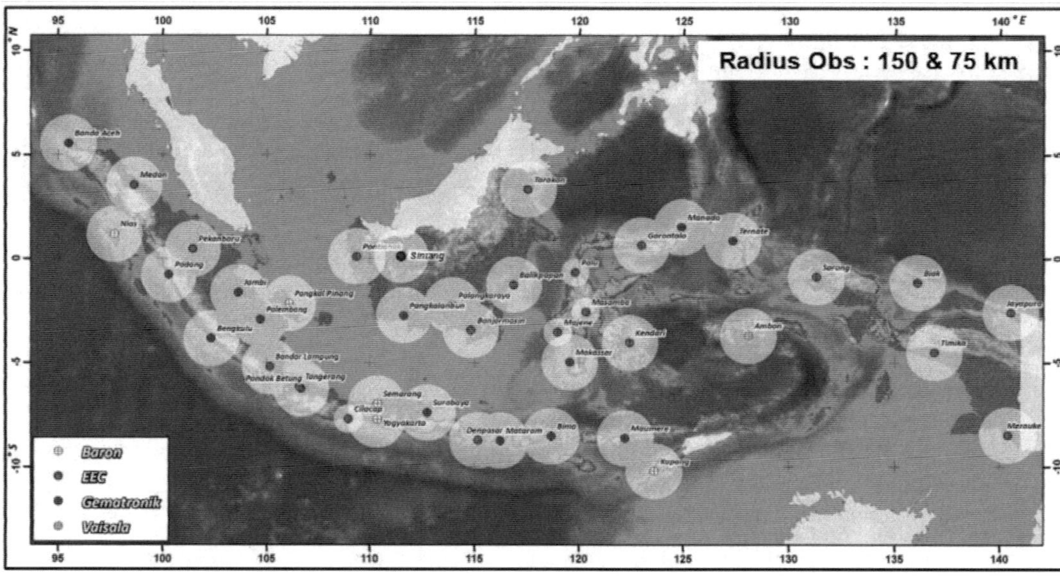

Figure 1. Map of BMKG weather radar locations in 2017. Colors represent different brands. The radius coverage is 75 km (X-Band) and 150 km (C-Band).

In recent years, the weather radar community has developed several open-source software systems to handle multiple radar file formats [2]. The open-source systems include BALTRAD [3-5], LROSE/TITAN (https://www.eol.ucar.edu/content/lidar-radar-open-software-environment), wradlib [6] and Py-ART [7]. Most of these software systems are using C/C++ and Python programming languages. In addition, another free and open library in the R language called radar.IRIS is also available for processing weather radar data (http://gaia.tru.ca/radar.IRIS) [8]. These software systems have been widely used in processing weather radar data and its applications in many countries. For instances, wradlib has been used to reconstruct monsoon rainfall using a radar network and evaluate the streamflow and flood simulations based on radar-derived rainfall in the Philippines [9-11]. In the meantime, BALTRAD software has been deployed throughout Europe and more recently in Canada [5].

Based on these considerations, Center for Research and Development (R&D), supported by Remote Sensing Management Division at Center for Public Meteorology of BMKG, have developed the radar integration system based on the open-source software called the Indonesia In-House Radar Integration System (InaRAISE) of BMKG. The core of InaRAISE has been developed using Python-based library wradlib since 2016 [1] and recently, Py-ART library was also used in 2017 for processing multi-radar file format. This study describes the development process, anticipated constraints and recent progress of InaRAISE. Some of InaRAISE products are also given including the composite radar maps and the future plan of InaRAISE is also discussed in this study.

2. Development Process

The radar data is typically stored in the polar coordinate format for each sweep and saved in the volumetric data file (.vol). The structure of the polar coordinate data mainly consists of (1) the reference point which is the radar center point, (2) the polar axis, (3) the radius or distance from the reference point and (4) the azimuth angle which is the angle from the polar axis (see figure in https://www.nssl.noaa.gov/publications/dopplerguide/images/2-1-1.gif). While the analysis and processing of meteorological data are typically conducted in Cartesian coordinate (latitude-longitude) format. Thus, the conversion from the polar to Cartesian coordinates is needed to facilitate the processing and analysis of radar data. Wradlib and Py-ART libraries in Python programming language are able to extract and convert the radar data into the Cartesian coordinate format and also save the data into NetCDF file format [1]. In addition, they also provide quality control tools (clutter identification and removal, and attenuation correction), the conversion from reflectivity into rain-rate and data visualization.

As shown in Figure 1, BMKG has four different radar brands which mean there are four different volumetric data. Each volumetric data have been directly extracted by wradlib, except the volumetric data from EEC. The EEC volumetric data must be converted into NetCDF (.nc) format using the *volconvert* package which only available in the EEC EDGE® software before being extracted by wradlib. For volumetric data from Vaisala, it also can be extracted by Py-ART library. The flowchart of InaRAISE is depicted in Figure 2.

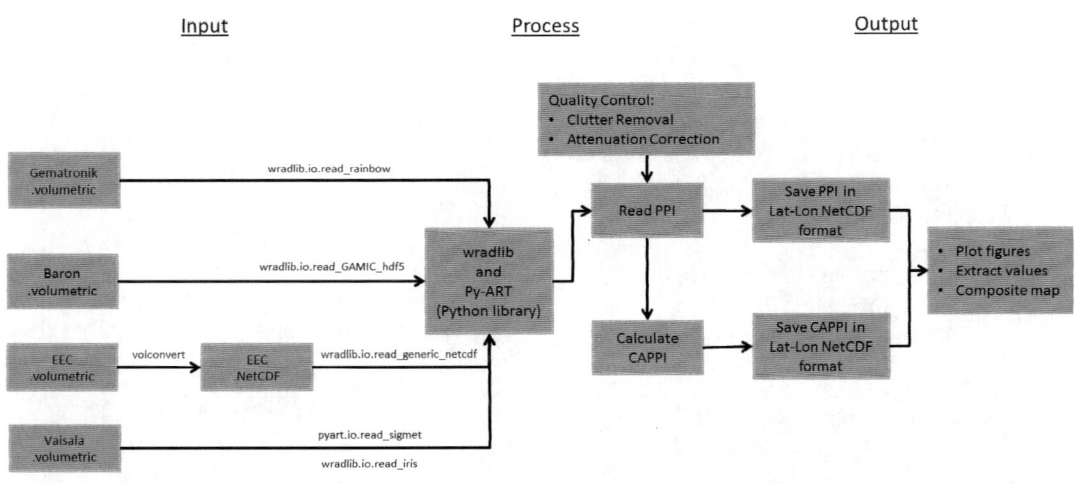

Figure 2. Flowchart of InaRAISE (after modification of Figure 3 in ref. 1).

InaRAISE automatically extracts the Plan Position Indicator (PPI) of reflectivity (Z) and radial velocity (V) from the radar data. While extracting the PPI, InaRAISE implements the quality control that utilizes the clutter identification and removal algorithms developed by Gabella and Notarpietro (2002) [12] and the attenuation correction method developed by Jacobi et al. (2009) [13] which both are provided by wradlib. Afterward, Constant Altitude PPI (CAPPI) is calculated with the following specifications: (1) the spatial resolution is 500 meters/pixel, (2) the CAPPI layers are generated from 1 to 10 km height with 1 km vertical resolution, and (3) the maximum CAPPI (CAPPI-MAX) layer is produced. Then, both PPI and CAPPI are saved into latitude-longitude (Cartesian) NetCDF file format. In the end, plotting figures, extracting values and compositing radars are conducted.

IOP Conf. Series: Earth and Environmental Science **303** (2019) 012051 doi:10.1088/1755-1315/303/1/012051

Currently, the InaRAISE near real-time web-based has been experimentally operated, but not officially launched. The web page links are http://www.bmkg.go.id/cuaca/citra-radar.bmkg (Center for Public Meteorology) and http://puslitbang.bmkg.go.id/radar (R&D Center). There are at least 23 radar sites have been integrated into InaRAISE, while the other radars still have network communication issues in transferring the real-time data to BMKG headquarters. Some examples of InaRAISE products are given in Figure 3. The map of Jabodetabek area is specifically cropped from Tangerang radar coverage with district boundaries to monitor the rainfall distribution over Jabodetabek (top left panel of Figure 3). The Ambon radar (Vaisala) is plotted with a radius of 25, 50 and ~200 km (top right panel of Figure 3). InaRAISE was also able to capture the movement of tropical cyclone Cempaka, both of rainfall intensity and wind speed on 27-28 November 2018 in the southern part of Central Java from Yogyakarta radar (bottom panel of Figure 3).

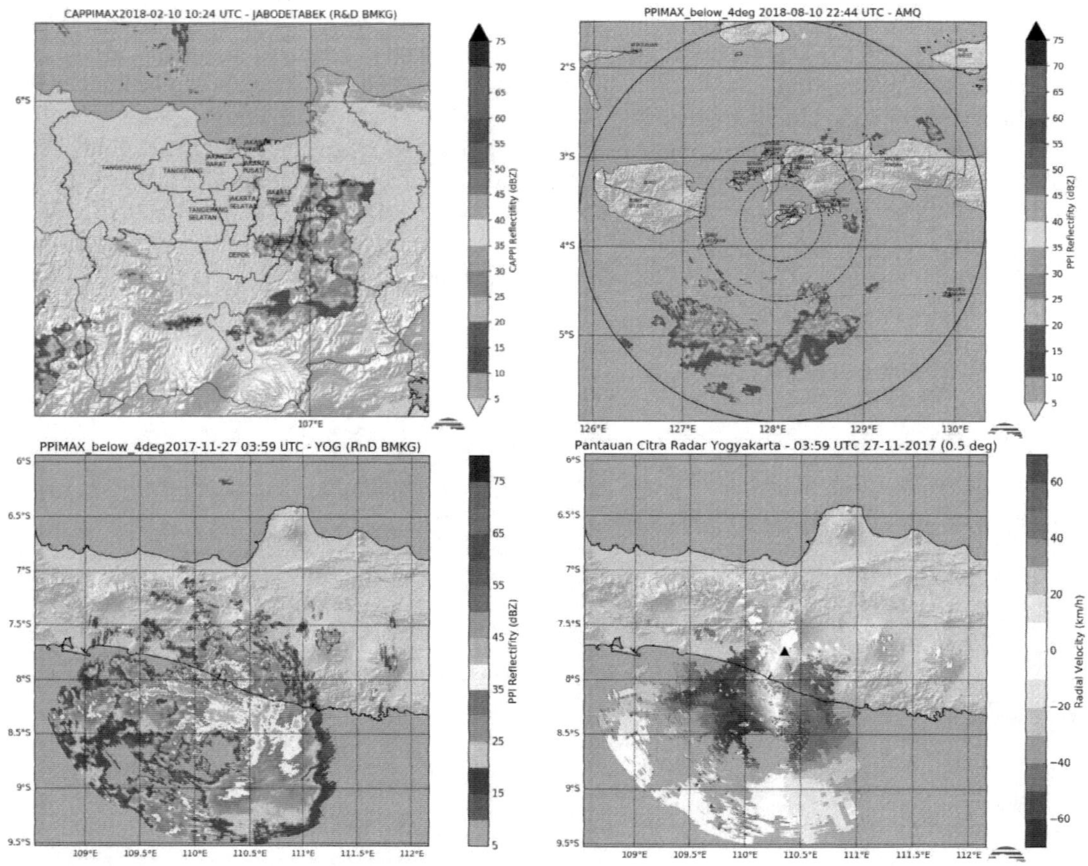

Figure 3. Examples of InaRAISE products. (Top left) the maximum CAPPI reflectivity layer over Jabodetabek area as a part of coverage of Tangerang radar (EEC) on 10 February 2018. (Top right) the maximum PPI reflectivity layer over Ambon (Vaisala) on 10 July 2018. (Bottom left) the maximum PPI reflectivity and (bottom right) 0.5 degree radial velocity layers over Yogyakarta (Baron) during the occurrence of tropical cyclone Cempaka on 27 November 2017.

3. Anticipated Constraints

The main anticipated constraint in the InaRAISE development is the susceptibility of the communication network between radar sites and BMKG headquarter which cause real-time data transfer issues. This issue is more pronounced for radar sites located in the eastern part of Indonesia. The other constraints are the quality control of radar data. Although clutter removal and attenuation correction have been applied, the method of removing intermittent interferences by signal communication is not implemented yet. This interference issue mainly occurs in the city area such as Jakarta. In addition, the bright band filter is also not implemented in the current state of INARAISE. Another important feature that is currently not implemented in INARAISE is the quality control with external data including comparison with rain gauge data to derive the reflectivity-rainrate (Z-R) relationship and comparison with satellite-derived rainfall data. However, R&D center has been working on the comparison of radar reflectivity and observed rainfall data to obtain a new Z-R relationship on several radar locations in Medan, Surabaya, Bali, and Palembang.

4. Recent Progress

INARAISE has been able to generate reflectivity composite maps from multiple radars by using column-maximum reflectivity comparison method. The idea of this method is basically comparing the reflectivity values over the intersection coverage areas of multiple radars and the obtained maximum reflectivity values over that area would be considered as the composite reflectivity values. For instance, radars in Tangerang and Lampung have an intersection coverage area over 105 – 107 °E and 7 – 4.5 °S (Figure 4). On 14 February 2017 13:10 UTC, both radars captured rainfall over the intersection area but its spatial distribution and intensity were slightly different possibly due to the distance difference from the radar center point. By using this method, the composite map indicates that the intersection area was filled with the maximum reflectivity values between the two radars (Figure 4). This method is also applied for multiple radars (more than two radars).

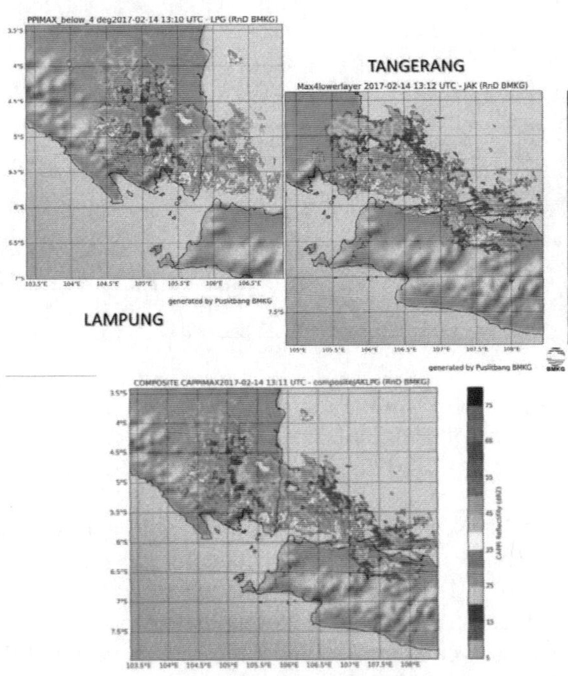

Figure 4. (Top panel) Individual radars in Tangerang and Lampung captured rainfall on 14 February 2017 13:10 UTC. (Bottom panel) Composite radar reflectivity map between Tangerang and Lampung radars.

For multiple radars, InaRAISE has generated composite CAPPIMAX reflectivity maps for main islands in Indonesia. For instance, the composite map for Java and Bali are composed of four radars. While for Sumatera, it consists of seven radars and for Kalimantan, it consists of four radars (Figure 5). The composite map clearly shows the spatial rainfall distribution over a larger area which is very important in understanding current weather condition and possibly to predict the rainfall cloud movements.

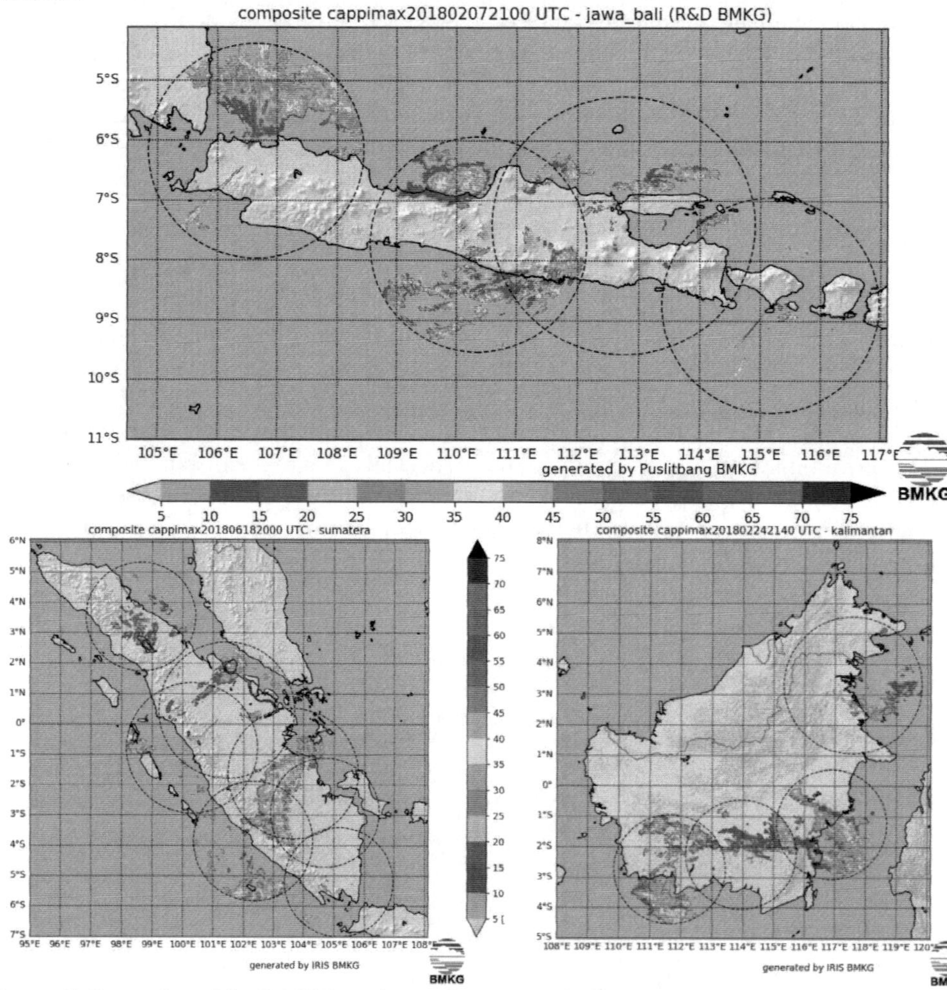

Figure 5. Examples of InaRAISE products. Composite CAPPIMAX reflectivity maps on Java and Bali (top panel), Sumatera (bottom left panel) and Kalimantan (bottom right panel).

In addition to composite maps, InaRAISE serves a feasible radar data extraction for data assimilation in numerical weather prediction (NWP) model. The R&D center has developed the R-language scripts to extract the radar data (reflectivity and radial velocity) values from PPI NetCDF output file and convert it into the format of data assimilation in Weather Research Forecast (WRF) model. However, this module is not automatically running yet in InaRAISE, but it is experimentally used for simulations of extreme events and study cases.

5. Summary

BMKG has 41 weather radars in 2017 consisting of four different brands with its specific volumetric data format and software. In recent years, the weather radar community has developed open-source software systems to process multiple radar data formats. Based on this, BMKG has initiated to develop the Indonesia In-House Radar Integration System (InaRAISE) utilizing the open-source software systems based on wradlib and Py-ART libraries in Python language. BMKG radar data have been successfully extracted and saved into a latitude-longitude NetCDF file format for post-processing. The multiple radars have been successfully composited by column-maximum reflectivity comparison method. The InaRAISE main constraint is the susceptible communication network between radar sites and BMKG headquarter causing real-time data transfer issues. InaRAISE serves feasible data radar extraction for data assimilation in the numerical weather prediction model. InaRAISE could serve as a supporting of the existing radar integration system or possibly as a replacement in the future.

Acknowledgments
Authors wishing to acknowledge assistance or encouragement from BMKG colleagues in Meteorology Research Division of R&D Center, Remote Sensing Management Division of Center for Public Meteorology and Database Center. This research project is financially supported by R&D Center and Center for Public Meteorology BMKG.

References

[1] Permana, D. S., Hutapea, T. D., Praja, A. S., Fatkhuroyan, F., and Muzayanah, L. F. (2016) Pengolahan multi data format radar cuaca menggunakan wradlib berbasis Python. *Jurnal Meteorologi dan Geofisika*, 17(3) (in Indonesian).
[2] M. Heistermann, S. Collis, M. J. Dixon, S. Giangrande, J. J. Helmus, B. Kelley, J. Koistinen, D. B. Michelson, M. Peura, T. Pfaff, and D. B. Wolff. (2015) The Emergence of Open-Source Software for the Weather Radar Community. *Bull. Amer. Meteor. Soc.* 96, 117–128, doi: 10.1175/BAMS-D-13-00240.1.
[3] Michelson, D., R. S. Gill, M. Peura, and J. Szturc. (2010) Community-based weather radar networking with BALTRAD. Proc. Sixth European Conf. on Radar in Meteorology and Hydrology, Sibiu, Romania, Meteo-Romania. [Available online at www.erad2010.org/pdf/oral /wednesday/dataex/01_ERAD2010_0170.pdf.] .
[4] Michelson, D., J. Koistinen, T. Peltonen, J. Szturc, and M. R. Rasmussen. (2012) Advanced weather radar networking with BALTRAD+. Seventh European Conf. on Radar in Meteorology and Hydrology, Toulouse, France, Meteo France. [Available online at www. meteo.fr/cic/meetings/2012/ERAD/extended_abs /NET_073_ext_abs.pdf.].
[5] Michelson, D., Henja, A., Ernes, S., Haase, G., Koistinen, J., Ośródka, K., Peltonen, T., Szewczykowski, M. and Szturc, J. (2018) BALTRAD Advanced Weather Radar Networking. *Journal of Open Research Software*, (3).
[6] Heistermann, M., Jacobi, S., and Pfaff, T. (2013) Technical Note: An open source library for processing weather radar data (wradlib), *Hydrol. Earth Syst. Sci.*, 17, 863-871, doi:10.5194/hess-17-863-2013.
[7] Helmus, J.J. and Collis, S.M. (2016) The Python ARM Radar Toolkit (Py-ART), a Library for Working with Weather Radar Data in the Python Programming Language. *Journal of Open Research Software*. 4(1), p.e25. doi: http://doi.org/10.5334/jors.119.
[8] Hill, D. J., and Baron, J. (2015) radar. IRIS: A free, open and transparent R library for processing Canada's weather radar data. *Canadian Water Resources Journal/Revue canadienne des ressources hydriques*, *40*(4), 409-422.
[9] Heistermann, M., Crisologo, I., Abon, C. C., Racoma, B. A., Jacobi, S., Servando, N. T., David, C. P. C., and Bronstert, A. (2013) Brief communication" Using the new Philippine radar network to reconstruct the Habagat of August 2012 monsoon event around Metropolitan Manila". *Natural Hazards and Earth System Sciences*, *13*(3), 653-657.
[10] Abon, C. C., Kneis, D., Crisologo, I., Bronstert, A., David, C. P. C., and Heistermann, M. (2016) Evaluating the potential of radar-based rainfall estimates for streamflow and flood simulations in the Philippines. *Geomatics, Natural Hazards and Risk*, 7(4), 1390-1405.
[11] Kneis, D., Abon, C. C., Bronstert, A., and Heistermann, M. (2017) Verification of short-term runoff forecasts for a small Philippine basin (Marikina), *Hydrological Sciences Journal*, 62:2, 205-216, doi: 10.1080/02626667.2016.1183773.
[12] Gabella, M. and Notarpietro, R. (2002) Ground clutter characterization and elimination in mountainous terrain. Proceedings of ERAD 2002, Delft, The Netherlands, 305–311, available at http://www.copernicus.org/erad/online/erad-305.pdf.
[13] Jacobi, S., Heistermann, M., and Pfaff, T. (2011) Evaluation and improvement of C-band radar attenuation correction for operational flash flood forecasting. Proceedings of the weather radar and hydrology symposium, Exeter, UK, April 2011, IAHS Publ. 3XX.

IOP Conf. Series: Earth and Environmental Science **303** (2019) 012052 doi:10.1088/1755-1315/303/1/012052

WRF-3DVAR Radiance Data Assimilation Impact for Convective Rain Prediction on Jakarta

D Fatmasari, D M P Rosita Dewi and I Gustari

Indonesia Agency of Meteorology, Climatology and Geophysics. Angkasa I Street No. 2, Kemayoran, Capital Jakarta, 10610 Indonesia

fatmasaridevi@gmail.com

Abstract. Weather Research and Forecasting (WRF) is an open source numerical weather prediction model, but one of the main problems is the inaccuracy of initial condition that impacts the accuracy of weather predictions. Techniques that can be used to improve initial condition is to assimilate satellite radiance data using Three-Dimensional Variation (3DVAR) method. The Global Forecast System (GFS) data were used as initial conditions which were assimilated with several radiance data such as from AMSU-A sensors, MHS sensors, and ATMS sensors. The purpose of this study was to compare the effect of data assimilation on the initial condition model and the rainfall prediction for convective precipitation on February 24, 2016 in Jakarta. The results showed there is a clear flow of moisture to the northern part of Jabodetabek area that cause the model captured a high relative humidity which results in the heavy rain. The result of the study obtained different rainfall prediction between non-assimilation model and assimilation model. In general, both models have underestimated rainfall, but satellite radiance data assimilation is better able to describe the convective rain patterns and has better accuracy on the convective rainfall accumulation per hour.

Keyboard : WRF-3DVAR, radiance, assimilation, convective.

1. Introduction

Precipitation consists of two types which is stratiform and convective. Stratiform precipitation falls from stratus, stratocumulus, and nimbostratus clouds, while convective occurs from cumulus and cumulonimbus clouds [1]. This type of cloud may occur separately or in the same cloud complex. Convective precipitation occurs in relation to strong low-level convergence. Convective clouds cause local rain that occurs due to convection located in unstable air [2]. Rain from the convective cloud has a large rainfall intensity, from moderate to heavy rain (shower). Rainfall from convective clouds has a value of >10 mm / hour [3]. Because of the high rainfall intensity, it often causes flooding in the Jakarta area, so it is imperative to use a high-resolution heavy rainfall forecast as an important role to produce accurate forecast for warning information for operational purposes. One of the latest generations of the meso scale model of Numerical Weather Prediction (NWP) is Weather Research and Forecasting (WRF). WRF model has good potential in observing rainfall events such as rain time, location, and evolution [4]. But the accuracy of rainfall forecasts is also influenced by many factors in the dynamic part of the numerical model, especially the uncertainty of the initial conditions [5]. Improvements to the initial conditions can improve the accuracy of rainfall forecasts [6] which can be done through data assimilation on the WRF model. Data assimilation is a method of improving the initial conditions of the atmosphere as a model input by calculating the observation data into a grid model system [7]. The application of data assimilation in the NWP model consists of several techniques, one of which is The Three-Dimensional Variation (3DVAR). 3DVAR techniques have better performance in producing rational analysis of hydrometeorological events with greater computational efficiency than other techniques [8].

The data used for assimilation consists of various types of data, ranging from surface observation data, upper air observation data, remote sensing data, and the combination of these types of data. The lack of surface-based weather observation platforms creates problems in monitoring meso-scale weather systems and rainfall characteristics in an area. Although there have been already many papers about this topic, but only several of them that discuss the effect of satellite radiance data assimilation in the tropic region, especially in Jakarta area. Therefore, it is necessary to use satellite observation data which has a wider scope of observation to cover the shortcomings of the surface weather observation platform. Moreover, because the satellite data used in assimilation is polar orbit satellite to some extent, not all satellite data can be used in an event in an area. Even though the initial global model data has been assimilated by satellite radiation data, but when downscaling is finished, it is uncertain that the area and time of the event taken as a case have been assimilated or not. Consequently, it is very necessary to further assimilate each domain that has been downscaled to ensure that each domain has been influenced by the assimilation of satellite radiation data. Based on data from *rsl.out.0000* of the assimilation results, it shows the satellite radiance data has successfully assimilated hundreds of data from satellites that crossed the WRF model's first guess data. Based on the previous research that investigate Quantitative Precipitation Forecast (QPF) skill evaluating the influence of data assimilation for rainfall prediction in Java region showd that radiance data assimilation has slightly contribution on rainfall prediction, but it presented better accuracy on rainfall prediction for the heavy rainfall cases [9]. Moreover, the other results from WRF data assimilation that examine rainfall prediction in Jakarta region using satellite data assimilation provides the best effect compared to other assimilation models, but it still overestimates compared to the observational data [10]. The addition of satellite radiance assimilation data has an impact on improving the quality of forecasts, initial conditions, spatial patterns of rainfall and rainfall intensity [5, 11].

This proves that satellite data can have a positive impact on forecast results through the method of data assimilation in the WRF model. Direct radiance assimilation is theoretically superior to retrieval assimilation because the observational error statistics are more justified in direct radiance assimilation than in retrieval assimilation [12]. This approach requires an observation operator built into the data assimilation system to transform model variables into radiances. The linkage between forecast model state variables and observed radiances is expressed mathematically by a Community Radiative Transfer Model (CRTM) that was developed by the U.S. Joint Centre for Satellite Data Assimilation (JCSDA) to provide fast, accurate satellite radiance simulations, which calculates radiance from model state vertical profiles. In this study, the WRF-3DVAR system is used to explore the effect of the assimilation of NCAR AMSU-A and MHS microwave radiance data on forecasts of heavy precipitation on the Jakarta region. This study aims to see how far the results of predictions of numerical rainfall predictions have improved after assimilation of satellite radiance data when compared to models without data assimilation. This paper is organized as follows, Section 2 gives a concise overview of the method and experiment design. Section 3 provides information on the study area and the data. The results of the data assimilation experiments are presented and evaluated in Section 4, and conclusions are given in Section 5.

2. Method and experimental design

2.1. WRF model set up

The numerical data assimilation experiments in this study are conducted using the Advanced Research WRF model Version 3.8. WRF is a model of numerical weather prediction systems and atmospheric simulations designed for both research and operations [7]. WRF-ARW has been developed in recent years and is designed to be a sophisticated and efficient atmospheric simulation system on existing parallel computing platforms [13]. As shown in Figure 1, we use two nested model domains with 9 km

(100 × 100) and 3 km (75 × 75) horizontal spacing. Each domain has 28 vertical pressure levels with the top-level set at 50 hPa. The WRF physical parameterization schemes used in this study include the Purdue Lin microphysical parameterization, Rapid Radiative Transfer Model (RRTM) longwave radiation, Dudhia shortwave radiation, Monin Obukhov surface layer, Noah land surface, Yonsei University (YSU) planetary boundary layer scheme, and Kain Fritsch (KF) cumulus scheme. The projection method is Mercator.

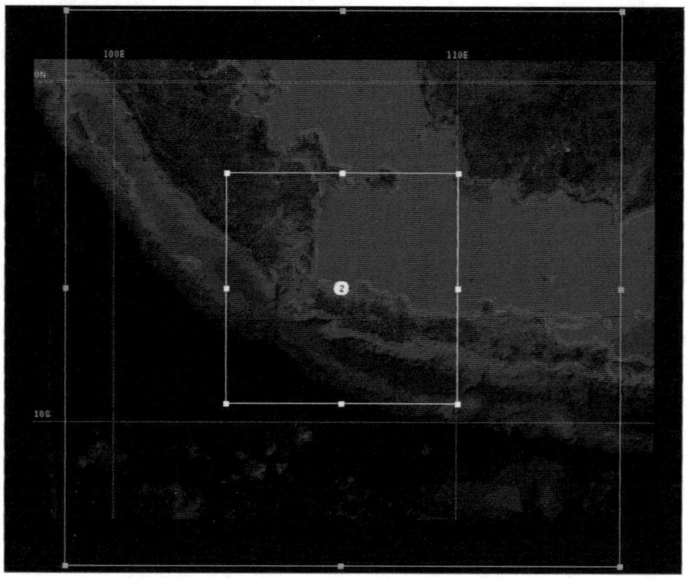

Figure 1. Domains for the WRF ARW. The outer box is the coarse grid with a resolution of 9 km (d01); the inner box is the nested grid (d02) with a resolution of 3 km.

2.2. Satellite radiance data

Satellite data is very important in the forecast process using the NWP model because assimilation of sounding satellite data can significantly improve temperature and humidity forecasts [14]. Data on Global Data Assimilation System (GDAS) Satellite Radiance Data (ds735.0) in PREBUFR format as data used in the process of assimilating radiance satellite data into WRFDA provided by NCEP. In this research, the Advanced Microwave Sounding Unit (AMSU-A) and Microwave Humidity Sounder (MHS) satellite sensors are found on NOAA-15 and NOAA-18 satellites. AMSU-A has spatial resolution at nadir points of 48 km and measures in 15 spectral bands and divided into two separate units: AMSU-A1 dan AMSU-A2 [15]. AMSU-A2 with two channels at 23.8 GHz and 31.4 GHz and AMSU-A1 with twelve channels in the range 50.3 GHz up to 57.290344 GHz plus one channel at 89.0 GHz. The 12 oxygen-band channels (channels 3–14) will provide microwave temperature sounding for regions from the Earth's near surface up to about 42 kilometres or from 1000 millibars to 2 millibars. The extreme spectral windows (channels 1, 2 and 15) allow correction of the other measurements for surface emissivity, atmospheric liquid water and total precipitable water. These channels also provide information concerning precipitation, sea ice and snow coverage. While MHS has spatial resolution at nadir of 16 km and has five-channel microwave radiometer, which complements the Advanced Microwave Sounding Unit-A (AMSU-A) channels [15]. Humidity profiles are planned to be derived from data in this frequency range. Information about cloud liquid water content, precipitation, and precipitation rates is also derived from these channels.

2.3. Data assimilation system

Data assimilation can be defined as a technique where observation is combined with NWP products and error statistics to provide a means to combine all available information to make the best estimates of atmospheric conditions [16]. The basic purpose of the variational data assimilation system is to produce optimal forecasts of the actual atmospheric state at the time of analysis through the iterative solution of the cost function equation. The following is the cost function formula for the 3DVAR assimilation method [17]:

$$J(x) = J_o(x) + J_b(x) = \frac{1}{2}(y^0 - H(x))^T R^{-1}(y^0 - H(x)) + \frac{1}{2}\left(x - x^b\right)^T B^{-1}\left(x - x^b\right) \quad (1)$$

where J_b and J_o represent background and observational terms respectively, $H(x)$ is an operator used to transform the model grid points analysis x to observational space, and B and R are the background and observation error covariance matrices respectively. The minimization procedure changes x iteratively in order to approach the (unique) minimum of $J(x)$ and as a result, the analysis x^a. There are three input data for the data assimilation process, namely first guess data (x^b), observation data (y^0), and background error (B) data [13].

2.4. Experimental Design

Two sets of experiments have been conducted using two domains and run for two days from February 24 to February 25, 2016 in Jakarta region. The model simulation without data assimilation will be referred to as the control experiment (CTRL). While assimilation experiments are designed using satellite radiance data in the SATA experiment which assimilated AMSU-A and MHS radiance data.

3. Study area and data

As shown in figure 2, the study area of Jabodetabek (Jakarta Bogor Depok Tangerang and Bekasi) which is on the north coast of Java Island. A total of 2 national observation stations have been used to verify the temporal distribution of the simulated precipitation every hour, they are Soekarno-Hatta station and Kemayoran station.

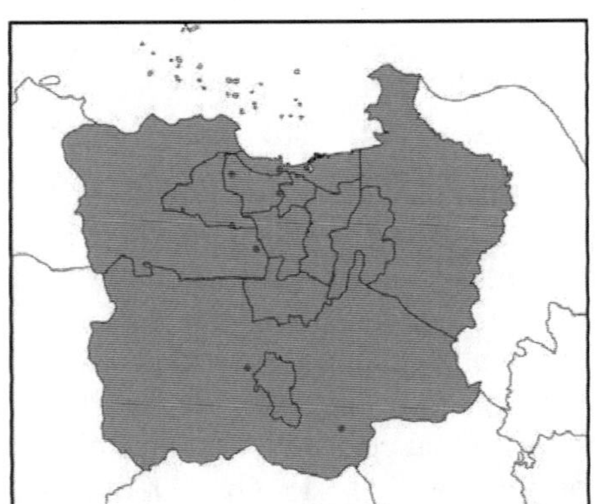

Figure 2. Location map showing study area and 3 observation stations

The initial and boundary conditions necessary to run the WRF are the GFS (*Global Forecasting System*) data at 0.25° x 0.25° grid resolution obtained from the National Centre for environmental Prediction (NCEP). This is because we want to simulate the precipitation forecast in the region of Jabodetabek. In the model integration, the coordinates of the central point are 6.117°S and 106.65°E. The WRF-3DVar experiment was conducted with the aim of modifying the initial condition data model obtained by assimilating other observation data. The assimilated data includes NCAR satellite radiance data from AMSU-A and MHS radiance data. The data are initially downloaded in PREPBUFR format and can be assimilated directly into WRFDA. The AMSU-A and MHS satellite radiance data (NOAA-15/18) from the NOAA ATOVS instruments can be read in WRFDA via CRTM2.0.2, which is in BUFR format.

4. Results and discussion

4.1. Impact of data assimilation on the initial fields

The purpose of data assimilation is to improve the initial condition data of the model. To find out the changes and improvements to the data of the initial condition of the model, an analysis of the initial conditions of the model will be carried out without assimilation of data and models with assimilation of satellite radiance data. The parameters used in the analysis are moisture flux, air temperature, and wind at an altitude of 850 hPa.

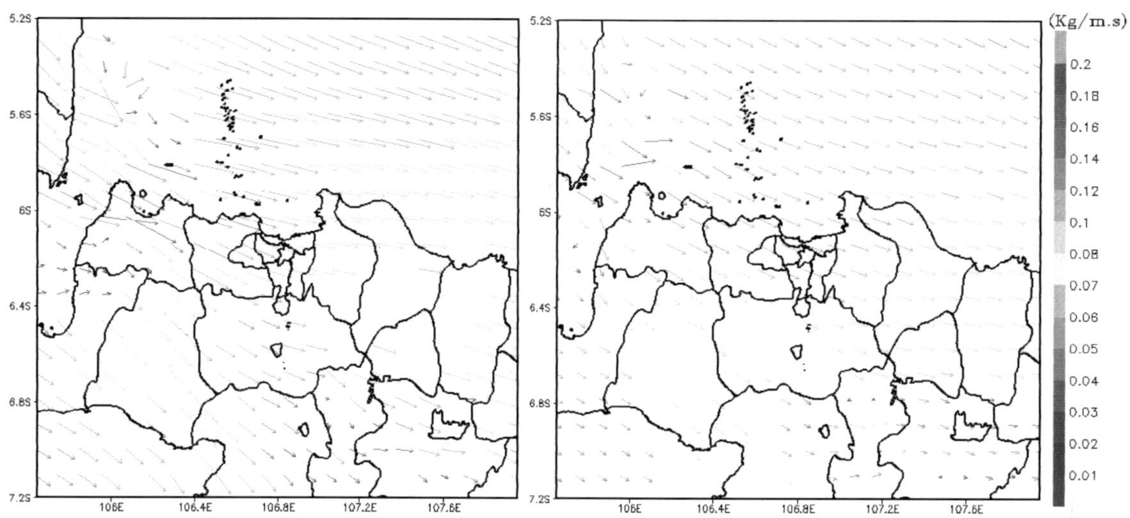

Figure 3. The 850-hPa moisture flux for initial condition for the (a) CTRL and (b) SATA

Figure 3 shows the moisture flux between control experiments and data assimilation. Compared with CTRL experiment (Figure 3(a)), there is only a slight change of moisture flux in SATA (Figure 3(b)). The moisture flux of SATA experiment is slightly increased in the south of the study area. It may bring more precipitation in the Jabodetabek region. The moisture flux of SATA experiment also shows marked increases in the northern part of the Java Island in comparison with the control experiment.

Based on the previous experiment that also stated that the use of satellite radiance assimilation had a large influence on the parameters of air temperature and atmospheric humidity in the tropics, while another assimilation such as radiosonde was more dominant in extratropical regions [18]. The 850-hPa circulations (figure 4) from both experiment over domain 2 show that an eastward moving wind occurs in the study area. In the Jabodetabek area, it is also seen that there is a slowing in the wind speed that allows the occurrence of air mass associations around the area. Besides that, based on the air temperature (shade) in the SATA experiment (figure 4 (b)) also shows that the Jabodetabek area is also warmer when compared to the CTRL experimental model (figure 4 (a)). The warmer the air temperature in the lower layers of the atmosphere can potentially affect the air flow where air flows into places with higher temperatures. This is in accordance with the existence of an air mass slowdown as indicated by the wind vector). However, there is a convergence zone located in Jakarta region at 1400 UTC 24 February in DA experiment.

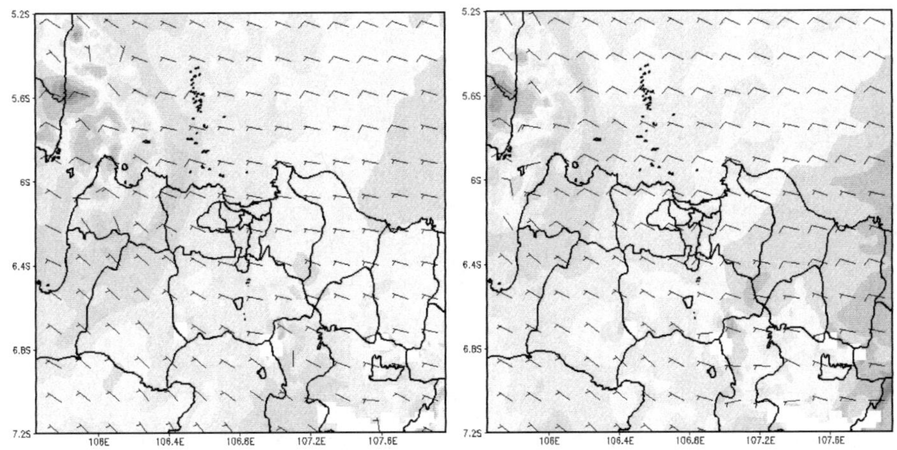

Figure 4. The 850-hPa temperature (shade; °C) and wind (barb; m s^{-1}) over domain 2 for (a) CTRL and (b) SATA

The 850-hPa humidity and circulations at 1200 UTC, 1800 UTC 24 February and 0000 UTC 25 February over domain 2 for CTRL and SATA are shown in figure 5. The initial relative humidity at 1200 UTC in the CTRL is at least 20% higher than that of SATA in the southern of Jabodetabek. The temperature is about 0.4 °C lower than that of SATA in the Jabodetabek. The temperature at 1200 UTC in SATA improved in the southeast of Jabodetabek compared with the CTRL. At 1800 UTC, a clear convergent flow and weak north-westerly wind flow occurred over the Jabodetabek area in both models, and the relative humidity of above 90% controlled the whole Jakarta area for SATA experiment (figure 5(e)). While in CTRL experiment, it only shows around 80% - 85% of relative humidity in the Jakarta area and above 95% over Bekasi. The wind flow in the SATA forming a convergence in the southern of Jabodetabek at 1800 UTC 24 February and 0000 UTC 25 February stronger than in CTRL experiment. Meanwhile, at 0000 UTC 25 February, the relative humidity of above 95% is mainly located over Jabodetabek area in model simulations. Differences of temperature and humidity from initial conditions lead to a displacement of the moisture flux in the CTRL and SATA. As the convergences move and enhance in the southern Jabodetabek, the relative humidity of SATA experiment seems to be increased at 1800 UTC 24 February and 0000 UTC 25 February over Jabodetabek. To summarize, the WRF model presented the favourable mesoscale synoptic conditions for convective precipitation over Jabodetabek. Inaccurate initial fields are particularly associated with the location errors and underestimated precipitation. Assimilation of the satellite radiance data has a positive impact on the initial fields, but the convergence still cannot be placed in the right position.

4.2. Impact of data assimilation on the precipitation forecast

From Figure 6 almost both simulations underestimated rainfall accumulation at all two observation points. Underestimation can be seen in the diagram where both all model simulations have a lower accumulation value of rainfall compared to the data of rainfall accumulation of observation stations. Although the intensity and accumulation of rainfall from some outputs of the model still underestimate when compared with rainfall observation data, but the SATA assimilation model can describe rain patterns similar to the data from observation points. The SATA experimental model has a rain pattern that is similar to the rain-gauge data at both observation points although it still shows an underestimate value for rainfall. While the CTRL experiment still cannot capture the rain pattern and is less accurate in simulating convective rain peaks. This can be caused by the influence of satellite radiance data assimilation in improving the initial conditions data on air temperature parameters and atmospheric humidity [5].

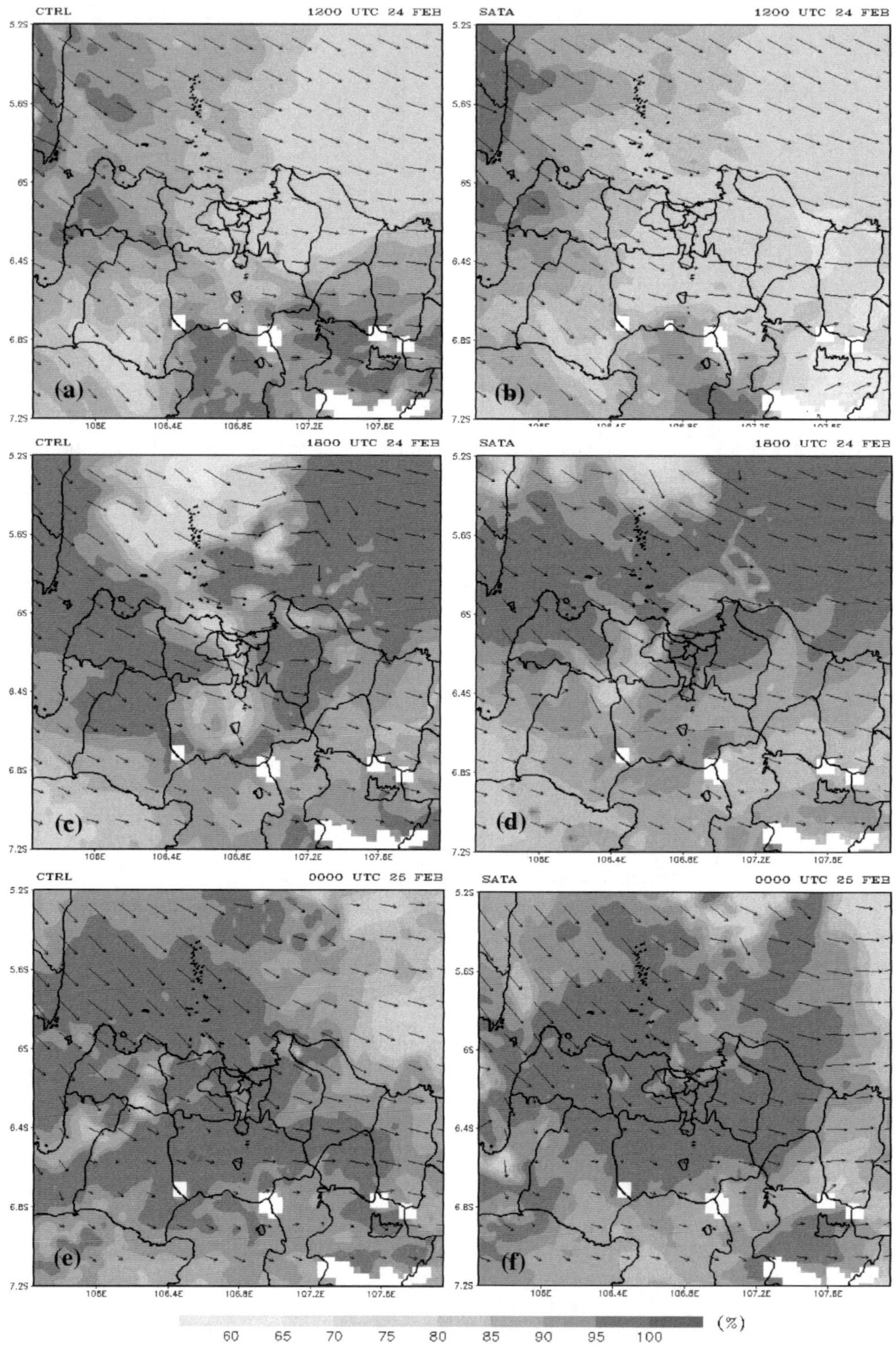

Figure 5. The 850-hPa relative humidity (shade; %) and wind (vector; m s^{-1}) over domain 2 for CTRL and SATA. (a-b) 1200 UTC 24 February, (c-d) 1800 UTC 24 February, (d-e) 0000 UTC 25 February

IOP Conf. Series: Earth and Environmental Science **303** (2019) 012052 doi:10.1088/1755-1315/303/1/012052

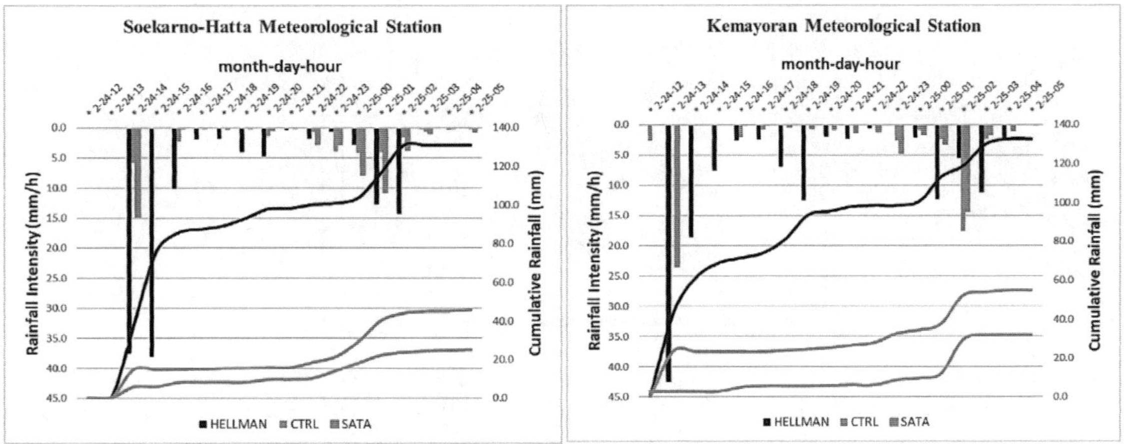

Figure 6. The intensities and cumulative rainfall of the values measured at the rain-gauge instrument and the simulated results from numerical experiments

The possibility of the number of rains forecast estimates according to predictions but not occuring is presented through POFD values, while the POD index is an index to show the probability of events that can be detected by the model (hit rate). Based on Table 1, the POD value of the model that has been assimilated by satellite radiance data is higher than the model that has not been assimilated, this shows that the level of truth from the prediction of rain events generated by the radiance data assimilation model is better than the model without assimilation. In addition, the model performance after data assimilation is also shown by the lower POFD value compared to the model without assimilation. This lower POFD value indicates that the model prediction error rate when there is no rain is also better when compared to the model before assimilation. So that based on the results of point verification, the satellite radiance data assimilation model has a better performance in predicting rain events compared to models without assimilation.

Table 1. POD and POFD of control and assimilation experiment

	CTRL	SATA
POD	0.66	0.76
POFD	0.50	0.44

The results of WRF rain prediction both WRF without assimilation and WRF output assimilation of satellite radiance data were also verified against GSMaP rain estimation data. Regridding process needs to be carried out first between WRF data and GSMaP data which aims to equalize the second grid data resolution. The results of daily rainfall accumulation in the GSMaP data show that the rainfall area is on the north coast of Java and Banten, while for the Jakarta area there is light rain with an intensity of 20 mm - 50 mm. Meanwhile, the WRF model of both assimilation and without assimilation also shows that the area of rain tends to be more widespread in the Java Sea which causes the model to overestimate the region. In addition, the model is still not able to capture the occurrence of heavy rain that occurred in the Banten region. Rain intensity shown by both models in the area of heavy rain is still very far from the actual intensity, this causes the model to underestimate to predict the occurrence of heavy rain to very heavy.

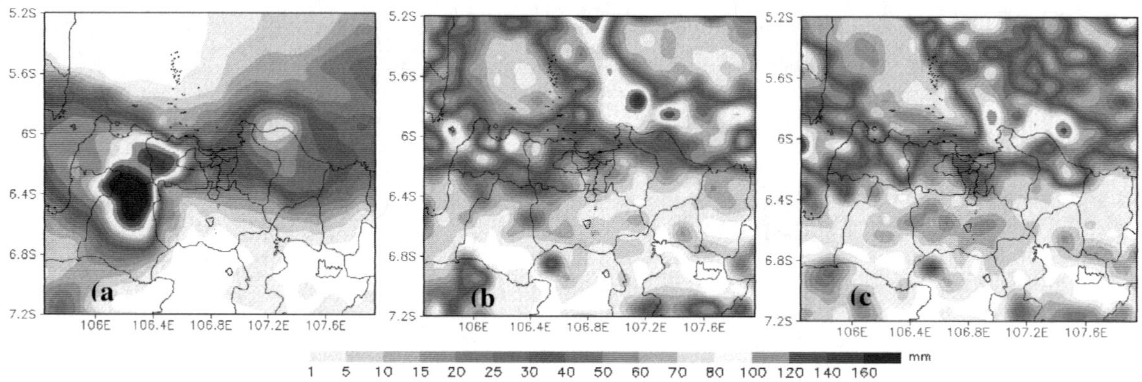

Figure 7. Simulated daily precipitation (mm) over the study area for the CTRL (b) and SATA (c), together with the observed (GSMaP) daily precipitation (a) during 24 February 2016

Hence, even though the two experimental models are less able to capture the occurrence of rain, the model without assimilation has more overestimated precipitation area compared to the assimilation model. As for the assimilation model, even though it is overestimated, the value is still smaller. The results of the analysis of the WRF prediction model show that the WRF prediction model with assimilation of satellite radiance data gives a better influence compared to the model without assimilation. This is due to an improvement using satellite radiance data which adds the radiation temperature of the cloud peak temperature in the calculation of the initial conditions of the WRF model [19]. As previously shown, there is a change in the value of the initial condition of the model after assimilating satellite radiance data even though the changes are small. These small changes are caused by ignoring small-scale features or local characteristics of a region when using 3DVAR assimilation techniques which produce a homogeneous contour [20] and also the number of satellite radiance data that assimilated is very low / limited because it uses polar orbit satellite. Using direct assimilation of satellite radiance data gives positive results on the initial condition data and a significant increase in the simulation of heavy rain events up to 24 hours [5]. Even though infrared and microwave radiances have been assimilated successfully in the original operational GFS forecast system since 1998 using a clear-sky approach [21] and the all-sky AMSU-A radiance assimilation and became operational in the 4D EnVar GFS system upgrade of 12 May 2016 [22], but GFS model is run using the previous model so data assimilation is very important to combine irregular observations to generate the initial conditions that are distributed on regular model grids [23], therefore it is neccesary to assimilate the instruments in downscaling process to update the initial condition data.

5. Conclusions

In this study, the 3DVAR assimilation system is used to improve the forecasting of heavy rainfall in Jabodetabek area. A control and satellite radiance data assimilation experiments were designed. These experiments demonstrate that the improvement of the rainfall prediction by the assimilation model is still not too significant, but in general, satellite radiance assimilation can reduce the distribution of the overestimated area and intensity of the precipitation, through enhancing the accuracy of the initial field. For the point verification, rain-gauge data from 2 sites were used to evaluate the impact after data assimilation, and resulted that data assimilation experiment has the same pattern as the measured rainfall intensity by the rain-gauge instrument. The result shows that SATA experiment has higher accuracy and lower error rate in predicting rain events compared the model without assimilation. And also, the cumulative curve in both experiments is below the curve of the observed rainfall, but SATA experiment has the closest value to the observed rainfall. In conclusion, based on the results of verification, the satellite radiance data assimilation experiment has a better performance in predicting rain events compared to models without assimilation. These results indicate that efforts are still needed to improve the model prediction results in the case of rain events. Utilizing more satellite radiometer data into the model as the initial conditions could specifically benefits the heavy rainfall nowcasting. In addition to improving the initial condition data model, improvements to static data on WRF models such as topography and land cover are also needed to improve the accuracy of rainfall predictions.

Reference

[1] Houze Jr R A 2014 *Cloud Dynamics 2nd ed* (Amsterdam: Elsevier Inc.)
[2] Tjasyono BHK 2007 *Mikrofisika Awan dan Hujan* (Jakarta: Badan Meteorologi dan Geofisika)
[3] Nzeukou A and Sauvageot H 2004 Raindrop size distribution and radar parameters at cape Verde *Journal of Applied Meteorology* **43** 90-105
[4] Hong S Y and Lee J W 2009 Assessment of the WRF model in reproducing a flash-flood heavy rainfall event over Korea *Atmospheric Research* **93** 818–31
[5] Xie Y, Xing J, Shi J, Dou Y and Lei Y 2016 Impacts of radiance data assimilation on the Beijing 7.21 heavy rainfall *Atmospheric Research* **169** 318-330
[6] Etherton B and Santos P 2008 Sensitivity of WRF forecasts for South Florida to initial conditions *Weather and Forecasting* **23** 725–40
[7] Skamarock W C, Klemp J B, Dudhia J, Barker D O, Duda M G, Huang X Y, Wang W and Powers J G 2008 A Description of the Advanced Research WRF Version 3, *NCAR Technical Note*, NCAR/TN–475+STR
[8] Barker D M, Huang W, Guo Y R, Boourgeous A J and Xiao Q N 2004 A three-dimensional variational data assimilation system for MM5: implementation and initial results *Monthly Weather Review* **132** 897-914
[9] Sagita N, Hidayati R, Hidayat R and Gustari I 2017 Satellite radiance data assimilation for rainfall prediction in Java Region *IOP Conference Series: Earth and Environmental Science* **54**
[10] Paski J A I 2017 Impact of remote sensing data assimilation (radar and satellite) on numerical weather prediction for rainfall estimation *Jurnal Penginderaan Jauh* **14** 77-86
[11] Yang J, Duan K, Wu J, Qin X, Shi P, Liu H, Xie X, Zhang X and Sun J 2015 Effect of data assimilation using WRF-3DVAR for heavy rain prediction on the North-Eastern Edge of the Tibetan Plateau *Advances in Meteorology* **2015**
[12] Eyre Jr A, *Fast-Radiative Transfer Model for Satellite Sounding Systems* (ECMWF Tech Memo 1991)
[13] Wang W, Bruyère C, Duda M, Dudhia J, Gill D, Kavulich M, Keene K, Chen M, Lin H C, Michalakes J, Rizvi S, Zhang X, Berner J, Ha S and Fossell K 2017 *WRF-ARW Version 3 Modeling System User's Guide* (United States: National Center for Atmospheric Research)
[14] Laing A, and Evans J L 2016 Introduction to Tropical Meteorology 2nd Edition [online], *http://www.meted.ucar.edu/tropical/textbook_2nd_edition/*
[15] EUMETSAT 2003 *AMSU-A Level 1 Product Generation Specification* (German: EUMETSAT)
[16] Lackmann G 2012 *Midlatitude Synoptic Meteorology: Dynamics, Analysis, And Forecasting* (Boston: American Meteorological Society)
[17] Kalnay E 2003 *Atmospheric Modeling, Data Assimilation, and Predictability* (Cambridge UK: Cambridge University Press)
[18] Kutty G and Wang X 2015 A Comparison of the impacts of radiosonde and AMSU radiance observations in GSI based 3DEnsVar and 3DVar data assimilation systems for NCEP GFS *Advances in Meteorology* **2015**
[19] Sagita N 2017 Asimilasi model Weather Research and Forecasting (WRF) dengan data observasi untuk prediksi curah hujan di wilayah Jawa (Bogor: *Thesis* of Applied Climatology in Bogor Agricultural Technology Institute)
[20] Barcons J, Folch A, Afif A S, Miró J R 2015 Assimilation of surface AWS using 3D-Var and LAPS and their effects on short-term high-resolution weather forecasts *Atmospheric Research* 160-173
[21] Derber J C and Wu W S 1998 The use of TOVS cloud-cleared radiances in the NCEP SSI analysis system *Monthly Weather Review* **126** 2287–99
[22] Zhu Y, Liu E, Mahajan R, Thomas C, Groff D, Van Delst P, Collard A, Kleist D, Treadon R and Derber J C 2016 All-sky microwave radiance assimilation in NCEP's GSI analysis System *Monthly Weather Review* **144** 4709–35
[23] Sahu, D K, Dash S K and Bhan S C 2014 Impact of surface observations on simulation of rainfall over NCR Delhi using regional background error statistics in WRF-3DVAR model *Meteorology and Atmospheric Physics* **125** 17- 42

IOP Conf. Series: Earth and Environmental Science **303** (2019) 012053 doi:10.1088/1755-1315/303/1/012053

Numerical Analysis using Empirical Orthogonal Function Based on Multivariate Singular Value Decomposition on Indonesian Forest Fire Signal

P Septiawan[1], S Nurdiati[1] and A Sopaheluwakan[2]

[1]Mathematics Department, Faculty of Mathematics and Natural Science, Bogor Agricultural University, Bogor, 16680, Indonesia
[2]Research and Development Center, Meteorology, Climatology and Geophysics Agency, Jakarta, 10720, Indonesia

Email: Septiawan.pandu@gmail.com

Abstract. Over the past years, fires have burn Indonesian forest annually. The strength of the fire and many of fire points is vary each year. In 1997 there was 9.75 million ha forest burned in one year with 206,6 million ton of carbon emissions. In 2015 there was 2.089 million ha forest and 805 million tons carbon emissions. Both years involved the biggest incidences of burned forest in Indonesia. The purpose of this research is to analyze the pattern of burned forest in Indonesia and to analyze linkages of variables that affect it. Datasets that used in this research is monthly data from 1997 until 2016 Global Fires Emissions Database (GFED) with variable burned area, GFEDs with variable carbon emissions, and Climate Hazards Group InfraRed Precipitation with Station data (CHIRPS). Using Singular Value Decomposition and Empirical Orthogonal Function analysis, for the overlapping period detected that there is annual signal that dominate the incidences of fires in Indonesia, especially in Riau, Palembang, South Kalimantan and Central Kalimantan. There was 4 annual signal that can explain 72% of fires incidences in Indonesia in 1997-2016. In Riau there is another signal that have 6 months period caused by 6 months rainfall period.

Keywords : fire, GFEDs, empirical orthogonal function, singular value decomposition.

1. Introduction

Forest fires are one of the impacts of increasing levels of pressure on forest resources. The impacts associated with forest or land fires are environmental damage and pollution, such as damage to flora and fauna, soil and water. There are several parameters that affect forest fires, including carbon emissions and rainfall. In this research focused on seasonal rainfall in order to see seasonal patterns of forest fire patterns in Indonesia.

In a previous research entitled "Precipitation-fire linkages in Indonesia (1997-2015)" by Thierry Fanin and Guido R. van der Werf examined the relationship between rainfall and forest fires in Indonesia. This research used 2 datasets namely the Global Fire Emissions Database (GFED) and the Tropical Rainfall Measurement Mission (TRMM) 3B42 in 1997-2015 and focused on the Sumatra and Kalimantan regions. The dataset used in the research is a daily dataset that focuses on night time to prove its consistency in the daily period. In this research, 5 patterns were obtained in 4 regions with different spatial and temporal patterns.

Global Forest Watch Fires shows that more than half of Indonesia's forest fires occur on peatlands, which are found in Central Sumatra, South Sumatra, Central Kalimantan, South Kalimantan and Papua.

IOP Conf. Series: Earth and Environmental Science **303** (2019) 012053 doi:10.1088/1755-1315/303/1/012053

In 2015, there were 96,937 hotspots detected in Indonesia, with total emissions in Indonesia of 1,403 million tons of carbon. Based on the number of hotspots and carbon emissions, we can predict or look for patterns about forest fires [6].

The aim of this research is to analyze burned land, carbon emissions and rainfall data in Indonesia using combined Empirical Orthogonal Function method based on Singular Value Decomposition (SVD). The next aim is to find the seasonal joint patterns between variables of forest fires, carbon emissions, and rainfall. Then the aim is to analyze the linkages of these three variables to the shared patterns that are formed. This research is expected to provide a very dominant signal that affects forest fires in Indonesia related to rainfall, so that it can be used for future purposes in determining the prediction of forest fires in Indonesia.

2. Types and Data Sources

The first data used in this research is the Global Fire Emissions Database (GFED) by taking the burned area variable. The second data is a small Global Fire Emissions Database (GFEDs) with a small fire fraction to get emissions variable. Both of these data have a spatial resolution of 0.250 x 0.250 and a monthly temporal resolution with HDF format. The third data related to rainfall variable is Climate Hazards Group Infrared Precipitation with Station data (CHIRPS). CHIRPS data has a spatial resolution of 0.050 x 0.050 and a monthly temporal resolution with bil format. Data reduction and transformation in this research were carried out in accordance with the Indonesian domain.

3. Methods

The methods that will be used in this research are combined empirical orthogonal function (EOF) method based on singular value decomposition (SVD) and Fourier transformation.

3.1. Combined Empirical Orthogonal Function based on Singular Value Decomposition

Combined EOF is a generalization of EOF which can be considered as an EOF with combined parameter. Combined EOF is a method that allows simultaneous consideration of modes of different variable variations. One way is to combine several different data by equating the variables owned by both data. Defined a new data set by combining variables in data. The data matrix is then arranged in such a way that the bounded fields at the same time are arranged into the same column [8].

For example:

$$\boldsymbol{Z} = [\boldsymbol{Z_1}, \boldsymbol{Z_2}, \boldsymbol{Z_3}, \dots, \boldsymbol{Z_n}]$$
$$S = [S_1, S_2, S_3, \dots, S_n]$$

Then:

$$Y_n = \begin{bmatrix} Z_1 & Z_2 & Z_3 & \dots & Z_n \\ S_1 & S_2 & S_3 & \dots & S_n \end{bmatrix}$$

The SVD-based EOF method is used to reduce the data matrix into computer programs using software. Suppose X is any matrix measuring $n \times p$ with rank (X) = r. Singular value decomposition (SVD) of X is a factorization in the form of:

$$X = U \sum V^T$$

with $U = [u_1, u_2, \dots, u_n]$ and $V = [v_1, v_2, \dots, v_n]$ are orthogonal matrix. U is $n \times n$ matrix, V is $p \times p$ matrix, and $\sum$ is diagonal matrix n × p with $\sigma = \text{diag}(\sigma_1, \sigma_2, \dots, \sigma_n)$ dan $\sigma_1 \geq \sigma_2 \geq, \dots \geq \sigma_n \geq 0$. Where σ_i is singular value matrix X. U and V are left singular vector and right singular vector X.

The variant of i main component i = 1,2, ..., r can be calculated by:

$$\mu_i = \frac{\sigma_i^2}{\sum_{i=1}^{r} \sigma_i^2}$$

with μ_i, i = 1,2, ..., r is the singular value of matrix X. In practice, k mode EOF1 or the first major component with k $\ll$ r explains the proportion of the largest variant. EOF2 mode is a linear combination of all observed variables that are orthogonal to EOF1 mode and have the second largest variant and so on. Therefore, the K EOF mode has a maximum variance of k and does not correlate with the previous EOF mode [5].

3.2. Fourier Transformation

Fourier transformation method is used to convert the signal from the time domain to a signal with a frequency domain. Fourier transform is defined as

$$F(v) = \int_{-\infty}^{\infty} f(x)e^{-i2\pi vx}dx \qquad f(x) = \int_{-\infty}^{\infty} F(v)e^{i2\pi vx}dv$$

$f(x)$	= signal with the time domain	v	= frequency,
$e^{-i2\pi vx}$	= kernel function,	x	= time.
$F(v)$	= signal with a frequency domain,		

The functions $f(x)$ and $F(v)$ are called the Fourier transform pairs. The first equation is called an advanced transformation (analysis) to transform time to frequency and second equation is called an inverse (synthesis) transformation to transform function from frequency to time domain [11].

4. Research Steps

This stage begins with a literature study related to the method that will be used. SVD-based combined EOF method is used to reduce data matrix into computer programs using software. This method can reduce large data without removing the main information from the data. Reductions were carried out on GFED, GFEDs and CHIRPS data in the Indonesian coverage area. This method aims to determine the dominant patterns in bulk data both spatially and temporally. After that, the patterns that are formed will be analyzed and find the linkages between the three variables used.

CHIRPS data has a smaller grid resolution of 0.05^0 x 0.05^0, so before being reduced, grid resolution is equalized by changing the resolution of CHIRPS data to 0.25^0 x 0.25^0. The resolution is equalized by taking the average of each 5x5 CHIRPS data in a resolution of 0.05^0 x 0.05^0, so that new data will be obtained that reflects the average of rainfall in a resolution of 0.25^0 x 0.25^0. After that, the data is removed for data located in the ocean. This is done because the data used is not measured in the sea area, so that the pattern formed can focus on the land area.

After analyzing the entire territory of Indonesia, the analysis continued on the Sumatra and Kalimantan regions. This is done to find out patterns more specifically in areas where frequent forest fires occur annually. Then, the temporal signal formed from each mode will be converted into a periodic signal using the Fourier transform, so that the period of each pattern can be determined from each mode. This period will be used to determine the number of years the forest fires occur in Indonesia.

5. Result and Analysis

Indonesia is a country that has varying periods of rain in its territory. This is because Indonesia is located between two oceans, and has a land topology area that has many mountains. Most parts of Indonesia have a rainy period of twelve months, with the dry season occurring after June. In addition to that, in some regions in Indonesia which are located around the Earth's equator, there is a rainy period with a shorter period of time with six months of rain. For example, at Riau in one year there are two rainy seasons and two dry seasons, from February to March, and July to September.

GFED and GFEDs data are global data with a spatial resolution of 0.25^0x0.25^0, while the CHIRPS extraction is a 0.05^0x0.05^0, so the resolution of CHIRPS data needs to be changed to fit GFED and GFEDs data. The data that has been changed in resolution will be used in the overall pattern analysis of the territory of Indonesia, and data that has an initial resolution will be used to observe patterns in the regions of Kalimantan and Sumatra.

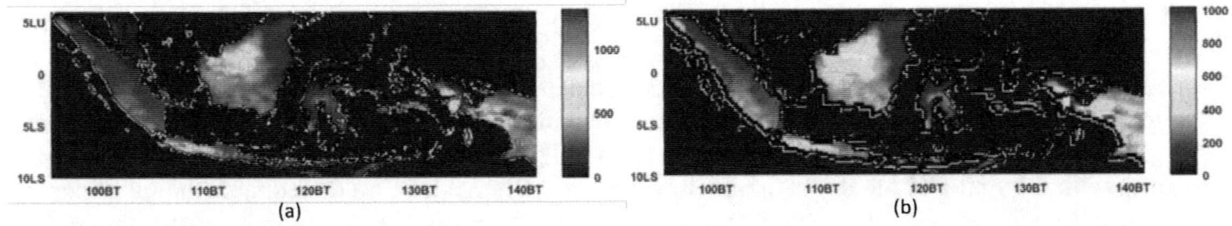

Figure 1. Indonesian rainfall distribution map for September 1997 with (a) data resolution of 0.05^0 x 0.05^0 and (b) data resolution of 0.25^0 x 0.25^0.

In this research the EOF mode obtained from each data matrix analysis is 240 modes. However, the fourth to 240th mode has a pattern similar to previous modes. The only difference lies in the strength of the signal. Therefore, this research only focused on three dominant modes in the analysis of each region. The first mode shows that there are patterns causing forest fires in Central Sumatra, South Sumatra, and South Kalimantan. When viewed from the spatial pattern of carbon emissions in Figure 2, many regions also have large carbon emissions but there are no forest fires. This is caused by the rainfall in the region is also high.

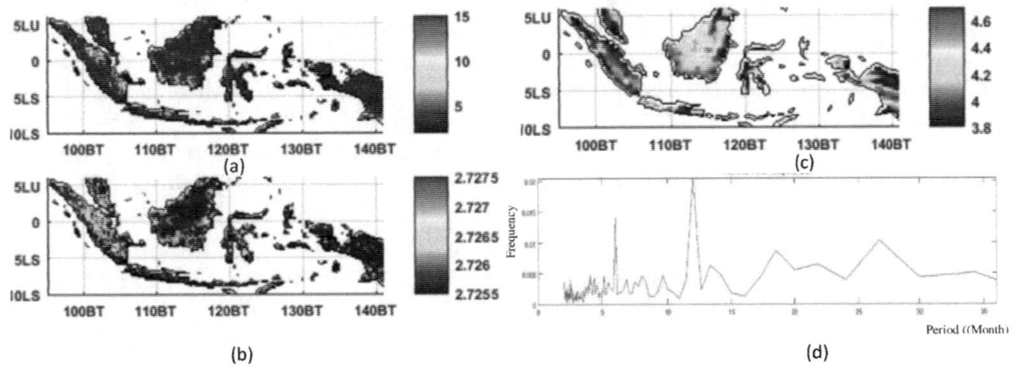

Figure 2. spatial pattern of variables (a) burned area, (b) carbon emission, (c) rainfall, and (d) period of the first temporal mode in Indonesia

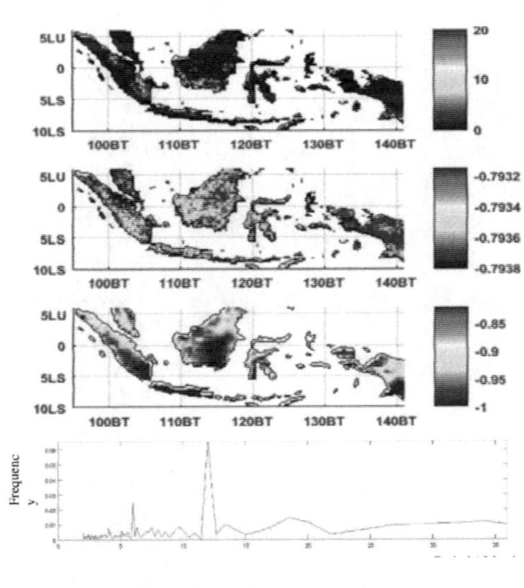

Figure 3. spatial pattern and second temporal mode of the Indonesia

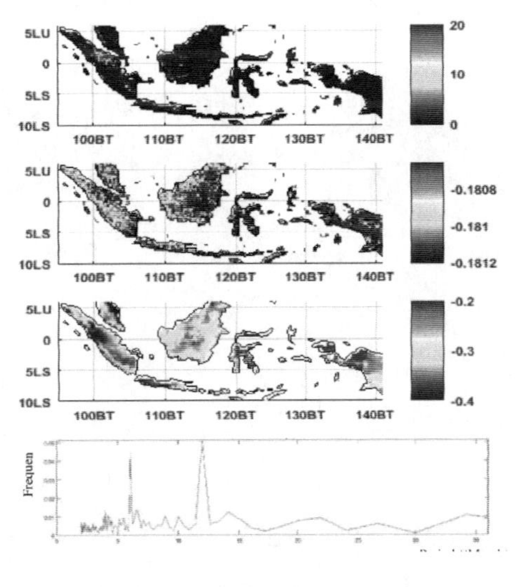

Figure 4. spatial pattern and third temporal mode pattern in Indonesia

In 1997-2017, this pattern is always occurred after July which was the time of the dry season in the territory of Indonesia which had a rainy period of twelve months. Based on this information it can be seen that the temporal pattern of forest patterns dominates variable was burned area. From Figure 2 it can be seen that the pattern has a period of six months and twelve months. This is influenced by the Indonesian rainy season which generally has a 12-month period, with 6 months of the rainy season alternating with 6 months of the dry season. However, for the area around the equator the rainy period of the region is 6 months, with 3 months of the rainy season alternating with 3 months of the dry season, resulting in 2 rainy periods in one year.

In the second and third mode observations that is shown in Figure 3 and 4, there were two quite different and contradictory modes in Sumatra and Kalimantan. The second mode of carbon emissions shows that most parts of Indonesia, especially in Sumatra and Kalimantan, have large emissions. However, large forest fires only occur in southern Sumatra and southern Kalimantan. This is because in Sumatra, Kalimantan and the other parts have high rainfall, which can prevent forest fires. The temporal pattern is the opposite of the first mode where in the second mode the pattern always occurs at the beginning of the year between January and mid-year. This pattern is a twelve-month period rainfall pattern in the area, so in the second mode the rainfall pattern dominates the third temporal pattern. From the third picture in Figure 3 it can be seen that the pattern has a twelve-month period, so this pattern only affects regions that have a period of twelve months of rainfall.

The carbon emission variable in the third mode shows the same as the second mode, where there is a large carbon emission in most parts of Sumatra and Kalimantan. However, only forest fires occurred in central Sumatra. This is because in Figure 4 almost all of Sumatra and Kalimantan have high rainfall except in Central Sumatra that shown in the third picture. Based on Figure 4 this pattern occurs within six-month period, so that only the area with six-month rainfall period that is affected by pattern in the third mode. This rainfall signal causes different time of dry season that only happened in equatorial area while twelve-month rainfall area has rain season. Consequently, this signal cause different pattern of carbon emission that does not occur in other area that only have twelve-month rainfall period. Based on the analysis of all regions of Indonesia, forest fires only occur in Sumatra and Kalimantan and separate in 2 patterns which has sixth and twelve period, therefore analysis will be carried out each of the two regions.

5.1. Kalimantan

Kalimantan is an area that has the second largest peatland area in Indonesia. The area of peatland in Kalimantan reaches 4.78 million hectares (ha). the area of peatland constitutes 32% of the total peatland in Indonesia. In fact, almost every year there are forest fires in Kalimantan, especially in peatland areas.

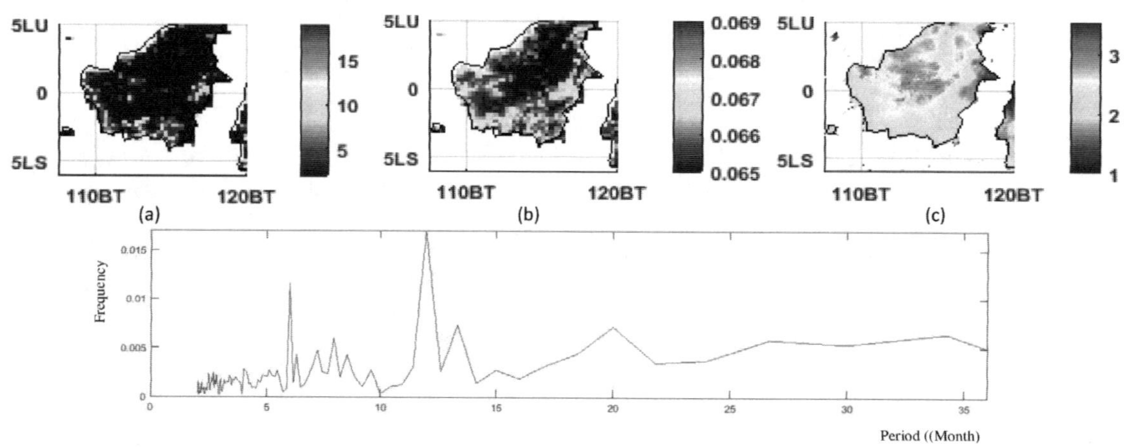

Figure 5. spatial patterns of variables (a) burned area (b) carbon emissions (c) rainfall, (d) period of temporal patterns in the first mode of Kalimantan

The first mode of analysis in the Kalimantan region shows that carbon emissions tend to be high in areas around the coast of Kalimantan or areas close to the ocean, while rainfall in the first mode is high in the regions of Central Kalimantan to North Kalimantan. The pattern that appears in the first mode has a dominant period of six months and twelve months. This is influenced by different rainfall period in some parts of Kalimantan. The equatorial area of Kalimantan has two rainfall signals, there are six-month and twelve-month period, while in South Kalimantan which is the area that forest fire occurred has twelve-month rainfall only. Interesting patterns are found in the second and third modes. The second and third modes of burned area variable have a similar pattern to the first mode. However, the variables of carbon emissions and rainfall in all modes have significant differences.

In the second mode, the temporal pattern that formed has a high frequency in the beginning of the year to mid-year, then low in the middle of the year until the end of the year. Based on the formed pattern, the pattern is the rainfall of the Kalimantan region with a period of twelve months, so it can be concluded that the second mode is dominated by the twelve-month rainfall pattern. High emissions in second mode appear in regions that have twelve-month rainfall period, but because of peatland area, emissions also always appear in almost of Kalimantan.

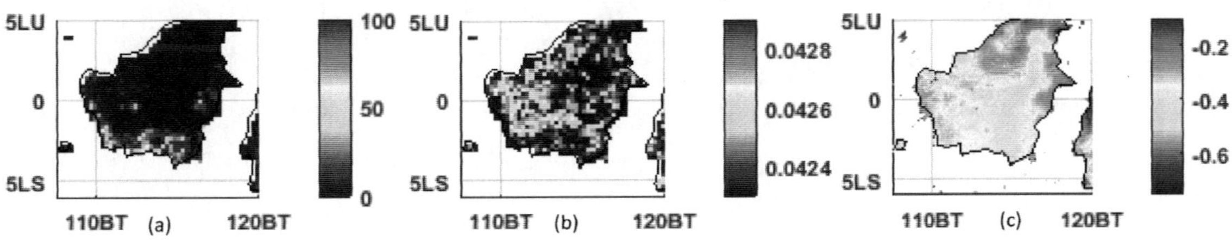

Figure 6. spatial patterns of variables (a) burned area (b) carbon emissions (c) rainfall in the second mode of Kalimantan

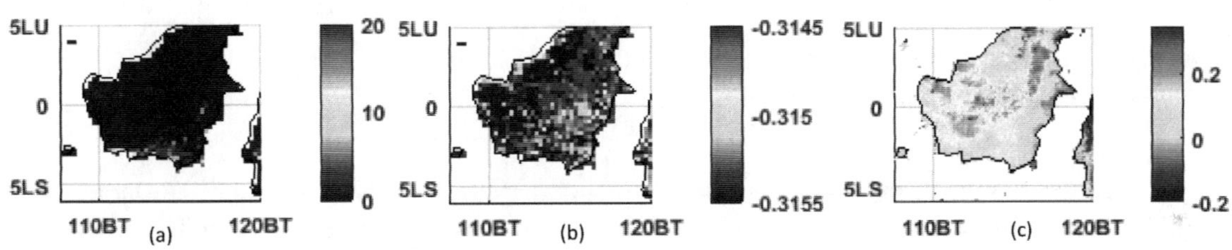

Figure 7. spatial patterns of variables (a) burned area (b) carbon emissions (c) rainfall in the third mode of Kalimantan

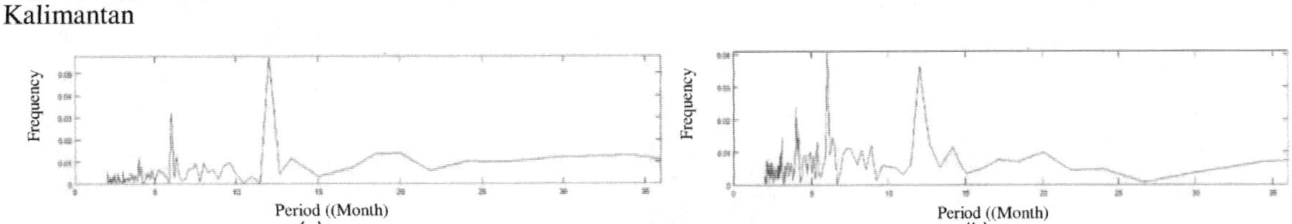

Figure 8. period of temporal patterns in the (a) second mode and (b) third mode of Kalimantan

The pattern formed in the third mode of the Kalimantan region has a high frequency of rainfall in the area around the equator and has a six-month temporal pattern period. Based on these two things, the third mode pattern is strongly influenced by rainfall patterns that have a six-month period. In this pattern the burned area was slightly compared to the previous mode, so it can be concluded that Kalimantan has a six-month forest fire pattern but the frequency is quite small and dominated by forest fire patterns with a period of twelve months. This also mean that forest fires in Kalimantan mostly occur on regions that has large of peatland area in South Kalimantan that produce high carbon emissions during dry seasons.

IOP Conf. Series: Earth and Environmental Science **303** (2019) 012053 doi:10.1088/1755-1315/303/1/012053

5.2. Sumatra

Further analysis was carried out on the Sumatra region. The first mode of burning area variable shows that fires occur in the area around Palembang and Riau. Carbon emission variables show that high carbon emissions occur in almost all parts of Sumatra but focuses on areas where there is a forest fire. Rainfall variables indicate that higher rainfall occur in areas close to the south coast or close to the Indian Ocean. Consequently, although some area has high emissions, there was no fires when rainfall on that regions was also high. The temporal pattern in the first mode shows that high frequency occurs two times in one year and has two dominant periods, six months and twelve months. This shows that the dominant variable in the temporal pattern is rainfall, with six-month rainfall signal in Riau and twelve-month rainfall period around Palembang. These two periods will be analyzed separately in the second and third mode.

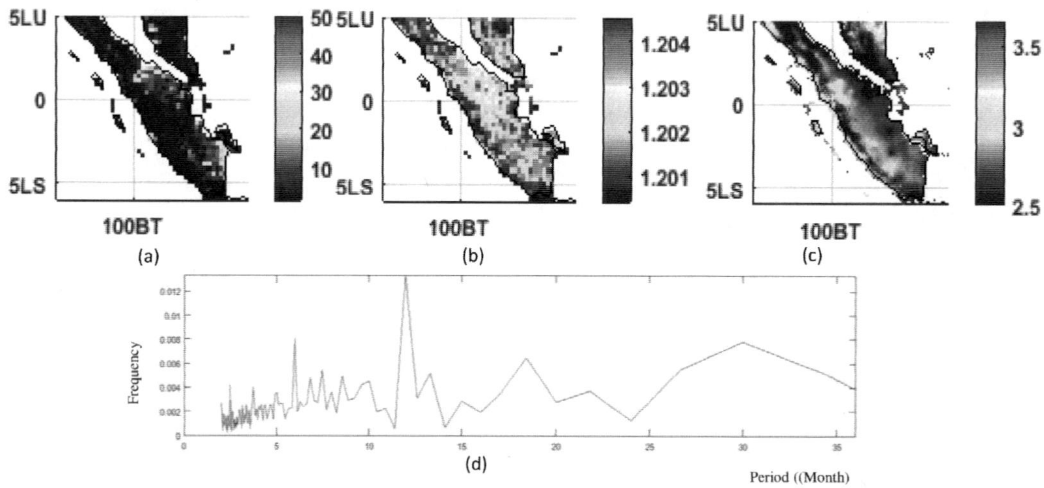

Figure 9. spatial patterns of variables (a) burned area (b) carbon emissions (c) rainfall, (d) period of temporal patterns in the first mode of Sumatra

The pattern in the second mode has similar burning area pattern to the first mode, but larger fires occur in the area around Palembang. While the carbon emission spreads in the southern Sumatra region and higher rainfall is in northern Sumatra. The temporal pattern that is formed has a high frequency around the beginning of the year and has twelve months period, so it can be concluded that the formed temporal pattern is strongly influenced by the South Sumatra rainfall pattern for a twelve-month period.

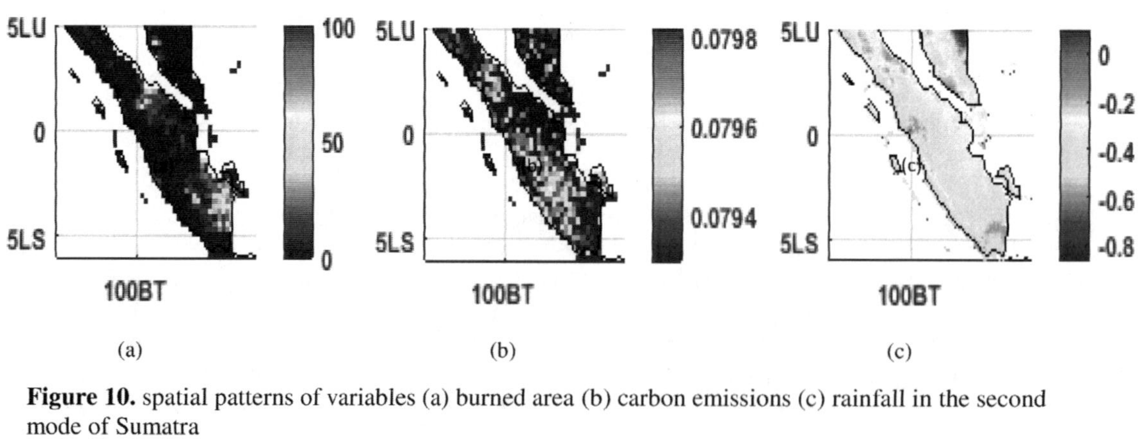

Figure 10. spatial patterns of variables (a) burned area (b) carbon emissions (c) rainfall in the second mode of Sumatra

IOP Conf. Series: Earth and Environmental Science **303** (2019) 012053 doi:10.1088/1755-1315/303/1/012053

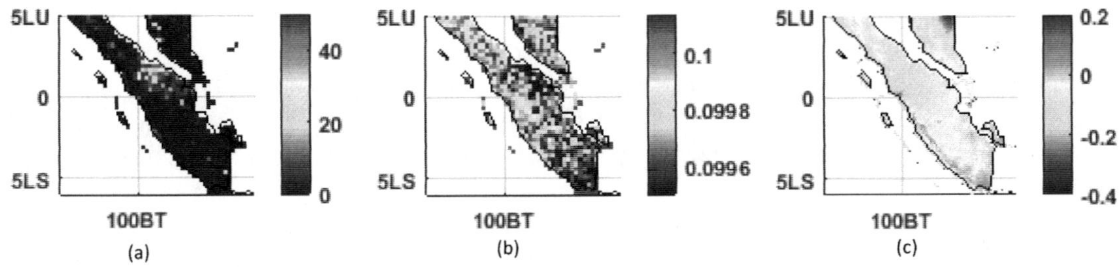

Figure 11. spatial patterns of variables (a) burned area (b) carbon emissions (c) rainfall in the third mode of Sumatra

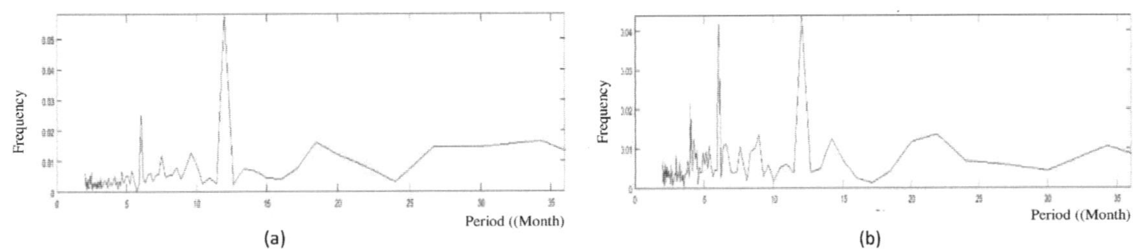

Figure 12. period of temporal patterns in the (a) second mode and (b) third mode of Sumatra

The pattern formed in the third mode is the opposite of the pattern in the second mode. Carbon emissions in the second mode are high in the northern Sumatran region around Riau and high rainfall is in the southern Sumatra region. This causes the third mode of burning area to occur only in areas around Riau. The temporal pattern formed has a six-month period and has a high frequency two times in one year, so it can be concluded that the formed temporal pattern is strongly influenced by rainfall which has a six-month period. Although there were high emissions in almost of Sumatra, this pattern only caused fires in Riau that has also six-month period of Rainfall.

From the analysis of these three modes, it can be concluded that the first mode is a combination of the second and third modes, while the second and third modes of forest fire patterns are influenced by the period of rainfall that occurs in areas where forest fires occur. Therefore, further analysis is carried out to see separately the patterns formed in northern Sumatra and southern Sumatra. Separation is carried out by dividing into two parts of Sumatra region transversely, so that a two region with the same matrix resolution is formed.

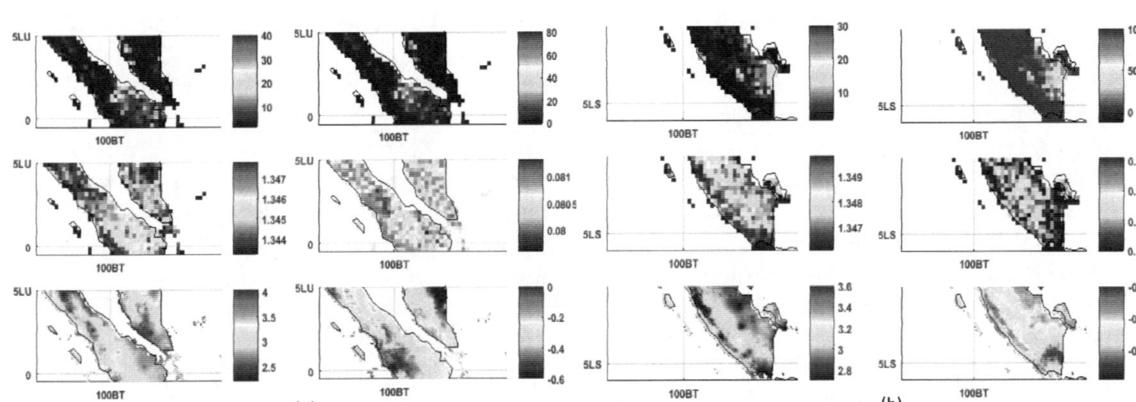

Figure 13. spatial patterns of variables burned area (top), carbon emissions (middle), rainfall (bottom) in the first (left) and second (right) mode of (a) North Sumatra and (b) South Sumatra

IOP Conf. Series: Earth and Environmental Science **303** (2019) 012053 doi:10.1088/1755-1315/303/1/012053

The discussion will then focus on separating the six-month and twelve-month periods. The spatial pattern that appears in the first mode shown in Figure 13(a) is a pattern similar to the first mode of the Sumatra region shown in Figure 9 and Figure 11. The difference lies in the temporal pattern and period. The formed temporal pattern is a pattern that is strongly influenced by the burned area. The next difference is the period of the pattern which has high frequency at six-month period. This causes two forest fires in one year in Riau.

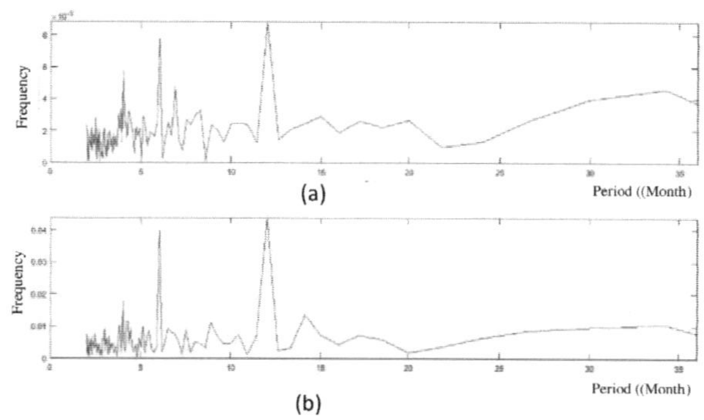

Figure 14. period of temporal patterns in the (a) first and (b) second mode in North Sumatra

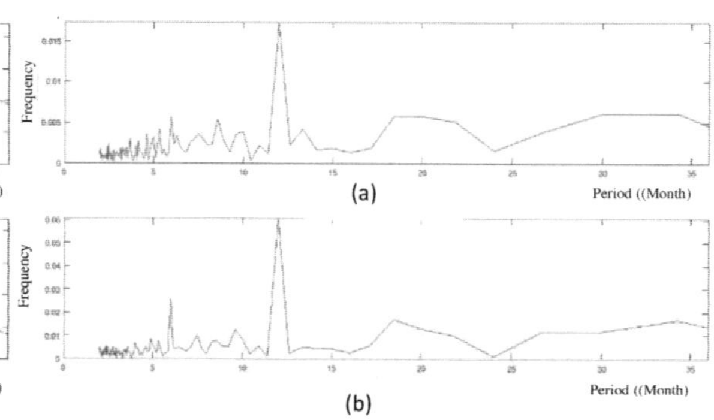

Figure 15. period of temporal patterns in the (a) first and (b) second mode in South Sumatra

In the spatial pattern of the second mode, formed burned area resembles the pattern in the first mode. While the temporal pattern in the second mode is a similar with the first mode. This pattern is a rainfall pattern that has a six-month period as shown in Figure 14. Significant differences occur in emission and rainfall where both of these variables provide a spatial pattern that is opposite to the first mode. This is because the second pattern is strongly influenced by rainfall, such as in the temporal pattern so that the spatial pattern is different with the pattern in the first mode which is dominated by burned area. Rainfall that dominated the third mode will make high frequency in temporal mode when there was high rainfall, different with burned area, second mode temporal signal frequency will high when there was large burned area.

Based on Figure 13(b) it can be seen that the spatial pattern in South Sumatra resembles the spatial pattern in the first mode of Sumatra which is shown in Figure 9 and Figure 10. In addition, the temporal pattern t is also almost the same, the difference lies in the period of the temporal pattern, where in Figure 8 (d) there are two dominant periods of six and twelve months, in Figure 15 there is only one dominant period of twelve months. The formed temporal pattern is a signal that strongly influenced by the burned area. This can be seen from the high signal frequency around the end of the year when the dry season occurs in Indonesia which has a period of twelve months of rainfall.

The spatial pattern in the second mode shown in Figure 13(b) is a pattern similar to the pattern that appears in the first mode, however the frequency is different. Significant differences are seen in the temporal pattern. The temporal pattern in the second mode is strongly influenced by the rainfall pattern of the region, so the pattern has a high frequency when the beginning of the year to mid-year. This is the time of the rainy season in the twelve-month rain period in Sumatra. In additions there are also periods that appear in the temporal pattern of six-month rainy periods. However, these signals tend to be small and are dominated by rainfall signals that have a twelve-month period.

From all of these modes, the third modes from all analysis give different patterns with the other modes significantly. The different appears on both spatial and temporal pattern. All third modes have pattern occur in six-month period. This happened because six-month rainfall signal that is happened around equator. This signal lead to be different in dry seasons to the other area that has twelve-month rainfall signal. Therefore, all three modes will only be affected on equatorial area. Furthermore, the six-month rainfall also leads to dry season that happened at the same time with twelve-month period dry season. This phenomenon was showed by first mode in all region. All first modes showed forest fire happened in areas where forest fires occur caused by both rainfall signal. However, the six-month period dry season was slightly weaker than the twelve-month dry season. Consequently, the frequency of six-month period was much lower than the twelve-month period.

6. Conclusion

Forest fires in Indonesia are phenomena that occur every year in Indonesia and are influenced by many factors. Most of carbon emissions variable in this research shows that most of region has high carbon emissions on all modes. It can be explained that in almost of region in Indonesia has a high temperature in afternoon and the rain will occur mostly after 2 pm. This fact make generally Indonesia will produce high emissions during afternoon, with different some of emissions depend on land type of each regions. The highest carbon emission was produced by peatland area. Patterns obtained from the analysis of burned area, carbon emissions, and rainfall the Indonesian region can be grouped into two. There are six-months period and twelve-months period patterns. This was same as rainfall in Indonesia that mostly can be grouped into sixth-month and twelve-month.

Forest fires in Indonesia occur annually in the regions of Central Sumatra, South Sumatra and South Kalimantan. In general, forest fires in Indonesia occur with a twelve-month period and occur when the dry season takes place in some of these areas. Besides that, number of peatland area in that regions produces high carbon emissions during dry season.

Besides having a twelve-month period, there are several areas that have a period of six months of forest fires. The area is a forest fire area whose rainfall is dominated by six months of rainfall such as the area around Riau. This is because when the rainfall is dominant for six months, there are two dry seasons which can cause forest fires twice a year. This six-month forest fire pattern does not appear in areas with little rainfall signal strength and is dominated by twelve months of rainfall. This can be seen in the analysis of the Kalimantan region that there are six-month forest fire patterns, but because of the dominant rainfall in Kalimantan twelve months, the small burning area is small.

Acknowledgments

This research was funded by the Directorate of Research and Community Service, Directorate General of Research and Development Strengthening, Ministry of Research, Technology, and Higher Education, in accordance with the Research Contract of the National Post-Graduate Team for Budget Year 2017 Number: 011 / SP2H / LT / DRPM / IV / 2017, April 20, 2017.

References

[1] Anton H, Rorres C. 2010. Elementary Linear Algebra. 10 Ed. New Jersey: J Wiley.
[2] Chen CC, Lin HW, Yu JY, Lo MH. 2016. The 2015 Borneo fires: What have we learned from the 1997 and 2006 El Niños? IOP Publishing Environ. Res. Lett 11 104003
[3] Fernandes K, Verchot L, Baethgen W, Gutierrez-Veles V, Pinedo-Vasquez M, Martius C. 2017. Heightened fire probability in Indonesia in non-drought conditions: the effect of increasing temperatures. IOP Publishing Environ. Res. Lett. 12 054002
[4] Fanin T, Van der Werf GR. 2017. Precipitation–fire linkages in Indonesia (1997-2015). Biogeosciences, 14, 3995–4008
[5] Hannachi A. 2004. A primer for EOF analysis of climate data: Department of Meteorology, University of Reading, Reading RG6 6BB, UK
[6] Harris N, Minnemeyer S, Stolle F, Aris OP. 2015. Indonesia's Fire Outbreaks Producing More Daily Emissions than Entire US Economy. World Resources Institute.
[7] Leon SJ. 2010. Linear Algebra with Applications. 8 ed. New Jersey: Prentice Hall.
[8] Navarra A, Simoncini V. 2010. A Guide to Empirical Orthogonal Function for Climate Data Analysis: Springer.
[9] Nicholson WK. 2001. Elementary Linear Algebra. Singapore: McGraw-Hill.
[10] Van dur Werf GR, Randerson JT, Giglio L, Collatz GJ, Mu M, Kasibhatla PS, Morton DC, DeFries RS, Jin Y, Van Leeuwen TT. 2010. Global fire emissions and the contribution of deforestation, savanna, forest, agricultural, and peat fires (1997–2009). Atmospheric Chemistry and Physics, 10 11707-11735
[11] Wiley E, Hansen. 2014. Fourier Transform: Principles and Applications. New Jersey: J Wiley.

Performance of Principal Component Analysis to Classify Precipitation Type from Raindrop Size Distribution Data at Kototabang, Indonesia

Marzuki[1], H Hashiguchi[2], M Vonnisa[1], Harmadi[1], and T Shimomai[3], M Yoseva[1] and E Saufina[1]

[1]Department of Physics, Andalas University, Padang, Indonesia.
[2]Research Institute for Sustainable Humanosphere (RISH), Kyoto University, Japan.
[3]Interdisciplinary Faculty of Science and Engineering, Shimane University, Japan.

marzuki@sci.unand.ac.id

Abstract. This study examines the use of principal component analysis (PCA) to classify the RDSD data at Kototabang, Indonesia. In addition to PCA with 6 attributes (hereinafter called PCA6) that had been developed by a previous researcher, this study also examines PCA with 7 attributes (PCA7) by adding radar reflectivity factor. The PCA is applied to the RDSD that had been classified by a wind profiler into Stratiform (S), deep convective (DC), shallow convective (SHC) and mixed stratiform/ convective (MSC). The number of unclassified data is much smaller than that reported by previous study in which it is about 33-47% with PCA6 and 29-44% with PCA7. While the PCA classifies the same group for different rain type from wind profiler, especially for Group I (moderate D_0 and large N_w) and II (small D_0 and N_w), some differences are observed. Each rain type classified by wind profiler has different dominant group in which Group II is dominant for S, Group I and V (large D_0 and low N_w) are dominant for DC, Group I, IV (small D_0 and large N_w) and VI (small D_0 and very large N_w) are dominant for SHC and Group I is also dominant for MSC type.

Keywords : principal component analysis, MSC, deep convective, raindrop.

1. Introduction

Rain type classification is very important because each rain type undergoes different physical processes. The most common type of rain is stratiform and convective. The cloud dynamics of these two rain types are different so that the microphysical processes that affect the growth of hydrometeor in both clouds are also different. Convective cloud is characterized by vertical air motion which is much stronger than that associated with stratiform rain. Hydrometeor in convective core grows through riming or accretion processes which generate large and dense raindrops. On the other hand, hydrometeor in stratiform rain grows through vapor deposition and aggregation processes, lead to smaller and less dense hydrometeors. Once melted, hydrometeor in stratiform produce more small-sized drops [1]. The difference in raindrop size distribution (RDSD) between convective and stratiform led to the importance of separation of these two types of rain for weather radar data conversion [2].

In general, classification of precipitation can be done in two ways, namely using radar and RDSD data. Precipitation can be classified using the texture of the radar reflectivity fields from weather radar

[3]. The precipitation type can be also deduced by analysing the vertical structure of reflectivity, velocity, and spectral width derived from the vertical beam of wind profiler observation [4]. Rain classification based on radar data is very accurate and it does not depend on rainfall intensity. However, radar observation particularly wind profiler is limited to several locations. As an alternative, classification can be done using the characteristics of RDSD. Rain classification deduced by analysing the RDSD can be based on the relationship between RDSD parameters and rainfall intensity. For example, the relationship between median mean diameter (D_0) and normalized intercept parameter of gamma distribution (N_w) given by the equation of $\log_{10}N_w = -1.6Do + 6.3$ separates RDSD into two clusters (convective-stratiform) [5]. Hereinafter, this method is called as the BR09 method. Recently, $N_w = 3.85$ is recommended as a threshold between convective and stratiform, and hereinafter this method is abbreviated as TH15 [6].

Rain classification which is based on the RDSD parameters depends on rainfall intensity. Precipitation with rainfall rate more than 10 m h^{-1} is always classified as convective rain and vice versa, precipitation with rainfall rate less than 5 m h^{-1} is classified as stratiform. This classification is not wrong because convective and stratiform rains are generally characterized by high and low rainfall intensity, respectively. However, not all rains with low intensity are stratiform type. Precipitation with low rainfall rate sometimes also occurs in convective rain [7]. To overcome this limitation, it has recently been developed rain classification using principle component analysis (PCA) [8].

The PCA is a technique which is commonly used to observe the variability of complex and large number of data. It uses linear regression to determine the main mode of variability in data. The PCA does not require the assumption of rainfall rate. However, previous study found that the PCA did not classify large number of data (70%) into any group [8]. Therefore, this work examined the application of PCA to classify the RDSD data in Indonesia. The performance of PCA was examined using the RDSD with known type of rain which has been classified using wind profiler observation [9].

2. Data and Method

This work applies the PCA to RDSD data collected by a 2D-Video Disdrometer (2DVD) at Kototabang, West Sumatra (100.32° E, 0.20° S), over eight consecutive years (2003-2010). An overview of 2DVD observation at Kototabang can be found in two publications [9,10]. The RDSD data have a temporal resolution of 2 minutes, adopting 0.2 bin interval from 0.4 to 10.25 mm. We only consider the RDSD with rainfall rate larger than 0.1 mm h^{-1} and having four consecutive bins with non-zero value. The RDSD data have been classified into stratiform (S), deep convective (DC), shallow convective (SHC) and mixed stratiform/convective (MSC) rains, using a 1.3 GHz wind profiler [9].

The PCA requires M x N array as input where M is the number of attributes that describe the characteristics of RDSD and N is the number of data points. In this study, we examined the performance of PCA with M = 6 and M = 7. For M = 6, the attributes consist of rainfall rate (R), liquid water content (LWC), mass-weighted mean diameter (D_m), total number of raindrops (N_t) , normalized intercept parameter of gamma distribution (N_w) and the normalized standard deviation of the mass spectrum (σ_m), which are expressed as:

$$R = 6\pi.10^{-4}\int_0^\infty D^3 v(D)N(D)dD \tag{1}$$

$$LWC = \frac{\pi.10^{-3}}{6}\int_0^\infty D^3 N(D)dD \tag{2}$$

$$D_m = \frac{\int_0^\infty D^4 N(D)dD}{\int_0^\infty D^3 N(D)dD} \tag{3}$$

IOP Conf. Series: Earth and Environmental Science **303** (2019) 012054 doi:10.1088/1755-1315/303/1/012054

$$N_w = \frac{4^4}{\pi \rho_w} \frac{LWC}{D_m^4} \tag{4}$$

$$\sigma_m = \sqrt{\frac{\int_0^{D_{max}} (D - D_m)^2 D^3 N(D) dD}{\int_0^{D_{max}} D^3 N(D) dD}} \tag{5}$$

The parameters of N_w, N_t, R, and LWC are included in the PCA in log scale. For M = 7, the value of Z is added into the array, which is expressed as

$$Z = \int_0^{\infty} D^6 N(D) dD . \tag{6}$$

The value of Z is the most widely used integral rainfall parameter because this parameter is directly provided by weather radar. The parameter of Z is also included in the PCA in log scale.

The PCA produces a set of vectors called empirical orthogonal functions (EOFs). This work only uses two EOFs, namely, EOF-1 and EOF-2, as in previous study [8]. EOF-1 is a vector that represents the largest amount of variance and EOF-2 describes the largest amount of variance after the first vector is removed. The PCA is only applied to the RDSD with shape parameter (μ) more than -3 and less than 15. The shape parameter is expressed by

$$\mu = \left(\frac{\sigma_m}{D_m} \right)^2 - 4 . \tag{7}$$

3. Result and Discussion

Figure 1 (a) shows the distribution of PC1 and PC2 for all 2DVD data at Kototabang obtained using 6 attributes (PCA6). The dashed line shows the 1.5 threshold as used in the previous study [8]. Data were grouped into 6 groups: Group I (positive PC1 but ambiguous PC2), Group II (negative PC1), Group III (positive PC2 but ambiguous PC1), Group IV (negative PC2 and ambiguous PC1), Group V (positive PC1 and positive PC2) and VI (positive PC1 and negative PC2). Of 8-year data, about 58% of the data are classified into these 6 groups and 48% data are unclassified. The unclassified data in the current study is smaller than that obtained by the previous study (~ 77%) [8]. EOF1 and EOF 2 explain about 59% and 30% of the variability, respectively.

Although the name of the group in this study is the same as in previous study [8], the physical meaning of each group is different. In this study, Group I means moderate D_0 and large N_w, Group II indicates small D_0 and N_w, Group III means moderate D_0 and small N_w, Group IV indicates small D_0 and large N_w, Group V means large D_0 and low N_w, and Group VI indicates small D_0 and very large N_w (figure 2). Kototabang RDSD has the largest number of data point in Group I and II about 16% and 20%, respectively. Group I is classified as convective rain and Group II is classified as stratiform by TH15 method.

Figures 1b-e show the distribution of 6 DSD groups in 4 types of rain classified by wind profiler. The stratiform rain has a large number of point with negative PC1 and deep convective rain has a large number of points in positive PC1. This characteristic is also observed in mixed stratiform/convective and shallow convective rains. Moreover, shallow convective has a larger number of points with negative PC2 than other rain types. Number of unclassified data for stratiform, deep, shallow and mixed stratiform/convective is 47, 45, 34, and 33%, respectively (table 1). In stratiform rain, the largest number of data is observed in Group II and the largest number of data is in Group I and V for deep convective rain. For shallow convective rain, the largest percentage of data points lie in Group I, followed by Group IV, II and VI. In mixed stratiform/convective, the data are mostly concentrated in Group I. Thus, the largest number of data point for convective rain lie in Group I, while it is in Group II for stratiform rain. This is consistent with the DSD classified by TH15 method.

IOP Conf. Series: Earth and Environmental Science **303** (2019) 012054 doi:10.1088/1755-1315/303/1/012054

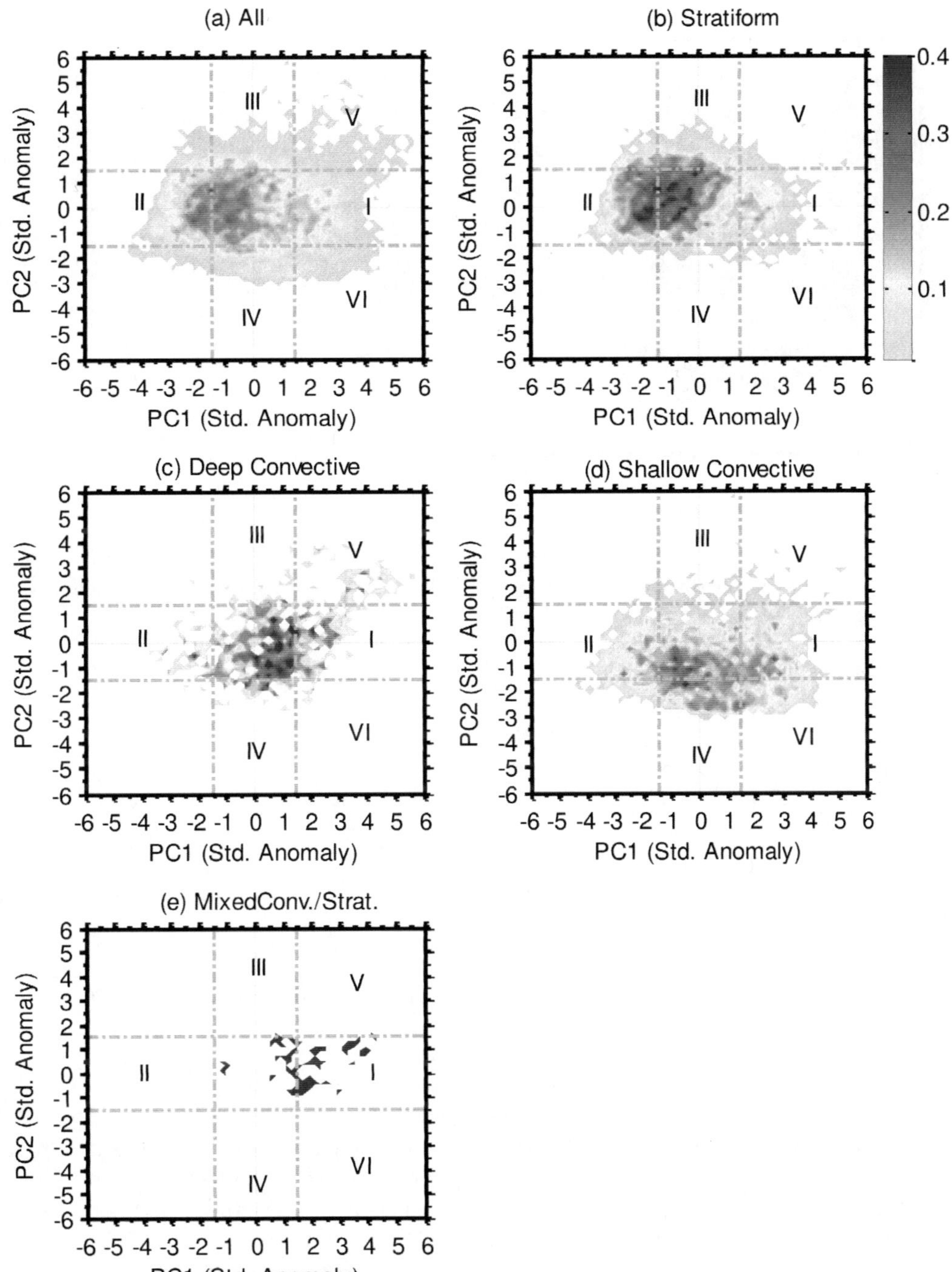

Figure 1. Distribution of normalized frequency of PC1 and PC2 estimated using 6 parameters. Dashed gray lines indicate $\sigma_{1.5}$ thresholds. All (a) indicates all data without rain classification, stratiform (b), deep convective (c), shallow convective (d) and mixed convective/stratiform (e) indicate raindrop spectra classified by wind profiler.

IOP Conf. Series: Earth and Environmental Science **303** (2019) 012054 doi:10.1088/1755-1315/303/1/012054

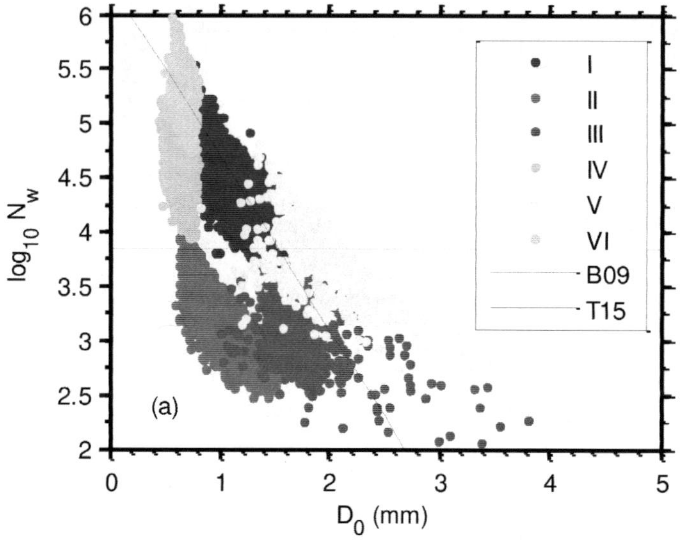

Figure 2. Two dimensional distribution of each group for $\log_{10}N_w - D_0$. Lines from BR09 and TH15 methods are also given.

Table 1. Percentage of data for each DSD group. U indicates unclassified data.

Type	Data	6 Parameters (%)							7 Parameters (%)						
	Number	I	II	III	IV	V	VI	U	I	II	III	IV	V	VI	U
All	13991	16	20	7	8	3	4	42	16	20	8	9	4	5	38
ST	7499	12	27	10	2	1	1	47	12	25	13	2	2	2	44
DC	1415	18	9	5	8	13	2	45	24	12	4	9	12	2	38
SHC	4828	20	13	3	17	3	10	34	17	17	3	19	3	10	31
Mix	249	43	8	7	0	7	2	33	49	8	5	1	6	3	29

Figure 3 shows the distribution of PC1 and PC2 for all 2DVD data classified by PCA with 7 attributes (PCA7). EOF1 and EOF2 explain about 61% and 30% variance, respectively. In general, the distribution of data is not much different from figure 1. However, we can see some differences. The number of unclassified data in figure 3 is smaller than Figure 1 and it is also observed for all rain types classified by wind profiler. The number of unclassified data decreases about 3-7% from figure 1. In deep convective and mixed stratiform/convective, there is an increase of data number in Group I. There is a decrease the data number in Group II and an increase in Group III for stratiform rain. The number of data in Group III that exceeds the BR09 line is also smaller in figure 3 than that in figure 1. In addition, the percentage of data in Group I distributed below the TH15 line also increases (figure 4). From figures 2 and 4, we can see some limitation of BR09 and TH15 in classifying the RDSD at Kototabang. BR09 does not provide a clear separation between stratiform and convective in which BR09 line lies in the middle of Group I, III, V and VI. TH15 method seems better because it only splits one group namely, Group V.

IOP Conf. Series: Earth and Environmental Science **303** (2019) 012054 doi:10.1088/1755-1315/303/1/012054

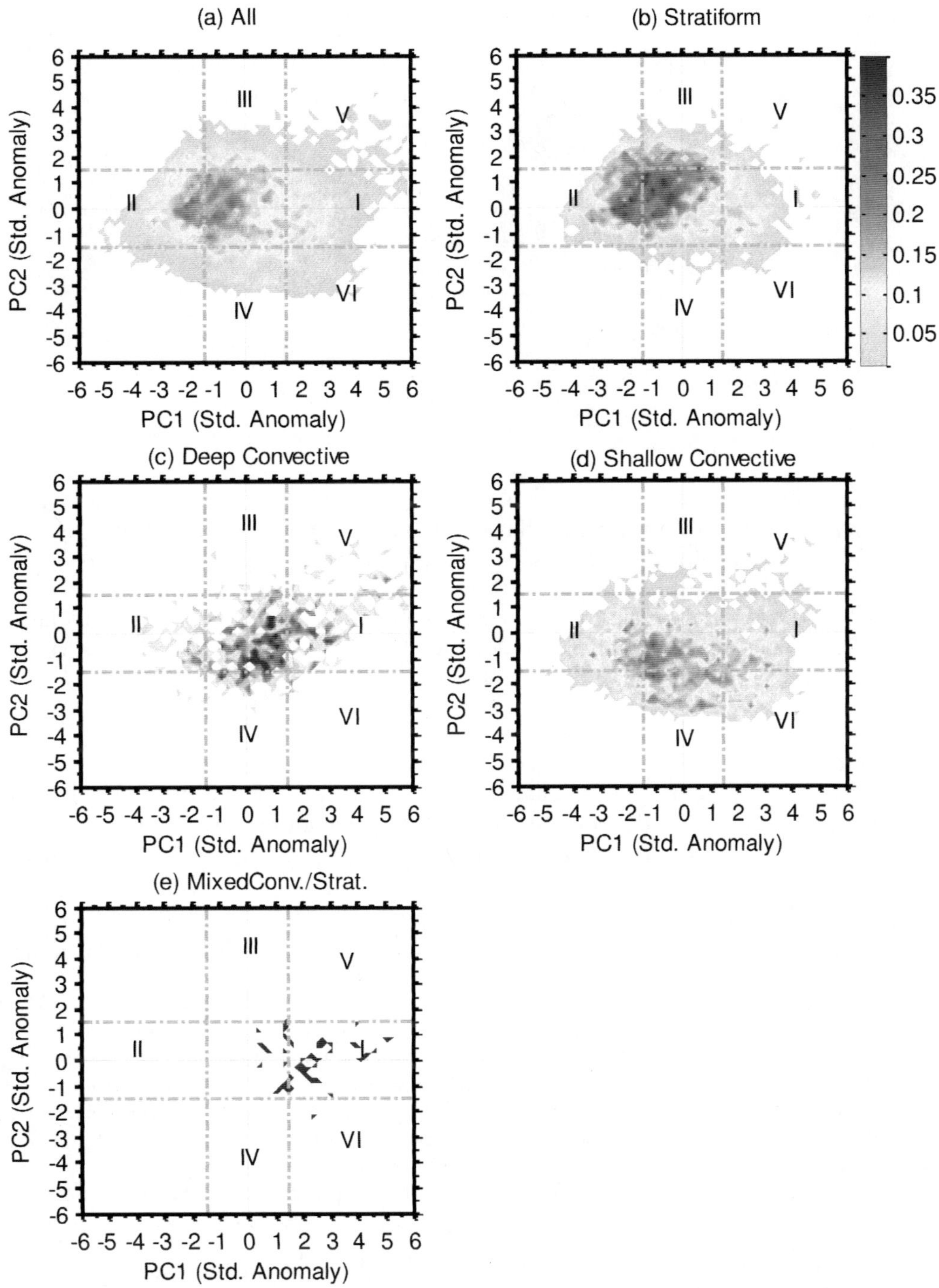

Figure 3. Same as figure 1 but for PCA with 7 attributes.

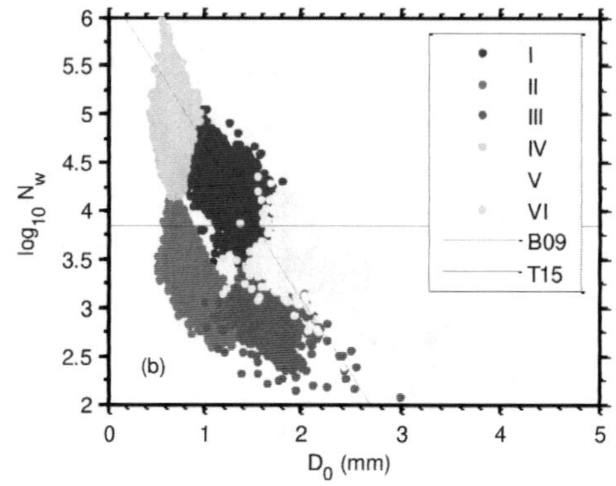

Figure 4. Same as figure 2 but for PCA with 7 attributes.

4. Conclusion

The results of this study show that the PCA can be used to classify precipitation types through RDSD observation in Indonesia. The PCA with 7 attributes provides a smaller number of unclassified data than that of 6 attributes. In general, characteristics of RDSD for each group which is classified by the PCA are consistent with the characteristics of DSD based on wind profiler classification scheme. The current study only analyzed data at one location. Data from several locations in Indonesia are being analyzed and will be published in another article. Moreover, RDSD in Indonesia has significant diurnal, intra-seasonal and regional variations. Application of PCA to observe such variability is also being examined.

Acknowledgments

This work was supported the Ministry of Research, Technology and Higher Education of the Republic of Indonesia under the 2017 and 2018 International Joint Collaboration and Scientific Publication grant (Contract no. 02/UN.16.1.17/PP.KLN/LPPM/2017 and 07/UN.16.17/PP.PKLN.K/LPPM/2018).

References

[1] Houze Jr R A 1993 *Cloud Dynamics* (California: Academic Press)

[2] Tokay A and Short D A 1996 Evidence from tropical raindrop spectra of the origin of rain from stratiform versus convective clouds *J. Appl. Meteor.* **35** pp 355–71

[3] Steiner M, Houze Jr R A and Yuter S E 1995 Climatological characterization of three-dimensional storm structure from operational radar and rain gauge data *J. Appl. Meteor.* **34** pp 1978–2007

[4] Williams C R., Ecklund W L and Gage K S 1995 Classification of precipitating clouds in the tropics using 915-MHz wind profilers *J. Atmos. Oceanic Technol.* **12** pp 996–1012

[5] Bringi V N, Williams C R, Thurai M and May P T 2009 Using dual-polarized radar and dual-frequency profiler for DSD characterization: a case study from Darwin, Australia *J. Atmos. Oceanic Technol.* **26** pp 2107–22

[6] Thompson E J, Rutledge S A, Dolan B and Thurai M 2015 Drop size distributions and radar observations of convective and stratiform rain over the equatorial indian and west pacific oceans *J. Atmos. Sci.* **72** pp 4091–25

[7] Tokay A, Short D A, Williams C R, Ecklund W L and Gage K S 1999 Tropical rainfall associated with convective and stratiform clouds: intercomparison of disdrometer and profiler measurements *J. Appl. Meteor.* **38** pp 302–20

[8] Dolan B, Fuchs B, Rutledge S A, Barnes E A and Thompson E J 2018 Primary modes of global drop size distributions *J. Atmos. Sci.* **75** pp 1453–76

[9] Marzuki, Hashiguchi H, Kozu T, Shimomai T, Shibagaki Y, Takahashi Y 2016 Precipitation microstructure in different Madden–Julian Oscillation phases over Sumatra *Atmos. Res.* **168** pp 121-38

[10] Marzuki, Randeu W L, Kozu T, Shimomai T, Hashiguchi H and Schoenhuber M 2013 Raindrop axis ratios, fall velocities and size distribution over Sumatra from 2D-video disdrometer measurement in advances in precipitation science ed Michaelides S *Atmos. Res.* **119** pp 23-37

IOP Conf. Series: Earth and Environmental Science **303** (2019) 012055 doi:10.1088/1755-1315/303/1/012055

Real Time Air Quality Monitoring System in Three Sites (Bogor, Cibeureum and Serpong)

G S Immanuel[1], A Rakhman[1], B Budianto[1], R Boer[1], M Ardiansyah[1], N Hamid[1], I Gumilang[1], M Nishihashi[2], H Mukai[2], Y Terao[2], S Hashimoto[2], Y Osonoi[2], R Nugroho[3], N Suwedi[3], A Rifai[3], I M Ihsan[3], A Sulaiman[3], D Gunawan[4], E Suharguniyawan[4], M S Nugraha[4] and R C Wattimena[4]

[1] Bogor Agricultural University (IPB), Kampus IPB Baranangsiang, Indonesia
[2] National Institute for Environmental Studies (NIES), 16-2 Onogawa, Tsukuba, Ibaraki 305-8506, Japan
[3] Agency for the Assessment and Application of Technology (BPPT), Indonesia
[4] Meteorological, Climatological, and Geophysical Agency (BMKG), Indonesia

gito.kribo@gmail.com

Abstract. Center for Climate Risk and Opportunity Management of Bogor Agricultural University in collaboration with National Institute for Environmental Studies-Japan developed an on-line system to monitor real time air quality and greenhouse gases for three sites surrounding Jakarta area, i.e. Bogor (Center of Bogor city), Serpong (Jakarta suburb) and Cibeureum (mountainous area, background-like site). The system uses standards air-quality monitoring equipped with an array of gaseous and meteorological sensors. The system records data every minute and stores in local storage and sent automatically to a Dropbox every 6 hours. The data in the Dropbox is accessed using API and being processed and analysed using several modules to be converted into useful information and presented in a website. The website will display real time air quality for ozone (O_3) and particulate matters ($PM_{2.5}$ and PM_{10}) and greenhouse gas data (CO_2, CH_4), Carbon monoxide (CO) Nitrogen Oxide (NO_x), and Sulfur Dioxide (SO_2), in the past 30-day including local weather data for the past 7 days. These data are then converted into an index indicating the status of the quality of the air surrounding the stations. The index is divided into five categories based on EPA air quality standard, namely good, moderate, unhealthy for sensitive group, unhealthy and very unhealthy. The system can provide the status of air quality index on a six hours basis and easy to be accessed by any stakeholder. The system allows local authorities, communities and industries to easily access and use the information for air quality management purposes and early warning.

Keywords : air quality, ozone, CO, SO_2, NO_x, CH_4, PM.

1. Introduction

The presence of air quality monitoring system is very important for air pollution management, particularly in the big cities. It provides information about the status of air quality for the city authorities and helps the cities' authorities in formulating the mitigation policies and developing air pollution management. It also provides information to warn people when pollution levels reach harmful level and help communities to do actions in limiting the impact of poor air quality on bad days. Jakarta, Bogor,

Depok, Tangerang and Bekasi (Jabodetabek), the growing and the biggest cities in Indonesia, have installed air quality monitoring system. However, many of them do not function well and only monitor a few air quality parameters [1]. In addition, the air quality information is also not processed and disseminated to public informatively.

Center for Climate Risk and Opportunity Management in Southeast Asia Pacific (CCROM-SEAP) of Bogor Agricultural University (IPB), Agency for the Assessment and Application of Technology (BPPT) and National Agency for Meteorology, Climatology, and Geophysics (BMKG) are implementing a joint monitoring of air pollutants and greenhouse gases with the National Institute for Environmental Studies (NIES) Japan. The monitoring systems are ground base system installed at three sites in Jabodetabek, i.e., Bogor (Center of Bogor City), Serpong (Jakarta suburb), and Cibeureum (mountainous area, background-like site) in 2016-2017. The system monitors many air quality parameters continuously consisting of GHGs and air pollutants (CO_2, CH_4, O_3, CO, NO_x, SO_2, aerosol concentrations $PM_{2.5}$, PM_{10}, Black Carbon, chemical components and flask sampling of air) [2]. Measurement of CO_2 and CH_4 uses G2301 from Picarro, CO using CO-30r from Los Gatos Research, NO_x using Model 42i-TL from Thermo, SO_2 using Model 43i-TLE from Thermo, O_3 using OA-787 from Kimoto Electric, aerosol concentrations ($PM_{2.5}$, PM_{10}, BC) and chemical components using ACSA-14 from Kimoto Electric, and automatic flask sampler of air from Koshin-RS. The system uses standards air-quality monitoring equipped with an array of gaseous and meteorological sensors.

We developed a website air quality monitoring system to display the processed air quality data measured by the monitoring system from the three sites in the form of air quality index information. This paper introduces the web-based air quality monitoring system along with its capabilities in analyzing the air quality data to produce real time air quality status along with the archive data in the past few hours and days.

2. Architecture of the Web-Based Air Quality Monitoring System

We developed the web-based air quality monitoring through two steps (Figure 1). The first is establishing a database of the air quality data from the local storage and the second is building a web service from the database that will display the real time and data archived from the past hours and days on the website.

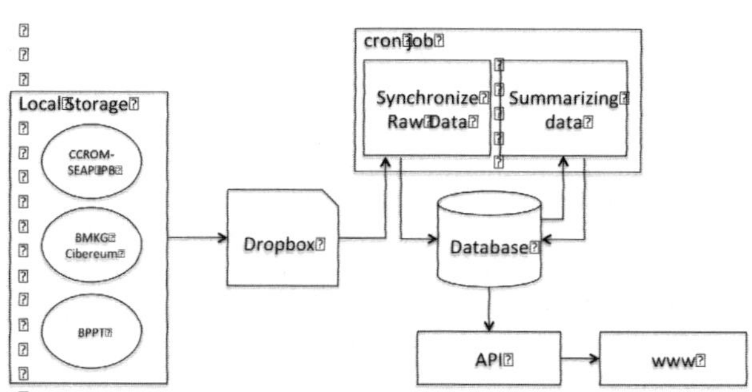

Figure 1. Architecture of the system

2.1. Database

The air quality monitoring system in the three sites records data every minute and stores the data in their local storage and sent automatically to a Dropbox every 6 hours. We developed modules using Application Programming Interface (API) to transfer the data from the Dropbox (cloud storage) to Database, The API is a protocol that forms an interface to allow for inter-programs to access and to analyze the data from the background. The Dropbox provides the v2 API through OAuth 2, in which

IOP Conf. Series: Earth and Environmental Science **303** (2019) 012055 doi:10.1088/1755-1315/303/1/012055

the access requires tokens generated from https://www.dropbox.com/developers site. Module for the database facilitate the process for receiving, storing, pre-processing and converting the data into useful information and presenting the information in the website. The module uses Data reader daemon that checks raw data (.csv) in a group files in Dropbox into the database (PostgreSQL) [4]. Data readers are built using python v3.6.4 with the following requirements:

- Dropbox library: Reads the structure and content of files stored in the Dropbox.
- Psycopg2 library: Facilitates interaction between the Dropbox and the database (PostgreSQL DBMS)
- Temporary folder (tmp): Temporary location for storing the data during the synchronization process.

During the synchronization process, the module scans all data in the Dropbox and retrieves the query list of files that have been synchronized in the database, so that they get a list of the data that is not in the database, and transfer the data to the database. The module can be run periodically using the scheduler on the operating system or on the website using 'cronjob feature' that is run every 6 hours.

2.2. *Web Service*

Modules for the web services run using PHP7.2.4 with Laravel Framework 5.6 requires the php-pgsql [5]. The web services accommodate the following data feature:

a) **Real time Air quality monitoring (O_3, $PM_{2.5}$ and PM_{10}) and Weather**, Air quality monitoring (O_3, $PM_{2.5}$ and PM_{10}) and weather are placed on the main page of the website as this information is considered as the main important information. Like the weather, the O3, PM2.5 and PM10 vary considerably between times (hour to hour, day to day etc.). The module will develop Air Quality Index (AQI) based on the condition of the concentration of the O3, PM2.5 and PM10 in the air using EPA standard [6]. AQI provides simple information about status of the air quality, whether the index is at the stage of 'unhealthy' or any status as presented in Table 1 and Figure 3, so that people can mitigate the impact on their health. The AQI is calculated using the last 6 hours data from eight time periods, i.e. from 01:00 to 07:00, 04:00 to 10:00, 07:00 to 13:00, 10:00 to 16:00, 13:00 to 19:00, 16:00 to 22:00, 19:00 to 01:00 and 22:00 to 04:00. Thus, if the page is accessed at 09:00, the information displayed is an average between 01:00 and 07:00 of the day before the data accessed. Each data has a flag indicating the status of the data, i.e. 'A' indicating the data obtained during adjustment mode of the instruments, 'X' indicating data error, and 'N' indicating normal (the data can be used for the calculation of the AQI).

Table 1. Classification of Air Quality Index.

	Classification	Color	O_3 (ppb)	$PM_{2.5}$ (ug/m3)	PM_{10} (ug/m3)	Health Concern
1	Good	Green	0-54	0-12	0-0.54	None
2	Moderate	Yellow	55-70	12.1-35.4	55-154	Unusually sensitive people should consider reducing prolonged or heavy exertion.
3	Unhealthy for Sensitive Group	Orange	71-85	35.5-55.4	155-254	People with heart or lung disease, older adults, and children should reduce prolonged or heavy exertion.
4	Unhealthy	Red	86-105	55.5-150.4	255-354	People with heart or lung disease, older adults, and children should avoid prolonged or heavy exertion. Everyone else should reduce prolonged or heavy exertion.
5	Very Unhealthy	Purple	>= 106	>= 150.5	>= 355	People with heart or lung disease, older adults, and children should avoid all physical activity outdoors. Everyone else should avoid prolonged or heavy exertion.

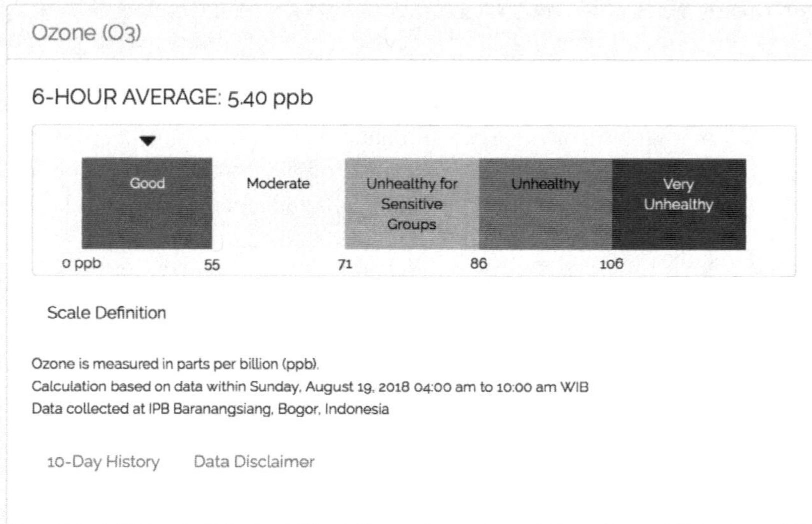

Figure 2. Bogor City AQI for Ozone

Similar with AQI, weather data is also grouped into eight time periods (segments), i.e. 01:00 to 07:00, 04:00 to 10:00, 07:00 to 13:00, 10:00 to 16:00, 13:00 to 19:00, 16:00 to 22:00, 19:00 to 01:00 and 22:00 to 04:00. The weather information presented in the front page of the web page consists of:

- Wind speed in m/s
- Wind direction in arc degree (DEG)
- Temperature (°C), average air temperature for the eight segments,
- Relative humidity (%), average relative humidity for the eight segments,
- Rainfall (mm), total rainfall for the eight segments.

Wind speed and wind direction are treated as vector which is calculated using the following formula:

$$\overline{ws} = \left(\left(\sum \frac{ws * \sin(wd)}{n} \right)^2 + \left(\sum \frac{ws * \cos(wd)}{n} \right)^2 \right)^{1/2}$$

$$\overline{wd} = atan2\left(\sum \frac{ws * \sin(wd)}{n}, \sum \frac{ws * \cos(wd)}{n} \right)$$

$\overline{ws}$ = Average value of wind speed in each 6-hour segments
$\overline{wd}$ = Average wind direction in each 6-hour segments.
ws = Average wind speed per minute.
wd = Average wind direction per minute.
n = Amount of data in each 6 hours segment.

b) **Archived Data** is displayed in two plots, line chart and scatter plot. The line chart shows an average for 8 hours and scatter plots show average for 1 hour but for $PM_{2.5}$ and PM_{10} the plot shows average for 24 hours and 3 hours respectively. Only rainfall is displayed in the form of a histogram and the data is using summation instead of the average. The average value presented in the graph is the moving average data (Figure 4). The definition of visualization of each parameter is presented in Table 2.

Table 2. Definition of visualization plot for each parameter

	Parameter	Scatter Plot (average of *n* hours)	Line Chart (average of *n* hours)
1	O_3	1	8
2	$PM_{2.5}$ and PM_{10}	3	24
3	Weather: Wind Speed, Wind Direction and Humidity	1	8
4	Weather: Temperature	1	-
5	Weather: Precipitation*	1	-
6	CO_2 and CH_4	1	8
7	CO	1	8
8	NO_X, NO, and NO_2	1	8
9	SO_2	1	8

* summation of hourly rainfall over certain periods (1 to 7 days)

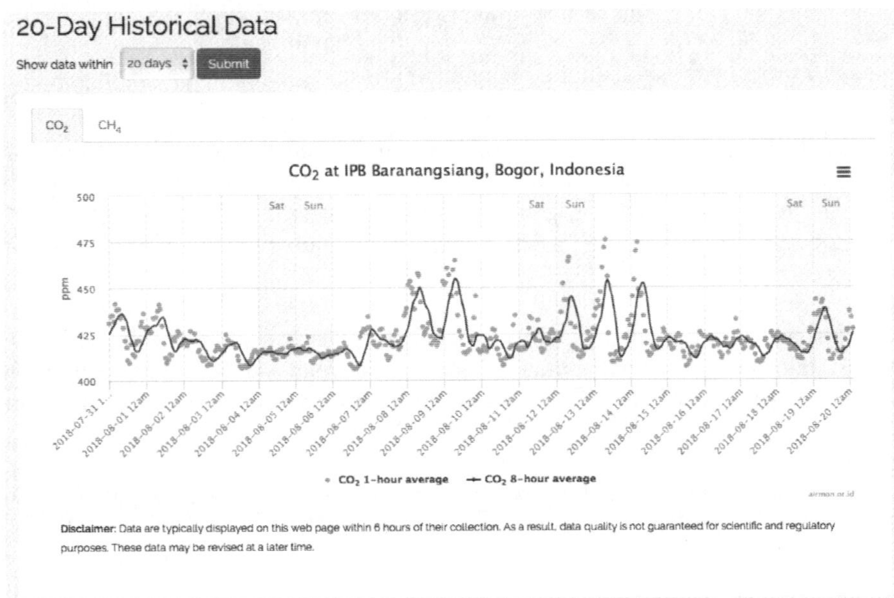

Figure 3. CO_2 archived data

c) **Website Widget** is a small application that can be installed and executed within a web page by the user. A widget has a role of a transient or auxiliary application, meaning that it just occupies a portion of a webpage and does something useful with information fetched from other websites and displayed in any position in the screen [8]. Summary of the information can be put on a page of affiliated sites using the iframe tag in HTML4 or the embed tag in HTML5. The url address used in the src attribute is: https://www.sample.com/plugin?f=fitur[&footer=footer&bg=warna].

The widget has two features namely *air* for air quality information, and *weather* for weather-related information. The users can also change the feature of the parameters and also background colors etc. The default color is white and the other colors can be chosen by using encoded value, for example, *&bg* =%23ffffff for the hex value #ffffff. Example of the visualization of the parameter can be seen in Figure 5.

IOP Conf. Series: Earth and Environmental Science **303** (2019) 012055 doi:10.1088/1755-1315/303/1/012055

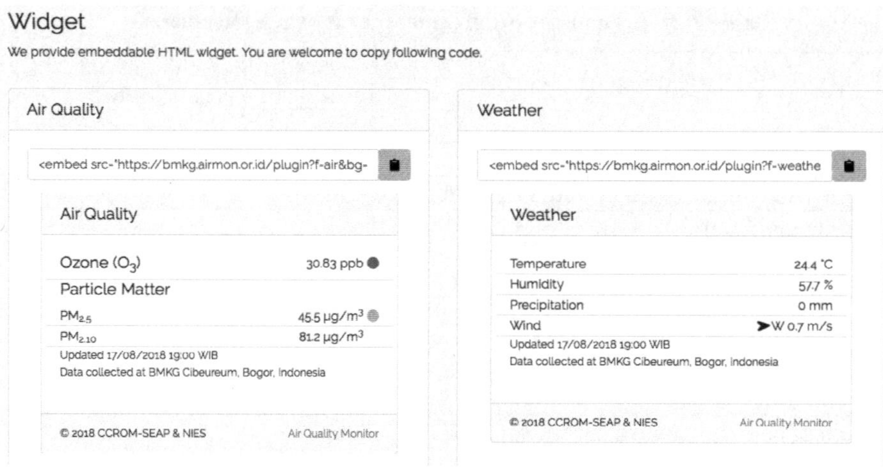

Figure 4. Website widget web service

3. Applications

The presence of the real time air quality monitoring system and also data archive in the last few hours and days allows the user to get the current status of air quality and also trend in the last few hours or days. This information will be useful for local authorities, communities and industries for air quality management purposes and early warning for anticipating the negative impact of the severe air pollution.

4. Conclusion

The system can present a summary of air quality data and weather periodically (updated every 6 hours). The use of flag for data status (normal, error, adjusted) will greatly affects the presentation of the air quality and weather in the form of graphics as the graphs only process and present valid data.

Acknowledgments

This joint monitoring project was funded by the Ministry of the Environment, Japan. We would like to thanks the Bogor City Authority for their inputs in the development of the real time air quality monitoring system.

References

[1] Greenpeace 2017 Briefing Paper - Kualitas Udara yang Buruk di Jabodetabek http://www.greenpeace.org/seasia/id/press/reports/
[2] Nishihashi M, Mukai H, Terao Y, Hashimoto S, Osonoi Y, Boer R, Ardiansyah M, Budianto B, Immanuel G.S, Rakhman A, Nugroho R, Suwedi N, Rifai A, Ihsan I.M, Sulaiman A, Gunawan D Suharguniyawan E, Nugraha M.S, Wattimena R.C and Ilahi A F 2018 Greenhouse gases and air pollutants monitoring project around Jakarta megacity *ICTMAS* 2018 Workshop Bandung
[3] PIP documentation 2018 https://pip.pypa.io/en/stable/
[4] PostgreSQL documentation 2018 https://www.postgresql.org/docs/
[5] Laravel Framework 5.6 documentation 2018 https://laravel.com/docs/5.6
[6] EPA 2018 NAAQS Table. https://www.epa.gov/criteria-air-pollutants/naaqs-table
[7] National Partk Service (NPS) 2018 Air Quality https://www.nps.gov/subjects/air/webcams.htm?site=yose
[8] Wikipedia 2018 Web widget. https://en.wikipedia.org/wiki/Web_widget
[9] AirNOW 2018 Sensitive to air pollution https://cfpub.epa.gov/airnow/index.cfm?action=aqikids.health2
[10] Lawrence K W; Jerry R T; Yung-Tse H and Kathleen H L 2004 Handbook of Environmental Engineering: Volume **1** Air Quality and Pollution Control Humana Press Inc Totowa
[11] Snehal S and Priya K 2015 Review Paper on Air Pollution Monitoring System *IJARCCE* Vol. **4**

Comparison and Analysis of The First Signal Detection of Two X-Band Radar with RGB Himawari Satellite Data before Localized Heavy Rain March 17 2017 Bandung Basin Indonesia

G A Nugroho, T Sinatra and Risyanto

National Institute of Aeronautics and Space, LAPAN

Email: ginaldi.ari@lapan.go.id

Abstract. Early detection of the localized heavy rainfall is necessary to avoid heavy losses and even casualties. Localized heavy rainfall occurred in complex topography Bandung Basin March 17, 2017. Previous research showed a distinctive v-shaped pattern (peak period of heavy rainfall) followed with heavy rainfall and strong wind in 11.30 LT. Early dataset from two x-band radar observation data (GMWR25SP and SANTANU) and composite RGB Himawari satellite are used to investigate the first signal before the v-shaped occurred. The period of the dataset is from 10.50 LT until 11.10 LT. Comparison from both time and visualization of the dataset are conducted to analyze the event. Time comparison from three dataset showed differences for the emergence of early indication. GMWR25SP is the first that could detect the early origin of convective cell. Visual comparison in three dataset showed slightly different pattern, especially for composite RGB Himawari. Similarity pattern are only showed in the period 11.00 LT. Analysis of each dataset are necessary to see more detailed on the generation process. SANTANU showed a rapid growth process from one cell into two convective cells spatially within 20 minutes period. The vertical growth of these two convective cells are persistent and reaching up to 10 km based on GMWR25SP. These strong convective cells are probably because of the vorticity couplet that occur in 11.00 LT until 11.10 LT. The dominant air mass movement based on the water vapor distribution RGB Himawari are showed towards the west side. This probably affect the position of the generated second convective cell (position on the left side of the previous cell). Data indication in 11.10 LT is the important period because it showed the generated second convective cell. Combination from three dataset give a more comprehensive analysis on the process before localized heavy rainfall occurred.

Keywords: localized heavy rainfall, v-shaped, GMWR25SP, SANTANU, RGB Himawari

1. Introduction

Localized severe weather in Bandung Basin have occurred in March 17,2017. Severe weather with the radar reflectivity area reach up to $\pm$ 30 km and an indicate maximum winds shear of 14.5 m/s have caused several damages in infrastructure. Strong wind speed has thrown away rooftop, billboard, traffic sign, etc. Preliminary study showed the location of this storm are located in the middle of Bandung urban area in a period between 10.50 LT until 12.00 LT [1]. This local convective storm reached its maximum in 11.20 LT until 11.30 LT and forming a v-shaped pattern and caused heavy rainfall and also strong wind speed. The strong wind speed is calculated reaching maximum of 12 – 15 m/s around 11.30 LT until 12.00 LT based on the ground based automatic weather station observation and Doppler velocity of x-band radar [2]. All of the reported location with heavily damaged were located inside this v-shaped.

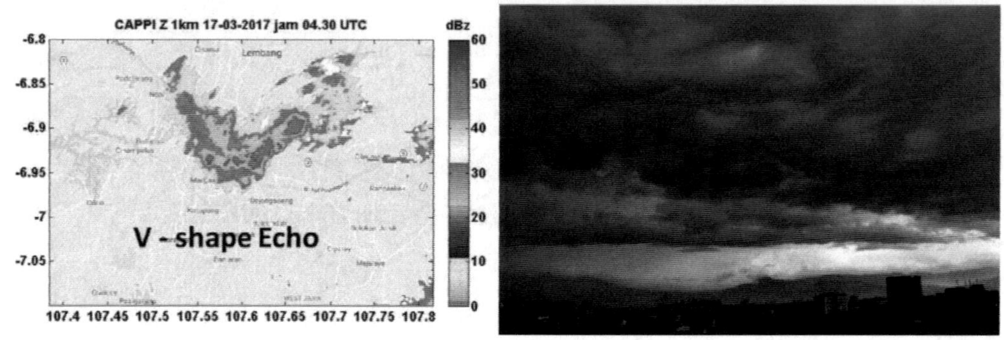

Figure 1. V shape echo pattern and condition of dark cloud in March 17, 2017 in Bandung [3]

Hail event are also occurred during this local convective storm based on visual report. Result from the x-band radar data processing for the estimated total amount of liquid water concentration using VIL (Vertically Integrated Liquid) showed strong potential of hail in one of flank of the v-shaped area [4]. Preliminary comparison between radar data with Himawari satellite data are studied from the dataset to analyze that is there any MCS contribution during the v-shaped period. Based on T_{BB} value the cloud activity is still low in 11 LT and increase several hours afterwards [1].

A localized heavy rainfall is generally formed by even an isolated cumulonimbus cloud which is suddenly generated aloft and develops rapidly in a very short time [5]. Due to short period of time, it is essential to detect the early detection before this heavy rainfall reaches the ground. Based on the previous study and combination with RGB composite Himawari data, this research is intended to analyze the dataset before 11.20 LT and what variables that can become a parameter for early warning system.

2. Data and Method

Two x-band radar with different type were used in this research. The first radar is the LAPAN Transportable GMWR25SP. It is a single polarization Doppler X-band radar. It is capable to conduct a volume scan with maximum radius of 83 km. The radar is mounted on a vehicle for mobile observation purpose. The second radar is the rain scanner SANTANU. This radar is a single polarization X-band non-doppler radar. This fixed-mounted radar can only scan in a single level of altitude and have a relatively large vertical beam width with maximum radius of 44 km. Both radars have its own advantage and disadvantage. GMWR25SP is able to scan spatially and vertically with a volume scan, but low temporal resolution since it needs time to achieve a whole volume scan. Meanwhile, SANTANU is able to scan only spatially, but high temporal resolution.

Red Green Blue Himawari (hereafter stated as RGB Himawari) composite imagery works based on the principle of color combination which consists of three primary colors, i.e. red, green and blue [6]. Combination of three image bands (individual or channel difference) of satellite data can be displayed as one image, showing new color information that can be used for various analysis. This technique is simple and creates image that relatively easy to be interpreted since the objects are able to be distinguished from their appearances. Furthermore, the use of high spectral resolution of Himawari-8 satellite data in this study enables us to generate more information from this RGB method. The water vapor classification from RGB combination of Himawari band (Table 1) are used to analyze the effect of the water vapor distribution with the process generation of the v-shaped.

Three early datasets (GMWR25SP, SANTANU, and Himawari) are used to see the first indication before the V-shaped echo in March 17, 2017. The period will be in the interval from 10.50 LT until 11.10 LT. Each dataset is extracted only from the desired area of the v-shaped. Time and coordinate location comparison from this dataset are used to see the similarity process for the early indicator as well as visual comparison.

IOP Conf. Series: Earth and Environmental Science **303** (2019) 012056 doi:10.1088/1755-1315/303/1/012056

Each dataset is then analyzed based on its characteristic to see the process before the v-shaped occur. Calculation of the object are used to estimate the area. Growth-decay rate of the reflectivity area from more high temporal resolution SANTANU data are used to calculate the rapid growth process [7]. Image processing are very useful tool to analyze rainfall distribution including growth-decay rate [8]. Meanwhile, the height profile in each CAPPI reflectivity and doppler velocity data from GMWR25SP are used to analyze the position and the generation of the convective cell.

Table 1. RGB combination band for water vapor

Classification	R	G	B
True color	0.64 μm (Band 3)	0.51 μm (Band 2)	0.47 μm (Band 1)
Water vapor	6.2 μm (Band 8)	6.9 μm (Band 9)	7.3 μm (Band 10)
Natural color	2.3 μm (Band 6)	0.86 μm (Band 4)	0.64 μm (Band 3)
Day microphysics	0.86 μm (Band 4)	3.9 μm (Band 7)	10.4 μm (Band 13)

3. Result and Discussion

Time comparison from the early data in each dataset are showed in Figure 2 and Table 2. There are differences for the emergence of early indication. Data from two x band radar is nearly in the same time (10.50-10.52 LT), but for Himawari data (RGB water vapor distribution) start to appear in the 11.00 LT. GMWR25SP is able to detect the first data compared with the other dataset. From this data showed there is one convective cell (> 40 dBZ). SANTANU are able to detect the first data in 10.52 LT (Figure 3) but with relatively small area compared to GMWR25SP. Meanwhile first data of RGB Himawari is in 11.00 LT that relatively showed low response than the other two datasets.

Visual comparison in three dataset showed slightly different pattern, especially for composite RGB Himawari. Similarity pattern are only showed in the period 11.00 LT. Initial stage of the second convective cell are detected in period 11.10 LT with both x-band radar. In this period, RGB Himawari only showed one cell but have a large area spread to the left from its origin. This showed that RGB Himawari is also detect the growth of the second cell, but could not distinguish both cells. Analysis of each dataset are necessary to see more detailed on the generation process.

Table 2. Comparison of the First data in each dataset

No	Dataset	Spatial Resolution	Temporal Resolution	Time (LT)
1	SANTANU	120x120 meters	2 minutes	10.52
2	GMWR25SP	250x250 meters	5 minutes	10.50
3	Himawari	2000 x 2000 meters	10 minutes	11.00

Growth reflectivity are used to analyze the growth process of reflectivity area. SANTANU image data are used for the growth rate calculation due to more high temporal resolution compared with the other dataset. Figure 3 showed the SANTANU data of accumulated and area reflectivity every 2 minutes from 10.50 LT to 11.10 LT. This figure showed rapid growth of the area and also the high slope of accumulated reflectivity from 10.52 LT until 11.00 LT. The average growth rate of the accumulated reflectivity within this period (10.52 LT – 11.00 LT) is 15.6% from total area, meanwhile from 11.04 until 11.10 only reach 1.2%. This rapid process showed a second convective cell appear in 10.58 LT. Since SANTANU only have data in single level, so we need a volume scan data to observe more on the convective cell.

Reflectivity profile ascending with time are the evidence of an updraft occur, in which the precipitation is generated and carried upward [9]. From reflectivity profile showed the first convective cell are already have a strong updraft with the echo top reaching up to 6 km (Figure 4a). The second convective cell then exist at 11.00 LT (more specific at 10.58 LT) in the left side from previous cell. Two reflectivity columns (Figure 4b) then exist in 11.00 LT with echo top in different height (first cell at 9 km, second cell at 6 km). Within the next 10 minutes (Figure 4c) the echo top change with stronger updraft in the second cell rather in the first cell. These two convective cells that appear in 11.10 LT become the bases of the v-shaped pattern.

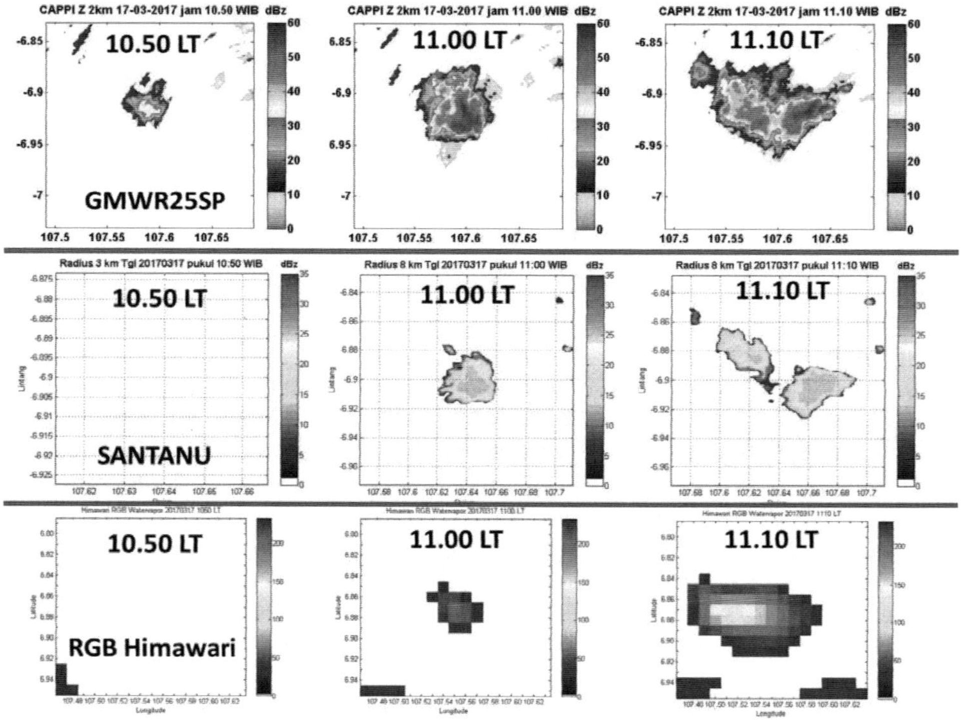

Figure 2. Comparison of GMWR25SP, SANTANU, and RGB Himawari in 10.50 LT,11.00 LT, and 11.10 LT.

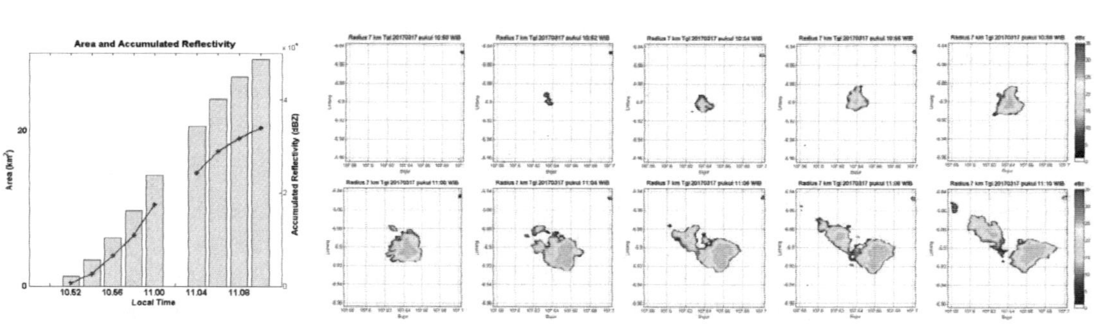

Figure 3. Area and Accumulated reflectivity from SANTANU data

It is shown by the radial velocity profile that there is single vorticity with clockwise rotation detected in 10.50 LT (Figure 4d). High activity (Figure 4e – 4f) are shown in the next 10 until 20 minutes (11.00 LT until 11.10 LT) where there are exhibited two vorticities, usually called vorticity couplet [4], which have a different movement (clockwise and anticlockwise) in the middle altitude (3 - 4 km). This vorticity couplet is also showed until certain altitude that probably related with the vorticity tube that need to be analyze further.

IOP Conf. Series: Earth and Environmental Science **303** (2019) 012056 doi:10.1088/1755-1315/303/1/012056

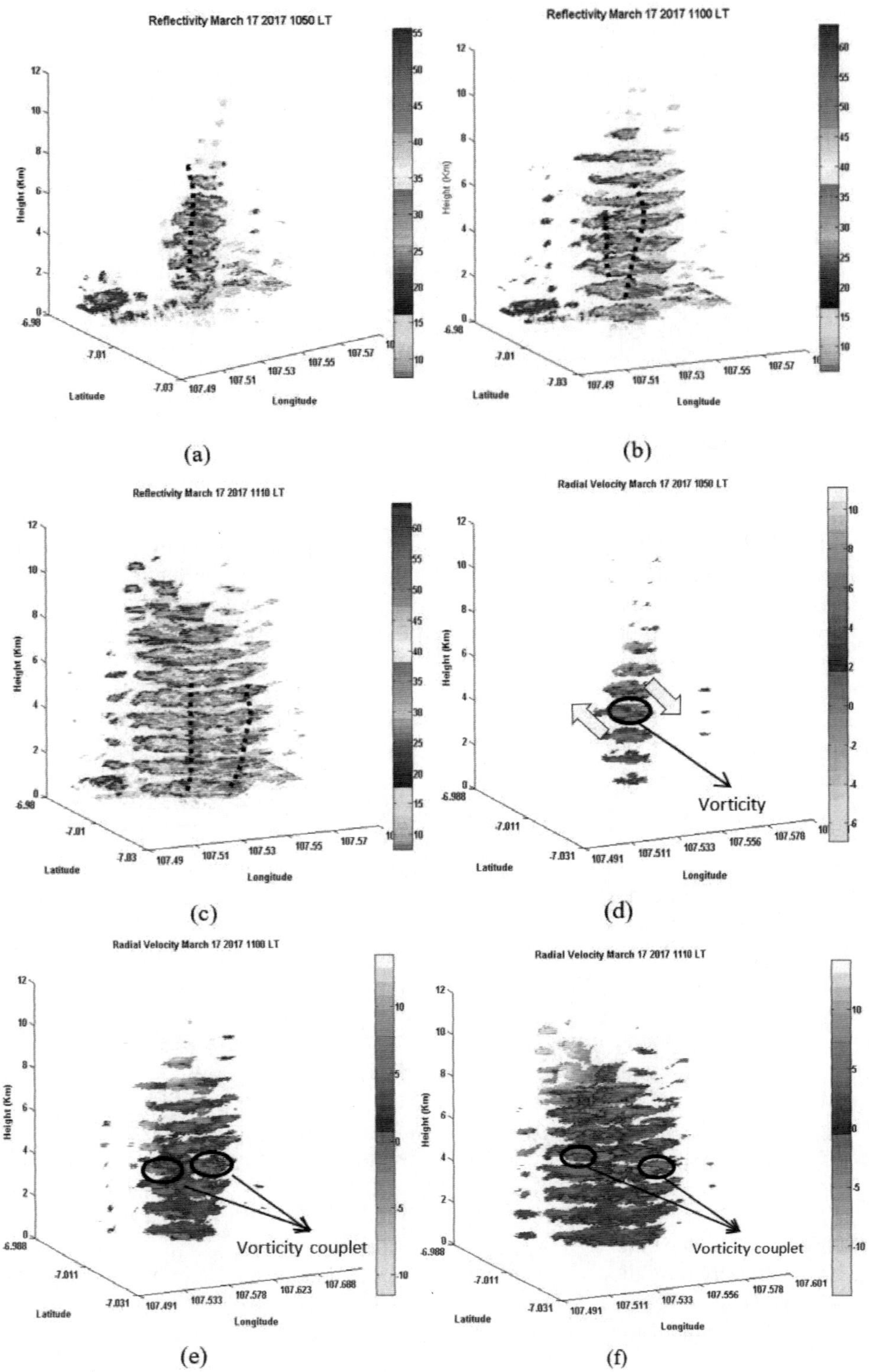

Figure 4. Reflectivity and Radial velocity profile from CAPPI data

Water vapor distribution from RGB Himawari sequence from 10.50 until 11.30 are shown in Figure 5. From this figure we can be estimated that ambient flow is characterized by water vapor moving towards the west side. This condition did not affect the updraft condition of the first convective cell in 10.50 LT.

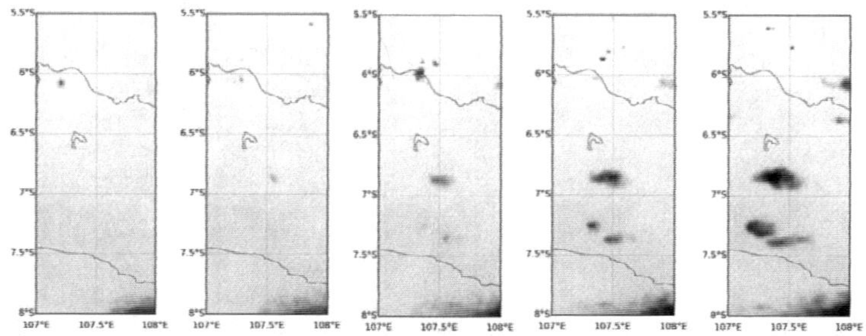

Figure 5. Water vapor distribution

However, for the newest updraft from second convective cell started tilting and move to the right side of the first cell. This ambient airflow possibly affects the location of the second convective cell (the left side from previous cell). Ambient flow started tilting the reflectivity to downward direction with height, with the second cell did not reach higher than the first cell [10]. This condition is the result of the v-shaped reflectivity pattern.

Three datasets give a more comprehensive analysis on the process before the localized severe weather with v-shaped pattern on March 17, 2017 happened. Data indication in 11.10 LT is the important period for the early indicator of this severe weather. In this period, the second convective cell was detected. Several indicators that presumably could become an early indicator of this severe weather are rapid reflectivity growth rate, second convective cell, and vorticity couplet.

4. Conclusion

Localized heavy rainfall occurred in Bandung Basin March 17, 2017 were observed from GMWR25SP, SANTANU, and also RGB Himawari Satellite. The early convective cell is detected in 10.50 LT. Then, it grows rapidly for the next 10 minutes and generate another convective cell on its left. Based on the water vapor distribution it showed the dominant movement towards the west part. This condition is probably the cause why the second cell generates on the left from previous cell. The strong convective activity is related with these two convective cell activities, with two vorticities (vorticity couplet) are detected in the 3 to 4 km altitude both in 11.00 LT until 11.10 LT. This vorticity couplet is also shown until certain altitude that probably related with the vorticity tube that need to be analyzed further. Data indication in 11.10 LT is the important period for the early indicator of localized heavy rainfall in March 17, 2017 followed by existence of rapid reflectivity growth rate, second convective cell, and vorticity couplet.

Acknowledgment

This research was supported by the Center of Atmospheric Science and Technology, National Institute of Aeronautics and Space - PSTA LAPAN, and also thank the Rancakasumba electric substation for support and providing the radar location site.

References

[1] Nugroho G A, Sinatra T, Trismidianto and Fathrio I 2018 Application of rain scanner SANTANU and transportable weather radar in analyze of Mesoscale Convective System (MCS) events over Bandung, West Java *Proc Int LISAT 2017 doi:* **10**.1088/1755-1315/149/1/012058

[2] Nugroho G A, Sinatra T and Trilaksono N J 2017 Post Processing and analysis using 2D combined shear method on X- band Transportable Weather Radar during Strong Wind in Bandung west Java Proc. International Symposium on Earth Hazard and Disaster Mitigation ISEDM 2017 doi: **10**.1063/1.5047369
[3] http://ayobandung.com/view/2017/03/17/17790/awan-hitam-menyelimuti-bandung
[4] Sinatra T and Nugroho G A 2018 Analysis of Heavy Rainfall with Hail using Reflectivity and Radial Velocity Derived from X-Band Radar in Bandung (Study Case: March 17, 2017) *Proc. International doi*: **10**.1088/1755-1315/166/1/012039
[5] Nakakita E, Sato H, Nishiwaki R, Yamabe H and Tamaguchi K 2017 Early Detection of Baby-Rain-Cell Aloft in a Severe Storm and Risk Projection for Urban Flash Flood Advances in Meteorology, Vol **2017** https://doi.org/10.1155/2017/5962356
[6] Bessho K, Date K, Hayashi M, Ikeda A, Imai T, Inoue H, Kumagai Y, Miyakawa T, Murata H, Ohno T, Okuyama A, Oyama R, Sasaki Y, Shimazu Y, Shimoji K, Sumida Y, Suzuki M, Taniguchi H, Tsuchiyama H, Uesawa D, Yokota H and Yoshida R 2016 An introduction to Himawari8/9 - Japan's new-generation geostationary meteorological satellites *J. Meteor. Soc. Japan* **94** doi:10.2151/jmsj.2016-009
[7] Nugroho G A, and Awaludin A 2015 Analisis Distribusi dan Pertumbuhan Area Hujan Hasil Observasi RDH (Radar Deteksi Hujan) pada Tanggal 19 Februari 2014 di Daerah Bandung dan Sekitarnya berdasarkan Metode Pengolahan Citra Buku Pengembangan Teknologi Atmosfer dan pemanfaatannya ISSN: 978-979-1458-92-4
[8] Wang P, Smeaton A, Lao S, O'Connor E, Ling Y and O'Connor N 2009 Short-term rainfall nowcasting using rainfall radar imaging The **9th** Irish workshop on computer graphics Eurographics Ireland
[9] Brown R A and Torgerson K L 2003 Interpretation of Single-Doppler Radar Signatures In a V-Shaped Hailstorm: Part I - Evolution of Reflectivity-Based Features National Weather Digest **27** 3-14
[10] Brown R A and Torgerson K L 2005 Interpretation of Single-Doppler Radar Signatures In a V-Shaped Hailstorm: Part II - Evolution of Updraft Interactions with Ambient Mid-Altitude Flow National Weather Digest **29** 65-80

Identification of rainfall area in Indonesia using infrared channels of Himawari-8 Advance Himawari Imager (AHI)

Risyanto, F Lasmono and GA Nugroho

Center for Science and Technology Atmosphere, Indonesian National Institute of Aeronautics and Space

risyanto@lapan.go.id

Abstract. The study aims to investigate the use of Infrared (IR) channels of Himawari-8/AHI for identification of rainfall area in Indonesia. The parameters used include the IR brightness temperature (BT) at 10.4 microns (T10.4) and seven IR BT differences (BTD), which were inferred as proxies for cloud properties. Identification of rainfall in this study is based on lookup table (LUT) approach, which is used to create probability of rainfall map. The LUTs were developed by combining the IR at 10.4 microns and IR BTD with the transportable X-band radar data, gathered during the campaign period on 15 March – 5 May 2017. Statistical skill scores were used in the study to determine the overall performance of the methods. The study indicated that the best IR and BTD combination to identify rainfall area is from the bands correlated to cloud-top height proxy (T10.4 and BTD 13.3- 10.4). In visual comparison with Global Satellite Mapping of Precipitation (GSMaP) hourly rainfall image, this IR-BTD method produced rain maps with high similarity. In general, almost all IR-BTD combinations could be used to identify rainfall area with comparable results. However combination of T10.4 and BTD at 6.2 & 7.3 micron generate high false alarm rates and underestimate the area of rainfall.

Keywords : Himawari-8, Brightness temperature, Cloud, Lookup Table, X-band radar.

1. Introduction

Indonesia hosts one of the most variable climate system in the world. Rainfall variability in this region is highly variable and convoluted due to its geographical features and complex topography conditions. Rainfall pattern in Indonesia is also incoherent spatially and temporally [1]. Scientific investigations on rainfall variability is important since it influences many aspects of life of million people living in the region. Moreover, the most frequently natural disasters occurred in Indonesia, such as floods and landslides are primarily triggered by extreme rainfall events. Study on rainfall in this region is necessary in order to improve knowledge and quality of weather/climate forecasting, and the Decision Support System (DSS) development to minimize the risk of disaster impact. Thus, accurate rainfall measurements are crucial and highly needed, which includes better methods of rainfall estimation from remotely-sensed instrument.

Over the past three decades, there have been numerous attempts to use satellite measurements for rainfall estimation, which is proven to be cost-effective, continuous and reliable sources over the remote regions where rainfall data is very limited [2,3]. A range of rainfall estimation techniques has been developed. The methods that utilized visible (VIS) and infrared (IR) measurements rely upon the

IOP Conf. Series: Earth and Environmental Science **303** (2019) 012057 doi:10.1088/1755-1315/303/1/012057

information provided by cloud-top characteristics. In spite of their indirect relationship between the cloud-top and rain-rate [4], they have the advantage of frequent observations compared to the techniques that use passive microwave (PM). In addition, the use of geostationary-orbit satellite data for this purpose offers very good temporal resolution that the dynamic of cloud and rainfall can be observed.

The newest generation of Japanese geostationary meteorological satellites, Himawari-8/9, equipped with an Advanced Himawari Imager (AHI) instrument, with significantly higher radiometric, spectral, and spatial resolution than those previously available in the geostationary orbit. The satellites have 16 observation bands, specifically: 3 VIS bands, 3 NIR bands, and 10 IR bands with high spatial resolution: 0.5 or 1 km for VIS and NIR bands and 2 km for IR bands [5]. Moreover, a temporal resolution of ten minutes provides real time monitoring of rainfall and cloud activities. The IR bands of AHI are currently not yet fully utilized for rainfall estimation researches. Therefore, this study seeks to investigate the use of IR channels of AHI brightness temperature (BT) and brightness temperature difference (BTD) to identify rainfall area specifically within Indonesia region.

Previous several studies pointed out that IR BTD can be a proxy of cloud properties, which can be associated with rainfall rate [6,7]. These cloud properties include cloud-top height, cloud optical thickness, cloud phase, and cloud water path. This study will examine the capability of the wide range Himawari-8 IR channels for rainfall area detection. Identification on the rainfall area is based on look-up table (LUT) approach, which is developed by utilizing the X-band radar data observed during a weather observation campaign in Bandung for about two months. The LUTs are used to produce probability of rain (POR) maps which are determined from the combination of IR and IR BTD channels of AHI observation.

2. Data dan area of study

The data used in this study includes the IR and IR BTD from Himawari-8 AHI (Table 1), the constant altitude plan-position indicator (CAPPI) product at 3km altitude of horizontal reflectivity from the transportable X-band radar, and the GSMaP MVK hourly rainfall rate product. The radar data were obtained during the observation campaign in Bandung on 15 March – 5 May 2017. The instrument was installed at Rancakasumba Substation (107.77 E, 7.04 S) located in Majalaya, south of Bandung city and covered the area within 100 km radius (Figure 1). For evaluation, the study area was expanded to the almost all region of Indonesia (10 S – 10 N; 95 W – 150 W) and the study period was extended to March until 31 December 2017.

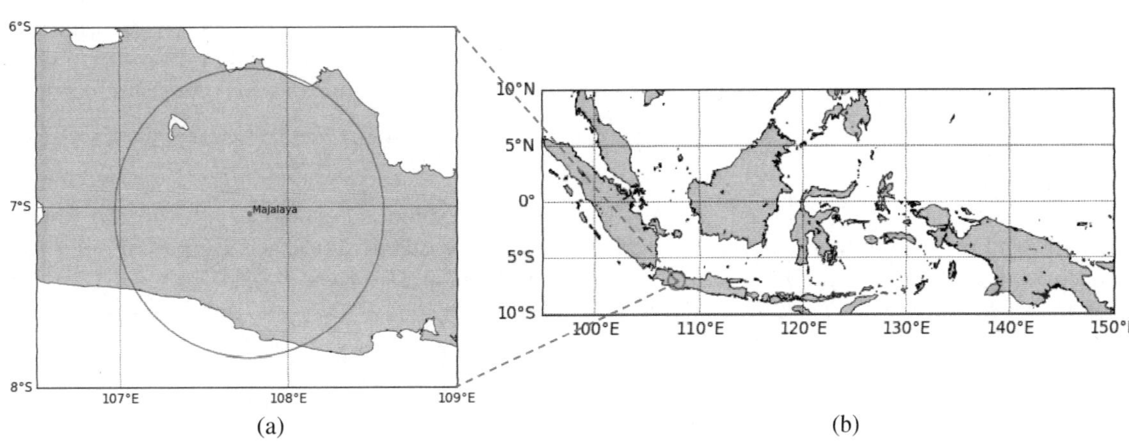

Figure 1. Location of the transportable radar at Rancakasumba substation, Majalaya (a) and the area of study for verification stage (b). Red circle shows the area coverage of radar.

Table 1. Selected BTD combinations used in creation of POR LUT

Related cloud property	Label	IR-BTD Combination
Cloud Top Height	CTH1	$T_{10.4}$ & $T_{6.2-10.4}$
	CTH2	$T_{10.4}$ & $T_{6.2-7.3}$
	CTH3	$T_{10.4}$ & $T_{13.3-10.4}$
	CTH4	$T_{10.4}$ & $T_{9.6-13.3}$
Cloud Optical Thickness	COT1	$T_{10.4}$ & $T_{10.4-12.3}$
	COT2	$T_{10.4}$ & $T_{8.6-12.3}$
Cloud Phase	CP	$T_{10.4}$ & $T_{8.6-10.4}$

3. Methods

Initial step to derive rainfall area from Himawari-8 is creating the POR LUT. These LUTs were developed based on relationship between IR and BTD with the collocated X-band radar data. Different BTD combinations in this process were utilized, thus, creating different LUTs. These POR LUTs were used to produce POR maps. Rainfall maps were determined and assessed to classify rain from no-rain based on POR threshold. Furthermore, a comparison of the resulting rainfall maps with another satellite rainfall data known as GSMaP hourly rainfall data was performed to validate the output maps within the region of Indonesia.

3.1. LUT Construction

POR is defined as the percentage value of rain event frequency over the total number of events. The equation is presented as [7]:

$$POR(x_1, \dots, x_n) = \frac{N_{rain}(x_1, \dots, x_n)}{N_{rain}(x_1, x_2 \dots, x_n) + N_{no-rain}(x_1, \dots, x_n)} \quad (1)$$

Where $(x_1, \dots, x_n)$ are the combination of IR-BTD as listed in Table 1, and $N_{rain}(x_1, \dots, x_n)$ and $N_{no-rain}(x_1, \dots, x_n)$ are the number of pixels with rain and with no–rain denoted by $(x_1, \dots, x_n)$ respectively. There are seven combinations of IR and BTD used in this study (Table 1).

POR LUTs are created through temporal and spatial matching of IR and BTD maps with the radar reflectivity maps. Rain and no-rain pixels were identified and assigned from radar reflectivity data to the equivalent IR and BTD pixels. The POR distribution as a function of IR and BTD combinations were generated by using Equation 1. Each IR and BTD combinations produced an LUT which in turn can be utilized to create maps with rain and no-rain areas. Finally, a map classifying rain from no-rain areas were derived by setting the threshold in POR distribution applying the LUT on satellite data images. This study used a threshold value of 0.32 [7].

3.2. Verification metrics

The categorical statistics [8], was used in this study to further evaluate and compare the performances of each IR and BTD combinations in Himawari-8 rainfall area detection. The categorical statistics applied in this study included: i) Probability of Detection (POD); ii) False Alarm Ratio (FAR); iii) Probability of False Detection (POFD); and iv) Critical Success Index (CSI), all calculated as follows [9]:

$$PD = H/(H + M) \quad (2)$$

$$FAR = F/(H + M) \quad (3)$$

$$POFD = F/(F + C) \quad (4)$$

$$CSI = H/(H + F + M) \quad (5)$$

Table 2. Contingency table used for computation of categorical statistics.

Forecasted or estimated	Observed	
	Rain	No-rain
Rain	Hits (H)	False Alarm (F)
No-rain	Misses (M)	Correct Negative (C)

The POD is the fraction, which the Himawari-8 algorithm correctly estimated rainfall events as measured in observed rain maps. It ranges from 0 to 1 with a perfect value of 1. FAR is the fraction, which the satellite algorithm has detected rainfall events, but there is no rain events in observed rain maps. It ranges from 0 to 1 with a perfect value of 0. POFD is almost similar to FAR, but the condition is on observations rather than the forecasts. It ranges from 0 to 1 with a perfect value of 1. CSI is a fraction, which satellite algorithm has correctly estimated rainfall but taking into account the false alarms in contrast to POD. It ranges from 0 to 1 with a perfect value of 1. In this study, observation data of rainfall events were obtained from GSMaP MVK hourly data product. The calculation of these statistical scores were applied for each pixel of rainfall area map derived from each LUT, which matched spatiotemporally with the GSMaP data.

4. Results and Discussion

4.1. Probability of Rain (POR) LUT

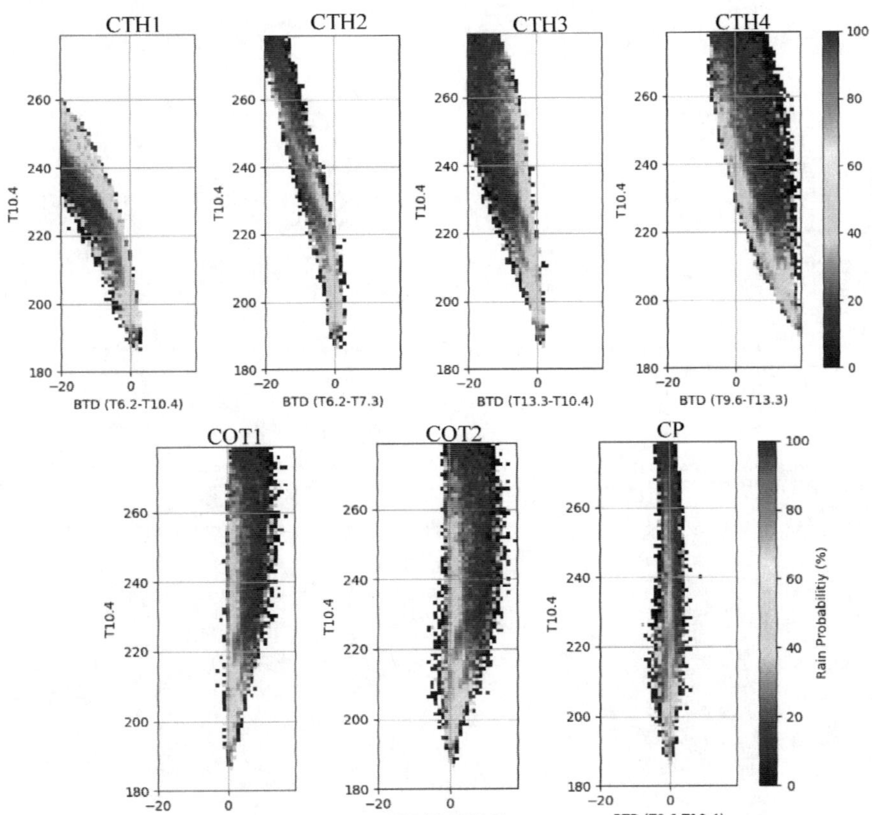

Figure 2. Probability of rain (POR) LUTs developed from IR-BTD combinations

IOP Conf. Series: Earth and Environmental Science **303** (2019) 012057 doi:10.1088/1755-1315/303/1/012057

Figure 2 shows the POR LUTs produced by comparing IR and BTD combinations to the X-band radar reflectivity which has been identified as rainfall. The POR was calculated by using Equation (1). To determine rainfall area from these LUTs, a threshold value of 0.32 have been used. IR and BTD values that have POR higher than or equal to the threshold will be assigned as rainfall area, and vice versa. This threshold was successfully used to define rain and no-rain area using the similar method in Phillipines [7]. Determining a proper threshold value in this method is really important since it significantly affects the rainfall area generated by the LUTs.

The POR LUTs indicate that rain areas (POR $\geq$ threshold) have brightness temperature at 10.4 microns (T10.4) generally lower than 260K. This is acceptable, since lower temperature is strongly associate with clouds, and colder clouds highly correlated with rain. The LUTs also show that there were no/very least POR values could reach 100%. This means that there are no exact values of IR-BTD combination could determine rainfall with high certainty. POR distribution mostly lies in the data range of 20-30%. This implies that the uncertainty of IR method to detect rainfall should be noticed.

4.2. Comparison with GSMaP Data

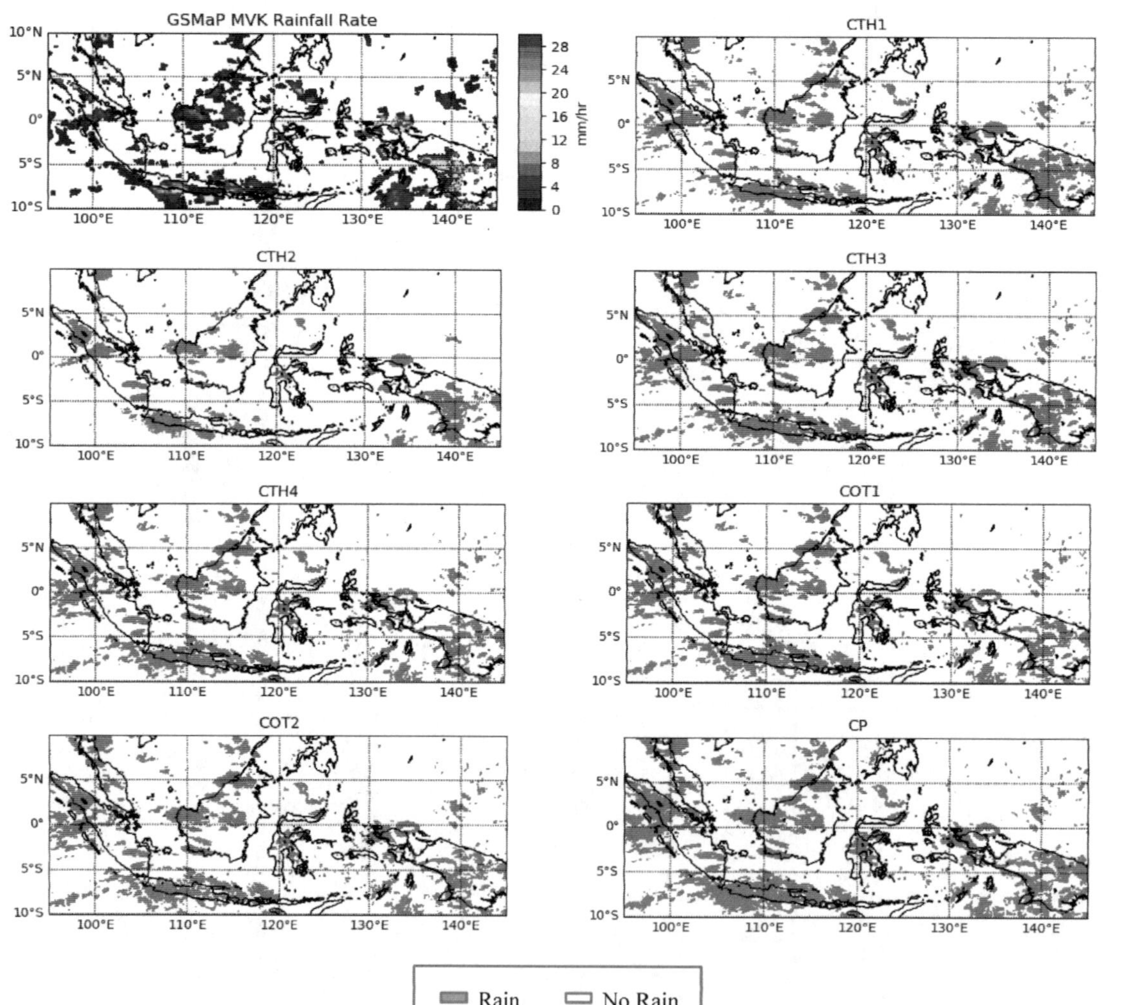

Figure 3. Comparison of GSMaP rain rate product and rain area maps produced by IR-BTD combinations taken at April 06, 2017 12:00 UTC

POR LUTs that have been created were used to generate rainfall area. The performance of the method of each IR-BTD combination will be evaluated by comparing them with rain rate products from GSMaP data. The intensity of GSMaP rainfall will not be further analysed since comparison was only done to investigate whether rain or no-rain occurred at the region of interest. Figure 3 shows one of example of the comparison results. In visual comparison, generally all IR-BTD combinations could be used to identify rainfall area with relatively high similarities to GSMaP product. There was only one exception, that is CTH2, which seems to generate smaller area of rain compared to the others. Almost all of IR-BTD combinations generated an overestimation compared to GSMaP, except CTH2 which is mostly underestimates. However, in several locations there were also the cases where rainfall occurrence in GSMaP were not appeared in the rain maps produced by LUTs. Almost all of combinations of BTD used in the method yielded the similar pattern of rainfall area. The differences amongst them were not significant.

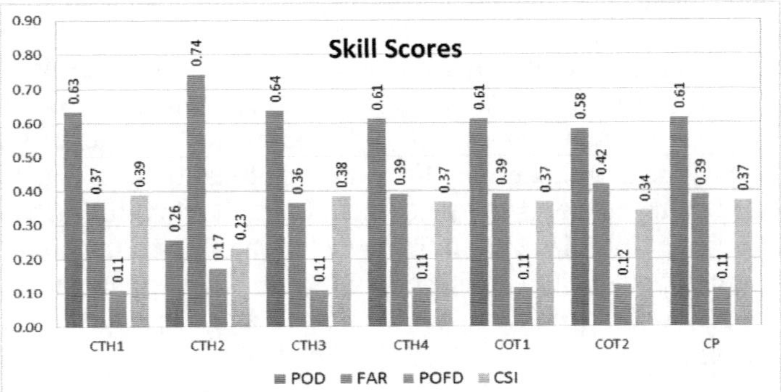

Figure 4. Some statistical scores used in the study to evaluate performance of the method.

Figure 4 shows the comparison of rainfall area produced by LUTs method and rain rate product of GSMaP qualitatively. This study used four categorical statistics (i.e. POD, FAR, POFD and CSI) to examine the performance of the methods. Based on their statistical scores, almost all IR-BTD combination have similar results, except CTH2. Very high FAR (0.74) obtained by CTH2, showing that this combination has the poorest accuracy compared to the others. The best combination of IR-BTD to detect rainfall area is from CTH3 (POD=0.64, FAR=0.36, POFD=0.11, CSI=0.38), which also has very close result to CTH1 (POD=0.63, FAR=0.37, POFD=0.11, CSI=0.39). It is interesting to see that the CTH3 and CTH2, that have the strongest and the poorest correlation with GSMaP, come from the same group that represent cloud top height properties.

Three channels (BT at 10.4 micron and BTD at 6.2 & 7.3 micron) were involved in CTH2 combination. The use of BT at 6.2 microns in CTH1 produced good results, which is acceptable since T6.2 can represents amount of water vapour in upper layer of atmosphere thus it can relate to clouds and rainfall. This channel usually used to extracting deep convective clouds with heavy rainfall [10]. However, the combination of T6.2 and T7.3 as used in CTH2 generated the lowest score. Brightness temperature at 7.3 micron is also sensitive to water vapour, but its capability to retrieve rainfall area when combined with other water vapour channel (T6.2) is poor.

Underestimation of rainfall area produced by CTH2 may also caused by inappropriate threshold value used to define rain and no-rain from POR LUT. Modification of this value, for example by lowering the threshold for CTH2, might generate larger areas of rainfall and expected to be more consistent to the GSMaP observation. However, even after changing this value for CTH2, the results were not too different. Figure 5 shows one example of the change of rainfall area produced by CTH2 by setting the threshold to become 10% and 20%. Based on the Figure 5, modification of threshold value for CTH2 have no impact to the rainfall area map generated.

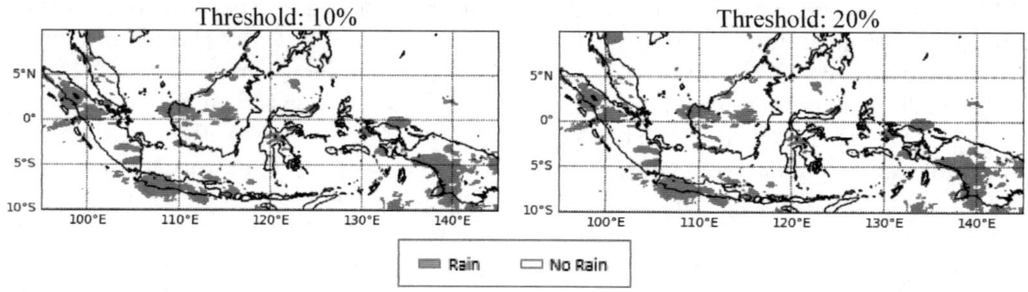

Figure 5. Rainfall area generated from CTH2 after changing the threshold value of rain/no-rain to 10% (left) and 20% (right)

5. Conclusion

This study investigated the application of Himawari-8 infrared channels for rainfall area identification through LUT method, which is developed based on relationship between IR-BTD of AHI and X-band radar reflectance. Almost all combinations of IR BT and BTD have generated rainfall area with comparable result to GSMaP data observation. However, combination of BT at 10.4 micron and BTD at 6.2 & 7.3 micron based on this study is not recommended to be used for rainfall area retrieval. The method used in the study can be utilized as initial step for satellite-based rainfall rate estimation or specifically for heavy rainfall potential area detection which is highly related to the natural disaster. In addition, further investigation on specific type of clouds and other different cases of rain like thunderstorms, is still needed.

Acknowledgments

This work is fully supported by Center for Atmospheric Science and Technology of National Institute of Aeronautics and Space. The authors would like to thank Rancakasumba electric substation staff for support and providing the radar location site. The authors also wish to acknowledge the JAXA Precipitation Measuring Mission (PMM) Science Team for providing GSMaP rain rate and dataset.

References

[1] Hidayat R 2016 *Procedia Environmental Sciences* ***33*** *167 – 77*

[2] Tapiador F J, Turk F J, Petersen W, Hou A Y, Garcia-Ortega E, Machado L A T, Angelis C F, Salio P, Kidd C, Huffman G J and de Castro M 2012 *Atmos. Res.* ***104–105****, 70–97*

[3] Collins M., Achuta R K, Ashok K, Bhandari S, Mitra A K, Prakash S, Srivastava R and Turner A 2013 *Nat. Clim. Change* ***3*** *(11) 940–41*

[4] Kidd C and Levizzani V 2011 *Hydrol. Earth Syst. Sci.* ***15*** *1109–16*

[5] Bessho K, Date K, Hayashi M, Ikeda A, Imai T, Inoue H, Kumagai Y, Miyakawa T, Murata H, Ohno T, Okuyama A, Oyama R, Sasaki Y, Shimazu Y, Shimoji K, Sumida Y, Suzuki M, Taniguchi H, Tsuchiyama H and Yoshida R 2016 *J. Meteorol. Soc. Jpn.* ***94*** *151-83. 10.2151/jmsj.2016-009.*

[6] Kuhnlein M, Thies B, Nauß T and Bendix J 2010 *Journal of Applied Meteorology and Climatology* ***49****(7, 1477–95*

[7] Nacario R G, Perez G J and Bagtasa G 2016 Rainfall detection using the Infrared bands of Himawari-8 Advanced Himawari Imager in the Philippines *37th Asian Conference on Remote Sensing (ACRS 2016) Proceeding*

[8] Wilks D S 2006 Statistical Methods in the Atmospheric Sciences, second ed *Academic press, Burlington, MA.*

[9] Lekula M, Lubczynski M W, Shemang E M and Verhoef W 2018 *Physics and Chemistry of the Earth 105 (2018) 84–97*

[10] Kurino T 1997 Rainfall Estimation with the GMS-5 Infrared Split-Window and Water Vapor Measurement *Meteorological Satellite Center Technical, JMA, Note No. 33, 91-101*

Interaction of MJO and diurnal cycle observed by radiosonde data during the first YMC campaign in Bengkulu (November 2017 - January 2018)

D S Permana[1*], T D Hutapea[1*], J A I Paski[1], A S Praja[1], Fatkhuroyan[1], H Harsa[1], D E Nuryanto[1], R Kurniawan[1], E E S Makmur[1] and D Ardiansyah[2]

1Center for Research and Development, BMKG, Kemayoran, Jakarta 10720, Indonesia
[2]BMKG Fatmawati Meteorological Station, Bengkulu 38213, Indonesia

donaldi.permana@bmkg.go.id; hutapea.df@gmail.com

Abstract. The interaction of MJO and diurnal cycle in the Maritime Continent has been identified from previous studies but more observations are needed to understand it clearly. The first campaign of Years of the Maritime Continent (YMC) was conducted in Bengkulu from 16 November 2017 to 15 January 2018. Among various kind of observations, the three-hourly radiosonde sounding was conducted in Bengkulu station during the campaign. This study analyses the radiosonde data to investigate the thermodynamic profile of the mean diurnal cycle (MDC) at different MJO phases. The MDC of potential temperature anomaly (PTA), relative humidity, winds, convective available potential energy (CAPE) and total precipitable water (TPW) in different MJO phases were compared to the MDC values during the campaign (average condition). The obtained MDC PTA has a similar pattern with the results from the Pre-YMC campaign (November - December 2015). During the active MJO phase (3-4), RH is much higher and the zonal westerly wind is up to 6 m/s faster than the average condition at most of the time and elevation below 15 km, and TPW is amplified about 1-4 mm. In contrast, the condition is reversed during inactive MJO phase (1-2). In addition, the CAPE peaks occurred at 06 - 09 UTC and followed by TPW peaks 6 hours later.

Keywords : MJO, YMC, radiosonde, CAPE

1. Introduction

The Maritime Continent (MC) is located within the Tropical Pacific Warm Pool and consists of many island and shallow oceans. The tropical MC has an important role in atmospheric circulation [1]. Due to the high sea surface temperature (SST) in the MC, there is a significant latent heat releases and accompanied by heavy precipitation over the MC. Because of releasing energy, the MC is known as an atmospheric 'boiler box' and one of the main drivers of the global atmospheric circulation [2, 3]. At intraseasonal timescale, the MC is influenced by the Madden Julian Oscillation (MJO) which is an atmosphere equatorial wave propagating from the Indian Ocean to the Pacific Ocean with a period of 30-90 days at speed of approximately 4-5 m/s [4, 5, 6]. MJO signals are observed to be strong in Austral summer (November – April) in the western Pacific [7] and systematically modulate the amplitude of the diurnal cycle of precipitation over the MC [8].

The diurnal cycle (DC) in the tropics is a principal mode of variability in the global climate system [9]. Activities of MJO have broad impacts on a wide range of global weather and climate phenomena, which is related to rainfall variability [10]. Active (wet) phases of the MJO correspond to enhanced rainfall in the tropical Western Pacific, with typical rainfall anomalies between the wet and dry phases in the range of 2 – 6 mm/day [11]. Though the diurnal variations of tropical deep convection largely depend on the region, it is generally accepted that the convective or precipitation maximum occurs over the continents in the late afternoon or early evening and over the ocean in the early morning [12, 13, 14, 15, 16].

Many studies have suggested that the MJO interacts with DC. The suppressed MJO to be associated with small, short-lived cloud system which form and decay in the afternoon and the active MJO to be characterized by larger, longer-lived systems which form during the afternoon but not peak until the early hours of the following morning, before decaying [17]. The DC is enhanced during the suppressed phase and diminished during the active phase of the MJO [18], whilst diurnal cycles of atmospheric variations are more pronounced during the wet phase [19]. The diurnal cycle of deep convection over both land and ocean areas to be enhanced during the active MJO compared with the suppressed MJO, but the diurnal phase to be similar in both regimes [20, 21]. However, more observations and research are needed to understand more details.

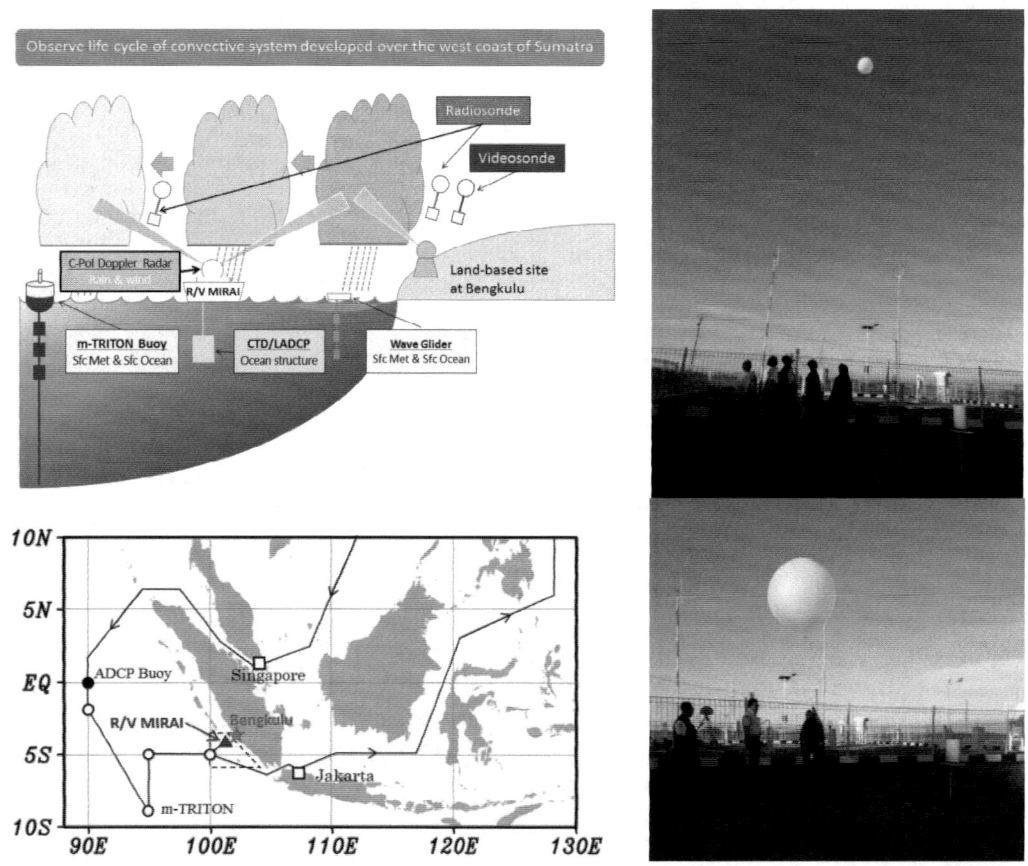

Figure 1. The first YMC campaign in the west coast of Sumatera from 16 November 2017 – 15 January 2018. (Left panel) activities conducted during the campaign including atmosphere and ocean observations (source: http://www.jamstec.go.jp/ymc/IOP_YMC-Sumatra_2017.html). (Right panel) Radiosonde launch in BMKG Fatmawati station with 3-hours intervals in a day (photos: BMKG).

Within the framework of "Years of the Maritime Continent" (YMC), a two-year (2017 – 2019) international project consisting of multiple intensive observation projects related to atmospheric and oceanic research focusing on the MC (http://bmkg.go.id/ymc; http://www.jamstec.go.jp/ymc). In the MC, BMKG is the host country of the YMC. Its overarching goal is to expedite the progress of improving understanding and prediction of local multi-scale variability of the MC weather-climate system and its global impact through observations and modeling exercises.

The first campaign of YMC was conducted by collaboration of the Japan Agency for Marine–Earth Science and Technology (JAMSTEC), the Indonesian Agency for the Assessment and Application of Technology (BPPT), and the Indonesian Agency for Meteorology, Climatology and Geophysics (BMKG) in the west coast of Sumatera from the mid-November 2017 to mid-January 2018 (Figure 1). Among various kind of observations, radiosonde observations were performed in Bengkulu station at 3-h intervals [01, 04, 07, 10, 13, 16, 19 and 22 local time (LT); LT = UTC + 7] with the use of Meisei-100 sensors manufactured by Meisei (Figure 1). Previously, the Pre-YMC intensive observation campaign was conducted in November-December 2015 by the same collaboration agencies at the same location for preliminary study [22].

This study investigates the atmospheric thermodynamic profile of the mean diurnal cycle (MDC) at different MJO phases using the radiosonde data collected from the first YMC campaign in Bengkulu station. The MDC of potential temperature anomaly, relative humidity, winds, freezing level height, convective available potential energy (CAPE), convective inhibition (CIN) and total precipitable water (TPW) at different MJO phases were compared to the MDC values during the whole campaign period to observe the interaction of MJO and MDC.

2. Data and Methods

In this study, we used radiosonde data collected at BMKG Fatmawati Station (102.337°E; 3.858°S) from 16 November 2017 to 15 January 2018. The recorded parameters consisted of pressure, temperature, relative humidity, wind speed, and wind direction in the vertical level. The highest recorded radiosonde height was 32430.9 meters (7.9 mb) on 5 January 2018 at 0600 UTC when the weather condition was clear. The lowest radiosonde height was recorded at 4587.5 meters (584.8 mb) on 22 December 2017 at 2100 UTC when the continuous rain was observed.

As the intensity of convective activity is influenced by the atmospheric thermodynamic profile before the convection begins, the potential temperature (θ) was derived from radiosonde data. The potential temperature of an air parcel at pressure P is defined as the temperature that the parcel would attain if adiabatically brought to a standard reference pressure (usually 1000 mb). By a given air parcel with temperature T (°C) and pressure P (mb), the potential temperature (Kelvin) is calculated by the following formula:

$$\theta = (T + 273.15) * \left(\frac{1000}{P}\right)^{0.286} \tag{1}$$

The vertical profiles of potential temperature anomaly (PTA), relative humidity (RH), zonal and meridional wind during the campaign at Bengkulu station are depicted in Figure 2 (Panel 1-4). The altitude is limited to 20.5 km (~ 20 mb) in this study. In order to obtain the MDC, the listed above parameters were averaged for eight particular times [00, 03, 06, 09, 12, 15, 18, and 21 UTC]. The MJO phase and intensity [23] during the campaign are shown in Figure 2 (Panel 5-6). In this study, the dates associated with strong MJO phases are defined as follows: phase 3-4 (23-30 November 2017), phase 5-6 (1-7 December 2017), phase 7-8 (8-26 December 2017), and phase 1-2 (27 December 2017-12 January 2018). The MDC at particular MJO phases were obtained by averaging the parameters in the associated dates. Furthermore, the CAPE, CIN and TPW for each radiosonde observation were calculated using SkewT library in Python language (https://pypi.org/project/SkewT). In addition, the changes of the freezing level height (FLH, the altitude at which the temperature is 0°C), CAPE, CINand TPW were obtained at particular MJO phases.

IOP Conf. Series: Earth and Environmental Science **303** (2019) 012058 doi:10.1088/1755-1315/303/1/012058

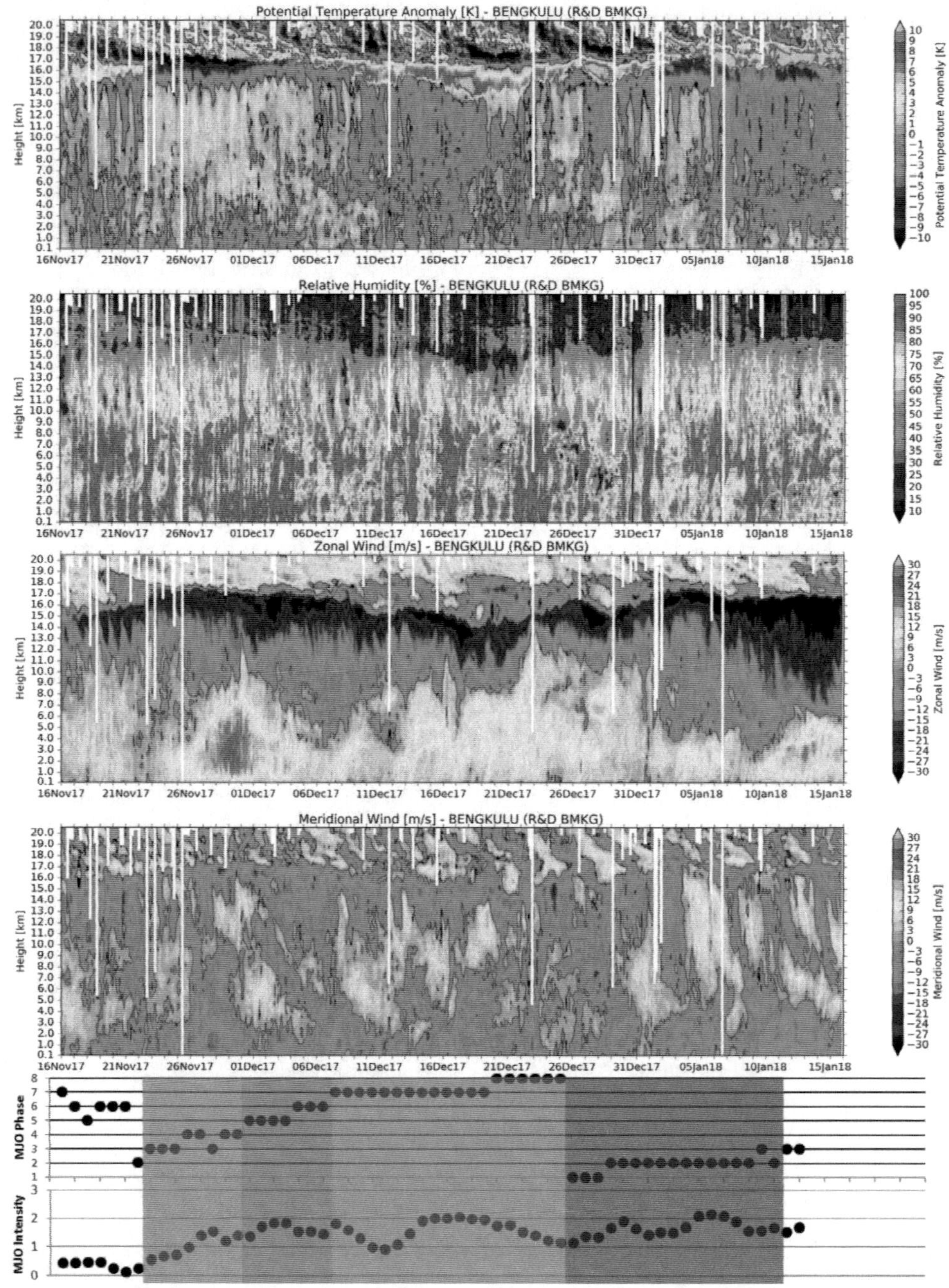

Figure 2. Radiosonde (Meisei-100) data collected from 16 November 2017 to 15 January 2018. The vertical profiles of PTA, RH, zonal and meridional winds are shown in Panel 1-4. The MJO phase and intensity are given in panel 5-6. Colored boxes indicate the dates associated with the MJO phases used in this study (blue: phase 3-4, red: phase 5-6, green: phase 7-8, purple: phase 1-2).

3. Results and Discussions

The MDC of PTA, RH, zonal and meridional winds during the whole campaign period (defined as "average condition") is shown in Figure 3. The obtained MDC of PTA has a similar pattern with the results from Pre-YMC campaign (November - December 2015) [23]. In the troposphere (< 10km), the positive maxima PTA are observed at 06 UTC (13 LT) from the surface up to the altitude of 500 meters, whereas negative maxima are observed at 18, 21 and 00 UTC (01-07 LT) at the surface. In average, the positive PTA is observed at 06-12 UTC (13-19 LT). At the altitude of 10-15 km, the positive maxima PTA are observed at 09-12 UTC (16-19 LT) from 11 to 14 km, whereas the negative maxima were observed at 21 and 00 UTC (04-07 LT) from 11 to 13 km. Relative humidity is generally lower during 03-09 UTC (daytime) than during nighttime, which represents the typical weather in Bengkulu that usually have rainfall (high humidity) during nighttime. The highest relative humidity is observed at 18-21 UTC in altitude of ~5 km and at 00 UTC in altitude of ~2 km. During the campaign, the westerly wind (monsoon) is dominant from the surface up to 5 km with maxima occurs at the altitude of 2-3km with the speed of 8-12 m/s. From 10 to 15 km, the tropical easterlies are dominant with a maximum speed of 24 m/s, whereas the winds are relatively calm all the time at 7 to 9 km (not shown) due to the meeting of monsoon westerlies and tropical easterlies. In the other hand, the northerly winds are strong below 2 km with a maximum speed of 5-6 m/s at 18, 21, and 00 UTC and a minimum speed of 1-2 m/s at 09 UTC below 500 meters.

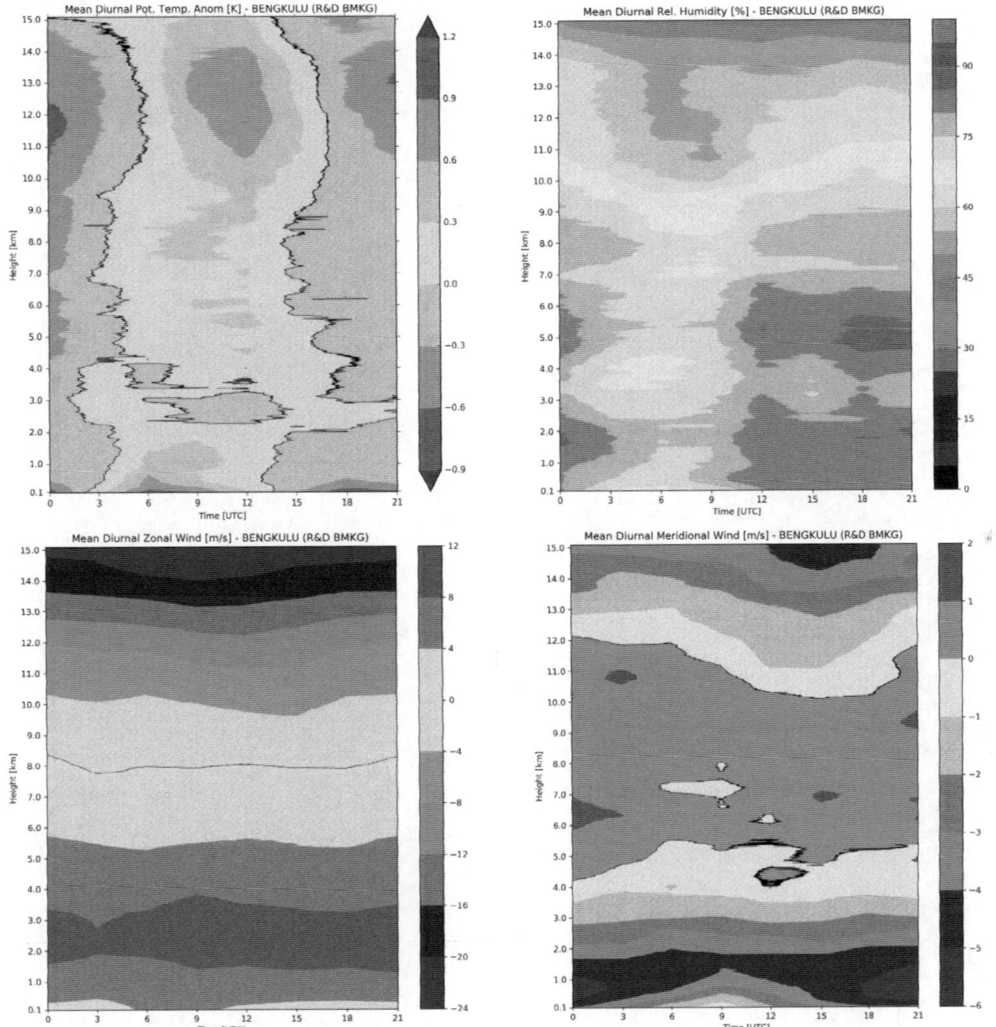

Figure 3. Mean diurnal cycle of potential temperature anomaly, relative humidity, zonal, and meridional winds at Bengkulu station.

IOP Conf. Series: Earth and Environmental Science **303** (2019) 012058 doi:10.1088/1755-1315/303/1/012058

In order to observe the interaction between MJO and MDC, the differences between the MDC at particular MJO phases and the average condition are obtained for PTA and RH (Figure 4), and for zonal and meridional winds (Figure 5). The PTA differences are mostly positive from 2 to 14 km during MJO phase 3-4 (active) and 5-6, and mostly negative during MJO phases 1-2 (inactive) and 7-8. During active phases (3-4) of MJO over Sumatera, RH is much higher (5-15%), suggesting additional moisture in the atmosphere and the zonal westerly wind is up to 6 m/s faster than the average condition at most of the time and elevation below 15 km. In contrary, the condition is reversed during inactive MJO phase (1-2). In addition, the meridional northerly wind is faster during the active MJO phase (3-4) below 4 km. The higher wind speed observed during the MJO phase 3-4 from 23-30 November 2017 could also contribute to the formation of tropical cyclone Dahlia on 29 November 2017 in the southwest of Bengkulu.

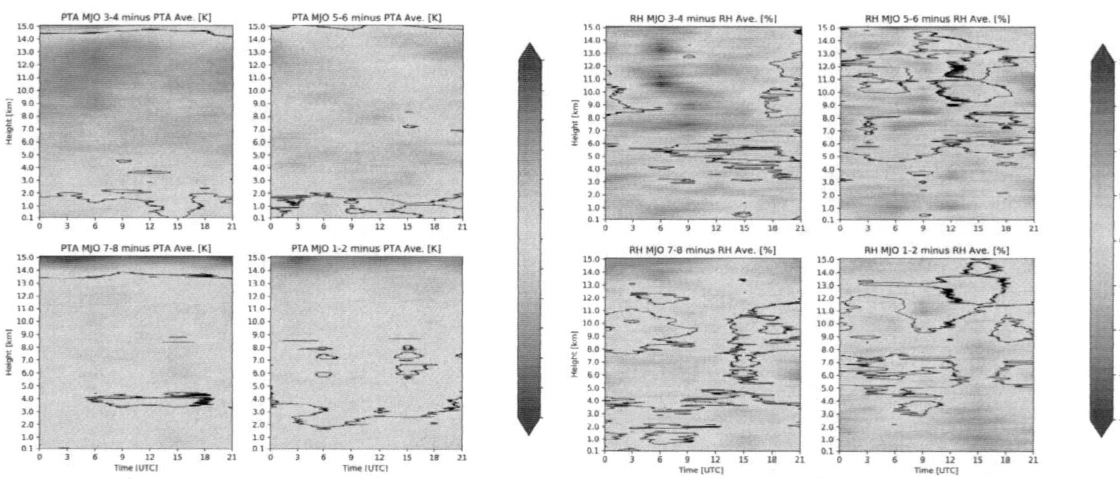

Figure 4. Differences of the MDC at particular MJO phases and the average condition for PTA (left) and RH (right).

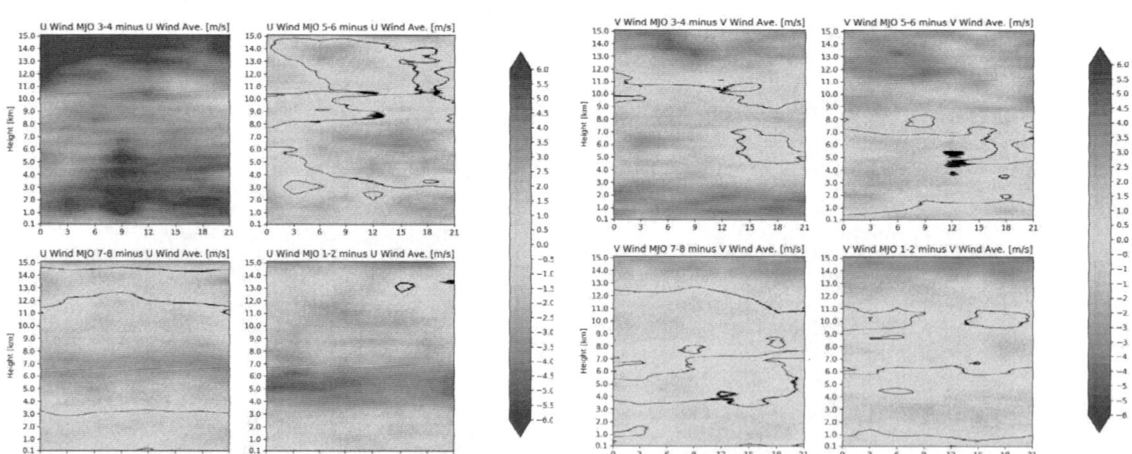

Figure 5. Differences between the MDC at particular MJO phases and the average condition for zonal (left) and meridional (right) winds.

Lastly, the MDC of FLH, CAPE, CIN and TPW at the average condition and at particular MJO phases in Bengkulu station are shown in Figure 6. The average FLH is about 4915 meters with maxima FLH of 4960 meters occurs at 09 UTC and minimum of 4840 meters at 00 UTC. During MJO phase 3-4 and 5-6, the FLH increased by 150 meters possibly due to rising warm air and strong convection, whereas during MJO phase 7-8 and 1-2, the FLH decreased by 30 meters. The CAPE peaks at 06 - 09 UTC (13-16 LT) with an energy of 3 KJ, describing strong convection during those times, while the CIN is near 0 KJ at most of the time. This condition suggests that precipitation would preferentially occur after 06-09 UTC, and it is shown by TPW peaks at 12-18 UTC (night-time rainfall). During MJO phase 3-4 and 5-6, the TPW is amplified by 1-4 mm, whereas during MJO phase 1-2, the TPW decreased by 2 mm.

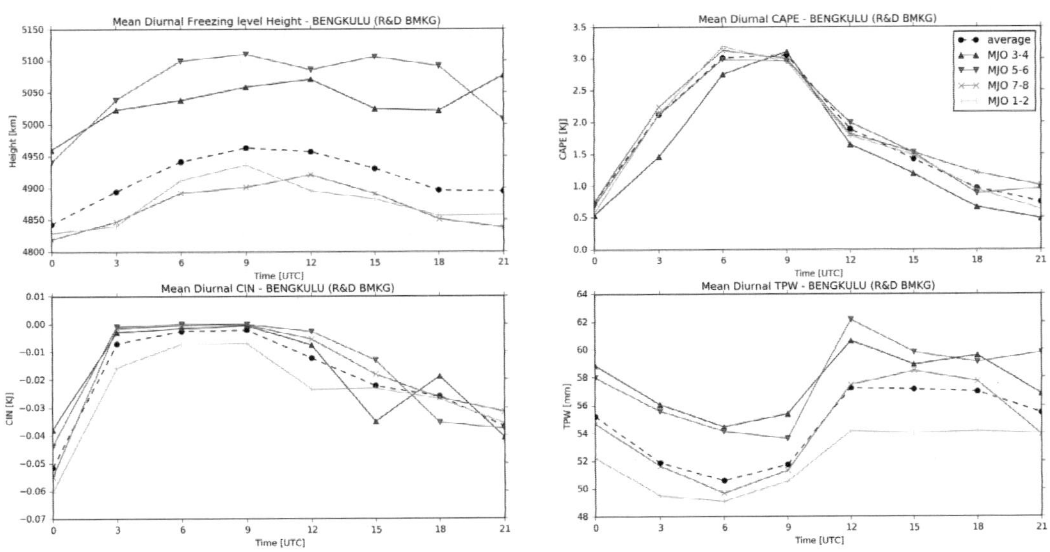

Figure 6. Mean diurnal cycle of freezing level height (FLH), CAPE, CIN and TPW during the campaign period (at average condition) and at particular MJO phases in Bengkulu station.

4. Conclusions

The first campaign of YMC project was conducted in Bengkulu from 16 November 2017 to 15 January 2018. Among various kinds of observation, radiosonde observations were performed at 3-h intervals in Bengkulu station. The results show that the MDC PTA has a similar pattern with the results from the Pre-YMC campaign. During active phases (3-4) of MJO over Sumatera, RH is much higher (5-15%), suggesting additional moisture in the atmosphere and the zonal westerly wind is up to 6 m/s faster than the average condition at most of the time and elevation below 15 km. In contrary, the condition is reversed during inactive MJO phase (1-2). During active MJO phases over the MC, the FLH is increased which possibly due to rising warm air and strong convection. The CAPE peaks at 06 - 09 UTC with an energy of 3 KJ, describing strong convection during those times, while the CIN are near 0 KJ at most of the time. The TPW peaks at 12-18 UTC, 6 hours after the peaks of CAPE. The TPW is amplified by 1-4 mm during active MJO phases and decreased by 2 mm during inactive MJO phases over Sumatera.

Acknowledgments
Authors wishing to acknowledge assistance or encouragement from BMKG colleagues in Center for Research and Development, Fatmawati Meteorology Station, Bengkulu Climatology Station, Kepahiang Geophysics Station, Centre for Public Meteorology, Centre for Aviation Meteorology and Database Centre. Authors also want to acknowledge Bengkulu local government and Ristekdikti for given

permission for the research project. This research project is financially supported by the collaboration of JAMSTEC, BMKG and BPPT under the YMC framework.

5. References

[1] Ramage C S 1968 Role of a tropical "Maritime Continent" in the atmospheric circulation *Mon. Wea. Rev.* **96** 365–370
[2] Simpson J, Keenan T D, Ferrier B, Simpson R H and Holland G J 1993 Cumulus mergers in the Maritime Continent region *Meteorol. Atmos. Phys.* **51** 73–99 doi: 10.1007/BF01080881
[3] Neale R and Slingo J M 2003 The Maritime Continent and its role in the global climate: A GCM study *J. Climate* **16** 834–848
[4] Madden R A and Julian P R 1972 Detection of a 40–50 day oscillation in the zonal wind in the tropical Pacific *J. Atmos. Sci.* **28** 702–708
[5] Hendon H, Liebmann B, Newman M E, Glick J D and Schemm J E 2000 Medium-range forecast errors associated with active episodes of the Madden-Julian Oscillation *Mon. Weather Rev.* **128** 69 – 86
[6] Waliser D E, Murtugudde R and Lucas L E 2003 Indo-Pacific Ocean response to atmospheric intraseasonal variability: 1. Austral summer and the Madden-Julian Oscillation *J. Geophys. Res.* **108**(C5) 3160 doi:10.1029/2002JC001620
[7] Zhang C and Dong M 2004 Seasonality in the Madden - Julian oscillation *J. Climate* **17(16)** 3169-80
[8] Peatman S C, Matthews A J and Stevens D P 2014 Propagation of the Madden-Julian Oscillation through the maritime continent and scale interaction with the diurnal cycle of precipitation *Q. J. R. Meteorol. Soc.* **140(680)** 814-825
[9] Yang G Y and Slingo J M 2001 The diurnal cycle in the Tropics *Mon. Wea. Rev.* **129** 784–801
[10] Seo H, Subramanian A C, Miller A J and Cavanaugh N R 2014 Coupled impacts of the diurnal cycle of sea surface temperature on the Madden-Julian oscillation *J. Climate* **27** 8422–43
[11] Matthews A J and Li H Y Y 2005 Modulation of station rainfall over the western Pacific by the Madden-Julian oscillation *Geophys. Res. Lett.* **32** L14827
[12] Nitta T and Sekine S 1994 Diurnal variation of convective activity over the tropical western Pacific *J. Meteor. Soc. Japan* **72** 627– 641
[13] Sui C-H, Lau K-M, Takayabu Y N and Short D A 1997 Diurnal variations in tropical oceanic cumulus convection during TOGA COARE *J. Atmos. Sci.* **54** 639 – 655
[14] Mori S, Hamada J-I, Tauhid Y I, Yamanaka M D, Okamoto N, Murata F, Sakurai N, Hashiguchi H and Sribimawati T 2004 Diurnal land-sea rainfall peak migration over Sumatera Island, Indonesian Maritime Continent, observed by TRMM satellite and intensive rawinsonde soundings *Mon. Wea. Rev.* **132** 2021-39
[15] Sakurai N, Murata F, Yamanaka M D, Mori S, Hamada J-I, Hashiguchi H, Tauhid Y I, Sribimawati T and Suhardi B 2005 Diurnal cycle of cloud system migration over Sumatra Island *J. Meteorol. Soc. Jpn.* **83** 835–850
[16] Bowman K P, Collier J C, North G R, Wu Q, Ha E and Hardin J 2005 Diurnal cycle of tropical precipitation in Tropical Rainfall Measuring Mission (TRMM) satellite and ocean buoy rain gauge data *J. Geophys. Res: Atm.* **110**(D21)
[17] Chen S S and Houze R A 1997 Diurnal variation and life-cycle of deep convective systems over the tropical Pacific warm pool *Q. J. R. Meteorol. Soc.* **123** 357–388
[18] Sui C-H and Lau K-M 1992 Multiscale phenomena in the tropical atmosphere over the western Pacific *Mon. Weather Rev.* **120** 407–430
[19] Sui C-H, Li X, Lau K-M and Adamec D 1997 Multiscale air-sea interactions during TOGA COARE *Mon. Weather Rev.* **125** 448–462
[20] Suzuki T 2009 Diurnal cycle of deep convection in super clusters embedded in the Madden–Julian Oscillation *J. Geophys. Res.* **114** D22102 doi: 10.1029/2008JD011303
[21] Tian B, Waliser D E and Fetzer E J 2006 Modulation of the diurnal cycle of tropical deep convective clouds by the MJO *Geophys. Res. Lett.* **33** L20704 doi: 10.1029/2006GL027752
[22] Yokoi S, Mori S, Katsumata M, Geng B, Yasunaga K, Syamsudin F, Nurhayati and Yoneyama K 2017 Diurnal cycle of precipitation observed in the western coastal area of Sumatra Island: offshore preconditioning by gravity waves *Mon. Weather Rev.* **145** 3745–3761 https://doi.org/10.1175/MWR-D-16-0468.1
[23] Wheleer M and Hendon H 2004 An All-Season Real-Time Multivariate MJO Index: Development of an Index for Monitoring and Prediction *Mon. Weather Rev.* **132** 1917-1932

IOP Conf. Series: Earth and Environmental Science **303** (2019) 012059 doi:10.1088/1755-1315/303/1/012059

The Characteristics of squall line over Indonesia and its vicinity based on Himawari-8 satellite imagery and radar data interpretation

A M Hidayat[1*], U Efendi[1], H N Rahmadini[1] and I R Nugraheni[2]

[1] Undergraduate Program of Meteorology, State College of Meteorology Climatology and Geophysics, 5 Perhubungan I Street, Banten, Indonesia 15221
[2] Lecturer, Department of Meteorology, State College of Meteorology Climatology and Geophysics, 5 Perhubungan I Street, Banten, Indonesia 15221

*anistia.malinda@gmail.com

Abstract. Squall line is a rare MCS phenomenon happened in Indonesia. In these recent years, radar showed the existence of squall line pattern on 31 December 2017, 25 and 27 January 2018 in disparate spot in Indonesia. This study aims to find the common characteristic of squall lines that occurs in tropical region of Indonesia. The result shows a persistent elongated convective pattern with the trailing stratiform region extends over 100 km in more than 6 hours in all squall line events observed in this study. The average cloud top temperature below -45ºC indicates an abundant amount of ice crystals. Radar data analysis shows that reflectivity value and surface rainfall intensity in the core part of the squall system always higher than the outer part, with maximum reflectivity value observed on the lower level of the convective cells. Analysis of the vertical structure of squall system shows that height of convective cells on the first squall system reached 17 km, while on the second squall system reached 12 km. The origin place where the squall line formed determines the structure and character of the squall system. These findings show that updraft mechanism on the continental area is generally stronger than around marine areas.

Keywords : Himawari, MCS, squall lines, convective.

1. Introduction

As a "mini" mesocyclonic thunderstorm, squall lines were assessed by many to be the primary host of tornadoes [1, 2]. Tropical squall line is a rare Mesoscale Convective Systems (MCS) phenomenon happened in Indonesia, a region with complexity on its atmosphere dynamics due to its geographic location. The tropical squall line is propagating mesoscale disturbance, commonly referred to as squall system, consists of cumulonimbus cloud elements and extensive precipitating anvil cloud [3] trails the squall line. Houze [4] and Zipser [5] further broadened the concept of the squall line to include the anvil and stratiform precipitation with a cool air often found behind the gust front as part of the squall line system. Thus, strong squall lines may contain supercells as building blocks, while weaker squall lines composed mainly of ordinary cells [6] or even precipitation elements forced by instabilities not exclusively due to gravity, such as conditional symmetric instability [7].

Further research on squall system was studied by Bluestein and Jain [8] and references therein. There are four classifications of squall line development: broken line squall lines, back-building squall lines, broken-areal squall lines, embedded-areal squall lines. The four-ways squall line formation as described by Bluestein and Jain [8] depicted in Figure 1. The broken-line and back-building formation processes are easily discerned in visible satellite images, but broken-areal and embedded-areal developments are not. All types of squall lines form in a conditionally and convectively unstable atmosphere which is characterized by strong vertical shear and turning of the shear vector with height at low levels, and weaker shear and only slight turning aloft. In addition, most lines formed within a range of 200 km, where the radar-beam resolution is fine enough to detect the appearance of new cells.

Fewer studies were conducted to examine the formation of squall line in tropical regions rather than in the midlatitude region. Houze [4] examined the structure and dynamics of tropical squall line system and found that the downpour of heavy precipitation below the sloping updraft of squall line system contained a convective-scale downdraft in the heavy rain zone which spread out at low levels. While, the precipitation falling from the anvil cloud was stratiform in character. Further studies of Gamache and Houze [9] showed that the squall line region was characterized by boundary layer convergence, which fed deep convection updrafts. The circulation in the anvil region, on the other hand, was characterized by mid-level convergence, which fed a mesoscale updraft within the anvil cloud and mesoscale downdraft below the anvil.

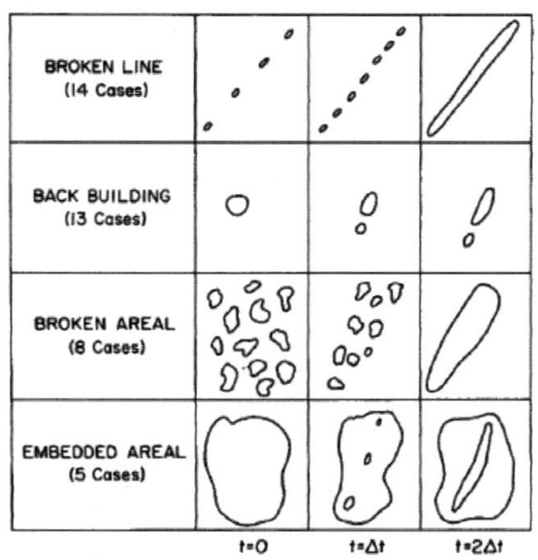

Figure 1. Idealized depiction of squall-line formation [8].

Although there have been several case studies about squall lines in the tropical region, there has been no systematic identification on its characteristics. Therefore, the purpose of this study is to examine the characteristics of tropical squall line by optimizing the use of infrared channel of satellite and radar data interpretation to identify the dynamic aspects. Combining satellite (upper-atmospheric observation) and radar (land-based observation) will give detail depiction of the squall system that occurred in Indonesia. The importance of understanding the characteristics of these disturbances and their dynamic aspects are to give wider insight on weather forecasting skills and provide a chance for doing further research of this rare MCS phenomenon.

2. Data and Methodology

2.1 Research sites

This study was conducted in two regions, both in land and water area in Indonesia, where squall lines system formed. The first squall system observed over South Sumatra region, located near Bukit Barisan mountains on 31 December 2017. While the second and third squall lines system observed almost in the same area around Java sea, located near Kalimantan Island. The second squall line system occurred on 25 January 2018, while the third squall system occurred on next two consecutive days of 26 – 27 January 2018. More information about research sites can be seen in Figure 2.

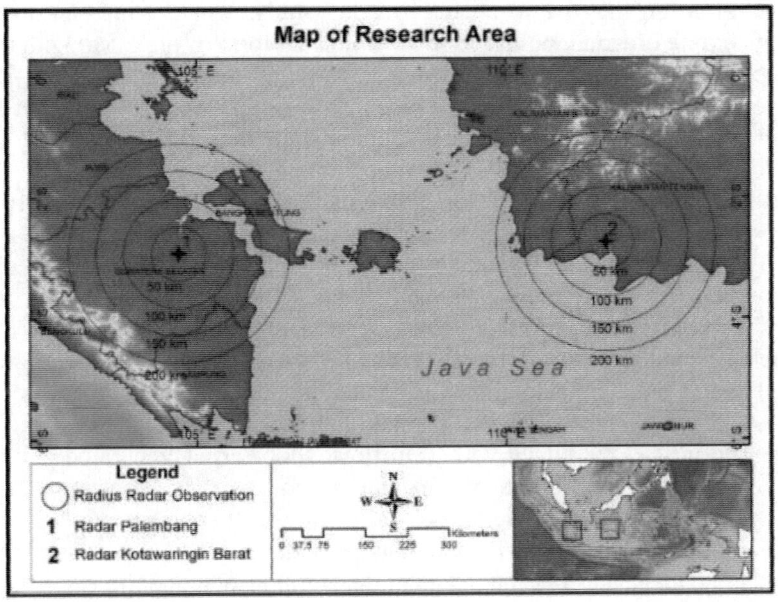

Figure 2. Research area awas conducted near Sumatra and Kalimantan

2.2 Satellite image data
This study focused on optimizing the use of visible (VS) and infrared (IR) channel via WMO Information System (WIS) portal. The other satellite product that used in this study is High-resolution Cloud Analysis Information (HCAI), derivation product from Himawari-8/9 satellite imagery. These data obtained from subdivision management of weather satellite image of Indonesian Agency for Meteorology Climatology and Geophysics (BMKG). Himawari WIS product has spatial resolution of 4 km and temporal resolution of 10 minutes. Satellite Animation and Interactive Diagnosis (SATAID) then used to process these satellite data. Dewita [10] stated that visible channel of the satellite was really effective to detect squall line events in daylight.

2.3 Radar data interpretation
Radar provides a good temporal and spatial resolution and able to give 3-dimension looks of a squall system. Another research using radar to study the convective cells in the mesoscale system and found that vertical profiles of radar reflectivity in Global Atmospheric Research Programme's Atlantic Tropical Experiment (GATE) cells exhibit generally modest reflectivity at low levels, decreasing rapidly with height above the freezing level [11]. Further studies showed that Gematronik Radar was competence enough for analysing the structure of squall line [10]. Radar data on 31 December 2017 over South Sumatra region observed by radar located in Palembang, while radar data on 26 and 27 January 2018 obtained C-Band Gematronik Weather Radar from meteorological station of Iskandar Kotawaringin Barat. Unfortunately, radar data on 25 January 2018 was missing due to technical issues. This data obtained using Gematronik radar then processed using Rainbow software. Gematronik radar used two range operational radar, 240 km for intensity observation and 120 km for observation resolution and better analysis [12].

IOP Conf. Series: Earth and Environmental Science **303** (2019) 012059 doi:10.1088/1755-1315/303/1/012059

3. Characterization of Squall Line System on Satellite Images

3.1 Squall Lines Observation

Hamilton and Archbold [13] defined squall lines as a typical tropical squall line consists of a row of cumulonimbus clouds forming at the edge of a downdraft region. This description was highly matched with satellite observation on the visible channel in this study, where cumulonimbus clouds, which depicted as a lumpy texture of cloud, observed in an elongated form of the cloud with trailing stratiform surround it. In some cases, the squall system forms a bow echo model, a special case when the squall line becomes bowed [10]. This occurs when the rear inflow jet is very strong, which can be identified on the radar as rear inflow notch, a weak echo region then shifted behind the core of the bowed convective line.

Based on the subjective interpretation of cloud types using the visible channel, a very thick and lumpy texture of cloud observed as an elongated convective line during the occurrence of the squall line system. The bright look showed that the cloud system consists of ice crystal on its top, reflected almost all of the light to the direction from which it comes. On the formation of the first squall line system on 31 December 2017, discrete convective cells observed for the first time at 09.10 UTC and continued to grow, until reached its mature stage at 12.30 UTC as shown in Figure 3. As it grows, the system development from time to time showing almost no movement, the movement only occurs is caused by the extend of clouds development, wider and spread [14]. The mature stage of this convective cloud indicated by a well-defined cloud edges with cauliflower looks that generated a line connection of convective cloud, with no gaps within the system. Interpretation of visible channel at 12.30 UTC was not possible since this channel strongly depends on the existence of solar radiation. The only movement caused by cloud development, the MCC system could persist in exact same location from growth phase until dissipating phase. This squall line might form due to monsoon Asia that hit the extent of Bukit Barisan mountains, resulting in the formation of elongated convective cloud over South Sumatra region. Another discrete cell started to form over the Java sea near Kalimantan on 25 January 2018, reached its mature squall systems stage at 02.00 UTC. While, the third squall line observed in this study, appeared as the convective cloud for the first time on 26 January 2018 at 23.00 UTC, reached its mature stage on 23.30 UTC and persisted until the next day, 27 January 2018.

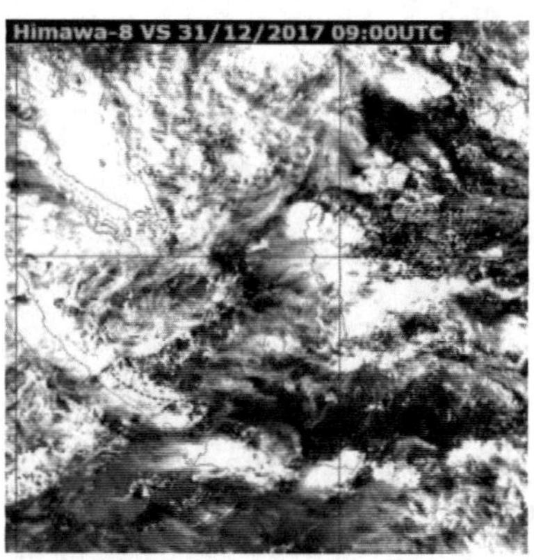

Figure 3. Visible channel of Himawari Satellite in the formation stage of squall lines over South Sumatera Island.

3.2 Objective Analysis of Squall Lines Structure

The objective analysis was done using HCAI products of Himawari satellite observation. This method used to ensure the type of clouds composing the whole structure of squall line formation, although the stratiform region was very hard to be observed using this method. The first squall lines showed that the convective cells located at the front edge of the system, surrounded by dense cloud (mixed cloud between cumulonimbus and high cloud region, hard to differentiate each type). This squall lines system was comprised of mostly individual convective elements. The stratiform region could not be detected. The second squall lines system showed that the squall system composed mostly by convective cells, broaden dense cloud area surrounding the convective area, and leading stratiform near the edge of the convective cloud. Houze et al. [15] identified the leading-line trailing stratiform (TS) structure as the most common organization of mature MCS. Further studies by Parker and Johnson [16] also recognized leading stratiform (LS) and parallel stratiform (PS) as a common arrangement of the convective line system. The third squall lines system showed that the squall system mostly composed by dense cloud and a small portion of convective cloud during the development of the squall lines system.

Another objective analysis on the IR channel used to understand the vertical wind profile of the squall system. Vertical wind analysis that used Numerical Weather Prediction (NWP) product, Global Spectral Model (GSM) on the first and second squall system, showed a change of wind speed and direction with the height in the atmosphere (commonly known as wind shear) observed at altitude about 300 – 400 mb, the wind speed increase along with the increase of the altitude. While the last squall line system in this study showed that wind shear occurred at altitude about 400 – 500 mb. This analysis showed that the three-squall line system has the same wind shear pattern at altitudes between 300 – 500 mb.

3.3 Squall Line Coverage Area

3.3.1 Typical length of squall lines system. On 31 December 2017, the typical length of the squall system formed over South Sumatra region was about 138 km. The typical length of the second squall systems was about 224 km, which is the longest linear type of squall lines formed during this study. The third squall systems formed around the same location with the second squall systems have a typical length of 109 km. It was the shortest squall system during this study. Thereby in general, the typical length of squall lines system form in Indonesia is about more than or equal to 100 km. However, these findings cannot be generalized because the sample size is too small to be statistically significant.

3.3.2 Squall line system total area. By adopting Maddox [17] research on characterizing the MCC as a convective system persisted for more than 6 hours, this study also defines the total coverage squall system area which has continuously low IR black-body temperature $T_{BB} \leq -52°C$, $T_{BB} \leq -32°C$, and $T_{BB} \leq -40°C$. When a large portion of squall systems have $T_{BB} \leq -52°C$, it indicates that the system is active and precipitation is falling over significant area, $T_{BB} \leq -32°C$ was a depiction of mature stage which indicates an average cold cloud shield area, and when most squall lines portion have $T_{BB} \leq -40°C$, all the water droplet content in that cloud system will suddenly change into an ice form [18]. Ahrens also emphasized that clouds which have a cloud top temperature of -40°C composed mainly from ice crystals [19].

The total area of cloud shield with $T_{BB} \leq -32°C$ on the first squall system formed on 31 December 2017 was 115296 km^2, $T_{BB} \leq -40°C$ was 79632 km^2, while the total cloud coverage with $T_{BB} \leq -52°C$ was 74016 km^2. The second squall system showed that the total area of cloud shield with $T_{BB} \leq -32°C$ was 113440 km^2, $T_{BB} \leq -40°C$ was 77136 km^2, while the total cloud coverage with $T_{BB} \leq -52°C$ was 46624 km^2. The mature stage of the third squall system showed that the total area of cloud shield with $T_{BB} \leq -32°C$ was 141648 km^2, $T_{BB} \leq -40°C$ was 87232 km^2, while the total cloud coverage with $T_{BB} \leq -52°C$ was 27680 km^2. From this observation, we can deduce that the squall lines system form in Indonesia has the total area of cloud shield of more than or equal to 100000 km^2 for $T_{BB} \leq -32°C$, more than or equal to 70000 km^2 for $T_{BB} \leq -40°C$, and more than or equal to 20000 km^2 for $T_{BB} \leq -52°C$.

Table 1. Total area coverage by squall lines system based upon analysis of IR satellite imagery data.

	31 December 2017	25 January 2018	27 January 2018
$T_{BB} \leq -32°C$	115296 km^2	113440 km^2	141648 km^2
$T_{BB} \leq -40°C$	79632 km^2	77136 km^2	87232 km^2
$T_{BB} \leq -52°C$	74016 km^2	46624 km^2	27680 km^2

3.3.3 Top cloud temperature and lifetime. This study identifies the top cloud temperature and the lifetime of three squall systems formed in Indonesia. The first squall systems formed on 31 December 2017 showed that the convective system persists for about 12 hours with the average cloud top temperature around -67.8°C and its minimum temperature is -86.8°C as shown in Figure 4.

The second squall systems observed on 25 January 2018 showed that the convective system persists for about 6 hours with the average cloud top temperature around -45°C and its minimum temperature is -74.5°C. The characteristics of the third squall system which formed on 27 December 2018 showed that the convective system persists for about 6 hours with the average cloud top temperature around -50.9°C and its minimum temperature is -85.1°C. If it is assumed that the cloud system only consists of the pure water droplet, freezing will occur at temperature about -40°C [18].

In fact, there is an abundant amount of cloud nuclei, ice nuclei, and aerosol in the atmosphere which accelerates the formation of a cloud system. The top cloud temperature of these three squall lines system showed a temperature more than -40°C, means that the top cloud layer mostly consists of ice crystal. The thick cloud which inherit the height of high clouds with its top cloud temperature is less than -40°C was a brief evidence to deduce it as a convective, cumulonimbus cloud.

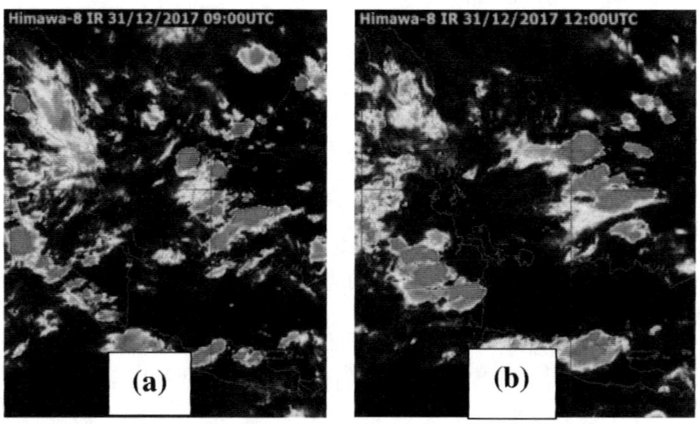

Figure 4. IR channel from Himawari-8 Satellite, (a) shows the growth stage of system and (b) shows the mature stage of the system. Red color show the convective cloud and very cold temperature.

Table 2. Identification of the lifetime and top cloud temperature of squall line systems based upon analysis of IR satellite imagery data.

	31 December 2017	25 January 2018	27 January 2018
Lifetime (hours)	12	6	6
Top cloud average temperature (°C)	-67.8 °C	-45 °C	-50.9 °C
Minimum top cloud temperature (°C)	-86.8 °C	-74.5 °C	-85.1 °C

4. Dynamic Aspects of the Squall Line System Using Radar Data Interpretation

4.1 Reflectivity Pattern of the Squall Line System

The appearance of discrete convective cells forming at nearly the same time on the first squall system started to appear at 09.10 UTC. Initially, this cell created a linear pattern of the convective line moved from south-west to north-east and reached its mature stage at 12.30 UTC in the form of bow echo, as shown in Fig 2a. The front edge of its mature element, as illustrated by Houze [4], had maximum reflectivity value around 40 – 55 dBZ and thereby, identified as the convective region of the squall system (region of heavy convective showers), showed the storm motion as well. This cloud system also consisted of the region of trailing stratiform, some defined it as anvil region, right behind the convective region indicated by lower reflectivity value around 10 – 30 dBZ. As the squall system moved, the trailing stratiform region became wider. Reflectivity pattern of the third squall system initially created a linear form. This pattern started to appear at 22.00 UTC on 26 January 2018 and continued to grow until the next day.

This system became wider and started to move to north-east. This squall system had the same pattern with the first squall line formed over South Sumatra, where the reflectivity of the convective region was around 35 – 50 dBZ and the trailing stratiform was around 10 – 30 dBZ. As time went by, this squall system showing a bow echo form at 23.30 UTC, a circular shape of tropical squall line illustrated by Maddox [17].

According to squall lines classification identified by Bluestein and Jain [8], the first squall lines categorized as a broken line formation. These squall systems appeared as a line of discrete cells, each cell forming at nearly the same time and transformed the line of convective cells into a solid line as the area of each existing cell expands and new cells develop in between the older cells. This type of squall line formation has been documented in Figure 5a. The third squall line system observed in this study was a back-building squall line formation. This type of squall system consists of the periodic appearance of a new cell upstream, relative to cell motion, from an old cell, and the resulting merger of the new cell with the old cell as the former expands in the area and moves into the latter. Although this process is usually initiated from a single cell, it can also occur to a group of widely spaced (much longer than the cell length) cells, each of which can back build to form a line or line segment. This type of squall line formation has been well-documented in Figure 5b. Besides using CMAX data, squall line types can be determined by interpreting visible channel data. These two-data inherit the same pattern that showed the best performance on examining the ways in which the "building block" that formed the mature squall line system initially become organized into a line. Due to the absence of radar data on 25 January 2018, the type of the second squall line was determined solely by analysing visible channel. This analysing showed that the second squall line systems categorized as a broken line type.

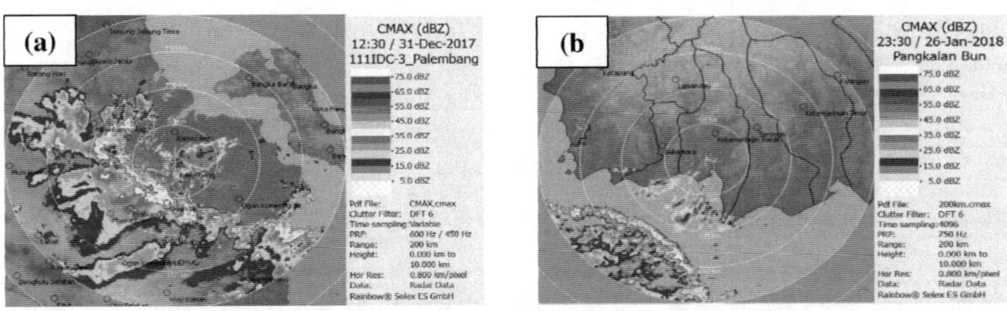

Figure 5. CMAX product analysis showing the range of reflectivity between the first and third squall line system observed in this study; (a) reflectivity pattern of the first squall line system formed over South Sumatra at 12.30 UTC, (b) reflectivity pattern of the third squall line system formed over Java sea near Kalimantan at 23.30 UTC.

4.2 Vertical Profile of the Squall Line System

Analysis of the vertical structure of the first squall system showed that the base height of reflectivity was around 1 – 4 km above the ground. This height then became smaller, reached 3 km at 08.00 UTC and less than 1 km at 13.00 UTC. While the top height of this reflectivity reached 17 km which can be indicated the top of cumulonimbus where the anvil region spread out around this height (Figure 6a). The maximum reflectivity value, around 50 – 55 dBZ, located at the height of 2 – 6 km. Szoke and Zipster [11] stated that this could happen because a weak updraft caused the maximum reflectivity value located at low levels because water droplets could not be lifted up. However, VCUT analysis at 09.00 UTC showed a different result, where the maximum reflectivity pattern located a height of 19 km. This change indicated that the updraft force became stronger. While, the reflectivity of the third squall, indicated its vertical profile, was in the range of 45 – 50 dBZ at 4 – 6 km, where the lowest reflectivity was 5 – 20 dBZ. If compare to the first squall line that formed over land area, this third squall had lower reflectivity.

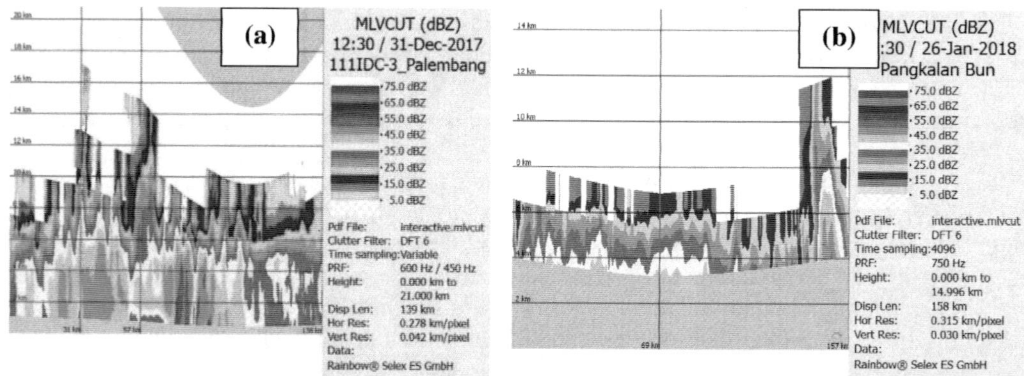

Figure 6. Vertical structure or cross section of the first and third squall line system can be identified using VCUT product of a radar data; (a) VCUT product along the front edge of the squall system in its mature stage, (b) Multi Line VCUT (MLVCUT) product along the front edge of the third squall system. This vertical section has taken from two lines following the bowed form of the squall lines system.

In general, the base height of the third and first squall system nearly the same, but the top height of this third squall system was lower than the first squall system, it was about 12 km at 22.30 UTC as shown on Figure 3b.

4.3 Surface Rainfall Intensity Analysis

Surface Rainfall Intensity (SRI) radar product indicated rainfall intensity at a given altitude above the earth's surface. In this study, analysis on SRI was set at an altitude of 3.5 km above the surface so that it could reach a distance at radius of 180 km, where the squall system began to form. The value of rainfall intensity in the initial phase of squall formation ranges from 0.2 – 0.8 mm/h on the outer side of the system, while it reached 12.6 – 25.1 mm/h in the core part of the system.

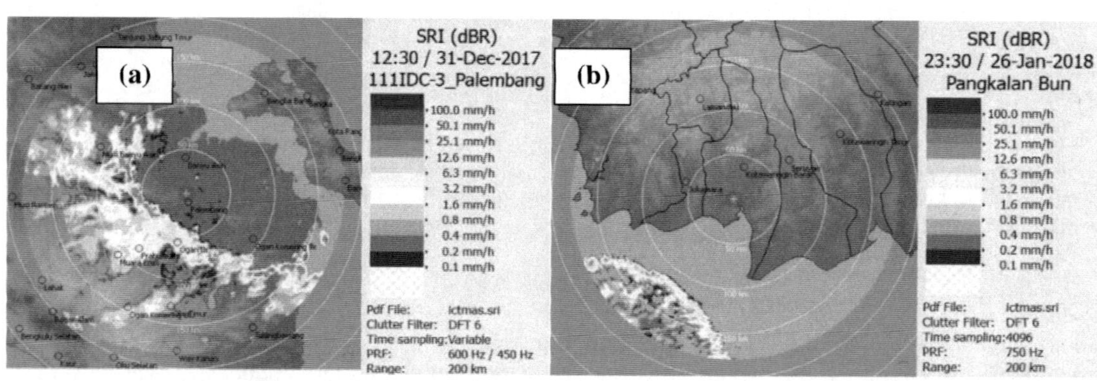

Figure 7. Surface rainfall intensity analysis along the squall system identified using SRI radar data. This product can estimate the amount of water that will precipitate at a constant height above the ground; (a) SRI product showing the maximum value of surface rainfall intensity of the first squall line system over South Sumatra region at 12.30 UTC, (b) SRI product showing the maximum value of surface rainfall intensity of the third squall line system over Java sea, near Kalimantan at 22.30 UTC.

IOP Conf. Series: Earth and Environmental Science **303** (2019) 012059 doi:10.1088/1755-1315/303/1/012059

The mature stage of first squall system occurred at 12.30 UTC with rainfall intensity in the core part of the system reaching 12.6 – 50.1 mm/h, while it reached 0.2 – 12.6 mm/h on the outer part (Figure 7a). The rainfall intensity of this squall system started to decrease afterward at 12.50 UTC. Surface rainfall intensity in the core part of the squall system seemed to decrease, but the coverage area increased. The decrease in rainfall intensity continued to occur until the system was extinct at around 13.40 UTC.

Analysis of the SRI of the third squall system (Figure 7b) showed that the rain intensity of the system reached 0.2 – 25.1 mm/h at 22.30 UTC. The SRI value seemed to increase along with the growth of the squall system. The highest rainfall intensity in this squall system occurred at 23.30 UTC, where the rainfall intensity on its core part reached 6.3 – 25.1 mm/h and the outer side of squall system reached 0.2 – 6.3 mm/h.

5. Discussions

Based on the analysis using satellite and radar data, the origin place where the squall line system formed to influence the formation structure of the system. Analysis on the cloud temperature using IR channel of Himawari-8 satellite data showed that the average cloud top temperature of the three-squall line system in this study was less than -45°C, while their average minimum temperature was less than -70°C. The squall system which had the lowest cloud top temperature is the first squall line system formed over South Sumatra region, basically categorized as continental area.

Moreover, analysis of the vertical structure of squall system showed that the height of convective cells on the first squall line system over South Sumatra reached 17 km, while the height of convective cells on the second squall system reached 12 km. Based on this result, we can know that the height of the convective cloud in a squall line system that formed over the continental region was higher than those which formed over marine areas. Analysis of the CMAX product showed that the cloud convective region in the first squall line system that formed over South Sumatra had higher reflectivity value than the third squall line system that formed over the Java Sea, basically categorized as marine areas. The reflectivity value in the core part of the first squall system was about 40 – 55 dBZ, while it was about 35 – 50 dBZ in the core part of the third squall system. The maximum reflectivity value of the first squall system was observed at 2 – 6 km above the base of the cloud, while the maximum reflectivity value on the third squall system was observed at 4 – 6 km. Generally, the maximum reflectivity value was observed at a lower level of a convective cloud system. This could happen because the liquid water content on the lower level of convective cloud was commonly the highest. When the updraft force became stronger the maximum reflectivity will be shifted on a higher level along with the more water droplet concentration that lifted up into a higher level. Furthermore, analysis on the surface rainfall intensity showed that the rainfall intensity of the first squall line system which formed over the continental region was higher than rainfall intensity of the third squall line system which basically formed over marine areas. The maximum rainfall intensity on the first squall line system reached up to 50.1 mm/h, while on the third squall system was only about 25.1 mm/h.

These findings showed that the updraft mechanism on the continental area is generally stronger than around marine areas. This could happen because there is no topographic factor involved in the cloud formation processes over marine areas, while the topographic complex over continental area caused frequent turbulences that induce stronger updraft. Furthermore, most of the marine clouds have droplet concentrations less than 100 cm^{-3}, and none has a droplet concentration greater than 200 cm^{-3}. In contrast, some of the continental cumulus clouds have droplet concentrations in excess of 900 cm^{-3}, and most have concentrations of a few hundred per cubic centimeter [20]. Thereby, many more droplets compete for the available moisture in continental clouds than in maritime clouds, and they are smaller consequently [18]. Ryan et al. [21] confirmed that maritime clouds usually have a broader size spectrum than continental clouds. This broader size caused condensation process became more intensive, resulting in the formation of narrower convective cloud with smaller updraft compared to the formation of convective cloud over the continental area. In addition, the condensation nuclei concentration above continental area was higher and diverse in size due to human activities, while the size and type of this

IOP Conf. Series: Earth and Environmental Science **303** (2019) 012059 doi:10.1088/1755-1315/303/1/012059

nuclei commonly the same above marine areas. This is why the cloud formation processes that form the squall line system over continental and marine areas inherit different structure and character.

6. Summary

Tropical squall line system that formed over Indonesia, both in continental and marine area, inherit commonly the same pattern, where the system comprised of the elongated convective line (core part of the squall system) and trailing stratiform region and the anvil region (outer part of the squall system). However, the type of squall line formation was diverse, the first and second squall line formation categorized as a broken line, while the third squall system indicated a back-building squall line formation. Generally, the wind shear pattern on these three-squall line system occurred at height 300 – 500 mb.

In general, the typical length of the squall line system in this study was more than 100 km and the total coverage squall system area with $T_{BB} \leq$ -32°C ranged from 113440 – 141648 km^2, with $T_{BB} \leq$ -40°C ranged from 77136 – 87232 km^2, and with $T_{BB} \leq$ -52°C ranged from 27680 – 74016 km^2. The average of cloud top temperature of the three-squall line's system observed in this study was -67,8°C, -45°C, and -50.9°C, consecutively. Based on CMAX radar data analysis, the reflectivity value in the core part of the first squall system was around 40 – 55 dBZ, while on the third squall system was around 35 – 50 dBZ. The reflectivity value on the outer part of the squall system on this two-squall system was around 10 – 30 dBZ, lower than the core part of the system. VCUT product analysis showed that the peak height of the convective cloud on the first squall reached up to 17 km with the maximum reflectivity of 50 – 55 dBZ at 2 – 6 km, while on the third squall system was lower at about 12 km, with maximum reflectivity of 35 – 50 dBZ at 4 – 6 km. Furthermore, analysis on the surface rainfall intensity showed that the rainfall intensity in the core part of the squall system was always higher than the outer part of the system. Surface rainfall intensity on the core part of the first squall system was about 12.6 – 50.1 mm/h, while the surface rainfall intensity of the third squall line system was lower in value around 6.3 – 25.1 mm/h.

The origin place where the squall line formed determines the structure and character of the squall system. Convective cloud formation was strongly developed above continental areas rather than marine areas, thereby the squall line system formed over the continental region had higher cloud top with more intense liquid cloud water content. This happened because of several factors, such as the more complex topographic structure of continental region, higher concentration of cloud condensation nuclei and aerosol, and larger average size droplet spectrum which composed a convective cloud system.

Acknowledgments

We acknowledge Indonesian Agency for Meteorology Climatology and Geophysics and Meteorological Station of Iskandar Kotawaringin Barat for providing radar data information and subdivision management of weather satellite image of Indonesian Agency for Meteorology Climatology and Geophysics (BMKG) for providing satellite data.

7. References

[1] Lloyd J R 1942 The development and trajectories of tornadoes *Mon. Wea. Rev.* **70** 65-75
[2] Fulks J R and Smith C D Jr. 1950 A December storm accompanied by tornadoes *Mon. Wea. Rev.* **78** 220-225
[3] Brown J M 1979 Mesoscale unsaturated downdrafts driven by rainfall evaporation: a numerical study *J. Atmos. Sci.* **36** 313-338
[4] Houze R A Jr 1977 Structure and dynamics of a tropical squall-line system *Mon. Wea. Rev.* **105** 1540-67
[5] Zipser E J 1977 Mesoscale and convective-scale downdraughts as distinct components of squall-line circulation *Mon. Wea. Rev.* **105** 1568-89
[6] Browning K A, Fankhauser J C, Chalon J P, Eccles P J, Strauch R C, Merrem F H, Musil D J, May E L and Sand W R 1976 Structure of an evolving hailstorm. Part V: Synthesis and implication for hail growth and hail suppression *Mon. Wea. Rev.* **104** 603-10
[7] Emanuel K A 1983 The lagrangian parcel dynamics of moist symmetric instability *J. of the Atmos. Sci.* **40** 2368-76

[8] Bluestein H B and Jain M H 1985 Formation of mesoscale lines of precipitations: Severe squall lines in Oklahoma during the spring *J. of the Atmos. Sci.* **42(16)** 1711-32
[9] Gamache J F and Houze R A Jr. 1981 Mesoscale air motion associated with a tropical squall line *Mon. Wea. Rev.* **110** 118-135
[10] Anggi D, Zulqisthi H T and Dwicahyo K 2015 Identifikasi squall line menggunakan radar gematronik dan satelit (studi kasus Lombok 21 Pebruari 2015) *Seminar Nasional Penginderaan Jauh 2015* 305-312
[11] Szoke E J, Mueller C K, Wilson J W and Zipser E J 1985 Development of convection and severe weather along outflow boundaries in Northeast Colorado: Diagnosing the above-surface environment with a mobile sounding system Preprints, 14th Conf. on Severe Local storms, Indianapolis *Amer. Meteor. Soc.* 386-389
[12] Eko W 2012 *Pengantar III Modul Radar Cuaca* (Jakarta: BMKG)
[13] Hamilton R A, Archbold J W and Douglas C K M 1945 Meteorology of Nigeria and adjacent territory *Quaterly J. of the Roy. Meteor. Soc.* **71** 231-264
[14] Ahmad M and Mulsandi A 2017 Distribusi spasial dan temporal *Mesoscale Convective Complex (MCC)* di Indonesia selama periode MJO menggunakan citra satelit *Jurnal Meteorologi Klimatologi dan Geofisika* **4**(1) (Jakarta: STMKG)
[15] Houze R A Jr, Smull B F and Dodge P 1990 Mesoscale organization of springtime rainstorms in Oklahoma *Mon. Wea. Rev.* **118(3)**
[16] Parker M D and Johnson R H 2000 Organizational modes of midlatitude mesoscale convective system *Mon. Wea. Rev.* **128** 3413-36
[17] Maddox R A 1980 Mesoscale convective complexes *Amer. Meteor. Soc.* **61** 1374-87
[18] Roger R R and Yau M K 1989 *A Short Course in Cloud Physics 3rd Edition* (USA: Butterworth Heinemann)
[19] Ahrens C D 2009 *Meteorology Today 9th Edition* (USA: Brooks/Cole, Cengage Learning)
[20] Wallace J M and Hobbs P V 2006 *Atmospheric Science 2nd Edition* (Canada: British Library Cataloging-in-Publication Data)
[21] Ryan R T, Blau, Van Thuena P C, Cohen M L and Roberts G D 1972 Cloud Microstructure as determined by optical clous particle spectrometer *J. Appl. Meteor.* **11** 149-156

IOP Conf. Series: Earth and Environmental Science **303** (2019) 012060 doi:10.1088/1755-1315/303/1/012060

Unveiling of the SST gradient over the South China Sea and its possible relationship to the Asian precipitation

B H Vaid
School of Marine Sciences, Nanjing University of Information Science and Technology (NUIST), Nanjing, China.

bakshi32@gmail.com

Abstract. The present study examines the Sea Surface Temperature (SST) variations over the South China Sea (SCS) using the National Oceanic and Atmospheric Administration (NOAA) SST daily datasets for the period 1988–2009. In the SST over the South China Sea (SCS), a north-west versus south-east SST gradient (hereafter gra-SCS SST) is revealed. This SST gradient is seen to be induced by the distinct wind anomaly along with fresh water flux changes. The slope difference between the gra-SCS SST before and after 1999 is observed to be statistically significantly different than zero at a 95% confidence level. A decreasing/increasing trend in the gra-SCS SST is revealed before/after 1999. Besides, it is found that the gra-SCS SST characterizes the Asian monsoon variability by inducing cross equatorial monsoon flow towards the Asian Continent. In addition, we established the role of the gra-SCS SST in the Asian monsoon by computing causation based on the information flow concept between the gra-SCS SST and traditional monsoon indices, which pronounces that former impact the later. Based on the present corollary, we suggest that the gra-SCS SST should be considered as an important parameter in the climate studies as it is meticulously related to the precipitation over Asia.

Keywords : SST, South China Sea, monsoon, precipitation.

1. Introduction

The South China Sea (SCS) is one of the largest semi-enclosed marginal seas in the world ocean but the SCS remains poorly observed in terms of climate variability. It is connected with the East China Sea to the northeast, the Pacific Ocean and the Sulu Sea to the east, and the Java Sea and the Indian Ocean to the southwest. It is a very important region of interest for the scientific community because 1) the onset of the SCS summer monsoon (SM) indicates the onset of the East Asian SM [1], 2) Indian and East Asian monsoon systems exchange heat, momentum and moisture with the sea over the SCS [2], 3) the changes in Sea Surface Temperature (SST) over the SCS because of these fluxes during the monsoon affect chlorophyll concentration and therefore affects half billion people live along the coast of the SCS with its fisheries and other important resources [3], and 4) the anomalous state of the SCS can affect East and Southeast Asian climate [4] [5] [6] [7] hence have a profound influence on the social and economic condition of over 60% of the earth's population.

Moreover, the origin, development, and evolution of the SCS processes can greatly influence the weather and climate in East and Southeast Asia [8] [9], so it always necessary to comprehend the SST variability over the SCS. The SST over the SCS is considered as an important parameter in many operational and research activities, ranging from weather forecasting to climate research and plays a crucial role in maintaining the monsoon precipitation over the SCS and the surrounding area [7] [10]. Thus investigating the SST variability over the SCS region can help to figure out the influences of SCS

SST on rainfall. Moreover, it's worth to study SST s it is an important surface boundary conditions which are decisively important for the establishment of the monsoon circulation and rainfall [11] by modifying surface heat fluxes and the deep convection in the tropical atmosphere [12] [13] [14]. For example, in the recent study [14], the explicit prominence of SST (over the Bay of Bengal) in facilitates deep convection was remarkably identified. Besides, SST is considered to be the most important representative quantity of the ocean, which communicates the ocean's relatively large thermal inertia to the atmosphere, through an exchange of the surface fluxes [15] and indeed plays an important component in the precipitation variability. Therefore, in the view of the importance of the SST in the monsoon and because of the peculiar location of SCS where the most intense atmospheric convection associated with East Asia occurs, it is desirable to study the climate variability of SST over the SCS. The climate variability has become an important topic of scientific pursuit during the past few decades, intimately linking the economy of a nation with its climate-resources [16] [6] [17]. For example, [16] found that SST gradients across the equatorial Pacific undergo a regime change in 1998/99 due to a significant cooling (warming) over tropical eastern (western) Pacific in the later period. Recall that, SCS SST is largely influenced by one of the largest source of Earth's climate variability commonly known by El Niño–Southern Oscillation [18] [3] [19]. We intended to comprehend the climate variability of SST over the SCS and its association with Asian precipitation. To best of our knowledge, none of the previous studies so far brought out the SST gradient over the SCS, its associated oblivious trends and its association to Asian monsoon. In the following, first, list the data source and methodology, and then give a brief presentation of the results. This study is summarized in section 4.

2. Data and Methodology

The National Oceanic and Atmospheric Administration (NOAA) SST product from 1988 to 2009 (hereafter NOAA SST) based on the Advanced Very High-Resolution Radiometer (AVHRR) infrared observations merged with the Advanced Microwave Scanning Radiometer-Earth Observing System (AMSR E) is used (please refer to [20] [21] for more details) in the present study. Besides, National Center for Environmental Prediction (NCEP) - DOE AMIP 2 reanalysis daily datasets for wind fields [22], which are provided by the NOAA/OAR/ESRL PSD, Boulder, Colorado, USA, and are available from their website at http://www.esrl.noaa.gov/psd/ is also used for dynamical explanation. The precipitation data is taken from Asian Precipitation - Highly-Resolved Observational Data Integration Towards Evaluation (APHRODITE's) of Water Resources [23]. Surface salt flux observations are obtained from the Global Ocean Data Assimilation System (GODAS) developed at NCEP [24]. Notably, to obtain statistical significance, we followed two-tailed student t-test.

In methodologies, we will be using the traditional ones (e.g., regression analysis) and a newly developed tool for identifying the causality between time series. The causality from X_2 to X_1 (units: nats per unit time) is:

$$T_{2\rightarrow 1} = \frac{C_{11}C_{12}C_{2,d1} - C_{12}^2 C_{1,d1}}{C_{11}^2 C_{22} - C_{11}C_{12}^2}, \quad (1)$$

where C_{ij} is the sample covariance between X_i and X_j (i,j=1,2), and $C_{i,dj}$ the covariance between X_i and $[X_j(t+k\Delta t) - X_j(t)] / (k\Delta t)$, with Δt being the time step size and $k \geq 1$ some integer. Ideally, when $T_{2\rightarrow 1}$ is nonzero, then X_2 is causal to X_1, and vice versa (please refer [25] [26] for more details for computing causation).

Besides, to understand the underlying oceanography. we also computed the wind-driven vertical velocity at the bottom of Ekman layer using stress according to [27] and formulae is given below.
The vertical velocity at the bottom of the Ekman layer:

$$W_E = (1/\rho f)\,(curl_z\tau + \beta\tau_x/f) \quad (2)$$

IOP Conf. Series: Earth and Environmental Science **303** (2019) 012060 doi:10.1088/1755-1315/303/1/012060

here the Coriolis parameter f is equal to 2Ω *sin* φ, where φ is the latitude, Ω is the rotation rate of Earth (7.29×10^{-5} rad s^{-1}), and β, the rate of change of the Carioles parameter with latitude, is equal to 2Ω *cos* φ/*R*, with *R* equal to the radius of Earth (6.37×10^{6} m). $curl_z\,\tau$, the vertical component of wind stress curl, is defined by,

$$curl_z\tau = \partial\tau_y/\partial x - \partial\tau_x/\partial y \tag{3}$$

Wind stress and its components (τ , τ_x and τ_y) were computed using the bulk aerodynamic formulae. The meridional Ekman transport per unit zonal width computed as,

$$-\tau_x/\rho f \tag{4}$$

3. Results

The SST changes over SCS for the period 1988-1998 (PRE99) and 1999-2009 (POST99) is analyzed using the NOAA SST daily data sets. It is known that SCS SST is largely influenced by one of the largest sources of Earth's climate variability commonly known by El Niño–Southern Oscillation [1] [3] [19]. And since after the late 1990's, significant changes were seen over the Pacific Ocean due to the anomalous wind divergence in the central Pacific that cause a shift in the anomalous atmospheric convection westward, leading to a westward shift of the anomalous westerly response, thereby preventing the eastward propagation of the SST anomaly after the late 1990s [28]. For example, [28] and [16] found that the first leading empirical orthogonal function (EOF) mode of the POST99 tropical Pacific SSTA has shown maximum warming is the central Pacific (around 150°W), whereas the PRE99 EOF1 mode shows a warming in the eastern Pacific. So, therefore, it is of great scientific interest to comprehend the SST changes over the SCS during these two different periods; PRE99 and POST99. Analysis of the NOAA SST data reveals that mean SST during June, July, August, and September (hereafter JJAS) of PRE99 and POST99 is above 27°C (Figure 1a and 1b), which is conducive for enhanced convective precipitation [29] and is therefore of paramount importance.

Figure 1c represents JJAS SST differences, POST99 minus PRE99 which vividly shows that there exist substantial differences in the variability of SST during JJAS of PRE99 and POST99. It is found that these SST differences within the marked rectangular boxes in Figure 1(b) are statistically distinct from zero at a 99% level from that in 1(a). The shaded region in Figure 1d shows the statistical significant region. The substantial difference in SST is observed in the period PRE99 and POST99, particularly over regions 106°E; 115°E, 17°N; 22°N (hereafter region I) and 117°E; 120°E, 12°N; 17°N (hereafter region II) with the statistical significance of 99%. Interestingly, in the SST over SCS, a gradient in the SST with one center over north-west and another over south-east is found (Figure 1). To comprehend the SST gradient, the SST over region II is subtracted from that over region I, and we referred to it as gra-SCS SST.

The time series (solid line) and linear trends (dash line) of the gra-SCS JJAS SST from 1988 to 2009 are plotted in Figure 2. Clearly, we see that during POST99 the gra-SCS SST is significantly increased in comparison to PRE99. As seen from Figure 2, the gra-SCS SST has experienced a change in the pattern since PRE99. It is found that before 1999, the gra-SCS SST has a decreasing trend and whereas after 1999 the gra-SCS SST shows an increasing trend. Further, it is to be noted that the two slopes (gra-SCS SST for PRE99 and POST99) do not overlap and indeed observed to be statistical significance at a 95% confidence level. Essentially, we find that, at a 95% confidence level, the two slopes are -3.64 ± 1.03 and 0.63 ± 1.00 (both are multiplied by 10E-4) respectively and they lie within a confidence interval of [-4.65, -2.61] and [-0.37, 1.63]. And these do not overlap because it is clearly evident from the interval that during PRE99 the interval is ending at -2.61 and while during the POST99 interval is starting from -0.37, thus we can say that the two slopes indeed completely differ at a 95% confidence level. The least squares method applied to estimate the slope is already being used in our recent study [14]. This distinguishes the role of the gra-SCS SST during the two different climate phases PRE99 and POST99.

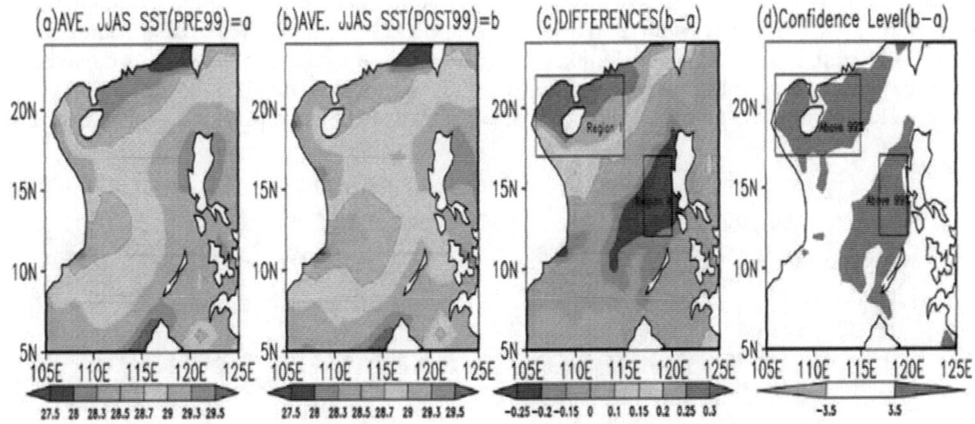

Figure 1. Average JJAS SST gra-SCS (in °C) (a) PRE99 (b) POST99 (c) POST99 minus PRE99 and (d) represent (b) is statistically distinct from zero at a 99% level from that in (a).

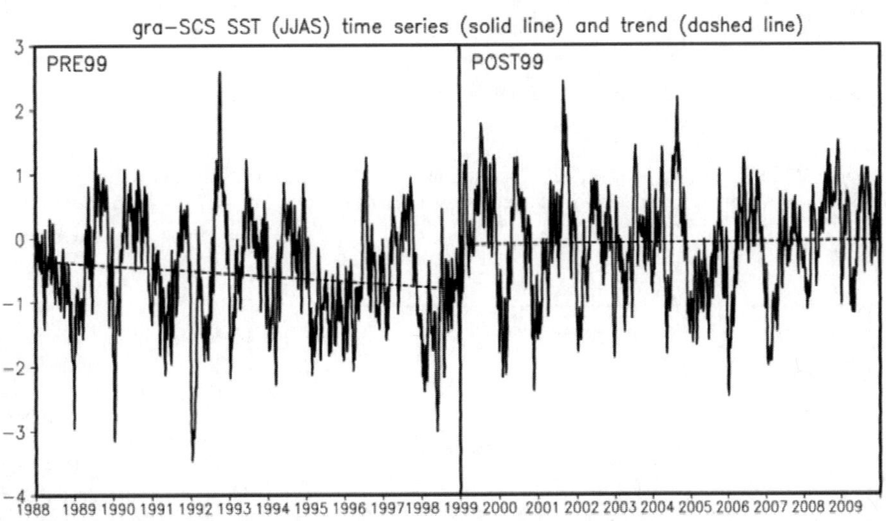

Figure 2. gra-SCS SST (JJAS) time series (solid line) and trend (dashed line). The trend plotted is statistically significant at 99% confidence level.

In the next, the possible dynamical reason behind the gra-SCS SST change during the study period is explored. It is identified that the gra-SCS SST variation during the POST99 seems to be associated with the distinctive wind patterns (Figure 3a). The distinctive wind pattern is clearly seen over the gra-SCS SST region. The decreased/increased wind speed (Figure 3a) along with the presence of high/low freshwater influx (Figure 3b) is observed during POST99. And, this seems to be the cause of SST increase/decrease during POST99 over the region I/region II respectively. Because it is well known that the presence freshwater in flux and low-intensity winds can break down the stable stratification that forms due to the flux and results in leaving the ocean warmer than most of the other parts of the Ocean, this is indeed natural and known fact [30]. Apart from this, an attempt has also been made to understand the underlying oceanography and hence vertical velocity at the bottom of Ekman layer and meridional Ekman transport per unit zonal width along 17°N during POST99 minus PRE99 also been computed (Figure 3c and Figure 3d). A clear contribution of Ekman transport to coastal upwelling is depicted from the analysis. It reveals that another possible reason to the reduction of SST over the region II is the coastal upwelling along with the maximum positive (or northward) Ekman transport (which can be seen between 116°E and 122°E).

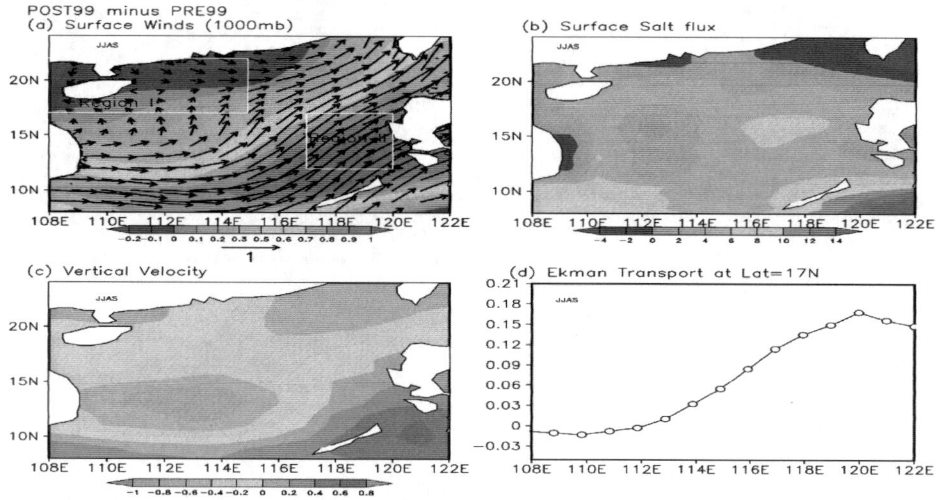

Figure 3. POST99 minus PRE99 differences during JJAS (a) surface (1000hpa level) wind magnitude/vector difference represent by shaded/arrow (b) surface salt flux (in x10^{-8}g/cm^2/s) (c) vertical velocity at the bottom of Ekman layer (in x 10^{-6} m/s) (d) meridional Ekman transport per unit zonal width along 17°N.

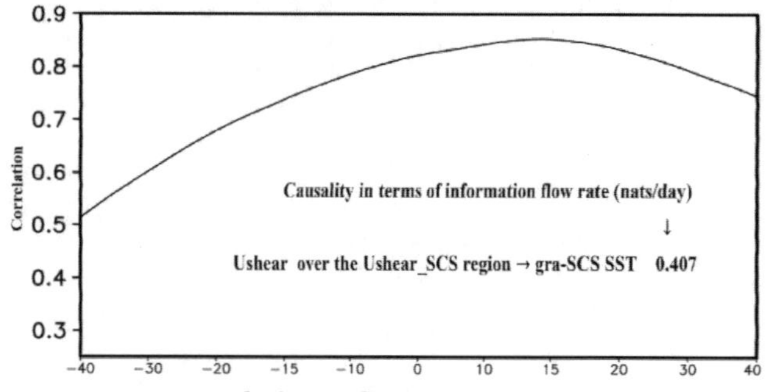

Figure 4. Lead-lag correlation between the gra-SCS SST and the Ushear over the Ushear_SCS region; a positive number means that the gra-SCS SST lags Ushear over the Ushear_SCS region (in days). The correlation coefficient is statistically significant at 99% confidence level.

Moreover, in order to strengthen our claim, especially the role of wind fields in the gra-SCS SST, the association of wind fields with gra-SCS SST is further explored by computing the lead-lad correlation followed by examining the causation in a quantitative way (please refer [25] [26] for more details for computing causation) between the gra-SCS SST and the U shear (U850-U200) over the SCS region (105°;120°E, 5°-20°N; hereafter called as Ushear_SCS region). It is to be noted that the Ushear region chosen here is similar to Liang et al., (2002) and it is of crucial importance in terms of SCS, as it characterizes the broad-scale feature of the SCS [31]. The results reveal a significantly high correlation

IOP Conf. Series: Earth and Environmental Science **303** (2019) 012060 doi:10.1088/1755-1315/303/1/012060

between the Ushear_SCS region and gra-SCS SST (Figure 4). Further correlation analysis reveals the possible contribution of Ushear_SCS (Ushear_SCS seems to lead the gra-SCS SST) to gra-SCS SST. To corroborate the findings, information flow rates from the Ushear_SCS region to gra-SCS SST is computed and it is found to be 0.407 nats/day at 95% significant level. This indeed says that there is causation from Ushear_SCS region to gra-SCS SST. Thus, in other words, we can say that the former could be one of the causes for the latter and this is in accordance with aforementioned findings.

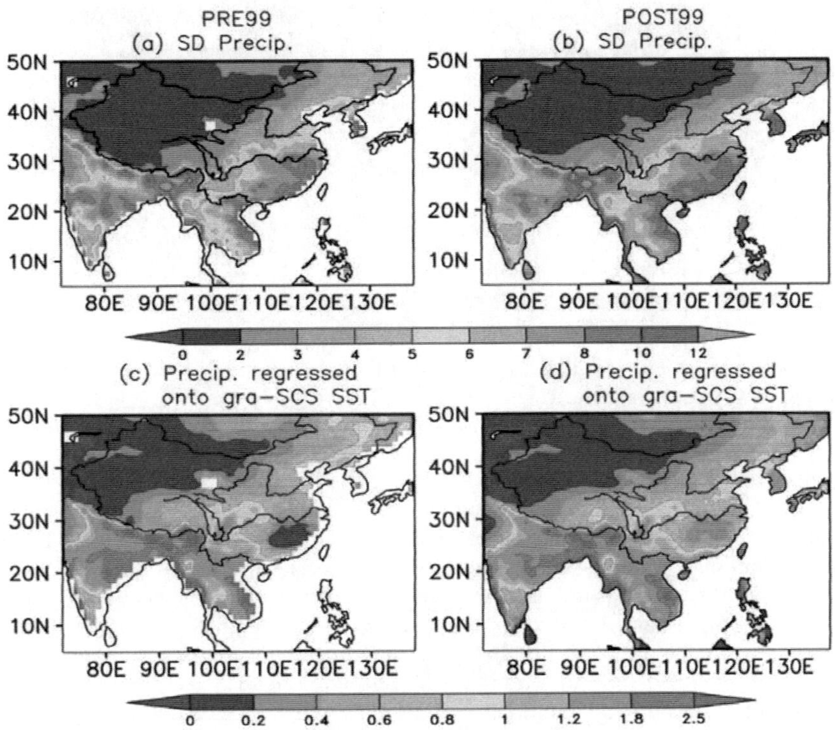

Figure 5. (a) and (b) are the standard deviation of the precipitation, (c) and (d) are precipitation regressed onto the gra-SCS SST for PRE99 (left panel) and POST99 (right panel). The regression coefficient is statistically significant at 99% confidence level.

Seeing at the SCS SST potent and knowledge which states that Asian monsoon system (which includes the South Asian monsoon and the East Asian monsoon) begins over the SCS [18], it is highly recommended to talk over the relationship between the SCS SST gradient and precipitation over the Asian region. Therefore, we analyzed the precipitation variability and its association with the SCS gradient in SST (Figure 5). To do so, we regressed the precipitation onto the gra-SCS SST for PRE99 (Figure 5c) and POST99 (Figure 5d) using the least square regression method. The regression equation has the form Y=a+bX, where X and Y are two variables, *b* is the slope of the line, and *a* is the y-intercept. In this case, X is the gra-SCS SST time series and Y the series of gridded precipitation values on a 1° x 1° latitude-longitude grid. Both the series have a time resolution of 1 day, and the regressed precipitation is plotted in Figure 5c (during PRE99) and 5d (during POST99). Interestingly, the comparison between Figure 5 (a and c) and Figure 5(b and d) shows that the precipitation variability over the vast region of East and South Asia can be well explained by the gra-SCS SST. Besides, we also have drawn the differences (POST99 minus PRE99) in standard deviation in precipitation and regressed values of precipitation in Figure 6. If we compare Figure 6a and Figure 6b, the similarity can be clearly evident. In other words, the analysis reveals that gra-SCS SST bears a strong relationship with the precipitation variability over Asia. Thus, we may say that gra-SCS SST must be considered as an important parameter in the climate studies as it is closely related to precipitation over Asia.

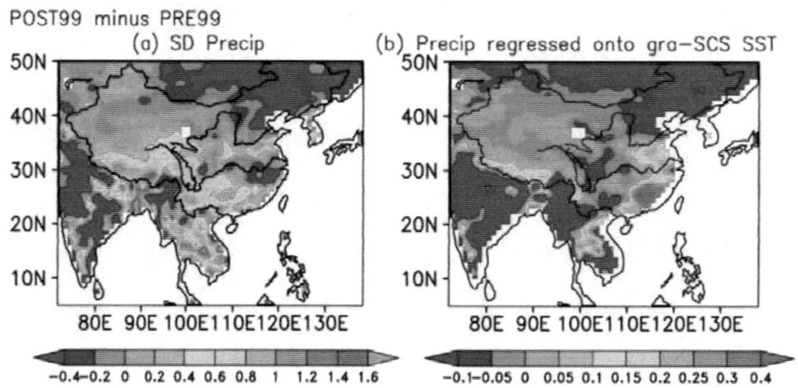

Figure 6. (a) alike Figure 5a but for POST99 minus PRE99 differences, and (b) alike Figure 5c but for POST99 minus PRE99 differences.

Since it is evident, that wind shear is the deep-rooted component that drives the monsoon precipitation over Asia [32] [33]. So to describe the gra-SCS SST association with precipitation over Asia, the wind shear over the Asian region is examined. To investigate this, we regressed the wind shear over the Asian region onto gra-SCS SST (Figure 7). It is clearly evident that gra-SCS SST induces a cross-equatorial shear flow towards the Asian Continent in order to participate actively in the dynamics of Asian monsoon (Figure 7c and 7f). Further, to make corroboration on the role of gra-SCS SST in the Asian monsoon, we examine the causation (by applying aforementioned discussed rigorously causality analysis) between the gra-SCS SST and traditional monsoon indices which are used to define Asian monsoon in a quantitative way. The tradition monsoon indices on which causation has been applied are Webster and Yang circulation index (hereafter USHEAR index) and Goswami monsoon Hadley circulation index (hereafter VSHEAR index); please refer [32] and [33] respectively for more details regarding tradition monsoon indices. The USHEAR index is defined as a time-mean zonal wind (U) shear between 850 hPa and 200 hPa, written U850 − U200, averaged over South Asia from the equator to 20°N and from 40° to 110°E (region is marked in Figure 7d) and the VSHEAR index is defined by the meridional wind (V) shear between 850 hPa and 200 hPa (V850–V200) averaged over the region 70°; 110°E, 10°; 30°N (region is marked in Figure 7e). We found a significant (at 95% level) causation between 1) gra-SCS SST and USHEAR index and 2) gra-SCS SST and VSHEAR index. In case (1), the flow rate is 0.242 nats/day. In case of (2), the flow rate is 0.144 nats/day. The significant causation between the gra-SCS SST and the monsoon index (USHEAR and VSHEAR) is found, i.e., we can say that the former could be one of the causes for the latter, which is indeed overwhelming because this reveals the interaction of SCS SST with Asian monsoon quantitatively. Also drawn are the POST99 minus PRE99 differences of the regressed U shear, V shear and UV shear vectors (Figure 8). Interestingly, an unambiguous role of the gra-SCS can be depicted in the differences as shown in Figure 8c. It shows a precise association between the gra-SCS and weakening of the cross-equatorial flow in the recent decade. The weakening of cross-equatorial monsoon flow and its association to the Indian Ocean is already been reported in the recent study [34]. Nevertheless, the exact cause of the weakening of large-scale cross-equatorial monsoon flow is still an open debate for the scientific community. Based on Figure 8 findings, we suggest that the gra-SCS is one of the reasons for the weakening of the large-scale cross-equatorial monsoon flow in the recent decade. However, we keep it as debatable until a detailed study using an adequate model with the gra-SCS SST forcing data is done because we feel that it would be necessary before a conclusive statement can be made. We leave this concern for future study.

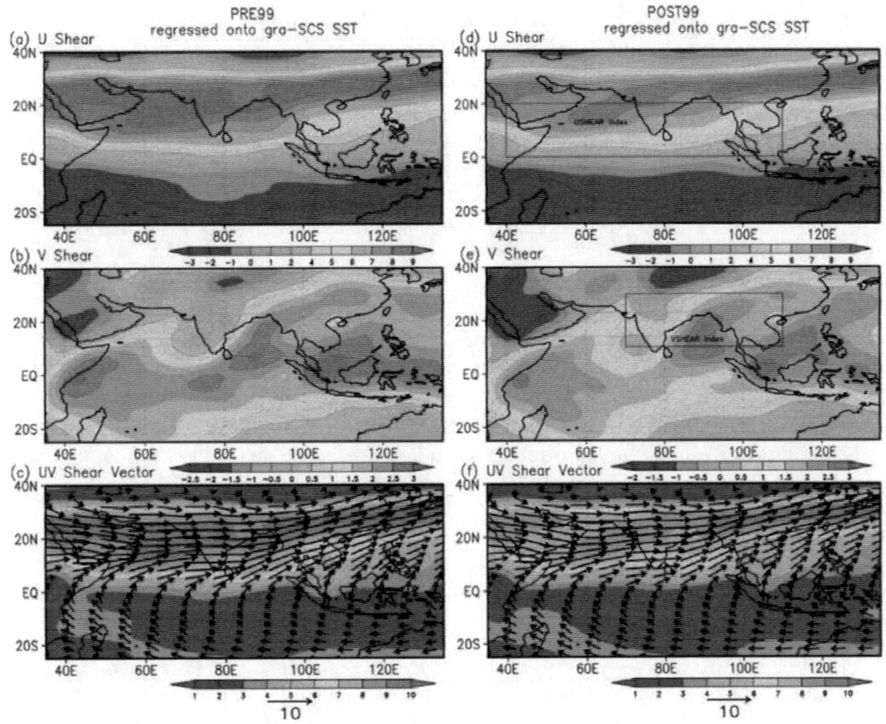

Figure 7. (a) and (d) are U shear regressed onto gra-SCS SST, (b) and (e) are V shear regressed onto gra-SCS SST, (c) and (f) are UV shear regressed onto gra-SCS SST for PRE99 (left panel) and POST99 (right panel). The regression coefficient is statistically significant at 99% confidence level.

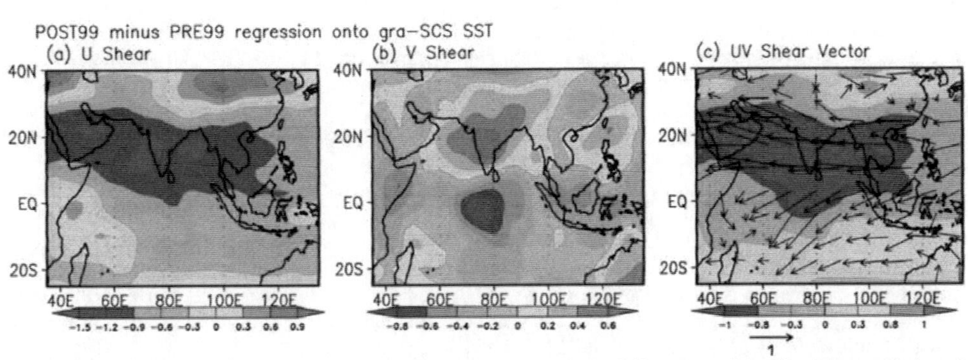

Figure 8. (a) alike Figure 7a but for POST99 minus PRE99 differences, (b) alike Figure 7b but for POST99 minus PRE99 differences, and (c) alike Figure 7c but for POST99 minus PRE99 differences.

4. Summary

Our analysis with the daily mean NOAA SST data has shown the gra-SCS SST has distinctly different climatic patterns before and after 1999, which is referred to as PRE99 and POST99 respectively. In the SST over the SCS, a north-west versus south-east SST gradient is revealed. The gra-SCS SST is identified to have decreasing slope before 1999 and increasing since 1999. This gra-SCS SST is seen to induced by the distinct wind anomaly along with fresh water flux changes and underlying oceanic processes. Interestingly, it is found that gra-SCS SST shares a strong relationship with precipitation over Asia by inducing the cross-equatorial monsoon shear flow towards the Asian Continent. Further corroboration made on the role of gra-SCS SST in the Asian monsoon by using established information flow concept between gra-SCS SST and traditional monsoon indices. It reveals explicit causation between gra-SCS SST to the monsoon indices. Thus with all due fairness, the present study proposed that the gra-SCS SST must be considered as important in climate change signals in describing the precipitation variability over Asia.

Acknowledgments

The comment from an anonymous referee is appreciated. The author thanks NOAA High-Resolution SST, NCEP-DOE AMIP 2 reanalysis and the APHRODITE's team for providing the data and ICTMAS's organizing committee for generous support. The Key Laboratory of Meteorological Disaster of Ministry of Education (KLME1603) and the Jiangsu Provincial Government through the 2015 Jiangsu Program for Innovation Research and Entrepreneurship Groups for the financial support. The figures were prepared using GrADS.

5. References

[1] Wang B and Lin H 2002 Rainy season of the Asian-Pacific summer monsoon *J. Clim.* **15** 386–396

[2] Mao J Y and Chan J C L 2005 Intraseasonal variability of the SCS SM *J. Clim.* **18(13)** 2388–2402

[3] Xie S P, Xie Q, Wang D and Liu W T 2003 Summer upwelling in the South China Sea and its role in regional climate variations *J. Geophys. Res.* **108(C8)** 3261. doi:10.1029/2003JC001867

[4] Zhou L, Tam CY, Zhou W and Chan J C L 2010 Influence of SCS SST and the ENSO on winter rainfall over South China *Adv. Atmos. Sci.* **27** 832-844. doi: 10.1007/s00376-009-9102-7

[5] Roxy M and Tanimoto Y 2011 Influence of SST on the intraseasonal variability of the South China Sea Summer Monsoon *Clim. Dyn.* **39 (5)** 1209 – 1218. DOI: 10.1007/s00382 – 011 – 1118 – x

[6] Vaid B H and Polito P S 2016 Influence of the SCS biweekly SST on SCS SM especially during Indian Ocean Dipole *Atmosphere-Ocean* **54(1)** 48-59 DOI:10.1080/07055900.2015.1130682

[7] Vaid B H 2017 Biweekly SST over the SCS and its association with the Western North Pacific SM Pure *Appl. Geophys.* **174** 463–475. DOI: 10.1007/s00024-015-1198-3

[8] Scsmex Project Office 1995 Scientific Plan of the South China Sea Monsoon Experiment (SCSMEX) Published by SCSMEX Project Office pp 42

[9] Ding Y H 2004 Seasonal march of the East Asian summer monsoon In*: The East Asian Monsoon* (Chang CP, ed). Singapore. World Scientific Publisher pp 560

[10] Vaid B H, Preethi B and Kripalani R H 2018 The Asymmetric Influence of the South China Sea Biweekly SST on the Abnormal Indian Monsoon Rainfall of 2002 *Pure Appl. Geophys.* 175(12) 4625–4642. doi.org/10.1007/s00024-018-1934-6

[11] Webster P J, Magana V O, Palmer T N, Shukla J, Tomas RA, Yanai M and Yasunari T 1998 Monsoons: process, predictability and the prospects for prediction *J. Geophys. Res.* **103(C7)** 14451–14510

[12] Thum N, Esbensen S K, Chelton D B and Mcphaden M J 2002 Air-sea heat exchange across the northern equatorial SST front in the eastern tropical Pacific *J. Clim.* **15** 3361–3378

[13] Raymond D J, Esbensen S K, Paulson C, Gregg M, Bretherton C S, Petersen W A, Cifelli R, Shay L K, Ohlmann C, and Zuidema P 2004 EPIC2001 and the coupled ocean-atmosphere system of the tropical east Pacific *Bull. Am. Meteorol. Soc.* **85** 1341–1354

[14] Vaid B H and Liang X S 2018a The changing relationship between the convection over the western Tibetan Plateau and the sea surface temperature in the northern Bay of Bengal *Tellus A: Dynamic Meteorology and Oceanography* **70**(1) 1440869

[15] Deser C, Alexander M A, Xie SP and Phillips A S 2010 Sea surface temperature variability: patterns and mechanisms *Ann. Rev. Mar. Sci.* **2** 115–143

[16] Chung P H and Li T 2013 Interdecadal relationship between the mean state and El Nino types *J. Clim.* **26** (2) 361-379
[17] Vaid B H and Liang X S 2018b An Abrupt Change in Tropospheric Temperature Gradient and Moisture Transport Over East Asia in the Late 1990s *Atmosphere-Ocean* DOI: 10.1080/07055900.2018.1429381
[18] Wang Q, Liu Q, Hu R and Xie Q 2002 A possible role of the South China Sea in ENSO cycle *Acta Oceanol. Sin.* **21** 217–226
[19] Wang C, Wang W, Wang D and Wang Q 2006 Interannual variability of the South China Sea associated with El Niño *J. Geophys. Res.* **111** C03023. doi:10.1029/2005JC003333
[20] Reynolds RW, Smith T M, Liu C, Chelton D B, Casey K S and Schlax M G 2007 Daily high-resolution blended analyses for sea surface temperature *J. Clim.* **20** 5473-5496
[21] Vaid B H, Gnanaseelan C, and Kumar J 2011 Intraseasonal oscillation in Reynold SST over the tropical Indian Ocean and their validation *Int. J. Rem. Sens.* 09/2011, **32(17-17)** 4835-4856. DOI:10.1080/01431161.2010.489585
[22] Kanamitsu M, Ebisuzaki W, Woollen J, Yang S K, Hnilo J, Fiorino M and Potter G L 2002 NCEP-DOE AMIP-II Reanalysis (R-2) *Bull. Am. Meteorol. Soc.* **83** 1631-1643
[23] Yatagai AK, Kamiguchi A O, Hamada A, Yasutomi N and Kitoh A 2012 APHRODITE: Constructing a Long-term Daily Gridded Precipitation dataset for Asia based on a Dense Network of Rain Gauges *Bulletin of American Meteorological Society* **93(9)** 1401-1415 doi:10.1175/BAMS-D-11-00122.1
[24] Saha S, et al. 2006 The NCEP climate forecast system *J Clim.* **19** 3483–3517
[25] Liang X S 2014 Unraveling the cause-effect relation between time series *Phys. Rev. E.* **90**:052150
[26] Liang X S 2016 Information flow and causality as rigorous notions ab initio *Phys. Rev. E.* **94** 52201
[27] Stommel H M 1965 *The Gulf Stream Berkeley*: University of California Press. pp 248
[28] Xiang B, Wang B and Li T 2013 A new paradigm for the predominance of standing Central Pacific Warming after the late 1990s *Clim. Dyn.* **41** (2) 327-340
[29] Lau KM, Wu H T and Bony S 1997 The role of large-scale atmospheric circulation in the relationship between tropical convection and sea surface temperature *J. Clim.* **10**(3) 381–392
[30] Shenoi S S C, Shankar D and Shetye S R 2002 Differences in heat budgets of the near-surface Arabian Sea and BOB: implications for the summer monsoon *J. Geophys. Res.* **107** 1-14. doi:10.1029/2000JC000679
[31] Liang J and Shang-Sen W U 2002 A Study of Southwest Monsoon onset date over the South China Sea and its impact factors Chin *J. Atmos. Sci.* **26**(6) 829-844
[32] Webster P J and Yang S 1992 Monsoon and ENSO: Selectively interactive systems *Quart. J. Roy. Meteor. Soc.* **118** 877–926
[33] Goswami B N, Krishnamurthy V and Annamalai H 1999 A broad scale circulation index for interannual variability of the Indian summer monsoon *Quart. J. Roy. Meteor. Soc.* **125** 611–633
[34] Swapna P, Krishnan R and Wallace J M 2014 Indian Ocean and monsoon coupled interactions in a warming environment *Clim. Dyn.* **42** 2439–2454

IOP Conf. Series: Earth and Environmental Science **303** (2019) 012061 doi:10.1088/1755-1315/303/1/012061

The activity of Borneo Vortex as a trigger for extreme rain in West Borneo (case study: 24th – 25th January 2017)

K F N Isnoor[1], P U Firdianto[2] and H A Rejeki[2]

[1]Indonesian Agency for Meteorology, Climatology and Geophysics, JL. Angkasa I, No.2, Kemayoran, Centre Jakarta, Indonesia
[2]Departement of Meteorology, School of Meteorology, Climatology and Geophysics, South Tangerang, Indonesia

khalidfikri16@gmail.com

Abstract. On January 24th – 25th, 2017, Borneo Vortex has occurred that started to form at 00.00 UTC and caused the extreme rain in some areas of West Borneo. It recorded 154.1 mm/day of heavy rain in the Ketapang regency, part of West Borneo region that indicated as the effect of Borneo Vortex. The objective of this study was to determine the activity of the Borneo Vortex phenomenon that triggers the extreme rain in some areas of West Borneo. Conducted data for this study obtained from ECMWF reanalysis data for some parameters such as vorticity, divergence, vertical velocity, and humidity transport. Those data used for providing an overview of atmospheric conditions during the Borneo Vortex. Rainfall data from six meteorological stations in West Borneo also used to identify the extreme rain. Regional and local scale analysis conducted to determine changes in atmospheric conditions that occur as triggers of extreme rain in the West Borneo region. The result showed that Borneo Vortex has a significant impact to the changing of atmospheric conditions at the area where close to the forming zone of Borneo Vortex and trigger to increase the convective cloud growing up which affects the increase of rainfall in some areas of West Borneo.

Keywords : ECMWF, Borneo Vortex, vorticity, extreme rain.

1. Introduction

Kalimantan Island (Borneo) is one of the islands directly adjacent to the South China Sea. In the northwestern coast of the Borneo Island to the south of South China Sea, in the Riau Islands region, there is a low-level cyclonic circulation characterized by the Asian monsoon [1]. The Previous study stated that when the Asian monsoon occurred around early November and ended at the end of February there was a vortex phenomenon over Indonesian waters [2]. This vortex circulation is formed continuously on the northwest coast of Borneo Island and called Borneo Vortex [1].

Monsoon is a pattern of wind circulation that blows periodically in a period (at least 3 months) and other periods, the pattern will change to the opposite direction [3]. Indonesia has two types of monsoon. They are the Asian monsoon and the Australian monsoon. In the Indonesian Maritime Continent (IMC), rainy season happened when the Asian monsoon active and dry season happens at the time of Australian monsoon occurred [4]. Borneo Vortex is one of the phenomena that occurs while Asia winter monsoon is active and occur in the northwest of the island of Borneo and often associated with heavy rain and flood disasters [7].

On January 24th – 25th, 2017, the Borneo Vortex phenomenon occurred in Borneo Island. This phenomenon has led to convective cloud growing up which had the effect of causing heavy rain and strong winds in some areas of West Borneo [1]. As a result, it recorded on January 24th and there was an increasing of the rainfall intensity in several regions of West Borneo. Some part of West Borneo regions such as Ketapang, Pontianak, Paloh, and Putusibau had a high rainfall intensity, and the highest rainfall happened in Ketapang region that reached up to 154.1 mm/ day. When Borneo Vortex occurred, there was a significant change in atmospheric conditions in West Borneo and triggered extreme rain in several regions. Therefore, this research will analyse the activity of Borneo Vortex as a trigger for extreme precipitation in the West Borneo region [6].

2. Data and Method

The data used in this research were the daily data from Era Interim ECMWF (European Center for Medium-Range Weather Forecast) in netCDF file format at 00.00 Coordinated Universal Time with 0.125° x 0.125° spatial resolution [8]. The data include of wind component parameters U (east-west), wind component V (north south), divergence, relative vorticity, vertical velocity, specific humidity, and daily rainfall data from BMKG Meteorological Station observations. The research area is in West Borneo which located in 2.46° N - 3.70° S and 108.03° E -114.90° E (Figure 1).

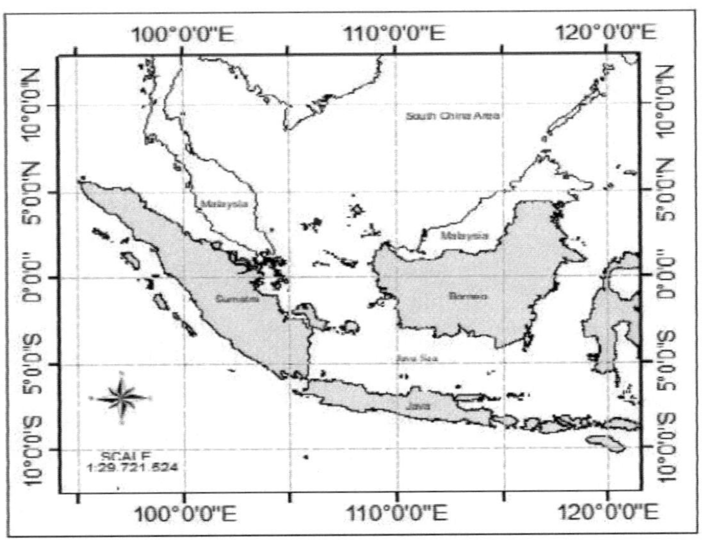

Figure 1. Research Area. Include identified area of Borneo Vortex on 2.5^0 S-7.5^0N and 102.5^0-117.5^0E [2].

The composite method was used in this research. Borneo vortex was identified by seeing a counter-clockwise circulation in the 925hPa layer in the area of 2.50 S -7.50 N and 102.50 - 117.50 E and at least one wind exceeds 2 ms-1 [2]. The interaction between Borneo Vortex and the selected parameter were described by using composite analysis. Then the atmospheric conditions when Borneo Vortex occurred were analyzed to identify the effect of Borneo Vortex to each parameter such as vorticity, vertical velocity, divergence, and humidity transport during the occurrence of Borneo Vortex.

3. Result and Discussion

3.1 Identification of The Borneo Vortex

Formation of Borneo vortex happened due to the vorticity which reproduced by wind shear and reinforced by the northeast monsoon wind convergence with the topography of the island of Borneo [11]. Figure 2 shows the streamline of wind in level 925hPa which can be used to identify the existence

IOP Conf. Series: Earth and Environmental Science **303** (2019) 012061 doi:10.1088/1755-1315/303/1/012061

of Borneo Vortex on January 24th and 25th, 2017 (green box). It observed wind vortex around West Borneo that has closed circulation and counter-clockwise direction. The wind speed at that vortex reaches 4 - 6 ms^{-1}. This condition agrees with [2], as the Borneo Vortex has closed circulation and counter-clockwise direction at the level 925hPa which identified at 2.5°S - 7.5°N and 102.5°E – 117.5°E and there is at least one wind velocity exceeds 2 m/s. Also, the vortex seen to move to the southwest which on 24th was in the West Borneo region, and on the 25th was in the northern part of Borneo Island, as well as caused the formed of rain clouds in the region.

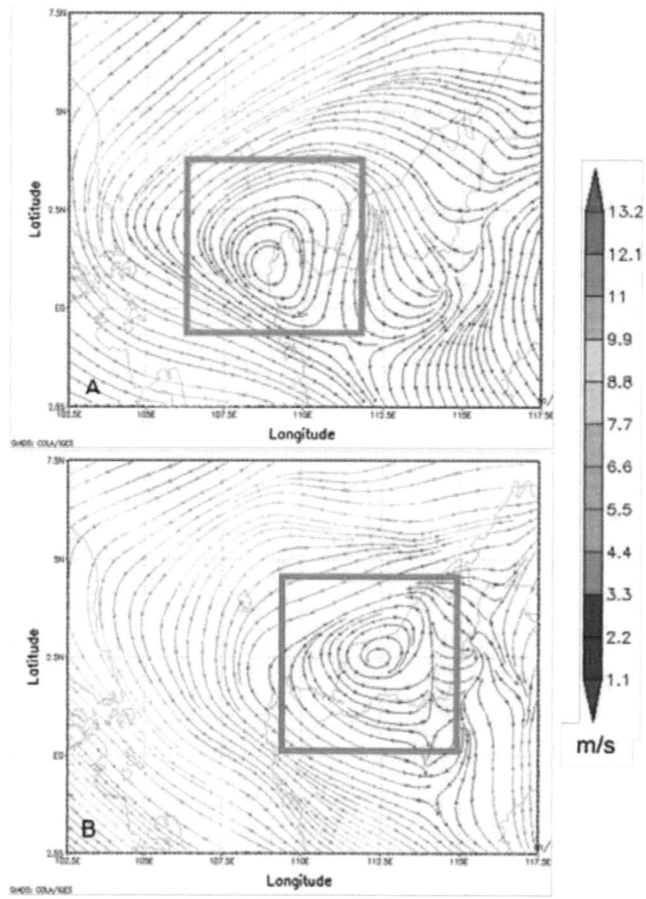

Figure 2. Streamline Wind Layer 925hPa on January 2017 (A) 24th 2017 and (B) 25th 2017.

3.2 Regional Scale Analysis

3.2.1 Analysis of Sea Surface Temperature

Figure 3 explains the daily mean sea surface temperature around the West Borneo region. It is shown that sea surface temperature is high in the north and south of Borneo Island with a high anomaly value. While the cold sea surface temperature with a low anomaly in the southwest part of the island of Borneo, caused by the flow of cold air masses from the South China Sea [1]. It is the cause of the air flow from the South China Sea to Borneo. The Asian winter monsoon brings a lot of moisture from the Pacific Ocean through the South China Sea, and meets winds from the southeast in the northwest region of Borneo and then forms cyclonic circulation [4]. The warming sea surface temperatures caused by the influence of Borneo Vortex, can have an impact on increasing convection and increasing the intensity of rainfall in the region [12].

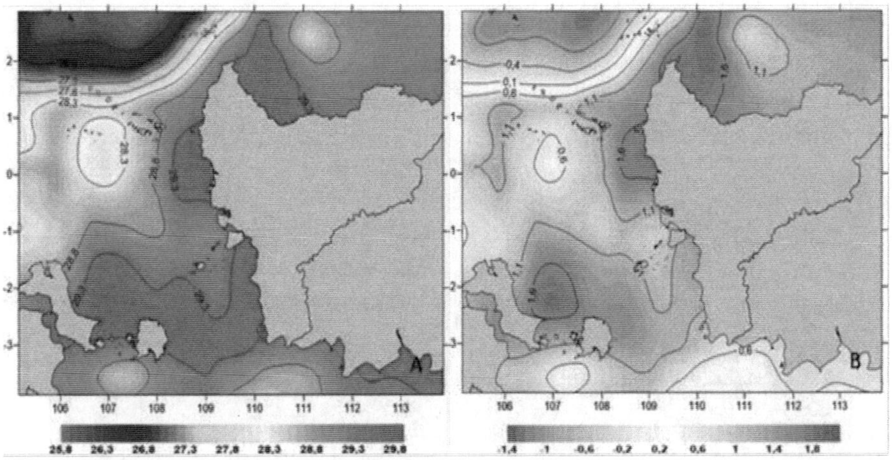

Figure 3. Daily Mean Sea Surface Temperature (A) and Sea Surface Temperature Anomaly (B) on January 24th and 25th 2017.

3.3 Local Scale Analysis

3.3.1 Surface Air Pressure Analysis

The result from observation showed there is a low surface air pressure in the northern part of Borneo Island reaches 920hPa (figure 4.). The low pressure increased from 24th until 25th of January 2017, indicating that the vortex formed caused a pressure drop in the Borneo Island region and supported the formation of rain clouds formation produced [12] and evenly distributed rain in the West Borneo region on 24th January, 2017.

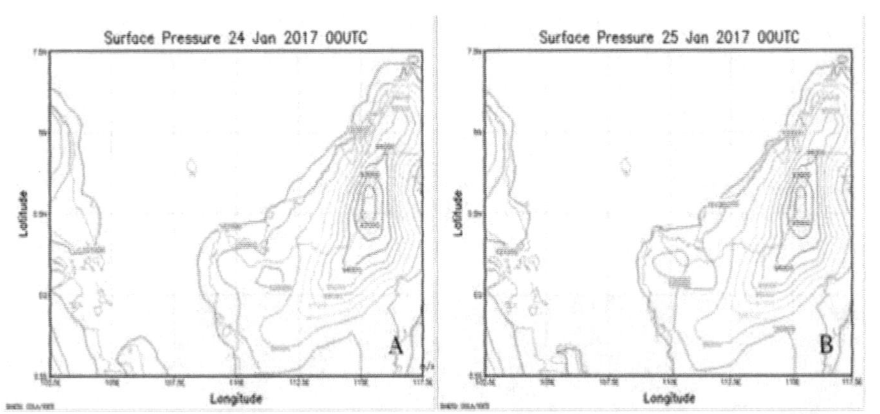

Figure 4. Surface Air Pressure (hPa) on January 24th 2017 (A) and (B) 25th 2017

3.3.2 Atmosphere Parameter Analysis

3.3.2.1 *Vorticity Analysis*

The vorticity values (Figure 5) on January 24th and 25th 2017, has average vorticity value of 1.5 x 10-5 s^{-1} to 4.5 x 10-5 s^{-1} [6]. It shows the distribution of vortices value. High vortices showed in the Northwest of Borneo Island waters, the South China Sea to northern Borneo, western and southern West Borneo. The center of vorticity was identified on northwest Borneo Island waters and South China Sea then the area of vortex spread to the mainland of West Borneo. Propagations of vorticity are quite high from the South China Sea to the equator and concentrated in the northwest waters of Borneo Island and it is expanding into the South China Sea due to the influence of vortex [10]. Then it caused the formation of rain clouds in the West Borneo region affected by vortex and due to horizontal differences in wind speeds formed in that region but the wind speeds in the Vortex center has been decreased [5].

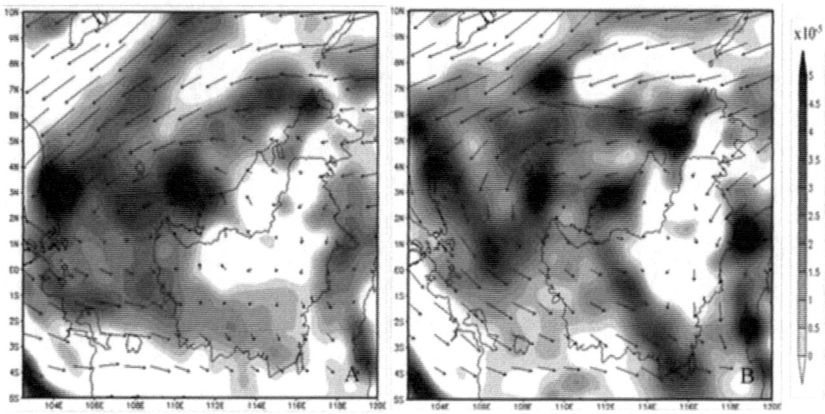

Figure 5. Vorticity (s^{-1}) on January (A) 24th 2017 and (B) 25th 2017.

3.3.2.2 Divergence Analysis

The divergence value (Figure 6) on January 24th and 25th 2017. Indicated that there is a negative or small divergence with an average value of -7.0 x 10-6 s^{-1} to -1.0 x 10-6 s^{-1} [6], with high values of areas that have high convergence appear and concentrated in almost all the South China Sea to the Java Sea, the east of waters Borneo Island to the northwest of Borneo Island. While in the west Borneo region the divergence looks more dominant which is characteristic by very high divergences that are almost throughout the West Borneo region reached value 7.0 x 10-6 s^{-1}, except in the southern part of West Borneo, there is a convergence that causes the air mass gather and formed rain clouds in this area. On January 25th, 2017, it seems that convergence tends to spread from east to south of West Borneo. Wind speed turns slow in the north-western of Borneo Island's waters and it causes cold air masses that is moving from the Pacific Ocean to the equator gathered on this region, which is influenced by high convergence values and the presence of Vortex vorticity.

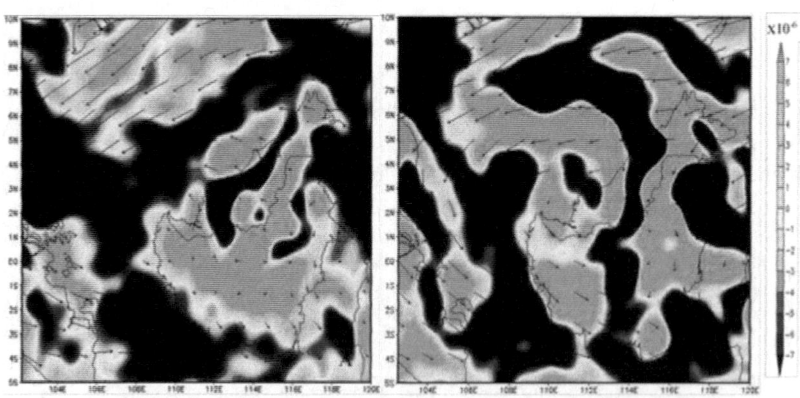

Figure 6. Divergence (s^{-1}) on January (A) 24th 2017 and (B) 25th 2017.

3.3.2.3 Vertical Velocity Analysis

The vertical velocity value (Figure 7) on January 24th and 25th, 2017 shows that there is a negative divergence which means that there is an air mass that rises with an average value of -0.1 Pa s^{-1} to -0.06 Pa s^{-1}. The areas with the highest vertical velocity are seen in the South China Sea to the Java Sea, and in the East to Northwest of Borneo waters, this indicates that there is an increase on air mass around Borneo islands and the South China Sea [6]. The atmosphere that is considered as a fluid and is considered (compressible) is very easily influenced by its environmental conditions [9], then the air masses this is strengthened by the cold air masses flows from the Pacific Ocean to the equator, so that the mass of air collects because the cold air masses tend to have a higher density so that it can push warmer air masses up and form rain clouds [12]. Then, the vertical velocity in the West Borneo region have value 0 Pa s-1, this happened because the influence of the cold air mass on the equator only affected the territorial waters of Indonesia, so that the West Borneo region did not experience an increase in air mass compared to the Riau Islands region [6]. The influence of the movement of the wind and the topography of the islands in IMC is a major factor in the formation of Borneo Vortex, due to the interaction between continents and oceans [11].

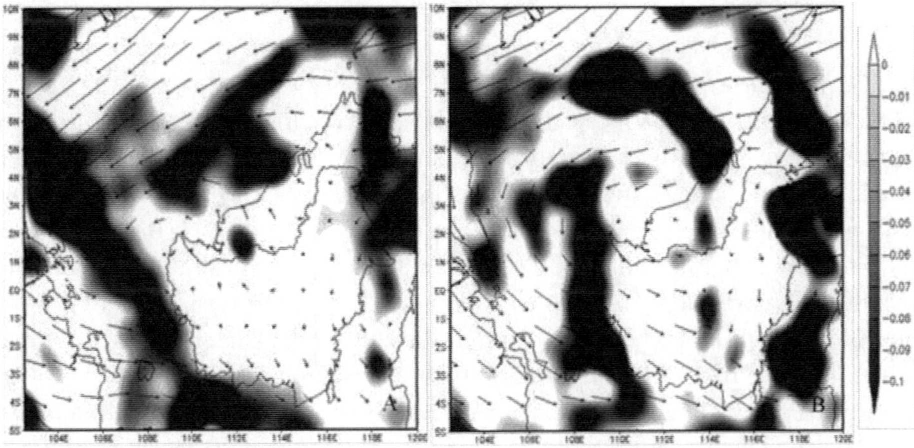

Figure 7. Vertical Velocity (Pa s^{-1}) on January (A) 24th 2017 and (B) 25th 2017.

3.3.2.4 Moisture Transport Analysis

Moisture transport value (Figure 8) on January 24th and 25th , 2017 shows that there is a movement of water vapor supply that enters to Indonesian waters that spread through the South China Sea into the waters of Bangka Belitung. It can be seen that the moisture transport value reaches an average value of 600 to 1400 $kgms^{-1}$, but on January 25th 2017, it can be seen that the humidity transport value is decreased because there are weakening effect of the vortex, which is 600 to 1200 $kgms^{-1}$. Indicates that the availability of moisture in on January 25th, 2017 is less than on January 24th, 2017. Showed the area that has high humidity transport moves from the Pacific Ocean to the equator and looks dominant in the South China Sea up to 2.5° N. This high humidity transport values has caused by the influences of moisture vapor by Borneo Vortex formed in the Northwest waters of Borneo Island [6]. There is indicating that there sufficient supply of water vapor to make the air conditions in the region wet, moist, and can support the process of large convective clouds growth with high top clouds, so that the chance of rain with moderate to heavy intensity increase [7]. Wind velocity in the event of Borneo Vortex shows a strong influence to support the increase of wind movement from the Pacific Ocean to the equator carrying the supply of water vapor, later than the air mass is concentrated in the Northwest waters of Borneo Island due to the influence of the Vortex formed [10].

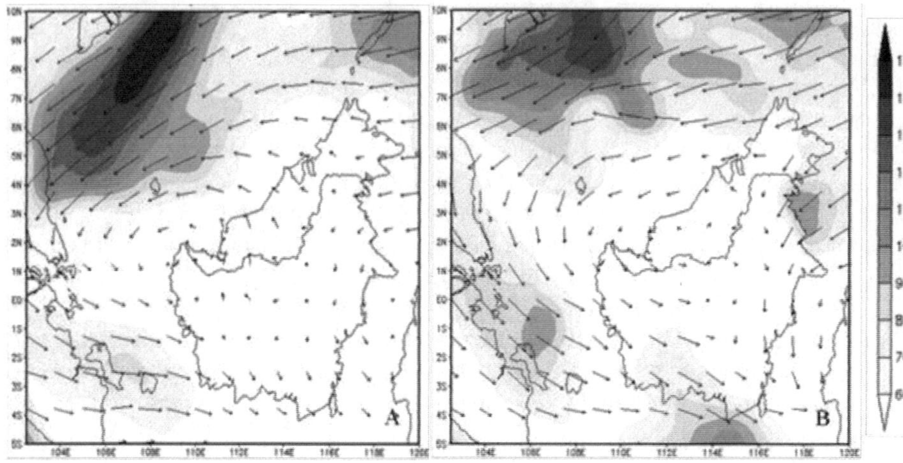

Figure 8. Moisture Transport (kg ms^{-1}) on January (A) 24th 2017 and (B) 25th 2017.

3.3.3 Rainfall Distribution Analysis

Rainfall distribution values (Figure 9) when Borneo Vortex occurred generally shows that there is a high distribution of rainfall in several research areas. It can be seen that there is a distribution of rainfall reaching a value of more than 27 mm/day on the South China Sea to the coast of Borneo Island, northwest of Borneo Island waters, and the western and southern parts of West Borneo. The distribution of rainfall indicates that there is a high rainfall concentration in the West Borneo region, the north, west, and south, marked by a large supply of water vapour on the coast of Borneo Island to the waters of the southern island of Borneo [6]. This confirms that heavy until very thick intensity rain in several areas of the research area has occurred. Then based on rainfall distribution shown at the time of the Borneo Vortex incident on January 24th 2017, indicated the tendency to have a higher distribution value more than caused on January 25th 2017 in all research locations. It shows that the most powerful influence of Borneo Vortex occurred on January 24th, 2017, because on January 24th was the strongest occurred of Borneo Vortex intensity.

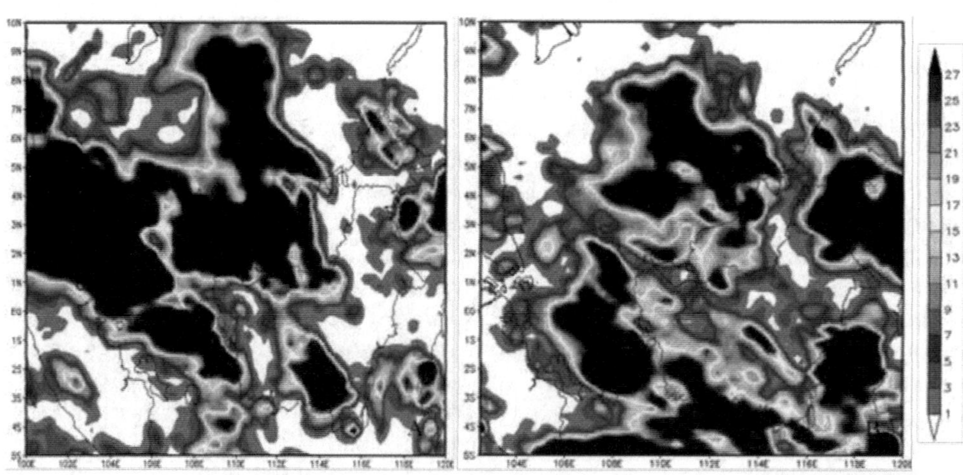

Figure 9. Rainfall Distribution of TRMM (mm) on January (A) 24th 2017 and (B) 25th 2017.

Figure 10 shows the chart of rainfall observation values of each station. When Borneo Vortex occurred on 24 until 25 January 2017, the incident caused an increasing rainfall intensity on 24 January 2017, recorded in the region of West Borneo the highest rainfall was in the area of Paloh, Pontianak and Ketapang. It recorded rainfall ranged from 73.1 until 154.1 mm/day, where the region represents the northern, western and southern regions. The central and eastern regions have lower rainfall there are Sintang, Nangapinoh, and Putusibau regions ranging from 10.9 - 52.0 mm/day, it shows that the rainfall occurs in almost all regions with intensity range from slight to very heavy.

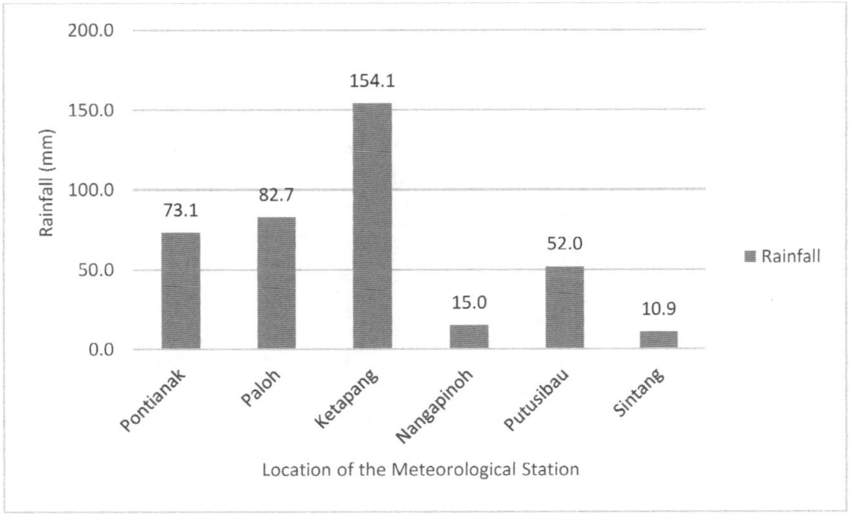

Figure 10. Rainfall Observation Chart on Meteorological Station at January 24th until 25th, 2017 (Source: *www.ogimet.com/synop report*).

4. Conclusion

Based on regional scale analysis sea surface temperatures in the South China Sea cooler than the heat of around West Borneo. This causes air flowed from the Pacific Ocean travel through the South China Sea and convergence the winds from the southeast in the northwest region of Borneo Island, and then formed the cyclonic circulation. Based on a local scale analysis of atmospheric conditions during the Borneo Vortex event, there was a significant increase on surface pressure, vorticity, divergence, vertical velocity, and moisture transport. It was observed that in the event of Borneo vortex, the value of atmospheric parameters as the indicator of atmospheric conditions can change become higher than the normal value. Based on the results of the analysis of rainfall distribution during the Borneo Vortex on January 24th until 25th, 2017, the incidence of vortex triggered the increase of rainfall in several regions of West Borneo. Recorded on January 24th, 2017 in the West Borneo region the highest rainfall was affected Paloh, Pontianak, and Ketapang regions, which recorded rainfall ranges from 73.1 - 154.1 mm/day. This was significantly affected the region represents the northern, western, and southern regions of West Borneo, while the central and eastern regions have lower rainfall in Sintang, Nangapinoh and Putusibau regions ranging from 10, 9 - 52.0 mm/day.

Acknowledgments

Thanks to the authors conveyed to the European Center for Medium-Range Weather Forecast (ECMWF) which has provided data reanalysis in the form of netCDF that can be accessed easily in general. Ms. Hasti Amrih Rejeki, SST, M.Si as the mentor of Meteorology, Indonesian State College of Meteorology, Climatology and Geophysics, South Tangerang, Indonesia

5. References

[1] Chang C P, Harr P A and Chen H J 2005 Synoptic disturbances over the Equatorial South China Sea and western maritime continent during boreal winter *J. Am. Meteorology Society* **133** 489

[2] Anip M H M and Lupo A 2012 Interannual and interdecadal variabillity of the Borneo Vortex during Boreal Winter monsoon *Thesis University of Missouri Columbia, USA* p 206

[3] Zakir A, Sulistya W and Khotimah M K 2010 *Perspektif Operasional Cuaca Tropis* (Research Centre and Development BMKG Jakarta)

[4] Tangang F T, Juneng L, Salimun E, Vinayachandran P N, Seng Y K, Reason C J C, Behera S K and Yasunari T 2008 Long-Term trends of winter monsoon synoptic circulations over The Maritime Continent in 1962 - 2007 *J. Atmospheric Sci. Lett.* **11** p 199-203

[5] Chang C P, Liu C H and Kuo H C 2003 Typhoon Vamei: An equatorial tropical cyclone formation *J. Geophys* **30(50)** p 1 – 4

[6] Isnoor K F N and Rejeki H A 2018 Analisis kondisi atmosfer saat terjadi Borneo Vortex dan pengaruhnya terhadap curah hujan di wilayah Kalimantan Barat dan Kepulauan Riau (Periode 2006 - 2016) *Prosiding J. National Remote Sensing Conference 2018.* p 615

[7] Ardianto R 2014 *Kajian dampak Borneo Vortex terhadap curah hujan di wilayah Borneo Barat* Skripsi School of Meteorology Climatology and Geophysics of Indonesia, Jakarta

[8] Boccara G, Hertzog A, Basedevant C and Vial F 2008 Accuracy of NCEP/NCAR reanalysis and ECMWF analysis in the lower stratosphere over Antarctica in 2005 *J. Geophys. Res.* **133**

[9] Holton J R 2004 *An Introduction to Dynamic Meteorology* 4th ed. (USA: Elsevier Publication)

[10] Paulus A W and Prayuda S S 2017 Atmospheric study of the impact of cold surges and Borneo Vortex over Western Indonesia Maritime Continent area *J. Clim. Weather Forecast* **5(1)** p 189

[11] Ooi S H, Samah A A and Braesicke P 2011 A Case Study of the *Borneo Vortex* Genesis and Its Interaction with the Global Circulation *J. Geophys. Rev.* **116** D21116

[12] Johnson R H dan Houze, Jr R A 1987 Precipitating cloud systems of the Asian monsoon *Journal of Monsoon Meteorology* p 298 - 353 Departement of Atmospheric Science Fort Collins, Colorado

Revisiting the impact of mid-latitude cold air outbreaks on the Maritime Continent weather

M R Abdillah[1] and T Iwasaki[2]

[1]Research Group of Atmospheric Science, Faculty of Earth Sciences and Technology, Institut Teknologi Bandung, West Java, 40132, Indonesia
[2]Department of Geophysics, Graduate School of Science, Tohoku University, Sendai 980-8578, Japan

m.rais@itb.ac.id

Abstract. The East Asian winter exhibits frequent cold air outbreaks (CAOs), or cold surges, which often cause severe cold waves in the extratropics. CAOs are also of importance for forecasters in the Maritime Continent because the associated equatorward outflows can influence tropical weather. Various CAO definitions have been documented, but a definition based on a quantitative approach was only recently proposed. Studies of the CAO impact often linked a surge index in the subtropics instead of an index in the midlatitude where CAOs basically occur. Here, we investigate the impact by using a quantitative CAO index, which is defined by integrating cold air mass flux below a threshold potential temperature over midlatitude East Asia (45°N, 90°-135°E). From a climatological analysis, tropical convections are significantly observed two-to-four days following CAO event (i.e., the peak of CAO index), indicating CAO impacts. However, case studies show that the impacts vary among independent CAO cases due to influences of associated synoptic conditions, which affect the pathways of northerlies propagation and consequently hinder or amplify the impacts. Several impact patterns and their possible causes are discussed. A better understanding of East Asian CAO variability can improve the predictability of weather over the Maritime Continent.

Keywords : cold air outbreaks, weather, maritime continent.

1. Introduction

Cold air outbreak (CAO) is a wintertime phenomenon characterized by the outflow of cold air mass from high latitudes to lower latitudes. CAOs, often referred to as cold surges, basically occur at a synoptic time scale and greatly influence extratropical weather [1, 2, 3]. In East Asia, the CAOs are typically strong—its associated northerly flow sometimes penetrates far to tropical regions and plays a role on the weather over the Maritime Continent [1, 4]. Previous studies investigated CAOs by using various CAO definitions [4, 5, 6, 7]. However, the studies may be inaccurate in defining CAO magnitude because the mass of cold air was not quantitatively measured [3]. Some proxies, such as surface temperature or wind, are used. Moreover, studies focused on the tropical impact of CAOs tend to rely on surge indices over subtropical/tropical regions [8], not in the midlatitudes where CAOs occur.

Here, we aim to revisit the impact of East Asian CAOs by using a quantitative CAO definition. This paper follows a recent isentropic method as documented in [3, 9, 10], which used CAO index defined by equatorward cold air mass fluxes below a threshold potential temperature of 280 K at 45°N. Section 2 explains the methodology in details. Section 3 discusses the mean impact of CAO from lagged

IOP Conf. Series: Earth and Environmental Science **303** (2019) 012062 doi:10.1088/1755-1315/303/1/012062

regressions. Section 4 shows variability among individual CAO events by conducting several case studies. Section 5 concludes the result and stresses important future studies.

2. Data and method

2.1. Data

We use the JRA-55 reanalysis as the main atmospheric dataset, which has 6-hr temporal and 1.25° horizontal resolutions [11]. To depict tropical convections, we use daily NOAA outgoing longwave radiation (OLR) [12] and GPCP precipitation [13]. The OLR and precipitation have 2.5° and 1° horizontal resolutions, respectively. The analysis period is December-February from 1997/98 to 2014/15.

2.2. Isentropic cold air mass and cold air outbreak index

A recent study proposed an isentropic method to diagnose polar cold air mass (CAM) below a threshold potential temperature [14]. This approach has quickly inspired other studies to develop CAO indices [3, 9, 10] because the method allows us to estimate CAM fluxes that regulate CAO events. Before defining the CAO index, let us start with the formulation of CAM below a threshold potential temperature (θ_T) [14]. The amount of CAM at each grid point (DP) is defined as

$$DP \equiv p_s - p(\theta_T) \tag{1}$$

where p_s and $p(\theta_T)$ denote pressure at the ground surface and pressure at θ_T, respectively. We can obtain the flux of cold air mass (**F**) by integrating horizontal wind (**v**) from θ_T to the surface, so that

$$\mathbf{F} \equiv \int_{p(\theta_T)}^{p_s} \mathbf{v}\, dp \tag{2}$$

The Tendency of DP is controlled by the convergence of **F** and genesis/loss rate due to diabatic heating (G),

$$\frac{\partial}{\partial t} DP = -\nabla \cdot \mathbf{F} + G \tag{3}$$

DP and **F** are calculated from 6-hr reanalysis, and G is estimated from residual in equation (3). Consistent with previous studies, we use 280 K as θ_T. This is justified by the fact that the zonal-mean equatorward mass flux below 280 K represents the lower part of extratropical direct circulation. We can say that the mass flux below this threshold clearly denotes "hemispheric" cold air outflow [14,15]. The hemispheric outflow is greatly attributed to a cold air stream in East Asia, which elongates from inland Asia to the central North Pacific (see [3,14] for climatologies of DP, **F** and G).

The East Asian cold stream is a manifestation of many CAO events occurring in there [3]. The stream exhibits two dominant modes of variability (western and eastern modes), which simply indicate two distinct CAO pathways and suggest two types of CAOs in East Asia: western CAO and eastern CAO [9]. While both CAOs cause robust impacts over the mid-latitudes, only western CAO seems to be capable of triggering remote influence over the tropics because the outflow of eastern CAO does not reach low latitudes [10]. Thus, here we only discuss the impact caused by western CAO. The magnitude of western CAO (hereafter referred to as CAO or East Asian CAO) can be measured by defining a CAO index at 45°N [3, 9, 10]:

$$\mathrm{CAOI} \equiv \int_{90°\mathrm{E}}^{135°\mathrm{E}} -F_v\, d\lambda \tag{4}$$

where $-F_v$ is the equatorward component of CAM flux and λ is the longitude. We use daily CAOI. The climate mean of CAOI is removed and then, CAOI magnitude is normalized by standard deviation.

In section 3, to depict mean temporal evolution associated with CAO events, we perform day-lagged regressions with CAO index for all days in the 18 winters. The statistical significances are assessed by using the two-sided student's *t*-test, where the numbers of degree of freedom are reduced following a decorrelation method [16]. In section 4, we perform case studies on several unique CAO events aiming to highlight variability among the events and its possible causes. The identification method of CAO events is shown in that section.

3. Mean synoptic evolution of CAO event and its impact

Figure 1 shows the synoptic evolution of CAO event in East Asia constructed from lagged regressions with CAOI. The CAO is led by an increase of CAM amount north of Tibetan Plateau and by the strengthening of Siberian High from day -4. The anomalous cold air and high pressure become larger and propagate southeastward. The equatorward CAM flux then intensifies, reaching its maximum at day 0. The CAO is also associated with anomalous low pressure over the northwestern Pacific, which provides eastward pressure gradient force and thus facilitates southward geostrophic flow. The cold air spreads up to southern China, Japan, and the western Pacific. The spread over the ocean and low latitudes causes intense surface heat fluxes and leads to massive disappearance of CAM. These synoptic conditions are discussed thoroughly in [3, 4, 7].

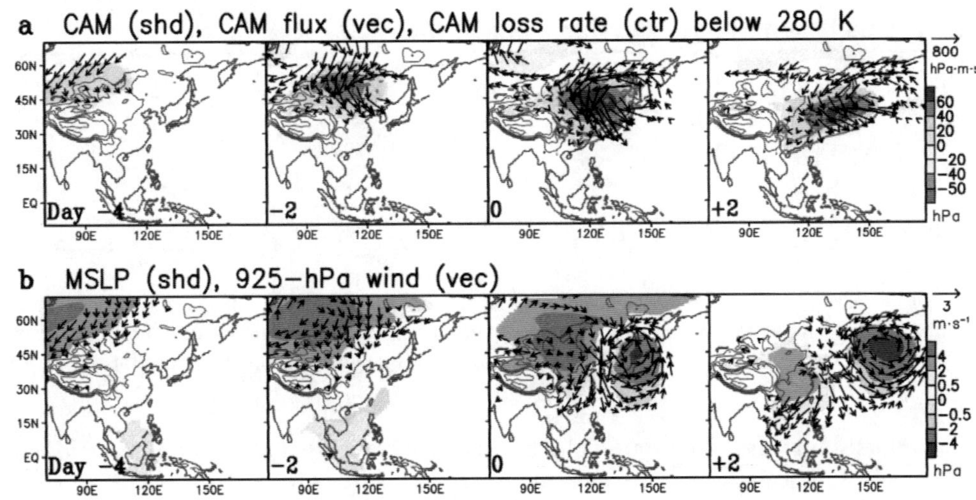

Figure 1. Synoptic evolution of an East Asian cold air outbreak (CAO) constructed by using lagged regressions with the isentropic CAO index (day -4 to day +2). (**a**), anomalies of 280-K cold air mass (CAM) amount (*DP*; shaded), CAM flux (**F**; vector), and diabatic CAM loss rate (*G*; red contour). For loss rate, the contours enclose areas of significant CAM losses regardless the amplitude. (**b**), anomalies of mean sea level pressure (MSLP) (shaded) and 925-hPa wind (vector). Only anomalies exceeding statistical significance (99.99%) are shown. Brown contours denote topography with 1000 m interval.

Although the CAM amount and its equatorward flux rapidly disappear following the peak of CAO, the low-level northerly wind anomalies persist and even propagate to low latitudes, potentially affecting tropical weather (Figure 1b). The northerly momentum is guided by the southward expansion of Siberian High. The mean impacts of the enhanced northerlies are shown in Figure 2. Despite the cold- and dry-features of CAO, the outflow gradually gains heat and moistures when they enter low latitudes and sea surface. Therefore, we observe anomalous southward moisture fluxes approaching the Maritime Continent (Fig. 2a), strengthening the northeasterly monsoon flow. The enhanced northerlies trigger tropical convections when they interact with islands and mountains over the Maritime Continent. The locations of tropical convections agree well with the locations of moisture flux convergence and precipitation. The statistically significant precipitation signals appear over the Philippines, South China

Sea, and north of Borneo at day +2 and day +4. This result is consistent with a previous study showing that regressions using an East Asian pressure index at midlatitude yield significant signals only over the north of equator [4]. Prior to day 0, we observe some convective and precipitation anomalies leading the peak of CAO; these indicate signatures of tropical CAO precursor associated with Madden-Julian Oscillation (MJO), whose regression signals would be more significant in intraseasonal temporal bands [10]. MJO can control the occurrence of CAO by deepening a low-pressure trough over the northwestern Pacific through poleward upper-level wave train [10].

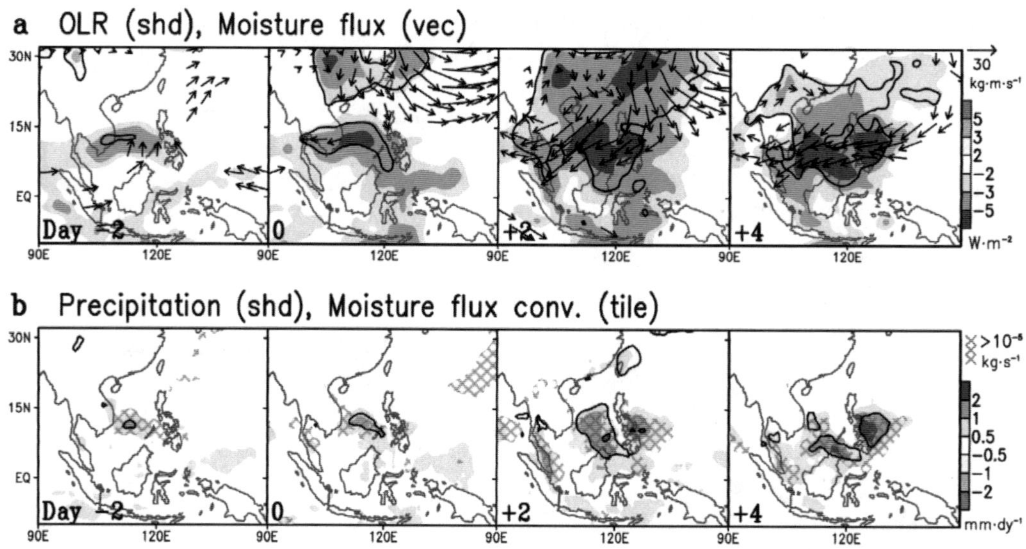

Figure 2. As in Figure 1, but from day -2 to day +4. Regressions with (a) outgoing longwave radiation (OLR) (shaded) and moisture flux (1000-200 hPa) (vector); (b) precipitation (shaded) and moisture flux convergence (red tiled). Black contours enclose statistically significant OLR/precipitation (99%).

Figure 3 exhibits southward propagation characteristics of several parameters associated with CAO along East Asia and Southeast Asia. It is captured by tracing anomaly peaks from 90° to 135°E of mean sea level pressure (MSLP), meridional wind at 925 hPa, moisture flux convergence, and precipitation. The high pressure and northerly wind propagate at the approximately same speed. It suggests that their southward movements are guided by pressure-wind imbalance, where the high pressure drives the northerlies and cold advection, and the cold northerlies give feedback to high pressure [3,4,8]. Following the path of northerlies, the anomalies of moisture flux convergence develop with a lag of 1-2 days, favoring instability and initiation of precipitation over the tropics. The peak of precipitation appears at day +3 over 10°N, but the convective signals vary regionally over the tropics possibly due to complex topography over the Maritime Continent.

4. Case studies of three selected CAO events

The precipitation anomalies obtained from the regressions seem limited in area and rather weak. Several kinds of literatures have shown that cold surges over the South China Sea can exert wide and strong impacts on precipitation over the Maritime Continent, including regions south of equator [8,18,19]. Our regression uses a base location of CAOI at midlatitude, thus the downstream impact of CAO may be subject to varying independent synoptic conditions over the subtropics and the tropics, which weaken average propagation of northerlies at low latitudes.

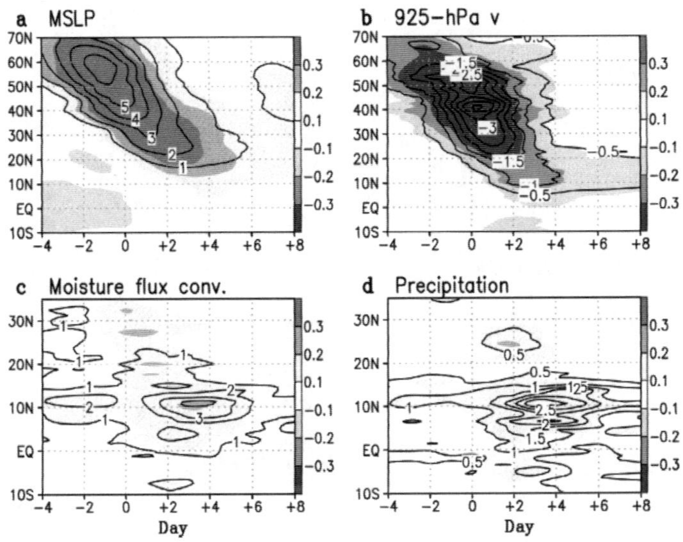

Figure 3. Hovmöller diagrams (latitude vs time) along East Asia and the Maritime Continent (tracing anomaly peaks from 100° to 135°E) obtained from lagged correlations (shading) and regressions (contour) with CAO index. Panels show (**a**) MSLP (hPa), (**b**) 925-hPa meridional wind (m s^{-1}), (**c**) moisture flux convergence (10^{-5} kg s^{-1}), and (**d**) precipitation (mm day^{-1}); zero contours are omitted. Correlation coefficients of ±0.1 denote confidence level greater than 95%.

To investigate possible variations among CAO impacts, we perform case studies of a few independent CAO events. Any CAO events can be collected by identifying local maxima of daily CAOI that satisfy a given threshold [9,10]. The daily CAOI is filtered by 6-day low-pass filter before CAO identification to ensure synoptic signals. We select three CAO events by considering different southward propagation characteristics. They occurred on 4 January 2006 (hereafter Case 1), 3 February 2010 (Case 2), and 30 December 2007 (Case 3). Although the events exhibit strong outbreaks, as indicated by CAOI greater than 1.5, their equatorward intrusions significantly differ of each other (Figure 4). Case 1 shows a clear propagation of northerly along East Asia and the South China Sea. Case 2, however, shows a lack of equatorward propagation despite having large CAO amplitude (CAOI=2.26). Case 3 has relatively weaker CAO amplitude but it exhibits very strong equatorward intrusion of northerlies—even crossing the equatorial line. This nonlinearity leads to weak regressions shown in Figure 2. We examine probable causes for these differing cases in the following sub-sections.

4.1. Case 1: Clear southward propagation

The magnitude of CAO occurred on 4 January 2006 is typically strong (CAOI>2). Two days before the outbreak, there is large CAM accumulation from 45°-60°N in the north of the Tibetan Plateau (Figure 5a). During the outbreak, the equatorward flux of CAM prevails in East Asia, transporting abundant cold air that spans from inland China to Japan. A portion of cold air travels further southward and evolves into warmer air; and the rest moves eastward toward the North Pacific following the propagation of midlatitude waves. The southward movement of high pressure is also evident (Figure 5b). It propagates along the eastern boundary of Tibetan Plateau and controls the circulation in the lower troposphere. In the northwestern Pacific, there is a strong cyclonic center, which strengthens the equatorward outflow at midlatitude.

IOP Conf. Series: Earth and Environmental Science **303** (2019) 012062 doi:10.1088/1755-1315/303/1/012062

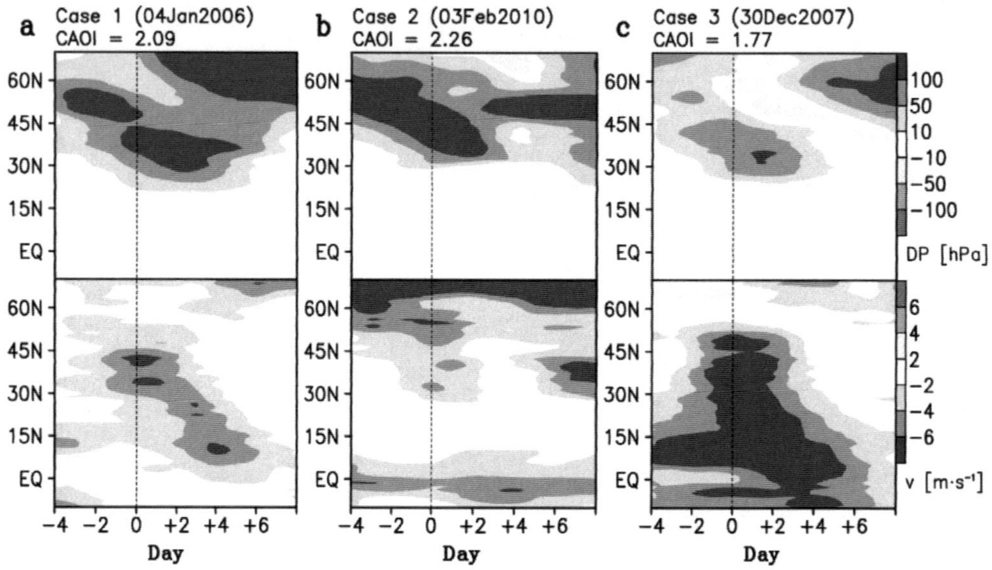

Figure 4. Propagation characteristics of three selected CAO cases. Top and bottom panels show CAM below 280 K and meridional wind at 925 hPa, respectively (tracing anomaly peaks from 100° to 135°E). The magnitude of CAOI is shown in the top.

In the tropics, the changes in wind direction and magnitude are robust (Figure 5c). Strong northerlies over the South China Sea and vicinity appear at day +2 and day +4 and enhance the northeasterly monsoon flow. The influence of northerly flow significantly affects the western Maritime Continent (Peninsular Malaysia, Singapore, and Sumatra Island). It confirms that the midlatitude CAO can cause substantial impact over a larger area of the Maritime Continent, which is not apparent in the regression analysis. The precipitation signals over there are quite strong and seem to be associated with a meso-to-synoptic scale disturbance. It shows a common pattern of South China Sea surges where the precipitation is induced by the interactions of incoming flow and terrains, which have been reported in previous studies [17, 18]. In the Philippine Sea, the precipitation band is well-aligned with a frontal line of northerly cold flow, suggesting front-like precipitation due to confluence of cold air with easterly Pacific warm air.

The associated northerly flow is well produced but it does not significantly cross the equatorial line. In this case, the anomalous high pressure only reaches subtropical regions and spreads to the east, consequently favor to development of easterly rather than northerly around the equator. We argue that the additional pressure gradient over the tropics is necessary to maintain further intrusion of the northerlies (see a discussion of Case 3).

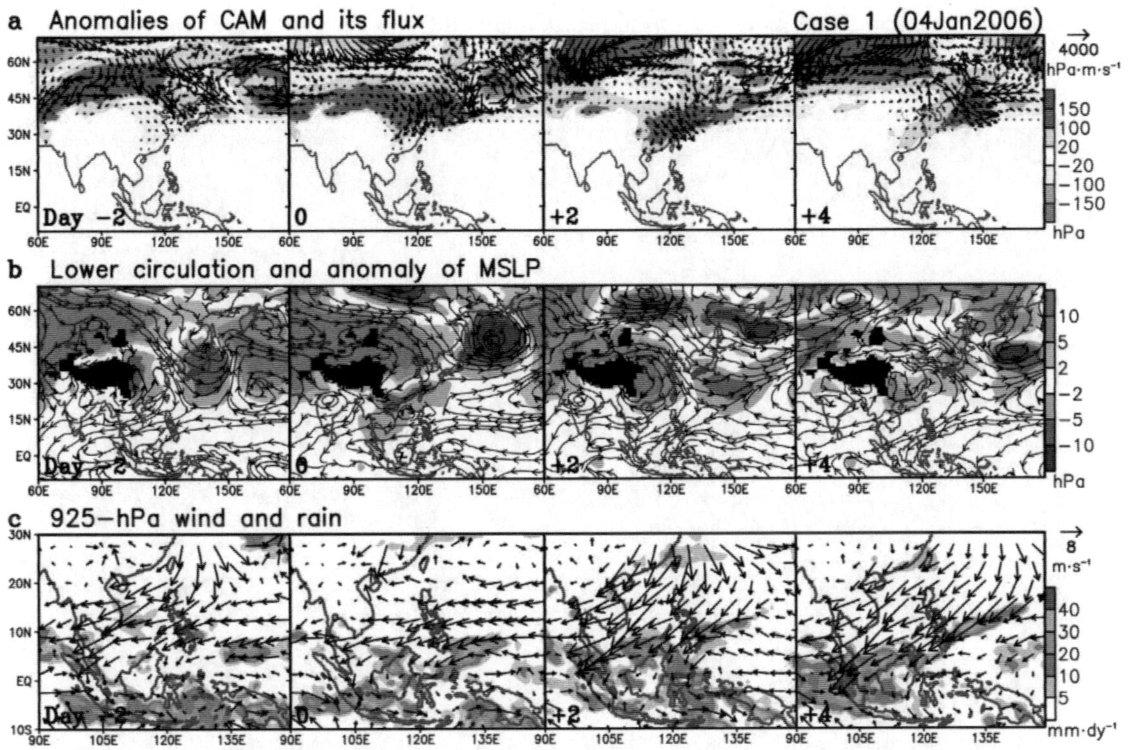

Figure 5. Synoptic conditions during a CAO case on 4 January 2006 (Case 1). Columns denote time sequence from 2 days prior to and 4 days after CAO. Top to bottom: (**a**) CAM and its flux below 280 K; (**b**) MSLP and 850-hPa streamlines (black areas: surface pressure < 850 hPa); and (**c**) 925-hPa wind and rain rate. Note that, parameters in (a) and MSLP in (b) are anomalies, while those in (c) and streamlines in (b) are obtained from actual daily values.

4.2. Case 2: Nonexistent subtropical surge

A CAO case occurred on 3 February 2010 shows a strong magnitude of CAOI, which is larger than Case 1, but this case has a unique synoptic pattern in the extratropics. Prior to the peak of CAO, the CAM accumulation is centered over the eastern Siberia (Figure 6a). The outbreaks occur mainly over the East Asian coast, transporting CAM to Korea, Japan, and North Pacific. The CAM anomalies over China and lower latitudes are very small and temporary. Comparing with the previous case, this case clearly shows a weak southward propagation of cold air in East Asia (see also Figure 4). The outflow rapidly turns eastward toward the North Pacific and develops into westerly surge.

Figure 6b indicates weak southward extension of Siberian High. The MSLP pattern suggests that the driving force of this CAO comes from interactions between quasi-stationary high-pressure anomalies in high latitudes and cyclonic circulation north of Japan rather than traveling waves trapped over the Tibetan Plateau. One more important pattern is located over low latitudes. A large anticyclone appears in the north of the Philippines. This anticyclone seems to be associated with low-level vorticity response [19] due to suppression of tropical convection over the eastern Maritime Continent (supported by high pressure in Fig. 6b and positive OLR anomalies in Fig. 8b). The lack of Siberian High extension and the existence of subtropical anticyclone greatly hinder southward migration of cold air outflow along East Asian coast and the South China Sea; and therefore, the CAO does not excite precipitation over the Maritime Continent. Nevertheless, at day +4, we observe stronger northerlies propagating over the western Pacific around 140°E, enhancing easterly surge over the Philippines Sea. The flow resembles a Philippines Sea surge-type as reported by [4]. But, the existence of high pressure over the western Pacific is not favorable for convection development around the Philippines.

Case 2 is rather unusual but it highlights the fact that not all CAO events cause disturbances over the tropics. Examination on other CAO events reveals that there are few other cases resembling Case 2.

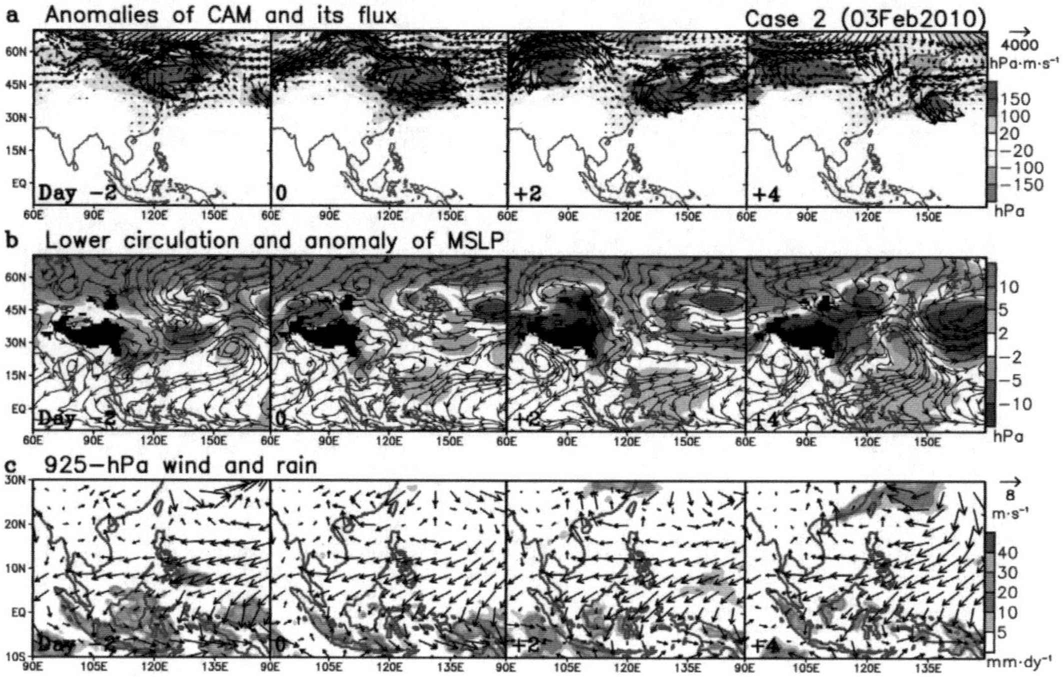

Figure 6. As in Figure 5, but during a CAO case on 3 February 2010 (Case 2).

4.3. Case 3: Strong subtropical surge and cross equatorial surge

Contrary to Case 2, Case 3 shows strong southward propagation of northerlies and even affects cross-equatorial flow (Figures 4c and 7). The patterns of CAM and high pressure evolve in a similar manner to those in Case 1, which shows the southward intrusion of high and northerlies along east of Tibetan Plateau. A unique feature occurred in this case is large scale negative pressure anomalies situated over the Maritime Continent, which elongates meridionally up to midlatitude East Asian coast (Figure 7b). The strong pressure contrasts along East Asia and Southeast Asia are believed to guide the southward momentum up to Southern Hemisphere. In the tropics, we observe the increase of northerlies from the South China Sea to the Java Sea, which leads to convections in both hemispheres (Figure 7c).

The accompanying low-pressure anomalies over the Maritime Continent appear to be induced by an active phase of MJO, which is indicated by eastward propagating intraseasonal OLR anomalies over the tropics that are coincided with the CAO (Figure 8c). The existences of precipitation signals *prior to* day 0 and strong westerlies in the south of equator support the MJO existence (Figure 7c). Furthermore, the MJO may affect the propagation speed of the northerlies. As shown in Figure 4, the propagation speed of northerlies in Case 3 is larger than that in Case 1.

Case 3 indicates that the midlatitude CAO does have an impact over the southern Maritime Continent, but with an influence of another forcing situated over the tropics. The role of active MJO is quite clear. It provides strong southward and eastward pressure gradients and maintains the northerly momentum to propagate further southward, which eventually enhances the precipitation. This finding is consistent with a previous study that shows the MJO influence on modulating precipitation associated with cross-equatorial northerly surge (CENS) [18]. Note that, tropical OLR anomalies in Case 1 does not show active MJO signal during its CAO event; and in Case 2, the tropical OLR anomalies are significant but they show *inactive* convection over the western Pacific (Figures 8a and 8b).

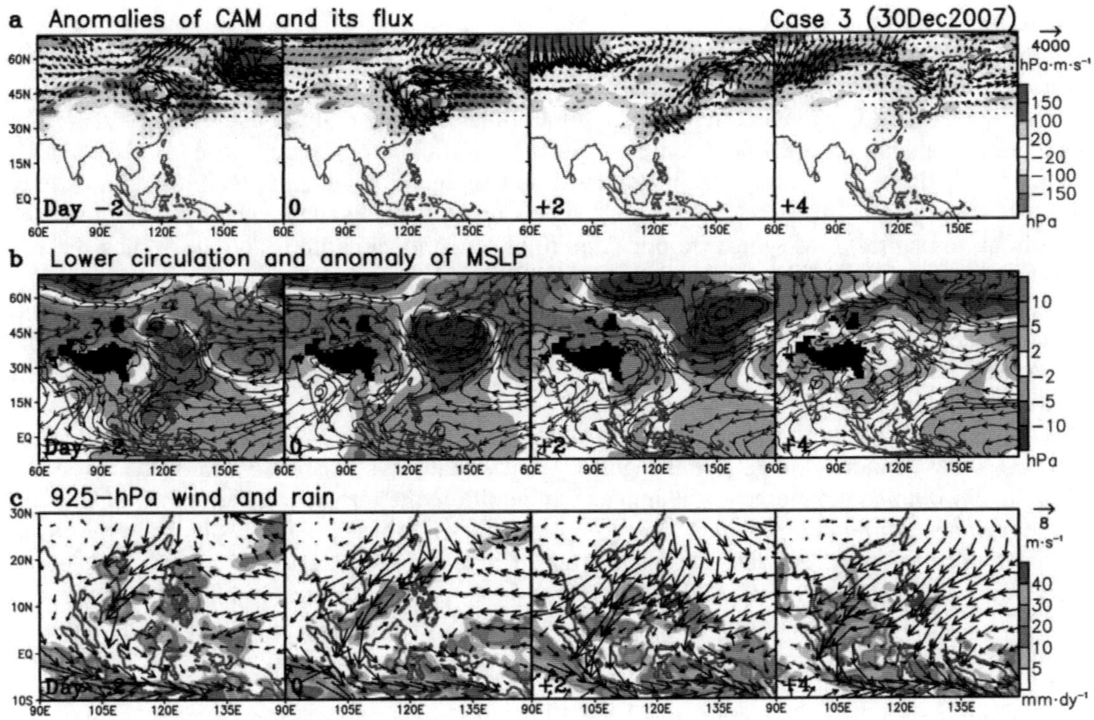

Figure 7. As in Figure 5, but during a CAO case on 30 December 2007 (Case 3).

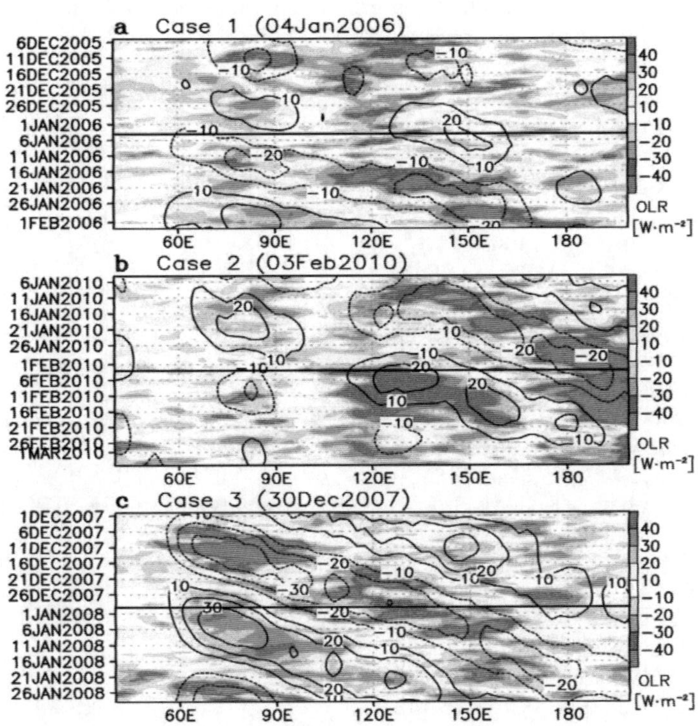

Figure 8. Hovmöller diagrams showing evolution of tropical OLR (averaged at 15°S-10°N) associated with the three CAO cases (**a-c**) from 30 day prior to and 30 day after CAOs. Shadings are daily OLR anomalies constructed by removing the first three annual harmonic. Contours denote 30-80 day bandpass filtered anomalies. The day of CAO event is shown by thick solid line.

5. Concluding remarks

This study revisits the impact of East Asian cold air outbreak (CAO) over the tropical regions. The CAO is defined on the basis of isentropic coordinate, which directly measures the quantity of cold air mass fluxes that regulate CAO events at midlatitude. In a climatological analysis using lagged regressions, we examine the linear linkages between East Asian CAO and tropical response. We find that the associated northerlies significantly reach the Maritime Continent, but excite precipitation mostly over the north of the equator. Investigations on three independent CAO events reveal that the downstream impact of CAO is subject to varying synoptic conditions over the midlatitude and the tropics. There is a case where the northerlies strongly propagate southward and cause substantial impacts (Case 1); but there is also a case that does not show northerlies propagation and thus results in non-existent impact (Case 2). In the tropics, we find a case that shows strong northerlies reaching Southern Hemisphere, which are partly affected by an active MJO (Case 3). The synoptic pattern suggests a role of active MJO in maintaining the northerlies to deeply propagate.

These findings highlight the variability of CAO impacts and call for next research. We investigate three CAO cases. Future studies are suggested to evaluate systematically all CAO events and identify/classify dominant patterns of the impacts. In addition, the interaction between CAO and other tropical phenomena should be explored. For example, some studies reported the roles of easterly waves and vortices can change the flow and the impact [e.g., 19].

Forecasting CAO impact by tropical weather advisories probably relies on a surge index over the South China Sea. We anticipate that developing a forecast system on the basis of midlatitude index would offer better preparedness for the surge impact over the tropics.

Acknowledgments

We thank Dr. Y. Kanno for providing a tool for isentropic calculations and constructive comments. This study is partly supported by the Japanese Ministry of Education, Culture, Sports, Science and Technology (MEXT) through Grant-in-Aid 15H02129 and the Program on Climate Change Adaptation Technology (SI-CAT).

6. References

[1] Chang C-P, Erickson J E and Lau K M 1979 *Mon. Weather Rev.* **107** 812–29
[2] Cellitti M P, Walsh J E, Rauber R M and Portis D H 2006 *J. Geophys. Res. Atmos.* **111** 1–14
[3] Shoji T, Kanno Y, Iwasaki T and Takaya K 2014 *J. Clim.* **27** 9337–48
[4] Compo G P, Kiladis G N and Webster P J 1999 *Q. J. R. Meteorol. Soc.* **125** 29–54
[5] Lau N-C and Lau K-M 1984 *Mon. Weather Rev.* **112** 1309–27
[6] Park T-W, Ho C-H, Jeong S-J, Choi Y-S, Park S K and Song C-K 2011 *J. Geophys. Res.* **116**
[7] Zhang Y, Sperber K R and Boyle J S 1997 *Mon. Weather Rev.* **125** 2605–19
[8] Chang C-P, Harr P A and Chen H-J 2005 *Mon. Weather Rev.* **133** 489–503
[9] Abdillah M R, Kanno Y and Iwasaki T 2017 *J. Clim.* **30** 2989–3007
[10] Abdillah M R, Kanno Y and Iwasaki T 2018 *J. Clim.* **31** 473–90
[11] Kobayashi S, Ota Y, Harada Y, Ebita A, Moriya M, Onoda H, Onogi K, Kamahori H, Kobayashi C, Endo H, Miyaoka K and Takahashi K 2015 *J. Meteorol. Soc. Japan*
[12] Liebmann B and Smith C a. 1996 *Bull. Amer. Meteor. Soc.* **77** 1275–77
[13] Huffman G J, Adler R F, Morrissey M M, Bolvin D T, Curtis S, Joyce R, McGavock B and Susskind J 2001 *J. Hydrometeorol.* **2** 36–50
[14] Iwasaki T, Shoji T, Kanno Y, Sawada M, Ujiie M and Takaya K 2014 *J. Atmos. Sci.* **71** 2230–43
[15] Iwasaki T and Mochizuki Y 2012 *SOLA* **8** 115–118
[16] Livezey R E and Chen W Y 1983 *Mon. Weather Rev.* **111** 46–59
[17] Lim S Y, Marzin C, Xavier P, Chang C-P and Timbal B 2017 *J. Clim.* **30** 4267–81
[18] Hattori M, Mori S and Matsumoto J 2011 *J. Meteorol. Soc. Japan* **89A** 27–47
[19] Chen T-C, Tsay J-D, Matsumoto J and Alpert J 2015 *J. Clim.* **28** 1417–43

Analysis of Mesoscale Convective Complex during Madden Julian Oscillation Phase 4 (Case Study: Heavy rain in Cilacap on Sept 16-17, 2016)

L F Aiqiu [1], A Zakir [2] and Trismidianto [3]

[1]Undergraduate program of Meteorology, State College of Meteorology, Climatology and Geophysics, Jakarta Indonesia
[2]Lecturer, Department of Meteorology, State College of Meteorology, Climatology and Geophysics, Jakarta Indonesia
[3]Center for Atmospheric Science and Technology, National Institute of Aeronautics and Space, Bandung Indonesia

fararetalavia@yahoo.com

Abstract. This research aims to analyze the heavy rain which accompanied by severe winds, caused floods and landslides in several of sub-districts in Cilacap, Central Java on 16-17 September 2016 that coincides with the phase 4 of Madden Jullian Oscillation (MJO) event. In the same time, also found a Mesoscale Convective Complex (MCC) in the around Central Java. MCCs were identified by infrared satellite imagery from Himawari 8 using an algorithm that combines criteria of cloud coverage, eccentricity, and cloud lifetime. The data used is a combination of observation data, satellite data and reanalysis data. This study results that there is an increase in vertical velocity accompanied by strong negative vorticity since the development of MCC until it reached its mature phase, then the vertical velocity decreases as the cloud systems disseminated. These results indicate there is the evolution of MCC convective cells from growth until dissipation. Throughout its lifetime the MCC system moves south-east into the Indian Ocean and is decapitated there. Based on GSMaP rainfall data it was observed that in Central Java area especially Cilacap area has extreme rainfall which caused by MCC activity. The influence of MCC on significant rainfall is also supported by the existence of MJO phase 4. The significant increase in rainfall when phase 4 active MJO is followed by the MCC. The value of rainfall intensity when MJO followed by the presence of MCC is greater than in the absence of MCC.

Keywords : MCC, MJO, IMC, GSMaP

1. Introduction

Ramage [1] have stated that Indonesian Maritime Continent (IMC) is one of the regions with the largest amount of rainfall, rainfall levels over the IMC are among the highest globally. Spatial and temporal variations in rainfall affect the global circulation through latent heat release, because of the IMC is composed of a large number of islands and surrounded by sea, both regional and temporal variations in rainfall may be quite complicated due to strong interactions between land and the adjacent sea. The precipitation organization mechanism over the IMC may involve several factors, such as the complex geographical variation mentioned above and the large-scale atmospheric circulation. Madden-Julian

Oscillation (MJO) is one of the dominant modes of tropical variability on the intraseasonal time scale [2] of the ocean-atmospheric pairing system which greatly influences atmospheric circulation throughout the global tropical including IMC. The influence of MJO on IMC is also very significant, especially in phase 4 where MJO is in the Maritime continent [3].

Besides MJO, Mesoscale Convective Complex (MCC) also has a considerable influence on rainfall in Indonesia. Trismidianto [4] has reported that the contribution of MCC to total rainfall in the Indonesia territory is between 18% -20% with the concentration of the most MCC formation region is in the coastal areas south of Sumatra, the sea of South China, Central Kalimantan, Central Sulawesi and the Indian Ocean, while the biggest influence of the MCC to the rainfall is in the region of Central Kalimantan which reaches 20%. Several studies regarding the interaction between deep convection and MJO global scale oscillation systems have been carried out. But previous research only focused on the interaction between MJO and deep convection in general such as MCS. Research that focuses on the interaction between MJO and MCC is still not much done that left many questions about the interaction between those systems. Therefore, this study aims to determine the interaction between the MCC and MJO, but is limited to phase 4 [3], so that the results of this study are expected to increase knowledge about interaction between the MCC and MJO and also help to improve forecast and early warning skills to the existence of a potential hydrometeorology disaster. This paper also discusses the evolution of cloud top temperature, precipitation cell evolution, environmental conditions around the occurrence of the MCC system and also thermodynamic conditions before the initiation of MCC, when the MCC initiate and mature also after the MCC dissipate which happened coincides with phase 4 of MJO.

2. Data and Method

Identification of the MCC system will be using Himawari 8 Infrared Channel (IR1/B13) satellite imagery data obtained from Kochi University which can be accessed freely at the website address http://weather.is.kochi-u.ac.jp/sat/ in the form of portable gray map (.pgm) data and its calibration in the form of (.dat) format. The satellite data has a horizontal resolution of 0.05° × 0.05° and temporal resolution of 60 minutes. The MCC identification technique used Ismanto's [5] identification method. This method selected cloud shield (SA) and interior cloud (IA) temperature as Maddox [6] definition and Machado technique (least square) to determine the quasi-circular shape of the MCC system. Next is to do a selection of the system by eliminating system that can't last up to 6 hours. The resulting output is the time data (date and time of occurrence of the system) and MCC identification location (latitude, longitude, pixels). The strength of the convective activity that occurs during MCC will be analysed using Convective Index (CI) analysis. This analysis will separate between surface temperatures and high convective clouds through threshold based on to the research of Adler and Negri [7]. The temperature of 253 K as the warmest brightness temperature (T_{BB}) to determine the clouds which allegedly result in the rain. The greater CI values, the stronger the convective activity occurs. If T_{BB} is smaller than threshold value so CI can be determined by reducing the threshold with the value of T_{BB} (T_{BB} <threshold, CI = threshold - T_{BB}) and if T_{BB} greater than or equal to the threshold value, CI is 0 ($T_{BB} \geq$ threshold, CI = 0).

Based on the results of this identification, analysis of evolution and propagation analysis of TBB and the precipitation area of the MCC system will be performed. This paper adapted Cotton et al. [8] and Trismidianto et al. [9] technique, the analysis will be focus on 4 phases of life of MCC, that's in the initiation phase, mature phase, decay phase and dissipation phase. The MCC initiation phase is the phase of the MCC system where the cloud system meets the MCC requirements defined by Maddox [6] that's having T_{BB} IA $\leq$ -52 °C, with IA coverage of $\geq$ 50,000 km^2 and T_{BB} SA $\leq$ -32 °C, with SA coverage $\geq$100,000 km^2 and eccentricity $\geq$ 0.7. While the pre-MCC phase is a phase that occurs 3 hours before MCC reaches the initiation phase. The decay phase is the phase where the system no longer qualifies as an MCC and begins to decay. Dissipation phase is the phase where the system has been dissipated. T_{BB} analysis will used Himawari 8 satellite channel IR1 data while rainfall analysis on the system is using Global Satellite Mapping of Precipitation (GSMaP) with a horizontal resolution of 0.1° × 0.1° and temporal resolution of 60 minutes and rainfall observation data obtained from Cilacap Meteorological Station on September 16 to September 17, 2016.

The thermodynamic analysis will be using upper air sounding observation data obtained from the Cilacap Meteorological Station on September 16 to September 17 at 0700 LT (Local Time) with LT is UTC + 7 and 1900 LT. In addition, to show the environmental conditions at the time of the MCC system, this study uses European Center for Medium-Range Weather Forecasts (ECMWF) data with a spatial resolution of 0.125° × 0.125° with parameters such as divergence, vertical velocity, u and v component of wind in the pressure level from 1000 mb to 100 mb. MJO monitoring data is obtained from Australian BoM which can be accessed through http://www.bom.gov.au/climate/mjo/ in the form of RMM1 and RMM2 data.

3. Result and Discussion

3.1. Analysis of the Evolution of MCC during MJO Phase 4

Heavy rains accompanied by strong winds that occurred in the Cilacap region on 16 - 17 September 2016 resulted in a number of losses, including the occurrence of flooding in some location of Sidareja District, such as Gunungreja, Sidareja, and Sidamulya. In addition, landslides also occurred in Bantar and Madura villages, Wanareja District [10]. Cilacap Meteorological Station recorded that the rain on 16 to 17 September 2016 reached 117.2 mm. The heaviest rainfall occurred on 17 September 2016 at 0300 LT until 17 September 2016 at 0700 LT. Based on the RMM MJO index shown in Figure 1 (a) and Figure 1 (b), the heavy rain was coincided with the MJO being active in phase 4. Many of the previous researchers has found that MJO has significant impact on generating heavy rain based on where it active [10, 11, 12].

The axes (RMM1 and RMM2) represent the daily values of the principal components from the two leading modes. The triangular areas indicate the location of the enhanced phase of the MJO, counter-clockwise motion is indicative of eastward propagation. The RMM index indicates a recent strengthening of the signal over the Maritime Continent in Phase 4 of MJO over Indian Ocean on 16 - 17 September 2016.

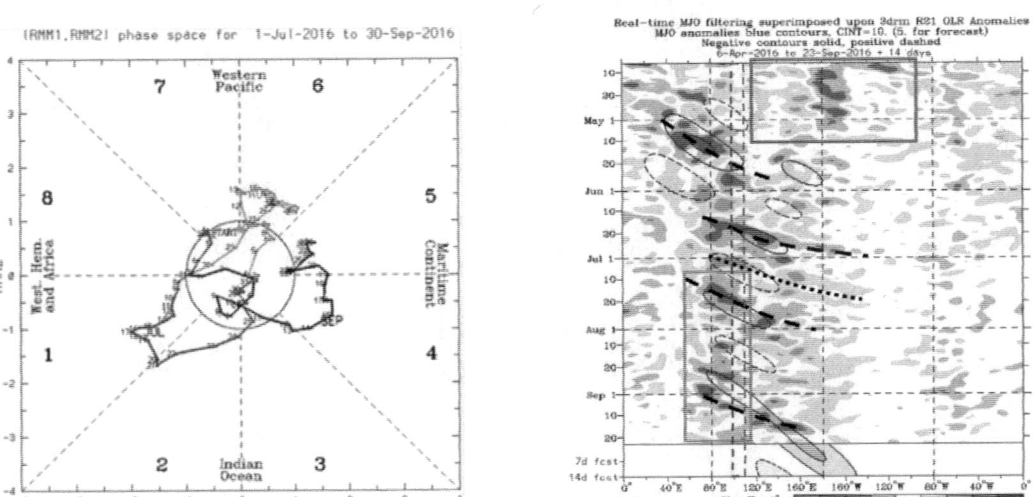

Figure 1. (a) RMM MJO index (left) (source: http://www.bom.gov.au/climate/mjo/#tabs=Time-longitude) and (b) Outgoing Longwave Radiation (OLR) Anomalies (7.5°S - 7.5°N) (source: http://www.cpc.ncep.noaa.gov/products/precip/CWlink/MJO/foregfs.shtml). The dashed red line is longitude of Cilacap.

Figure 1 (b) shows that the 2015-2016 El Niño background state is observed (red box) as a dipole of anomalous convection extending from the Maritime Continent to the East Pacific. The signal weakened steadily through boreal Spring. Several intra-seasonal events were observed during May through July, with other modes such as tropical cyclone activity also influence the pattern. A low-frequency state favoring enhanced convection over the eastern Indian Ocean has been evident since July (green box). This activity is likely related to a negative phase Indian Ocean Dipole event. During September, a fast-eastward moving envelope was evident, likely linked to intra-seasonal activity. More recently, the OLR pattern became increasingly incoherent. This reinforces that in September there was MJO phase 4, but it is interesting to see that from the beginning to mid-September there was significant convective activity. Satellite imagery shows that there is a convective system that last for more than 6 hours around Cilacap. Based on the identification, it is known that the system is the MCC system as shown in Figure 2.

Cotton et al. [8] have defined eight stages in the life cycle of an MCC: MCC-12 h, pre-MCC, initial, growth, mature, decay, dissipation, and post-MCC; however, the most important period for an MCC is from the initial to the dissipation stage. The MCC began to develop from the pre-MCC stage at 1800 LT on 16 September 2016. At that time, small-scale clouds were located over Java, as shown in Figure 1 (a). The system, which was initially monitored as a number of small convective cells, continues to develop into a system of large size and meets all criteria to be categorized as an MCC system before the system is finally dissipated. The MCC system initiated in the mainland at night local time. The system is at night, based on research conducted by Gray and Jacobson [13] in Laing and Fritsch [14] stated that the growth of the system at night is related to differences in solar radiative heating between the MCC system and the surrounding environment, which results in the subsidence of environmental peripherals and the convergence of the lower layers to the system. During the initial stage, deep convective seem over the Indian Ocean and several small clouds in the Java Island.

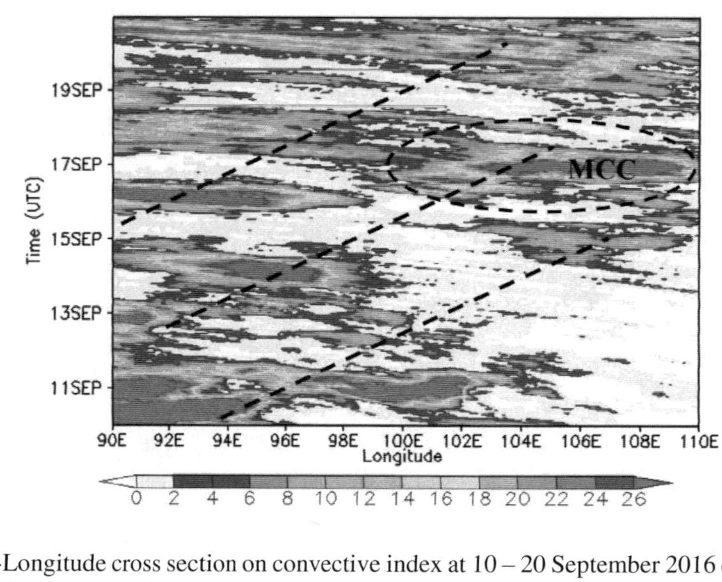

Figure 2. Time-Longitude cross section on convective index at 10 – 20 September 2016 during MJO phase 4 event and MCC system event.

During the growth stage of the MCC in the midnight, the sizes of the deep convective and small clouds had increased further, and they began merging with each other, such that the maximum extent of the MCC was attained at 0300 LT (Figure 3 (e)). At 0800 LT, during the decay stage in the morning, the MCC began to split and dissipated (Figure 3 (h)). During the dissipation stage in the late afternoon

(1400 LT) (Figure 3 (k)), the MCC had split into small-scale clouds that propagated southward toward the Indian Ocean near the Java Island. The system has been completely dissipated and in the surrounding environment, a new convective system begins to form. The MCC system has a long-life duration that's around 20 hours because the system has an extensive SA coverage area around 373,244 km^2 and the IA coverage area of the system is about 304,768.75 km^2.When the system reaches mature phase and the dissipation phase it formed a new convective system around the Java Sea region and in the southern area of Indian Ocean near East Java. The system is formed due to cold pool activity [4, 5, 9]. Based on the results of the analysis of the time series evolution of MCC which was formed on September 16, 2017, this study supports the results of previous studies conducted by Laing and Fritsch [14] on the global population of MCC. The results of research conducted by Laing and Fritsch [14] showed that the MCC system formed on land with large cold cloud cover has a long-life duration.

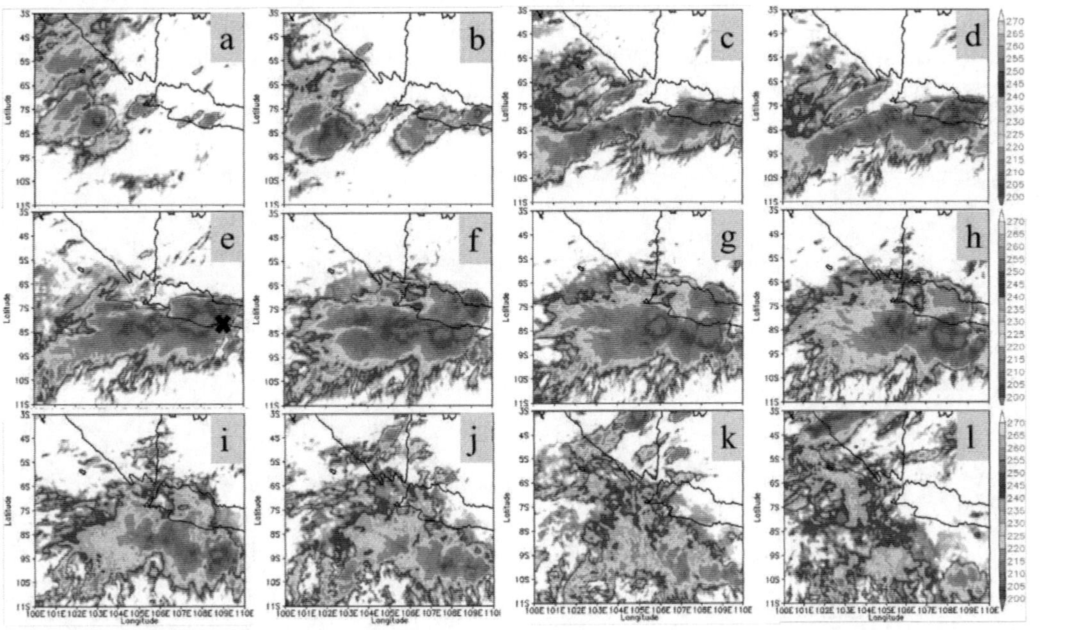

Figure 3. Horizontal distribution of T_{BB} for MCC criteria from infrared data obtained by Himawari 8 on 16 September 2016, (a) 1700 LT (initial phase) , (b) 2100 LT, (c) 0000 LT, (d) 0100 LT, (e) 0300 LT (maximum phase), (f) 0600 LT, (g) 0700 LT, (h) 0800 LT (decay phase), (i) 1000 LT, (j) 1200 LT, (k) 1400 LT (dissipation phase), (l) 1700 LT. Unit for T_{BB} is kelvin. Black cross sign in Figure 3 (e) is representative to location of Cilacap Meteorology Station.

3.2. *Influence of the MCC to the Rainfall*

Figure 4 shows the horizontal distribution of rainfall during the studied MCC events. Previous research has been conducted on the impact of the MCC system on rainfall in the surrounding area. In Reynolds [15], based on Fritsch [16] research shows that the MCC system is the most productive rainfall generation system. The comparison between Hurricane Alicia and the three MCC systems that occurred earlier shows that the MCC system was able to produce twice as much rainfall as Hurricane Alicia. In the dry season, the occurrence of the MCC system often results in inaccurate quantitative predictions of rainfall. To better understand the effect of the MCC system on precipitation in the surrounding area, this section will explain the evolution of the MCC system precipitation region using GPM GSMaP data.

The result of analysis of evolution precipitation cell is that the system was identified as the first storm on September 16, 2016 at 0000 LT, the system precipitation area is limited. The system, which was originally centered in the Sundanese Strait region to the south of the island of Java as propagate to the northeast and continues to grow. Maximum rainfall recorded by Cilacap Meteorological Station is about 108 mm/3 hours from September 16 at 0500 until September 0700 LT. This is support previous research conducted by Kane [17] which states that the heaviest precipitation occurs in the first half of the life phase of the MCC convective system.

After reaching maximum precipitation activity, the MCC system began to be decay. The precipitation area will become narrow, the system begins to propagate to the southern area of Indian Ocean region in the central Java and finally dissipated around there. System dissipation occurred on September 17, 2016 at 1000 LT. After the decay phase, then the system will reach the dissipation phase. In this phase the system has been dissipated and the environment has formed a new convective system. This phase occurs on September 17, 2016 at 1900 LT. Analysis of time series evolution of MCC precipitation cell also provides information that as the MCC system develops and propagates towards the Indian Ocean region, formed new precipitation cell in the waters southern area of the Indian Ocean south of eastern Java. The formation of these new convective cells is related to the cold pool activity [4, 5].

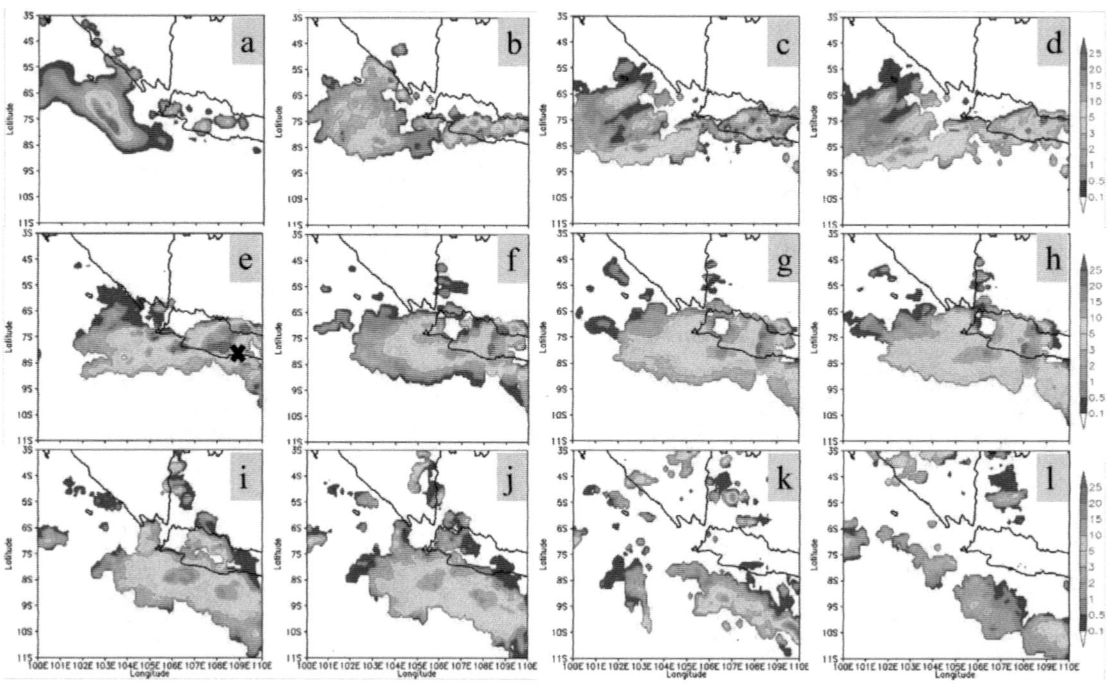

Figure 4. Horizontal distribution of rainfall from GSMaP data obtained during MCC event on 16 September 2016, a) 1700 LT (initial phase) , (b) 2100 LT, (c) 0000 LT, (d) 0100 LT, (e) 0300 LT (maximum phase), (f) 0600 LT, (g) 0700 LT, (h) 0800 LT (decay phase), (i) 1000 LT, (j) 1200 LT, (k) 1400 LT (dissipation phase), (l) 1700 LT. Unit for rainfall is mm/hr. Black cross sign in Figure 4 (e) is representative to location of Cilacap Meteorology Station.

Figure 5 (a) shows the rainfall accumulation during the MCC events from the Pre-MCC stage until 6 hours after the post-MCC stage for all of the case study. The greatest rainfall accumulation is presented in the center of MCC system area for the oceanic, continental and coastal MCCs. It indicated that the significant influence of MCC to the rainfall is during MCC reached maximum extent in the mature stage. However, the high rainfall accumulation also appears in several regions near the MCC system area. It indicated that the effect of MCCs are not only over the MCC system area but also influencing to the rainfall in the surrounding area of the MCC systems. In this research, the area in the surrounding of MCC systems that get affected by the MCC. The influence of MCC on significant rainfall is also supported by the existence of MJO phase 4. Figure 5 (b) shows that there is a significant increase in rainfall when phase 4 active MJO is followed by the MCC. The value of rainfall intensity when MJO followed by the presence of MCC is greater than in the absence of MCC.

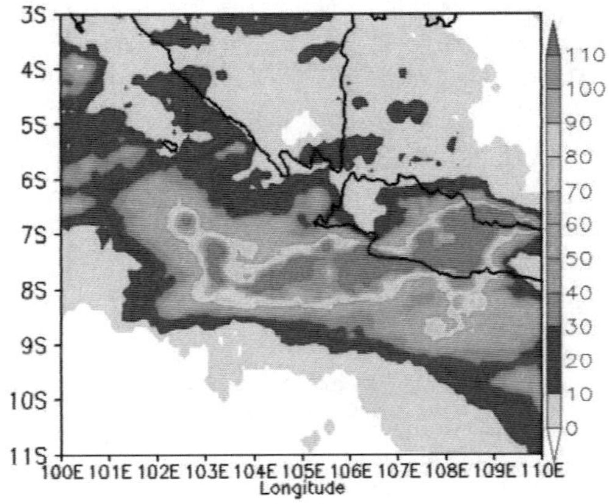

(a)

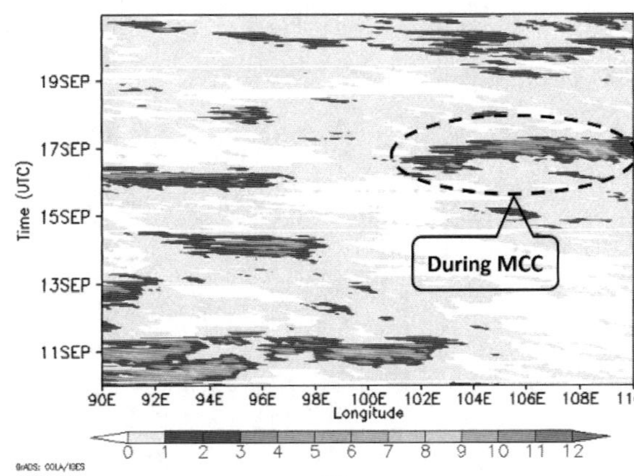

(b)

Figure 5. (a) Rainfall accumulation during MJO event on 16 – 17 September 2016, and (b) Time-Longitude cross section on rainfall at 10 – 20 September 2016 during MJO phase 4 event and MCC system event. Unit for rainfall in Figure 5(a) is mm/2 day, while in Figure 5(b) is mm/11day.

3.3. *Analysis of the Meteorological Condition during MCC*

The development of MCC in Cilacap region can't be separated from the influence of the environmental conditions around the system where it was formed. The MCC was formed on September 17, 2016 at 0000 LT. The wind comes from several directions, that is western winds and northeasterly winds that originate from the Indian Ocean region and eastern winds originating from the Pacific Ocean, the wind was convergence around south western area of Cilacap. The MCC initiation phase is also characterized by an increase in potential temperature around the IA of MCC. These environmental conditions will support the MCC system convective process.

In the mature phase, there was occurred a shear line and convergence of wind (Figure 6 (b)). In the western area of the MCC system there is a weak divergence pattern and the potential temperature around the system has decreased. This is related to the formation of cold pools during the mature phase of MCC. Another mature phase characteristic that can be captured in the MCC case in Cilacap is that there is a large area of precipitation and the development of new convective system around the Java Sea region and the southern area of Indian Ocean of near East Java. The existence of this cold pool is in accordance with the research by Fritsch and Brown [18] and Trismidianto et al. [4].

When entering the decay phase at 1000 LT, the system has propagated to the Indian Ocean region (Figure 6 (c)). The potential temperature around the MCC system is lower than the surrounding environment. The wind convergence pattern is still formed, but the system that has propagated from the convergence region. Another thing that marks the decay phase of the MCC system is that the convective cloud begins to decay so that the area of precipitation becomes smaller and the formation of a new convective system due to cold pool activity. The system propagated from the convergence region causes the system lose water vapour supply from the lower layers and the stability of the air of the region is more stable [6] so the system cannot evolve anymore. The dissipation phase of the MCC system occurred at 1900 LT on 17 September 2016. This phase is marked by the system that already dissipated and the convective system that is induced by cold pools starting to grow and propagate (Figure 6 (d)).

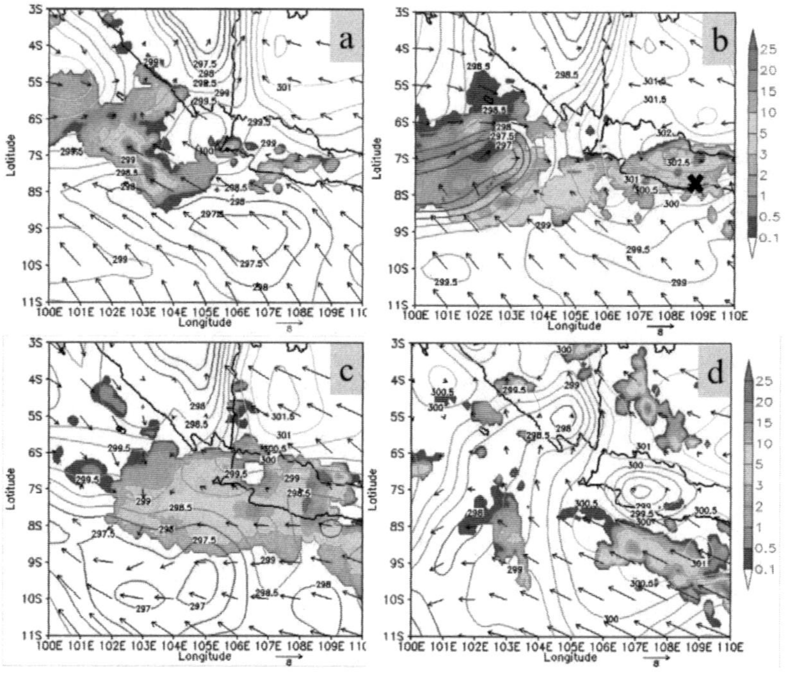

Figure 6. Horizontal distribution of rainfall from GSMaP data (mm/hr, shaded), potential temperature (K, contour), wind vector (m/s, vector) in the environment around MCC in each phase of life of the MCC system on September 16 - 17, 2016. (a) Initiation Phase (1900 LT) (b) Mature phase (0100 LT) (c) Decay phase (0700 LT) and (d) Dissipation phase (1300 LT). Black cross sign in Figure (e) is representative to the location of Cilacap Meteorology Station.

3.4. Thermodynamic Condition

In the research about convective cloud, knowing the condition of atmospheric stability is very important. Atmospheric stability conditions can provide information about the potential of thick and towering convective cloud formation. To find out more about the thermodynamic conditions at the time of the occurrence MCC, in this study will use upper air observations data from Cilacap Meteorological Station. Based on radiosonde observations dated September 16, 2016 and September 17 at 0700 LT and 1900 LT that resulted atmospheric stability index, some of them were listed below (Table 1).

The data shows the stability index value at the time of MCC occurring on September 16-17, 2016 at 0700 LT and 1900 LT, there is an increase in atmospheric lability from September 16 at 0700 LT until September 17, 2016 at 0700 LT. It can be seen by the increase of CAPE (Convective Available Potential Energy), SWEAT (Severe Weather Threat Index), and Total Totals Index and the decrease of Showalter Index, Lifted Index and CIN. This is also accompanied by the increase of possibility of Ts, that can be seen on the increase of K Index. When the MCC reach its mature stage upper air data can't get K Index data. On September 17, 2016 at 1900 LT it appeared that the value of atmospheric stability began to increase, it means that the atmosphere began to stabilize, that showed by the decrease of CAPE Index, SWEAT and Total Totals Index also the increase of Showalter Index, Lifted Index and CIN. Based on these data it can be seen that the most unstable condition of the atmosphere happened on September 16 at 1900 LT where the system is still in the form of a small convective cell. These atmospheric conditions support convective cell growth into a larger system. At 0700 UTC the

IOP Conf. Series: Earth and Environmental Science **303** (2019) 012063 doi:10.1088/1755-1315/303/1/012063

system begins to enter pre-MCC phase and at 1700 LT the system starts to begin to enter the initiation phase. On September 17 at 1900 LT the atmospheric stability index value showed that atmospheric conditions became more stable, at the same time the MCC system had been dissipated.

Table 1. Stability Index from Upper Air Sounding Cilacap Meteorological Station.

	16 September 2016 (J/kg)		17 September 2016 (J/kg)	
Stability Index	0700 LT	1900 LT	0700 LT	1900 LT
Showalter Index (SI)	-0.20[c]	-1.59 [c]	-1.59 [c]	1.01[b]
Lifted Index (LI)	-1.98[b]	-2.85[c]	-0.59[b]	-0.04[b]
SWEAT Index	196.5	221.9	255.4	218
K Index (KI)	32.60[c]	34.70[c]	-	21.90[b]
Totals Total Index (TT)	44.10[b]	45.70[b]	45.80[b]	42.00[a]
CAPE	464.4	794.6	58.61	89.04
CIN	-0.13	-5.69	-40.1	-6.52

[a] Stable.
[b] Possible Thunderstorm (Ts), strong trigger needed
[c] Unstable, Ts probability
[c] Extremely Unstable.

In the initiation phase of the MCC system (Figure 7 (a)), shows that there is convergence in the lower layers of the troposphere, ie at layers of 1000 hPa to 800 hPa, convergence with strong intensity is found in layers 1000 hPa to 900 hPa. there is a divergence pattern in the middle layer between 800 hPa and 300 hPa layer, however, a strong divergence is split into two parts, that's on layers of 800 hPa to 700 hPa and layers of 350 hPa to 250 hPa. While in the upper layer there is strong convergence and divergence, strong convergence in the upper layer occurs in layers of 200 hPa to 100 hPa. While moisture shows a moist layer (RH 80%) from layers 1000 hPa to layer 925 hPa. Then RH decreases with increasing altitude, up to 300 mb the humidity is 40%. Above this layer, RH will increase again to the 100 hPa layer.

When the MCC system reaches the mature phase (Figure 7 (b)), the convergence of the lower layer becomes wider and stronger, namely between layers of 1000 hPa to 700 hPa. Strengthening the lower layer convergence is also followed by the increase of the lower layer of moisture. The RH layer 1000 hPa reaches 90% while the 850 hPa layer reaches 65%. While the layer above is 650 hPa to 100 hPa layer, the air is dominated by strong divergence. Strong divergence is in the 700 mb layer to 400 mb layer and in the high layer between 300 hPa to 100 hPa layer. In the divergence pattern, the humidity decreases to only 40%. While in the 700 hPa layer up to 500 hPa layer convergence is formed but not so strong.

When the MCC system reaches the mature phase (Figure 7 (b)), the convergence of the lower layer becomes wider and stronger, between layers of 1000 hPa to 700 hPa. Strengthening the lower layer convergence is also followed by the increase of the lower layer of moisture. The RH layer 1000 hPa reaches 90% while the 850 hPa layer reaches 65%. While the layer above is 650 hPa to 100 hPa layer, the air is dominated by strong divergence. Strong divergence is in the 700 hPa layer to 400 hPa layer and in the high layer between 300 hPa to 100 hPa layer. In the divergence pattern, the humidity decreases to only 40%. While in the 700 hPa layer up to 500 hPa layer convergence is formed but not so strong.

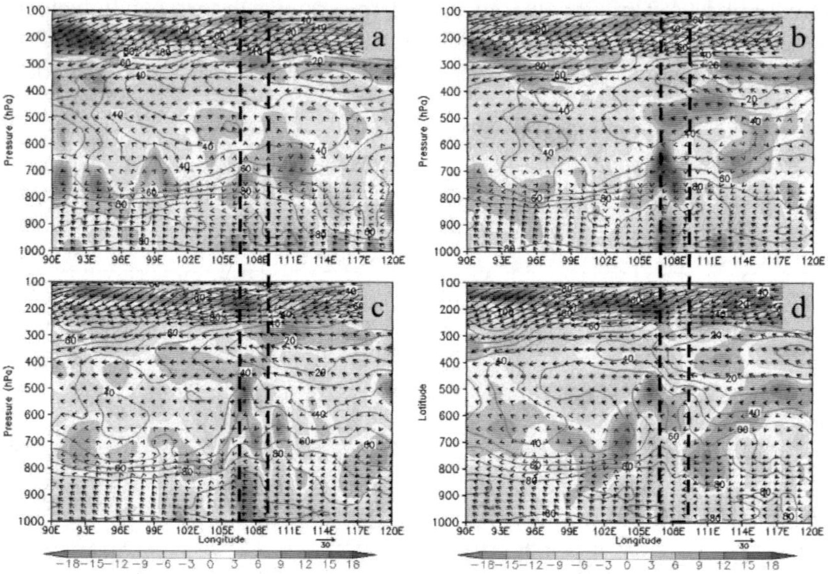

Figure 7. Pressure-Longitude cross section (by averaging latitude 11 ° S - 5 ° S) for divergence (10^{-5} s^{-1}, shaded), relative humidity (%, contour) and wind direction (m/s, vector) in the environment around MCC in each phase of life of the MCC system on September 16 - 17, 2016. (a) Initiation Phase (1900 LT), (b) Mature phase (0100 LT), (c) Decay phase (0700 LT) and (d) Dissipation phase (1300 LT). Black dashed box is representative to the location of MCC Occurrence.

The system reaches dissipation phase (Figure 7 (d)), it appears that the lower to middle layers are still dominated by divergence. The bottom layer convergence is only formed from layers of 1000 hPa to 900 hPa. In the middle layer there is convergence from layers of 900 hPa to 500 hPa. The next layer is dominated by convergence, which is from the layer 500 hPa to 300 hPa. While in the upper layer, that is at the 250 hPa layer to 100 hPa layer there is a strong divergence. RH layer is 1000 hPa to 800 hPa layer moist, but in the 750 hPa layer to 500 hPa layer has a low RH that is 40%. In the next layer, layer 400 hPa to 100 hPa of moisture will increase to 80%.

The next discussion is about the pressure longitude cross-section between vertical velocity, divergence and wind direction in each phase of the MCC system life. The discussion needs to be done in order to obtain an overview of vertical air motion conditions, divergence and wind direction in each phase of the life of the MCC system. By this analysis, it will give us a better picture of the condition of the MCC system in each phase. In the initiation phase (Figure 8 (a)), it appears that the air around the central region of the formation of IA MCC is dominated by the presence of upward. The upward motion last from 1000 hPa to 100 hPa layer, the strongest upward vertical motion can be in the layer 1000 hPa to 700 hPa where the air convergence also occurs. At layers 600 hPa to 100 hPa, there is air divergence with the strongest downward motion in the 400 hPa to 200 hPa layer. Upward vertical motion and air downward help the development of the MCC to maintain the system.

The mature phase of MCC (Figure 8 (b)), is marked by expanding upward from layers 1000 hPa to 100 hPa layer. Widespread upward is in accordance with the strengthening of the lower layer convergence and the weakening of the divergence in the upper layer. In the mature phase, the MCC system is dominated by strong upward air motion. In the decay phase (Figure 8(c)), it appears that there is a downward layer in the middle layer, ie at layers of 800 hPa to 400 hPa. In addition, there is an increase in air divergence activity. Strong divergence occurs in layers of 200 hPa to 100 hPa. Upward still survives from 1000 hPa to 150 hPa. The increase in downward and divergence can help the MCC system decay process. In the dissipation phase of the MCC system (Figure 8(d)), the air is dominated by the downward from 1000 hPa to 450 hPa. The strongest downward occurs in the middle layer, starting from 800 hPa to 450 hPa. Upward still occurs in the lower layer but occurs in very narrow layers, while in the upper layer between layers 400 hPa to 150 hPa occurs upward strong. Divergence in the lower layers and upper layers still occurs but in the middle layer, there is strong convergence.

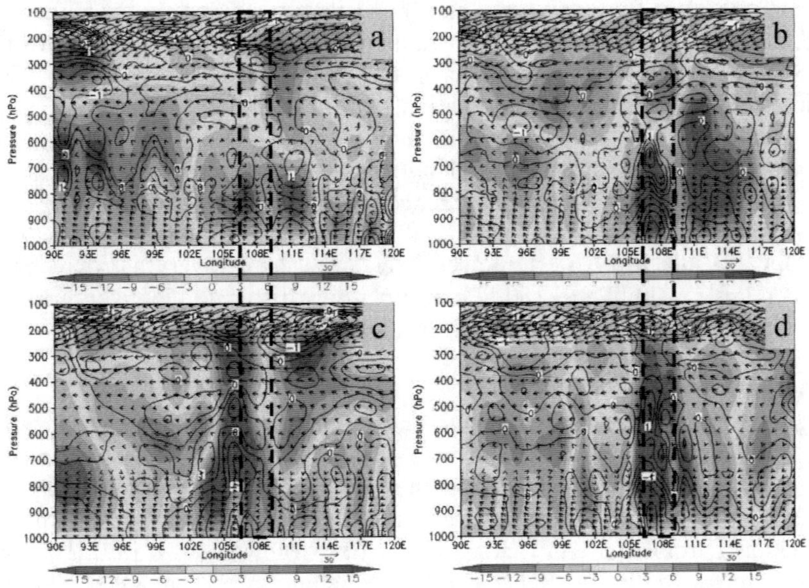

Figure 8. Pressure-Longitude cross section (by averaging latitude 11 - 5 ° S) between vertical velocity (10^{-2} Pa s^{-1}, shaded), divergence (10^{-5} s^{-1}, contour) and wind direction (m / s, vector) in the environment around MCC in each phase of life of the MCC system on September 16 - 17, 2016. (a) Initiation Phase (1900 LT) (b) Mature phase (0100 LT) (c) Decay phase (0700 LT) and (d) Dissipation phase (1300 LT). Black dashed box is representative to the location of MCC Occurrence.

4. Conclusion

The formation of MCC in the Cilacap on September 16-17, 2016 impacted to the occurrence of heavy rain accompanied by strong winds and a number of other hydrometeorological disasters. Based on the analysis that has been done it is known that the system is a land system which then propagates to the Indian Ocean region in southern Java. The MCC system begins to enter the initiation phase on September 16 at 1500 LT and reaches the mature phase on September 17 at 0700 LT. The MCC system reaches the decay phase at 1000 LT and dissipated at 1900 LT. Overall the MCC system has a duration of life about 20 hours.

The suitable atmospheric condition will support the growth of convective cell into larger convective system. The initiation phase of the MCC system begins with the formation of one or more small convective systems that can continue to evolve into larger systems because they are supported by the surrounding environment including unstable atmospheric conditions and allowing Ts. There is a disturbance in the form of convergence patterns around the northwest of the Cilacap region. Based on the pressure longitude cross section it is known that in the pre-MCC phase until the initiation phase there is a strong convergence in the lower layer. In general, the convergence of the lower layer is in the layer 1000 hPa to 800 hPa. In the mature phase of the MCC system, there is still support from the surrounding environment and the wind convergence area is still formed. The results of the analysis of the pressure longitude cross section show that in the mature phase the convergence of the lower layer strengthens, namely between the layers of 1000 hPa to 700 hPa. When the system enters the decay phase, the system no longer receives support from the surrounding environment. This is indicated by the propagation of the system towards the Indian Ocean region away from the convergence pattern with the stability of the atmosphere more stable.

The influence of MCC on significant rainfall is also supported by the existence of MJO phase 4. The significant increase in rainfall when phase 4 active MJO is followed by the MCC. The value of rainfall intensity when MJO followed by the presence of MCC is greater than in the absence of MCC. Environment conditions when the system begins to enter the decay phase is the formation of the lower layer divergence, that is at 1000 hPa to 800 hPa. In the upper layer, there is also a divergence between layers 300 hPa to 100 hPa. In this phase, it begins to form downward in the middle or lower layers. The last phase in MCC's life cycle is MCC dissipation phase. In this phase the MCC system has been dissipated and the environment has begun to form a new convective system. Pressure longitude cross-section analysis is known that in this phase around the location where the MCC system is formed, a lower layer convergence is formed. This convergence is able to withstand 1000 hPa to 900 hPa layers. Vertical velocity in this phase shows that there is still a downward vertical motion between layers 1000 mb to 450 hPa. Upward still occurs in the lower layer but occurs in very narrow layers, while in the upper layer between layers 400 hPa to 150 hPa occurs upward strong. Divergence in the lower layers and upper layers still occurs but in the middle layer there is strong convergence.

Acknowledgment

Authors wishing to acknowledge assistance or encouragement from colleagues, special work by technical staff or financial support from organizations should do so in an unnumbered Acknowledgments section immediately following the last numbered section of the paper. Thank you so much for the three sources data which we can get for free and the data is very helpful to this research. This research uses data from ECMWF ERA-Interim (http://www.ecmwf.int/), Himawari-8 satellite (http://weather.is.kochiu.ac.jp/sat/GAME) and GSMaP satellite (http://sharaku.eorc.jaxa.jp/GSMaP/).

5. References

[1] Ramage C S 1968 Role of a tropical "maritime continent" in the atmospheric circulation *Mon. Wea. Rev.* **96(6)** 365-70

[2] Madden R A and Julian P R 1972 Description of global-scale circulation cells in the tropics with a 40–50 day period *Journal of the atmospheric sciences* **29(6)** 1109-23

[3] Wheeler M C and Hendon H H 2004 An All-season real-time multivariate MJO Index: Development of an index for monitoring and prediction *Mon. Wea. Rev.* **132(8)** 1917-1932

[4] Trismidianto, Yulihastin E, Satyawardhana H, Nugroho J T and Ishida S 2017a The Contribution of The Mesoscale Convective Complexes (MCCs) to Total Rainfall Over Indonesian Maritime *Continent IOP Conference Series: Earth and Environmental Science* **54(1)** 012027

[5] Ismanto H 2011 *Characteristics of MCC in the Maritime Continent* Thesis of Earth Sciences Master's Degree Study Program. (Bandung: Bandung Institute of Technology)

[6] Maddox R A 1980 Mesocale Convective Complexes *Bulletin of the American Meteorological Society* **(1)** 1374-87

[7] Adler R F, Negri A J 1988 A Satellite infrared technique to estimate tropical convective and stratiform rainfall *Journal of Applied Meteorology* **27(1)** 30-51

[8] Cotton W R, Lin M S, McAnelly R L and Tremback C J 1989 Composite model of mesoscale convective complexes *Monthly Weather Review* **117(4)** 765-783

[9] Trismidianto, Yulihastin E, Satyawardhana H, and Ishida S 2017 A Composite analysis of the Mesoscale Convective Complexes (MCCs) development over the central kalimantan and its relation with the propagation of the rainfall systems *IOP Conference Series: Earth and Environmental Science* **54(1)** 012036

[10] Yulihastin E, Trismidianto, Satyawardhana H and Nugroho G A 2017 MJO modulation on diurnal rainfall over West Java during pre-monsoon and strong El Niño periods *IOP Conference Series: Earth and Environmental Science* **54(1)** 012029

[11] Trismidianto, Hadi T W, Ishida S, Moteki Q, Manda A and Iizuka S 2016 Development processes of oceanic convective systems inducing the heavy rainfall over the western coast of Sumatra on 28 October 2007 *Sola* **12** 6-11

[12] Ji-Hyun O, K-Y Kim, G-H Lim 2016 Impact of MJO on the diurnal cycle of rainfall over the western Maritime Continent Climate Dynamics **38** 5-6

[13] Gray W M and Jacobson Jr R W 1977 Diurnal variation of deep cumulus convection *Monthly Weather Review*. 105 **9** 1171-88

[14] Laing A G, Michael F J 1997 The global population of mesoscale convective complexes. *Quarterly Journal of the Royal Meteorological Society*. 123 **538** 389-405

[15] Reynolds H 1990 Mesoscale Convective Complexes: An Overview *Report*. Master of Science University of Toronto

[16] Fritsch J M, Kane R J, Chelius C R 1986 The contribution of mesoscale convective weather systems to the warm-season precipitation in the United States Journal of climate and applied meteorology. 25 **10** 1333-45.

[17] Kane R J 1985 The Temporal And Spatial Characteristics Of Precipitation From Mesoscale Convective Weather Systems, M.S. Thesis. The Pennsylvania State University, Philadelphia.

[18] Fritsch J M and Brown J M 1982 On the generation of convectively driven mesohighs aloft *Monthly Weather Review* **110(11)** 1554-1563.

Identification of The Influence of Long Winter Season in 2017 – 2018 to the Forming of Strong Northerly Cold Surge over West Indonesia

Y A Nugroho, R Fajarianti, A F R Imami and P A Winarso

Department of Meteorology, Indonesian State College of Meteorology, Climatology and Geophysics, Pondok Aren, South Tangerang, Indonesia

yoshua.meteo@gmail.com

Abstract. This study identifies the characteristic of strong Northerly Cold Surge (NCS) over West Indonesia in the period from November 2017 until April 2018. Usually the NCS occurs in the DJF period that indicates the rainy season that usually goes to trigger some serve weather in Indonesia. The severe weather accompanying the intense northerly cold surge often brings more major disruptions that often trigger natural disasters like floods and landslides to the affected regions in West Indonesia. Descriptive and statistical methods are used in this study by classifying the intensity of NCS in November 2017 until April 2018. The real-time data from Ogimet Weather Information Service, GSMaP Satellite, and restudy data from National Centers for Environmental Prediction (NCEP) are used to determine the 925 millibar meridional wind, relative humidity in 1000, 925, 850, and 700 millibar, mean sea level pressure, and rainfall estimation to identify the passage of NCS in West Indonesia region. The long winter season in 2017 – 2018 triggered the NCS to stay active even it comes to March and April 2018. It is proven by the natural disaster that comes from consecutive intense rainfall like floods and landslides that still intensively occurred in the March until April 2018 in West Indonesia.

Keywords : GSMaP, Northerly Cold Surge, meridional wind, landslides.

1. Introduction

The region of Indonesia is located geographically between two major continents (Asia and Australia) and two oceans (Indian and Pacific), this geographic position gives a huge impact to the weather dynamics in Indonesia [1]. That is why the weather dynamics in Indonesia are mostly influenced by monsoon. Monsoon is triggered by the surplus-deficit of energy between Asia and Australia. When the Asian monsoon that brings a very moist air masses occurs and passes above Indonesian continent, it could trigger the forming of convective clouds become more intense, that mostly lead to more intense rainfall (rainy season). Another hand, when the Australian monsoon that bring dryer air masses occurs and passes above Indonesian continent, it could reduce the concentration of air moisture, that means on that moment a number of convective clouds could be reduced. This deficit of convective clouds would lead into dry season in some region in Indonesia.

When it comes to the winter season in Northern Hemisphere in the South China Sea, usually a huge mass of cold air is pounded from Asia that originally comes from the isolated cold region in Siberia. This condition called Northerly Cold Surge (NCS) [2]. Despite the monsoon, the NCS is also has a significant impact on the weather dynamics in Indonesia, especially in West Indonesia. The huge cold

IOP Conf. Series: Earth and Environmental Science **303** (2019) 012064 doi:10.1088/1755-1315/303/1/012064

air mass pounds the warm air mass in West Indonesia, that forces the warm air mass to lift and leads to the forming of convective clouds that responsible to some intense rainfall events that occur in that region.

In the period between 2017-2018, the winter season in Northern Hemisphere expands until the end of April 2018 [3, 4, 5]. The unusual condition of winter season in 2017-2018 like the increasing of winter storms over Northern Hemisphere to affect aerodrome closed for certain airports in big city of the United States of America and Canada in Northern America Continent, and the freezing water Niagara for the first time during the last 50 years could be additional condition from coolest than normal and it might longer than usual based upon public information and collecting the global climatic chart from National Climate Centers under coordination from the World Meteorological Organization (WMO) [2]. This cold condition could expand over European Countries and North Asian Continent. With the expansion of the winter season, it suspected to be a big factor to the more intense energy of NCS that passes West Indonesia region.

The aim of this research is to identify the impact of the long winter season in 2017-2018 to the forming of strong NCS over Indonesia. We are using some major factors to determine the NCS in West Indonesia: the difference of mean sea level pressure of Gushi-Hongkong in November 2017 until April 2018, 925 millibar meridional wind (V component), 1000, 925, 850, and 700 millibar relative humidity and surface rainfall estimation in South China Sea and West Indonesia region (25°N to -10°S and 95°E to 122.5°E) in November 2017 until April 2018 to determine the passage of the NCS and its impact over West Indonesia.

2. Data and Methods

2.1. Data

We are using the daily mean sea level pressure of Gushi-Hongkong data in November 2017 until April 2018 from Ogimet Weather Information Service; 925 millibar restudy meridional wind (V component) data in South China Sea and West Indonesia region (25°N to -10°S and 95°E to 122.5°E) in November 2017 until April 2018 from National Centers of Environment Prediction (NCEP) with 2.5° × 2.5° data resolution; 1000, 925, 850, and 700 millibar relative humidity data in South China Sea and West Indonesia region (25°N to -10°S and 95°E to 122.5°E) in November 2017 until April 2018 from National Centers of Environment Prediction (NCEP) with 2.5° × 2.5° data resolution; and rainfall estimation data in South China Sea and West Indonesia region (25°N to -10°S and 95°E to 122.5°E) in November 2017 until April 2018 from Global Satellite Mapping of Precipitation (GSMaP) with 2.5° × 2.5° data resolution.

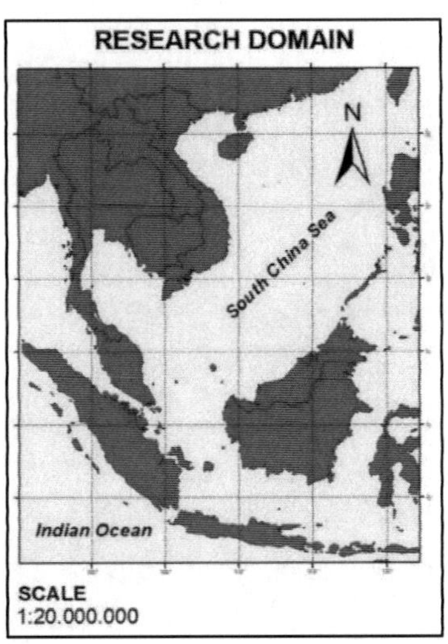

Figure 1. Research domain to determine the NCS in South China Sea and West Indonesia region (25°N to -10°S and 95°E to 122.5°E).

2.2. Method

The method used in the research is descriptive. The emergence of NCS from Siberia is identified from the cold surge index (the difference of mean sea level between Gushi and Hongkong), and then it would be classified in the range of weak (10,0 – 12,9 millibar), moderate (13,0 – 14,9 millibar), and strong (> 15 millibar) indexes [6]. The transport of cold surge from Siberia to South China Sea and finally would come to West Indonesia region can be identified from the meridional wind (V component) study, it would take at least 24 to 72 hours for the cold surge to come to West Indonesia region [6]. The passage of cold surge in West Indonesia would affect the increasing amount of air moisture (relative humidity) from the usual condition that would provide more chances to the forming of convective clouds (that could lead to hydro-meteorological hazards like floods and landslides). The relative humidity study in 1000, 925, 850, and 700 millibar is the key to identify the surplus of air moisture that could be determined as the trigger to the forming of convective clouds in the West Indonesia region when the NCS comes. The dynamics of moisture distribution in the research domain would be compared with the mapping of near real-time daily rainfall data based from GSMaP Satellite to strengthen the evidence of the increasing of the convective clouds formation in the research domain.

3. Result and Discussion

3.1. NCS Index Study (Gushi-Hongkong)

According to the mean sea level pressure difference between Gushi-Hongkong within 24 hours, the indication of NCS is determined when the difference of the pressure is equal or more than 10 millibar that implemented to the NCS index that shown on Figure 2 and 3 [7].

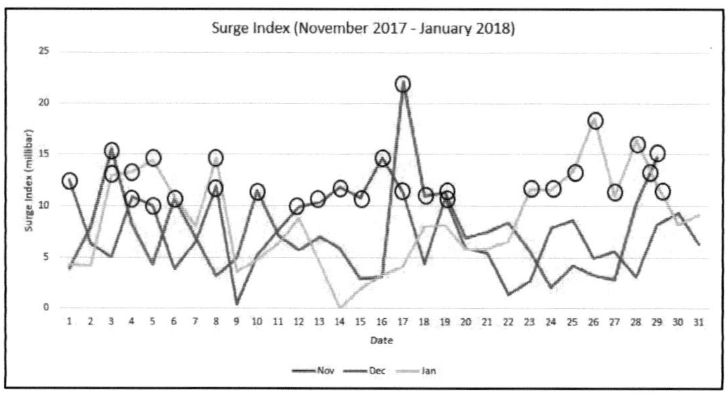

Figure 2. Daily NCS index on November 2017 – January 2018.

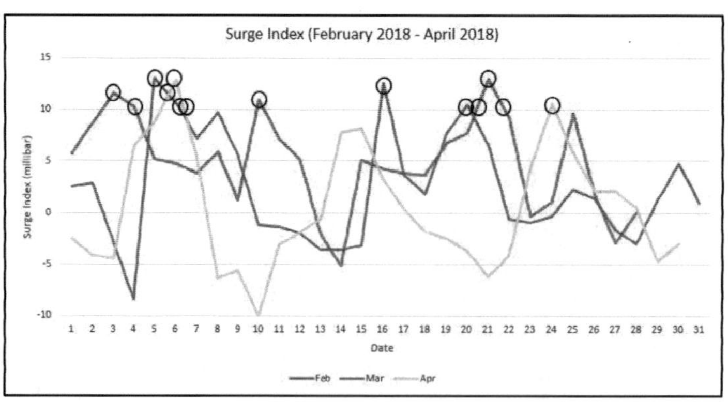

Figure 3. Daily NCS index on February 2018 – April 2018.

On Figure 2 and 3 we can see 8 cases of NCS occurred on November 2017, the month when the Asian Monsoon passes over Indonesia, then 10 cases of NCS occurred on December 2017, and 11 cases of NCS in January 2018. In February 2018 the amount of NCS cases started to decrease to 3 cases and it slightly increased to 4 cases in March 2018 and decreased again in April 2018 with 2 NCS cases. The result of the study is well-corresponded with recent research by Winarso and Prayuda, 2017 [8].

According to [6], in November 2017 there were 5 weak NCS, 1 moderate NCS and 2 strong NCS. In December 2017, there were 9 weak NCS and 1 moderate NCS. In January 2018, there were 5 weak NCS, 4 moderate NCS and 2 strong NCS. In February 2018, there were 3 weak NCS. In March 2018, there were 3 weak NCS and 1 moderate NCS. In April 2018, there were 2 weak NCS.

3.2. Meridional Wind Study

The significant pressure difference between Gushi and Hongkong causing an air that flows from high-pressure area to the low-pressure. The presence of cold advection that brought by the air flow causes the NCS. The NCS can also be identified by analyzing the 925 mb meridional wind [6, 9]. The 925 mb meridional wind is eligible to use for the study because on this level there is no interference to the air flow that caused by surface obstacles (mountains, buildings, etc.) [6].

Based on data study, on November 4th, 2017 the NCS passed Hongkong with the meridional wind speed on the range of 5 – 15 m/s from 15 N – 20 N down to the south, but there has not been any propagation to western Indonesia. This is identified with the meridional wind upon western Indonesia still went to the north with a speed of 0 – 5 m/s. On November 5th, 2017, the surge began to appear to the south with speed of 0 – 5 m/s reaching western Indonesia, especially in most parts of Kalimantan, but there was still a force of the wind towards the north in Indonesian waters (Malacca Strait and Strait Java). The NCS reached the territory of west Indonesia within 48 hours. On November 6th, 2017 ,the wind that carrying cold water vapor reached almost over all-region in western Indonesia.

The activation of NCS on November 7th, 2017 strengthened the spread of NCS until November 10th, 2017 as shown in green color on Figure 4. On November 18th, 2017 a strong NCS and followed by a weak NCS for the next 2 days strenghtening the head south wind over west Indonesia. This consecutive surge makes the propagation period spent only 24 hours to pass the west Indonesia region. After this case, weak to moderate NCS occurred consecutively until the end of the month that caused the wind towards the equator to reach 5 – 10 m/s pulled to the west of Sumatra and southern Java (shown on Figure 5) and the wind rose to a maximum of 15 m/s on November 26th – 28th, 2017 over western Sumatra.

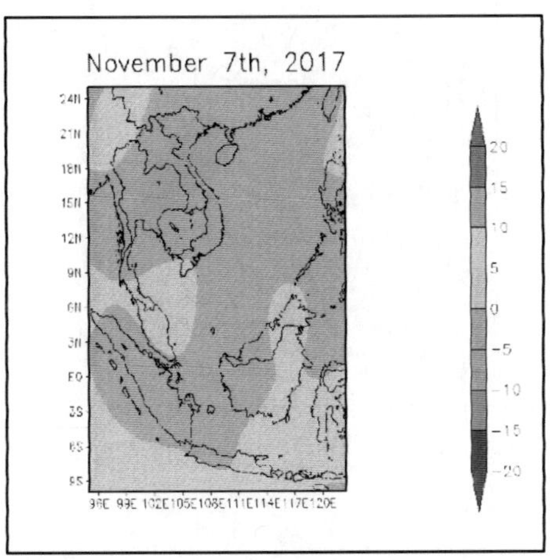

Figure 4. Meridional wind approached the equator.

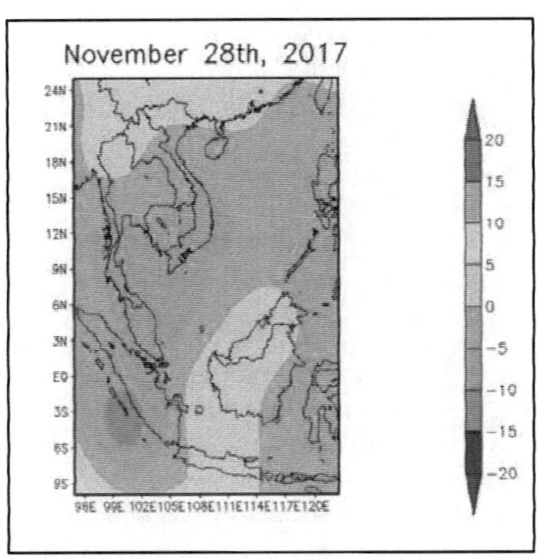

Figure 5. Meridional wind strengthened and pulled to the West Sumatra.

The active weak NCS still occurred on December 1st, 2017 with the surge reaching the South China Sea with meridional wind speed of 5 – 10 m/s but weakening to 0 – 5 m/s when it reached the south of the equator. The NCS was blowing constantly until December 4th, 2017. On December 13th, 2017 an active weak NCS consecutively occurred, this occurrence strengthened the head south wind until the end of December 2017 with the meridional wind speed of 15 – 20 m/s in South China Sea (Figure 6) and 5 – 15 m/s in the west Indonesia (Figure 7).

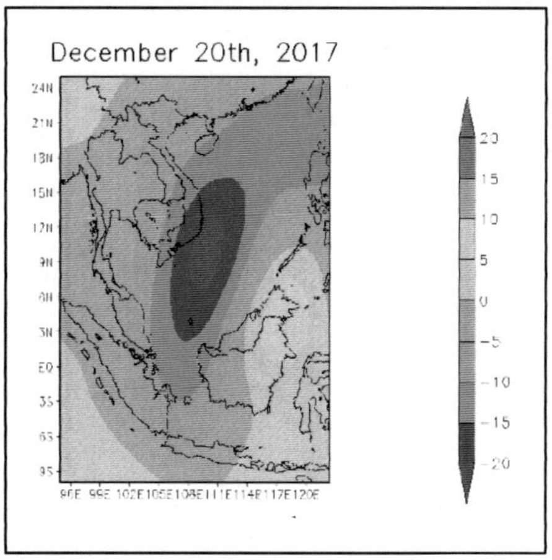

Figure 6. Meridional wind with wind speed of 15 – 20 m/s over South China Sea.

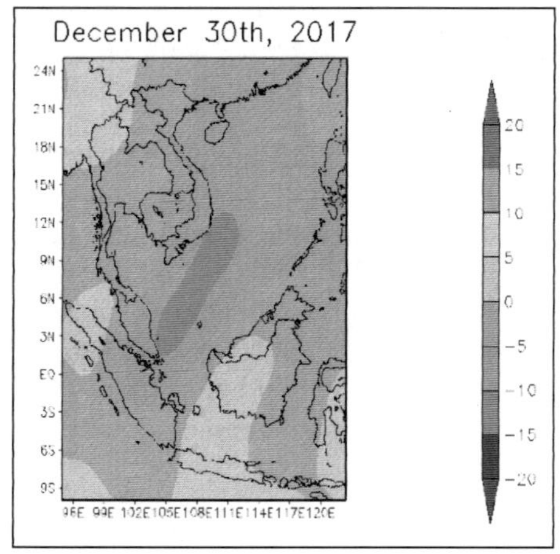

Figure 7. Meridional wind with wind speed of 5 – 15 m/s over west Indonesia.

On January 2018, the frequency of NCS reached its highest peak. The increasing frequency of NCS was supported by the Asian Monsoon is still active above Indonesia. This condition allows the NCS to have a larger influence on the weather dynamics especially over west Indonesia region.

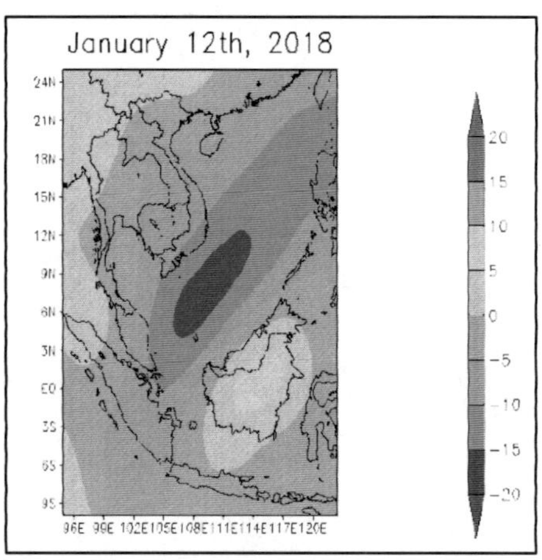

Figure 8. Meridional wind on January 12th, 2018.

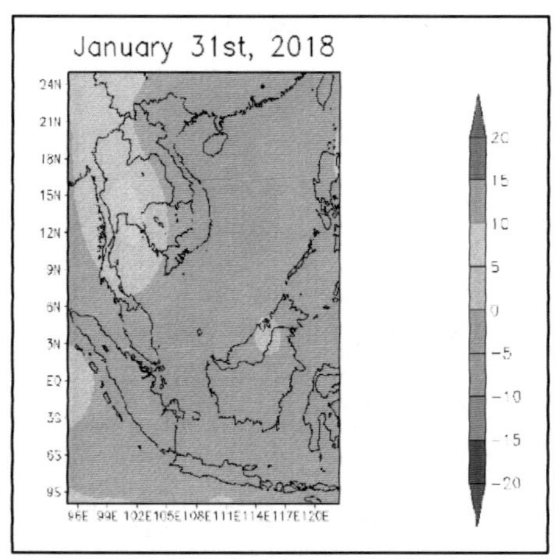

Figure 9. Meridional wind on January 31st, 2018.

On the first 3 days of January 2018, the winds with a speed of 5 – 10 m/s reached South China Sea and then the surge reached west Indonesia with a speed of 0 – 5 m/s. While on January 9th, 2018 the surge has reached the South China Sea and west Indonesia with an average meridional wind speed of 5

– 10 m/s due to the previous active NCS moderate intensity that occurred in those regions. The average wind speed strengthens for the next 3 days with average meridional wind speed of 10 – 20 m/s in the South China Sea. Meanwhile in the north of the equator reaches 5 – 15 m/s and south of the equator reaches 5 – 10 m/s as shown in Figure 8. However, on January 13th, 2018 the speed weakened to 5 – 10 m/s in the South China Sea and western Indonesia, this condition lasts until January 19th, 2018. On January 23rd, 2018 until January 29th, 2018 the active NCS occurred again with weak to strong intensity, the active NCS reaches South China Sea for 72 hours with meridional wind speed of 5 – 10 m/s and 0 – 5 m/s in west Indonesia (Figure 9).

Due to the active influence of NCS on January 31st, 2018 it makes the NCS stay active on February 3 and 4, 2018. This condition increased the meridional wind speed to 10 – 15 m/s in the South China Sea and it lasts until February 9, 2018 with meridional wind speed decreased to 5 – 10 m/s with the narrowing of the affected area (see Figure 10 and 11). On February 10, 2018 the weak NCS has activated again so that it re-establishes the wind speed to become stronger in the South China Sea and Indonesian territory with average meridional wind speed of 5 – 15 m/s with 48 hours of propagation time. The effect of the surge lasted until the end of February with an average meridional wind speed of 0 – 5 m/s, this condition was allegedly due to the synoptic conditions of the South China Sea which supported its effects to last for days.

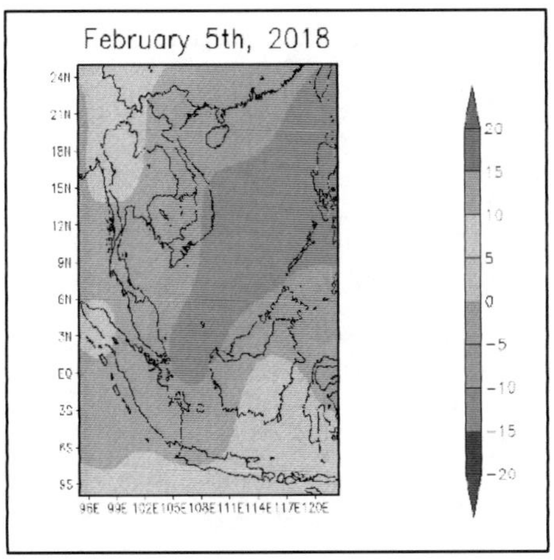

Figure 10. Meridional wind on February 5th, 2018.

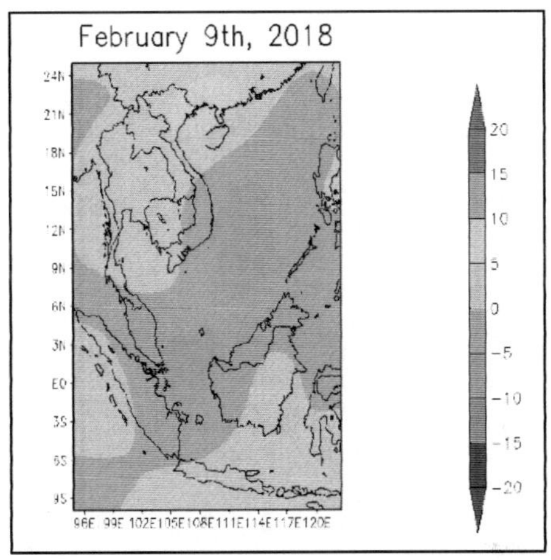

Figure 11. Meridional wind on February 9th, 2018 with the narrowing of the affected area.

On March 9th, 2018, the NCS has activated again and it took 72 hours propagation time (Figure 12) with an average meridional wind speed of 10 – 15 m/s in the South China Sea and 0 – 10 m/s in west Indonesia. The NCS in the area lasted until March 14th, 2018 with weaker wind speed (0 – 5 m/s). On March 20, 2018 the NCS was active and reached the South China sea for 24 hours with meridional wind speed of 5 – 10 m/s and in west Indonesia with a speed of 0 – 5 m/s with some intensity fluctuation until the end of the month.

The Asian Monsoon usually occurred on the December-January-February (DJF) period that usually accompanied by NCS in some certain condition.[10] But in the period of 2017 – 2018 the NCS remains active until April 2018. It could be proved by on April 1st, 2018 the propagation of NCS was still active in the South China Sea and the north of the equator with the meridional wind speed average between 10 – 15 m/s and in the south of the equator 0 – 5 m/s (Figure 13). On April 6th, 2018 the NCS appeared to be active with weak intensity. The propagation took 24 hours to come over in west Indonesia with the average meridional wind speed 0 – 10 m/s (Figure 14). The condition consecutively occurred until April 9th, 2018. On April 24th, 2018 the NCS spotted to be active but the propagation did not reach the region of The South China Sea and West Indonesia.

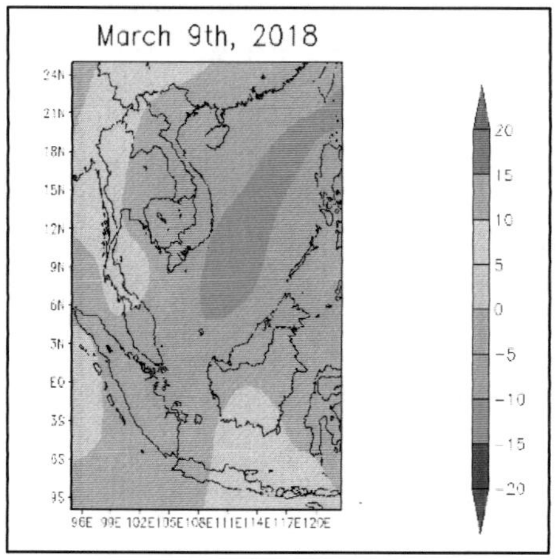

Figure 12. Meridional wind on March 9th, 2018.

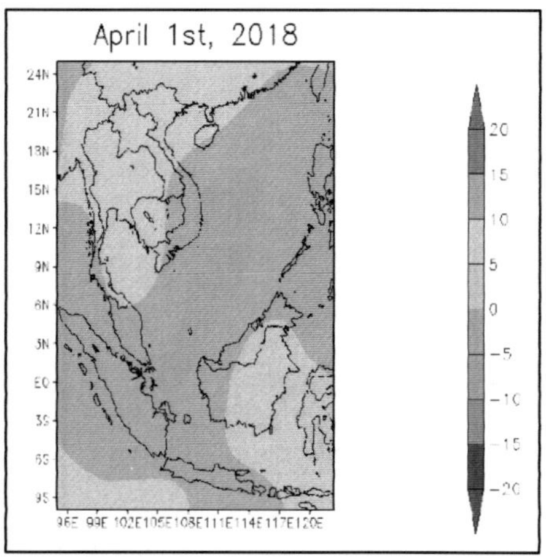

Figure 13. Meridional wind on April 1st, 2018.

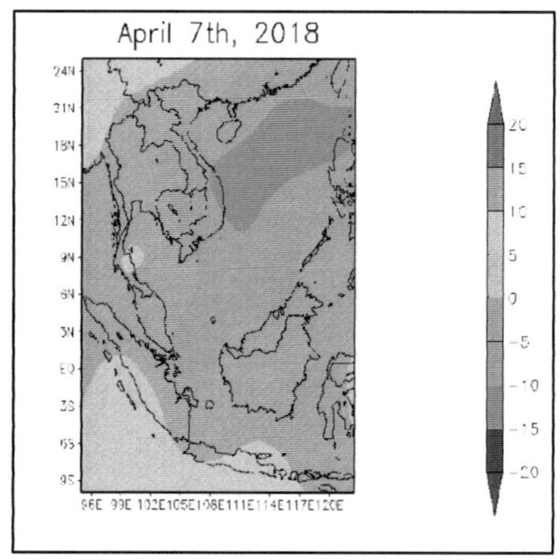

Figure 14. The 24 hours propagation of meridional wind on April 7th, 2018 over South China Sea and west Indonesia.

3.3. Relative Humidity and Surface Rainfall Study

The availability of water vapor is closely related to cloud formation. The availability of a lot of water vapor and an increase in unstable air parcels causes the air mass to continue to move up and carry moisture. The height achieved is associated with the formation of convective clouds that rise high and have the chance to cause sudden heavy rainfall [11]. From several flight paths through Cumulus clouds, it was found that on the outside of the cloud, air usually has a relative humidity between 95% and 100%, then swoops down as low as 70% near the ends of the clouds where turbulent mixing is responsible for the entry of dry air from outside the cloud. In the deeper parts of the cloud, relative humidity is found which is 100% to 107% [12].

Table 1. Relative humidity in west Indonesia on November 2017 – April 2018

Year	Month	Date	Relative Humidity (%)			
			1000 mb	925 mb	850 mb	700 mb
2017	November	4	80-100	70-90	70-90	60-100
		7	80-100	70-90	80-100	70-100
		11	80-100	70-90	80-90	70-90
		18	80-100	80-90	80-100	50-80
		19	80-100	80-90	80-90	70-90
		20	80-100	70-90	80-90	70-90
		29	80-100	80-100	80-100	70-90
		30	80-100	80-100	70-90	60-90
	December	1	80-100	80-100	80-100	60-90
		4	80-100	80-100	70-90	60-90
		5	80-100	70-90	70-90	60-90
		8	80-100	70-90	70-90	70-90
		13	80-100	80-100	80-100	70-100
		14	80-100	70-90	70-90	60-90
		15	80-100	70-90	70-90	50-80
		16	80-100	70-90	70-90	50-90
		17	70-100	70-90	70-90	60-90
		19	80-100	80-100	70-90	60-90
2018	January	4	90-100	70-90	70-90	60-90
		5	90-100	70-90	60-90	60-90
		6	90-100	60-90	60-90	50-90
		8	90-100	70-90	70-90	70-90
		23	90-100	80-100	70-90	60-90
		24	90-100	80-100	70-90	60-80
		25	90-100	70-90	60-80	50-80
		26	90-100	60-90	60-80	40-80
		27	80-100	60-80	60-80	40-80
		28	80-100	60-80	60-80	40-80
		29	80-100	60-90	60-80	40-80
	February	3	80-100	70-90	60-90	70-100
		4	80-100	80-100	70-100	70-100
		10	70-90	60-80	70-90	60-90
	March	5	80-100	70-90	80-100	70-100
		6	80-100	70-90	80-100	70-100
		16	80-100	60-80	60-80	40-80
		20	80-100	60-80	70-90	60-90
	April	6	80-100	70-90	70-90	60-100
		24	80-100	70-90	70-100	60-100

IOP Conf. Series: Earth and Environmental Science **303** (2019) 012064 doi:10.1088/1755-1315/303/1/012064

Based on the relative humidity data that has been processed, the relative humidity in the period between November 2017 until April 2018 is relatively high on the average of 70 – 90% on each layer. The surplus of water vapor in the atmosphere would make the production of convective clouds increased that usually leads to intense rainfall. It could be proven by the study from the surface rainfall study that shows the distribution of rainfall on the period from November 2017 to April 2018 with the estimation in west Indonesia is around 5 – 35 mm/day on November 2017. While on December 2017 the rainfall estimation is on between 5 – 25 mm/day but in some areas in the west coast of Sumatra the rainfall estimation reached 20 – 35 mm/day. In January 2018 the rainfall estimation is between 5 – 25 mm/day. On February 2018 the rainfall estimation is between 0 – 15 mm/day.

With the extension of the winter season in the Northern Hemisphere, the NCS occurred until early April 2018, NCS in March was identified on March 5th, 2018 with moderate intensity, followed on March 6th, 2018 with weak intensity. On March 5th, 2018 the relative humidity in the South China Sea of the 1000 mb layer was 90 – 100% and continued to expand to Natuna waters on March 6th, 2018, on March 8th – 11th, 2018 the relative humidity increased to 80 – 90%, on March 12th – 14th, 2018 humidity increased to 90 – 100% and extends to Kalimantan and Sumatra; at layer 925 mb the relative humidity is on 60 – 80%, then on March 9th, 2018 relative humidity increased to 80 – 90% in the South China Sea, and extended to Kalimantan, southern Sumatra and parts of Java until March 12th, 2018; layer 850 mb humidity 70 – 90%, but on March 7th – 10th, 2018 it drops to reach 40 – 60%, then rises with variations of 60 – 90% on March 12th, 2018; the 700 mb layer of relative humidity varies between 40 – 70% and decreases up to 20 – 50% until March 11th, 2018, and on March 12th, 2018 the humidity rises again to 40 – 70%. On this period the rainfall estimation is between 0 – 20 mm/day.

NCS in April was identified on April 6th, 2018 with a weak intensity. Relative humidity in the South China Sea at layer of 100 mb was 80 – 100%, and rising on April 10th, 2018 reached 90 – 100%, and then dropped to 80 – 90% on April 16th, 2018; at layers of 925 mb 50 – 80% relative humidity, increases with 80 – 100% variation on April 8th – 9th, 2018, and drops to 70 – 90% on April 10th, 2018; in the 850 mb layer the relative humidity varies between 40 – 70%, rises on April 8th, 2018 to 60 – 100%, and goes down again on April 10th, 2018 with relative humidity variations of 60 – 80%; while at the 700 mb layer the humidity reaches 30 – 50% going to April 9th, 2018 then rises to 40 – 70%. On this period the rainfall estimation is between 0 – 20 mm/day.

4. Conclusion

Based on the study, the NCS period in 2017 – 2018 has been expanded until early April 2018 due to the unusual winter season in the 2017 – 2018 that also expanded until April 2018 in the Northern Hemisphere. The extended passage of NCS dominantly affected the atmosphere dynamics over West Indonesia. It proven by the huge number of water vapor brought by meridional wind (NCS) that still spreading widely from north equator to the West Indonesia even though it comes to March – April period that in the normal condition on that period the water vapor concentration should be decreased due to the dry air from Australia that should flows from south equator started to pass over West Indonesia (Australian Monsoon).

5. References

[1] Tjasyono B, Gernowo R, Harijono S W B and Juaeni I 2008 The character of rainfall in the Indonesian monsoon *Symposium on Equatorial Monsoon System* Yogyakarta

[2] Winarso P A 2018 Impact long term winter season 2017/2018 over Indonesia Maritime Continent area *Recent Advances in Petrochemical Science* **4** Issue 4

[3] https://weather.com/storms/winter/news/2018-04-25-winter-storm-takeaways-2017-2018 (accessed on January 18th, 2019)

[4] https://www.noaa.gov/news/us-had-its-coldest-april-in-more-than-20-years

[5] (accessed on January 18th, 2019)

[6] https://www.washingtonpost.com/news/capital-weather-gang/wp/2018/04/18/its-the-coldest-and-snowiest-april-on-record-near-the-great-lakes-and-residents-are-fuming/?noredirect=on&utm_term=.d4a28bc56758 (accessed on January 18th, 2019)

[7] Aldrian E and Utama G S A 2007 Identifikasi dan karakteristik seruak dingin (cold surge) tahun 1995-2003 *Jurnal Sains Dirgantara* **4** **(2)**
[8] Setiawati F Z 2015 Kajian superposisi cold surge dan gangguan pusaran udara (vortex) di Kalimantan Barat : Studi kasus tahun 2009 *Jurnal Meteorologi Klimatologi dan Geofisika* **2(2)** 32-40
[9] Paulus AW and Shanas S P 2017 Atmospheric study of the impact of cold surge and borneo vortex over western indonesia maritime continent area *J. Climatol Weat Forcesat* **5** 1000189
[10] Fauzi R R and Hidayat R 2017 *Pengaruh Cold Surge dan MJO terhadap Peningkatan Hujan di Indonesia pada Periode East Asian Winter Monsoon* (Bogor:Institut Pertanian Bogor)
[11] Chen T-C, Huang W-R and Yoon J-H 2004 Interannual variation of the east asian cold surge activity *Journal Ametsoc* **17** Issue 2 401-413
[12] Dayantolis W 2006 *Analisa Kondisi Fisis Atmosfer Pada Saat Hujan Ekstrim dan Terjadinya Banjir Bulan Februari 2006 di Manado* (Depok:Universitas Indonesia)
[13] Tjasyono B 2012 *Mikrofisika Awan dan Hujan* BMKG

IOP Conf. Series: Earth and Environmental Science **303** (2019) 012065 doi:10.1088/1755-1315/303/1/012065

Analysis of the Meteorological Condition of Tropical Cyclone Cempaka and Its Effect on Heavy Rainfall in Java Island

F Samrin[1], I Irwana[1], Trismidianto[2] and N Hasanah[1]

[1]Geophysics Study Program, Physics Department, Hasanuddin University, Makassar, Indonesia
[2]Center for Atmospheric Science and Technology, National Institute of Aeronautics and Space, Bandung, Indonesia

samrinfajriani@gmail.com

Abstract. Indonesia, literally tropical cyclone free area, still gains some influence from the surrounding cyclones, such as high-intensity precipitation which causes flooding. Therefore, cyclones monitoring is important to observe its growth, movement and intensity. Using ECMWF ERA-Interim data, Himawari-8 channel infrared 1 (IR1), and GSMaP satellite data, this study attempts to spatially analyze the evolution of Cempaka tropical cyclones (occurred on November 2017), propagation cyclone and its effect to rainfall over Java Island based on the brightness temperature of the cloud. The results show that the Cempaka tropical cyclone has a three-day life span and moves from the southern ocean of Central Java to the northeast (East Java) and the peak occurs on November 27th, 2017 at 19:00 LT to 23:00 LT. Compared to rainfall data, the intensity of rainfall at 7 areas closed to Cempaka tropical cyclone began to increase ever since the cyclone peak and continue to rise despite the cyclone was slowly decaying. With this result, we hope in the future we can anticipate the impact of any tropical cyclones that arise near Indonesia.

Keywords : tropical cyclone, flood, ECMWF, rainfall.

1. Introduction

Tropical cyclones (TC) occur at the tropics or subtropics area in the form of winds storm. It began with the presence of tropical or low-pressure centers that are intensive over the oceans, thus triggering an intensive process of convection and cloud formation. The study of TC has been long introduced by Fasiq (1913) in his book that contains traces, the beginning and destructive action of TC. Many studies of TC are also carried out separately in various cyclone basins on this earth. However, based on historical events, generally TC is formed effectively above 10°LU and LS region. Therefore, Indonesia territory should be free from TC formation, but there is still some part of Indonesia that adjacent to the formation and trajectory of TC [1]. For example is the TC Cempaka which detected on November 27th, 2017, by the TC Warning Center (TCWC) of BMKG. The TC Cempaka grew very close to the southern coast of Java Island.

TC greatly affect the surrounding weather conditions. A number of studies have examined the contribution of TCs to precipitation at the regional scale [2, 3]. Villarini and Denniston [4] showed that there was a higher probability of TC-induced annual maximum rainfall events in Australia during La Niña. Yin et al [5] pointed out that the number of TC-heavy rain days is higher during El Niño years in eastern China, besides that the phenomenon of tropical cyclones can also be related to local, meso or other global phenomena [6]. TCs also not only affect the trajectory area directly but also affects/affected the surrounding weather conditions. Therefore, TCs affect the weather conditions in the Indonesian territory and can occur outside the appropriate period, the storm monitoring / tropical cyclone around Indonesia is important to do, so that information anticipates a possible impact on the form of bad weather such as floods and landslides [1]. The impact on the TC Cempaka is also expected to affect the surrounding weather patterns, but the trigger has not been clearly explained. For this reason, an in-depth study of the occurrence of this TC is needed. Based on this, this paper will explain in detail the analysis of meteorological conditions during TC Cempaka events and their effects on rainfall on the island of Java.

2. Data and Method

In this study, the three main data are brightness temperature (T_{BB}), rainfall estimation data, and meteorological parameter data. All data is processed using GrADS. The T_{BB} data was taken from the Himawari-8 satellite operated by the Japan Meteorological Agency (JMA) with a horizontal resolution of 0.05° x0.05° and temporal resolution of 1 hours for infrared (IR) imagery (http://weather.is.kochiu.ac.jp/sat/GAME). Analysis of the development of TC Cempaka was carried out by visual observation method on the results of Himawari-8 data plotting on IR1 channels by looking at the T_{BB} value. Some literature states that analyzing of Convective Index (CI) can be determined by using cloud peak temperatures from T_{BB} [7, 8, 9, 10]. The threshold value of brightness temperature or black body temperature is determined to separate the surface temperature from high convective clouds. This threshold value according to Chen and Houze [11] can describe diurnal variation in convective activity. Adler and Negri [7] use a threshold value of 253 K as the hottest brightness temperature to determine clouds that are thought to produce rain. CI is equal to threshold - T_{BB} if T_{BB} is smaller than the threshold value (CI = threshold - T_{BB}, for T_{BB} <threshold) and a convective index equal to zero for the T_{BB} greater than or equal to the threshold value, (CI = 0, for $T_{BB} \geq$ threshold).

Rainfall estimation data used in this study comes from GSMaP (Global Satellite Mapping of Precipitation) satellite data which is Near Real Time (NRT). GSMaP near real- time (GSMaP NRT) has been implemented since October 2008. Data sets are provided in real time (about four to five hours after observation) by file transfer protocol (ftp). This GSMaP_NRT data from JAXA has been validated and calibrated with both the incisive data and radar data in Japan with validation results likely to be quite good [12]. GSMaP data can be accessed at http://sharaku.eorc.jaxa.jp/GSMaP/ (JAXA, 2014). The effect of the TC Cempaka on the surrounding rainfall is seen by visual observations of rainfall estimates from GSMaP satellite data during the TC Cempaka. The movement of the TC Cempaka can be analyzed using a hovmoller diagram or cross section during the occurrence of the TC Cempaka.

The meteorological parameter data is obtained from ECMWF ERA-Interim. ERA-interim is a reanalysis of the global atmosphere covering the data-rich period since 1979 (originally, ERA-Interim ran from 1989, but the 10-year extension for 1979-1988 produced in 2011). ERA-Interim is the latest global atmospheric reanalysis produced by the ECMWF. This data has a time resolution of 6-hourly, a horizontal resolution of 1° x 1° in latitude and longitude in 27 pressure level that include 1000, 975, 950, 925, 900, 875, 850, 825, 800, 775, 750, 700, 650, 600, 550, 500, 450, 400, 350, 300, 250, 225, 200, 175, 150, 125 and 100 hPa. The environmental parameters that used in this study are the wind, relative humidity, temperature, specific humidity, divergence, vertical velocity, vorticity and geopotential height. However, there are some parameters should be derived by using equation from some variable of ECMWF ERA-Interim data due to is not available. These data can be downloaded at http://www.ecmwf.int/. The Meteorological conditions analysis is also carried out with spatial analysis of ECMWF ERA-Interim data which is also combined with other data.

3. Result and Discussion

3.1. Analysis of Evolution of TC Cempaka Development

The evolution of the development of the TC Cempaka can be analyzed using T_{BB} from the Himawari-8 satellite data shown in Figure 1. Since morning at 0900 LT (LT=Local Time = UTC+7 hours) on

November 26, 2017 a small-scale cloud had been around the south of East Java water around 9.5°S and 111.5°E or about 75 km south of Pacitan an area which effected of the TC Cempaka based on several news reports. From the small T_{BB} value below 253 K it can be indicated that the cloud is a convection cloud that produces rain [7]. After several hours, the cloud underground a decay around 1100 LT but because the thermodynamic process of the decaying atmosphere of the cloud was developing again to form a fairly large convection cloud. The maximum peak of the cloud size occurs at 2300 LT at midnight. Viewed from its peak, this cloud is indicated as a cloud of Mesoscale Convective Complex (MCC) because it has a physical characteristic of cloud temperature similar with the MCC characteristic according Maddox [13] where is the cloud temperature above 221 K and has an almost circle shape, it is also in accordance with the explanation from the previous paper [14] which also states that MCC is occurring in the oceans around Indonesia are more nocturnal (maximum peak occurring at night). Trismidianto et al. [14] also stated that MCC could be a trigger for tropical cyclones.

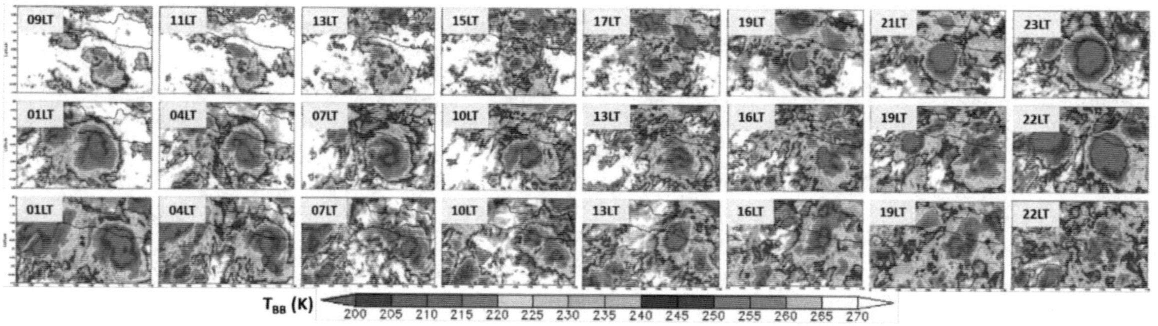

Figure 1. Horizontal distribution of brightness (T_{BB}) temperature on November 26, 2017 (top row), November 27, 2018 (middle row) and November 28, 2018 (bottom row) of Himawari-8 satellite data.

Since the morning of November 27, 2017 after the MCC decays, it starts to show a high convective cloud vortex as shown in Figure 1 in the middle row. Convective clouds that have enlarged at 1100 LT. (November 26, 2017) combine with other convective clouds around them forming a vortex of clouds in the waters of the Indian Ocean (south of Java). This high convective cloud vortex was very clearly visible at 0700 LT which indicated a cyclone eye with very low T_{BB} below 210 K. The vortex continued to spin and move closer to the mainland of Java Island, can be seen at 1500 LT convective clouds begin to be above the mainland of Java Island and continue to spread until convective clouds cover most of the area in Java. The peak of the TC Cempaka formation occurs at night around 1900 - 2300 LT. At 2300 LT, it was also seen that deep tropical or deep convective clouds were thought to influence the formation of the TC Cempaka. The core of the convective cloud of the TC Cempaka has also reached land at 2300 LT. Figure 1 on the bottom row shows that tropical cyclones began to decay from the morning of November 28, 2017. Several hours after decaying from 2200 LT began to notice the movement of convective clouds towards the mainland of Java which also joined with other convective clouds from the waters of the Java Sea making convective clouds over Java Island are intensified with low T_{BB} values.

Figure 2 to strong the analyzed of Figure 1 that convective activity has been seen since morning at 0900 LT. on November 26, 2017 which continued to strengthen until the formation of MCC and continued to become a TC Cempaka. Figure 3 shows that the growth potential of TC Cempaka seedlings has been detected since November 23, this can be seen in the presence of low-pressure centers on the south coast of Java Island. Geopotential height approximates the actual height of surface pressure (1000 hPa) above mean sea-level. Therefore, a geopotential height observation represents the height of the surface pressure on which the observation was taken. So that a low geopotential height value can also represent low pressure. The low geopotential height is followed by the presence of shortwave trough. A trough is a region with relatively lower heights tend to bring in cooler weather. Shortwaves may be

IOP Conf. Series: Earth and Environmental Science **303** (2019) 012065 doi:10.1088/1755-1315/303/1/012065

contained within or found ahead of long waves and range from the mesoscale to the synoptic scale. Shortwaves are most frequently caused by either a cold pool or an upper-level front. So, this also strengthens the discussion before that TC is triggered more by the presence of meso-scale convective clouds that are thought to be MCC.

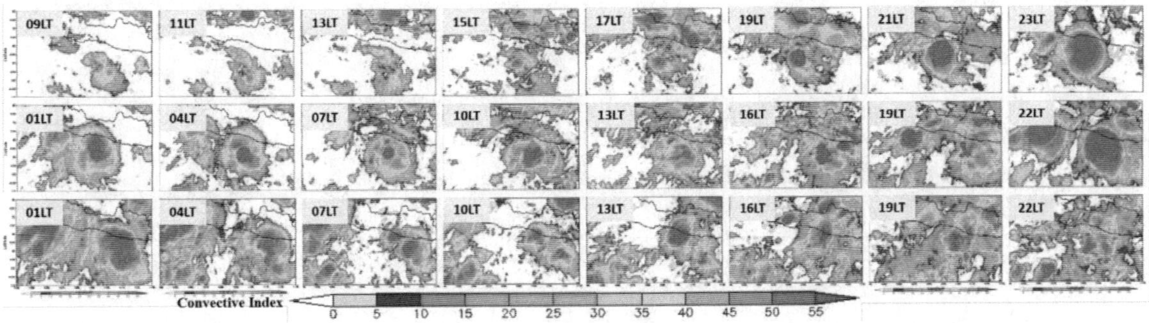

Figure 2. Convective index on November 26, 2017 (top row), November 27, 2018 (middle row) and November 28, 2018 (bottom row) of Himawari-8 satellite data.

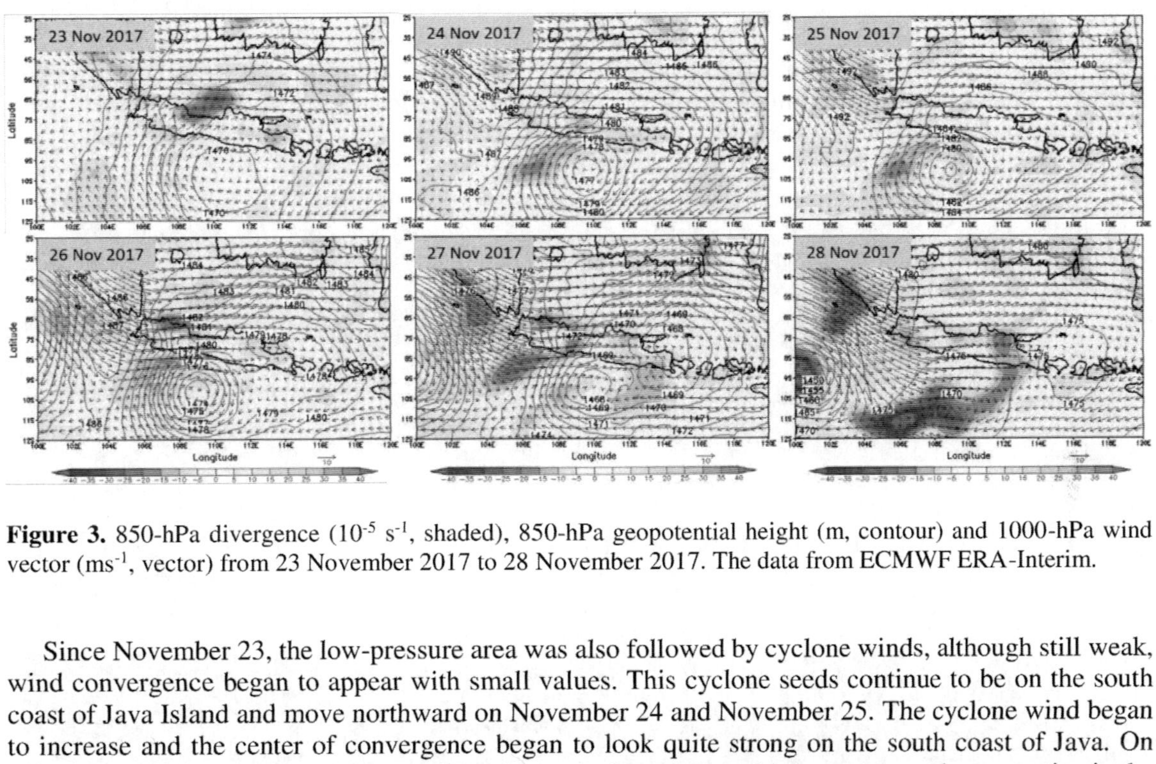

Figure 3. 850-hPa divergence (10^{-5} s^{-1}, shaded), 850-hPa geopotential height (m, contour) and 1000-hPa wind vector (ms^{-1}, vector) from 23 November 2017 to 28 November 2017. The data from ECMWF ERA-Interim.

Since November 23, the low-pressure area was also followed by cyclone winds, although still weak, wind convergence began to appear with small values. This cyclone seeds continue to be on the south coast of Java Island and move northward on November 24 and November 25. The cyclone wind began to increase and the center of convergence began to look quite strong on the south coast of Java. On November 26, the cyclone seeds moved southwest and strengthened into a meso-scale convective in the form of the tropical depression. On November 27, tropical depression became the TC Cempaka and moved northeast towards the southern coast of Java. On November 28, TC Cempaka experienced landfall on the southern coast of Java. Figure 3 also shows that the formation of tropical cyclones formed due to the Low-Level Jet (LLJ) moving from the west coast of Sumatra and the Indian Ocean to the southern coast of Java since November 23, 2017, this is also strengthened by the conversion of the convergence center from the Indian Ocean to the coast south of Java Island.

Figure 4 shows that since November 25, there has been a strong low-level convergence movement and a strong upper- level divergence from the Indian Ocean towards the southern coast of Java Island.

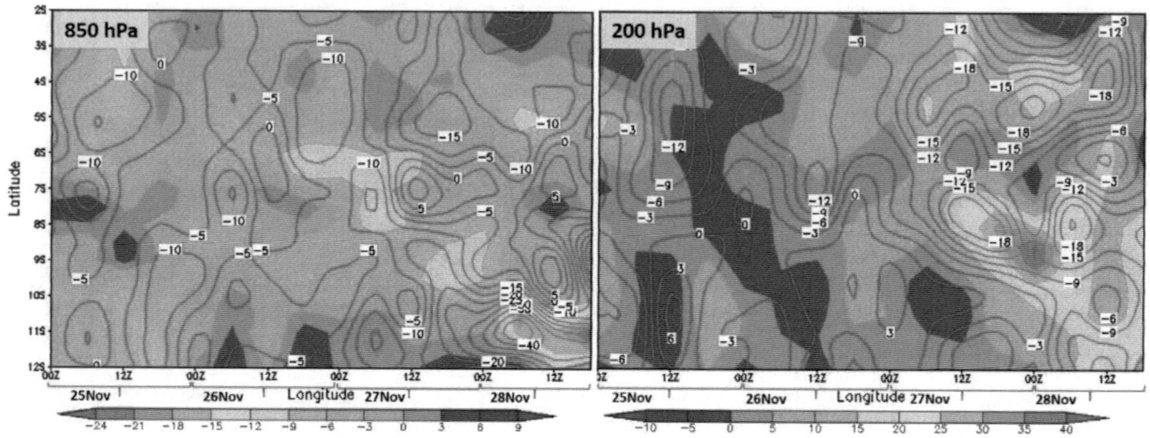

Figure 4. 850-hPa and 200 hPa divergence (10^{-5} s^{-1}, shaded) and vertical velocity (10^{-2} Pa s^{-1}, contour) from 25 November 2017 to 28 November 2017. The data from ECMWF ERA-Interim.

Figure 5 strengthens the analysis of figure 4, almost the same as the analysis of Figure 4, but in this picture, it focuses on the condition of the level, low level (850 hPa), middle level (500 hPa), upper level (200 hPa). The conditions are if there is convergence in the lower level, then in the upper-layer ,there will be a strong divergence. Focus on the location of the occurrence of TC Cempaka, between 110 - 111°E.

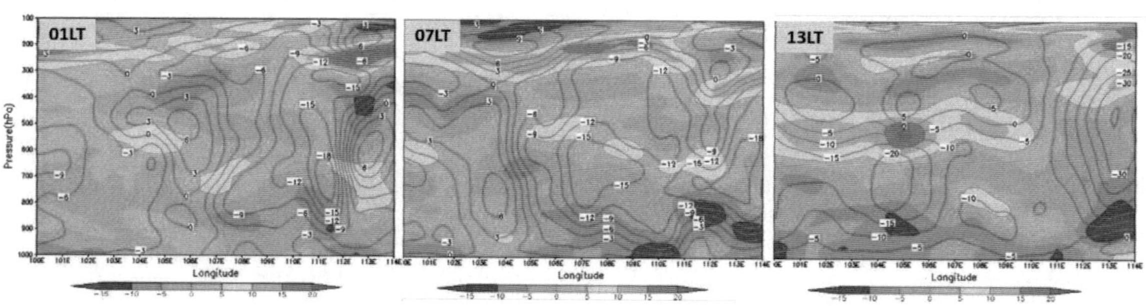

Figure 5. Pressure Longitude Cross Section of divergence (10^{-5} s^{-1}, shaded) and vertical velocity (10^{-2} Pa s^{-1}, contour) from 25 November 2017 to 28 November 2017. The data from ECMWF ERA-Interim.

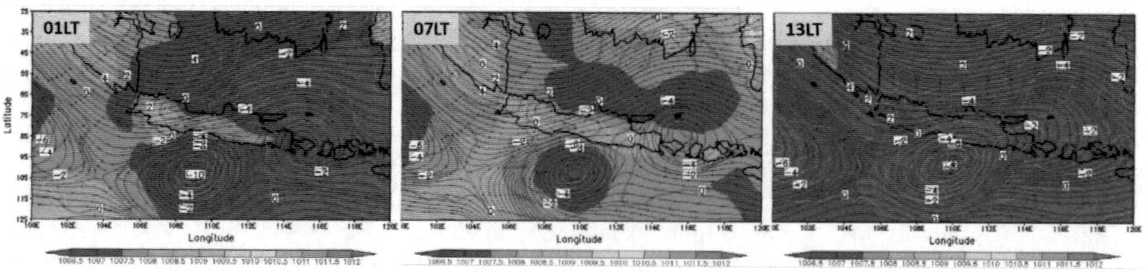

Figure 6. Horizontal distribution in 850 hPa for Mean Sea Level Pressure (m, shaded) and vorticity (10^{-6} s^{-1}, contour) on 27 November 2017. The data from ECMWF ERA-Interim.

IOP Conf. Series: Earth and Environmental Science **303** (2019) 012065 doi:10.1088/1755-1315/303/1/012065

Figure 6 shows that the surface pressure at the peak of cyclones is very low compared to others. And the vortices are also negative which indicates the presence of tropical cyclones which normally carry a mass of moisture around it.

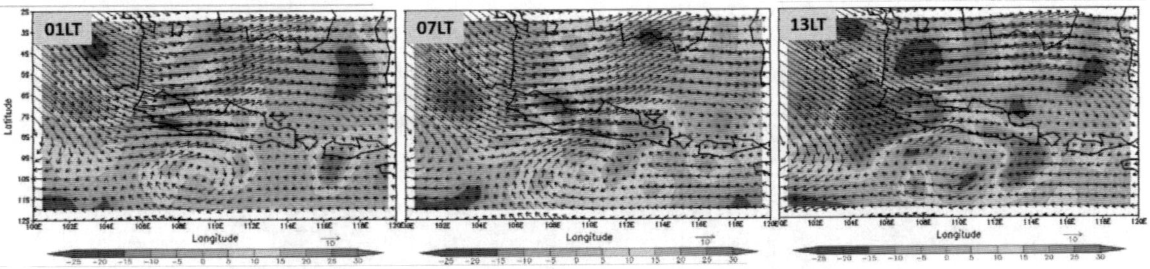

Figure 7. Horizontal distribution in 850 hPa for MFC (moisture flux convergence in 10^{-3} g kg^{-1} s^{-1}, shaded), and wind vector (ms^{-1}, vector) on 27 November 2017. The data from ECMWF ERA-Interim.

Figure 7 explains that the source of moisture in the vapor mass that causes cyclones is from the Indian Ocean near the coast of western Sumatra.

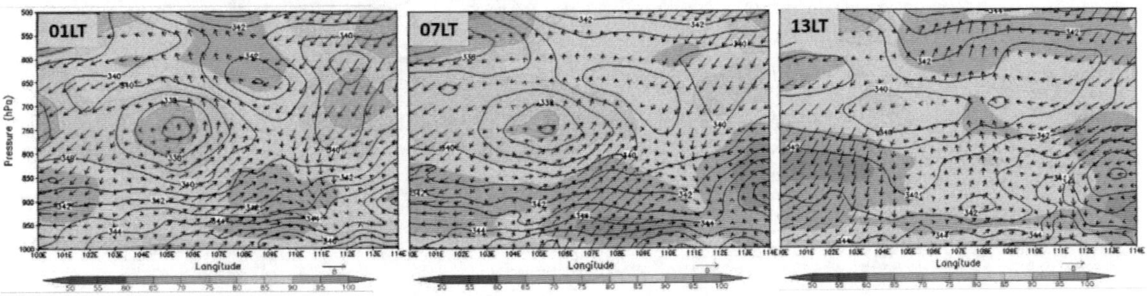

Figure 8. Pressure-longitude cross section of Equivalent Potential Temperature (θ_E, contour; K) and relative humidity (shaded; %) and horizontal wind (vector; ms^{-1}; upward represents northward) on 27 November 2017. The data from ECMWF ERA-Interim.

Figure 8 shows relative humidity and equivalent potential temperature, the high relative humidity values indicate that there is a process of precipitation or convective clouds. Whereas θ_E, equivalent potential temperature, is the temperature that results after all, the heat is released in a parcel of water and then brought to the 1000 hPa level. The increases θ_E as dewpoint and temperature increases. A region with a relatively high θ_E is often the region with the most instability. Low-level temperature and higher low-level dewpoints increase instability. The low level jet from values and thus increased instability. θ_E ridge is regions with higher θ_E. They are often the bursting point for convective activity.

3.2. Propagation Cyclone and Its effect on rainfall over Java Island

One way to show the movement of a convective cloud or tropical cyclone is to create and analyze a hovmoller diagram. In this study, a hovmoller diagram was carried out by averaging the longitude of the area of the TC Cempaka from 102° - 116°E. The hovmoller diagramming is carried out evenly averaged because the cyclone movement is estimated from top to bottom with time changes. Figure 4.9 shows that the formation of the TC Cempaka was triggered by the incorporation of several convective clouds from the Indian Ocean at a latitude of around 7.5° - 9°S with the presence of convective clouds from the waters of the Java Sea around latitude 3° - 4°S on November 26 around 2400 LT with high rainfall intensity. The convective cloud combination triggers the occurrence of the TC Cempaka with extreme rainfall intensity at its center at latitude 7.5°LS on November 27, 2017 at around 2100 LT. This high rainfall occurs after the peak of the TC Cempaka. The extreme rainfall then spreads in various directions, namely to latitude 8°S (Indian Ocean) and to latitude 5.5° - 6.2°S (around the mainland of Java Island) at around 1100 LT, November 28, 2017 shows the occurrence of extreme rainfall in most areas on the mainland of Java Island.

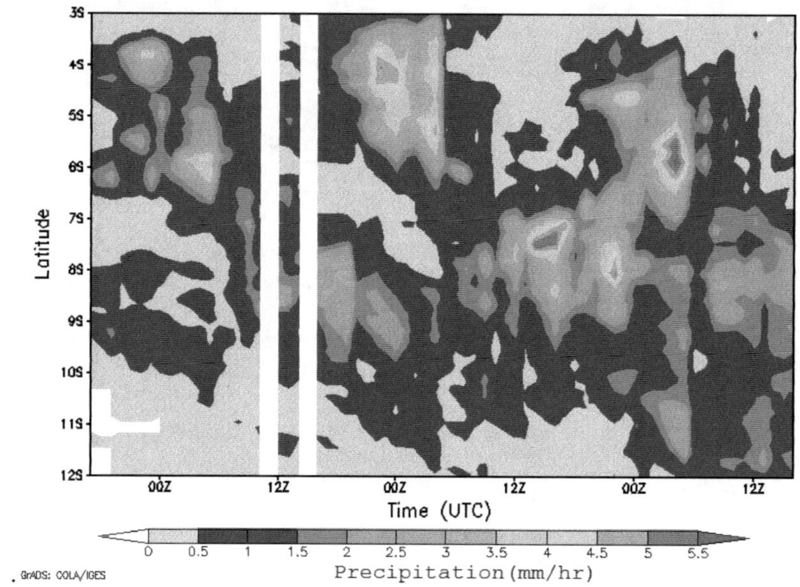

Figure 9. Hovmoller Time-Latitude for rainfall on November 26-28, 2017 from GSMaP satellite data (average longitude 102° - 116°BT)

As previously explained that the influence of TC Cempaka on the surrounding rainfall can be seen by visual observation of rainfall estimates from GSMaP satellite data during the TC Cempaka. Strong convection activity is closely related to the high intensity of rainfall in the area, and vice versa. As with Himawari-8 satellite data visualized by colors that represent T_{BB} values, GSMaP satellite data is also visualized by colors representing the value of precipitation in mm / day. Figure 9 shows that throughout the day on November 26, 2017 in the Indian Ocean waters (south of Java Island) there was significant rainfall with rainfall intensity of around 0.5 - 10 mm / day, even at 1600 LT rainfall ranges from 20 mm/hr (extreme) which means there is heavy rainfall in the area, and at 2300 LT the area with rainfall reaching 10 mm/day is increasingly widespread. This is in accordance with the Figure 9 which shows the distribution of convection clouds in the waters of the Indian Ocean (south of Java Island), even seen a large cloud of convection indicated as MCC at 2300 LT (cloud convection causes rain with high rainfall).

Figure 10 shows the distribution of rainfall on November 27, 2017, when the process formed until the peak of the tropical cyclone in the Indian Ocean (south of Java). From Figure 10, it is known that the cloud vortex in the Indian Ocean waters began to appear, at that time there was moderate to heavy rain in the area, even during the process of forming a tropical cyclone cyclone some areas in Java Island near the Indian Ocean waters had experienced rain with intensity extreme rainfall above 20 mm / day. Until the time of the peak of the cyclone, which is around 1900 - 2300 LT, the rain does not only occur in the waters of the Indian Ocean, but also rain in most areas of Java, starting at 1400 LT with rainfall intensity ranging from 1 - 10 mm / day up to 2300 LT. Rainfall in some areas exceeds 20 mm / day. Most likely the extreme rainfall occurred due to the movement of cloud swirls from the waters of the

Indian Ocean towards the mainland of Java which resulted in the spread of convective clouds in most areas in Java (Figure 10 at 1100 LT).

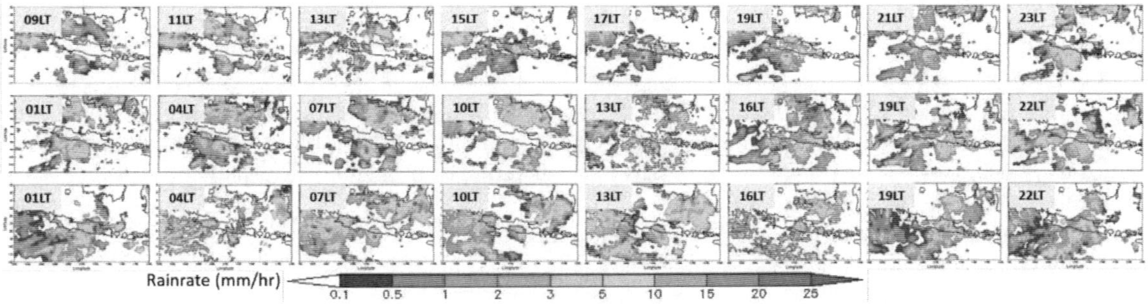

Figure 10. Horizontal distribution of rainfall on November 26, 2017 (above), November 27, 2017 (middle) and November 28, 2017 (bottom) of GSMaP satellite data.

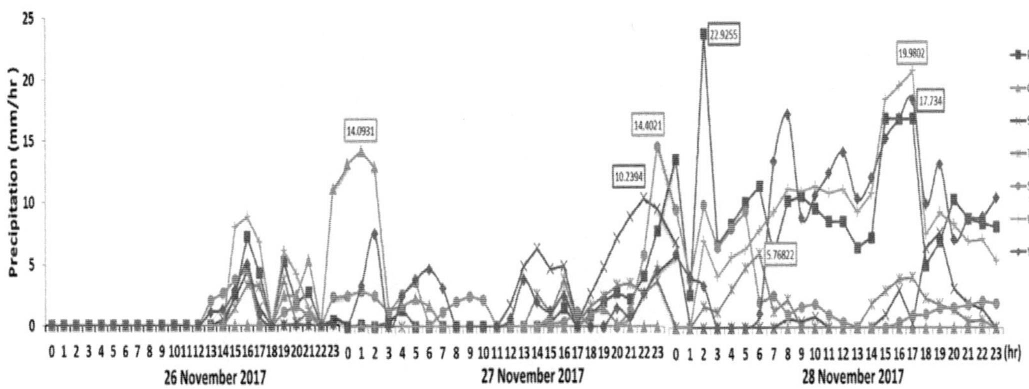

Figure 11. Rainfall graph in the affected area by the Cempaka tropical cyclone.

After it was discovered that the TC Cempaka affected rainfall in several areas of Java, several areas were chosen which were considered to have had a large influence from the occurrence of the TC Cempaka. These areas are Sukabumi, Cilacap, Yogyakarta, Pacitan, Wonogiri, Tulungagung, and Sidoarjo.

Figure 11 shows a graph of rainfall hourly for 3 days, which is November 26 -28, 2017. On November 26 , 2017 before the occurrence of the TC Cempaka, since 1300 LT., the rainfall in the Sukabumi and Yogyakarta regions has increased (not reaching the bulk extreme rain), a significant increase in rainfall occurred in Wonogiri at 1500 LT. with rainfall reaching 8 mm/hr which was then followed by increased rainfall in Pacitan at 1600 LT. with a rainfall of 7.2 mm/hr, and during the night a significant bulk increase occurred in Cilacap at 2300 LT the rainfall reached 11 mm/hr (extreme rainfall), it continued until at 0100 LT (November 27, 2017) the rainfall in Cilacap reached 14 mm/hr.

On November 27, 2017 since 0000 LT there was rain in Cilacap (extreme rainfall) and Sukabumi (rainfall 2 mm/hr), in Yogyakarta began to rain lightly at 0100 LT with rainfall of 3 mm/hr , in Pacitan it rained lightly at 0300 LT with a rainfall of 0.3 mm/hr, in Sidoarjo it started to rain at 1200 LT with rainfall of 1.7 mm/hr, in Tulungagung it began to rain at 1400 LT with rainfall of 2.6 mm/hr, and in Wonogiri began to rain in the afternoon at 1600 LT with rainfall of 2.9 mm / hr. Throughout 27 November 2017 the highest rainfall occurred in Sukabumi at 2300 LT with rainfall reaching 14.4 mm /

hr (extreme rainfall). This corresponds to Himawari-8 satellite data which shows that on November 27, 2017 at 2300 LT the TC Cempaka has enlarged so that some areas experience heavy rains at night.

On November 28, 2017, a swirl of clouds from the waters of the Indian Ocean moved towards the mainland of Java and then its convection clouds were scattered so that several areas experienced heavy rain throughout the day. The region with the highest rainfall is Pacitan with rainfall reaching 22.9 mm/hr (extreme rainfall) at 0200 LT, then the Wonogiri area with rainfall of 19.9 mm/hr at 1700 LT. At 1700 LT, in Yogyakarta also reached the peak of extreme rainfall, namely 17.7 mm/hour, while for the Sukabumi area heavy rain only lasted until 0500 LT., after that only mild rain - moderate to the night, even during the day at 1400 - 1500 LT there is no rain. In Tulungagung all day there is light rain - moderate with the highest rainfall value of 5.7 mm/hr at 0600 LT. In Sidoarjo there was only rain at 1500 LT with rainfall of 6.2 mm / hour, medium rain only lasted until 1900 LT, after that there was no rain. Whereas in Cilacap all day there was no rain at all.

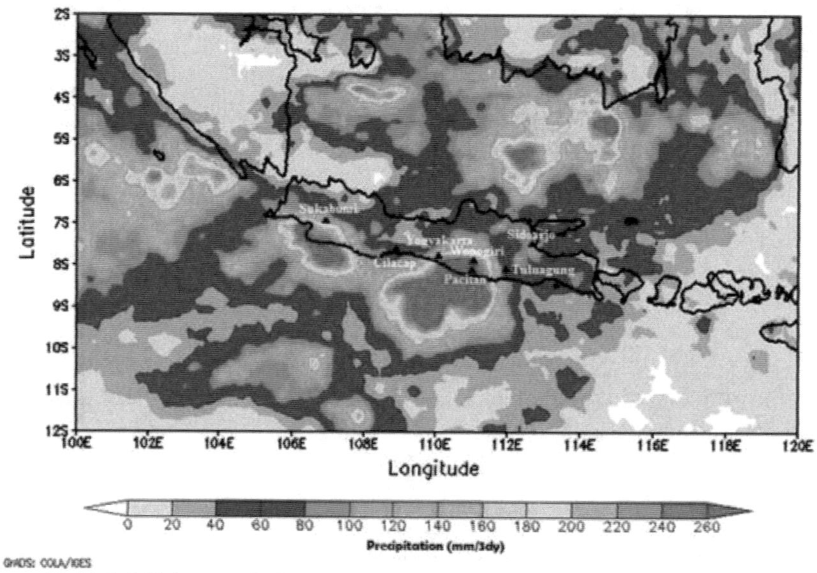

Figure 12. Rainfall Accumulation on November 26 -28 , 2017 from GSMaP satellite data

Rainfall accumulation for 3 days (November 26 -28, 2017) using GSMaP satellite data was conducted to see which areas had the most influence during the occurrence of the Cempaka tropical cyclone. Figure 12 shows that Pacitan and Wonogiri are areas with extreme rainfall during the Cempaka tropical cyclone, which is in accordance with Figure 11 which shows that during the 3 rainy days with the highest rainfall occurred in Pacitan and Wonogiri with 22 rainfall each, 9 mm / hour and 19.9 mm / hour, and rainfall accumulation reaches 260 mm/3dy.

4. Conclusion

Since November 26, 2017 small-scale clouds were formed around the waters south coast of East Java that trigger convective activities in the area, then the clouds that previously formed MCC and continued to become tropical cyclones. The peak is on November 27, 2017. Based on CI, the emergence of TC Cempaka due to the low-pressure center on the south coast of Java Island which triggered the emergence of cyclone winds. In addition, the formation of the devastating tropical cyclone is also caused by the presence of LLJs moving from the west coast of Sumatra and the Indian Ocean to the southern coast of Java since November 23, 2017, this is also strengthened by the conversion of the convergence center from the Indian Ocean to the south coast of the Island Java. TC Cempaka affects the weather system in most regions of Java, which is increasing the intensity of rainfall, starting from the growth stage until after the cyclone occurs. On November 26 - 28, 2017 rainfall in several regions in Java Island experienced a significant increase, Pacitan and Wonogiri were recorded as areas with the highest rainfall.

Acknowledgment
We are grateful to ECMWF ERA-Interim (http://www.ecmwf.int/), Himawari-8 satellite (http://weather.is.kochiu.ac.jp/sat/GAME) and GSMaP satellite (http://sharaku.eorc.jaxa.jp/GSMaP) for available the data to the public.

5. References

[1] Haryani, Suryo N and Any Z 2012 *Dinamika siklon tropis di Asia Tenggara menggunakan data pengindraan jauh Pusbangja LAPAN* p 324
[2] Chen F and Fu Y 2015 Contribution of tropical cyclone rainfall at categories to total precipitation over The Western North Pacific from 1998 to 2007 *China Earth Sci.* **58** 2015–25 doi:10.1007/s11430-015-5103-9
[3] Ng B, Walsh K and Lavender S 2015 The contribution of tropical cyclones to rainfall in Northwest Australia *Int. J. Climatol.* **35** pp 2689–97 doi:10.1002/joc.4148
[4] Villarini G and Denniston RF 2016 Contribution of tropical cyclones to extreme rainfall in Australia *Int. J. Climatol.* **36** pp 1019–1025 doi:10.1002/joc.4393
[5] Yin Y, Gemmer M, Luo Y and Wang Y 2010 Tropical cyclones and heavy rainfall in Fujian Province China *Quat. Int.* **226** pp 122–128 doi:10.1016/j.quaint.2010.03.015
[6] Curtis S and Adler R F 2003 Evolution of El Nino precipitation relationships from satellite and gauges *J. Geoph. Res.* **108**
[7] Adler R F and Negri A J 1988 A satellite infrared technique to estimate tropical convective and stratiform rainfall *J. Appl. Meteor.* **27** pp 30-51
[8] Mapes B E and Houze Jr R A 1995 Diabatic divergence profiles in western Pacific mesoscale convective systems *J. Atmos. Sci.* **52** 1807–1828
[9] Nitta T and S. Sekine 1994 Diurnal variation of convective activity over tropical western Pacific *J. Meteor. Soc. Japan* **72** 627-641
[10] Sakurai N, Murata F, Yamanaka M D, Hashiguchi H, Mori S, Hamada J-I, Tauhid Y-I, Sribimawati T and Suhardi B 2005 Diurnal cycle of migration of convective cloud systems over Sumatera Island *J. Meteorol. Soc. Japan* **83** 835–850
[11] Chen S S and Houze Jr R A 1997 Diurnal variation and life-cycle of deep convective systems over the tropical Pacific warm pool *Q. J. R. Met. Soc.* **123** 357-388
[12] Syaifullah M D 2014 Validasi data TRMM terhadap data curah hujan aktual di tiga DAS di Indonesia UPT Hujan Buatan – BPPT **15** 2:109-118
[13] Maddox R A 1980 Mesoscale convective complexes *Bull. Amer. Meteor. Soc.* **61** 1374 - 13
[14] Trismidianto, Yulihastin E, Satyawardhana H, and Ishida S 2017 A composite analysis of The Mesoscale Convective Complexes (MCCs) development over The Central Kalimantan and its relation with the propagation of the rainfall systems *IOP Publishing Ltd.* DOI 10.1088/1755-1315/54/1/012036.

AUTHOR INDEX

AUTHOR INDEX

AUTHOR INDEX